Selected Titles in This Series

9 **Brian Conrad and Karl Rubin, Editors,** Arithmetic Algebraic Geometry, 2001
8 **Jeffrey Adams and David Vogan, Editors,** Representation Theory of Lie Groups, 2000
7 **Yakov Eliashberg and Lisa Traynor, Editors,** Symplectic Geometry and Topology, 1999
6 **Elton P. Hsu and S. R. S. Varadhan, Editors,** Probability Theory and Applications, 1999
5 **Luis Caffarelli and Weinan E, Editors,** Hyperbolic Equations and Frequency Interactions, 1999
4 **Robert Friedman and John W. Morgan, Editors,** Gauge Theory and the Topology of Four-Manifolds, 1998
3 **János Kollár, Editor,** Complex Algebraic Geometry, 1997
2 **Robert Hardt and Michael Wolf, Editors,** Nonlinear Partial Differential Equations in Differential Geometry, 1996
1 **Daniel S. Freed and Karen K. Uhlenbeck, Editors,** Geometry and Quantum Field Theory, 1995

IAS/PARK CITY MATHEMATICS SERIES

Volume 9

Arithmetic Algebraic Geometry

Brian Conrad
Karl Rubin
Editors

American Mathematical Society
Institute for Advanced Study

IAS/Park City Mathematics Institute runs mathematics education programs that bring together high school mathematics teachers, researchers in mathematics and mathematics education, undergraduate mathematics faculty, graduate students, and undergraduates to participate in distinct but overlapping programs of research and education. This volume contains the lecture notes from the Graduate Summer School program on Arithmetic Algebraic Geometry held in Park City, Utah, on June 20–July 10, 1999.

2000 *Mathematics Subject Classification.* Primary 11F11, 11F33, 11F70, 11F80, 11G05, 11G07, 11G18, 11R23, 11S25.

Library of Congress Cataloging-in-Publication Data
Arithmetic algebraic geometry / Brian Conrad, Karl Rubin, editors.
  p. cm. — (IAS/Park City mathematics series, ISSN 1079-5634 ; v. 9)
  ISBN 0-8218-2173-3 (acid-free paper)
  1. Arithmetical algebraic geometry—Congresses. I. Conrad, Brian David, 1970– II. Rubin, Karl. III. Series.

QA242.5 .A73 2001
512′.7—dc21   2001035291

Copying and reprinting. Material in this book may be reproduced by any means for educational and scientific purposes without fee or permission with the exception of reproduction by services that collect fees for delivery of documents and provided that the customary acknowledgment of the source is given. This consent does not extend to other kinds of copying for general distribution, for advertising or promotional purposes, or for resale. Requests for permission for commercial use of material should be addressed to the Assistant to the Publisher, American Mathematical Society, P. O. Box 6248, Providence, Rhode Island 02940-6248. Requests can also be made by e-mail to reprint-permission@ams.org.

Excluded from these provisions is material in articles for which the author holds copyright. In such cases, requests for permission to use or reprint should be addressed directly to the author(s). (Copyright ownership is indicated in the notice in the lower right-hand corner of the first page of each article.)

© 2001 by the American Mathematical Society. All rights reserved.
The American Mathematical Society retains all rights
except those granted to the United States Government.
Printed in the United States of America.

∞ The paper used in this book is acid-free and falls within the guidelines
established to ensure permanence and durability.
Visit the AMS home page at URL: http://www.ams.org/
10 9 8 7 6 5 4 3 2 1   06 05 04 03 02 01

Contents

Preface	xiii
Introduction	1
Joe P. Buhler, Elliptic Curves, Modular Forms, and Applications	5
Preface	7
Lecture 1. Elliptic Curves	9
Curves in the affine plane	11
Projective curves	14
The group operation	17
Rational maps	18
Genus	19
Weierstrass form and the group law via divisors	21
Lecture 2. Points on Elliptic Curves	25
Finding points	25
Minimal discriminants	27
Reduction	28
The p-adic filtration	29
Lecture 3. Elliptic Curves Over **C**	33
Fundamental isomorphisms	33
The j invariant of an elliptic curve	34
Elliptic curves via lattices	36
Lattices via the upper half-plane	38
Lecture 4. Modular Forms of Level 1	41
Weak modular forms	41
Modular Forms	42
Dimensions of spaces of modular forms	44
Hecke operators	46
Lecture 5. L-series; Modular Forms of Higher Level	49
Dirichlet series	49
L-functions of modular forms of level 1	50
Forms of higher level	51
L-series of elliptic curves	54
Lecture 6. l-adic Representations	57

Galois representations 57
One-dimensional representations 58
Galois representations coming from elliptic curves 59
The conductor 60

Lecture 7. The Rank of Elliptic Curves Over $\mathbf{Q}$ 63
Mordell-Weil 64
The Weak Mordell-Weil Theorem 65
The analytic rank 69

Lecture 8. Applications of Elliptic Curves 71
Primality 71
Factoring 73
Cryptography 75

Bibliography 79

Alice Silverberg, Open Questions in Arithmetic Algebraic Geometry 83

Chapter 1. Overview 85
Introduction 85
Elliptic curves 86
Abelian varieties 87
Mordell-Weil groups 88
Torsion subgroups 89
Ranks 90
Looking ahead 90
Suggested reading 90

Chapter 2. Torsion Subgroups 91
Torsion subgroups for elliptic curves 91
A quick look at modular curves 92
Conjectures 93
Other results for elliptic curves 93
Abelian varieties over number fields 95
Abelian varieties over other fields 98
Digression on complex multiplication 98

Chapter 3. Ranks 99
Examples 99
Unboundedness/boundedness 100
Vary the field 101
Quadratic twists 101
Average ranks 103
Distribution of ranks 104

Chapter 4. Conjectures of Birch and Swinnerton-Dyer 107
The L-function of an elliptic curve 107
Modularity 109
BSD I 111

The Congruent Number Problem	112
Evidence for BSD I	112
BSD II	113
Evidence for BSD II	115
Selmer groups	116
Examples — descents	117
Chapter 5. ABC and Related Conjectures	**123**
ABC Conjectures	123
ABC Conjectures for function fields	125
Szpiro Conjecture	125
Goldfeld-Szpiro Conjecture	126
de Weger Conjectures	127
Additional remarks	128
Chapter 6. Some Other Conjectures	**129**
Integral points on elliptic curves	129
Bounds on numbers of rational points on curves	129
Conductors of elliptic curves	130
Abelian varieties, Shimura varieties, and open questions	130
Bibliography	**133**

Kenneth A. Ribet and William A. Stein, Lectures on Serre's Conjectures — **143**

Preface	145
Chapter 1. Introduction to Serre's Conjecture	**149**
Introduction	149
The weak conjecture of Serre	153
The strong conjecture	154
Representations arising from an elliptic curve	156
Background material	157
The cyclotomic character	157
Frobenius elements	158
Modular forms	159
Tate curves	160
Mod ℓ modular forms	161
Chapter 2. Optimizing the Weight	**163**
Representations arising from forms of low weight	164
The ordinary case	164
The supersingular case and fundamental characters	164
Representations of high weight	166
The supersingular case	168
Systems of mod ℓ eigenvalues	169
The supersingular case revisited	170
The ordinary case	172
Distinguishing between weights 2 and $\ell + 1$	172
Geometric construction of Galois representations	172

Representations arising from elliptic curves	174
Frey curves	175
Examples	175
Companion forms	176
Chapter 3. Optimizing the Level	177
Reduction to weight 2	177
Geometric realization of Galois representations	178
Multiplicity one	179
Multiplicity one representations	180
Multiplicity one theorems	181
Multiplicity one for mod 2 representations	181
The key case	182
Approaches to level optimization in the key case	183
Some commutative algebra	184
Aside: Examples in characteristic two	184
III applies but I and II do not	185
II applies but I and III do not	185
Aside: Sketching the spectrum of the Hecke algebra	186
Mazur's principle	187
Level optimization using a pivot	190
Shimura curves	191
Character groups	191
Proof	192
Level optimization with multiplicity one	193
Chapter 4. Exercises	197
Exercises	197
Solutions	200
Chapter 5. Appendix by Brian Conrad: The Shimura Construction in Weight 2	205
Analytic preparations	205
Algebraic preliminaries	212
Proof of Theorem 5.12	216
Chapter 6. Appendix by Kevin Buzzard: A Mod ℓ Multiplicity One Result	223
Bibliography	227
Fernando Q. Gouvêa, Deformations of Galois Representations	**233**
Introduction	235
Lecture 1. Galois Groups and Their Representations	237
Galois groups of infinite algebraic extensions	237
The galois group of $\mathbb{Q}$	241
Restricting the ramification	244
Galois representations	247
Complements to Lecture 1	249
Lecture 2. Deformations of Representations	253

Why deform Galois representations?	253
The deformation functor	255
Universal deformations: why representable functors are nice	261
Representable functors and fiber products	262
The tangent space	266
Complements to Lecture 2	268
Lecture 3. The Universal Deformation: Existence	271
Schlessinger's criteria	272
Universal Deformations exist	275
Absolutely irreducible representations	278
Example: the case $n = 1$	279
Extra Problems	281
Complements of Lecture 3	282
Lecture 4. The Universal Deformation: Properties	283
Functorial properties	283
Tangent spaces and cohomology groups	284
Tangent spaces and extensions of modules	285
Obstructed and unobstructed deformation problems	287
Galois representations	290
Lecture 5. Explicit Deformations	293
The basic setup	293
Group theory	295
Pro-p-extensions	296
Tame representations	298
Lecture 6. Deformations With Prescribed Properties	303
Deformation conditions	304
Deformations with fixed determinant	306
Categorical deformation conditions	308
Ordinary deformations	309
Deformation conditions for global Galois representations	311
Representations that are ordinary at p	313
Complements to Lecture 6	313
Lecture 7. Modular Deformations	317
Classical modular forms and their representations	318
p-adic modular forms	323
The ordinary case	327
Imposing deformation conditions	328
Lecture 8. p-adic Families and Infinite Ferns	331
The slope of an eigenform	332
p-old and p-new	333
p-adic families of modular forms	335
Infinite ferns	335
APPENDIX 1. A Criterion for Existence of a Universal Deformation Ring by Mark Dickinson	339

APPENDIX 2. An Overview of a Theorem of Flach by Tom Weston	345
Unobstructed deformation problems	346
Galois modules and the calculus of Tate twists	346
First reductions	348
Selmer groups	351
The L-function of $\operatorname{Sym}^2 T_l E$	354
Kolyvagin's theory of Euler systems	356
The Flach map	360
The geometry of modular curves	363
Some modular units	366
Local behavior of the c_r	367
Appendix A: On the local Galois invariants of $E[l] \otimes E[l]$	370
Appendix B: The definition of the Flach map	373
APPENDIX 3. An Introduction to the p-adic Geometry of Modular Curves by Matthew Emerton	377
The curve $X_0(2)$	378
Canonical subgroups of elliptic curves over $\mathbb{C}_2$	382
The parameter q	384
The Hasse invariant	386
Return to $X_0(2)$	387
The theory for arbitrary primes p	390
p-adic modular forms	393
Guide to the literature	395
Bibliography	399

Ralph Greenberg, Introduction to Iwasawa Theory for Elliptic Curves

	407
Preface	409
Chapter 1. Mordell-Weil Groups	411
Exercises	422
Chapter 2. Selmer Groups	425
Exercises	435
Chapter 3. Λ-Modules	439
Exercises	449
Chapter 4. Mazur's Control Theorem	453
Exercises	461
Bibliography	463

John Tate, Galois Cohomology

	465
Preface	467
Group modules	467
Cohomology	468
Examples	469

Characterization of $H^r(G,-)$	470
Kummer theory	470
Functor of pairs (G, A)	472
The Shafarevich group	473
The inflation-restriction sequence	474
Cup products	475
G-pairing	475
Duality for finite modules	476
Local Fields	476
Bibliography	479

Wen-Ching Winnie Li, The Arithmetic of Modular Forms — 481

Introduction	483
Lecture 1. Introduction to Elliptic Curves, Modular Forms, and Calabi-Yau Varieties	485
Classical modular forms	485
Elliptic curves	487
Zeta functions	488
Calabi-Yau varieties	489
Lecture 2. The arithmetic of modular forms	491
Eisenstein series	491
Hecke operators	492
The structure of $\mathcal{C}(k, N, \chi)$– The theory of newforms	493
Lecture 3. Connections Among Modular Forms, Elliptic Curves, and Representations of Galois Groups	497
Functional equations	497
Connections with elliptic curves	498
Connections with representations of the Galois group over $\mathbb{Q}$	500
Comparisons with automorphic forms of GL_2 over function fields	502
Bibliography	505

Noriko Yui, Arithmetic of Certain Calabi-Yau Varieties and Mirror Symmetry — 507

Introduction	509
Acknowledgements	511
Lecture 1. The Modularity Conjecture for Rigid Calabi-Yau Threefolds over the Field of Rational Numbers	513
Definition of Calabi-Yau varieties	513
The mirror symmetry conjecture	516
Rigid Calabi-Yau threefolds over $\mathbb{Q}$	517
The modularity conjecture for rigid Calabi-Yau threefolds over $\mathbb{Q}$	517
Evidence for the modularity conjecture	519
Strategy for establishing the modularity conjecture	522
The intermediate Jacobian of a Calabi-Yau threefold	524

The order of vanishing of L-series at $s=2$	526
The conjecture of Beilinson and Bloch	527
Lecture 2. Arithmetic of Orbifold Calabi-Yau Varieties Over Number Fields	531
Fermat hypersurfaces and their deformations	531
Orbifold Calabi-Yau varieties	532
Sketch of proof of Theorem 2.2	535
The L-series of orbifold Calabi-Yau varieties	538
Sketch of proof of Theorem 4.2	539
The Tate conjecture	540
The conjecture of Beilinson and Bloch	541
Construction of Mirror Calabi-Yau varieties	543
Problems	546
Lecture 3. $K3$ Surfaces, Mirror Moonshine Phenomenon	549
Mirror Moonshine phenomena	549
$K3$ surfaces	550
The $K3$ lattice	552
Lattice polarized $K3$ surfaces	553
Moduli space of lattice polarized $K3$ surfaces	554
Picard-Fuchs differential equations for one-parameter families of $K3$ surfaces	557
Mirror maps	558
Generalizations and open problems	562
Bibliography	565

Preface

The IAS/Park City Mathematics Institute (PCMI) was founded in 1991 as part of the "Regional Geometry Institute" initiative of the National Science Foundation. In mid 1993 the program found an institutional home at the Institute for Advanced Study (IAS) in Princeton, New Jersey. The PCMI will continue to hold summer programs alternately in Park City and in Princeton.

The IAS/Park City Mathematics Institute encourages both research and education in mathematics and fosters interaction between the two. The three-week summer institute offers programs for researchers and postdoctoral scholars, graduate students, undergraduate students, high school teachers, mathematics education researchers, and undergraduate faculty. The summer institute is preceded by the Mentoring Program for Women in Mathematics. One of PCMI's main goals is to make all of the participants aware of the total spectrum of activities that occur in mathematics education and research: we wish to involve professional mathematicians in education and to bring modern concepts in mathematics to the attention of educators. To that end the summer institute features general sessions designed to encourage interaction among the various groups. In-year activities at sites around the country form an integral part of the High School Teacher Program.

Each summer a different topic is chosen as the focus of the Research Program and Graduate Summer School. Activities in the Undergraduate Program, and at the Mentoring Program for Women in Mathematics, deal with this topic as well. Lecture notes from the Graduate Summer School are being published each year in this series, in some cases combined with notes from the graduate lectures at the Mentoring Program for Women in Mathematics. The first nine volumes are:

Volume 1: *Geometry and Quantum Field Theory* (1991)
Volume 2: *Nonlinear Partial Differential Equations in Differential Geometry* (1992)
Volume 3: *Complex Algebraic Geometry* (1993)
Volume 4: *Gauge Theory and the Topology of Four-Manifolds* (1994)
Volume 5: *Hyperbolic Equations and Frequency Interactions* (1995)
Volume 6: *Probability Theory and Applications* (1996)
Volume 7: *Symplectic Geometry and Topology* (1997)
Volume 8: *Representation Theory of Lie Groups* (1998)
Volume 9: *Arithmetic Algebraic Geometry* (1999)

A future volume from the 2000 Summer School on *Computational Complexity Theory* is in preparation. The 2001 Research Program and Graduate Summer School topic is *Quantum Field Theory, Supersymmetry, and Enumerative Geometry*.

Some material from the Undergraduate Program is published as part of the Student Mathematical Library series of the American Mathematical Society. We hope to publish material from other parts of the IAS/Park City Mathematics Institute in the future. This will include material from the High School Teacher Program and publications documenting the interactive activities which are a primary focus of the PCMI. At the summer institute late afternoons are devoted to seminars of common interest to all participants. Many deal with current issues in education; others treat mathematical topics at a level which encourages broad participation. The PCMI has also spawned interactions between universities and high schools at a local level. We hope to share these activities with a wider audience in future volumes.

David R. Morrison, Series Editor
October, 2000

Introduction

The 1999 Graduate Summer School of the Institute for Advanced Study/Park City Mathematics Institute was held in Park City, Utah from June 20 to July 10, 1999. The topic was arithmetic algebraic geometry. Joe Buhler, Fernando Gouvêa, Ralph Greenberg, Ken Ribet, and Alice Silverberg each gave a lecture course, and there were two special lectures by John Tate.

In general terms, arithmetic algebraic geometry is the use of methods from algebraic geometry to study problems in number theory. This is both a very active and very broad area. In deciding which topics to choose for the lectures, we were heavily (but not exclusively) influenced by the recent spectacular work of Wiles on modular elliptic curves and Fermat's Last Theorem.

Many problems in number theory involve finding solutions of polynomial equations (or systems of polynomial equations) in integers, in rational numbers, or in other number fields. In studying such questions one is quickly led to try to understand the collection of all number fields, or equivalently, to understand the absolute Galois group $G_\mathbf{Q} = \mathrm{Gal}(\bar{\mathbf{Q}}/\mathbf{Q})$ of the rational number field $\mathbf{Q}$.

How can one try to understand $G_\mathbf{Q}$? One of the most productive ways has been by studying its continuous representations (over the complex numbers, or over p-adic or finite fields). For example, the one-dimensional representations of $G_\mathbf{Q}$ are described by the Kronecker-Weber theorem, which says that there is a natural bijection between the set of representations of $G_\mathbf{Q}$ into $\mathbf{C}^\times$ and the set of Dirichlet characters. More generally, if F is a number field then the one-dimensional representations of G_F are described by class field theory, and have been understood since the early 20th century.

The situation with higher-dimensional representations is much more difficult, and a great deal remains to be understood, even for two-dimensional representations. It is generally expected that all "well-behaved" representations should "come from geometry" in a certain precise sense.

What does this mean? First of all, one can use algebraic geometry to produce examples of higher dimensional representations of $G_\mathbf{Q}$. For example, if E is an elliptic curve defined over $\mathbf{Q}$, then the torsion points on E with coordinates in $\bar{\mathbf{Q}}$ form a group isomorphic to $(\mathbf{Q}/\mathbf{Z})^2$. For every prime p, $G_\mathbf{Q}$ acts linearly on the p-part of this group, and thus we get a representation $\rho_{E,p}$ of $G_\mathbf{Q}$ into $\mathrm{GL}_2(\mathbf{Q}_p)$. In a similar but more complicated way, certain modular forms f (cusp forms which are eigenfunctions of the Hecke operators) give rise to two-dimensional representations $\rho_{f,\mathfrak{p}}$ of $G_\mathbf{Q}$ into $\mathrm{GL}_2(K_{f,\mathfrak{p}})$ where K_f is a number field determined by f and $K_{f,\mathfrak{p}}$ is its completion at a prime $\mathfrak{p}$.

The lectures of Buhler give an introduction to these topics, including elliptic curves, modular forms, and their L-functions and Galois representations.

The lectures of Silverberg discuss some of the current frontiers of research in arithmetic algebraic geometry by focusing on important open problems that are guiding the development of the field. For example: what is known, and what is not known, about torsion points on elliptic curves and abelian varieties? What is known, and what is conjectured, about ranks of elliptic curves? What is the Birch and Swinnerton-Dyer conjecture, and how much of it has been proved? What does it mean for an elliptic curve to be modular?

One formulation of the modularity conjecture of Shimura and Taniyama[1] states that for every elliptic curve E defined over $\mathbf{Q}$ there is a cusp form f with $K_f = \mathbf{Q}$ such that $\rho_{E,p}$ is isomorphic to $\rho_{f,p}$ (in which case we say that E is modular).

Beginning in the early 1970's Serre developed conjectures which (among other things) give a "first approximation" to the Shimura-Taniyama conjecture. Namely, Serre conjectured (the "weak" conjecture) that if $\bar{\rho}$ is an irreducible representation of $G_\mathbf{Q}$ into $\mathrm{GL}_2(\bar{\mathbf{F}}_p)$, and $\det(\bar{\rho}(\tau)) = -1$ where τ is a complex conjugation, then $\bar{\rho}$ is modular in the sense that there is a cusp form f such that for some prime $\mathfrak{p}$ of K_f dividing p, the reduction of $\rho_{f,\mathfrak{p}}$ modulo $\mathfrak{p}$ is isomorphic to $\bar{\rho}$ (after a suitable embedding of the residue field at $\mathfrak{p}$ into $\bar{\mathbf{F}}_p$). Serre also gave a recipe (the "strong" conjecture) for the minimal weight, level, and character of such an f in terms of $\bar{\rho}$.

Not much progress has been made in proving Serre's weak conjecture. A notable exception is when $\mathbf{F} = \mathbf{F}_3$, in which case work of Langlands and Tunnell shows that such a form f does exist. This case is crucial for Wiles' program.

On the other hand, when one such f exists, a great deal of progress has been made in showing that an f of the prescribed weight, level and character exists. The lectures of Ribet (written for this volume jointly by Ribet and William Stein) introduce Serre's conjectures and give some details of the methods and results which enable one to go from Serre's weak conjecture to the strong conjecture.

Suppose now that one has a representation $\rho : G_\mathbf{Q} \to \mathrm{GL}_2(\mathbf{Q}_p)$ which one wants to show is modular, and let $\bar{\rho} : G_\mathbf{Q} \to \mathrm{GL}_2(\mathbf{F}_p)$ be the reduction of ρ modulo p. In the 1980's Mazur introduced his deformation theory of Galois representations to study deformations of $\bar{\rho}$, i.e., two-dimensional representations of $G_\mathbf{Q}$ which reduce to $\bar{\rho}$. With appropriate restrictions on the deformations, Mazur showed that there is a "universal deformation ring" R, and a "universal deformation" $\rho_R : G_\mathbf{Q} \to \mathrm{GL}_2(R)$, which parametrizes all deformations in the sense that every representation of $G_\mathbf{Q}$ into $\mathrm{GL}_2(S)$ (of the appropriate type) whose reduction is $\bar{\rho}$ is the composition of ρ_R with a unique homomorphism $R \to S$.

Deformation theory is the subject of the lectures of Gouvêa. The goal is to get information about all deformations by studying the universal deformation ring. For example, if Serre's conjecture holds for $\bar{\rho}$, then one can also define a "universal modular deformation" $\mathbb{T}$ which parametrizes all deformations of $\bar{\rho}$ which are modular (i.e., isomorphic to some $\rho_{f,\mathfrak{p}}$). Then (since R is universal) there is a homomorphism $R \to \mathbb{T}$, and if (after placing appropriate restrictions on the deformation problem) one can show that this map is an isomorphism, then one can conclude that "every deformation is modular", including the original ρ. This was Wiles' successful strategy.

In a different direction, not so directly related to the modularity problem, are Greenberg's lectures on Iwasawa theory and elliptic curves. One can attach an

---

[1] The senior organizer would like to thank the junior organizer and his colleagues Breuil, Diamond, and Taylor for completing their work on the program begun by Wiles just in time to announce a complete proof of this conjecture in Park City.

L-function $L(E,s)$ to an elliptic curve E over $\mathbf{Q}$. This is a function of a complex variable s, initially defined by an Euler product in a right half-plane. It follows from the fact that E is modular that $L(E,s)$ has an analytic continuation to all of $\mathbf{C}$ and satisfies a functional equation relating $L(E,s)$ and $L(E,2-s)$. The Birch and Swinnerton-Dyer conjecture predicts the order of vanishing of $L(E,s)$ at $s=1$, as well as the value of the first non-vanishing derivative of $L(E,s)$ at $s=1$, in terms of important invariants of E. One can think of $L(E,s)$ as a generalization of Dirichlet L-functions, and the Birch and Swinnerton-Dyer conjecture as a generalization of classical analytic class number formulas.

Iwasawa theory was introduced by Iwasawa beginning in the late 1950's as a tool to study ideal class groups of cyclotomic fields. He observed that the group algebra $\varprojlim \mathbf{Z}_p[\mathrm{Gal}(\mathbf{Q}(\boldsymbol{\mu}_{p^n})/\mathbf{Q})]$ is in many ways easier to work with than the individual group rings $\mathbf{Z}_p[\mathrm{Gal}(\mathbf{Q}(\boldsymbol{\mu}_{p^n})/\mathbf{Q})]$. In particular, he found that the ideal class groups of the fields $\mathbf{Q}(\boldsymbol{\mu}_{p^n})$ for $n \geq 1$ can be much better understood if one studies them all at once rather than for each n separately. In the early 1970's, Mazur used Iwasawa theory to study elliptic curves, and since then it has been extended by Greenberg and others to study a large class of problems.

Iwasawa theory for elliptic curves has two sides, analytic and algebraic. On the analytic side one can attach to E an analytic p-adic L-function, defined by interpolating values of classical L-functions. On the algebraic side one can study the arithmetic of E over the infinite abelian extension $\mathbf{Q}(\boldsymbol{\mu}_{p^\infty})$ of $\mathbf{Q}$, and define an algebraic p-adic L-function which encodes some of the arithmetic of E. If one can show that these two p-adic L-functions are the same (the "main conjecture"), then one has begun to find the link between $L(E,s)$ and arithmetic invariants of E which was predicted by Birch and Swinnerton-Dyer. Greenberg's lectures deal with the algebraic side of the theory.

Galois cohomology is an essential tool used throughout arithmetic algebraic geometry, and Tate kindly agreed to give two introductory lectures which are included in this volume.

Also in this volume are the lectures which were delivered at the Mentoring Program for Women in Mathematics portion of PCMI at the Institute for Advanced Study in May 1999. The three lecturers were Winnie Li, Noriko Yui, and Alice Silverberg. For details about that program see the introduction to Li's lectures.

Finally we would like to express our thanks to all the lecturers for their lectures and their written contributions to this volume. Their efforts to keep the Park City lectures accessible to as many of the participants as possible were successful and greatly appreciated. We would also like to thank all the teaching assistants, the audiences, and Catherine Jordan and the IAS/PCMI staff for making the program a success.

Karl Rubin
Brian Conrad

Elliptic Curves, Modular Forms, and Applications

Joe P. Buhler

Elliptic Curves, Modular Forms, and Applications

Joe P. Buhler[1]

Preface

The second half of the twentieth century has seen many exciting developments in number theory, and this continuing vitality was especially obvious during the summer 1999 Park City Mathematics Institute on Arithmetic Algebraic Geometry. Several major new results were announced at approximately the time of the meeting. Perhaps the most notable was the announcement by Breuil, Conrad, Diamond, and Taylor of the completion of the Wiles/Taylor program for proving that all elliptic curves are modular. In addition, important results were announced (or hinted) on icosahedral galois representations, function field versions of Langlands' conjectures, and admissible (in the sense of Fontaine) Galois representations.

As part of the Graduate Summer School connected with the program, the organizers asked me to give eight introductory lectures that would introduce some of the basic objects of arithmetic algebraic geometry: elliptic curves, modular forms, ℓ-adic representations associated thereto, etc. There are many excellent book-length treatments of these subjects, and in a few lectures it is possible to give only a taste of the ideas with an eye to motivating the student to learn the details and proofs from those books. The first few lectures covered elliptic curves and modular forms, and the last few gave brief treatments of applications to the construction of ℓ-adic representations, the solution of Diophantine equations, and algorithms for primality testing, factoring, and cryptography.

In preparing these notes for publication, an attempt was made to retain the basic feel of the lecture notes, so that the order of topics, varying levels of completeness, and overall flavor reflect the sense of the lectures.

The PCMI session was a pleasant experience, and I'd like to thank the organizers for inviting me, and also thank Lily Khadjavi and Amod Agashe for being the TA's during the session, and Amod Agashe, Brian Conrad, Vinay Deolalikar, Michael Drinen, Lily Khadjavi, and Karl Rubin for providing suggestions on portions of the notes.

---

[1] MSRI, Berkeley, CA 94720
E-mail address: jpb@msri.org

LECTURE 1
Elliptic Curves

Two fundamental facts about elliptic curves are that their equations can be put into a standard form, and that the solutions to those equations form a commutative group under a natural geometric operation.

Theorem 1.1 (Weierstrass Form and Group Law): Every elliptic curve over a field K is isomorphic to the projective curve corresponding to a nonsingular affine cubic of the form
$$E: \quad y^2 + a_1 xy + a_3 y = x^3 + a_2 x^2 + a_4 x + a_6.$$
The set $E(K)$ of points on an elliptic curve has the structure of a commutative group in which the identity is the point at infinity and three points sum to the identity if and only if they are colinear.

There are a number of terms in this theorem that need to be defined, including "projective," "elliptic curve," "nonsingular," and "isomorphism." It is tempting (but unsatisfying) to make short work of the first part of the theorem by defining elliptic curves to be equations of the indicated form!

The equation in the theorem is said to be in "Weierstrass form" and will be fundamental for our future investigations. However, elliptic curves often arise in other guises in nature, and Fermat's last theorem provides two examples.

Example 1.2. The Fermat cubic $x^3 + y^3 = z^3$ isn't in Weierstrass form, but if we substitute
$$x = Y + 36, \quad y = -Y + 36, \quad z = 6X$$
we arrive at the equation
$$Y^2 = X^3 - 432$$
which is in Weierstrass form.

Example 1.3. Fermat's Last Theorem for $n = 4$ asserts that there is no solution to $x^4 + y^4 = z^4$ in nonzero integers. This is usually proved by showing that the equation
$$x^4 + y^4 = z^2$$
has no nonzero solution in rational numbers. Substituting
$$x = 2XY, \quad y = 4X^2, \quad z = 4X^2Y^2 + 32X^3$$

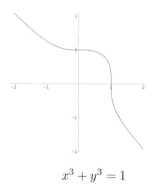

$$x^3 + y^3 = 1$$

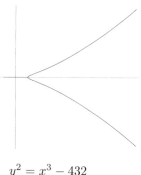

$$y^2 = x^3 - 432$$

leads to an equation that is (a power of X times)

$$Y^2 = X^3 - 4X,$$

which is an equation in Weierstrass form. More generally, an equation of the form

$$y^2 = \text{a quartic polynomial in } x$$

can be transformed to an elliptic curve in Weierstrass form if the polynomial on the right hand side has distinct roots.

Example 1.4. (see [**Bre**]) An article in the 1795 Gentleman's Diary exhibited a pair of integers whose sum, sum of squares, and sum of cubes are all perfect squares. In other words, there are integers x, y, u, v, w such that

$$\begin{aligned} x + y &= u^2 \\ x^2 + y^2 &= v^2 \\ x^3 + y^3 &= w^2 \end{aligned}$$

(Puzzle: find the smallest solution in positive integers!). By dividing the third equation by the first, we see that any solution to this system, in rational numbers, gives rise to a rational solution to

$$\begin{aligned} x^2 + y^2 &= z^2 \\ x^2 - xy + y^2 &= t^2 \end{aligned}.$$

Conversely, any rational solution to the second system gives rise to an integer solution to the first equation after scaling by a suitable integer so that the sum is a perfect square. Although apparently unrelated to the Weierstrass equation (there are two homogeneous equations of degree 2 in 4 variables rather than a single inhomogeneous equation of degree 3 in 2 variables), in fact this is another common disguised form of an elliptic curve: the intersection of two general "quadric surfaces" (zero sets of homogeneous equations in 4 variables of degree 2). In this specific case, if we set

$$X = -3tz^2, \quad Y = 3(y^2 - x^2)z, \quad Z = t^3$$

we find, after some algebraic juggling, that

$$Y^2 = X^3 + 8X^2Z + 12XZ^2 = X(X + 2Z)(X + 6Z)$$

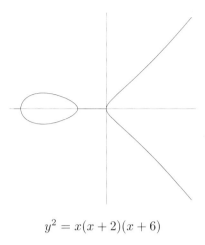

$$y^2 = x(x+2)(x+6)$$

which is in (the projective version of) Weierstrass form. This curve will be used as an illustrative example at several later junctures, and the reader will be able to use those discussions to describe all (x, y) that answer the original question.

Remark 1.5. If the characteristic of K isn't equal to 2 then in the Weierstrass form
$$E: \quad y^2 + a_1 xy + a_3 y = x^3 + a_2 x^2 + a_4 x + a_6$$
we can complete the square on the left hand side to get
$$Y^2 = (y + a_1 x/2 + a_3/2)^2 = \text{ a cubic in } x.$$

(As we will see, this equation is the equation of an elliptic curve if and only if the cubic polynomial on the right hand side does not have repeated roots.) If the characteristic of the field isn't equal to 3 then we can further simplify the equation by shifting the variable x to eliminate the x^2 term to end up with an equation of the form
$$Y^2 = X^3 + aX + b.$$

For many purposes it suffices to consider this simpler Weierstrass form rather than the general equation in the theorem, and we will tend to prefer this form merely because of its simplicity.

The goal of this lecture is to introduce enough algebraic geometry to be able to define the terms in the theorem and give an overview of its proof. There are by now a large number of treatments of elliptic curves, from several different perspectives. The semi-canonical text for graduate students is [**Si1**], and the reader may also enjoy consulting [**Cas**], [**Hus**], [**Kna**], [**Kob**], [**La2**], [**Mi1**], [**ST**], [**Si2**], etc.

1. Curves in the affine plane

We start with curves in the plane as solution sets to polynomials in two variables.

Fix a field K. When convenient we will assume that its characteristic is not equal to 2 or 3. On occasion we will use the fact that K is contained in a field that is **algebraically closed** in the sense that every polynomial in one variable with coefficients in the field factors into linear factors. An **algebraic closure** $\overline{K}$ of K is an algebraically closed field containing K such that any element of $\overline{K}$ is algebraic over (i.e., satisfies a polynomial equation with coefficients in) k. Any two such algebraic closures are isomorphic (though there are usually many such isomorphisms).

The **affine plane** over K is the set $\mathbf{A}^2 = \mathbf{A}^2(K) := K^2$.

An **affine plane curve** C defined over K is a nonzero polynomial $f(x,y)$ in two variables with coefficients in K. (A multiple of f by a nonzero scalar in K is regarded as the same curve.) The zero set of this polynomial is what we usually visualize as the curve, and if L is any field containing K then solutions to $f(x,y) = 0$ with coordinates in L is denoted

$$C(L) := \{(x,y) \in L^2 : f(x,y) = 0\}.$$

A K-rational point on the curve is a point in $C(K)$, and a **point** on the curve means an element in $C(\overline{K})$ for a fixed choice of algebraic closure $\overline{K}$ of K.

Remark 1.6. If $f(x,y) = 0$ is a curve over $\mathbf{Q}$ then an explicit solution (x,y) to the equation with coordinates that are transcendental numbers is not a point on the curve. Modern algebraic geometry gives a more satisfactory and conceptual definition of a "point," but the direct notion above suffices for our purposes.

We say that a curve is **irreducible** if its defining polynomial $f(x,y)$ is absolutely irreducible in the sense that it is irreducible as a polynomial in $\overline{K}[x,y]$.

Example 1.7. The curve $x^2 + y^2 = 0$ has one $\mathbf{Q}$-rational point, but is reducible since it factors $(x+iy)(x-iy)$ over $\mathbf{Q}(i)$; the curve is the union of two lines, each of which is defined over $\mathbf{Q}(i)$ but not over $\mathbf{Q}$.

From now on we make the convention that curves are irreducible unless we make an explicit remark to the contrary.

Assume that $P = (0,0)$ is a point on a curve $C : f(x,y) = 0$. the Taylor expansion of the curve at P is

$$0 = f(x,y) = f_1(x,y) + f_2(x,y) + \cdots$$

where $f_d(x,y)$ is a homogeneous polynomial of degree d. If d is the smallest integer such that $f_d(x,y)$ is nonzero then the (not necessarily irreducible) curve $f_d(x,y) = 0$ is an approximation to C in a sufficiently small vicinity of $(0,0)$.

The curve is **smooth** or **nonsingular** at P if $f_1(x,y)$ is nonzero, i.e., if the curve can be approximated by a straight line in a small neighborhood of P. The linear form f_1 has the form

$$f_1(x,y) = Ax + By, \quad A = \partial_x f(0,0), \quad B = \partial_y f(0,0),$$

so the point P is nonsingular if at least one of the two partial derivatives is nonzero at P. In this case we say that $f_1 = 0$ is the tangent line at P.

More generally, if P is singular and if d is the least integer such that f_d is nonzero then the homogeneous form f_d factors into linear forms

$$f_d(x,y) = \prod_{i=1}^{d}(\alpha_i x + \beta_i y),$$

over an algebraic closure of K, and the d lines $\alpha_i x + \beta_i y = 0$ are said to be the tangent lines to the curve at P (counted with multiplicity if some of the factors are repeated).

All of this generalizes to arbitrary points P by considering the Taylor expansion at P (or changing coordinates to put P at the origin).

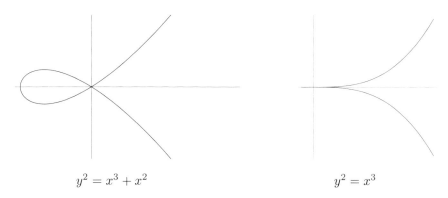

$y^2 = x^3 + x^2$ $\qquad\qquad\qquad\qquad$ $y^2 = x^3$

Example 1.8. (a node) The cubic curve $y^2 - x^3 - x^2 = 0$ is singular at the origin. The leading term of the Taylor expansion is

$$f_2(x,y) = y^2 - x^2 = (y-x)(y+x)$$

so that there are two tangent lines at the origin: $y = x$ and $y = -x$. Any singular point where f_2 factors into distinct lines is said to be a **node**, and it is easy to see how these singular points can be thought of as "double points."

Example 1.9. (a cusp) The cubic curve $y^2 - x^3 = 0$ is also singular at the origin and the leading term of the Taylor expansion is $y^2 = 0$. This means that the x-axis is tangent line of multiplicity two. Singularities of this type are called **cusps**.

The natural coordinate changes in the affine plane are nonsingular linear transformations followed by translations, i.e., maps of the form

$$v \mapsto Mv + w, \qquad M \in GL_2(K), \quad w \in \mathbf{A}^2(K).$$

These are sometimes called affine changes of coordinates, or linear changes of coordinates, and they are benign; all of our definitions should be essentially invariant under such changes of coordinates.

If a line L in the plane intersects an affine curve C at $P = (x_0, y_0)$ then the **intersection multiplicity**

$$I_P(C \cap L)$$

of the point P is the order of vanishing of the polynomial $g(t) := f(x_0 + at, y_0 + bt)$ at $t = 0$, where C is the curve $f = 0$ and the line has been parametrized:

$$L = \{(x_0 + at, y_0 + bt) : t \in \overline{K}\}.$$

Another way to describe this is to say that one obtains the intersections (and corresponding multiplicities) of a line and curve as follows: choose a variable that occurs in the equation of the line (with a nonzero coefficient) and solve that equation for that variable and substitute into the equation of the curve. The resulting polynomial (in one variable) factors according to the intersections.

Example 1.10. The intersections of the line $x + y + 1 = 0$ and the parabola $y = x^2 + x$ can be obtained by solving for y and substituting, leading to the equation $x^2 + 2x + 1 = 0$. Thus a line and parabola intersect with multiplicity two at the point $(-1, 0)$.

Example 1.11. The line $y = x$ and the circle $x^2 + y^2 = 1$ intersect in two points, both $\mathbf{Q}(\sqrt{2})$-rational, but not $\mathbf{Q}$-rational.

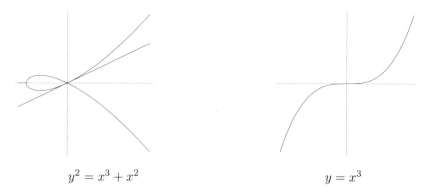

$$y^2 = x^3 + x^2 \qquad\qquad y = x^3$$

The intersection multiplicity $I_P(C \cap L)$ is equal to 0 if the curve and line don't intersect at P, and it is equal to 1 if they intersect transversally at P.

Example 1.12. The nodal cubic $y^2 = x^3 + x^2 = x^2(x+1)$ intersects the line $y = 0$ twice at $x = 0$ and once at $x = -1$. Indeed, if we use the equation $y = 0$ to eliminate y then we get the equation $x^2(x+1) = 0$ in one variable.

One of the tangent lines at the $(0,0)$ on this curve is the line $y = x$. If we use the equation $y = x$ to eliminate the variable y we get $x^3 = 0$, so that that line intersects the curve with multiplicity 3 at $(0,0)$.

Example 1.13. (a point of inflection) The origin is nonsingular on the curve $y = x^3$, and the tangent line is the x-axis $y = 0$. If we eliminate y we get $x^3 = 0$ so that the curve meets its tangent line three times at the origin. Tangent lines ordinarily intersect the curve twice at the point of tangency; if the intersection multiplicity is strictly greater than two, then the point is said to be a **point of inflection**.

If C is a curve, not necessarily irreducible, then the definition of the intersection multiplicity of a line and the curve still makes sense. One thing that can happen is that the result of using the equation of the line to eliminate a variable is that the equation of the curve becomes zero; this means that the equation of the curve is divisible by the equation of the line, in which case the line is said to be a **component** of C.

2. Projective curves

Although we will mostly be concerned with curves in the projective plane in later lectures, it is useful to begin with a brief sketch of curves in projective n-space. Details can be found in many different texts on algebraic geometry including, for instance, [**Sha**] and [**Mum**].

Projective n-space $\mathbf{P}^n(K)$ is defined to be the set of equivalence classes of nonzero $n+1$-tuples $x = (x_1, \cdots, x_{n+1})$ under the relation "scalar multiple" relation

$$x \simeq \lambda x, \qquad \lambda \in K, \lambda \neq 0.$$

Equivalence classes will be written

$$x_1 : x_2 : \cdots : x_{n+1}.$$

We usually think of affine space $\mathbf{A}^n(K) := K^n$ as embedded in $\mathbf{P}^n(K)$

$$\begin{aligned} \mathbf{A}^n(K) &\to \mathbf{P}^n(K) \\ x = (x_1, \cdots, x_n) &\mapsto x_1 : \cdots : x_n : 1. \end{aligned}$$

Any point in projective space with nonzero last coordinate is in the image of this map, and the complement of the image, consisting of all equivalence classes of points with $x_{n+1} = 0$ is isomorphic to $\mathbf{P}^{n-1}(K)$ and is sometimes said to be the hyperplane at infinity.

The projective line $\mathbf{P}^1(K)$ consists of the affine line $\mathbf{A}^1(K) = \{x : 1\}$ together with a single point.

A point in $\mathbf{P}^n(K)$ is defined over a subfield k if some representative of its equivalence class has components lying in the subfield, i.e., if the components lie in k after being multiplied by a suitable nonzero scalar in K. Note the this means that if we dehomogenize (i.e., divide by a nonzero variable), then the resulting $(n+1)$-typle has coordinates in k.

A **change of coordinates** (sometimes called a projective change of coordinates) in projective space is a map of the form

$$x \mapsto Mx$$

where M is a nonsingular $n+1$ by $n+1$ matrix; two such matrices that differ by a scalar determine the same change of coordinates on $\mathbf{P}^n(K)$. If a change of coordinates maps the hyperplane at infinity to itself, then one can check that the restriction of the mapping to affine space $\mathbf{A}^n$ is an affine change of coordinates in affine space.

A nonzero polynomial $f(x)$ in $n+1$ variables is homogeneous of degree d if it is a linear combination of monomials of total degree d, which is equivalent to $f(\lambda x) = \lambda^d f(x)$ for all $\lambda \in \overline{K}$. Homogeneous polynomials are sometimes called **forms**. The zero set of such a polynomial is a well-defined subset of $\mathbf{P}^n(\overline{K})$ since $f(x) = 0$ is equivalent to $f(\lambda x) = 0$. A (projective) **algebraic variety** is, loosely speaking, the zero set in projective n-space over $\overline{K}$ of a collection of homogeneous polynomials. Algebraic varieties can be written as a finite union of **irreducible** varieties which can not be written as finite unions of proper subvarieties. A **projective curve** is an irreducible 1-dimensional algebraic variety. The dimension of an algebraic variety can be defined rigorously either topologically or algebraically; either method requires careful work, and proving their equivalence (over appropriate fields) requires further work.

The concept of a tangent line can be extended to projective curves in n-space, and a projective curve is **nonsingular** or **smooth** if every point has a unique tangent line.

To simplify the notation we now restrict to the case of most interest to us, namely the projective plane, i.e., $n = 2$. It is worth remarking that curves in higher-dimensional projective spaces can be mapped to the plane by "projecting from a point" and, if this is done carefully, many of the properties of the curve are retained. Moreover, curves in the projective plane are the zero sets of a single polynomial so that curves can be defined as forms in 3 variables (up to multiplication by a nonzero scalar). As above, we will assume irreducibility unless the contrary is explicitly allowed.

If $f(x, y) = 0$ is a polynomial in 2 variables of degree d then it can be **homogenized** to a polynomial

$$f^*(x, y, z) := z^d f(x/z, y/z).$$

Any homogeneous polynomial $g(x, y, z)$ not divisible by z is the homogenization of $g(x, y, 1)$.

Thus any affine curve $f = 0$ determines a projective curve $f^* = 0$, and this curve consists of the points in $\mathbf{A}^2$ with $f(x, y) = f^*(x, y, 1) = 0$ as well as the points "at infinity," i.e., on the line $z = 0$, which are found by solving $f^*(x, y, 0) = 0$.

Example 1.14. Often when an affine equation is written, the implicit object of study is the associated projective curve. Thus the equation
$$y^2 = x^3 + ax + b$$
will be taken here to be just shorthand for
$$y^2 z = x^3 + axz^2 + bz^3.$$
The points "at infinity" are those with $z = 0$ and the only such point on this curve is $0:1:0$. Thus the points on the curve consist of the affine solutions $x:y:1$ to $y^2 = x^3 + ax + b$ together with a single point $0:1:0$ at infinity.

The notions of smooth point, and intersection multiplicity with a line, for a point P on a (projective plane) curve $f(x, y, z) = 0$ can be reduced to the case considered earlier by a change of coordinates (i.e., by an invertible 3 by 3 matrix) to move P to $0:0:1$. Then the definitions in the affine case apply to the affine curve $f(x, y, 1) = 0$.

The projective version of the tangent line is straightforward. A point P on $f(x, y, z) = 0$ is smooth if at least one of the partial derivatives of f at P is nonzero, and in that case the tangent line is
$$Ax + By + Cz = 0, \quad A = \partial_x f(P), \quad B = \partial_y f(P), \quad C = \partial_z f(P).$$
The intersections between a line and a curve can be found by eliminating one of the variables (that occurs with a nonzero coefficient in the equation for the line) and factoring the resulting form in two variables.

Example 1.15. Consider the general cubic curve in 2 variables
$$ax^3 + bx^2 y + cxy^2 + dy^3 + ex^2 + fxy + gy^2 + hx + iy + j = 0.$$
Implicitly this means that we are looking at the zero set, in projective space, of the general homogeneous polynomial of degree 3
$$F(x, y, z) = ax^3 + bx^2 y + cxy^2 + dy^3 + ex^2 z + fxyz + gy^2 z + hxz^2 + iyz^2 + jz^3.$$

- When does the point $O = 0:1:0$ lie on this curve? By direction substitution, this happens exactly when $d = 0$.
- What does it mean for the point O to be nonsingular on the curve? The degree 1 part of the Taylor expansion at O is $cx+gz$, so the point is nonsingular if at least one of c or g is nonzero.
- What does it mean to say that the tangent line at O is the line at infinity? The line at infinity is the line $z = 0$ so this is the tangent line if $c = 0$ and g is nonzero.
- Finally, what does it mean for O to be a point of inflection, given that $z = 0$ is the tangent line? Eliminating z between the equations z and the cubic gives $ax^3 + bx^2 z = 0$. This has three equal roots only if $b = 0$.

These remarks give the following result.

Lemma 1.16: A plane cubic curve has the form
$$E: \quad y^2 + a_1 xy + a_3 y = x^3 + a_2 x^2 + a_4 x + a_6$$

if and only if the point $O := 0\!:\!1\!:\!0$ is a smooth point on the curve, is an inflection point, and has the line $z = 0$ at infinity as the tangent line to the curve at that point.

An equation of this form does not necessarily define an elliptic curve since it might have a singular point P, $P \neq O$.

3. The group operation

Let $E(K)$ be the solution set to a Weierstrass equation

$$y^2 + a_1 xy + a_3 y = x^3 + a_2 x^2 + a_4 x + a_6.$$

There are three intersections (counting multiplicity) of a line and E in the projective plane (over an algebraic closure of K). If two of the three intersections are defined over K then so is the third. (If a cubic form has two factors with coefficients in K, then its third factor does also.)

This suggests that we try to define an operation on the points $E(K)$

$$(P, Q) \mapsto P * Q := \text{the third intersection of the line } PQ \text{ with } E(K).$$

Here we take PQ to be the line joining P and Q if they are distinct, and the tangent line to E at P if $P = Q$. Since this requires the existence of a unique tangent line at every point, we see that E has to be nonsingular in order for this operation to be defined everywhere.

Anytime there is a binary operation on a set it is natural to inquire whether or not this operation satisfies the axioms of a group. It is easy to see that the above operation does not make $E(K)$ into a group. The set $E(K)$ might have more than one inflection point, so that there would have to be more than one identity, and the operation isn't associative. However, $E(K)$ does turn out to be an abelian group under the operation

$$P + Q = O * (P * Q)$$

where $O = 0\!:\!1\!:\!0$. Commutative is obvious, and O is the identity (O is a point of inflection, so $O * O = O$). Moreover, it is easy to find an inverse of P, namely $P * O$. Indeed,

$$P + (P * O) = P * (P * O) - O.$$

The only thing remaining in order to show that $E(K)$ is a group is the associative law. One (unappetizing) way to verify the associative law is to convert the geometric definition into a purely algebraic one and then verify the associative law directly from the explicit formulas. This can be done, but requires the consideration of various special cases. A more conceptual method, relying on deeper results in algebraic geometry, will be sketched at the end of this lecture.

In any event, this so-called "chord-tangent" addition law can be used to generate new points on the curve, and is at the base of the numerous applications of elliptic curves. If $K = \mathbf{Q}$ is the field of rational numbers, then a celebrated theorem, implicitly conjectured by Poincaré, later proved by Mordell and then vastly generalized by Weil, says that for any elliptic curve over the rational numbers there is a finite set of points from which all others can be derived by this chord-tangent process.

4. Rational maps

In order to sketch the standard proof of Theorem 1.1 (giving canonical equations and the group law) it is useful to work with the field of rational functions on a curve, and to use (part of) the Riemann-Roch theorem. We give a brief sketch of some of the general definitions, and then consider rational functions on a curve in detail; the reader may find it useful to keep the case rational functions on the line in mind. Thorough developments of this material can be found in (among many other sources), such as [**Ful**], [**Sha**], [**Mum**], [**Lor**], etc.

If V is an irreducible variety defined over K then a rational mapping from V to $\mathbf{P}^m$ is an equivalence class of m forms (with coefficients in $\overline{K}$) in $n+1$ variables

$$g = g_0 : \cdots : g_m$$

of the same degree, such that at least one of them does not vanish identically on V. Two such collections are equivalent if they give the same result where they are both defined; i.e., g is equivalent to (defines the same rational map as) G if

$$g_i(x) G_j(x) = g_j(x) G_i(x)$$

for all i, j and $x \in V$. The **domain of definition** of the rational map is the set of points x such that there is some collection of forms g for which $g(x)$ is defined in the sense that $g(x)$ is a nonzero $(m+1)$-tuple. In other words, the domain of definition is the maximal set on which g defines a mapping from $V(\overline{K})$ to $\mathbf{P}^m(\overline{K})$ in the set-theoretic sense. A rational mapping is a **morphism** if its domain of definition is all of V. Later we will need the nontrivial fact that a non-constant rational map defined on a smooth curve is actually a morphism.

A rational map is **defined over** a subfield k of K if some collection g of forms has coefficients in the subfield k. A rational map defined over k is also defined over any field containing k.

Example 1.17. The rational map

$$g(x_0 : x_1 : x_2) = x_0 x_1 : x_0 x_2 : x_1 x_2$$

from $\mathbf{P}^2$ to $\mathbf{P}^2$ is defined everywhere except at the three points $1:0:0$, $0:1:0$, and $0:0:1$.

A rational mapping is **birational** if there is a rational mapping that is inverse to it, wherever both are defined, e.g., $f(g(x)) = x$ if x is in the domain of g and $g(x)$ is in the domain of f.

A **rational function** on a variety V defined over K is a rational map from V to $\mathbf{P}^1$. Usually $f = f_0 : f_1$ is written as $f(x) = f_0(x)/f_1(x)$ and f has a **zero** at x if $f_0(x) = 0$, $f_1(x) \neq 0$, and has a **pole** if $f_0(x) = 1$, $f_1(x) = 0$. The set of all rational functions on V is a field (if we include the zero map as an honorary rational map) called the function field of V and is denoted $K(V)$. Here the irreducibility of V is needed in order for $K(V)$ to be a field.

If f is a rational function on a curve V and P is a smooth point of V then there is a well-defined "order of vanishing" of f at P

$$\operatorname{ord}_P(f) \in \mathbf{Z} \cup \{\infty\}$$

where we set $\operatorname{ord}_P(0) = \infty$ by convention. More precisely there is a rational function u_P (called a uniformizer at P) such that $u_P(P) = 0:1$, and every nonzero rational function can be written

$$f = u_P^n g$$

where n is an integer, and g is a rational function whose value at P is neither 0 nor infinity (i.e., neither $0:1$ nor $1:0$). The integer n is the order of vanishing $n = \text{ord}_P(f)$ of f at P. If $\text{ord}_P(f) < 0$ then f has a **pole** of order $-\text{ord}_P(f)$ at P, and if $\text{ord}_P(f) > 0$ then f has a **zero** of order $\text{ord}_P(f)$ at P.

Example 1.18. Rational functions on the projective line are familiar:
$$f(x:y) = F(x,y):G(x,y) = \frac{F(x,y)}{G(x,y)}.$$

In order for this to give a well-defined function on $\mathbf{P}^1$ the forms F and G have to have the same degree; without loss of generality we can also assume that they are relatively prime. These forms factor into linear forms (over an algebraically closed field), that correspond to points on the line; the zeros of f correspond to roots of the numerator F and the poles of f correspond to the roots of the numerator G.

5. Genus

The natural topological definition of genus of a curve C (when C is defined over a subfield K of the complex numbers) follows from the fact that the points of C with complex coordinates form a one-dimensional complex manifold, and hence a compact orientable surface. Any compact orientable surface is homeomorphic to a sphere with some number g of handles attached, and g is called the **genus** of the curve. The genus can be calculated by taking a triangulation of a surface and applying Euler's formula which says that the "Euler characteristic" of the surface is
$$2 - 2g = v - e + f$$
if there are v vertices, e edges, and f faces in the triangulation.

This topological definition is natural, especially when algebraic curves are viewed as Riemann surfaces. But it is not ultimately sufficient for our purposes since we will want to consider, for example, finite fields. It is possible to give a purely algebraic definition of the genus and, with quite a bit of work, it can be shown that the algebraic and topological definitions are equivalent for curves defined over (a subfield of) the complex numbers.

The genus of a curve can be defined algebraically by projecting it into the plane, and analyzing the singularities in the plane carefully (see [**BK**]). It can be shown that it is possible to find a birational map from a curve to a plane curve whose only singularities (if any) are nodes, i.e., singular points with two distinct tangent lines.

Def 1.19: If C is a plane projective curve of degree d and there are n nodes and no other singular points, then the **genus** of C is
$$g := \frac{(d-1)(d-2)}{2} - n.$$

For instance, the Fermat cubic $x^3 + y^3 = z^3$ is a smooth curve of degree 3, and by the definition has genus equal to 1.

The genus is invariant under birational mappings. Moreover, if C is any smooth projective curve then there is a rational map from C to $\mathbf{P}^2$ whose image is as in the theorem.

Another algebraic formulation of the genus, more suitable to our current purposes, is implicit in (part of) the Riemann-Roch theorem. Assume for the rest of this section that K is an algebraically closed field and that C is a smooth projective curve.

A **divisor** is a finite formal integral linear combination of points

$$D = \sum_{P \in C(K)} a_P P, \qquad a_P \in \mathbf{Z}.$$

We are interested in the vector space of functions whose poles are "no worse than D" in the sense that $\mathrm{ord}_P(f) \geq -a_P$ (of course, if a_P is negative this requires that f have a zero of order at least a_P at P). We denote this vector space by $L(D)$, i.e.,

$$L(D) = \{f \in K(C) : \mathrm{ord}_P(f) \geq -a_P\}.$$

(We take $f = 0$ to be in $L(D)$ by convention.) It is a basic fact that this is a finite dimensional vector space. For instance, if $D = 0$ then $L(D)$ consists of rational functions that have no poles, and it is known that the only such functions are constant functions, i.e., $\dim L(0) = 1$.

The **degree** of a divisor D is

$$\deg(D) = \sum_P a_P.$$

Every rational function f determines its divisor $\mathrm{div}(f)$ of its zeros and poles,

$$\mathrm{div}(f) := \sum_P \mathrm{ord}_P(f) P,$$

and a fundamental theorem says the degree of this divisor is 0. The only functions whose divisors are trivial are the constant functions.

Two divisors are said to be **linearly equivalent** if they differ by the divisor of a function, i.e.,

$$D = D' + (f)$$

where (f) denotes the divisor of zeros and poles of a rational function f. Linearly equivalent divisors have the same degree.

Example 1.20. ($\mathbf{P}^1$) Let $P = 0:1 \in \mathbf{P}^1$, and consider $D = nP$, so that $L(nP)$ consists of rational functions f on the projective line whose only pole is at $P = 0:1$ and the order of the pole at most n. Any rational function in $L(nP)$ can be written

$$f(x:y) = \frac{F(x:y)}{x^n}$$

where F is an arbitrary form of degree n. Forms (in two variables) of degree n have $n+1$ coefficients, so the dimension of the space of such forms is $n+1$ and

$$\dim L(nP) = n + 1.$$

More generally, the Riemann-Roch theorem has the following theorem/definition as a special case.

> **Theorem/Definition 1.21:** If C is a nonsingular curve then there is an integer g, called the genus $g = g(C)$ of C, such that if $\deg(D) > 2g - 2$ then
> $$\dim L(D) = \deg(D) + 1 - g.$$

Remark 1.22. The projective line has genus 0, by the calculation above.

Remark 1.23. The full Riemann-Roch theorem gives a formula for the dimension of $L(D)$, for arbitrary D, in terms of the degree, genus, and dimension of $L(D_{can} - D)$, where D_{can} is the "canonical divisor" on the curve.

6. Weierstrass form and the group law via divisors

The preceding results and definitions can be used to give a definition of an elliptic curve, and a sketch of the proof that elliptic curves can be put into Weierstrass form, and the proof that the points form an abelian group.

Def 1.24: An elliptic curve over a field K is a smooth projective curve E over K of genus 1, together with a specified rational point O. Two elliptic curves are isomorphic if there is a non-constant birational map from one curve to the other that maps the specified point on one curve to the specified point on the other curve.

According to the Riemann-Roch Theorem we have
$$\dim L(2O) = 2, \quad \dim L(3O) = 3.$$
This theorem applies to E as a curve over the algebraic closure $\overline{K}$, but one finds that the vector spaces $L(nO)$ are defined over K if O is, in the sense that they have a basis of rational functions F/G where the coefficients of F an G are in K.

Let x be a non-constant element of $L(2O)$, defined over K, so that $\{1, x\}$ is a basis of $L(2O)$.

The vector space $L(3O)$ contains 1 and x and, since it is 3-dimensional, an additional element y, also defined over K. Thus $L(3O)$ has basis $\{1, x, y\}$.

There are 7 obvious elements of the 6-dimensional vector space $L(6O)$:
$$1, x, y, x^2, xy, x^3, y^2.$$
These elements must be linearly dependent; moreover, the coefficients of x^3 and y^2 in a linear dependence relation must be nonzero since those are the only two elements with poles of order 6. By suitable scaling we can assume that the two coefficients are ± 1 so as to put the equation into Weierstrass form. One then has to check that the map
$$P \mapsto (x(P), y(P))$$
gives an embedding of E into the projective plane such that the image is a nonsingular plane cubic in canonical form (see [**Si1**] for details).

Remark 1.25. In practice, if one starts with an elliptic curve in nonstandard form, such as some of the examples at the beginning of this lecture, it is usually best to either (a) use one of the numerous software packages for elliptic curves to compute the corresponding Weierstrass equation, or (b) consult the explicit algebraic reductions described in [**Cas**].

What about the group law? Again, the basic idea is easy, though there are numerous details that have to be checked. Assume that K is algebraically closed. The degree 0 **Picard** group of E, $\text{Pic}^0(E)$, is defined to be the equivalence classes of divisors of degree 0, under the relation of being linearly equivalent. We define a map from $E(K)$ to $\text{Pic}^0(E)$ by
$$P \mapsto [P - O].$$
This map is injective: if $P - O = \text{div}(f) + Q - O$ then $P = \text{div}(f) + Q$ and $f \in L(Q)$. However, $L(Q)$ is one-dimensional, by the preceding theorem, and it contains the constant functions. So f is constant and $P = Q$. On the other hand, this map is surjective: if D is a divisor of degree 0 then the divisor $D + O$ has degree 1, and $L(D + O)$ has dimension 1. If f is a nonzero element of that vector space, then

$\operatorname{div}(f) \geq -D - O$ so there is a point P such that $\operatorname{div}(f) = -D - O + P$ which says that the class of D is the image of the point P.

It can be shown that if k is a subfield of K then this bijection takes k-rational points to divisors that are "defined over k" in the sense that they are fixed under all automorphisms of K that fix k.

This gives a bijection
$$E(K) \to \operatorname{Pic}^0(E).$$
The right hand side is a group in a natural way, and we can pull back the group structure to $E(K)$. Since O is mapped to the zero divisor class, the identity of this group operation on $E(K)$ is O. To be explicit about the group operation on $E(K)$ we need to assume that E is given in Weierstrass form. Suppose that P, Q, and R are colinear on $E(K)$, and let L be the linear form corresponding to the line, defined over K, that intersects the curve in those three points. Then L/Z is a rational function on E whose divisor is
$$\operatorname{div}(L/Z) = P + Q + R - 3O$$
(since O is a point of inflection and the tangent line $Z = 0$ intersects the curve with multiplicity 3 at O). This means that
$$[P - O] + [Q - O] + [R - O] = 0 \in \operatorname{Pic}^0(E)$$
and we see that the natural group structure on $\operatorname{Pic}^0(E)$ pulls back to the unique group structure on $E(K)$ in which O is the identity, and three points sum to 0 if and only if they are colinear. More generally, one can supply $E \times E$ with the structure of a projective variety, such that there is a morphism of varieties
$$m \colon E \times E \to E$$
defined over K that realizes the above group law, which in turn reduces to the chord-tangent law on the points of E in any extension field of K.

Ex 1.26. Use a symbolic computation program to plot graphs of elliptic curves $y^2 = x^3 + ax + b$, for various choices of $a, b \in \mathbf{Z}$, in the real coordinate plane.

When does this graph have two components?

Prove that in the case that there are two components, that the bounded one is *not* an ellipse.

When does the unbounded component of an elliptic curve $y^2 = x^3 + ax + b$ have a "pinch"? (i.e., the height y of points (x, y) on the unbounded component, $y > 0$, is not a monotonically increasing function of x).

Ex 1.27. Find all singular points on the (projectivization of the) curve $y^2 = x^4 + 1$.

Ex 1.28. Roughly sketch a graph of $x^3 + y^3 = a$, for a positive rational number a. Find a linear change of coordinates that converts this elliptic curve to canonical form.

Ex 1.29. Show that a curve $y^2 = f(x)$ is irreducible over a field K if $f(x)$ has distinct roots in $\overline{K}$, and the characteristic of K isn't equal to 2.

Ex 1.30. Express the addition law on $y^2 = x^3 + ax + b$ completely in algebraic form.

Ex 1.31. Find several points on $y^2 = x^3 - 4x + 4$ that do *not* have integer coordinates. (Later we will see that points of finite order have integer coordinates, so the points that you find in fact all have infinite order.)

Ex 1.32. Write a Mathematica or Maple or Pari (or ...) script that verifies the associative law for elliptic curve addition; you can assume that all points that arise are distinct.

Ex 1.33. (suggested by Hendrik Lenstra) Prove "Fermat's First Theorem": The equation $x^3 - y^3 = a$ has a solution in positive rational numbers x, y if and only if $x^3 + y^3 = a$ has a solution in positive rational numbers.

LECTURE 2
Points on Elliptic Curves

The study of the group of points $E(\mathbf{Q})$ of an elliptic curve over the field of rational numbers is one of the fundamental objects of study in arithmetic algebraic geometry.

In order to construct points it is natural to begin by searching for points with small integer coordinates; additional points can be constructed by using the group operation. One finds that all non-integer points have the form $(u/d^2, w/d^3)$ where u, v, d are integers. In an attempt to prove that this is generally true we are led to looking at the equations for E modulo primes or powers of primes. This leads to a "p-adic" analysis of elliptic curves. The goal of this lecture is to outline the basic facts, and point out their consequences for points of finite order in $E(\mathbf{Q})$. Again, this is an entirely standard story that can be found in almost all of the general references on elliptic curves given in the first lecture.

1. Finding points

Consider the elliptic curve E over $\mathbf{Q}$ defined by the equation
$$y^2 = x^3 - 9x + 9.$$

How can we go about finding points on this curve? The only idea that comes to mind immediately is brute-force search.

In fact, if we plug in integer values of x into the cubic polynomial on the right hand side to see when we get a perfect square, we find the points

$$P_{-3}^\pm = (-3, \pm 3), \quad P_0^\pm = (0, \pm 3), \quad P_1^\pm = (1, \pm 1)$$
$$P_3^\pm = (3, \pm 3), \quad P_7^\pm = (7, \pm 17), \quad P_{15}^\pm = (15, \pm 57).$$

Further searching doesn't seem to yield any further points with integer x. Of course, points in $E(\mathbf{Q})$ might have rational, but non-integral, coordinates.

In any event, we can generate further points by applying the group law, and it is convenient to have explicit formulas for the algebra underlying the group operation.

To find $2P$ where $P = (x, y)$ is a point on the curve $y^2 = x^3 - 9x + 9$ we must start by finding the slope of the tangent line at P. Differentiating $y^2 = x^3 - 9x + 9$

implicitly, we find that the slope of the tangent line at $P = (x, y)$ is

$$m = \frac{dy}{dx} = \frac{3x^2 - 9}{2y}$$

and the equation of the tangent line is

$$Y - y = m(X - x).$$

The three intersections of this line with E have x-coordinates that are roots of the cubic

$$(y + m(X - x))^2 = X^3 - 9X + 9.$$

This cubic equation has the form

$$0 = X^3 - m^2 X^2 + \cdots.$$

The sum of the three roots of a monic cubic polynomial is the negative of the next-to-leading coefficient. Taking into account that x is a root of multiplicity two (since the tangent line intersects E at least twice) we have

$$2x + x' = m^2$$

where x' is the x-coordinate of $2P$. Thus

$$x' = m^2 - 2x.$$

Substituting into the equation for the tangent line gives

$$y' = -(y + m(x' - x)) = mx - y - mx'$$

(remembering that the y-coordinate of $2P$ is the negative of the y-coordinate of $P * P$).

To add two distinct points we follow the same basic procedure, except that the slope m is just given by the usual formula for the slope of the line between two points.

Using this algebra we can produce many new points:

$$2P^+_{-3} = P^+_{15}, \quad 2P^+_0 = (9/4, 3/8), \quad 2P^+_3 = P^-_3,$$
$$2P_7 = (715/289, 6731/4913), \quad 2P^+_{15} = (1491/361, 44601/6859)$$
$$3P^+_{-3} = (-8/9, 109/27), \quad P^+_{-3} + P^+_7 = (-51/25, -543/125).$$

In addition the points P^+_{-3}, P^+_0, P^+_3 obviously add to O since they lie on the line $y = 3$. The point $P_{3,3}$ is of order 3.

If we add a point to itself repeatedly we either get O, if the point is a point of finite order (called a torsion point), or go on forever to get infinitely many points. Later in this lecture we will learn that points of finite order on this curve have integer coordinates, so all of the points on the above list have infinite order except $P^\pm_3$, since the other points, or some multiple of them, have non-integer coordinates.

A careful study of the list shows that in fact every point can be written

$$aP^+_3 + bP^+_0, \quad a \in \{0, 1, 2\}, \quad b \in \mathbf{Z}.$$

Thus we have discovered that

$$\mathbf{Z}/3\mathbf{Z} \times \mathbf{Z}$$

is a subgroup of $E(\mathbf{Q})$.

An even closer look at the shape of the points that have been produced shows that every point has the shape $(u/d^2, v/d^3)$ for integers u, v, d. To prove this, we

consider the power of p dividing the denominators of coordinates, for each prime p. If x is a nonzero rational number then define the "power of p dividing x" by
$$\mathrm{ord}_p(x) = n, \quad \text{if } x = p^n \frac{r}{s}, \quad n \in \mathbf{Z}$$
where p does not divide either r or s.

Theorem 2.1: Let p be a prime, and let $E : y^2 = x^3 + ax + b$ be an elliptic curve with $a, b \in \mathbf{Z}$. If $P = (x, y)$ is a non-integer point and p divides one of the denominators then there is a positive integer n such that
$$\mathrm{ord}_p(x) = -2n, \quad \mathrm{ord}_p(y) = -3n.$$

Proof. We write the point projectively as $P = X : Y : Z$ where X, Y, Z are integers, not all divisible by p. The projective equation
$$Y^2 Z = X^3 + aXZ^2 + bZ^3$$
shows that p divides X since it divides Z by the assumption that one of the denominators of the inhomogeneous coordinates is divisible by p. Therefore p does not divide Y. This equation can be true only if there are two terms divisible by the same power of p. Since the last two terms, aXZ^2 and bZ^3 are divisible by strictly higher powers of p than $Y^2 Z$ it follows that the two terms divisible by equal powers of p must be $Y^2 Z$ and X^3; therefore, using $\mathrm{ord}_p(Y) = 0$, we get
$$\mathrm{ord}_p(Y^2 Z) = \mathrm{ord}_p(Z) = \mathrm{ord}_p(X^3) = 3\,\mathrm{ord}_p(X).$$
Let $n = \mathrm{ord}_p(X)$. Then
$$\mathrm{ord}_p(x) = \mathrm{ord}_p(X/Z) = -2n, \text{ and } \mathrm{ord}_p(y) = \mathrm{ord}_p(Y/Z) = -3n.$$
$\square$

Since this is true for every prime, it follows that a rational point (x, y) has the form $(u/d^2, v/d^3)$ as we observed experimentally for the particular curve above.

For a fixed prime p and positive integer n let E_n be the set of points where the denominator of x is divisible by (at least) p^n:
$$E_n := \{(x, y) \in E(\mathbf{Q}) : \mathrm{ord}_p(x) \leq -2n\}.$$
These groups are subgroups, and a careful analysis of their structure reveals that all points of E_1 have infinite order for curves $y^2 = x^3 + ax + b$.

2. Minimal discriminants

In considering elliptic curves over $\mathbf{Q}$ it is sometimes convenient to start with an equation that is "as nice as possible" among all equations with integral coefficients for a given curve.

It can be shown that two elliptic curves over $\mathbf{Q}$ in Weierstrass form
$$E : y^2 + a_1 xy + a_3 y = x^3 + a_2 x^2 + a_4 x + a_6, \quad a_i \in \mathbf{Q}$$
are isomorphic by a birational map defined over $\mathbf{Q}$ if and only if there is a birational map between them of a very simple form:
$$x = u^2 x' + r, \quad y = u^3 y' + s u^2 x' + t$$
where r, s, t, u are in $\mathbf{Q}$ and u is nonzero. One checks easily that coordinate changes of this form turn one Weierstrass equation into another.

Following standard conventions, we define auxiliary quantities

$$b_2 = a_1^2 + 4a_2, \quad b_4 = 2a_4 + a_1 a_3, \quad b_6 = a_3^2 + 4a_6,$$

$$b_8 = a_1^2 a_6 + 4a_2 a_6 - a_1 a_3 a_4 + a_2 a_3^2 - a_4^2,$$

$$c_4 = b_2^3 - 24 b_4, \quad c_6 = b_2^3 + 36 b_2 b_4 - 216 b_6,$$

and define the **discriminant** of E to be

$$\Delta(E) = -b_2^2 b_8 - 8 b_4^3 - 27 b_6^3 + 9 b_2 b_4 b_6 = (c_4^3 - c_6^2)/1728.$$

Remark 2.2. The discriminant of $y^2 = x^3 + a_4 x + a_6$ is $\Delta = -16(4a_4^3 + 27 a_6^2)$.

Under the change of coordinates above, one finds (using obvious notation) that Δ, c_4, and c_6 change as

$$c_4 = u^4 c_4', \quad c_6 = u^6 c_6', \quad \Delta = u^{12} \Delta'.$$

The **j-invariant** of the curve is defined to be

$$j(E) = c_4^3 / \Delta(E).$$

Unlike the discriminant, the j-invariant is invariant under coordinate changes. In other words, if one uses the above blizzard of formulas to compute $j(E)$, then isomorphic curves give the same result.

Any elliptic curve over $\mathbf{Q}$ is isomorphic to a curve with integral coefficients a_i, by scaling if necessary to get rid of denominators. Among all equations with integer coefficients for E, ones with the smallest possible power of p dividing the discriminant are said to be a **minimal model for E at** p, and we often prefer to work with these models when discussing the "reduction" of elliptic curves.

Remark 2.3. It isn't too hard to prove that there is a model that is minimal for all primes [**Kna**, p. 290], [**Si1**], although the analogous assertion is not true over arbitrary number fields. This global minimal model is unique if we also require $a_1 \in \{0, 1\}$, $a_2 \in \{0, 1, 2\}$.

Remark 2.4. If $p > 3$ it is very easy to find a model that is minimal at p: an integral model is minimal at p if and only if the discriminant is not divisible by p^{12}. More generally, an algorithm due to Laska finds the minimal model easily ([**Lsk**], [**Cre**]).

Remark 2.5. The global minimal model should not be confused with the *Néron model* [**Si1, Si2**] which is a more elaborate object.

3. Reduction

Now return to the situation where we have a fixed elliptic curve E over $\mathbf{Q}$, and a fixed prime p. We assume that we have a specific equation for E that is of minimal discriminant at p.

If x is an integer then we let $\bar{x}$ denote the image of x in the field $\mathbf{F}_p :=$ $\mathbf{Z}/p\mathbf{Z}$ of p elements. If $P = x\colon y\colon z$ is a point in $\mathbf{P}^2(\mathbf{Q})$ then by multiplying by a suitable nonzero rational number we can assume, without loss of generality, that all components are integers and at least one of them is not divisible by p. If $x\colon y\colon z$ is such a representative of $P \in \mathbf{P}^2(\mathbf{Q})$ then the map

$$x\colon y\colon z \mapsto \bar{x}\colon \bar{y}\colon \bar{z}$$

is a well-defined mapping from $\mathbf{P}^2(\mathbf{Q})$ to $\mathbf{P}^2(\mathbf{F}_p)$, called the reduction map modulo p.

If E is an elliptic curve over $\mathbf{Q}$ in Weierstrass form then we can then consider the curve $\overline{E}$ defined over $\mathbf{F}_p$ whose coefficients are $\overline{a_i} := a_i \bmod p$. Thus we obtain $\overline{E}$ by "reading the equation for E modulo p." This curve is called the **reduction** of E modulo p.

If the discriminant of E is not divisible by p then the discriminant of $\overline{E}$ is nonzero (as an element of $\mathbf{F}_p$) and $\overline{E}$ is an elliptic curve over $\mathbf{F}_p$. If the discriminant of E is divisible by p then $\overline{E}$ is singular, and it has a unique singular point (that can be checked to be $\mathbf{F}_p$-rational).

Let $\overline{E}_{ns}$ denote the $\mathbf{F}_p$-rational points which are nonsingular on $\overline{E}$ (as a curve over $\mathbf{F}_p$). Then the chord-tangent law can be used exactly as before to define a structure of a finite abelian group on $\overline{E}_{ns}$ (recall that on an irreducible cubic the only thing that we required was a unique tangent line in order to define $P * Q$).

Theorem/Definition 2.6 (Reduction Types): Let E be the equation of an elliptic curve which is minimal at p. Then one of the following cases holds:

1. The discriminant $\Delta(E)$ is not divisible by p. Then $\overline{E}$ is an elliptic curve defined over $\mathbf{F}_p$, and we say that E is has **good** reduction at p. In this case, the elliptic curve $\overline{E}$ is independent (up to isomorphism) of the choice of the minimal Weierstrass equation that was reduced.
2. The curve $\overline{E}$ is singular and has a cusp; the group $\overline{E}_{ns}$ is isomorphic to the additive group of $\mathbf{F}_p$ and is therefore cyclic of order p. In this case, E is said to have **additive** reduction at p.
3. The curve $\overline{E}$ is singular and has a node at which the two tangent lines are both defined over $\mathbf{F}_p$; the group $\overline{E}_{ns}$ is isomorphic to the multiplicative group of $\mathbf{F}_p$, so that it is cyclic of order $p-1$. In this case, the curve is said to have **split multiplicative** reduction at p.
4. The curve $\overline{E}$ is singular and has a node at which the two tangent lines are not defined over $\mathbf{F}_p$; the group $\overline{E}_{ns}$ is cyclic of order $p+1$. In this case, the curve is said to have **non-split multiplicative** reduction at p.

The proof that the reduced groups have the given form is an easy exercise in the chord-tangent law group operation over finite fields.

4. The p-adic filtration

Fix an elliptic curve E over $\mathbf{Q}$, given in minimal form, and fix a prime p. Let E_0 denote the set of points in $E(\mathbf{Q})$ whose image under reduction is nonsingular.

If $L : rx + sy + tz = 0$ is a line defined by an equation with rational coefficients r, s, t then by multiplying the equation by a suitable nonzero rational number we can assume that the coefficients r, s, t are integers and that at least one of them is not divisible by p. The equation $\overline{L} : \overline{r}x + \overline{s}y + \overline{t}z = 0$ determines a line in the projective plane over $\mathbf{F}_p$, defined over $\mathbf{F}_p$, called the reduction of L modulo p.

The algebra involved in computing an intersection multiplicity $I_P(E \cap L)$ reduces modulo p to the algebra involved in computing $I_{\overline{P}}(\overline{E} \cap \overline{L})$. (For simplicity, we assume that all intersections are $\mathbf{Q}$-rational, which is the case for intersections needed for computing the group law over $\mathbf{Q}$.) Therefore

$$I_P(E \cap L) \leq I_{\overline{P}}(\overline{E} \cap \overline{L})$$

since the multiplicity of a root (of a polynomial with integer roots) can only increase when the coefficients are reduced modulo p. However, the sum of the multiplicities is 3 ($\overline{E}$ is irreducible and therefore does not contain a line). Therefore, the multiplicities don't change under reduction!

The group law on both E_0 and $\overline{E}_{ns}$ is defined in terms of intersection multiplicities. So E_0 is a subgroup, since $P + Q$ reduces to $\overline{P} + \overline{Q}$, and the reduction map
$$E_0 \to \overline{E}_{ns}$$
is a group homomorphism. With considerable work, one can prove more.

Theorem 2.7: The subgroup E_0 is of finite index in the full group $E(\mathbf{Q})$ of rational points.

The kernel of the reduction map
$$E_0 \to \overline{E}_{ns}$$
is just the subgroup E_1 referred to earlier. Indeed, if a point $x\!:\!y\!:\!z$ maps to $0\!:\!1\!:\!0$ then z is divisible by p, and p divides the denominator of the point $(x/z, y/z)$ when expressed in inhomogeneous coordinates.

A detailed analysis of the filtration
$$E_1 \supset E_2 \supset E_3 \supset \cdots$$
can be used to show the following.

Theorem 2.8: If E has the form $y^2 = x^3 + ax + b$ then the group E_1 is torsion-free (i.e., has no points of finite order).

The most natural context for proving this arises in looking at the "formal group" associated to the elliptic curve ([**Si1**]), but more elementary proofs can be found in many of the standard introductions to elliptic curves.

The following corollary enables us to find torsion points easily, at least on curves of the form $y^2 = x^3 + ax + b$.

Corollary 2.9 (Nagell-Lutz): If $P = (x, y)$ is a torsion point on an elliptic curve
$$y^2 = x^3 + ax + b, \qquad a, b \in \mathbf{Z}$$
then x and y are integers, and either $y = 0$ (so that P is a point of order 2) or else y^2 divides $D = 4a^3 + 27b^2$. If p does not divide the discriminant of the curve then the reduction map from the group $E(\mathbf{Q})_{\text{tors}}$ to $E(\mathbf{F}_p)$ is injective.

Proof. For all primes p the theorem says that p does not divide the denominator of x, and y, so they are integers.

If P has order 2 then $y = 0$. Otherwise $2P = (x', y')$ is a torsion point so x' and y' are integers. From our formula for the double of a point we have $2x + x' = m^2$ where $m = (3x^2 + a)/2y$. Thus
$$2x + x' = \frac{(3x^2 + a)^2}{4y^2} = \frac{(3x^2 + a)^2}{4(x^3 + ax + b)}$$
and y^2 divides $(3x^2 + a)^2$. However the (apparently) magical identity
$$(-27x^3 - 27ax + 27b)(x^3 + ax + b) + (3x^2 + 4a)(3x^2 + a)^2 = 4a^3 + 27b^2$$
shows that $y^2 = x^3 + ax + b$ divides D as claimed.

The last assertion follows from the observation that if E has good reduction at p then $E(\mathbf{Q}) = E_0$ so that the reduction map is defined on all torsion points, and no non-trivial torsion point is in the kernel of reduction. $\square$

Remark 2.10. If the curve has the more general Weierstrass form then it is possible to have torsion points with non-integral coordinates, but a slightly more general result holds and it is easy to use it to find all torsion points in a straightforward way ([**Si1**], [**Kna**]).

Using the full force of modern Grothendieck-style algebraic geometry, Mazur was able to completely classify the possible torsion groups over the rational numbers.

Theorem 2.11 (Mazur's Theorem): If E is an elliptic curve over $\mathbf{Q}$ then the group $E(\mathbf{Q})_{\text{tors}}$ of points of finite order on E is one of the following groups:

$$\mathbf{Z}/n\mathbf{Z}, \quad 1 \leq n \leq 10, n = 12$$
$$\mathbf{Z}/n\mathbf{Z} \times \mathbf{Z}/2\mathbf{Z}, \quad n = 2, 4, 6, 8.$$

Ex 2.12. Prove that the curve $y^2 = x^3 + ax + b$, for $a, b \in \mathbf{Z}$, is singular if and only if the discriminant $\Delta(E) = -16(4a^3 - 27b^2)$ is zero.

Ex 2.13. Describe all points P, Q, R on an elliptic curve such that the "star" operation is associative, i.e.,

$$P * (Q * R) = (P * Q) * R.$$

Ex 2.14. Find all $\mathbf{Q}$-rational torsion points on $y^2 = x^3 + 1$.

Ex 2.15. Find all $\mathbf{Q}$-rational torsion points on $y^2 = x(x+2)(x+6)$.

Ex 2.16. Find an elliptic curve over $\mathbf{Q}$ and a prime p where E has split multiplicative reduction at p. Find an elliptic curve and a prime where the reduction is non-split multiplicative.

Ex 2.17. Prove that if E is an elliptic curve over the complex numbers $\mathbf{C}$ then the group

$$E[3] := \{P \in E(\mathbf{C}) : 3P = O\}$$

is of order 9. Find an elliptic curve E over $\mathbf{Q}$ such that $E[3]$ contains $\mathbf{Q}$-rational points, other than the identity. (Hint: a point P is of order 3 if $2P = -P$.)

Ex 2.18. Find an elliptic curve over $\mathbf{Q}$ with a point of order 4. Find one with a point of order 5.

Ex 2.19. Find a $\mathbf{Q}$-rational point P of infinite order on $y^2 = x(x+2)(x+6)$. Show that a point nP determines a solution to the "Gentleman's Diary" problem if n is even. Use this to find the first few terms of an infinite sequence $G_n = (a_n, b_n)$ solutions to the "Gentleman's Diary" problem introduced in the first lecture, where G_n corresponds to $2nP$. Find a recursion for G_{n+1} in terms of G_n.

LECTURE 3
Elliptic Curves over C

In studying mathematical objects of a given type up to isomorphism it is often fruitful in algebraic geometry to endow the set, or "moduli space", of all isomorphism classes with a mathematical structure. The goal of this lecture is to consider the set of isomorphism classes of elliptic curves over the complex numbers, and to give several different descriptions of this set.

The various tools that arise will be useful when we look at modular forms in subsequent lectures. The story is very well-known (dating from the nineteenth century) and expositions can be found in many places. An especially concise treatment can be found in [**Se1**], and similar or more elaborate treatments can be found, for instance, in [**Si1**], [**Si2**], [**Shi**], [**Kna**], [**Kob**], [**La1**], [**La2**], [**Ogg**],[**Mi1**], [**Mi2**], [**Miy**], etc.

1. Fundamental isomorphisms

In order to state the various realizations of the moduli space referred to above, it is useful to recall lattices in the complex plane, the upper half plane, and the action of the full modular group on the upper half plane.

A **lattice** in $\mathbf{C}$ is the additive subgroup of $\mathbf{C}$ generated by two complex numbers ω_1, ω_2 that are linearly independent over $\mathbf{R}$. In other words, if ω_1 and ω_2 are nonzero, and don't lie on a line through the origin in the complex plane, then the set of linear combinations $a\omega_1 + b\omega_2$, where a and b range through all integers, is a lattice. Perhaps the most familiar example is the set of Gaussian integers

$$\mathbf{Z}[i] := \{a + bi : a, b \in \mathbf{Z}\}$$

form a square lattice consisting of the points in the complex plane whose real and imaginary parts are integers.

Two lattices Λ and Λ' are **homothetic** if one is a scalar multiple of the other, i.e., $\Lambda' = u\Lambda$, $u \in \mathbf{C}^*$. Geometrically, this means that one lattice is similar to the other (i.e., one can be rotated and scaled so as to coincide with the other).

The **upper half plane** consists of all complex numbers whose imaginary part is positive

$$\mathbf{H} := \{z \in \mathbf{C} : \mathrm{Im}(z) > 0\}.$$

If z is in the upper half-plane and a, b, c, d are real numbers such that $ad - bc$ is positive then a simple calculation shows that $(az + b)/(cz + d)$ is in the upper half-plane. We will be interested in the group $\mathrm{SL}_2(\mathbf{Z})$ of integer 2-by-2 matrices of determinant 1. Following convention (which will be clearer only later, when we consider congruence subgroups), we sometimes denote $\mathrm{SL}_2(\mathbf{Z})$ by $\Gamma(1)$.

Thus the group $\mathrm{SL}_2(\mathbf{Z}) = \Gamma(1)$ acts on the upper half plane in the sense that we have a map
$$\Gamma(1) \times \mathbf{H} \to \mathbf{H}$$
which is defined here by
$$\begin{bmatrix} a & b \\ c & d \end{bmatrix} z := \frac{az + b}{cz + d}.$$
Saying that this is a group action means that if z is in the upper half-plane and γ and γ' are in $\Gamma(1)$ then
$$(\gamma\gamma')z = \gamma(\gamma' z).$$
We often say that if $\gamma z = z'$ then z and z' are equivalent under $\mathrm{SL}_2(\mathbf{Z})$, or we just say that they are equivalent.

We remark that since $-I$ acts trivially on the upper half-plane, where I is the 2 by 2 identity matrix, the group
$$\mathrm{PSL}_2(\mathbf{Z}) := \mathrm{SL}_2(\mathbf{Z})/\{\pm I\} = \Gamma(1)/\pm I$$
acts on $\mathbf{H}$ in a natural way. In some circumstances it is more convenient to think of this as the basic group that acts on the upper half plane.

The set $\mathbf{H}/\Gamma(1)$ denotes the set of orbits of this action. This can be constructed as a set by choosing a collection X of elements of $\mathbf{H}$ such that any element of $\mathbf{H}$ is equivalent to an element of X and no two elements of X are equivalent. As will be outlined below, in order for this to be useful one has to also put a topology on X.

With this terminology in hand we can state our fundamental identifications that are the object of this lecture: there are natural bijective maps
$$\frac{\mathbf{H}}{\mathrm{SL}_2(\mathbf{Z})} \longrightarrow \frac{\text{lattices in } \mathbf{C}}{\text{homothety}} \longrightarrow \frac{\text{elliptic curves over } \mathbf{C}}{\text{isomorphism}} \longrightarrow \mathbf{C}.$$
In subsequent sections we consider each of these identifications in turn, starting with the one on the right.

2. The j invariant of an elliptic curve

Two elliptic curves over $\mathbf{C}$ of the form $y^2 = x^3 + ax + b$ are isomorphic if and only if there is a change of coordinates
$$x = u^2 x', \qquad y = u^3 y'$$
between them, for some nonzero $u \in \mathbf{C}$. If we are only interested in elliptic curves up to isomorphism it is natural to try to find invariants — functions of a and b that are invariant under coordinate changes. For reasons partly of convention and partly related to the more general formulas given below, we consider the "j-invariant" of the curve, defined by
$$j(E) = 1728 \cdot \frac{4a^3}{4a^3 + 27b^2}.$$
Note the the denominator must be nonzero since, by earlier calculations, E is nonsingular if and only if $4a^3 + 27b^2 \neq 0$.

Theorem 3.1: Two elliptic curves over $\mathbf{C}$ of the form given above are isomorphic if and only if they have the same j-invariant. If $j \in \mathbf{C}$ then there is an elliptic curve with $j(E) = j$.

Proof. The coordinate change $x = u^2 x'$, $y = u^3 y'$ produces an elliptic curve $E': y'^2 = x'^3 + a'x' + b'$ with $a' = u^{-4}a$, $b' = u^{-6}b$, and an immediate calculation shows that $j(E) = j(E')$. Conversely, if E and E' have the same j-invariant then simple algebra shows that

$$\frac{a^3}{b^2} = \frac{a'^3}{b'^2}$$

and that, at least if a and b are nonzero, then an isomorphism can be defined by taking

$$u = (a'/a)^{1/4} = (b'/b)^{1/6}.$$

To finish the proof, one has to check that if j is given, and if $j \neq 0, 1728$, then the curve with $a = b = -27j/(4(j-1728))$ has j-invariant equal to j. The special values $j = 0$ and $j = 1728$ can be handled immediately by considering the curves with $a = 0$ and $b = 0$ respectively. $\square$

The map

$$j: \text{elliptic curves/isomorphism} \to \mathbf{C}$$

is the first of the identifications promised above.

Of course all of these extends to the general Weierstrass model. In terms of the formulas in the previous lecture, we define

$$j(E) = c_4^3/\Delta(E)$$

and one can then prove:

- The curve E is nonsingular if and only if $\Delta(E) \neq 0$.
- If the ground field is algebraically closed then two curves are isomorphic if and only if their j-invariants are equal.
- If j is in the ground field then there is an elliptic curve defined over the field whose j-invariant is equal to j.

With this general Weierstrass form, one has to consider coordinate changes

$$x = u^2 x' + r, \qquad y = u^3 y' + u^2 s x' + t$$

and the algebra needed for the proof is more elaborate.

As an additional remark, we note that this classification can be used, together with some elementary Galois cohomology, to classify curves up to isomorphism over arbitrary fields K. For instance, if $E: y^2 = x^3 + ax + b$ is an elliptic curve over $\mathbf{Q}$ with $j(E) = j$, with $j \neq 0, 1728$ (for simplicity), then any other curve with the same j-invariant is of the form $y^2 = x^3 + d^2 ax + d^3 b$ for some nonzero $d \in \mathbf{Q}$, and that this curve is isomorphic to E over $\mathbf{Q}$ if and only if d is a square. Thus in this case the set of $\mathbf{Q}$-isomorphism classes of curves that have a given j-invariant (i.e., are isomorphic over the algebraic closure) is isomorphic to $\mathbf{Q}^*/(\mathbf{Q}^*)^2$. Usually one says that the second curve above is a **twist** of the first, and we note that we can scale the variables so that its equation is $dy^2 = x^3 + ax + b$.

3. Elliptic curves via lattices

The next fundamental isomorphism relates lattices and elliptic curves. A lattice $\Lambda = \{a\omega_1 + b\omega_2\}$ is a subgroup of the additive group of $\mathbf{C}$, and the quotient group $\mathbf{C}/\Lambda$ can be visualized as the "fundamental parallelogram"

$$\{x\omega_1 + y\omega_2 : 0 \leq x < 1,\ 0 \leq y < 1\}.$$

Algebraically, this is isomorphic to the product $\mathbf{S}^1 \times \mathbf{S}^1$ of the circle group with itself. Topologically, one can check that $\mathbf{C}/\Lambda$ inherits a topology induced the topology on $\mathbf{C}$, and is homeomorphic to a torus (obtained from the fundamental parallelogram by identifying opposite sides). Analytically, one can check that $\mathbf{C}/\Lambda$ inherits the structure of a Riemann surface (one-dimensional complex manifold) from the obvious Riemann surface structure on $\mathbf{C}$, and that it has genus one. Moreover, two such Riemann surfaces $\mathbf{C}/\Lambda$ and $\mathbf{C}/\Lambda'$ are isomorphic as Riemann surfaces if and only if the lattices are homothetic in the sense that there is a complex number u with $\Lambda' = u\Lambda$.

Meromorphic functions on $\mathbf{C}/\Lambda$ can be identified with meromorphic functions on $\mathbf{C}$ that are **periodic** with respect to Λ in the sense that

$$f(z+\omega) = f(z), \quad z \in \mathbf{C}, \omega \in \Lambda.$$

There are several ways to construct such functions; the most useful for our purposes starts with the Weierstrass $\wp$ function

$$\wp(z) = \frac{1}{z^2} + \sum_{\omega \in \Lambda \setminus \{0\}} \left(\frac{1}{(z-\omega)^2} - \frac{1}{\omega^2} \right).$$

Using basic complex analysis, one shows that this series converges absolutely to a periodic meromorphic function on $\mathbf{C}$ and, using some elegant contour integration arguments, that this function's only poles are double poles at all $\omega \in \Lambda$. As a function on $\mathbf{C}/\Lambda$ its only pole is a double pole at the origin.

In addition, the function

$$\wp'(z) = -2 \sum_{\omega \in \Lambda} \frac{1}{(z-\omega)^3}$$

has poles only at lattice points, and those poles are triple poles. On the Riemann surface $\mathbf{C}/\Lambda$ we have, using the earlier language of divisors, a non-constant function $\wp$ in $L(2O)$ and another function $\wp'$ in $L(3O)$, where $O = [0] \in \mathbf{C}/\Lambda$. A formal calculation with the expansions at O shows that the function

$$\wp'(z)^2 - 4\wp(z)^3 + g_2 \wp(z) + g_3$$

has no poles. Therefore this function is a constant function. Since it takes the value 0 at the specified point O, the function is identically 0. Here the numbers $g_2 = g_2(\Lambda)$ and $g_3 = g_3(\Lambda)$ are defined by

$$g_2 = 60 G_4, \quad g_3 = 140 G_6, \quad G_{2k} = \sum_{\omega \in \Lambda \setminus \{0\}} \frac{1}{\omega^{2k}}$$

and they depend only on the underlying lattice. The sums G_{2k} converge absolutely for $k > 1$.

The main results about the Weierstrass $\wp$-function are as follows.

Theorem 3.2: Fix a lattice Λ as above.

- Any double periodic function is a rational function of $\wp$ and $\wp'$. The field $\mathbf{C}(\wp, \wp')$ is a quadratic extension of $\mathbf{C}(\wp)$.
- The number $g_2^3 - 27g_3^2$ is nonzero, and the mapping
$$z \mapsto (\wp(z), \wp'(z))$$
induces an analytic isomorphism between $\mathbf{C}/\Lambda$ and the points $E(\mathbf{C})$ of the curves $y^2 = 4x^3 - g_2 x - g_3$. Homothetic lattices give rise to isomorphic curves.
- This bijection is a group homomorphism with respect to the addition on $\mathbf{C}/\Lambda$ and the chord-tangent law on $E(\mathbf{C})$.
- If E is an elliptic curve over $\mathbf{C}$ then it can be written in the form $y^2 = 4x^3 - g_2 x - g_3$, and there exists a lattice Λ such that $g_2(\Lambda) = g_2$ and $g_3(\Lambda) = g_3$.

This is the middle identification given in the first section, between lattices up to homothety and elliptic curves up to isomorphism. The proof requires work, and can be found in the sources mentioned at the beginning of the lecture.

Remark 3.3. The homomorphism property in the theorem implies several addition laws for the Weierstrass function; for instance, by carefully consulting the formula for the x-coordinate of the sum of two points on $E(\mathbf{C})$ we see that if $z \neq z'$ in $\mathbf{C}/\Lambda$ then $\wp(z + z') = m^2/4 - \wp(z) - \wp(z')$ where
$$m = \frac{\wp'(z) - \wp'(z')}{\wp(z) - \wp(z')}.$$

Remark 3.4. The fact that every elliptic curve comes from a lattice is a special case of the Uniformization Theorem. In this case, the proof reduces to some interesting calculations very roughly as follows. To invert the $\wp$ function we would like to define
$$\wp^{-1}(w) = \int_{w_0}^{w} \frac{dx}{y}$$
where dx/y is the unique (up to scalar multiple) holomorphic differential on the Riemann surface $E(\mathbf{C})$. However, this integral is only well-defined up to a basis for the one-homology $H_1(E(\mathbf{C}), \mathbf{R})$, i.e., it is only well-defined up to contour integrals of the differential around the torus in two different directions. These period integrals are precisely the generators of the lattice Λ that produces the elliptic curve that we started with.

Remark 3.5. The group structure $\mathbf{C}/\Lambda$ lets us understand the points of finite order very easily. For elliptic curves over an arbitrary field K, let
$$E[n] := \{P \in E(\overline{K}) : nP = O\}$$
be the group of points (over the algebraic closure) of order dividing n. If $K = \mathbf{C}$ then the points of order n in $\mathbf{C}/\Lambda$ are obviously of the form
$$\frac{a_1}{n}\omega_1 + \frac{a_2}{n}\omega_2, \quad a_i \in \mathbf{Z}, \quad 0 \leq a_i < n$$
so that
$$E[n] \simeq \mathbf{Z}/n\mathbf{Z} \times \mathbf{Z}/n\mathbf{Z}.$$
In fact, this is true over an algebraically closed field of characteristic not dividing n. If $n = p^k$ is a power of p and K has characteristic p then there are two cases that

occur: either $E[n]$ has order n, in which case E is said to be **ordinary**, or else $E[n]$ is trivial, in which case E is said to be **supersingular**.

Remark 3.6. Homomorphisms (isogenies) of elliptic curves are easy to think about in terms of the model $E = \mathbf{C}/\Lambda$. Namely, homomorphisms correspond to multiplication by complex numbers that preserve the lattices, i.e.,

$$\mathrm{Hom}(E, E') = \mathrm{Hom}(\mathbf{C}/\Lambda, \mathbf{C}/\Lambda') = \{u \in \mathbf{C} : u\Lambda \subset \Lambda'\}.$$

4. Lattices via the upper half-plane

Finally, we consider the identification between lattices in $\mathbf{C}$, up to homothety, and the upper half plane modulo the full modular group $\Gamma(1)$.

If we have a lattice Λ with basis $\langle \omega_1, \omega_2 \rangle$ it is natural to make two restrictions to begin with. First, we order the basis so that ω_2/ω_1 has positive imaginary part. (Since the ω_i are independent over $\mathbf{R}$, the imaginary part of the quotient is nonzero, and interchanging the ω_i reverses the sign of the imaginary part.) Then by replacing Λ by the homothetic lattice $(1/\omega_1)\Lambda$ we can assume, without loss of generality, that our lattice has a basis of the form $\langle 1, z \rangle$ were z is in the upper half plane $\mathbf{H}$.

When are the lattices $\langle 1, z \rangle$ and $\langle 1, z' \rangle$ homothetic? If the lattices $\langle u, uz \rangle$ and $\langle 1, z' \rangle$ are equal then there are integers a, b, c, d such that

$$\begin{aligned} uz' &= az + b \\ u &= cz + d \end{aligned}$$

and the 2-by-2 matrix on the right hand side is invertible over the integers so that its determinant is ± 1. Moreover, if the basis on the left hand side is oriented then the determinant is 1. These equations imply that

$$z' = \frac{az + b}{cz + d}.$$

In other words, a number in the upper half plane determines a lattice in an obvious way, and numbers that are equivalent under the action of $\Gamma(1)$ determine homothetic lattices.

There is a well-known description of the orbit space $\mathbf{H}/\Gamma(1)$ as a subset of the upper half-plane $\mathbf{H}$ that is a "fundamental domain" in the sense that every element of $\mathbf{H}$ is equivalent to a unique element of the set. Consider the orbit of some $z \in \mathbf{H}$, i.e., the set of all elements of the upper half-plane that are equivalent to z. Among all of these, choose the element z' of largest imaginary part. Let

$$T := \begin{bmatrix} 1 & 1 \\ 0 & 1 \end{bmatrix}$$

and then note T corresponds to translation by a distance 1, so that for some integer n so that $1/2 < \mathrm{Re}(T^n z') \leq 1/2$. One checks that absolute value of this element is larger than 1 (using the fact that $\mathrm{Im}(z')$ was maximal), so that any orbit contains an element of

$$F := \{z \in \mathbf{H} : -1/2 < \mathrm{Re}(z) \leq 1/2, |z| \geq 1, |z| > 1 \text{ if } \mathrm{Re}(z) < 0\}.$$

The details of this argument show that the element of the orbit in F is unique, so that the orbit space $\mathbf{H}/\Gamma(1)$ can be identified with F (though one has to think carefully about the Riemann surface structure on this quotient space).

LECTURE 3. ELLIPTIC CURVES OVER C

Remark 3.7. The details of the argument also show that T and
$$S := \begin{bmatrix} 0 & -1 \\ 1 & 0 \end{bmatrix}$$
generate $\mathrm{PSL}_2(\mathbf{Z})$, and in fact show that $\mathrm{PSL}_2(\mathbf{Z})$ is a free product of S of order 2 and ST of order 3.

If we compose all of our fundamental identifications we get a bijection
$$j\colon \mathbf{H}/\Gamma(1) \to \mathbf{C}.$$
It turns out that the set $\mathbf{H}/\Gamma(1)$ has a natural Riemann surface structure and that it is isomorphic to a compact Riemann surface with one point missing. It is possible to add the missing point "at infinity" in the complex plane and produce appropriate coordinate charts so that the slightly larger space is isomorphic to the sphere, and the map j extends to an isomorphism
$$j\colon \mathbf{H}/\Gamma(1) \cup \{\infty\} \to \mathbf{P}^1(\mathbf{C})$$
that is an isomorphism of compact Riemann surfaces.

This can be described slightly more precisely as follows. Let
$$\mathbf{H}^* := \mathbf{H} \cup \mathbf{P}^1(\mathbf{Q})$$
and let $\Gamma(1)$ act on $\mathbf{P}^1(\mathbf{Q})$ in a fairly obvious way:
$$\gamma x = \begin{bmatrix} a & b \\ c & d \end{bmatrix} (x_0 : x_1) = (ax_0 + bx_1 : cx_0 + dx_1).$$
We let
$$Y = Y(1) = \mathbf{H}/\Gamma(1)$$
and
$$X = X(1) = \mathbf{H}^*/\Gamma(1).$$
Then $X = Y \cup \{\infty\}$ and X is a compact Riemann surface.

Ex 3.8. What are the zeros of $\wp'(z)$?

Ex 3.9. Prove that if $g_2(\Lambda) = g_2(\Lambda')$ and $g_3(\Lambda) = g_3(\Lambda')$ then $\Lambda = \Lambda'$.

Ex 3.10. Find $j(i)$ and $j(e^{2\pi i/3})$.

Ex 3.11. Prove that $G_8 = (3/7)G_4^2$.

Ex 3.12. Show that the real points of an elliptic curve are isomorphic, as a topological group, either to a circle or the product of a circle and a group of order 2.

Ex 3.13. Show that $\wp'(z)^2 - \wp(z)^3 - g_2\wp(z)$ is a constant function.

Ex 3.14. Let E be an elliptic curve defined over $\mathbf{C}$. Find the size of the automorphism group $\mathrm{Aut}(E)$ of the curve. (Hint: the group has size 2 if $j(E) \neq 0, 1728$.)

LECTURE 4
Modular Forms of Level 1

From the identifications given in the previous lecture we know that functions defined on elliptic curves over **C** (up to isomorphism) correspond to functions on lattices (up to homothety) or functions defined on the upper half plane (up the action of $\Gamma(1)$). The j-function is an example of such a function, and it is defined as a quotient of functions that aren't themselves invariant under $\Gamma(1)$. However, these homogeneous "forms" transform with a given weight under the action of $\Gamma(1)$. As functions on lattices, they transform with a given weight when the lattice is scaled.

In many circumstances it is useful to consider these more general functions for their own sake, and the goal of this lecture is to outline the basic theory of these modular forms for the full modular group $\Gamma(1)$. This is again a very standard story, since all subsequent results about modular forms are extensions in one way or another of results in this basic case. Treatments can be found in [**Se1**], [**Si1**], [**Si2**], [**Shi**] [**Kna**], [**Kob**], [**La2**], [**Ogg**],[**Mi1**], [**Mi2**], [**Miy**], etc.

1. Weak modular forms

A function on the upper half plane is a modular form if it transforms suitably under certain linear fractional transformations, and behaves appropriately at infinity. We start by focusing on the first part of the definition.

Def 4.1: Let k be a nonnegative integer. A holomorphic function f on the upper half plane is a **weak modular form** of weight k if

$$f(\gamma z) = (cz+d)^k f(z)$$

for all $z \in \mathbf{H}$, where

$$\gamma = \begin{bmatrix} a & b \\ c & d \end{bmatrix} \in \Gamma(1), \qquad \gamma z = \frac{az+b}{cz+d}.$$

If $f \neq 0$ then then expression $(cz+d)^k$ depends only on the image of γ in $\text{PSL}_2(\mathbf{Z})$ so we must have

$$(cz+d)^k = (-cz-d)^k$$

for all z, i.e., k must be even. So the weight of any nonzero weak modular form is even. Unless explicit mention to the contrary is made, we will assume in this lecture that the weight of any weak modular form, usually denoted k, is a nonnegative even integer.

The formula
$$f^*(\Lambda) = \omega_2^k f(\omega_2/\omega_1), \qquad \Lambda = \langle \omega_1, \omega_2 \rangle$$
sets up a correspondence between weak modular forms f of weight k and certain functions f^* of lattices that have weight $-k$, with respect to homotheties, in the sense that
$$f^*(u\Lambda) = u^{-k} f^*(\Lambda).$$
(We are assuming, as in the previous lecture, that we only consider oriented lattice basis in the sense that we require that $\omega_2/\omega_1 \in \mathbf{H}$.) Thus we can think of modular forms as functions on lattices, and in some situations this is useful.

In order to check that $f(z)$ is a weak modular form it suffices to check that
$$f(z+1) = f(z), \qquad f(-1/z) = z^k f(z)$$
since $\Gamma(1)$ is generated by the matrices
$$T = \begin{bmatrix} 1 & 1 \\ 0 & 1 \end{bmatrix}, \qquad S = \begin{bmatrix} 0 & -1 \\ 1 & 0 \end{bmatrix}.$$

Remark 4.2. Modular forms can be interpreted as "higher order differentials;" suitably interpreted, $f(z)(dz)^k$ is invariant under $\Gamma(1)$. Less obviously one can also realize them as elements of the vector space $L(D)$ for suitable divisors D on $X(1)$.

The functions G_k were originally defined for lattices, but by the identifications in the last lecture, can also be thought of as functions defined on the upper half-plane. We therefore define
$$G_k(z) = \sum \frac{1}{(mz+n)^k}, \qquad k \geq 4$$
where the sum is over all pairs (m, n) of integers, not both zero, and k is of course an even integer. These sums converge absolutely, and uniformly on compact subsets of the upper half plane, so that in particular they define analytic functions.

These functions are called **Eisenstein** series. It is easily seen that
$$G_k(z+1) = G_k(z), \qquad G_k(-1/z) = z^k G_k(z)$$
so G_k is indeed weak modular of weight k, at least for $k \geq 4$ when the series converges to an analytic function.

2. Modular forms

Any periodic function on the real line has a Fourier expansion. More generally any weak modular form f satisfies $f(z+1) = f(z)$ so that f is periodic of period 1 as a function of $\mathrm{Re}(z)$ and therefore has a Fourier series. One can show either using complex analysis, or by a direct proof (e.g., [**Kna**]p. 224) that the coefficients are independent of the imaginary part of z and that the series converges for all z in the upper half-plane; thus we have
$$f(z) = \sum_{n \in \mathbf{Z}} a_n e^{2\pi i n z}.$$

It is an (almost) universal convention to write this as

$$f(z) = \sum a_n q^n$$

where $q = e^{2\pi i z}$. The Fourier expansion is usually called a "q-expansion" in this context.

If $a_n = 0$ for $n < 0$ then we say that f is "holomorphic at infinity."

Def 4.3: A weak modular form $f(z)$ of weight k is said to be a **modular form** of weight k if it is holomorphic at infinity. If in addition the constant term in its q-expansion a_0 is zero then f is said to be a **cusp form**.

The function $G_k(z)$ is weak modular of weight k. If it is a modular form then its constant term a_0 is the limit of $G_k(it)$ as t goes to infinity. Since

$$\lim_{t \to \infty} G_k(it) = 2 \sum_{n=1}^{\infty} \frac{1}{n^k}$$

we see that the constant term has to be $2\zeta(k)$ where

$$\zeta(s) = \sum_{n=1}^{\infty} \frac{1}{n^s}, \qquad \text{Re}(s) > 1$$

is the usual Riemann zeta function.

It turns out to be convenient to normalize the Eisenstein series so that the constant term is equal to 1. Thus we define the normalized Eisenstein series E_k by

$$E_k(z) = \frac{G_k(z)}{2\zeta(k)}$$

for even integers $k \geq 4$. Somewhat surprisingly, it turns out that *all* of the coefficients of the q-expansion of the normalized Eisenstein series are rational numbers.

To state this result it is convenient to recall the Bernoulli numbers, and their relationship to the zeta function. The **Bernoulli numbers** are defined by the generating function

$$\frac{x}{e^x - 1} = \sum_{n=0}^{\infty} B_n \frac{x^n}{n!}$$

so that

$$B_0 = 1, \quad B_1 = -1/2, \quad B_2 = 1/6, \quad B_4 = -1/30, \quad B_6 = 1/42$$
$$B_8 = -1/30, \quad B_{10} = 5/66, \quad B_{12} = -691/2730, \quad B_{2n+1} = 0, n > 0.$$

For even k,

$$2\zeta(k) = \sum_{n \in \mathbf{Z}, n \neq 0} \frac{1}{n^k} = \frac{(-1)^{k-1} B_k (2\pi)^k}{k!}.$$

Special cases of this include the well-known formulas

$$\zeta(2) = \frac{\pi^2}{6}, \qquad \zeta(4) = \frac{\pi^4}{90}.$$

Theorem 4.4: The Eisenstein series E_k is holomorphic at infinity and is therefore a modular form of weight k. Its q-expansion is

$$E_k(z) = 1 - \frac{4k}{B_k} \sum_{n=1}^{\infty} \sigma_{k-1}(n) q^n$$

where
$$\sigma_{k-1}(n) := \sum_{d|n} d^{k-1}$$
and the sum is over positive divisors d of the integer n.

One natural starting place both for the proof of the formula for $\zeta(k)$ given above, and the q-expansion given in the theorem, is the cotangent expansion
$$\pi \cot(\pi z) = \sum_{n \in \mathbf{Z}} \frac{1}{z+n} = \frac{1}{z} + \sum_{n=1}^{\infty} \left(\frac{1}{z+n} + \frac{1}{z-n} \right)$$
usually derived in a complex analysis course. The sum of $1/(z+n)$ over all integers n may not converge nicely unless the terms are ordered so that the n and $-n$ terms are combined as in the second form of the sum.

We leave these results as elaborate exercises for the reader. To prove the formula for $\zeta(k)$, combine the $1/(z+n)$ and $1/(z-n)$ terms and expand the result as a geometric series to show that
$$\pi \cot(\pi z) = \frac{1}{z} - 2 \sum_{n=1}^{\infty} \zeta(2n) z^{2n-1};$$
rewrite the left hand side in terms of the generating function for the Bernoulli numbers.

To prove the q-expansion formula for E_k, start by expressing $\pi \cot(\pi z)$ in terms of $q = \exp(2\pi i z)$ to show that
$$\pi \cot(\pi z) = \pi i - \frac{2\pi i}{1-q}.$$

3. Dimensions of spaces of modular forms

The goal of this section is to state a theorem that gives the dimension of the vector space of modular forms of weight k. This result has some surprising consequences.

For instance, the product of modular forms of weights k and k' is a modular form of weight $k + k'$. Thus E_8 and E_4^2 have the same weight. We will see that the vector space of modular forms of weight 8 has dimension 1. Since the q-expansions both start with 1, we conclude that $E_8 = E_4^2$. This identity implies an unexpected relation between the sum-of-divisors function σ_3 and the sum-of-divisors function σ_7:
$$\sigma_7(n) = \sigma_3(n) + 120 \sum_{k=1}^{n-k} \sigma_3(k) \sigma_3(n-1).$$

Let k be a nonnegative even integer, let $M_k(\Gamma(1))$ denote the vector space of modular forms of weight k for the full modular group, and let $S_k(\Gamma(1))$ denote the corresponding space of cusp forms. For $k \geq 4$ the complex vector space $M_k(\Gamma(1))$ is nontrivial since it contains the corresponding Eisenstein series; moreover, for $k \geq 4$,
$$\dim M_k(\Gamma(1)) = \dim S_k(\Gamma(1)) + 1$$
since the map $f \mapsto a_0$ taking a modular form to the constant term of its q-expansion is a linear mapping whose kernel consists exactly of the cusp forms.

One fancy way to express the fact that the product of modular forms of weights k and k' is a form of weight $k + k'$ is to say that the space

$$\oplus_{k=0}^{\infty} M_k(\Gamma(1))$$

of all modular forms for $\Gamma(1)$ is a graded algebra. The key facts about the spaces $M_k(\Gamma(1))$ can be encapsulated in two statements about this algebra.

Theorem 4.5: The space of modular forms is generated, as an algebra, by E_4 and E_6. Moreover, these elements are algebraically independent.

To say that modular forms are generated by E_4 and E_6 means that any modular form is a linear combination of products of the form

$$E_4^a E_6^b.$$

(The weight of this monomial is $4a + 6b$.) To say that E_4 and E_6 are algebraically independent merely means that these monomials are linearly independent.

Corollary 4.6: The dimension of $M_k(\Gamma(1))$ is equal to the number of pairs (a, b) of nonnegative integers such that

$$4a + 6b = k.$$

By considering the equation $4a + 6b = k$ carefully, one shows that

$$\dim M_k(\Gamma(1)) = 1 + [k/12] - \delta_{k-2}$$

where $[x]$ denotes the integer part of x and δ_m is 1 (resp. 0) if m is (resp. is not) divisible by 12.

Since $\dim S_k(\Gamma(1)) = \dim M_k(\Gamma(1)) - 1$, for $k \geq 4$, the smallest weight of a nonzero cusp form will be the smallest weight for which $\dim M_k > 1$. A simple check for small k shows that $k = 12$ is the smallest example, where E_4^3 and E_6^2 are linearly independent. It is natural to construct a cusp form by subtracting these two modular forms, and we define

$$\Delta(z) = E_4(z)^3 - E_6(z)^2.$$

The q-expansion of E_4 and E_6 have rational coefficients, and the coefficients $\tau(n)$ of the q-expansion of $\Delta(z)$

$$\Delta(z) = \sum_{n=1}^{\infty} \tau(n) q^n$$

satisfy an even stronger property: they are integers. The function τ is the so-called Ramanujan τ function, and turns out to be multiplicative in the sense that $\tau(mn) = \tau(m)\tau(n)$ for all relatively prime positive integers m and n.

In a similar vein, the q-expansion of the j-function has the form

$$j(z) = \frac{1}{q} + \sum_{n=0}^{\infty} c_n q^n$$

where the c_n are integers; one finds

$$c_0 = 744, c_1 = 196884, \ldots.$$

Finally, we mention that the modular form Δ has a product expansion of the form

$$\Delta(z) = q \prod_{n=1}^{\infty} (1 - q^n)^{24}.$$

4. Hecke operators

The basis of M_k consisting of suitable $E_4^a E_6^b$ was useful for determining the dimension of the space. However, there is a much more elegant and useful basis, consisting of the Eisenstein series E_k together with cusp forms

$$f = \sum_{n=1}^{\infty} a_n q^n$$

whose coefficients have $a_1 = 1$ and are multiplicative in the sense that $a_{mn} = a_m a_n$ for relatively prime m and n. This idea is at the root of many further developments in the subject.

In order to have a glimpse how this more natural basis arises we have to consider the **Hecke operators** which are linear operators on the spaces M_k. Their definition is superficially somewhat abstruse. Roughly speaking, if f is a modular form of weight k then $T_n f$ is the modular form of weight k defined by

$$(T_n f)(z) = \sum f(z')$$

where z' runs over representatives in the upper half-plane of all sublattices of index n in the lattice corresponding to $z \in \mathbf{H}$.

Although this is reasonably natural when expressed in terms of lattices, the expression in terms of functions on the upper half-plane, or q-expansions, plane is more complicated. The simplest case is when $n = p$ is a prime, where we have

$$T_p f(z) = f(z/p) + \sum_{b=0}^{p-1} f(pz + b)$$

expressed in terms of functions of z, and

$$T_p f = \sum (a_{np} + p^{k-1} a_{n/p}) q^n$$

in terms of q-expansions, where $f = \sum a_n q^n$, and we let $a_x = 0$ if x isn't an integer. The basic facts about these Hecke operators are as follows.

Theorem 4.7: Fix a weight k as above. The Hecke operators T_n are linear operators on $M_k(\Gamma(1))$ that map cusp forms to cusp forms. If $f = \sum a_m q^m$ then the coefficient of q in the q expansion of $T_n f$ is a_n.

There is an inner product on $S_k(\Gamma(1))$ with respect to which the T_n are self-adjoint (hermitian). The algebra generated by the T_n is generated by the operators T_p for primes p, and the T_n can be expressed in terms of the T_p by (repeatedly) using the identities

$$T_{mn} = T_m T_n \quad \text{if} \quad \gcd(m, n) = 1$$

$$T_{p^r} = T_p T_{p^r} - p^{k-1} T_{p^{r-1}}, \quad r \geq 1.$$

Remark 4.8. The inner product in the theorem is called the **Petersson inner product** and, for cusp forms f and g in $S_k(\Gamma(1))$ is defined by

$$\langle f, g \rangle = \int_F f(z) \overline{g(z)} \mathrm{Re}(z)^{k-2} \, dz \, d\bar{z}$$

where F is a fundamental domain for the full modular group acting on the upper half-plane.

From the formulas for the Hecke operators one sees that the Hecke operators commute. A basic result from linear algebra says that a self-adjoint operator is diagonalizable in the sense that there is a basis of eigenvectors. A commuting family of self-adjoint operators is simultaneously diagonalizable in the sense that there is a basis consisting of eigenvectors for all of the operators.

If we apply this to the T_n (with the weight k fixed), we see that $S_k(\Gamma(1))$ has a basis of modular forms that are eigenvectors for all T_n; we will call these forms eigenforms.

Suppose that $f = \sum a_n q^n$ is an eigenform. By comparing coefficients of q in the equation
$$T_n f = \lambda f$$
we see, using the Theorem, that $\lambda a_1 = a_n$. Since some Fourier coefficient is nonzero it follows that $a_1 \neq 0$. We say that an eigenform is **normalized** if $a_1 = 1$, and clearly there is a basis of normalized eigenforms.

From the formula $T_n f = a_n f$ we see that the collection of modular forms with a given set of eigenvalues for the Hecke operators is one-dimensional. Moreover, the relations for the Hecke operators given in the theorem imply that if $f = \sum a_n q^n$ is an eigenform of weight k then
$$a_{mn} = a_m a_n, \qquad \gcd(m,n) = 1$$
$$a_{p^r} = a_p a_{p^{r-1}} - p^{k-1} a_{p^{r-2}}, \qquad r \geq 2.$$

The function
$$L_f(s) = \sum a_n n^{-s}$$
associated to a normalized eigenform f is called the L-series of the modular form, and the preceding identities will play a crucial role in the next lecture when studying this function.

LECTURE 5
L-series; Modular Forms of Higher Level

The idea of associating a Dirichlet series $\sum a_n n^{-s}$ to objects of geometric and arithmetic interest has proved to be remarkably fruitful in number theory and algebraic geometry. The goal of this lecture is to define "L-series" associated to modular forms and to elliptic curves. Along the way the extension of the earlier results on modular forms to forms of level N will be sketched. This material can be found in almost all of the references listed in the previous chapter.

1. Dirichlet series

A Dirichlet series is a generating function of the shape

$$L(s) = \sum_{n=1}^{\infty} a_n n^{-s}.$$

If the coefficients a_n are bounded by a constant times n^k then a simple comparison argument shows that the series converges absolutely if $\mathrm{Re}(s) > k+1$.

The archetypical example is the **Riemann zeta function**

$$\zeta(s) = \sum_{n=1}^{\infty} \frac{1}{n^s}, \qquad \mathrm{Re}(s) > 1.$$

As with many Dirichlet series of interest in number theory, the zeta function has a meromorphic continuation and a functional equation. This means that there is a meromorphic function on the complex plane, also denoted $\zeta(s)$, whose values agree with the above sum on the half-plane $\mathrm{Re}(s) > 1$; in this case, the function is analytic, except for a simple pole at $s = 1$, and satisfies the functional equation

$$\Lambda(s) := \pi^{-s/2} \Gamma(s/2) \zeta(s) = \Lambda(1-s)$$

that relates the values of ζ at s and $1 - s$.

The Riemann hypothesis is perhaps the most famous unsolved problem in mathematics: it asserts that if $\zeta(s) = 0$ and $0 \leq \mathrm{Re}(s) \leq 1$ then $\mathrm{Re}(s) = 1/2$.

Another standard example of a Dirichlet series of number-theoretic interest is the L-series associated to a Dirichlet character (homomorphism) $\chi\colon (\mathbf{Z}/N\mathbf{Z})^* \to \mathbf{C}^*$. If N' is a divisor of N then Dirichlet characters on $(\mathbf{Z}/N'\mathbf{Z})^*$ give rise to Dirichlet characters on $(\mathbf{Z}/N\mathbf{Z})^*$ by composition with the canonical mapping $\mathbf{Z}/N\mathbf{Z} \to \mathbf{Z}/N'\mathbf{Z}$; a Dirichlet character on $\mathbf{Z}/N\mathbf{Z}$ has **conductor** N if it does not arise in this way from a Dirichlet character defined modulo a proper divisor N' of N.

The L-series of a Dirichlet character $\chi\colon (\mathbf{Z}/N\mathbf{Z})^* \to \mathbf{C}^*$ is defined to be

$$L(s,\chi) = \sum_{n=1}^{\infty} \chi(n) n^{-s}.$$

Thus the Riemann zeta function is the L-series of the trivial character.

If χ has conductor N then its L-series satisfies a functional equation

$$\Lambda(s,\chi) := \pi^{-s'/2} N^{s'/2} \Gamma(s'/2) L(s,\chi) = w \Lambda(1-s, \overline{\chi})$$

where w is a complex number of absolute value one (defined in terms of the Gauss sums of χ) and

$$s' = \begin{cases} s & \chi(-1) = 1 \\ s+1 & \chi(-1) = -1. \end{cases}$$

These Dirichlet series have an **Euler product** of the form

$$L(s,\chi) = \sum_{n=1}^{\infty} \chi(n) n^{-s} = \prod_p \frac{1}{1 - \chi(p) p^{-s}}$$

where the product is over all primes p. This formula can be proved formally by expanding each term on the right hand side as a geometric series and then multiplying these terms; it is not hard to justify these manipulations where the series converges absolutely.

In these functional equations one can think of the exponential factors and gamma functions as arising from an extra term in the Euler product corresponding the the "prime" or "place" at infinity. It is customary to define a place to be (equivalence class of) an embedding of $\mathbf{Q}$ into a complete field, and to prove that there is a place for each prime p (coming from the obvious embedding of $\mathbf{Q}$ into $\mathbf{Q}_p$) and one place "at infinity" corresponding to the obvious inclusion of $\mathbf{Q}$ into the real numbers. Here $\mathbf{Q}_p$ is the completion of the rational number with respect to the p-adic absolute value used earlier, and the real numbers are of course the completion of the rational numbers with respect to the usual absolute value.

2. L-functions of modular forms of level 1

If $f \in S_k(\Gamma(1))$ is a cusp form for the group $\Gamma(1)$ with q-expansion $\sum a_n q^n$ then its L-series is defined to be

$$L(s,f) = \sum_{n=1}^{\infty} a_n n^{-s}.$$

A formal calculation shows that

$$\int_0^{\infty} f(it) t^s \frac{dt}{t} = (2\pi)^{-s} \int_0^{\infty} e^{-t} t^s \frac{dt}{t} \sum \frac{a_n}{n^s} = (2\pi)^{-s} \Gamma(s) L(s,f).$$

Straightforward complex analysis, using the fact that $f(x+iy)$ is periodic of period 1 with respect to x, can be used to show that the Fourier coefficients a_n are bounded by a constant times $n^{k/2}$. This in turn shows that the L-series for f converges absolutely and uniformly on $\mathrm{Re}(s) > 1 + k/2 + \epsilon$, for any $\epsilon > 0$, and that the formal calculation is valid on that half-plane.

The same estimates that bound the a_n also show that

$$\int_1^\infty f(it)t^{s-1}dt$$

exists for all complex numbers s, and is an entire function of s. This leads to the following idea: write

$$(2\pi)^{-s}\Gamma(s)L(s,f) = \int_0^1 f(it)t^{s-1}dt + \int_1^\infty f(it)t^{s-1}dt$$

and use the transformation property of a modular form to rewrite the first integral. Since f is a modular form we find that $f(-1/z) = z^k f(z)$ so that $f(i/t) = i^k t^k f(it)$. Using this to change variables by $t \to t^{-1}$ in the first integral above, we arrive at

$$(2\pi)^{-s}\Gamma(s)L(s,f) = i^k \int_1^\infty f(it)t^{k-s-1}dt + \int_1^\infty f(it)t^{s-1}dt.$$

This shows that the left hand side is an entire function of s, and therefore so is $L(s,f)$ since $\Gamma(s)$ is everywhere nonzero. The right hand side is multiplied by i^k if s is replaced by $k-s$, and this proves the functional equation for $L(s,f)$, namely that

$$\Lambda(s,f) := (2\pi)^{-s}\Gamma(s)L(s,f) = (-1)^{k/2}\Lambda(k-s,f).$$

It turns out that the L-series of cusp forms that are eigenforms for the Hecke operators are particularly nice in that they have an Euler product, for which the denominator of the factor corresponding to a prime p is quadratic polynomial in p^{-s}.

More precisely, the relations for the Hecke operators given in the previous lecture can be written as an identity relating formal Dirichlet series (whose coefficients are linear operators on $S_k(\Gamma(1))$)

$$\sum T_n n^{-s} = \prod_p \frac{1}{1 - T_p p^{-s} + p^{k-1}p^{-2s}}$$

where the denominator on the right hand side is written in that form to emphasize that it should be thought of as a quadratic polynomial in p^{-s}.

As was discussed earlier, the space $S_k(\Gamma(1))$ has a basis consisting of normalized eigenforms for the Hecke operators, i.e, modular forms $f = \sum a_n q^n$ such that $a_1 = 1$ and $T_n f = a_n f$. If we apply the formal identity above to f we get

$$L(s,f) = \sum_{n=1}^\infty a_n n^{-s} = \prod_p \frac{1}{1 - a_p p^{-s} + p^{k-1-2s}}.$$

3. Forms of higher level

Although modular forms of level 1 that we have been considering are sufficient for many purposes, the later Park City sessions will definitely need to modular forms of "level N." The definitions are somewhat technical at first glance, but have natural modular interpretations along the lines of the fundamental isomorphisms given in the third lecture.

Recall that in that context we found that $Y = \mathbf{H}/\Gamma(1)$ can be thought of as the set of all isomorphism classes of elliptic curves over $\mathbf{C}$, and that it has a natural Riemann surface structure which can be viewed as a Riemann sphere missing one point. Modular forms of weight k are "k-differentials" on this surface.

In order to generalize this idea we "add some additional structure" to the elliptic curve. This corresponds to considering the action of proper subgroups of the full modular group on the upper half-plane. Specifically, let N be a positive integer and define subgroups of $\Gamma(1)$ as follows.

$$\Gamma_0(N) = \{\gamma \in \Gamma(1) : \gamma \equiv \begin{bmatrix} * & * \\ 0 & * \end{bmatrix} \bmod N\}$$

$$\Gamma_1(N) = \{\gamma \in \Gamma(1) : \gamma \equiv \begin{bmatrix} 1 & * \\ 0 & 1 \end{bmatrix} \bmod N\}$$

$$\Gamma(N) = \{\gamma \in \Gamma(1) : \gamma \equiv \begin{bmatrix} 1 & 0 \\ 0 & 1 \end{bmatrix} \bmod N\}.$$

The notation is intended to mean, for instance, that $\Gamma_0(N)$ consists of all matrices whose lower left-hand entry is divisible by N. If $N > 1$ these groups are proper subgroups of $\Gamma(1)$ and their orbits on the upper half-plane will be proper subsets of the orbits of $\Gamma(1)$, so that the orbit spaces will be larger than the orbit space for $\Gamma(1)$.

If Γ is one of these three groups then we can consider

$$Y = \mathbf{H}/\Gamma \quad \subset \quad X = \mathbf{H}^*/\Gamma.$$

In order to verify that Y and X are Riemann surfaces, and that the complement of Y in X consists of finitely many points (called cusps) one has to construct local coordinate charts around each of these points. For the respective groups above this all works out nicely and the corresponding surfaces are denoted $Y_0(N), Y_1(N), Y(N)$ and $X_0(N), X_1(N), X(N)$ for $\Gamma = \Gamma_0(N), \Gamma_1(N), \Gamma(N)$.

In each case, the Riemann surface X actually has the structure of a projective curve, defined over some finite extension of $\mathbf{Q}$. The modular interpretation of points on these curves is as follows (in characteristic 0):

$$Y_0(N)(K) = \{(E,C) : E \text{ is an elliptic curve defined over } K,$$
$$\text{and } C \text{ is a cyclic subgroup of order } N \text{ defined over } K\}$$

$$Y_1(N)(K) = \{(E,P) : E \text{ is an elliptic curve defined over } K,$$
$$\text{and } P \text{ is a point of order } N \text{ defined over } K\}$$

$$Y(N)(K) = \{(E,\alpha) : E \text{ is an elliptic curve defined over } K, \text{ and } \alpha$$
$$\text{is a symplectic isomorphism of } E[N] \text{ with } (\mathbf{Z}/N\mathbf{Z})^2\}.$$

In the last example "symplectic" means that α carries a natural alternating form on $E[N]$, called the Weil pairing, to a canonical inner product on $(\mathbf{Z}/N\mathbf{Z})^2$; these pairings require that the field K contain a primitive N-th root of unity. Details on this pairing and on modular curves can be found in [**DeR**], [**KaM**], [**Roh**], [**Dia**], [**Dar**], and [**Shi**].

A modular form of weight k for one of these three groups Γ is a holomorphic function on $\mathbf{H}$ that satisfies

$$f(\gamma z) = (cz+d)^k f(z), \qquad \gamma = \begin{bmatrix} a & b \\ c & d \end{bmatrix} \in \Gamma$$

and is holomorphic at the cusps. Here k can be any nonnegative integer. (Since $\Gamma_0(N)$ contains $-I$, k actually has to be even in that case if $f \neq 0$.)

Being holomorphic at infinity means that the Fourier expansion
$$f(z) = \sum_n a_n e^{2\pi i n z/w} = \sum_n a_n q^{n/w}$$
has $a_n = 0$ for $n < 0$ (where w is the smallest positive integer such that
$$\begin{bmatrix} 1 & w \\ 0 & 1 \end{bmatrix}$$
is in Γ). If $x \in \mathbf{Q}$ is another cusp and $\gamma \infty = x$ for $\gamma \in \Gamma(1)$ then f is holomorphic at x if $(cz+d)^{-k} f(\gamma z)$ is holomorphic at infinity; this is independent of the choice of γ.

The group $(\mathbf{Z}/N\mathbf{Z})^*$ acts on modular forms for $\Gamma_1(N)$, and in many contexts it is useful to consider eigenspaces for that action. We say that a modular form f for $\Gamma_1(N)$ has level N, weight k, and character χ if
$$f(\gamma z) = \chi(d)(cz+d)^k f(z), \qquad \gamma \in \Gamma_0(N)$$
where $\chi \colon (\mathbf{Z}/N\mathbf{Z})^* \to \mathbf{C}^*$ is a Dirichlet character. By substituting $-I$ one finds that if f is nonzero then $\chi(-1) = (-1)^k$ so that the parity of the character χ must be the same as the parity of the weight. We let $S_k(N,\chi)$ denote the space of cusp forms of level N, weight k, and character χ. If $\chi = 1$ is the trivial character then $S_k(N,1)$ is just the space of cusp forms of weight k for the group $\Gamma_0(N)$.

Hecke operators can be defined much along the lines as before. We consider consider sub-lattices of index N that have the corresponding structure, and translate this into an operator on modular forms. This changes the formula for T_p for p dividing the level, and in fact Hecke operators for primes dividing the level N are often denoted U_p (for simplicity of notation we won't adopt this here). For forms $f \in S_k(N,\chi)$ the formulas are as follows. If $f = \sum a_n q^n$ we have
$$T_p f(z) = \sum_{n=1}^\infty (a_{np} + p^{k-1} \chi(p) a_{n/p}) q^n$$
where we define $\chi(p) = 0$ for p dividing N. The connection between the Hecke operators for prime n and all of the Hecke operators can be summarized in an identity of formal Dirichlet series:
$$\sum_{n=1}^\infty T_n n^{-s} = \prod_p (1 - T_p p^{-s} + \chi(p) p^{k-1-2s})^{-1}.$$

Although modular forms of level N are spanned by eigenvectors for the Hecke operators, it turns out that the eigenspaces do not have to be one-dimensional. However, this arises because forms of level N can (just as for Dirichlet characters) arise from forms of level strictly dividing N. If N' is a divisor of N and f is in $S_k(N',\chi)$ then it turns out that $f(mz)$ is in $S_k(N,\chi)$ where m is any divisor of N/N'. And in fact the $f(mz)$, for various choices of m, give eigenforms for T_p, p not dividing N, whose eigenvalues under T_p are identical.

The problem is that these modular forms $f(mz)$ do not "really" have level N. We let $S_k^{\text{old}}(N,\chi)$ denote the subspace of $S_k(N,\chi)$ spanned by all forms arising in this way modular forms of smaller level. The orthogonal complement (under the Petersson inner product) consists of forms that have "genuine" level N, and is sometimes called the space of newforms at that level. However, we will call a form in this space a **newform** only if it is normalized (has $a_1 = 1$) and is an eigenform for all of the Hecke operators. A key fact (sometimes called the Multiplicity One theorem) says the eigenspaces among newforms are one-dimensional: if f and g are distinct newforms then there is an n such that their eigenvalues for T_n are distinct.

Applying the formal Dirichlet series for Hecke operators given above to a newform $f = \sum a_n q^n \in S_k(N, \chi)$ leads to

$$L(s, f) = \sum_{n=1}^{\infty} a_n q^n = \prod_p (1 - a_p p^{-s} + \chi(p) p^{k-1} p^{-2s})^{-1}.$$

Finally, the outline given earlier for proving the functional equation follows very similarly: if f is a newform then there is a functional equation

$$\Lambda(s, f) := N^{s/2}(2\pi)^{-s}\Gamma(s)L(s, f) = w\Lambda(s, \overline{f})$$

where w is a complex number of absolute value 1 that can be specified precisely.

4. L-series of elliptic curves

Let E be an elliptic curve over $\mathbf{Q}$. Define a_p for primes p as follows.

$$a_p = \begin{cases} p + 1 - \#\overline{E}(\mathbf{F}_p) & \text{if } E \text{ has good reduction at } p \\ 1 & \text{if } E \text{ has split multiplicative reduction at } p \\ -1 & \text{if } E \text{ has non-split multiplicative reduction at } p \\ 0 & \text{if } E \text{ has additive reduction at } p. \end{cases}$$

In the case of good reduction the formula for a_p measures the difference between the number of points on the curve over $\mathbf{F}_p$ and the number of points on the projective line over $\mathbf{F}_p$.

In the next lecture we will learn that a positive integer N can be associated to E, called the conductor of E. The primes dividing N are exactly the primes where E has bad reduction, and for primes $p > 3$ the power of p dividing N is 0, 1, or 2 according to whether the reduction of E at p is good, multiplicative, or additive. (For primes $p = 2$ and $p = 3$ the exponent is more subtle.)

Now define the L-series associated to E by an Euler product:

$$L(E, s) = \prod_{p \nmid N} (1 - a_p p^{-s} + p^{1-2s})^{-1} \prod_{p | N} (1 - a_p p^{-s})^{-1}.$$

In the last lecture a famous theorem of Hasse will be stated which states that a_p is bounded by a constant times $p^{1/2}$; it follows from this and from general results on Dirichlet series that $L(E, s)$ converges for $\text{Re}(s) > 3/2$. One might hope that this L-series has an analytic continuation to the entire complex plane. In fact, it has long been known that for certain E, this L-series is a "Hecke L-series" and that it therefore satisfies the functional equation

$$\Lambda(s, E) := N^{s/2}(2\pi)^{-s}\Gamma(s)L(s, E) = w\Lambda(2 - s, E)$$

where $w = \pm 1$ can be specified precisely.

This L-series has exactly the same general shape as the L-series attached to a newform of weight 2, level N, and trivial character. In fact, Eichler and Shimura were able to carefully analyze certain modular forms and show that they give rise to elliptic curves. More precisely, if f is a newform of level N whose Fourier coefficients

a_n are integers then Eichler and Shimura construct an elliptic curve as follows. Fix $z \in \mathbf{H}$ and set
$$\Lambda := \{\int_z^{\gamma(z)} f(z)dz : \gamma \in \Gamma_0(N)\}.$$
This set is independent of the choice of z, and is a lattice in the complex plane. The quotient $\mathbf{C}/\Lambda$ is isomorphic to an elliptic curve with rational coefficients, explicitly given as an equation whose coefficients are (multiples of) the rational numbers $G_4(\Lambda)$ and $G_6(\Lambda)$. Moreover, the L-series of this elliptic curve coincides with the L-series of the modular form f, and the construction provides a natural surjective mapping from the curve $X_0(N)$ to the elliptic curve.

This circle of ideas led to a conjecture that says that any elliptic curve over $\mathbf{Q}$ arises in this way. This idea was expressed in a very fragmentary form by Taniyama in the 1950's, and then in a more precise form by Shimura in the 1960's. Weil published a version of the conjecture and connected the idea to the existence of functional equations. This conjecture has had various names over the last 30 years, including the Taniyama-Shimura-Weil conjecture.

More recently, Wiles [**Wil**], with assistance from Taylor [**TW**], proved the conjecture for curves without additive reduction, and used this to prove Fermat's Last Theorem. As announced about at the time of the PCMI conference, Breuil, Conrad, Diamond, and Taylor [**Brl**], using similar techniques, extended to proof to all remaining curves. These long string of developments is surely one of the major successes in 20-th century mathematics, and the continuing fecundity of these ideas is a major motivation for the PCMI conference.

Here are three forms of this marvelous theorem.

Theorem 5.1: Let E be an elliptic curve over $\mathbf{Q}$.
- **(a):** The L-series $L(f, s)$ has an analytic continuation, and satisfies the functional equation given above.
- **(b):** There is a newform $f \in S_2(N, 1)$ such that $L(f, s) = L(E, s)$.
- **(c):** There is a non-constant map of curves
$$X_0(N) \to E.$$

For more discussion of this modularity conjecture, see [**Sil**] in these notes, or the preprint [**Brl**].

LECTURE 6
l-adic Representations

One of the most effective ways to study a group is to study its representations, and much effort has been expended on studying the representations of the Galois group

$$G_{\mathbf{Q}} := \mathrm{Gal}(\overline{\mathbf{Q}}/\mathbf{Q})$$

of the algebraic closure of the rational numbers over the rational numbers. The goal of this lecture is to introduce important Galois representations associated to elliptic curves and modular forms.

Various basic facts about Galois representations can be found at the beginning of [**Gou**] in these proceedings, and further information can be found in [**Se2**], [**Dar**], and [**Dia**].

1. Galois representations

The group $G_{\mathbf{Q}}$ is an inverse limit group of the groups $\mathrm{Gal}(K/\mathbf{Q})$ as K ranges over all finite extensions of $\mathbf{Q}$ in $\overline{Q}$; thus an element σ of $G_{\mathbf{Q}}$ determines a compatible family of automorphisms $\sigma_K \in \mathrm{Gal}(K/\mathbf{Q})$ for each such finite Galois extension K of $\mathbf{Q}$. Here, a family of automorphisms is compatible if σ_L restricts to σ_K if $K \subset L$ inside the fixed algebraic closure $\overline{\mathbf{Q}}$. The group $G_{\mathbf{Q}}$ is a profinite group and has a natural topology in which it is a compact totally disconnected group. Roughly, this means that the group can be understood completely by understanding all of its finite quotients.

The large goal is to describe the n-dimensional representations

$$\rho \colon G_{\mathbf{Q}} \to \mathrm{GL}_n(F)$$

where F is a topological field and ρ is a continuous homomorphism of groups. Typically, one has $F = \mathbf{C}$, F finite, or F an extension of $\mathbf{Q}_p$.

We remind the reader that if $K/\mathbf{Q}$ is a Galois extension, p is a rational prime, and P is a prime of the ring of integers that contains p, then the **decomposition group** D_P is the subgroup of the Galois group that preserves P set-wise:

$$D_P := \{\sigma \in \mathrm{Gal}(K/\mathbf{Q}) : \sigma(P) = P\}.$$

Elements of the decomposition group determine automorphisms of the residue field $\mathcal{O}_K/P$. The **inertia group** I_P consists of those σ that induce the identity automorphism, i.e.,

$$I_P := \{\sigma \in D_P : \sigma(x) \equiv x \bmod P, \quad \forall x \in \mathcal{O}_K\}.$$

The extension $K/\mathbf{Q}$ is said to be **unramified** if I_P is trivial.

The decomposition and inertia groups associated to different primes lying over p are conjugate in $\mathrm{Gal}(K/\mathbf{Q})$.

Finally, the residue field is a Galois extension of the field $\mathbf{Z}/p\mathbf{Z}$, and the Galois group of this extension has a canonical generator, namely, in this case the automorphism that takes x to x^p. We let Frob_p denote an element of D_P that maps to this canonical generator. Thus Frob_p is only well defined modulo the inertia group, and we will usually only use it in the case in which p is unramified. The Frobenius map also depends on the choice of the prime ideal P, so that perhaps Frob_P would be better notation. However, the Galois group acts transitively on ideals of K lying over p, and one checks that the different Frobenius elements in $\mathrm{Gal}(K/\mathbf{Q})$ are conjugate. Therefore Frob_p could be thought of as a well-defined conjugacy class. Typically, one "knows" a representation ρ by knowing the elements $\rho(\mathrm{Frob}_p)$.

2. One-dimensional representations

With these preliminaries, we start by considering one-dimensional representations, which is perhaps the only case that is fully understood. These representations can be determined using class field theory. For instance, one-dimensional representations

$$\rho \colon G_\mathbf{Q} \to \mathbf{C}^* = \mathrm{GL}_1(\mathbf{C})$$

correspond to Dirichlet characters $\chi \colon (\mathbf{Z}/N\mathbf{Z})^* \to \mathbf{C}^*$ via the identification of abelian extensions of $\mathbf{Q}$ with cyclotomic extensions obtained by adjoining N-th roots of unity. Specifically, there is an isomorphism

$$(\mathbf{Z}/N\mathbf{Z})^* \to \mathrm{Gal}(\mathbf{Q}(\mu[N])/\mathbf{Q})$$

that takes an integer k relatively prime to N to the automorphism that takes roots of unity to their k-th power. Here, $\mu[N]$ denotes the set of N-th roots of unity in the algebraic closure of $\mathbf{Q}$.

This implies that the maximal abelian quotient of $G_\mathbf{Q}$ is

$$G_\mathbf{Q}^{ab} \simeq \mathrm{Gal}(\mathbf{Q}(\mu)/\mathbf{Q})$$

where μ denotes the group of all roots of unity in $\overline{\mathbf{Q}}$.

In order to sharpen this picture, we focus on ℓ-power roots of unity, where ℓ is a prime. We get representations

$$\chi[\ell^n] \colon G_\mathbf{Q} \to \mathrm{Aut}(\mu[\ell^n]) \simeq (\mathbf{Z}/\ell^n\mathbf{Z})^*$$

that are "compatible" for different n in the sense that $\chi[\ell^{n+1}]$ reduces to $\chi[\ell^n]$ modulo ℓ^n. Let $T_\ell(\mu)$ denote the **Tate module** that is the inverse limit of the $\mu[\ell^n]$, with respect to the "raising to the ℓ-th power" map from $\mu[\ell^{n+1}]$ to $\mu[\ell^n]$. One can think of an element of this inverse limit as a sequence

$$(u_1, u_2, \cdots)$$

where $u_n \in \mu[\ell^n]$ and $u_{n+1}^\ell = u_n$.

The Galois group acts on this inverse limit in the obvious way and we get a representation
$$\chi_\ell \colon G_{\mathbf{Q}} \to \operatorname{Aut}(T_\ell(\mu)) \simeq \mathbf{Z}_\ell^*.$$
We note that $T_\ell(\mu)$ is isomorphic to the ℓ-adic integers $\mathbf{Z}_\ell$ as an abstract group, but that it has further structure in the sense that the group $G_{\mathbf{Q}}$ acts on it via the representation χ_ℓ. This representation is called the **cyclotomic character** and it can be concretely defined by observing that
$$\sigma(x) = x^{\chi_\ell(\sigma)}$$
where $x \in \mu[\ell^n]$ for some n, and $\chi_\ell(\sigma)$ is in $\mathbf{Z}_\ell^*$. Note that for the action on $x \in \mu[\ell^n]$ we only need $\chi_\ell(\sigma) \bmod \ell^n$.

3. Galois representations coming from elliptic curves

Let E be an elliptic curve over $\mathbf{Q}$ and let $n > 1$. The points in $E(\overline{\mathbf{Q}})$ that are of order n form a group that $E[n]$, called the group of n-division points on E, that is isomorphic to the direct product of cyclic groups of order n:
$$E[n] \simeq \mathbf{Z}/n\mathbf{Z} \times \mathbf{Z}/n\mathbf{Z}.$$
In addition, the Galois group $G_{\mathbf{Q}}$ acts on this group: if $\sigma \in G_{\mathbf{Q}}$ and P is a point of order n then so is σP.

Here if $P = (x, y)$ then $\sigma P = (\sigma x, \sigma y)$ is obtained from P by applying the automorphism σ to both components of P. The basic fact that
$$\sigma(P + Q) = \sigma P + \sigma Q$$
can be proved from the explicit formulas for the group operation, or in terms of the definition of the group operation in terms of $\operatorname{Pic}^0(E)$.

All in all, this result gives us a representation
$$\rho_E[n] \colon G_{\mathbf{Q}} \to \operatorname{Aut}(E[n]) \simeq \operatorname{GL}_2(\mathbf{Z}/n\mathbf{Z})$$
where the group on the right is the group of all 2-by-2 invertible matrices over the ring $\mathbf{Z}/n\mathbf{Z}$. We will denote $\rho_E[n]$ by $\rho[n]$ if the elliptic curve E is understood.

Theorem 6.1: The x and y coordinates of nontrivial n-division points on E have algebraic coordinates. If K_n is the field extension of $\mathbf{Q}$ obtained by adjoining all coordinates of those points, then K_n is a Galois extension and the representation $\rho[n]$ factors through $\operatorname{Gal}(K_n/\mathbf{Q})$, i.e., there is a diagram
$$G_{\mathbf{Q}} \to \operatorname{Gal}(K_n/\mathbf{Q}) \hookrightarrow \operatorname{GL}_2(\mathbf{Z}/n\mathbf{Z})$$
where the map on the left is the canonical surjection and the map on the right is injective.

A difficult theorem of Serre ([**Se3**]) says that this Galois representation on points of order n is in fact usually as large as possible for n not divisible by a fixed finite set of primes. To state this more precisely we need to make a definition. The "multiply by m" map is an endomorphism for any elliptic curve E, so $\mathbf{Z} \subset \operatorname{End}(E)$ in a natural way. If the group of endomorphisms is strictly larger then E is said to have **complex multiplication**. Although elliptic curves with complex multiplication are very special, the theory of elliptic curves with complex multiplication is a rich source of ideas, and leads to an explicit class field theory (classification of abelian extensions) of imaginary quadratic fields.

Theorem 6.2 (Serre): If E is an elliptic curve defined over $\mathbf{Q}$ that does not have complex multiplication, then there is an integer m such that if $\gcd(m,n) = 1$ then the map $\rho[n]$ is surjective.

In other words, the coordinates of n-division points on an elliptic curve "usually" generate Galois extensions of $\mathbf{Q}$ whose Galois group is $\mathrm{GL}_2(\mathbf{Z}/n\mathbf{Z})$.

As in the case of one-dimensional representations, we now recast these ideas by focusing on powers of a fixed prime ℓ.

The representations
$$\rho[\ell^n]\colon G_{\mathbf{Q}} \to \mathrm{Aut}(E[\ell^n])$$
are compatible for different n. The inverse limit of $E[\ell^n]$ consists of all $(P_1, P_2, \cdots)$ where $P_n \in E[\ell^n]$ and $\ell P_{n+1} = P_n$. This group is denoted $T_\ell(E)$ and is the Tate module of the elliptic curve. It is isomorphic to $\mathbf{Z}_\ell^2$ as a group, and has a natural $\mathbf{Z}_\ell$-module structure and a natural action of $G_{\mathbf{Q}}$. By passing to the inverse limit of the $\rho[\ell^n]$ one gets a two-dimensional Galois representation
$$\rho_\ell\colon G_{\mathbf{Q}} \to \mathrm{Aut}(T_\ell(E)) \simeq \mathrm{GL}_2(\mathbf{Z}_\ell).$$

We state several facts about this Galois representation.

Theorem 6.3:
- $\det(\rho_\ell) = \chi_\ell$.
- If E has good reduction at p, and $p \neq \ell$, then
$$\mathrm{tr}(\rho_\ell(\mathrm{Frob}_p)) = 1 + p - \#\overline{E}(\mathbf{F}_p).$$
- The reduction mod ℓ of ρ_ℓ, i.e.,
$$\rho[\ell] = \rho_\ell \bmod \ell\colon G_{\mathbf{Q}} \to \mathrm{GL}_2(\mathbf{F}_\ell),$$
is irreducible for almost all primes ℓ.

Notice that the trace of Frobenius is independent of the choice of the prime ℓ (so long as $\ell \neq p$ and E has good reduction at p).

4. The conductor

The conductor $N = N_E$ of an elliptic curve E can be defined in terms of the Galois representations associated to the curve. The conductor of an elliptic curve E over $\mathbf{Q}$ is
$$N = N_E = \prod_p p^{e_p},$$
where the exponents e_p are nonzero only for the primes where E has bad reduction. To define the exponent e_p for p we choose a prime $\ell \neq p$ (the formula is independent of this choice) and set
$$e_p = \mathrm{tame}_p + \mathrm{wild}_p$$
where
$$\mathrm{tame}_p = \mathrm{codim}(V_\ell^{I(\overline{\mathbf{Q}}/\mathbf{Q})})$$
where $V_\ell = T_\ell(E) \otimes \mathbf{Q}_\ell$, $I(\overline{\mathbf{Q}}/\mathbf{Q})$ is the inertia group at p, and $V_\ell^{I(\overline{\mathbf{Q}}/\mathbf{Q})}$ denotes the vector space fixed by the inertia subgroup.

Since V_ℓ is a 2-dimensional vector space over $\mathbf{Q}_\ell$, the tame contribution to the exponent is either 0, 1, or 2 and the respective cases are easy to describe:

Theorem 6.4:

$$\text{tame}_p = \begin{cases} 0 & \text{if } E \text{ has good reduction at } p \\ 1 & \text{if } E \text{ has multiplicative reduction at } p \\ 2 & \text{if } E \text{ has additive reduction at } p \end{cases}$$

The wild part is a harder to describe, though as a (minor) compensation it depends only on the representation $\rho[\ell]$ on the field K_ℓ of ℓ-division points. Specifically, we define the higher ramification groups by

$$G_i = \{\sigma \in \text{Gal}(K_\ell/\mathbf{Q}) : \sigma(x) \equiv x \bmod P^{i+1}\}$$

where P is some prime over p in K_ℓ and x ranges over all elements of the ring of integers in that field. Then the wild part of the conductor is

$$\text{wild}_p = \sum_{i=1}^{\infty} \frac{\text{codim}(E[\ell]^{G_i})}{[G_0 : G_i]}.$$

Here G_0 is the inertia group in the finite extension K_ℓ. Also, the sum is finite since sooner or later the G_i are trivial and the codimension is zero.

This formula can be difficult to compute. Fortunately, if you have access to a program that implements Tate's algorithm for finding the minimal Weierstrass equation and the **Néron model** of the elliptic curve, then the computer can do the hard work for you.

Let

$c_p := $ the number of components of the special fiber of the Néron model at p.

Theorem 6.5: The wild part of the exponent of the conductor is zero unless $p = 2$ or $p = 3$ and E has additive reduction at p. In that case,

$$\text{wild}_p = \text{ord}_p(\Delta(E)) + 1 - c_p$$

where $\text{ord}_p(\Delta(E))$ is the power of p in the minimal discriminant of E.

LECTURE 7
The Rank of Elliptic Curves over Q

The earlier "Gentleman's Diary" problem of finding two integers whose sum, sum of squares, and sum of cubes were all perfect squares leads to the problem of finding rational points on the elliptic curve

$$E: \quad y^2 = x(x+2)(x+6).$$

The torsion points are easy to determine: there are three obvious points of order two, and the Nagell-Lutz Theorem can be used to verify that there are no other points of finite order. However, these points do not produce solutions to the original problem, so we need to determine full group of rational points $E(\mathbf{Q})$.

The Mordell-Weil theorem says that $E(\mathbf{Q})$ is finitely generated. (This was originally proved by Mordell, and then generalized, to elliptic curves over larger fields, and to abelian varieties, by Weil.)

For the curve above, the torsion is elementary abelian of order four, and the finitely generated abelian group $E(\mathbf{Q})$ has the shape

$$E(\mathbf{Q}) \simeq \mathbf{Z}^r \times E(\mathbf{Q})_{\text{tors}} \simeq \mathbf{Z}^r \times \mathbf{Z}/2\mathbf{Z} \times \mathbf{Z}/2\mathbf{Z}$$

for some integer r, called the rank of the curve. We would like to find the rank, and find an explicit set of r generators of infinite order.

The natural way to bound the rank from below is to search for points. In this case, one immediately finds, for instance, the point $P = (2, 8)$. We find $2P = (1/4, -15/8)$, so this point is of infinite order. Thus the rank r is at least 1 and suitable multiples of P can be used to construct solutions to the original problem.

This lecture gives an overview of the proof of the Mordell-Weil theorem, and develops some of the ideas, implicit in that proof, for actually finding the rank. This algorithmic problem is interesting, and can be approached from several directions depending on resources and goals: one might want to solve problems by hand, write computer programs to solve the problem, learn to use existing software, or to analyze the asymptotic complexity of the problem.

There are a number of open problems concerning the rank of elliptic curves; e.g., finding an algorithm (i.e., a procedure guaranteed to terminate) to find the rank, finding elliptic curves over $\mathbf{Q}$ of arbitrarily large rank, or determining the density of curves of a given rank (see [**Sil**]).

1. Mordell-Weil

There are two key ingredients in proof of the Mordell-Weil Theorem. One is a simple assertion that there are only finitely many points in the group "modulo m-th powers."

Theorem 7.1 (Weak Mordell-Weil): If E is an elliptic curve over $\mathbf{Q}$ and m is a positive integer then
$$E(\mathbf{Q})/mE(\mathbf{Q}) \quad \text{is finite.}$$

This result will be considered in more detail later in the lecture. It suffices for the sequel to use the special case $m = 2$, and we will fix stick to that value for the remainder of this lecture.

The second ingredient in the proof is a method for measuring the size or "height" of points in the group. If there are only finitely many points of height at most c, and there is a generating set among those points, then the Mordell-Weil group $E(\mathbf{Q})$ is finitely generated.

There are a number of different "naive" notions of the height of a point. One natural notion of the height of a rational number is (proportional to) the maximum number of digits in the numerator and denominator of that number (expressed in least terms), i.e.,
$$\text{height}(a/b) := \max(\log(|a|), \log(|b|)), \qquad \gcd(a,b) = 1.$$

We could then define the height of a rational point $P = (x, y)$ on an elliptic curve to be, for instance, the height of its x-coordinate:
$$\text{height}(P) = \text{height}(x).$$

Although this is a sufficient notion for the proof of the Mordell-Weil Theorem, it turns out to be far more elegant to consider the **canonical** (or Néron-Tate) height function
$$\hat{h}(P) = \frac{1}{2} \lim_{n \to \infty} \frac{\text{height}(2^n P)}{4^n}.$$

Theorem 7.2 (Canonical Height Properties):
1. For all c, the set of points $\{P \in E(\mathbf{Q}) : \hat{h}(P) \leq c\}$ is finite.
2. The parallelogram law holds: For all point P and Q,
$$\hat{h}(P + Q) + \hat{h}(Q - P) = 2\hat{h}(P) + 2\hat{h}(Q).$$
3. The canonical height function is quadratic in the sense that $\hat{h}(rP) = r^2 \hat{h}(P)$.
4. The canonical height is nonnegative, and is zero only on torsion points:
$$\hat{h}(P) \geq 0, \qquad \hat{h}(P) = 0 \iff P \in E(\mathbf{Q})_{\text{tors}}.$$

This result says that $\hat{h}$ is a quadratic form on the group $E(\mathbf{Q})$. It turns out that the canonical height function is essentially unique, so that if one starts with a different notion of the naive (logarithmic) height of a point, it refines to the same canonical height function. Careful treatments of the canonical height can be found in all of the standard texts on elliptic curves.

The Weak Mordell-Weil theorem, and the basic facts about canonical heights enable an easy proof of the Mordell-Weil Theorem. The proof is a descendant of Fermat's technique of "infinite descent" that he used successfully for several Diophantine equations.

Theorem 7.3 (Mordell-Weil): If E is an elliptic curve over $\mathbf{Q}$ then $E(\mathbf{Q})$ is finitely generated.

Proof. Let
$$P_1, \cdots, P_k$$
be elements of $E(\mathbf{Q})$ whose cosets in $E(\mathbf{Q})/2E(\mathbf{Q})$ cover the entire group. Let
$$c := \max(\hat{h}(P_j)).$$
The set
$$S = \{P \in E(\mathbf{Q}) : \hat{h}(P) \leq c\}$$
is finite, by the Canonical Height Theorem, and we finish the proof by showing that this set generates $E(\mathbf{Q})$. The idea is to show that by taking an arbitrary point and successively adding or subtracting elements of S one ends up at the identity. This shows that S generates $E(\mathbf{Q})$ as desired.

Suppose that P is any point in $E(\mathbf{Q})$. Then
$$P = P_j + 2Q$$
for some $P_j \in S$ and some point $Q \in E(\mathbf{Q})$. If $\hat{h}(Q) \leq c$ we are done (both points on the right hand side are in S) we are done. If $\hat{h}(Q) > c$ then the parallelogram inequality in the Canonical Height Theorem gives
$$\hat{h}(P + P_j) + \hat{h}(P - P_j) = 2\hat{h}(P) + 2\hat{h}(P_j).$$
Either $\hat{h}(P - P_j)$ or $\hat{h}(P + P_j)$ is less than or equal to $\hat{h}(P) + \hat{h}(P_j)$.

In the former case we have
$$\hat{h}(2Q) = \hat{h}(P - P_j) \leq \hat{h}(P) + \hat{h}(P_j) \leq \hat{h}(P) + c < 2\hat{h}(P).$$
Thus $\hat{h}(Q) \leq \hat{h}(P)/2$.

In the case $\hat{h}(Q + P_j) \leq \hat{h}(Q) + \hat{h}(P_j)$ then we apply the same argument to $2(Q + P_j)$ to arrive at $\hat{h}(Q + P_j) \leq \hat{h}(P)/2$.

In either case, we halve the height; continuing in this way we quickly arrive at a point of height less than c, i.e., a point in S. Reversing the steps shows that the original point P is in the group generated by S, as desired. □

2. The Weak Mordell-Weil Theorem

There are several methods for proving the weak Mordell-Weil theorem. We work through an elementary proof for of curves of the form
$$E : y^2 = (x - a)(x - b)(x - c), \quad a, b, c \in \mathbf{Z}$$
and then sketch what is involved in the general case.

The first step is to show that the somewhat unlikely map taking a point $(x, y) \in E(\mathbf{Q})$ to the rational number $x - a$ modulo squares is a group homomorphism
$$\phi_a : E(\mathbf{Q}) \to \mathbf{Q}^*/\mathbf{Q}^{*2}.$$
Actually, we need a slightly more precise definition, since the formula does not make sense for the point O or the point $(a, 0)$ of order 2.

We note the multiplicative group $\mathbf{Q}^*/\mathbf{Q}^{*2}$ of nonzero rational numbers modulo squares has exponent 2, i.e., the square of any element is the identity.

Lemma 7.4: If we define $\phi_a \colon E(\mathbf{Q}) \to \mathbf{Q}^*/\mathbf{Q}^{*2}$ by
$$\phi_a(P) = \begin{cases} [1] & P = O \\ [(a-b)(a-c)] & P = (a,0) \\ [x-a] & P = (x,y), x \neq a \end{cases}$$
then ϕ_a is a homomorphism.

Proof. Since $\phi_a(0) = 1$ and $\phi_a(-P) = \phi_a(P)$ it suffices to show that if
$$P_1 + P_2 + P_3 = 0 \quad \text{then} \quad \phi_a(P_1)\phi_a(P_2)\phi_a(P_3) = 1.$$
If none of the P_i are the identity or the point $(a,0)$ of order 2, then there is a line $y = rx + s$ that intersects the curve in the $P_i = (x_i, y_i)$. By calculating the intersection between the line and the curve by eliminating y we see that we must have an equation
$$(X - x_1)(X - x_2)(X - x_3) = (X-a)(X-b)(X-c) - (rX+s)^2$$
which says that
$$(x_1 - a)(x_2 - a)(x_3 - a) = (ra + s)^2$$
which means that $\phi_a(P_1)\phi_a(P_2)\phi_a(P_3) = 1$ as desired.

The result is immediate if any of the P_i are the identity, and is an easy calculation if one or more of the P_i are equal to $(a,0)$. $\square$

Now we consider all three roots at once and define
$$\phi \colon E(\mathbf{Q}) \to \mathbf{Q}^*/\mathbf{Q}^{*2} \times \mathbf{Q}^*/\mathbf{Q}^{*2} \times \mathbf{Q}^*/\mathbf{Q}^{*2}, \qquad \phi(P) = (\phi_a(P), \phi_b(P), \phi_c(P)).$$
By the previous lemma this is a homomorphism. Note that the image is contained in the set of triples (u, v, w) such that $uvw = 1$. Indeed, if $P = (x,y), x \neq a$, then
$$\phi_a(P)\phi_b(P)\phi_c(P) = [(x-a)(x-b)(x-c)] = [y^2] = 1$$
and there are similarly immediate calculations for $P = O$ and $P = (a,0)$.

One can say more about the image of ϕ.

Lemma 7.5: The image of ϕ is finite.

Proof. The natural choice of coset representatives for $\mathbf{Q}^*/\mathbf{Q}^{*2}$ is the set of square-free integers. To say that a prime p occurs to an odd power (positive or negative) in $x \in \mathbf{Q}^*/\mathbf{Q}^{*2}$ is to say that its natural coset representative is divisible by p. We will show that the only primes that can occur to an odd power in the components of $\phi(P)$ are primes that divide
$$d := (a-b)(b-c)(c-a).$$
Since there are only finitely many primes, there are only finitely many elements in the image of ϕ.

From results in the second lecture we know that if $P = (x,y)$ is a non-identity point then the denominator of x is a perfect square, and the denominator of y is a cube, i.e., $(x,y) = (u/w^2, v/w^3)$ where the integer w is relatively prime to u and v. Thus we get an equation in integers
$$v^2 = (u - aw^2)(u - bw^2)(u - cw^2).$$

If p divides the first two factors on the right hand side then it divides their difference
$$(u - aw^2) - (u - bw^2) = (a-b)w^2.$$

LECTURE 7. THE RANK OF ELLIPTIC CURVES OVER Q

Since p does not divide w (since it would then also divide u), it follows that p divides $a - b$ and therefore d.

If p divides only one of the factors on the right hand side then it divides it to an even power (since the left hand side is a perfect square).

All in all, we conclude the the only primes that divide components of $\phi(P)$ to odd powers are primes dividing d, and this proves the lemma. $\square$

The last result needed to prove the weak Mordell-Weil Theorem is a description of the kernel of ϕ.

Lemma 7.6: $\ker(\phi) = 2E(\mathbf{Q})$.

Note that this lemma finishes the proof that $E(\mathbf{Q})/2E(\mathbf{Q})$ is finite, since it shows that $E(\mathbf{Q})/2E(\mathbf{Q})$ maps injectively to a finite group.

Proof. Suppose that $\phi(P) = 1$, i.e., that there are rational numbers u, v, and w such that
$$x - a = u^2, \quad x - b = v^2, \quad x - c = w^2$$
where $P = (x, y)$. (The case that $P = (a, 0)$ is similar, and simpler, and will be left to the reader.) We want to show that there is a P' in $E(\mathbf{Q})$ such that $2P' = P$. This will be done by showing that there is a line that intersects the curve once at $-P$ and twice at P'.

To simplify the algebra, we can assume by translation that $x = 0$. This means that $a = -u^2$, $b = -v^2$, and $c = -w^2$.

Define e_i as elementary symmetric functions of u, v, w by
$$g(x) := (x - u)(x - v)(x - w) = x^3 - e_1 x^2 + e_2 x - e_3.$$

We now need a straightforward (if somewhat unmotivated) algebraic result.

Sublemma 7.7:
$$(X - a)(X - b)(X - c) = X(X - e_2)^2 + (e_1 X - e_3)^2.$$

The Sublemma gives an immediate proof of the Lemma; indeed it implies that the line
$$Y = e_1 X - e_3$$
intersects the cubic once at a point with $X = 0$ and twice at a point with $X = e_2$. $\square$

Proof. (of the Sublemma) This is straightforward high-school algebra, and can be proved directly. In an attempt to use at least a little bit of pure thought, we note that the polynomial $g(Z)g(-Z)$ is a monic cubic polynomial in $X = Z^2$ with root $X = u^2$, and similarly for v and w. Thus
$$g(Z)g(-Z) = (X - u^2)(X - v^2)(X - w^2) = (X + a)(X + b)(X + c).$$
Also
$$\begin{aligned} g(Z)g(-Z) &= (Z^3 - e_1 Z^2 + e_2 Z - e_3)(-Z^3 - e_1 Z^2 - e_2 Z - e_3) \\ &= -(Z^3 + e_2 Z - e_1 Z^2 - e_3)(Z^3 + e_2 Z - e_1 Z^2 - e_3) \\ &= -(Z^3 + e_2 Z)^2 + (e_1 Z^2 + e_3)^2 \\ &= -X(X + e_2)^2 + (e_1 X + e_3)^2. \end{aligned}$$

Replacing X by $-X$ and doing the algebra carefully gives the formula in the sublemma. $\square$

Example 7.8. The Gentleman's Diary problem considered in the first lecture gives rise to the elliptic curve
$$y^2 = x(x+2)(x+6).$$
Here we have $d = \pm 36$. Thus the image of ϕ lies in the multiplicative group of triples whose product is 1, and where each component has the form
$$(-1)^r 2^s 3^t, \qquad r, s, t \in \{0, 1\}.$$
This group is of order 2^6.

The image of ϕ has order at least 2^3 (from the two points of order 2 and the easily found point $(2, 8)$ of infinite order. Thus we conclude that the rank r satisfies
$$1 \leq r \leq 4$$
where the lower bound comes from a brute force search and the upper bound comes from the "descent."

In fact, one can refine the idea of the proof obtain sharper bounds on the rank (see, e.g., [**Kna**, p. 108]). For instance, if (x, y) is on the curve then $x \geq 0$ so $\phi_0(x)$ is never negative. A more careful analysis in this case shows that $r \leq 1$ so that this curve has rank exactly equal to 1.

This situation is typical: one finds lower bounds by searches and upper bounds by carrying out a constructive version of the proof. If these bounds agree we are happy. If not, our options are to search further or think more carefully about the upper bound argument, or do an "n-descent" by attempting to calculate bounds on $E(\mathbf{Q})/nE(\mathbf{Q})$ for $n > 2$ (which is hard!).

If the curve has the form $y^2 = f(x)$ where the cubic $f(x)$ does not have three rational roots then the calculations are more complicated, but the ideas are basically the same. The starting point is a map
$$\phi \colon E(\mathbf{Q})/2E(\mathbf{Q}) \to A^*/A^{*2}$$
that is an injective homomorphism with finite image. Here $A = \mathbf{Q}[X]/(f(X))$ is a ring that is a product of fields, according to how $f(X)$ factors over $\mathbf{Q}$, and A^* are the units of that ring. If $f(X)$ has three rational roots then A is the direct product of three copies of $\mathbf{Q}$ (the case considered above), if $f(X)$ has one rational root then A is the direct product of $\mathbf{Q}$ and a quadratic field, and A is a field if $f(X)$ is irreducible. The image of ϕ consists of elements such that each component satisfies the condition that adjoining the square root gives an extension of that component field that is unramified outside primes dividing the discriminant of $f(X)$. Standard results in algebraic number theory enable one to limit this group (see [**Sma**], [**Cas**], [**Si1**], or [**Kna**] for examples).

More generally, one can systematize this idea by using Galois cohomology [**Ta2**] to describe an explicitly computable finite "Selmer group" $S^n(E)$ that contains $E(\mathbf{Q})/nE(\mathbf{Q})$ as a subgroup. The calculation of the Selmer group, together with an attempt to identify $E(\mathbf{Q})/nE(\mathbf{Q})$ via these calculations, is referred to as "doing an n-descent." General cohomological machinations produce a group $\text{III}(E)$, called the Tate-Shafarevich group, such that the points $\text{III}(E)[n]$ of order n in the group are the quotient of the Selmer group by $E(\mathbf{Q})/nE(\mathbf{Q})$, i.e., there is an exact sequence
$$0 \to E(\mathbf{Q})/nE(\mathbf{Q}) \to S^n(E) \to \text{III}(E)[n] \to 0.$$
If the Tate-Shafarevich group is finite, then n-descent will prove successful for some n. Since the Tate-Shafarevich group is not known to be finite in general,

we do not have an algorithm (i.e., a procedure that is guaranteed to terminate) that determines $E(\mathbf{Q})$. Performing n-descents often works in practice, but the calculations can be truly arduous.

3. The analytic rank

Let $L(E,s)$ be the L-function of the elliptic curve E over $\mathbf{Q}$. By the theorem stated in Lecture 5 (i.e., the completion of the [**Wil**] and [**TW**] in [**Brl**]) the curve is modular and there is a holomorphic continuation of $L(E,s)$ to the entire complex plane, and a functional equation

$$\Lambda(E,s) = N^{s/2}(2\pi)^{-s}\Gamma(s)L(E,s) = w\Lambda(2-s), \qquad w = \pm 1.$$

The **analytic rank** of E is defined to be the order of vanishing of the L-function at $s=1$. One part of the famous conjecture of Birch and Swinnerton-Dyer asserts that the rank $r(E)$ is equal to the analytic rank, i.e., that

$$r(E) = \mathrm{ord}_{s=1} L(E,s).$$

The full conjecture also gives a conjectured value for the leading term of the L-function at $s=1$.

If the conductor isn't too large the analytic rank can be computed fairly easily by re-fashioning the L-series, and its derivatives, into a series with exponentially decreasing terms. The first step is to give a formula for the value of the L-series.

Theorem 7.9: Let t be a positive real number. Let E be an elliptic curve over $\mathbf{Q}$ with L-series

$$L(E,s) = \sum_{n=1}^{\infty} \frac{a_n}{n^s}, \qquad \mathrm{Re}(s) > 2.$$

Let N denote the conductor, and w the sign in the functional equation. Then

$$L(s,1) = \sum_{n=1}^{\infty} \frac{a_n}{n} e^{-2\pi nt/\sqrt{N}} + w \sum_{n=1}^{\infty} \frac{a_n}{n} e^{-2\pi n/(t\sqrt{N})}.$$

Remark 7.10. The theorem implies that the right hand side of the formula for $L(s,1)$ is independent of the choice of t. The formula is simplest when we choose t to be equal to 1 (for instance, if $w=-1$ then $L(E,1)$ is obviously zero). However, in practice it is often convenient to vary the parameter numerically; the extent to which the value changes is a good indication of the correctness of the program and the numerical stability of the results.

One straightforward way to prove this (see [**Cre**, section 2.13]) is to let $f = \sum a_n q^n$ be the cusp form of weight two corresponding to E, to calculate that

$$\Lambda(E,s) = N^{s/2}(2\pi)^{-s} L(f,s) = \int_0^{\infty} f\left(\frac{iy}{\sqrt{N}}\right) y^{s-1} dy.$$

One the finds that

$$\int_t^{\infty} f\left(\frac{iy}{\sqrt{N}}\right) dy = \frac{\sqrt{N}}{\pi} \sum_{n=1}^{\infty} \frac{a_n}{n} e^{-2\pi nt/\sqrt{N}}.$$

Using the modularity property of f (as in the proof of the functional equation) the integral from 0 to t can be recast as an almost identical integral to the previous integral, except that the interval of integration is $[t^{-1}, \infty)$. The formula in the theorem results.

Remark 7.11. Analogous results for much more general L-series of number theoretic interest are due to Lavrik, Friedman and others, and can be found in, for example, [**Co2**, p. 508]. In addition, it is shown there that higher derivatives of the L-series can be re-written similarly so that the order of vanishing can thereby be determined numerically. Both of these generalizations typically have more complicated exponentially decreasing special functions in the expansions, and evaluating these can be tricky.

The upshot of this analysis is that the analytic rank is relatively easy to compute. After trying to find the rank with a simple search for points and a simple 2-descent, it is probably best, if those calculations haven't settled the question, to ask your elliptic curve software package to calculate the analytic rank.

LECTURE 8
Applications of Elliptic Curves

Elliptic curves probably first arose in the context of Diophantine equations, but by the nineteenth century their study had spread in many guises to other areas of mathematics, including differential equations, algebraic geometry, and complex analysis. During the past twenty years elliptic curves have emerged as important objects of study in surprising new arenas including algorithms for primality, algorithms for factorization, and cryptography. The goal of this lecture is to give a small taste of these developments by briefly sketching two algorithms, and then describing a cryptosystem based on elliptic curves.

1. Primality

If a prime is needed in the real world, e.g., for cryptographic purposes, one starts by performing *pseudoprimality* tests. For instance, one could attempt to verify that p is prime by checking that

$$a^{p-1} \equiv 1 \bmod p$$

for a few values of a (mild generalizations of this idea are often used in practice).

These tests don't prove the primality of p, but they provide compelling evidence. Indeed, in many programs, numbers that pass these tests (called pseudoprimes) are assumed to be primes. (One argument cited in favor of this is that the probability of coming up with a composite pseudoprime is far less than the probability of a software bug, or hardware glitch, an errant cosmic ray, or mistake in a proof.)

What if one actually wants to *prove* that p is prime? Then more work is required. In recent years two different approaches have become popular for large numbers, and it is still a matter of debate as to which is better. One uses cyclotomy, due to Lenstra and others ([**Mih**]), and another, due to Goldwasser, Kilian, Atkin, Morain, and others, uses elliptic curves ([**Mor**]).

The Goldwasser-Kilian primality test is a precursor of the Atkin and Morain test. Although not as practical as the latter, it is easy describe, and was the starting point for the proof of Adleman and Huang that there is a probabilistic primality algorithm that takes expected polynomial time. The Goldwasser-Kilian test says, roughly, that if one can find an elliptic curve E such that the group of points on that curve modulo an integer p behaves like the group of points $E(\mathbf{F}_p)$, then p is a prime.

The basic result on points of elliptic curves over finite fields, which colors all of the results in this lecture, is Hasse's Theorem which gives an estimate for the number of points. This estimate says that the number of points is approximately equal to the number of points on the projective line, with an error term that is $O(\sqrt{p})$.

Theorem 8.1 (Hasse): If E is an elliptic curve over the field $\mathbf{F}_p$ of p elements, where p is a prime, then the number of points $\#E(\mathbf{F}_p)$ satisfies
$$|\#E(\mathbf{F}_p) - (p+1)| < 2\sqrt{p}.$$

The most straightforward algorithm to find $\#E(\mathbf{F}_p)$ is to count the number of points on the curve with each possible x-coordinate. Suppose that p is odd, and E has the form $y^2 = x^3 + ax + b$. Let
$$\left(\frac{u}{p}\right) \in \{-1, 0, 1\}$$
be the usual Legendre symbol. Then the number of points on the curve with first coordinate x is 0, 1, or 2 according to whether $x^3 + ax + b$ is a non-square, zero, or a nonzero square. Thus the number of points with first coordinate equal to x is
$$1 + \left(\frac{x^3 + ax + b}{p}\right).$$
Summing over all x (and remembering the point at infinity) gives
$$\#E(\mathbf{F}_p) = 1 + p + \sum_{x=0}^{p-1}\left(\frac{x^3 + ax + b}{p}\right).$$

It clearly takes time $O(p)$ to compute this sum, where as usual $O(p)$ denotes a function whose absolute value is bounded by a constant times p.

Is this fast or slow? For small p it is fast, but for large p it is excruciatingly slow. The standard way to quantify this is to say that the running time is an exponential function of the input length. Indeed, the input, p, takes $O(\log(p))$ bits to represent in any fixed base, so the input is of length proportional to $\log(p)$ and the running time is therefore exponential in the input length.

More recently Schoof ([**Scf**]) devised a polynomial-time algorithm to find $\#E(\mathbf{F}_p)$. This idea was later improved by Atkin, Elkies, and others (see [**BSS**] and references therein), and is now reasonably efficient, e.g., for primes of the size that is currently interesting in the cryptographic community. Interest in this problem has been an important focus of research, and the techniques use a surprisingly large amount of the theory of elliptic curves.

The Goldwasser-Kilian algorithm for proving the primality of a pseudoprime p is a probabilistic procedure (with "side-exits" and "re-tries" as explained below). First we give the algorithm, and then explain how it makes sense even if we don't know that p is prime. The steps in the Goldwasser-Kilian algorithm are:

1. Choose a random elliptic curve $E: y^2 = x^3 + ax + b$ over $\mathbf{F}_p$.
2. Use Schoof's algorithm to find $n := \#E(\mathbf{F}_p)$.
3. Attempt to factor n into $n = qn'$ where q is a pseudoprime and
$$q > (p^{1/4} + 1)^2.$$

4. Choose a random point $P \in E(\mathbf{F}_p)$ and verify that $n'P \neq O$, and $nP = O$.
5. Verify the primality of q.

Given that p is not known to be prime, what sense does this make? One way to proceed is to define $E(\mathbf{F}_p)$ when p is not necessarily prime. Instead, we will proceed at a more naive level and look at what happens when we work "as if" p was a prime until we have evidence to the contrary.

The algorithm can then fail in several ways:

- For instance, in the second step it might turn out that the group operation is undefined (since we might be asked to invert a nonzero element that does not have an inverse if p isn't a prime).
- In the third step, the factorization might not exist, or it might not have the desired form.
- The fourth step the group operation might again fail to be defined.
- The fourth step might fail because $n'P = O$ or because $nP \neq O$.
- Finally, the (recursive) call in the last step might fail.

If the group operation fails, or if $nP \neq 0$ in step 4, then we have a "side-exit" in the sense that we have an explicit and constructive disproof of the primality of p.

In the other cases, we "re-try" by choosing another curve (for failures in steps 3 or 5) or a new point P. The crucial point is that if everything does work, then p is a prime.

Theorem 8.2: If no failures are encountered in the above procedure then p is a prime.

Proof. We are given an equation modulo p for E, a solution P of the equation, an integer n such that $nP = O$ when we perform additions on the curve "as if p were prime", and a prime factor q of n with $q > (p^{1/4} + 1)$ such that $(n/q)P \neq O$.

Let r be the (unknown) smallest prime factor of p. The additions modulo p on E in the algorithm are valid computations in $E(\mathbf{F}_r)$ in the sense that the coordinates, modulo r, represent points on $E(\mathbf{F}_r)$ that are added according to the group operation on that group. The one additional remark that is needed here is that if we have an arbitrary point, say in homogeneous coordinates $x:y:z$ where the components are integers, we can detect whether it is O modulo an unknown prime divisor by checking that the coordinates have no common divisor, and z is relatively prime to p. (This is needed in step 4 when we test that $n'P \neq O$.)

The point $P' := n'P$ has the property that $P' \neq 0$ in $E(\mathbf{F}_r)$, but $qP' = 0$. This means that $\#E(\mathbf{F}_r)$ is divisible by q. Therefore

$$(r^{1/2} + 1)^2 = r + 2\sqrt{r} + 1 > \#E(\mathbf{F}_r) \geq q > (p^{1/4} + 1)^2.$$

The outer terms give $r > p^{1/2}$. The smallest prime factor of p is either equal to p, or less than $\sqrt{p}$. Thus $r = p$, and p is prime as claimed. $\square$

2. Factoring

Suppose that n is a large composite integer that is divisible by at least two primes (i.e., is not a prime power.) One could attempt to find a prime factor of n by trial division by successively larger integers. If n is at all large this is likely to be totally impractical, since this requires as many as $O(n^{1/2})$ divisions. For numbers of relevance in modern cryptography (say, 300 digits) this process would require, using any known or plausible computer, more time than has elapsed since (current guesses at) the beginning of the universe.

This is an exponential time algorithm since the running time $O(n^{1/2})$ is exponential in the input size $O(\log n)$ of the integer n.

Many algorithms have been proposed for factoring integers. Although at the moment no algorithm is known whose running time is bounded by a polynomial in the input size, some sub-exponential algorithms are known. To describe one of these, namely the "Elliptic Curve Method" of Lenstra ([**Len**]), it is useful to recall the Pollard $p-1$ method, which succeeds (in reasonable time) if n has a special property.

Suppose that we are trying to factor an integer n, and assume that there is a prime divisor p of the number n that has the property that $p-1$ factors into a product of distinct primes less than an integer M. (For the purposes of this lecture, M can be taken to be one million.) We don't know what p is, but are merely assuming that such a prime exists. If b is a multiple of $p-1$ and x is random then $x^b - 1$ is a multiple of p, and $\gcd(x^b - 1, n)$ is divisible by p and, if we are lucky, is a proper factor of n.

A number is said to be "M-smooth" if all of its prime divisors are at most M. Our assumption on n is that it has a prime divisor p such that $p-1$ is M-smooth.

This leads us to try the following randomized procedure.

Algorithm 8.3 ($p-1$ algorithm). Let $b = M!$. Choose a random integer x modulo n and then compute $\gcd(x^b - 1, n)$.

The choice of b is a little more sophisticated in practice; one could choose it to be the product of the primes less than M, or even

$$b = \operatorname{lcm}(1, 2, \cdots, M)$$

in order to increase the chance that $p-1$ divides b. The parameter M can be varied as is appropriate (for the capabilities of the computing machinery that will be used).

Greatest common divisors can be computed efficiently by Euclid's algorithm. More surprisingly, the computation of the large integer b is unnecessary, and $x^b - 1$ can be computed efficiently. Namely, we start with x and repeatedly calculate powers by all of the integers up to M. And we don't actually compute x^b but rather $x^b \bmod n$. Exponentiating modulo n can be computed efficiently by the "binary method," which takes at most $2\log(b)$ multiplications, by the recursive procedure implicit in the observation that

$$x^k = \begin{cases} (x^{k/2})^2 & \text{if } k \text{ is even} \\ x(x^{k-1}) & \text{if } k \text{ is odd.} \end{cases}$$

Algorithm 8.3 is probabilistic, but if an M-smooth prime divisor p of n exists then it is likely to work. Indeed it is reasonable to hope that $p-1$ should divide b, and x^b should be 1 modulo p, so that the final gcd in the algorithm is divisible by p. There are of course a number of things than can go wrong, e.g., all primes dividing n might have this property so that the gcd is n, or none of them might have the property in which case the final gcd will almost always be equal to 1. Or, $p-1$ might be divisible only by primes less than a million, but not be a divisor of b.

Nonetheless, this a reasonable strategy in practice; the ingenious idea is due to Pollard.

In practice, one takes b to be the product of all primes up to some limit (such as one million) but throws in higher powers of the smaller primes. If the algorithm fails, we can always increase the limit and try again. For arbitrary p, this is unfortunately an exponential time algorithm.

Lenstra's marvelous idea is to replace the multiplicative group $(\mathbf{Z}/p\mathbf{Z})^*$ with $E(\mathbf{F}_p)$ for a randomly chosen elliptic curve over $\mathbf{F}_p$. One crucial benefit of this is that there are many elliptic curves to choose from, so that if one curve fails we can try another one. The expected complexity of this Lenstra's algorithm is

$$O\left(\exp((1+o(1))\sqrt{\log n \log \log n})\right)$$

which is sub-exponential as a function of the input size $\log(n)$, though not polynomial time. In fact, this ECM (elliptic curve method) algorithm shares, with trial division, the property that it finds smaller primes faster, i.e., that the running time of algorithm depends on the size of the smallest prime factor and not only on the size of n. For this reason, the ECM is now the method of choice for initial factoring attempts for numbers of unknown character. (For numbers known to be of the form $n = pq$ for roughly equal primes, the quadratic sieve or number field sieve will be superior.)

In any event, the algorithm is:

Algorithm 8.4 (ECM).
1. Let $b = M!$.
2. Choose a random elliptic curve E modulo n, and a point $P = x:y:z$ on E modulo n.
3. Compute $bP = X:Y:Z$ and compute $\gcd(Z, n)$.

As indicated above, b is usually chosen in a slightly different way, and the parameter M can be varied as is useful. The choice $M = 1000000$ gives a very rough indication of what is possible on current machines.

Also, we remark the multiple bP is easy to compute by successive multiplication by small integers, since in any event the factorization of b is completely known.

The astute reader will note that $\mathbf{Z}/n\mathbf{Z}$ isn't a field, so that it's not clear what is meant by choosing an elliptic curve over $\mathbf{Z}/n\mathbf{Z}$; so far all of our elliptic curves have been defined over fields. In fact it is possible to define elliptic curves over rings. However, in the above procedure this isn't really needed: we compute "as if" N was a prime, and if this naive assumption fails, then it fails constructively in the sense that the process produces an explicit prime factor of N, which was our goal all along.

If p is an unknown prime divisor of n then the calculations in $E(\mathbf{Z}/n\mathbf{Z})$ reduce modulo p to the same calculations in $E(\mathbf{Z}/p\mathbf{Z})$. The order $\#E(\mathbf{F}_p)$ is an integer in the interval $[p+1-2\sqrt{p}, p+1+2\sqrt{p}]$ and if it is smooth enough to be a divisor of b then bP will be the identity and the last coordinate of bP will be divisible by p. As in the Pollard algorithm we could be extraordinarily unlucky and all primes dividing n might have this property, in which case the final gcd will be equal to n.

3. Cryptography

Until 1970 or so, systems for exchanging secret messages in the presence of possible eavesdroppers tended to have the following form. If Alice and Bob knew that they were going to need to exchange messages in the future, they arranged to share a

secret "key" K, typically in the form of a few words, or a string of bits, etc. Later, when Alice wanted to send a message she would take her plaintext P and encrypt it
$$P \mapsto C = E(P, K)$$
using an encryption function E to produce ciphertext C from the key and the plaintext. When Bob received the message he would use the inverse function D to decrypt the plaintext from the ciphertext
$$C \mapsto P = D(C, K).$$
This general scheme is known as a "symmetric" (or private-key, or secret key) cryptosystem, and relies on the property that $D(E(P, K), K) = P$.

One famous example is the "one-time pad" in which case P and K have the same length and are usually represented as bit strings, i.e., elements of the vector space $\mathbf{F}_2^n$ over the field with two elements. Then
$$E(x, y) = D(x, y) = x + y$$
is the so-called "one-time pad." Here the addition operation is just the operation of adding vectors in $\mathbf{F}_2^n$, which is realized in computer hardware as the the bitwise "exclusive-or" operation.

One can formulate a theorem that asserts that if K is random then this system is unbreakable; roughly, the idea is that an eavesdropper who intercepts $C = P + K$ can't deduce anything about C since an arbitrary plaintext P' gives rise to the same ciphertext C if the key is $K' = P' + C'$.

Another famous example of a symmetric cryptosystem is the Data Encryption Standard (DES), introduced 25 years ago, which has a key size of 56 bits. Currently there is an effort underway to replace the DES cryptosystem with a new system, which will be called the Advanced Encryption Standard (AES) that will be chosen among various competitors, and will allow keys up to 256 bits in length.

In most practical circumstances it is important for the key K to be of manageable size, e.g., to be much shorter than typical messages. For instance, the DES cryptosystem requires a 56-bit key in order to exchange messages, so that the key is about 8 symbols long. Even with this short key, the problem of managing the distribution of keys in the real world is very demanding. For every pair of entities that needs to exchange messages there has to be a key, and it is prudent to update these frequently.

This key-management nightmare was greatly ameliorated by the introduction of public-key cryptography. The fundamental idea is that Alice and Bob no longer have to share information before exchanging a message. In order for Alice to send a message to Bob, all that she has to do is to look up Bob's public key K_{Bob} (in some sort of public phone book) and use it to send a message. These "asymmetric" cryptosystems present vastly easier key management issues than symmetric cryptosystems and are in constant use nowadays.

The most famous such scheme is the Rivest-Shamir-Adleman cryptosystem in which the encryption function is
$$E(P) = P^e \bmod n$$
where Bob's public key consists of a pair (n, e) consisting of an integer n (that is the product of two roughly equal primes) and an exponent e. The decryption function is $D(C) = C^d \bmod n$ for suitable d. It is easy to compute the secret key d knowing e

and the factorization of n (a good exercise for the reader!), but no one knows how to find d without knowing the factorization. It is conjectured that breaking the RSA public-key scheme is equivalent to factoring n.

In order to give a simple illustration of elliptic curves in cryptography, we consider an alternate (and very famous) public-key protocol due to Diffie and Hellman. Suppose that Alice and Bob want to exchange public messages so that somehow they end up with mutual information K that is "random" and secret (i.e., not knowable by any eavesdropper). The Diffie-Hellman scheme allows them to do just this, despite the fact that Alice and Bob have never met. This public key-exchange process could be used as a precursor to exchanging messages with a classical symmetric cryptosystem using the secret key K.

As a minor digression, it should be noted that the RSA cryptosystem is significantly slower than symmetric systems by a factor of as much as 1000 or more. So in practice, it is in fact very common for messages to be exchanged by first conducting a public-key exchange session of some kind (e.g., with RSA or with Diffie-Hellman) and using the resulting key for a (much faster) dialogue that uses a symmetric cryptosystem.

At first glance the key-exchange problem sounds impossible (in fact, in an information-theoretic sense it is). Alice and Bob have to communicate entirely in public and end up jointly knowing some secret information K that no eavesdropper knows.

Here is a way to do it, at least under appropriate assumptions. A finite cyclic group G and a generator g of G are known publicly. For convenience, we write the group additively so that if m is an integer then $mg \in G$ is the element of G obtained by adding g to itself m times.

Algorithm 8.5 (Diffie-Hellman Key Exchange Algorithm).
1. Alice chooses a private key m that is a random integer (roughly of the order of magnitude of the order $\#G$ of the group G).
2. Bob chooses a similar private key m'.
3. Alice sends mg to Bob.
4. Bob sends $m'g$ to Alice.
5. Alice computes $K = m(m'g) \in G$ and Bob computes $K = m'(mg) \in G$.

At the end of the algorithm both Alice and Bob share the key K and we will argue that it is likely to be difficult for an eavesdropper to obtain this information!

What cyclic group G should we use? This might seem to be a silly question, since all cyclic groups of the same order are isomorphic. However, cyclic groups can be represented in many different ways, and this turns out to be crucial.

If G is the additive group $\mathbf{Z}/n\mathbf{Z}$ and elements x are represented in the usual way as integers $0 \leq x < n$, then it is an easy exercise to see that from the knowledge of n, g, and $mg \bmod n$ it is easy to use the Euclidean algorithm to find m.

On the other hand, if G is the multiplicative group $(\mathbf{Z}/p\mathbf{Z})^*$ (where it is customary, for simplicity, to choose the modulus to be prime) and elements are represented as integers $1 \leq x < p$, then it is not so clear that it is easy to find k if one is given p, g, and $g^k \bmod p$. Solving this problem is called the **discrete logarithm** problem modulo p.

For numbers of cryptographic size there are no known algorithms to solve the discrete logarithm problem modulo primes in a reasonable amount of time. The difficulty of the the discrete logarithm problem modulo p seems to be roughly equivalent to the problem of factoring integers of the size of p, and so the current state of the art is that numbers of, say, 300 digits should be reasonably secure, at least for a few years.

The alert reader will notice that to break the Diffie-Hellman Key Exchange algorithm it actually suffices to solve the following weaker problem, sometimes called the Computational Diffie-Hellman problem: Given a group G and elements g, mg, and $m'g$ (where m and m' are unknown integers), find $mm'g$.

It is clear that an algorithm for the discrete logarithm problem gives an algorithm for the Computational Diffie-Hellman problem, but the converse is not known. In recent years an even weaker version of the problem has proved to be important in theoretical cryptography: the Decision Diffie-Hellman problem which, very roughly, asks whether it is possible to distinguish in any useful way between triples of the form $(mg, m'g, mm'g)$ and arbitrary triples of group elements.

Elliptic curves were introduced into cryptography by Neal Koblitz and Victor Miller (independently). Although there are many ways to exploit elliptic curves over finite fields, the most straightforward application is to take the cyclic group G, in the Diffie-Hellman algorithm, to be a large cyclic subgroup of the group of points $E(k)$ of an elliptic curve over a finite field k.

Why should the elliptic curve version of the Diffie-Hellman protocol be more secure that the original version?

Arguments about cryptographic strength usually fall far short of mathematical proof, since it is extremely difficult to prove security for any practical cryptosystem. But one can make several observations. The elliptic curve algebra has the feeling of a black box, whereby the structure of the group $E(\mathbf{F}_p)$ is sufficiently murky that all that one really knows is how to compute $P + Q$. With $(\mathbf{Z}/p\mathbf{Z})^*$ the elements x are written as integers $1 \leq x < p$ and this additional structure (e.g., the size of x and its prime factorization) allow various attacks on the discrete logarithm problem. Several of these are based on an "index calculus" idea that gives algorithms whose complexity is similar to that of integer factorization algorithms; no such algorithms are known for elliptic curves.

If this argument is correct, elliptic curve Diffie-Hellman systems can use smaller primes than systems based on $(\mathbf{Z}/p\mathbf{Z})^*$, and still give equivalent security. This is potentially of considerable importance in practice, since such systems might be embedded inside "smart cards" where space is at a premium.

Needless to say, these brief remarks merely scratch the surface of a rapidly growing field. For a picture of the extent of the practical side of the field (together with extensive references to practical and theoretical works), you could consult Schneier's tome [**Sch**]. The use of elliptic curves in cryptography is described in many places, including [**BSS**].

The potential importance of cryptography will only continue to grow, and it is sobering to realize that, by now, many more people have probably heard of elliptic curves in connection with cryptosystems than Diophantine equations.

BIBLIOGRAPHY

[BSS] Blake, I., Seroussi, G., and Smart, N., *Elliptic Curves in Cryptography*, London Mathematical Society Lecture Note Series, vol. 265, Cambridge University Press, Cambridge, UK, 1999.

[Brl] C.Breuil, B.Conrad, F.Diamond and R.Taylor, On the modularity of elliptic curves over **Q**: Wild 3-adic exercises, preprint, **http://www.math.harvard.edu/ rtaylor**

[Bre] Bremner, Andrew, *A Diophantine System*, Internat. J. Math. & Math. Sci. **9** (1986) 413-415

[BK] Brieskorn, Egbert, and Knörrer, Horst, *Plane Algebraic Curves*, Birkhäuser, Boston, 1986

[Cas] Cassels, J. W. S., *Lectures on Elliptic Curves*, Cambridge University Press, Cambridge, 1991

[Cre] Cremona, J. E., *Algorithms for Modular Elliptic Curves* (2nd ed), Cambridge University Press, Cambridge, 1997

[Co1] Cohen, Henri, *A Course in Computational Algebraic Number Theory*, Springer Verlag, New York, 2000

[Co2] Cohen, Henri, *Advanced Topics in Computational Number Theory*, Springer Verlag, New York, 2000

[DeR] Deligne, P. and Rapoport, M., *Les schémas de modules de courbes elliptiques* in *Modular functions of one variable, II*, 143–316, Lecture Notes in Math., Vol. 349, Springer, Berlin, 1973.

[Gou] Gouvêa, Fernando, *Deformations of Galois Representations*, these proceedings.

[Dar] Darmon, Henri, Diamond, Fred, and Taylor, Richard *Elliptic curves, modular forms and Fermat's last theorem* in "Fermat's Last Theorem", edited by J. Coates and S.-T. Yau, 1997, Internat. Press, Cambridge, MA.

[Dia] Diamond, F., and Im, J. *Modular Forms and Modular Curves* 39-104 in "Seminar on Fermat's Last Theorem", edited by V. Kumar Murty, 1995, American Mathematical Society, Providence, RI.

[Ful] Fulton, William, *Algebraic curves. An introduction to algebraic geometry*, reprint of 1969 original, Addison-Wesley Publishing Company, Redwood City, CA, 1989.

[Har] Hartshorne, Robin, *Algebraic geometry*, Graduate Texts in Mathematics, No. 52. Springer-Verlag, New York-Heidelberg, 1977.

[Hus] Husemoller, Dale, *Elliptic Curves*, Springer Verlag, New York, 1987
[Kna] Knapp, Anthony W., *Elliptic Curves*, Princeton University Press, Princeton, 1992
[KaM] Katz, Nicholas, and Mazur, Barry, *Arithmetic moduli of elliptic curves*, Annals of Mathematics Studies, 108. Princeton University Press, Princeton, N.J., 1985.
[Kob] Koblitz, Neal, *Introduction to Elliptic Curves and Modular Forms*, Springer-Verlag, New York, 1984
[La1] Lang, Serge, *Introduction to Modular Forms* Springer-Verlag, New York, 1976
[La2] Lang, Serge, *Elliptic Functions*, Springer-Verlag, New York, 1987
[Lsk] Laska, Michael, *An algorithm for finding a minimal Weierstrass equation for an elliptic curve*, Math. Comp. **38** (1982) 257-260.
[Len] Lenstra, Hendrik W., Jr., *Factoring integers with elliptic curves*, Annals of Math. **126** (1987) 649-673
[Lor] Lorenzini, Dino, *An invitation to arithmetic geometry*, Graduate Studies in Mathematics, 9. American Mathematical Society, Providence, RI, 1996.
[Men] Menezes, Alfred J., *Elliptic Curve Public Key Cryptosystems*. Kluwer Academic Publishers, 1993
[Mih] Mihăilescu, Preda, *Cyclotomy Primality Proving — Recent Developments*, 1998, 95-110, in Springer Lecture Notes in Computer Science v. 1423.
[Mi1] Milne, J.S. ,*Elliptic Curves*, **http://www.jmilne.org/math**
[Mi2] Milne, J.S. ,*Modular Functions and Modular Forms*, **http://www.jmilne.org/math**
[Miy] Miyake, T., *Modular Forms*, Springer-Verlag, New York, 1989
[Mor] Morain, François, *Primality Proving Using Elliptic Curves: An Update*, 1998, 111-127, in Springer Lecture Notes in Computer Science v. 1423.
[Mum] Mumford, David, *Algebraic geometry. I. Complex projective varieties*, reprint of the 1976 edition. Classics in Mathematics. Springer-Verlag, Berlin, 1995.
[Ogg] Ogg, Andrew, *Modular Forms and Dirichlet Series*, W. A. Benjamin Inc., 1969
[Roh] Rohrlich, David *Modular Curves, Hecke Correspondences, and L-functions* in "Modular Forms and Fermat's Last Theorem", edited by G. Cornell, G. Stephens, and J. Silverman, 1997 Springer-Verlag, NY.
[Sch] Schneier, Bruce *Applied Cryptography*, second edition, 1996, John Wiley and Sons.
[Scf] Schoof, René, *Elliptic curves over finite fields and the computation of square roots mod p*, Math. Comp. 44 (1985), 483–494.
[Se1] Serre, J.-P., *A course in Arithmetic* Springer-Verlag, New York, 1973
[Se2] Serre, Jean-Pierre *Abelian l-adic representations and elliptic curves* Research Notes in Mathematics, **7** 1998, A K Peters, Ltd., Wellesley, MA.
[Se3] Serre, J.-P., *Properiétés galoisiennes des points d'order fini des courbes elliptique*, Invent. Math. **15** (1972), 259-331. Springer-Verlag, New York, 1973
[Se4] Serre, Jean-Pierre *Travaux de Wiles (et Taylor, ...)*, Séminaire Bourbaki, 1994/95. Astérisque **237** (1996), 5, 319–332.
[Sha] Shafarevich, Igor R. *Basic algebraic geometry. 1. Varieties in projective space*, second edition, Springer-Verlag, Berlin, 1994.

BIBLIOGRAPHY

[Shi] Shimura, Goro *Introduction to the Arithmetic Theory of Automorphic Functions*, Princeton Univ. Press, 1971, Princeton, NJ.
[Sil] Silverberg, Alice, *Open Questions in Arithmetic Algebraic Geometry*, these proceedings.
[Si1] Silverman, Joseph H., *The Arithmetic of Elliptic Curves* Springer Verlag, New York, 1986
[Si2] Silverman, Joseph H., *Advanced Topics in the Arithmetic of Elliptic Curves*, Springer Verlag, New York, 1994
[ST]   Silverman, Joseph H., and Tate, John, *Rational Points on Elliptic Curves*, Springer Verlag, New York, 1992
[Sma] Smart, Nigel P., *The Algorithmic Resolution of Diophantine Equations*, Cambridge University Press, Cambridge, 1998
[Ta1] Tate, John, *The arithmetic of elliptic curves* Invent. Math. **23** (1974) 179-206
[Ta2] Tate, John, *Galois Cohomology*, these proceedings.
[TW]   Taylor, Richard, and Wiles, Andrew, *Ring-theoretic properties of certain Hecke algebras*, Ann. of Math. **141** (1995), 553–572.
[Wil] Wiles, Andrew, *Modular elliptic curves and Fermat's last theorem*, Ann. of Math. **141** (1995), 443–551.

Open Questions in Arithmetic Algebraic Geometry

Alice Silverberg

Open Questions in Arithmetic Algebraic Geometry

Alice Silverberg[1]

CHAPTER 1
Overview

1.1. Introduction

This series of lectures introduces some important conjectures in Arithmetic Algebraic Geometry. Arithmetic Algebraic Geometry can be viewed as a branch of Number Theory where one solves number theoretic questions using powerful techniques from Algebraic Geometry. Number Theory and Arithmetic Algebraic Geometry grew out of studying classical questions concerned with finding whole number solutions to polynomial equations (with whole number coefficients). Fermat's Last Theorem is one such classical question. Starting with the simplest polynomial equations, one first encounters linear and quadratic equations, which are easy to solve. Next one is naturally led to consider cubic equations, and here one is already in an area with deep mathematical ideas, results, and conjectures. Elliptic curves are smooth projective cubics in the plane, and their study is part of a rich and beautiful theory, with many as yet unanswered questions.

Our central objects of study will be elliptic curves and (their higher-dimensional generalizations) abelian varieties, along with modular curves, Shimura curves, modular forms, automorphic forms, L-functions, and ℓ-adic representations.

---

[1]Department of Mathematics, Ohio State University, 231 West 18 Avenue, Columbus, Ohio 43210-1174, USA

E-mail address: silver@math.ohio-state.edu

Key words and phrases. Arithmetic algebraic geometry, elliptic curves, abelian varieties, conjectures.

The author was supported in part by the NSF, the NSA, and the Alexander von Humboldt Foundation. She would like to thank B. Birch, A. Brumer, B. Conrad, N. Elkies, H. Lange, K. Rubin, J-P. Serre, G. Shimura, and B. de Weger for helpful comments on earlier versions of the manuscript, the audience at the lectures for helpful feedback, A. Agashe for help collecting the references, and the Mathematics Institute of the University of Erlangen for its hospitality. Chapter 5 owes a lot to de Weger's paper [**183**].

©2001 American Mathematical Society

I hope that at least the first few chapters will be accessible to a wide audience (including undergraduates, teachers,...).

The exercises are designed to help the reader explore and become more at home with the topics, ideas, and objects. They are of varying difficulties, and can sometimes be explored from several directions or on several levels (computationally, theoretically, etc.). The exercises are not necessary for following the course.

I do not claim to be complete in these notes, nor do I claim to always focus on the most interesting or deepest problems or results (sometimes I have chosen what's fun or easy to explain, rather than what's cutting edge).

The first chapter gives an overview and introduction. We introduce elliptic curves, abelian varieties, and Mordell-Weil groups, and ask some questions about them. In later chapters we examine these questions and others in more detail, discussing some of the progress that has been made, and identifying which questions are still open.

This write-up combines my talks in Princeton and Park City. The chapters do not precisely correspond to the lectures. In Park City I gave (approximately) one lecture of overview, one lecture on torsion subgroups, two lectures on ranks, two lectures on the conjectures of Birch and Swinnerton-Dyer, and one lecture on ABC and other conjectures. In Princeton I gave an overview and a more compact treatment of the topics, focusing on Mordell-Weil groups, ranks, and the conjectures of Birch and Swinnerton-Dyer.

1.2. Elliptic curves

Definition 1.2.1. An *elliptic curve* over the rational numbers $\mathbf{Q}$ is the projective curve associated to an affine equation of the form

$$(1) \qquad y^2 = x^3 + ax + b$$

where a and b are integers such that

$$4a^3 + 27b^2 \neq 0.$$

The *discriminant* of the above elliptic curve E is

$$\Delta(E) = -16(4a^3 + 27b^2).$$

The condition $4a^3 + 27b^2 \neq 0$ is equivalent to the curve being smooth.

Exercise 1.2.2. *Show that if a and b are complex numbers, then the following are equivalent:*
  (i) $4a^3 + 27b^2 \neq 0$,
  (ii) *the cubic $x^3 + ax + b$ has three distinct complex roots,*
  (iii) *there are no complex values of x and y for which both partial derivatives of $x^3 + ax + b - y^2$ vanish.*

Example 1.2.3. Consider the elliptic curve

$$E : y^2 = x^3 - x.$$

This curve was considered by Fermat. He showed that the only rational solutions are $(0,0), (1,0), (-1,0)$.

Definition 1.2.4. An *elliptic curve* E over a field K is a smooth projective curve of genus one with a point $P \in E(K)$. Equivalently, it is a smooth projective plane cubic over K with a point over K. Equivalently, it is a smooth projective curve with an affine equation of the form

(2) $$y^2 + a_1 xy + a_3 y = x^3 + a_2 x^2 + a_4 x + a_6$$

where the a_i are in K. If the characteristic of K is not 2 or 3, then E has an equation (called a *Weierstrass form* for E) of the form (1), with $a, b \in K$. Equation (2) is called a *generalized Weierstrass form*.

The K-rational points of E form an abelian group. The group law (since it's abelian we will call it an addition law) is based on the principle that a line meets a cubic in 3 points, and is defined by the rule that the 3 points of intersection of a line with the cubic sum to O in the addition law. The identity element for the group law, O, is the point with projective coordinates $(x, y, z) = (0, 1, 0)$, where we are viewing the curve as a projective curve with homogeneous equation

$$y^2 z + a_1 xyz + a_3 yz^2 = x^3 + a_2 x^2 z + a_4 xz^2 + a_6 z^3.$$

In affine coordinates we would call this point (∞, ∞). This point is often called the *point at infinity*, and can be viewed as the point where all vertical lines meet (if we graph the affine equation in the plane).

Exercise 1.2.5. *Verify that when the elliptic curve E has equation $y^2 = x^3 + ax + b$, then the points of order 2 for the group law are exactly the points with $y = 0$.*

Exercise 1.2.6. *Verify that if P is a point on an elliptic curve with an equation as in (2), then the x-coordinate of P is the same as the x-coordinate of $-P$ (where $-P$ is the inverse of P with respect to the addition law).*

Definition 1.2.7. A *homomorphism* of elliptic curves over a field K is a map which is both a morphism over K (i.e., is defined by polynomial equations with coefficients in K) and a group homomorphism.

Exercise 1.2.8. *If E is an elliptic curve over K in generalized Weierstrass form, show that every isomorphism of E to another elliptic curve over K in generalized Weierstrass form can be given by a change of variables of the form $x \mapsto u^2 x + r$, $y \mapsto u^3 y + u^2 sx + t$ with $r, s, t, u \in K$ and $u \neq 0$. Further, show that every such change of variables produces an isomorphism.*

1.3. Abelian varieties

In these lectures I will sometimes state results and conjectures for abelian varieties over number fields. If you prefer a more concrete approach, you can substitute the phrase "elliptic curve over $\mathbf{Q}$" for the phrase "abelian variety over a number field" throughout the lectures. However, some of the conjectures will be theorems in the case of elliptic curves.

Definition 1.3.1. An *abelian variety* is a connected projective variety which is also an algebraic group.

Examples 1.3.2. The one-dimensional abelian varieties are exactly the elliptic curves. Products of abelian varieties are abelian varieties. Jacobian varieties of curves of genus g are abelian varieties of dimension g. For example, if C is a curve given by a smooth (i.e., there is no point at which all partial derivatives vanish) homogeneous polynomial in x, y, and z of degree N (for example, the Fermat equation $x^N + y^N = z^N$), then C has genus $(N-1)(N-2)/2$, and its Jacobian variety $J(C)$ has dimension $(N-1)(N-2)/2$.

1.4. Mordell-Weil groups

If A is an abelian variety over a field K, then the group of K-rational points of A is denoted $A(K)$. This group is called the *Mordell-Weil group* of A over K if K is a number field (because of the following theorem, proved for elliptic curves over $\mathbf{Q}$ by Mordell, and generalized by Weil).

Theorem 1.4.1 (Mordell-Weil Theorem). *If A is an abelian variety over a number field K, then $A(K)$ is a finitely generated abelian group.*

In other words, if A is an abelian variety over a number field K, then

$$(3) \qquad A(K) \cong \mathbf{Z}^r \oplus A(K)_{\mathrm{tors}}$$

where r is a non-negative integer, called the *rank* of $A(K)$, and $A(K)_{\mathrm{tors}}$ is a finite group, called the *torsion subgroup* of $A(K)$. The elements of $A(K)_{\mathrm{tors}}$ (i.e., the points of finite order with respect to the group law on A) are called the *torsion points* of $A(K)$.

In these notes we will discuss what is known and what is not yet known about Mordell-Weil groups of abelian varieties. We begin with some examples (without proof). Later, we will learn more about how to compute Mordell-Weil groups.

Example 1.4.2. Consider
$$E : y^2 = x^3 - x.$$
Then
$$E(\mathbf{Q}) = \{O, (0,0), (1,0), (-1,0)\} \cong \mathbf{Z}/2\mathbf{Z} \oplus \mathbf{Z}/2\mathbf{Z},$$
as shown by Fermat. Further,
$$E(\mathbf{Q}(i)) = \{O, (0,0), (1,0), (-1,0), (i, i-1), (i, 1-i), (-i, 1+i), (-i, -1-i)\}$$
$$\cong \mathbf{Z}/2\mathbf{Z} \oplus \mathbf{Z}/4\mathbf{Z},$$
where i is a square root of -1.

Exercise 1.4.3. *Calculate directly the orders of the points in $E(\mathbf{Q}(i))$ (by using the group law) in the previous example.*

Example 1.4.4. Consider
$$E : y^2 = x^3 - 2x.$$
Then
$$E(\mathbf{Q}) \cong \mathbf{Z} \oplus \mathbf{Z}/2\mathbf{Z},$$
and $E(\mathbf{Q})$ is generated by the point of infinite order $(-1,1)$ and the point of order two $(0,0)$.

Example 1.4.5. Consider
$$E : y^2 = x^3 + 17.$$
Then
$$E(\mathbf{Q}) \cong \mathbf{Z}^2,$$
and $E(\mathbf{Q}) = \mathbf{Z}P_1 + \mathbf{Z}P_2$ where $P_1 = (-2, 3)$ and $P_2 = (2, 5)$.

Exercise 1.4.6. *The equation*

$$y^2 = x^3 + 17$$

has exactly 16 *solutions with* x *and* y *both integers. Find them.*

Example 1.4.7. Consider

$$E : y^2 = x^3 - 17x.$$

Then

$$E(\mathbf{Q}) \cong \mathbf{Z}^2 \oplus \mathbf{Z}/2\mathbf{Z},$$

and $E(\mathbf{Q})/E(\mathbf{Q})_{\text{tors}}$ is generated by $(-1, 4)$ and $(9, 24)$.

1.5. Torsion subgroups

By the following result, given an elliptic curve E over $\mathbf{Q}$ (and clearing denominators to make the coefficients integers), the torsion subgroup $E(\mathbf{Q})_{\text{tors}}$ can be computed in a finite amount of time.

Theorem 1.5.1 (Nagell [**110**], Lutz [**84**]). *If* $E : y^2 = x^3 + ax + b$ *is an elliptic curve with* $a, b \in \mathbf{Z}$, *and* (x, y) *is a non-zero rational point on* E *of finite order, then* x *and* y *are integers, and either* $y = 0$ *or* y^2 *is a divisor of* $4a^3 + 27b^2$.

Exercise 1.5.2. *Compute* $E(\mathbf{Q})_{\text{tors}}$, *for*

$$E : y^2 = x^3 + 4x.$$

Find the orders of all the torsion points.

Exercise 1.5.3. *Compute* $E(\mathbf{Q})_{\text{tors}}$, *for*

$$E : y^2 = x^3 + 1.$$

Find the orders of all the torsion points.

There are many questions one can ask about torsion points and torsion subgroups, for elliptic curves and for abelian varieties.

Which positive integers occur as orders of torsion points on elliptic curves over $\mathbf{Q}$? (In the examples in §1.4, we have only seen the integers 2 and 1. There are more in the exercises.) Do they all occur? Do infinitely many occur? What about over number fields other than $\mathbf{Q}$? For example, can one find an integer d and an elliptic curve over $\mathbf{Q}(\sqrt{d})$ with a point over $\mathbf{Q}(\sqrt{d})$ of order 17?

Which finite abelian groups can occur as $E(\mathbf{Q})_{\text{tors}}$'s, with E an elliptic curve over $\mathbf{Q}$? (We have seen $\mathbf{Z}/2\mathbf{Z} \oplus \mathbf{Z}/2\mathbf{Z}$, $\mathbf{Z}/2\mathbf{Z}$, and the trivial group. Other groups appear in the above exercises.) Do all finite abelian groups occur? Do infinitely many occur? What about over number fields other than $\mathbf{Q}$? (We have seen $\mathbf{Z}/2\mathbf{Z} \oplus \mathbf{Z}/4\mathbf{Z}$ over $\mathbf{Q}(i)$.) What about abelian varieties over $\mathbf{Q}$ (of a fixed dimension, say)? Or abelian varieties (of a fixed dimension) over a fixed number field? Or abelian varieties (of a fixed dimension) over all number fields of a fixed degree over $\mathbf{Q}$?

We will discuss these questions in Chapter 2.

1.6. Ranks

We have seen (above) examples of elliptic curves over $\mathbf{Q}$ of ranks 0, 1, and 2. There are many interesting questions about ranks. For example, what integers can occur as the ranks of Mordell-Weil groups of elliptic curves over $\mathbf{Q}$? Do all non-negative integers occur? Is the set of non-negative integers which occur unbounded? If r is a non-negative integer, how "often" does r occur as the rank of the Mordell-Weil group of an elliptic curve over $\mathbf{Q}$?

What is the "average rank" as one runs over elliptic curves over $\mathbf{Q}$? This depends on how one defines "average rank". Once we define it, what is its numerical value? Is it unbounded? Is it zero?

Suppose we restrict to a family of elliptic curves over $\mathbf{Q}$ (or a family of abelian varieties of fixed dimension over a fixed number field, or a family of abelian varieties of fixed dimension over all number fields of a fixed degree over $\mathbf{Q}$). For this family, we can ask the above questions about ranks. Are they bounded or unbounded in the family? What is the average rank in the family? An example of a family is the family of quadratic twists of a given elliptic curve (i.e, the family $dy^2 = x^3 + ax + b$, where a and b are fixed and d ranges over square-free integers). Another example is the family of cubic twists of $y^2 = x^3 - 432$. This latter family is usually written as the family $x^3 + y^3 = d$, where d is a cube-free integer (a Weierstrass model for the Fermat cubic $x^3 + y^3 = 1$ is $y^2 = x^3 - 432$).

We can ask the same questions for ranks of Mordell-Weil groups of elliptic curves over number fields other than $\mathbf{Q}$. Moreover, we can consider ranks of Mordell-Weil groups of abelian varieties over $\mathbf{Q}$, or over other number fields.

We will discuss some of these questions in Chapters 3 and 4.

1.7. Looking ahead

Here are some of the major topics we will discuss in these lectures:
- **Torsion** Conjectures (for elliptic curves and abelian varieties);
- Conjectures about **ranks** (mostly for elliptic curves: average ranks in families, boundedness/unboundedness, density of curves of a given rank, ...);
- **Birch and Swinnerton-Dyer** Conjectures (These conjectures relate arithmetic properties or invariants of the variety to properties of its L-function, an analytic object. This includes a conjecture about ranks of Mordell-Weil groups of elliptic curves and abelian varieties, and leads to questions about the Tate-Shafarevich group III.);
- **ABC** Conjectures, and relations with other conjectures.

1.8. Suggested reading

For introductions to elliptic curves, see the books written by Husemöller [**64**], Koblitz [**73**], Silverman [**156**], and Silverman and Tate [**160**]. Cremona's book [**35**] gives elliptic curve algorithms and data. See Cohen's book [**33**] for number theory algorithms, including a chapter on elliptic curves. See Smart's book [**162**] for algorithms for solving Diophantine equations.

Useful elliptic curve software packages include PARI [**4**], mwrank [**36**], and the Maple package Apecs [**34**].

CHAPTER 2
Torsion subgroups

The torsion subgroup is the group of points of finite order in the Mordell-Weil group of an abelian variety or elliptic curve. For elliptic curves, the torsion subgroups are now fairly well understood, as we shall see below. For higher-dimensional abelian varieties there is still much work to be done.

2.1. Torsion subgroups for elliptic curves

Recall the following theorem stated in the last chapter:

Theorem 2.1.1 (Nagell-Lutz). *If $E : y^2 = x^3 + ax + b$ is an elliptic curve with $a, b \in \mathbf{Z}$, and (x, y) is a non-zero rational point on E of finite order, then x and y are integers, and either $y = 0$ or y^2 is a divisor of $4a^3 + 27b^2$.*

Example 2.1.2. Consider
$$E : y^2 = x^3 + 4x.$$
Then
$$E(\mathbf{Q}) = \{O, (0,0), (2,4), (2,-4)\} \cong \mathbf{Z}/4\mathbf{Z}.$$

Example 2.1.3. Consider
$$E : y^2 = x^3 + 1.$$
Then
$$E(\mathbf{Q}) = \{O, (-1,0), (0,1), (0,-1), (2,3), (2,-3)\} \cong \mathbf{Z}/6\mathbf{Z}.$$

Theorem 2.1.4 (Mazur [96]). *If E is an elliptic curve over $\mathbf{Q}$, then the torsion subgroup $E(\mathbf{Q})_{\text{tors}}$ is one of the following 15 groups:*

$$\begin{array}{ll} \mathbf{Z}/N\mathbf{Z} & \text{for } N = 1, \ldots, 10 \text{ or } 12, \\ \mathbf{Z}/2\mathbf{Z} \times \mathbf{Z}/2N\mathbf{Z} & \text{for } N = 1, 2, 3, \text{ or } 4. \end{array}$$

Further, it is known that each of these groups occurs infinitely often (i.e., for infinitely many $\mathbf{Q}$-isomorphism classes).

2.2. A quick look at modular curves

If K is a subfield of the field $\mathbf{C}$ of complex numbers, and $E : y^2 = x^3 + ax + b$ is an elliptic curve over K, then the points of order 2 are the points where $y = 0$. Therefore the subgroup $E(K)[2]$ of $E(K)$ of points of order 1 or 2 either is the trivial group, or is isomorphic to $\mathbf{Z}/2\mathbf{Z}$ or to $\mathbf{Z}/2\mathbf{Z} \times \mathbf{Z}/2\mathbf{Z}$.

Exercise 2.2.1. *Show that if E is an elliptic curve over K with $E[2](K) \cong \mathbf{Z}/2\mathbf{Z} \times \mathbf{Z}/2\mathbf{Z}$, then E is isomorphic over K to*

(4) $$E^{d,\lambda} : dy^2 = x(x-1)(x-\lambda)$$

for some $\lambda \in K - \{0, 1\}$ and some $0 \neq d \in K$.

Note that conversely, if $\lambda \in K - \{0, 1\}$ and $0 \neq d \in K$, then $E^{d,\lambda}$ is an elliptic curve over K with the property that $E^{d,\lambda}(K)[2] \cong \mathbf{Z}/2\mathbf{Z} \times \mathbf{Z}/2\mathbf{Z}$.

We can view the "λ-line" as parametrizing elliptic curves E over K with $E(K)[2] \cong \mathbf{Z}/2\mathbf{Z} \times \mathbf{Z}/2\mathbf{Z}$ (up to isomorphisms and twists). The equation

$$y^2 = x(x-1)(x-\lambda)$$

can be viewed as the equation of a surface "over" the λ-line (the fibers are the elliptic curves $E^{1,\lambda}$).

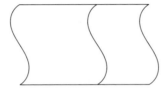

λ-line

Suppose now that we want to write down all the elliptic curves over $\mathbf{Q}$ with a rational point of order 5. Here's what to do.

Suppose more generally that K is a subfield of $\mathbf{C}$, E is an elliptic curve over K, and $P \in E(K)$ is a point of finite order ≥ 4.

By doing a change of variables over K, we can translate P to the point $(0, 0)$. This puts the equation for E in the form

$$y^2 + a_1 xy + a_3 y = x^3 + a_2 x^2 + a_4 x.$$

Note that $a_3 \neq 0$, since otherwise $(0, 0)$ would be a point of order 2 (check this!). After the change of variables $y \mapsto y + a_4 x/a_3$ we obtain an equation of the above form where the coefficient of x is 0. The coefficient of x^2 must be non-zero, since otherwise $(0, 0)$ would be a point of order 3 (check this!). The change of variables $x \mapsto (a_3/a_2)^2 x$, $y \mapsto (a_3/a_2)^3 y$ now gives an equation of the form

$$y^2 + \alpha xy + \beta y = x^3 + \beta x^2.$$

A computation (for example, using the program PARI) shows that

$$2P = (-\beta, \beta(\alpha - 1)) \quad \text{and} \quad 3P = (1 - \alpha, \alpha - \beta - 1).$$

Now suppose that P has order 5. Then $2P = -3P$, so the x-coordinates of $2P$ and $3P$ are the same (by Exercise 1.2.6). Therefore, $\alpha = 1 + \beta$, and E is isomorphic to

(5) $$y^2 + (1+\beta)xy + \beta y = x^3 + \beta x^2.$$

The discriminant of the elliptic curve (5) is $-\beta^5(\beta^2+11\beta-1)$. We can view (5) as the equation of a surface over the β-line, where the "bad", i.e., singular, fibers lie over the points 0, $\frac{-11\pm 5\sqrt{5}}{2}$, and infinity on the projective (β-)line $\mathbf{P}^1$. With $\beta \in \mathbf{Q}-\{0\}$, we obtain the elliptic curves over $\mathbf{Q}$ with a rational point (normalized to be $(0,0)$) of order 5. If K is a subfield of $\mathbf{C}$, then for $\beta \in K$ with $\beta \notin \{0, \frac{-11\pm 5\sqrt{5}}{2}\}$, we obtain the elliptic curves over K with a K-rational point of order 5.

Exercise 2.2.2. *Do the above with 4 instead of 5.*

Exercise 2.2.3. *Do the above with 7 instead of 5.*

Exercise 2.2.4. *In the above, check that $(0,0)$ is a point of order 5 on every elliptic curve given by (5).*

We write $Y_1(N)$ for the ("moduli" or "parameter") space of pairs (E, P) where E is an elliptic curve and P is a point of order N on E. (This only works nicely for $N \geq 3$ — think about why this is so! — but it can be fudged for $N = 1, 2$.) Then $Y_1(5)$ is the β-line minus the "bad" points. We write $X_1(N)$ for the compactification of $Y_1(N)$ (so $X_1(5)$ is $\mathbf{P}^1$). Then $X_1(N)$ is a curve defined over $\mathbf{Q}$, and is an example of what is called a modular curve. If K is a subfield of $\mathbf{C}$, then the elements of $Y_1(N)(K)$ correspond to isomorphism classes of pairs (E, P) where E is an elliptic curve over K and $P \in E(K)$ is a point of order N. The genus of $X_1(N)$ tends to grow as N grows.

Fact 2.2.5. *The genus of $X_1(N)$ is zero (and $X_1(N)$ is $\mathbf{P}^1$) if and only if either $N \leq 10$ or $N = 12$.*

Therefore, if either $1 \leq N \leq 10$ or $N = 12$, then there are infinitely many elliptic curves over $\mathbf{Q}$ with a rational point of order N (we wrote down these elliptic curves when $N = 5$).

B. Levi, and later A. Ogg, conjectured which groups can be torsion subgroups of Mordell-Weil groups of elliptic curves over $\mathbf{Q}$ (see [131]). Mazur (Theorem 2.1.4 above) proved that the conjectured set of finite groups was the correct one. Mazur's proof required a deep study of modular curves (see [96], [95]).

2.3. Conjectures

Conjecture 2.3.1 (Torsion Conjecture for Abelian Varieties). *If A is an abelian variety of dimension d defined over a number field K, then $\#A(K)_{\text{tors}}$ is bounded above by a constant depending only on d and K.*

Conjecture 2.3.2 (Strong Torsion Conjecture for Abelian Varieties). *If A is an abelian variety of dimension d defined over a number field K of degree m, then $\#A(K)_{\text{tors}}$ is bounded above by a constant depending only on d and m.*

The Strong Torsion Conjecture is sometimes called the Uniform Boundedness Conjecture. In [29], Cassels said that the Torsion Conjecture for elliptic curves was "part of the folklore".

Mazur's theorem is the same as the (Strong) Torsion Conjecture for $d = 1$ and $m = 1$.

2.4. Other results for elliptic curves

We begin by stating the famous result of Faltings:

Theorem 2.4.1 (Faltings [**43**]; formerly the Mordell Conjecture). *If C is a (smooth projective) curve of genus > 1 defined over a number field K, then $C(K)$ is finite.*

Faltings' Theorem (from the 1980's) implies in particular that for N sufficiently large there are only finitely many rational points on $X_1(N)$. By the "modular interpretation" of $X_1(N)$, there are therefore only finitely many elliptic curves over $\mathbf{Q}$ with a rational point of order N (up to isomorphism). Mazur (in the 1970's) proved more, namely that there are actually no rational points on $Y_1(N)$, for $N > 12$ and $N = 11$.

Theorem 2.4.2 (Manin [**88**]). *If K is a number field and p is a prime number, then there exists a constant $N(p, K)$ so that if E is an elliptic curve over K, then the number of torsion points in $E(K)$ whose order is a power of p is at most $p^{N(p,K)}$.*

Therefore to prove the (weak) Torsion Conjecture for elliptic curves, it suffices to bound the primes p that can occur as orders of torsion points. (Note that Manin's $N(p, K)$ depends on the field K, not just its degree.)

Exercise 2.4.3. Show that Faltings' Theorem implies Manin's Theorem. Hint: Use that there are only finitely many N for which the genus of $X_1(N)$ is ≤ 1. (In fact, what Manin proved is that for all n sufficiently large, $X_0(p^n)(K)$ is finite. Here $X_0(N)$ is (the compactification of) the moduli space of pairs (E, C) where C is a subgroup of order N on the elliptic curve E.)

Theorem 2.4.4 (Kamienny [**65**], using Kenku-Momose [**71**]). *If E is an elliptic curve defined over a quadratic number field K, then $E(K)_{\mathrm{tors}}$ is isomorphic to one of the following* 26 *groups:*

$$\begin{array}{ll} \mathbf{Z}/N\mathbf{Z} & \text{for } N = 1, \ldots, 16, \text{ or } 18, \\ \mathbf{Z}/2\mathbf{Z} \times \mathbf{Z}/2N\mathbf{Z} & \text{for } N = 1, \ldots, 6, \\ \mathbf{Z}/3\mathbf{Z} \times \mathbf{Z}/3N\mathbf{Z} & \text{for } N = 1 \text{ or } 2, \\ \mathbf{Z}/4\mathbf{Z} \times \mathbf{Z}/4\mathbf{Z}. & \end{array}$$

This proves the Strong Torsion Conjecture for $d = 1$ and $m = 2$, i.e., for elliptic curves over quadratic number fields (uniformly). The result was conjectured by Kenku and Momose in [**71**]. Kamienny proved that, for K quadratic, the order of a torsion point in $E(K)$ must divide $2^4 \cdot 3^2 \cdot 5 \cdot 7 \cdot 11 \cdot 13$, and in [**71**] it had already been determined exactly which groups can then occur. Note that not all of these groups will occur, for a given quadratic number field.

Kamienny and Mazur proved the Strong Torsion Conjecture for $d = 1$ and $m \leq 8$ (see [**66**] and [**67**]), and Abramovich proved it for $d = 1$ and $m \leq 14$ (see [**1**]). Then Merel proved it for $d = 1$ and all m, that is, he settled it for elliptic curves (see [**99**]).

Theorem 2.4.5 (Merel). *If E is an elliptic curve defined over a number field K of degree m, then $\#E(K)_{\mathrm{tors}}$ is bounded above by a constant depending only on d and m. If $m > 1$ and $E(K)$ has a point of prime order p, then $p < m^{3m^2}$.*

Next we state a strengthening of Merel's result, due to Oesterlé.

Theorem 2.4.6 (Oesterlé). *If E is an elliptic curve defined over a number field K of degree m, and $E(K)$ has a point of prime order p, then*

$$p \leq (1 + 3^{\frac{m}{2}})^2.$$

(Note: To get from this to the Strong Torsion Conjecture for Elliptic Curves, one needs to use Faltings' Theorem....)

The next result gives an effective version of the Strong Torsion Conjecture for Elliptic Curves.

Theorem 2.4.7 (Parent [121]). *If E is an elliptic curve defined over a number field K of degree m, and $E(K)$ has a point of order p^n where p is a prime number, then*
$$p^n \leq 129(5^m - 1)(3m)^6.$$

Oesterlé's result (for $m = 3$) shows that if E is an elliptic curve over a number field K of degree 3, and $E(K)$ has a point of prime order p, then $p \leq 37$. In a recent preprint Parent has a result which gives us reason to believe that the correct bound is 17 or 13.

Theorem 2.4.8 (Parent preprint [122]). *Suppose E is an elliptic curve defined over a number field K of degree 3, suppose $E(K)$ has a point of prime order p, and assume that a certain condition which would follow from the Birch and Swinnerton-Dyer Conjecture holds[1]. Then $p \leq 17$, and there are at most 49 elliptic curves "up to isomorphism" having a point of order 17 over a cubic number field.*

Exercise 2.4.9. *Show that there exists an elliptic curve with a point of order 11 over a cubic number field, by showing that the modular curve $X_1(11)$ has infinitely many points defined over cubic number fields. Hint: Use Hilbert's Irreducibility Theorem and the fact that $X_1(11)$ is an elliptic curve — an equation for it is $y^2 + y = x^3 - x^2$.*

2.5. Abelian varieties over number fields

For abelian varieties of dimension greater than one, there are no number fields for which the Torsion Conjecture is known to hold. However, one can give explicit bounds for torsion for certain families of abelian varieties.

First, let's consider a result that says that the torsion subgroup is "large". Flynn constructed sequences of curves over $\mathbf{Q}$ whose Jacobian varieties have $\mathbf{Q}$-rational torsion points whose orders grow quadratically in the genus of the curve.

Theorem 2.5.1 (Flynn [44]). *If g is even, $t \neq 0$, and $0 \leq r \leq g$, then the curves*
$$y^2 + t(x-1)y = \left(\sum_{i=1}^{g-r+1} x^{r+i}\right)^2 - t(x^{g+2} + x^{r+1})$$

have genus g, and their Jacobian varieties have rational torsion points of exact order $g^2 + 3g + 1 - r$. Moreover, there is a sequence $A(n)$ of abelian varieties over $\mathbf{Q}$ of strictly increasing dimension $d(n)$, each containing a $\mathbf{Q}$-rational torsion point of order $N(n)$, such that
$$N(n) > (1.6)^{(d(n) \log d(n))^{2/3}}.$$

The following result (see Theorem 3.3 of [154]) follows from results of Serre-Tate on abelian varieties with potentially good reduction. We will give the definitions and proof below.

---

[1] Namely, that the winding quotient of $J_1(q)$ has zero rank over $\mathbf{Q}$ for all primes q between 17 and 37 inclusive.

Theorem 2.5.2. *If A is a d-dimensional abelian variety defined over a number field K of degree m, and A has everywhere potentially good reduction, then*

$$\#A(K)_{\text{tors}} \leq [(1 + 2^{\#\text{GL}_{2d}(\mathbf{Z}/3\mathbf{Z})m/2})(1 + 3^{\#\text{GL}_{2d}(\mathbf{Z}/4\mathbf{Z})m/2})]^{2d}$$
$$< \left(1 + \frac{1}{10^{11}}\right)(2^{3^{4d^2}} \cdot 3^{4^{4d^2}})^{md} < 6^{md \cdot 4^{4d^2}}.$$

Definition 2.5.3. An abelian variety A over a number field K has *good reduction* at a prime ideal $\mathfrak{p}$ of the ring of integers $\mathcal{O}$ of K if (for some appropriate defining equations for A over $\mathcal{O}$) the reduction of A modulo $\mathfrak{p}$ is an abelian variety (i.e., is non-singular) over the (finite) field $\mathcal{O}/\mathfrak{p}$. (See [14] for a more precise discussion of the requirements on the "defining equations" and a detailed study of their existence.) It has *potentially good reduction* at $\mathfrak{p}$ if it has good reduction at all the primes dividing $\mathfrak{p}$ in some finite extension of the field K. It has *everywhere potentially good reduction* if it has potentially good reduction at all primes of K (in which case there will exist a finite extension of K over which A has good reduction at all prime ideals).

Example 2.5.4. The elliptic curve $y^2 = x^3 - x$ has good reduction at all odd prime numbers (this is easy to see directly from the definition), and has everywhere potentially good reduction (this is less easy).

Example 2.5.5. Let E be the elliptic curve

$$y^2 = x^3 - 35x - 98$$

(this is $X_0(49)$). By the change of variables $x \mapsto 4x - 1$, $y \mapsto 8y + 4x$, we have the equation

$$y^2 + xy = x^3 - x^2 - 2x - 1,$$

which has good reduction at all prime numbers except 7 (in fact, the discriminant is -7^3). However, by the change of variables $x \mapsto \sqrt{-7}x$, $y \mapsto (-7)^{3/4}y$ we have the equation

$$y^2 = x^3 + 5x - 2\sqrt{-7}$$

which has good reduction at all prime ideals of the field $\mathbf{Q}(\sqrt[4]{-7})$ (check that the discriminant is -1). Therefore E has everywhere potentially good reduction.

Definition 2.5.6. If E is an elliptic curve of the form $y^2 = x^3 + ax + b$, then its j-invariant is

$$j(E) = 1728 \frac{4a^3}{4a^3 + 27b^2}.$$

Fact 2.5.7. If $E : y^2 = x^3 + ax + b$ is an elliptic curve with $a, b \in \mathbf{Q}$, then E has everywhere potentially good reduction if and only if $j(E) \in \mathbf{Z}$.

Fact 2.5.8. Elliptic curves with complex multiplication (see §2.7 below) have everywhere potentially good reduction (and have j-invariants which are algebraic integers). More generally, abelian varieties of CM-type have everywhere potentially good reduction (see [149] and [139]).

Definition 2.5.9. Suppose A is an abelian variety over a number field K, and N is a positive integer. Let $A[N]$ denote the subgroup of points of $A(\mathbf{C})$ of order dividing N. Let K_N (or in more standard notation, $K(A[N])$) denote the smallest

field extension of K over which all the points on A of order dividing N are defined. Also, let
$$A(K)[N] = A[N] \cap A(K).$$

Example 2.5.10. For $E : y^2 = x^3 - x$ we have
$$E[2] = \{O, (0,0), (1,0), (-1,0)\} \quad \text{and} \quad \mathbf{Q}(E[2]) = \mathbf{Q}.$$

Example 2.5.11. For $E : y^2 = x^3 - 2x$ we have
$$E[2] = \{O, (0,0), (\sqrt{2},0), (-\sqrt{2},0)\} \quad \text{and} \quad \mathbf{Q}(E[2]) = \mathbf{Q}(\sqrt{2}).$$

Facts 2.5.12. Suppose A is a d-dimensional abelian variety over a number field K. Then
$$A[N] \cong (\mathbf{Z}/N\mathbf{Z})^{2d},$$
the field K_N is a finite Galois extension of K, and the action of $\mathrm{Gal}(K_N/K)$ on $A[N]$ gives a group homomorphism
$$\mathrm{Gal}(K_N/K) \to \mathrm{GL}_{2d}(\mathbf{Z}/N\mathbf{Z}).$$
This homomorphism is injective by the definition of K_N. Therefore,
$$[K_N : K] \leq \#\mathrm{GL}_{2d}(\mathbf{Z}/N\mathbf{Z}) < N^{4d^2}.$$

Fact 2.5.13 (Serre-Tate [139]). Suppose A is an abelian variety over a number field K, with everywhere potentially good reduction. If $N \geq 3$, then A has good reduction at all the prime ideals of K_N which do not divide N.

Definition 2.5.14. If A is an abelian variety over a number field M, and p is a prime number, let $A(M)_{p'-\mathrm{tors}}$ denote the subgroup of $A(M)_{\mathrm{tors}}$ of points of order prime to p.

Fact 2.5.15. Suppose A is an abelian variety over a number field M, suppose $\mathfrak{p}$ is a prime ideal of the ring of integers $\mathcal{O}$ of M, and suppose p is the characteristic of the field $\mathcal{O}/\mathfrak{p}$ (so $\mathfrak{p}$ divides p). If A has good reduction at $\mathfrak{p}$, then reduction modulo $\mathfrak{p}$ gives an injective homomorphism
$$A(M)_{p'-\mathrm{tors}} \hookrightarrow A(\mathcal{O}/\mathfrak{p}).$$

Fact 2.5.16 (Weil's bound). If A is a d-dimensional abelian variety over the finite field $\mathbf{F}_q$, then
$$\#A(\mathbf{F}_q) \leq (1 + \sqrt{q})^{2d}.$$

Proof of Theorem 2.5.2. With notation as in the statement of Theorem 2.5.2, let $\mathcal{O}$ denote the ring of integers of K_3, let $\mathfrak{p}$ be a prime ideal of $\mathcal{O}$ dividing 2, and let
$$q = 2^{\#\mathrm{GL}_{2d}(\mathbf{Z}/3\mathbf{Z})m}.$$
Then $\#(\mathcal{O}/\mathfrak{p})$ divides $2^{[K_3:\mathbf{Q}]}$ which divides q. By Fact 2.5.13, A has good reduction at $\mathfrak{p}$. By Facts 2.5.15 and 2.5.16 we have
$$\#A(K_3)_{2'-\mathrm{tors}} \leq \#A(\mathcal{O}/\mathfrak{p}) \leq \#A(\mathbf{F}_q)$$
$$\leq (1 + \sqrt{q})^{2d} = (1 + 2^{\#\mathrm{GL}_{2d}(\mathbf{Z}/3\mathbf{Z})m/2})^{2d}.$$
Similarly,
$$\#A(K_4)_{3'-\mathrm{tors}} \leq (1 + 3^{\#\mathrm{GL}_{2d}(\mathbf{Z}/4\mathbf{Z})m/2})^{2d}.$$
Since
$$\#A(K)_{\mathrm{tors}} \leq \#A(K_N)_{\mathrm{tors}} \quad \text{for every } N,$$

we have
$$\#A(K)_{\text{tors}} \leq \#A(K_3)_{2'-\text{tors}} \#A(K_4)_{3'-\text{tors}}.$$
The rest of the proof is left as an exercise. $\square$

2.6. Abelian varieties over other fields

As remarked by Cassels, the Torsion Conjecture would be false if the phrase "number field" were replaced by "local field" (Lemma 17.1 and p. 264 of [29]).

However, for abelian varieties (with no constant part) over function fields, one expects a Torsion Conjecture to hold. For a non-constant elliptic curve E defined over a function field in one variable K, Theorem 1 of [83] gives an explicit upper bound on the order of the torsion subgroup of $E(K)$, depending only on the genus of K. See also Theorem 7.2 of [62], which shows that if K has genus g and characteristic zero, then
$$\#E(K)_{\text{tors}} \leq 144(g+1)^{2/3}.$$

For surveys on Torsion Conjectures see [40] (for elliptic curves) and [154].

2.7. Digression on complex multiplication

If E is an elliptic curve over $\mathbf{C}$, and $m \in \mathbf{Z}$, then the map on $E(\mathbf{C})$ that takes a point P and sends it to mP defines an endomorphism of E. In this way we can view $\mathbf{Z}$ as a subring of $\text{End}(E)$.

Definition 2.7.1. If E is an elliptic curve over $\mathbf{C}$, we say that E has *complex multiplication* or *CM* if E has endomorphisms that are not multiplication by an integer (i.e., if $\text{End}(E) - \mathbf{Z} \neq \emptyset$).

If E has CM, then $\text{End}(E)$ is an order in an imaginary quadratic field K (so that $\text{End}(E) \otimes_{\mathbf{Z}} \mathbf{Q} = K$), and we say that E has *CM by* K. See [148] and [149] for more about complex multiplication of elliptic curves and abelian varieties.

Example 2.7.2. Consider
$$E : y^2 = x^3 - x.$$
The map $(x, y) \mapsto (-x, iy)$ is an automorphism of E. It is not multiplication by an integer, since it has order 4. Therefore E has complex multiplication.

Example 2.7.3. More generally, consider
$$E : y^2 = x^3 + ax,$$
with a a non-zero integer. As above, the map $(x, y) \mapsto (-x, iy)$ is in $\text{End}(E) - \mathbf{Z}$. Therefore E has complex multiplication (by $\mathbf{Q}(i)$).

Example 2.7.4. Consider
$$E : y^2 = x^3 + b,$$
with b a non-zero integer. The map $(x, y) \mapsto (\omega x, y)$, where ω is a primitive cube root of unity, is an automorphism of E. It is not multiplication by an integer, since it has order 3. Therefore E has complex multiplication (by $\mathbf{Q}(\omega)$).

See [149] for the theory of abelian varieties of CM-type. See [153], [154], [179], and [2] for bounds on torsion subgroups of Mordell-Weil groups of abelian varieties of CM-type.

CHAPTER 3
Ranks

Even for elliptic curves, the rank is still very mysterious. We will see in this chapter and the next some deep conjectures about the ranks of abelian varieties and elliptic curves.

3.1. Examples

By the Mordell-Weil Theorem, if A is an abelian variety over a number field K, then
$$A(K) \cong \mathbf{Z}^r \oplus A(K)_{\text{tors}}$$
for some non-negative integer r. We call r the *rank* of A over K.

Example 3.1.1.
$$E: y^2 = x^3 - x$$
$$E(\mathbf{Q}) \cong \mathbf{Z}/2\mathbf{Z} \times \mathbf{Z}/2\mathbf{Z}, \quad r = 0$$

Example 3.1.2.
$$E: y^2 = x^3 - 5^2 x$$
$$E(\mathbf{Q}) \cong \mathbf{Z} \times \mathbf{Z}/2\mathbf{Z} \times \mathbf{Z}/2\mathbf{Z}, \quad r = 1$$

The point $(-4, 6)$ has infinite order (this follows from the Nagell-Lutz Theorem, since $6 \neq 0$ and 36 does not divide $4(25)^3$).

Example 3.1.3.
$$E: y^2 = x^3 - (2 \cdot 7 \cdot 11)^2 x$$
$$E(\mathbf{Q}) \cong \mathbf{Z}^2 \times \mathbf{Z}/2\mathbf{Z} \times \mathbf{Z}/2\mathbf{Z}, \quad r = 2$$

The points $(-98, 1176)$ and $(350, 5880)$ are independent points of infinite order. (This can be shown by computing the determinant of the associated Néron-Tate pairing matrix — see §4.6 below — and showing it is non-zero.)

Example 3.1.4 (Kramarz [76]).
$$E: y^2 = x^3 - (2 \cdot 3 \cdot 11 \cdot 19)^2 x$$
$$E(\mathbf{Q}) \cong \mathbf{Z}^3 \times \mathbf{Z}/2\mathbf{Z} \times \mathbf{Z}/2\mathbf{Z}, \quad r = 3$$

The points $(-98, 12376)$, $(1650, 43560)$, and $(109554, 36258840)$ are independent points of infinite order.

Example 3.1.5.
$$E : y^2 = x^3 - (2 \cdot 3 \cdot 7 \cdot 17 \cdot 41)^2 x$$
$$E(\mathbf{Q}) \cong \mathbf{Z}^4 \times \mathbf{Z}/2\mathbf{Z} \times \mathbf{Z}/2\mathbf{Z}, \quad r = 4$$

Example 3.1.6 (Rogers [**126**]).
$$E : y^2 = x^3 - (2 \cdot 3 \cdot 11 \cdot 17 \cdot 19 \cdot 59 \cdot 163)^2 x$$
$$E(\mathbf{Q}) \cong \mathbf{Z}^5 \times \mathbf{Z}/2\mathbf{Z} \times \mathbf{Z}/2\mathbf{Z}, \quad r = 5$$

Example 3.1.7 (Rogers [**126**]).
$$E : y^2 = x^3 - (2 \cdot 3 \cdot 5 \cdot 11 \cdot 19 \cdot 41 \cdot 43 \cdot 67 \cdot 83)^2 x$$
$$E(\mathbf{Q}) \cong \mathbf{Z}^6 \times \mathbf{Z}/2\mathbf{Z} \times \mathbf{Z}/2\mathbf{Z}, \quad r = 6$$

Example 3.1.8. Roland Martin and William McMillen [**92**] of NSA found 24 independent points of infinite order on

$$y^2 + xy + y = x^3 - 120039822036992245303534619191166796374x +$$
$$504224992484910670010801799168082726759443756222911415116.$$

They (and a number of other people who have constructed elliptic curves of large rank) used methods of Mestre [**100**]. They had found earlier [**91**] an elliptic curve over **Q** with rank at least 23.

Example 3.1.9. Nagao [**109**] found an infinite family of elliptic curves over **Q** of rank at least 13, by finding an elliptic curve over $\mathbf{Q}(t)$ of rank at least 13. Kihara [**72**] found an infinite family of elliptic curves over **Q** of rank at least 14, by considering Nagao's $\mathbf{Q}(t)$-curve over the function field of a certain elliptic curve of positive rank.

Example 3.1.10. D. Zagier and G. Kramarz [**188**] considered the curves $x^3 + y^3 = d$, i.e., the elliptic curves $y^2 = x^3 - 432d^2$. These are the "cubic twists" of $x^3 + y^3 = 1$. They performed numerical calculations for d up to 70,000, and found that 23.3% of the even ranks are at least 2, and 2.2% of the odd ranks are at least 3, with these percentages remaining constant as d grows.

Example 3.1.11. Shioda gives $y^2 = x^3 + t^{360} + 1$ as an example of an elliptic curve over $\mathbf{C}(t)$ of rank 68 (see [**151**] and [**30**]; see also [**168**] and [**150**] for examples with rank 56).

3.2. Unboundedness/boundedness

The following conjecture is part of the folklore.

Conjecture 3.2.1. *Ranks of elliptic curves over* **Q** *are unbounded.*

One reason for believing the conjecture is that the analogous result over a function field in one variable over a finite field $\mathbf{F}_q$ is true:

Theorem 3.2.2 (Tate-Shafarevich [**175**]). *Ranks of elliptic curves over* $\mathbf{F}_q(t)$ *are unbounded.*

The examples of Tate-Shafarevich are of quadratic twists. In [**175**] it is stated that Lapin ([**81**], [**82**]) constructed analogous examples over $\mathbf{C}(t)$. However, I have been told that this does not seem to be the case.

3.3. Vary the field

One can fix the elliptic curve or abelian variety, increase the field, and see how the rank behaves. Such questions are considered, for example, in [**6**], [**112**], [**94**], [**57**], and [**120**]. Honda stated a controversial conjecture:

Conjecture 3.3.1 (Honda [**63**], see also Question 3 on p. 162 of [**137**]). *If A is an abelian variety over a number field L, then there exists a constant C_A, depending only on A, such that for all number fields $K \supseteq L$,*
$$\mathrm{rank}(A(K)) \leq C_A[K : \mathbf{Q}].$$

Honda based his conjecture on the following two theorems he proved about ranks of abelian varieties, and an analogy with the Dirichlet unit theorem.

Theorem 3.3.2 (Honda [**63**], Corollary of Theorem 5). *If A is an abelian variety of dimension d over a number field, then there exists a constant C_A, depending only on A, such that for every prime ℓ and every number field K over which A is defined and such that $A[\ell] \subset A(K)$,*
$$\mathrm{rank}(A(K)) \leq C_A[K : \mathbf{Q}] + 2dh_K(\ell),$$
where $h_K(\ell)$ is the ℓ-rank of the ideal class group of K.

Note that the condition $A[\ell] \subset A(K)$ is equivalent to $K(A[\ell]) = K$, which is equivalent to $A(K)[\ell] \cong (\mathbf{Z}/\ell\mathbf{Z})^{2d}$.

Theorem 3.3.3 (Honda [**63**], Theorem 6). *If A is a simple abelian variety over a number field K, and A is of CM-type with reflex CM-field F, then*
$$\mathrm{rank}(A(KF)) = [KF : K]\mathrm{rank}(A(K)).$$

See [**149**] or [**148**] for the definition of the reflex CM-field. In the case of CM elliptic curves, the reflex CM-field is the same as the imaginary quadratic field by which the curve has complex multiplication.

Honda's Conjecture is not widely believed, and would imply statements that many people believe to be false. For example, it would imply that ranks of elliptic curves are bounded in families of quadratic twists (see Exercise 3.4.7 below). However, in [**133**] Schneiders and Zimmer give data which they say provide numerical evidence for Honda's Conjecture.

3.4. Quadratic twists

From now until the end of the chapter, unless otherwise stated, E is an elliptic curve over $\mathbf{Q}$.

Definition 3.4.1. If $E : y^2 = x^3 + ax + b$ is an elliptic curve over $\mathbf{Q}$ and $0 \neq D \in \mathbf{Z}$, define the *quadratic twist of E by D* to be the elliptic curve
$$E^{(D)} : Dy^2 = x^3 + ax + b.$$

Exercise 3.4.2. *Check that the map $(x, y) \mapsto (Dx, D^2y)$ is an isomorphism from $E^{(D)}$ onto $y^2 = x^3 + D^2ax + D^3b$.*

Exercise 3.4.3. *Show that if $D_1 = n^2 D_2$ with $n \in \mathbf{Q}$, then $E^{(D_1)} \cong E^{(D_2)}$.*

Exercise 3.4.4. *Show that $E^{(D)}$ and E are isomorphic over $\mathbf{Q}(\sqrt{D})$.*

Definition 3.4.5. Let
$$r_E(D) = \mathrm{rank}(E^{(D)}(\mathbf{Q})).$$

Exercise 3.4.6. *Show that*
$$\mathrm{rank}(E(\mathbf{Q})) + r_E(D) = \mathrm{rank}(E(\mathbf{Q}(\sqrt{D}))).$$

Exercise 3.4.7. *Use Exercise 3.4.6 to show that (Honda's) Conjecture 3.3.1 implies that there exists a positive constant C_E depending only on E such that $r_E(D) \leq C_E$ for all D.*

Question 3.4.8. *Are ranks unbounded in (all, some) families of quadratic twists?*

Note that the result of Tate and Shafarevich was for a family of quadratic twists (i.e., they found quadratic twists of an elliptic curve over $\mathbf{F}_q(t)$ of unbounded rank). Honda's Conjecture would be false if number fields were replaced by function fields over finite fields.

Write $\omega(n)$ for the number of prime divisors of n.

Fact 3.4.9. (See Exercise 3.4.12 below.) If $E(\mathbf{Q})[2] \cong \mathbf{Z}/2\mathbf{Z} \times \mathbf{Z}/2\mathbf{Z}$, then
$$\mathrm{rank}(E(\mathbf{Q})) \leq 2\omega(2\Delta(E))$$
where $\Delta(E)$ is the discriminant of E.

Remark 3.4.10. For some positive constant C,
$$\omega(n) \leq C \frac{\log n}{\log \log n}$$
if $n > 2$. Therefore if $E(\mathbf{Q})[2] \cong \mathbf{Z}/2\mathbf{Z} \times \mathbf{Z}/2\mathbf{Z}$, then
$$\mathrm{rank}(E(\mathbf{Q})) \leq C \frac{\log |\Delta(E)|}{\log \log |\Delta(E)|}.$$

Exercise 3.4.11. *Use Fact 3.4.9 to show that if $E(\mathbf{Q})[2] \cong \mathbf{Z}/2\mathbf{Z} \times \mathbf{Z}/2\mathbf{Z}$ and $|D| > 2$, then*
$$r_E(D) \leq 2\omega(2D \cdot \Delta(E)) \leq C_E \frac{\log |D|}{\log \log |D|},$$
where C_E is a constant depending only on E.

Therefore, to get large rank, $|D|$ must be large. The bound

(6) $$r_E(D) \leq C_E \frac{\log |D|}{\log \log |D|}$$

is called the "trivial bound".

For $y^2 = x^3 - x$, compare the bounds given by the first inequality of Exercise 3.4.11 with the examples in §3.1.

Exercise 3.4.12 (Advanced exercise). *Suppose E is an elliptic curve over a number field K, suppose ℓ is a prime number, and suppose $E[\ell] \subset E(K)$. Let M denote the maximal Galois extension of K whose Galois group is an elementary abelian ℓ-group and which is unramified at all primes that do not divide $\ell \cdot \Delta(E)$. Define φ by $\varphi(P)(g) = g(Q) - Q$ for $P \in E(K)$ and $g \in \mathrm{Gal}(\bar{K}/K)$, where Q is a point of $E(\bar{K})$ such that $\ell Q = P$. Show that φ induces a well-defined injective group homomorphism*
$$E(K)/\ell E(K) \hookrightarrow \mathrm{Hom}(\mathrm{Gal}(M/K), E[\ell]).$$
Then let $\ell = 2$ and $K = \mathbf{Q}$ and use Class Field Theory to prove Fact 3.4.9, thereby proving the "trivial bound" (6).

The technique of proof in Exercise 3.4.12 can be generalized to lead to a proof of (Honda's) Theorem 3.3.2. The same technique is also used in proving the Mordell-Weil Theorem. We will see more of this later when we define Tate-Shafarevich groups.

3.5. Average ranks

The average rank will depend on how we define the average. Let's first consider the average rank in a family of quadratic twists, where we order the curves by $|D|$. Goldfeld conjectured that this average is always $\frac{1}{2}$:

Conjecture 3.5.1 (Goldfeld [51]).
$$\lim_{X \to \infty} \frac{\sum_{|D|<X} r_E(D)}{\sum_{|D|<X} 1} = \frac{1}{2}$$

(where the D's in the sums are square-free).

Theorem 3.5.2 (Goldfeld [51]). *Assuming the Birch and Swinnerton-Dycr Conjecture, the Modularity Conjecture, and a form of the Riemann Hypothesis, then*
$$\limsup\nolimits_{X \to \infty} \frac{\sum_{|D|<X} r_E(D)}{\sum_{|D|<X} 1} \leq 3.25$$

(where the D's in the sums are square-free).

In [17] Brumer pointed out that Goldfeld's 3.25 can be replaced by 2.5. Brumer tells me that Heath-Brown can prove that it can in fact be replaced by 1.5.

Heath-Brown has determined an upper bound for the average rank of the odd quadratic twists of the curve $y^2 = x^3 - x$:

Theorem 3.5.3 (Heath-Brown [60]). *If* $E : y^2 = x^3 - x$, *then*
$$\limsup\nolimits_{X \to \infty} \frac{\sum_{D \, odd, |D|<X} r_E(D)}{\sum_{D \, odd, |D|<X} 1} \leq 1.2645\ldots$$

(where the D's are square-free).

Exercise 3.5.4. *Show that every elliptic curve over* $\mathbf{Q}$ *has a unique model of the form*
$$E_{r,s} : y^2 = x^3 + rx + s$$
where r and s are integers with the property that if p^4 *divides r then* p^6 *does not divide s.*

Rather than looking at families of twists, Brumer looked at all elliptic curves, ordered by the size of the coefficients for a model as in Exercise 3.5.4.

Theorem 3.5.5 (Brumer [17]). *Assuming the Birch and Swinnerton-Dyer Conjecture, the Modularity Conjecture, and the Riemann Hypothesis for L-functions of elliptic curves, then*
$$\limsup\nolimits_{X \to \infty} \frac{1}{\#\mathcal{C}(X)} \sum_{E \in \mathcal{C}(X)} \mathrm{rank}(E(\mathbf{Q})) \leq 2.3,$$

where
$$\mathcal{C}(X) = \{E_{r,s} : |r| \leq X^{1/3}, |s| \leq X^{1/2}\}.$$

Brumer tells me that Heath-Brown can reduce the 2.3 to 2.

In [17] Brumer points out that we can't yet prove unconditionally that the average rank is > 0, or that it is bounded.

Brumer and McGuinness [19] did numerical calculations for elliptic curves over $\mathbf{Q}$ of prime conductor up to 10^8, for which the average rank seems to increase slowly to around 1.

Brumer proved the following unconditional result over $\mathbf{F}_q(t)$.

Theorem 3.5.6 (Brumer [17]). *If q is a power of a prime p, and $p \geq 5$, then the average rank of elliptic curves over $\mathbf{F}_q(t)$ is at most 2.3.*

Brumer tells me that he can now reduce this 2.3 to 2, using methods analogous to those Heath-Brown used to obtain the improvements stated above.

3.6. Distribution of ranks

The following conjecture is a consequence of the Conjecture of Birch and Swinnerton-Dyer, which we will discuss later.

Conjecture 3.6.1 (Weak Parity Conjecture). *For square-free D, the density of D with $r_E(D)$ even is $1/2$.*

Note that the Weak Parity Conjecture implies that the average rank in a family of quadratic twists is at least $1/2$. We will state a stronger Parity Conjecture later, when we discuss the Conjecture of Birch and Swinnerton-Dyer. Most experts believe the Parity Conjecture, since most believe the Conjecture of Birch and Swinnerton-Dyer. Many (but fewer) also believe the following conjecture.

Conjecture 3.6.2 (Density Conjecture). *For square-free D, the density of D with $r_E(D) = 0$ (respectively $= 1$, respectively ≥ 2) is $1/2$ (respectively $1/2$, respectively 0).*

Exercise 3.6.3. *Show that the Weak Parity Conjecture and Goldfeld's Conjecture together imply the Density Conjecture.*

Katz and Sarnak ([68], [69]) have a philosophy for why the Density Conjecture should be true.

Theorem 3.6.4 (Heath-Brown [60]). *If $E : y^2 = x^3 - x$, then*

(i) *the density of odd D with $r_E(D) = 0$ is greater than zero, and*

(ii)

$$\limsup_{X \to \infty} \frac{\#\{ \text{ odd } D : |D| \leq X, r_E(D) \geq R\}}{\#\{ \text{ odd } D : |D| \leq X\}} \leq (1.7313\ldots)2^{-(R^2-R)/2}$$

(where the D's are always square-free).

Part (ii) of Heath-Brown's Theorem implies that the density of odd square-free D with $r_E(D) \geq 2$ is at most .866, the density with $r_E(D) \geq 3$ is at most .213, etc.

Theorem 3.6.5 (Ono-Skinner [119]). *If E is a (modular) elliptic curve over $\mathbf{Q}$, then*

$$\#\{ \text{ square-free } D : |D| \leq X, r_E(D) = 0\} \geq C_E \frac{X}{\log X}.$$

This result does not give a positive density with rank 0 the way Heath-Brown's result did, but it holds for all (modular) elliptic curves over $\mathbf{Q}$.

In Definition 4.3.1 below we will define the analytic rank, which conjecturally is equal to the rank. There are many results in the literature about analytic ranks. For average and density results for analytic ranks in the family of cubic twists $x^3 + y^3 = d$, see [85] and [86].

Example 3.6.6 (Silverman [159]). Let E be the elliptic curve $y^2 = x^3 - x$, and for $0 \neq d \in \mathbf{Q}(i)$ let $E^{(d)}$ be the elliptic curve $dy^2 = x^3 - x$. Then $\text{rank}(E^{(d)}(\mathbf{Q}(i)))$ is even (since $\text{End}(E^{(d)}) = \mathbf{Z}[i]$, so $E^{(d)}(\mathbf{Q}(i))$ is a $\mathbf{Z}[i]$-module). Viewing $ty^2 = x^3 - x$ as an elliptic curve over the function field $\mathbf{C}(t)$, it has rank 0 (a Mordell-Weil Theorem holds here). Under the general philosophy that ranks are "no bigger than they have to be", one expects that for $d \in \mathbf{Q}(i)$, the rank of $E^{(d)}(\mathbf{Q}(i))$ is 0 for d in a set of density 1.

Theorem 3.6.7 (Gouvêa-Mazur [53]). *If E is an elliptic curve over $\mathbf{Q}$, and the Parity Conjecture holds, then for every $\varepsilon > 0$ there exists a constant $C_{E,\varepsilon} > 0$ such that for every $X > C_{E,\varepsilon}$,*

$$\#\{\text{square-free } D : |D| \leq X, r_E(D) \geq 2, \text{ and } r_E(D) \text{ is even}\} \geq X^{1/2-\varepsilon}.$$

Here is a quick sketch of what Gouvêa and Mazur did. Suppose E is given by the equation $y^2 = x^3 + ax + b$ with $a, b \in \mathbf{Z}$. If n is a non-zero rational number, write $s(n)$ for the square-free part of n, i.e., the unique integer s such that $n = sy^2$ for some $y \in \mathbf{Q}$ (check that this is well-defined). Take a rational number x and compute $x^3 + ax + b$. If it is non-zero, then write $x^3 + ax + b$ as Dy^2 where D is the square-free part and $y \in \mathbf{Q}$. Then (x, y) is a point in $E^{(D)}(\mathbf{Q})$. (For example, if E is $y^2 = x^3 - x$, and $x = 2$, then $D = 6$ and $y = 1$.) For all but finitely many $x \in \mathbf{Q}$, (x, y) is a point of infinite order. This gives a way of (usually) obtaining D's so that $r_E(D) \geq 1$. To deal with integers rather than rational numbers, write x as a fraction in lowest terms, u/v. Then

$$Dy^2 = \left(\frac{u}{v}\right)^3 + a\left(\frac{u}{v}\right) + b.$$

Therefore, $D(yv^2)^2 = u^3v + auv^3 + bv^4$. Let

$$F(u, v) = u^3v + auv^3 + bv^4.$$

Then $D = s(F(u, v))$. Proving Theorem 3.6.7 basically reduces to finding a (sufficiently good) lower bound for

$$\#\{D \in \mathbf{Z} : |D| \leq X, r_E(D) \text{ is even}, D = s(F(u, v)) \text{ for some } u, v \in \mathbf{Z}\}.$$

The Parity Conjecture gives a criterion for when $r_E(D)$ is even. Gouvêa and Mazur show that a positive proportion of $F(u, v)$'s are square-free (this is the hardest part of their proof). If $|u|, |v| \leq X^{1/4}$, then $|F(u, v)| \leq C_E X$. Further, they show that

$$\#\{F(u, v) : |u|, |v| \leq X^{1/4}\} \geq C'_{E,\delta} X^{1/2-\delta}$$

for every $\delta > 0$.

Theorem 3.6.8 (Stewart-Top [166]). *If $E : y^2 = x^3 + ax + b$ with $a, b \in \mathbf{Q}$ and $ab \neq 0$, then there exist positive constants C_0 and C_1 such that for every $X > C_0$,*

$$\#\{\text{square-free } D : |D| \leq X, r_E(D) \geq 2\} \geq C_1 \frac{X^{1/7}}{(\log X)^2}.$$

The result of Stewart and Top has a weaker conclusion than that of the theorem of Gouvêa and Mazur, but does not rely on the Parity Conjecture.

Karl Rubin and I [**130**] modified the idea of Gouvêa and Mazur, where instead of counting how many D's occur as square-free parts of $F(u,v)$'s, we count how often each D occurs as the square-free part of an $F(u,v)$. The D's that occur more often tend to correspond to higher rank curves. Using our method, Nick Rogers [**126**] carried out computations which led to the rank 5 and 6 curves in Examples 3.1.6 and 3.1.7.

CHAPTER 4
Conjectures of Birch and Swinnerton-Dyer

Birch and Swinnerton-Dyer based their conjectures ([**10**], [**11**], see also [**7**] and [**25**]) on extensive numerical evidence obtained from computer calculations they did in the 1950's and 1960's. As Birch recalls, "Back in the late 1950's there was a lot of excitement about Tamagawa's work on Siegel's formulae". Birch and Swinnerton-Dyer were trying to generalize Siegel's results on quadratic forms to the setting of elliptic curves, and wanted to understand the analogue of the Tamagawa numbers. Their conjectures remain among the major unsolved problems in number theory. The original conjectures were made for elliptic curves over $\mathbf{Q}$. They have been generalized by Tate (to abelian varieties over global fields) and others. See in particular the conjectures of Beilinson ([**124**], [**111**]) and Bloch-Kato [**12**].

4.1. The L-function of an elliptic curve

Let E be an elliptic curve over $\mathbf{Q}$.

Let N_E denote the conductor of E (defined in Joe Buhler's lectures). Then N_E divides $\Delta(E)$. Further, if p is a prime, then p divides N_E if and only if E has bad reduction at p.

There exists a generalized Weierstrass equation for E with coefficients in $\mathbf{Z}$ for which $|\Delta(E)|$ is minimal. Call this a *minimal Weierstrass equation* for E. In this chapter from now on we will only consider minimal Weierstrass equations. Then p divides N_E if and only if p divides $\Delta(E)$. Given an elliptic curve over $\mathbf{Q}$, the computer program PARI will tell you how to get a minimal Weierstrass equation for it.

If p is a prime, let
$$a_p = p + 1 - \#E(\mathbf{F}_p).$$

Exercise 4.1.1. *Show that if p divides $\Delta(E)$, then*
$$a_p = \begin{cases} 0 & \text{if } E \text{ has additive reduction at } p, \\ 1 & \text{if } E \text{ has split multiplicative reduction at } p, \\ -1 & \text{if } E \text{ has non-split multiplicative reduction at } p. \end{cases}$$

(See [**156**] for definitions of additive, split multiplicative, and non-split multiplicative reduction.)

Example 4.1.2. The equation $E : y^2 = x^3 - x$ is a minimal Weierstrass equation. Here, $\Delta(E) = 2^6$ and $N_E = 2^5$. The point $(1,0)$ is the singular point on the reduction of E modulo 2. We have (see §2 of [**11**])

$$a_p = \begin{cases} 0 & \text{if } p \equiv 3 \pmod 4 \text{ or } p = 2, \\ 2\alpha & \text{if } p \equiv 1 \pmod 4, \end{cases}$$

where $p = \alpha^2 + \beta^2$ with $\alpha, \beta \in \mathbf{Z}$, normalized so that β is even and 4 divides $\alpha - 1 - \beta$.

Let

(7) $$L(E, s) = \prod_{p \nmid \Delta(E)} \frac{1}{1 - \frac{a_p}{p^s} + \frac{p}{p^{2s}}} \prod_{p | \Delta(E)} \frac{1}{1 - \frac{a_p}{p^s}}.$$

The function $L(E, s)$ is called the *Hasse-Weil L-function* for E.

The next result is a special case of Weil's bound (see Fact 2.5.16 above).

Theorem 4.1.3 (Hasse).
$$|a_p| < 2\sqrt{p}.$$

Corollary 4.1.4 (Exercise). For $\mathbf{Re}(s) > 3/2$, the Euler product (7) defining $L(E, s)$ converges absolutely.

Multiplying out the Euler product, we have the Dirichlet series

$$L(E, s) = \sum_{n=1}^{\infty} \frac{a_n}{n^s},$$

which converges absolutely for $\mathbf{Re}(s) > 3/2$, and defines the a_n's for composite n.

Definition 4.1.5. Suppose A and A' are abelian varieties over a number field K. We say that *A is isogenous to A' over K* if there exists a surjective homomorphism $A \to A'$ over K with finite kernel.

Two elliptic curves E and E' are isogenous over K if and only if there exists a non-zero homomorphism $E \to E'$ over K.

Exercise 4.1.6. *Show that if E and E' are elliptic curves over $\mathbf{Q}$ which are isogenous over $\mathbf{Q}$, then*

$$\mathrm{rank}(E(\mathbf{Q})) = \mathrm{rank}(E'(\mathbf{Q})) \text{ and } L(E, s) = L(E', s).$$

In other words, the rank and L-function are both isogeny invariants.

Theorem 4.1.7 (Faltings [**43**]). *If E and E' are elliptic curves over $\mathbf{Q}$ and $L(E, s) = L(E', s)$, then E and E' are isogenous over $\mathbf{Q}$.*

Remark 4.1.8. Faltings proved more generally that if two abelian varieties over a number field have isomorphic ℓ-adic representations for some prime ℓ, then they are isogenous. This result was formerly known as Tate's Isogeny Conjecture.

Warning 4.1.9. Tate's Isogeny Conjecture is sometimes incorrectly stated in the literature as saying that two abelian varieties over a number field K are isogenous over K if and only if their L-functions over K are equal. However, this statement is false even for elliptic curves, i.e., Theorem 4.1.7 above is *false* if $\mathbf{Q}$ is replaced by an arbitrary number field (see Remark 3.4 of [**155**] for a counterexample over $\mathbf{Q}(i)$).

4.2. Modularity

Definition 4.2.2 below is the quickest definition of modularity that I know.

Let $\mathbf{H}$ denote the complex upper half plane, and let
$$\Gamma_0(N) = \left\{ \begin{pmatrix} a & b \\ c & d \end{pmatrix} \in \mathrm{SL}_2(\mathbf{Z}) : N \text{ divides } c \right\}.$$

Exercise 4.2.1. *Show that if $z \in \mathbf{H}$ and $\begin{pmatrix} a & b \\ c & d \end{pmatrix} \in \mathrm{SL}_2(\mathbf{Z})$, then $\frac{az+b}{cz+d} \in \mathbf{H}$.*

It follows that the group $\Gamma_0(N)$ acts on $\mathbf{H}$ by linear fractional transformations. For $N \geq 3$, let $Y_0(N)$ denote the moduli space of pairs (E, C) where E is an elliptic curve and C is a cyclic subgroup of E of order N. Then $Y_0(N)(\mathbf{C}) \cong \mathbf{H}/\Gamma_0(N)$. The compactification $X_0(N)$ of $Y_0(N)$ is an algebraic curve over $\mathbf{Q}$. We have
$$X_0(N)(\mathbf{C}) \cong \mathbf{H}/\Gamma_0(N) \cup (\mathbf{Q} \cup i\infty)/\Gamma_0(N),$$
and $X_0(N)(\mathbf{C})$ can be viewed as a compact Riemann surface. The elements of the finite set $(\mathbf{Q} \cup i\infty)/\Gamma_0(N)$ are called the *cusps* of the modular curve $X_0(N)$.

Definition 4.2.2. If E is an elliptic curve over $\mathbf{Q}$, we say that E is *modular* if there is a non-constant analytic map of Riemann surfaces $X_0(N)(\mathbf{C}) \to E(\mathbf{C})$ for some N.

Example 4.2.3. If E is the elliptic curve $y^2 = x^3 - x$, then there is a non-constant analytic map $X_0(32)(\mathbf{C}) \to E(\mathbf{C})$.

The following result of Shimura (which he proved much more generally, in the setting of abelian varieties; see Chapter 7, especially Proposition 7.19, of [**148**], and [**144**], especially Theorem 1 and Proposition 1) was a groundbreaking result that led to much work on modularity.

Theorem 4.2.4 (Shimura).
(i) *If f is a normalized Hecke eigenform of weight 2 for $\Gamma_0(N)$ with rational Fourier coefficients, then there is an elliptic curve E_f over $\mathbf{Q}$ such that:*
  (a) *there is a non-constant morphism $X_0(N) \to E_f$ defined over $\mathbf{Q}$, and*
  (b) *$L(E_f, s) = L(f, s)$ up to finitely many Euler factors.*
(ii) *If E is an elliptic curve over $\mathbf{Q}$, and for some N there is a non-constant morphism $X_0(N) \to E$ defined over $\mathbf{Q}$, then there is an f as in (i) such that E is isogenous to E_f over $\mathbf{Q}$ (and thus $L(E, s) = L(E_f, s)$).*

(For definitions of modular forms, Hecke eigenforms, etc., see [**148**].) Building on Shimura's work, Carayol (Corollaire on p. 411 of [**24**]) proved:

Theorem 4.2.5 (Carayol). *Under the hypotheses in Theorem 4.2.4(i), if f is a newform, then $L(E_f, s) = L(f, s)$ and $N = N_{E_f}$ (where N_{E_f} is the conductor of E_f).*

Exercise 4.2.6. *Use Theorems 4.2.4 and 4.2.5 (along with Faltings' result, Theorem 4.1.7 above) to deduce Theorem 4.2.7 below (see the appendix to* **[97]** *for the implication (a) $\Rightarrow$ (b)).*

Theorem 4.2.7. *If E is an elliptic curve over $\mathbf{Q}$, then the following statements are equivalent:*
  (a) *E is modular in the sense of Definition 4.2.2;*
  (b) *there is a non-constant morphism $X_0(N) \to E$ defined over $\mathbf{Q}$, for some N;*
  (c) *for some N, $g(z) := \sum_{n=1}^{\infty} a_n e^{2\pi i n z}$ satisfies*
$$g(\gamma(z)) = (cz+d)^2 g(z)$$
  for every $\gamma = \begin{pmatrix} a & b \\ c & d \end{pmatrix} \in \Gamma_0(N);$
  (d) *for some N, there exists a weight two cusp form f for $\Gamma_0(N)$ such that $L(E, s) = L(f, s)$ (if so, then f will be a normalized eigenform, and $f = g$);*
  (e) *there exists a normalized Hecke eigenform f of weight 2 for $\Gamma_0(N)$ with rational Fourier coefficients such that E is isogenous over $\mathbf{Q}$ to E_f;*
  (f) *the same as any of the above, with $N = N_E$, the conductor of E.*

If E is a CM elliptic curve over $\mathbf{Q}$, then E is modular (see **[143]** for the proof, as well as references to relevant papers of Hecke and Deuring).

Taniyama **[147]** conjectured that every elliptic curve is "automorphic", and Shimura conjectured that every elliptic curve over $\mathbf{Q}$ is modular. Breuil, Conrad, Diamond, and Taylor have recently announced that they can complete the work of Wiles to prove Shimura's Modularity Conjecture. (See **[186]** and **[176]** for the work of Wiles and Taylor-Wiles.)

Theorem 4.2.8 (Breuil-Conrad-Diamond-Taylor-Wiles). *Every elliptic curve over $\mathbf{Q}$ is modular*[1].

This result implies the following long-standing conjecture of Hasse and Weil (see **[184]** and the Commentary to [1952d] in **[185]**), which had been proved earlier when E is modular (see **[144]** and **[148]**; see also **[41]** and **[141]**).

Theorem 4.2.9. *If E is an elliptic curve over $\mathbf{Q}$, then $L(E, s)$ has an analytic continuation to the complex plane $\mathbf{C}$, and has a functional equation:*

(8)
$$\Lambda(E, s) = w_E \Lambda(E, 2-s)$$

with $w_E \in \{\pm 1\}$ and with

$$\Lambda(E, s) = N_E^{s/2} (2\pi)^{-s} \Gamma(s) L(E, s).$$

See §2.2 of **[108]** and Theorem 4.1 of **[31]** for the following result of Chandrasekharan and Narasimhan.

Theorem 4.2.10. *If Theorem 4.2.9 holds for E, then $\sum_{n=1}^{\infty} \frac{a_n}{n^s}$ converges (conditionally) to $L(E, s)$ for $\mathbf{Re}(s) > 5/6$.*

---

[1]Since this result is so new, when a result in this text relies on Theorem 4.2.8 or 4.2.9, this dependence is explicitly stated.

4.3. BSD I

If Theorem 4.2.9 holds for E, then taking the Taylor expansion of $L(E,s)$ around $s=1$ gives

$$(9) \qquad L(E,s) = \sum_{i=r}^{\infty} b_i(s-1)^i = b_r(s-1)^r + b_{r+1}(s-1)^{r+1} + \ldots$$

for some non-negative integer r, with $b_i \in \mathbf{C}$ and with $b_r \neq 0$.

Definition 4.3.1. This r is called the *analytic rank* of E (over $\mathbf{Q}$).

We have
$$r = \operatorname{ord}_{s=1} L(E,s).$$
Note that
$$b_r = \lim_{s \to 1} \frac{L(E,s)}{(s-1)^r} = \frac{L^{(r)}(E,1)}{r!},$$
where $L^{(r)}$ is the r-th derivative of L.

Exercise 4.3.2. *Show that all $b_i \in \mathbf{R}$.*

Birch and Swinnerton-Dyer made the following conjecture (see [11]).

Conjecture of BSD I. *If E is an elliptic curve over $\mathbf{Q}$, then*
$$\text{rank} = \text{analytic rank}.$$

A heuristic justification for this conjecture goes as follows. At $s=1$, the Euler product for $L(E,s)$ is

$$\prod_{p \nmid \Delta(E)} \frac{p}{\#E(\mathbf{F}_p)} \prod_{p \mid \Delta(E)} \left(1 - \frac{a_p}{p}\right)^{-1},$$

where the second (finite) product is a non-zero number. Therefore, $L(E,s)$ should have a high order zero at $s=1$ if and only if E has "many" points modulo primes, and the latter should be true if and only if E has "many" rational points.

Exercise 4.3.3. *Use the functional equation (8) to show that*
$$w_E = (-1)^r$$
where w_E is the "root number" defined in Theorem 4.2.9 and r is the analytic rank.

It follows that BSD I implies:

Parity Conjecture. *If E is an elliptic curve over $\mathbf{Q}$, then*
$$w_E = (-1)^{\operatorname{rank}(E(\mathbf{Q}))},$$
i.e., $\operatorname{rank}(E(\mathbf{Q}))$ is even if $w_E = 1$ and is odd if $w_E = -1$.

Suppose D is a square-free integer. Let ψ_D denote the Dirichlet character associated to $\mathbf{Q}(\sqrt{D})$. Define the conductor
$$N_{\psi_D} = \begin{cases} D & \text{if } D \equiv 1 \pmod 4, \\ 4D & \text{if } D \equiv 2,3 \pmod 4. \end{cases}$$

If $(N_E, N_{\psi_D}) = 1$, then
$$(10) \qquad w_{E^{(D)}} = \psi_D(-N_E) w_E.$$

Example 4.3.4. Suppose E is the elliptic curve $y^2 = x^3 - x$. Fermat showed that the rank (over $\mathbf{Q}$) is 0. Since E has complex multiplication, $L(E, s)$ is entire. Using that $L(E, s) = L(f_E, s)$ for an appropriate modular form f_E, one can compute that $L(E, 1)$ is approximately $0.655514\ldots \neq 0$. Therefore, the analytic rank $r = 0$, and thus $w_E = 1$ and BSD I holds for this E. Further,

$$(11) \quad w_{E^{(D)}} = \begin{cases} 1 & \text{if } |D| \equiv 1, 2, 3 \pmod 8, \\ -1 & \text{if } |D| \equiv 5, 6, 7 \pmod 8. \end{cases}$$

Exercise 4.3.5. *Use (10) to show that (11) holds when D is odd.*

4.4. The Congruent Number Problem

Definition 4.4.1. A *congruent number* is a positive integer D which is the area of a right triangle with three rational sides.

For example, 6 is a congruent number since the 3–4–5 right triangle has area 6.

The Congruent Number Problem is the problem of determining which positive integers are congruent numbers.

Throughout this section, let E be the elliptic curve $y^2 = x^3 - x$. Then $E^{(D)}$ is the curve $y^2 = x^3 - D^2 x$.

Fact 4.4.2 (see [**73**]). *A positive integer D is a congruent number if and only if $r_E(D) \geq 1$.*

If D is a square-free positive integer, then the Parity Conjecture (or BSD I for all the curves $E^{(D)}$) would imply that $D \equiv 5, 6, 7 \pmod 8$ if and only if $r_E(D)$ is odd. If $r_E(D)$ is odd, then in particular it is non-zero, and therefore D is a congruent number. In any case, if one knew BSD I and computed $L(E^{(D)}, 1)$, then one would know whether D is a congruent number. Thus, the Conjecture of Birch and Swinnerton-Dyer would give an algorithm for answering the Congruent Number Problem, a classical problem in number theory. Writing p_i for primes congruent to $i \pmod 8$, then $1 = r_E(p_5) = r_E(p_7) = r_E(2p_3) = r_E(2p_7)$ (see [**105**], [**61**], [**8**], [**9**]). In particular, if p is a prime and $p \equiv 5$ or $7 \pmod 8$, then p is a congruent number. For more on the Congruent Number Problem, see [**73**], [**177**], and [**135**].

4.5. Evidence for BSD I

What is known in the direction of BSD I? There is extensive numerical evidence (see especially [**11**] and [**164**]). There is also theoretical evidence. We give below some of the more recent results.

Theorem 4.5.1 (Kolyvagin, et al.). *If E is a (modular) elliptic curve over $\mathbf{Q}$ with analytic rank 0, then $\operatorname{rank}(E(\mathbf{Q})) = 0$.*

The result was proved in the case of complex multiplication by Coates-Wiles [**32**]. When E is modular, the result follows from work of Kolyvagin [**74**], combined with the work of Gross-Zagier [**56**] and an analytic number theory result proved independently by Bump-Friedberg-Hoffstein [**21**] and Murty-Murty [**107**].

Theorem 4.5.2 (Gross-Zagier, Kolyvagin, et al.). *If E is a (modular) elliptic curve over $\mathbf{Q}$ with analytic rank 1, then $\operatorname{rank}(E(\mathbf{Q})) = 1$.*

Gross and Zagier [**56**] showed that if E is modular, and the analytic rank is 1, then rank$(E(\mathbf{Q})) \geq 1$. Rubin [**128**] showed that if E has complex multiplication, and the analytic rank is 1, then rank$(E(\mathbf{Q})) \leq 1$. Work of Kolyvagin [**74**] (when combined with work of Waldspurger [**182**]) shows that if E is modular, and the analytic rank is 1, then rank$(E(\mathbf{Q})) \leq 1$.

More precisely, suppose that E is a modular elliptic curve over $\mathbf{Q}$. Let $K = \mathbf{Q}(\sqrt{D})$ be an imaginary quadratic field with the property that every prime of bad reduction for E splits into two distinct prime ideals for K.

Exercise 4.5.3. *Use (10) to show that this condition on K implies that $w_{E^{(D)}} = -w_E$.*

The concept of L-function can be defined more generally for elliptic curves (or abelian varieties) over number fields other than $\mathbf{Q}$. In particular, the L-function of E over K satisfies
$$L(E/K, s) = L(E, s)L(E^{(D)}, s)$$
where the L-functions on the right side are the L-functions for E and $E^{(D)}$ over $\mathbf{Q}$ as defined above (this is true for any quadratic number field $K = \mathbf{Q}(\sqrt{D})$, not just those as above). Gross and Zagier [**56**] showed that there is a point $P_K \in E(K)$ (closely related to the Heegner point on E corresponding to K), lying in $E(\mathbf{Q})$ when $w_E = -1$ and lying in $E^{(D)}(\mathbf{Q})$ when $w_E = 1$, with the property that P_K has infinite order if and only if $L'(E/K, 1) \neq 0$. Waldspurger [**182**] showed that if $w_E = -1$, then there exists a field K as above with $L(E^{(D)}, 1) \neq 0$. Bump-Friedberg-Hoffstein [**21**] and Murty-Murty [**107**] showed that if $w_E = 1$, then there exists a field K as above with $L'(E^{(D)}, 1) \neq 0$. Kolyvagin [**74**], [**75**] showed that if P_K has infinite order, then rank$(E(K)) = 1$.

Exercise 4.5.4. *Use Exercise 3.4.6 above to deduce Theorems 4.5.1 and 4.5.2 from the above information.*

4.6. BSD II

For the remainder of this chapter, E is an elliptic curve over $\mathbf{Q}$. Assuming BSD I, then Birch and Swinnerton-Dyer made the following conjecture (the new terms are defined below; note that they assumed the "folklore conjecture" that the group Ш is finite).

Conjecture of BSD II.
$$\lim_{s \to 1} \frac{L(E, s)}{(s-1)^r} = \frac{\Omega R \# \text{Ш} \prod_{p | \Delta(E)} c_p}{\#(E(\mathbf{Q})_{\text{tors}})^2}.$$

Note that the left side is just b_r defined in (9).

John Tate (p. 198 of [**174**]) wrote "This remarkable conjecture relates the behavior of a function L at a point where it is not at present known to be defined to the order of a group Ш which is not known to be finite!" Thanks to Theorem 4.2.9, $L(E, s)$ is now known to be defined at $s = 1$.

Tate generalized the conjecture to abelian varieties over global fields [**173**].

Definition 4.6.1. Let
$$\Omega = \int_{E(\mathbf{R})} \frac{dx}{|2y + a_1 x + a_3|} \in \mathbf{R},$$

where a_1 and a_3 are coefficients of a generalized Weierstrass model for E as in (2) in Chapter 1.

Like b_r, Ω is transcendental. The expression $\frac{dx}{2y+a_1x+a_3}$ is a holomorphic differential on E, and is invariant (under translation on E). All others are multiples of this one.

Tate found that the "fudge factors" c_p of Birch and Swinnerton-Dyer were the same as the following:

Definition 4.6.2. Let
$$c_p = \#(E(\mathbf{Q}_p)/E_0(\mathbf{Q}_p))$$
where $\mathbf{Q}_p$ is the field of p-adic numbers, and
$$E_0(\mathbf{Q}_p) = \{P \in E(\mathbf{Q}_p) : P \text{ reduces to a non-singular}$$
$$\text{point on the reduction of } E \text{ mod } p\}.$$

Note that $c_p = 1$ if E has good reduction at p.

Exercise 4.6.3. *Show that $E_0(\mathbf{Q}_p)$ is a subgroup of $E(\mathbf{Q}_p)$.*

Theorem 4.6.4 (Theorem 3 of [183]). *There exists an absolute constant κ such that if E is an elliptic curve over $\mathbf{Q}$, then*
$$\prod_{p|\Delta(E)} c_p < \Delta(E)^{\kappa/\log\log \Delta(E)}.$$

Next we define a height function on $E(\mathbf{Q})$, in order to define the regulator R. For $P = (x,y) \in E(\mathbf{Q})$, write $x = \frac{u}{v}$ with $u,v \in \mathbf{Z}$ in lowest terms, and define the *naive height*
$$h(P) = \log\max(|u|,|v|).$$
Define the *height*
$$\hat{h}(P) = \frac{1}{2}\lim_{n\to\infty}\frac{h(2^n P)}{4^n}$$
(and define $\hat{h}(O) = 0$). (For the existence of the limit, see for example Proposition 9.1 in Chapter VIII of [156].) Note that $\hat{h}$ is a function from $E(\mathbf{Q})$ to the set of non-negative real numbers, and $\hat{h}(P) = 0$ if and only if P is a point of finite order. Further, for $m \in \mathbf{Z}$,
$$\hat{h}(mP) = m^2\hat{h}(P).$$
Other properties of $\hat{h}$ were given in Joe Buhler's lectures (see also §9 in Chapter VIII of [156]). Define the *Néron-Tate pairing*, a bilinear form on $E(\mathbf{Q})$, by
$$<P,Q> = \hat{h}(P+Q) - \hat{h}(P) - \hat{h}(Q).$$
Let $\{P_1,\ldots,P_r\}$ be a $\mathbf{Z}$-basis for $E(\mathbf{Q})/E(\mathbf{Q})_{\text{tors}}$.

Definition 4.6.5. Define the *regulator*
$$R = \det(<P_i,P_j>)_{1\leq i\leq r, 1\leq j\leq r}$$
if the rank $r \geq 1$, and let $R = 1$ if $r = 0$.

Note that $R = 2\hat{h}(P_1)$ if $r = 1$.

Exercise 4.6.6. *Show that R is independent of the choice of basis $\{P_1,\ldots,P_r\}$.*

Warning 4.6.7. Be careful to use correct definitions for the "period" Ω and the height $\hat{h}$. They can be normalized differently, and this can lead to confusion. BSD II is sometimes misstated in the literature, due to incompatible normalizations of the period and height. A good rule to follow is to use the papers of Tate [**173**], [**174**], especially for BSD II for abelian varieties over number fields or, more generally, global fields.

If v is a prime number p (respectively, is ∞), then write $\mathbf{Q}_v$ for $\mathbf{Q}_p$ (respectively, $\mathbf{R}$), and write $\bar{\mathbf{Q}}_v$ for $\bar{\mathbf{Q}}_p$ (respectively, $\mathbf{C}$).

Definition 4.6.8. Define the *Tate-Shafarevich group*

$$\text{III} = \text{III}(E/\mathbf{Q}) = \ker\left[H^1(\mathbf{Q}, E(\bar{\mathbf{Q}})) \to \prod_v H^1(\mathbf{Q}_v, E(\bar{\mathbf{Q}}_v))\right]$$

where $H^1(K, E(\bar{K})) = H^1(\text{Gal}(\bar{K}/K), E(\bar{K}))$, and where the map is induced from the inclusions $\text{Gal}(\bar{\mathbf{Q}}_v/\mathbf{Q}_v) \hookrightarrow \text{Gal}(\bar{\mathbf{Q}}/\mathbf{Q})$.

The group III consists of the global cohomology classes which are locally trivial everywhere.

BSD II assumes:

Conjecture 4.6.9 (see pp. 239–240 of [**25**]). *III is finite.*

Theorem 4.6.10 (Cassels [**26**]). *If III is finite, then #III is a square.*

The Tate-Shafarevich group III for an abelian variety can be defined similarly.

Theorem 4.6.11 (Tate, [**173**]). *For principally polarized abelian varieties, if III is finite, then #III is a square or twice a square.*

In [**123**] Poonen and Stoll constructed interesting examples, including an example of a principally polarized abelian surface over $\mathbf{Q}$ where #III $= 2$.

Note that Tate-Shafarevich groups for abelian varieties over function fields can be infinite (see [**118**], [**140**]).

4.7. Evidence for BSD II

The first examples of elliptic curves for which III could be proved to be finite were obtained by Rubin [**128**].

If III is finite, then both sides of the equation in BSD II are isogeny invariant (although the individual factors on the right side are not). Cassels [**28**] showed this for elliptic curves over number fields, and Tate [**173**] showed it for abelian varieties over global fields, if the degree of the isogeny is relatively prime to the characteristic of the field.

For the function field case, Tate wrote (p. 199 of [**174**]): "The situation is encouraging." He proved that for abelian varieties over a function field over a finite field of characteristic p, the rank is bounded above by the analytic rank. Further, in the same setting, he showed that the prime-to-p part of III is finite if and only if the rank equals the analytic rank (i.e., if and only if BSD I holds), and that BSD I holds if and only if BSD II holds up to a power of p. Milne [**103**] (see also [**102**], [**104**], [**3**], [**132**], and [**5**] for related results) showed that in fact everything is also OK at p, if p is odd.

We return to the case of elliptic curves over $\mathbf{Q}$.

Theorem 4.7.1.
$$\frac{L(E,1)}{\Omega} \in \mathbf{Q}$$
and the denominator of this rational number can be effectively bounded.

For the family $y^2 = x^3 - Dx$ this was proved by Birch and Swinnerton-Dyer [**11**]. See [**37**] for the general CM case. In the modular case it follows from work of Shimura [**142**], [**145**], [**146**] and Manin [**89**], [**90**] (by writing the expression in terms of modular symbols that live in rational homology).

Theorem 4.7.2. *If the analytic rank is 1, then $E(\mathbf{Q})$ has a point P of infinite order such that*
$$\frac{L'(E,1)}{\Omega \hat{h}(P)} \in \mathbf{Q}.$$

When E is modular, this was proved by Gross-Zagier (see Theorem 7.3 of [**56**]).

If K is an imaginary quadratic field, let $w(K)$ denote the number of roots of unity in K. Then $w(K) = 2, 4,$ or 6, with $w(K) = 4$ if and only if $K = \mathbf{Q}(i)$, and $w(K) = 6$ if and only if $K = \mathbf{Q}(\sqrt{-3})$.

Theorem 4.7.3 (Rubin [**129**]). *If E is an elliptic curve over $\mathbf{Q}$ with CM by K, and $L(E,1) \neq 0$ (i.e., the analytic rank is 0), then*
  (i) Ш *is finite,*
  (ii) $\#Ш = \frac{L(E,1)}{\Omega}$ *up to powers of primes that divide $w(K)$,*
  (iii) *BSD II is true up to powers of primes that divide $w(K)$.*

Theorem 4.7.4 ([**55**], [**127**], [**129**]). *Suppose E is an elliptic curve over $\mathbf{Q}$ with CM, and the analytic rank is greater than 0. Then either the rank is greater than 0, or the p-primary part of Ш is infinite for all primes $p > 3$.*

When E has CM and odd analytic rank, Greenberg [**55**] (respectively, Rubin [**127**]) showed that either $\mathrm{rank}(E(\mathbf{Q})) > 0$ or the p-primary part of Ш is infinite for all odd primes p of good ordinary (respectively, good supersingular) reduction. For the full Theorem 4.7.4, see the Theorem on p. 25 of [**129**].

Theorem 4.7.5 (Kolyvagin [**74**]). *Suppose E is a modular elliptic curve over $\mathbf{Q}$, and the analytic rank is 0 or 1. Then Ш is finite.*

4.8. Selmer groups

To compute Mordell-Weil groups, we need to define Selmer groups.

If X is an abelian group, and m is a positive integer, let $X[m]$ denote the group of elements in X of order dividing m. As before, if E is an elliptic curve over $\mathbf{Q}$ we write $E[m]$ for $E(\mathbf{C})[m]$ $(= E(\bar{\mathbf{Q}})[m])$.

The short exact sequence given by multiplication by m on E
$$0 \to E[m] \to E(\bar{\mathbf{Q}}) \xrightarrow{m} E(\bar{\mathbf{Q}}) \to 0$$
induces cohomology sequences

$$\begin{array}{ccccccccc}
0 \to & E(\mathbf{Q})/mE(\mathbf{Q}) & \xrightarrow{\kappa} & H^1(\mathbf{Q}, E[m]) & \xrightarrow{\lambda} & H^1(\mathbf{Q}, E(\bar{\mathbf{Q}}))[m] & \to 0 \\
& \downarrow & & \downarrow \mathrm{res}_v & & \downarrow & \\
0 \to & E(\mathbf{Q}_v)/mE(\mathbf{Q}_v) & \xrightarrow{\kappa_v} & H^1(\mathbf{Q}_v, E[m]) & \xrightarrow{\lambda_v} & H^1(\mathbf{Q}_v, E(\bar{\mathbf{Q}}_v))[m] & \to 0
\end{array}$$

where the first vertical map is induced by inclusion, and the other two are restriction maps (restricting from $\text{Gal}(\bar{\mathbf{Q}}/\mathbf{Q})$ to its subgroup $\text{Gal}(\bar{\mathbf{Q}}_v/\mathbf{Q}_v)$). The map κ is defined to take a point $P \pmod{mE(\mathbf{Q})}$ and send it to the cocycle that takes $\sigma \in \text{Gal}(\bar{\mathbf{Q}}/\mathbf{Q})$ to $\sigma Q - Q$ where $Q \in E(\bar{\mathbf{Q}})$ and $mQ = P$. Note that $E(\mathbf{Q})/mE(\mathbf{Q})$ and $E(\mathbf{Q}_v)/mE(\mathbf{Q}_v)$ are finite groups.

Definition 4.8.1. Define the *m-Selmer group*
$$S_m = \bigcap_v \text{res}_v^{-1}\left(\kappa_v(E(\mathbf{Q}_v)/mE(\mathbf{Q}_v))\right) \subseteq H^1(\mathbf{Q}, E[m]).$$

The "local condition at v" that S_m satisfies is
$$\text{res}_v(S_m) \subseteq \kappa_v(E(\mathbf{Q}_v)/mE(\mathbf{Q}_v)).$$

The Selmer group consists of global cohomology classes that everywhere locally come from local points. We want to understand the Mordell-Weil group by understanding the image of κ, i.e., the global cohomology classes that come from global points. But we don't know how to do that, so we study the cohomology classes that everywhere locally come from local points.

Note that what Ralph Greenberg calls $\text{Sel}_E(\mathbf{Q})_p$ in his lectures is the direct limit of the S_{p^n}'s. (The maps might not be injective, so Ralph's $\text{Sel}_E(\mathbf{Q})_p[p^n]$ is not necessarily my S_{p^n}. They're the same if and only if there is no p-torsion in $E(\mathbf{Q})$.)

Exercise 4.8.2. *Show that S_m is finite. (Hint: This is very similar to the proof in Exercise 3.4.12. Show that the image of S_m in $H^1(\mathbf{Q}(E[m]), E[m])$ is contained in $\text{Hom}(\text{Gal}(M/\mathbf{Q}(E[m])), E[m])$, where M is the maximal abelian extension of $\mathbf{Q}(E[m])$ of exponent m unramified outside the primes dividing $m\Delta(E)$.)*

Exercise 4.8.3. *Show that we have a short exact sequence:*
$$0 \to E(\mathbf{Q})/mE(\mathbf{Q}) \xrightarrow{\kappa} S_m \xrightarrow{\lambda} \text{Ш}[m] \to 0.$$

Therefore, Ш can be viewed as the obstruction to knowing the Mordell-Weil group, once you know the Selmer group. The Tate-Shafarevich group can therefore be viewed as an "error term".

It follows from Exercises 4.8.2 and 4.8.3 that Ш$[m]$ is finite.

Birch points out that the Parity Conjecture is very close to an earlier conjecture of Selmer [**134**] that for primes p, the p-rank of Ш ($= \dim_{\mathbf{F}_p} \text{Ш}[p]$) is even (see [**106**]). By Theorem 4.6.10, Selmer's Conjecture would follow from the finiteness of Ш.

Tate (p. 193 of [**174**]) quotes Barry Mazur as saying that to compute $E(\mathbf{Q})$, by day one makes descents, calculating Selmer groups. By night one computes points in the Mordell-Weil group. If, some happy day or night, this process converges, then one has computed $E(\mathbf{Q})$. On the other hand, if Ш is infinite, then we are doomed to continue computing through all eternity.

Let's start computing.

4.9. Examples — descents

Example 4.9.1. Let E be the elliptic curve $y^2 = x^3 - x$. This equation is minimal, and we have $N_E = 2^5$, $\Delta(E) = 2^6$. One can compute that $L(E, 1) \approx 0.655514\ldots$.

Further,
$$\Omega = \int_{E(\mathbf{R})} \frac{dx}{2|y|} = \int_{-1}^{0} + \int_{1}^{\infty} \frac{dx}{\sqrt{x^3 - x}} \approx 5.2441\ldots.$$

Therefore, $L(E,1)/\Omega \approx \frac{1}{8}$. In fact, Birch and Swinnerton-Dyer [11] proved that $L(E,1)/\Omega = \frac{1}{8}$. Further, we have $c_p = 1$ if $p \neq 2$. To compute
$$c_2 = \#(E(\mathbf{Q}_2)/E_0(\mathbf{Q}_2)),$$
note that $(1,0) \in E(\mathbf{Q}_2) - E_0(\mathbf{Q}_2)$, since $(1,0)$ is the singular point modulo 2. Suppose $P = (x,y) \in E(\mathbf{Q}_2) - E_0(\mathbf{Q}_2)$. We will show that $P \in (1,0) + E_0(\mathbf{Q}_2)$, thereby proving that $c_2 = 2$. Letting $\tilde{P}$ denote the reduction of P modulo 2, we have $\tilde{P} = (1,0)$, so that $x \equiv 1 \pmod 2$. Thus,
$$P + (1,0) = (x,y) + (1,0) = \left(\frac{x+1}{x-1}, \frac{-2y}{(x-1)^2}\right).$$

If $x \equiv 1 \pmod 4$ then $\widetilde{P + (1,0)} = O$. If $x \equiv 3 \pmod 4$ then $\widetilde{P + (1,0)} = (0,0)$. In both cases, we can conclude that $P + (1,0) \in E_0(\mathbf{Q}_2)$, and therefore $P \in (1,0) + E_0(\mathbf{Q}_2)$. We will consider this curve again in Example 4.9.3 below.

Exercise 4.9.2. For $E^{(D)} : y^2 = x^3 - D^2 x$, show that
$$\Omega_{E^{(D)}} = \frac{\Omega_{E^{(1)}}}{\sqrt{D}}.$$

To perform additional computations in some examples, we need to do some more work. Suppose that E is an elliptic curve over $\mathbf{Q}$, and
$$E[2] \subseteq E(\mathbf{Q}).$$
Then $\mathrm{Gal}(\bar{\mathbf{Q}}/\mathbf{Q})$ and $\mathrm{Gal}(\bar{\mathbf{Q}}_v/\mathbf{Q}_v)$ act as the identity on $E[2]$. Therefore,
$$H^1(\mathbf{Q}, E[2]) = \mathrm{Hom}(\mathrm{Gal}(\bar{\mathbf{Q}}/\mathbf{Q}), E[2]),$$
$$H^1(\mathbf{Q}_v, E[2]) = \mathrm{Hom}(\mathrm{Gal}(\bar{\mathbf{Q}}_v/\mathbf{Q}_v), E[2]).$$
Let M be the maximal Galois extension of $\mathbf{Q}$ unramified at all primes not dividing $2\Delta(E)$ and with $\mathrm{Gal}(M/\mathbf{Q}) \cong (\mathbf{Z}/2\mathbf{Z})^n$ for some n. Using the hint in Exercise 4.8.2, we have
$$E(\mathbf{Q})/2E(\mathbf{Q}) \cong \mathbf{Z}/2\mathbf{Z} \times \mathbf{Z}/2\mathbf{Z} \times (\mathbf{Z}/2\mathbf{Z})^{\mathrm{rank}} \xrightarrow{\kappa} S_2$$
$$\hookrightarrow \mathrm{Hom}(\mathrm{Gal}(M/\mathbf{Q}), E[2]) \hookrightarrow \mathrm{Hom}(\mathrm{Gal}(\bar{\mathbf{Q}}/\mathbf{Q}), E[2]).$$
Further,
$$\mathrm{Hom}(\mathrm{Gal}(M/\mathbf{Q}), E[2]) \cong \mathrm{Hom}((\mathbf{Z}/2\mathbf{Z})^n, (\mathbf{Z}/2\mathbf{Z})^2) \cong (\mathbf{Z}/2\mathbf{Z})^{2n}.$$
As Ralph Greenberg told us in his lectures,
$$E(\mathbf{Q}_p) \cong \mathbf{Z}_p \times H$$
where H is a finite subgroup of $(\mathbf{Q}/\mathbf{Z})^2$. For $p = 2$, we have
$$E(\mathbf{Q}_2)/2E(\mathbf{Q}_2) \cong (\mathbf{Z}/2\mathbf{Z})^3$$

since $E[2] \subseteq E(\mathbf{Q})$. We therefore have

$$(\mathbf{Z}/2\mathbf{Z})^{\text{rank}+2} \xhookrightarrow{\kappa} S_2 \subset \text{Hom}(\text{Gal}(M/\mathbf{Q}), E[2]) \cong (\mathbf{Z}/2\mathbf{Z})^{2n}$$

$$\downarrow$$

$$\text{Hom}(\text{Gal}(M_\mathfrak{p}/\mathbf{Q}_2), E[2])$$

$$\uparrow$$

$$(\mathbf{Z}/2\mathbf{Z})^3 \cong E(\mathbf{Q}_2)/2E(\mathbf{Q}_2) \xhookrightarrow{\kappa_2} \text{Hom}(\text{Gal}(\bar{\mathbf{Q}}_2/\mathbf{Q}_2), E[2]) \cong (\mathbf{Z}/2\mathbf{Z})^6$$

where $\mathfrak{p}$ is a prime ideal of M dividing 2, and $M_\mathfrak{p}$ is the $\mathfrak{p}$-adic completion of M. Note that

$$\text{Hom}(\text{Gal}(\bar{\mathbf{Q}}_2/\mathbf{Q}_2), E[2]) \cong (\mathbf{Z}/2\mathbf{Z})^6$$

since the inertia subgroup of $\text{Gal}(\mathbf{Q}_2^{\text{ab}}/\mathbf{Q}_2)$ is $\mathbf{Z}_2^\times$, and thus

$$\text{Gal}(\mathbf{Q}_2^{\text{ab}}/\mathbf{Q}_2) \cong \mathbf{Z}_2^\times \times \text{Gal}(\bar{\mathbf{F}}_2/\mathbf{F}_2) \cong \mathbf{Z}_2 \times \mathbf{Z}/2\mathbf{Z} \times \hat{\mathbf{Z}},$$

where $\hat{\mathbf{Z}} = \prod \mathbf{Z}_p \cong \text{Gal}(\bar{\mathbf{F}}_2/\mathbf{F}_2)$.

Example 4.9.3. Let E be the elliptic curve $y^2 = x^3 - x$. Then one can check that M is now $\mathbf{Q}(\sqrt{2}, i)$ and that $\mathfrak{p}$ is totally ramified. The above diagram now is:

$$(\mathbf{Z}/2\mathbf{Z})^{\text{rank}+2} \xhookrightarrow{} S_2 \subset \text{Hom}(\text{Gal}(M/\mathbf{Q}), E[2]) \cong (\mathbf{Z}/2\mathbf{Z})^4$$

$$\|$$

$$\text{Hom}(\text{Gal}(M_\mathfrak{p}/\mathbf{Q}_2), E[2])$$

$$\uparrow$$

$$(\mathbf{Z}/2\mathbf{Z})^3 \cong E(\mathbf{Q}_2)/2E(\mathbf{Q}_2) \xhookrightarrow{\kappa_2} \text{Hom}(\text{Gal}(\bar{\mathbf{Q}}_2/\mathbf{Q}_2), E[2])$$

with a res_2 arrow from S_2 down to $\text{Hom}(\text{Gal}(\bar{\mathbf{Q}}_2/\mathbf{Q}_2), E[2])$.

Let $P = (\frac{1}{4}, \frac{\sqrt{-15}}{8}) \in E(\mathbf{Q}_2)$ and let $Q = (\alpha, \alpha^2 + 1)$ where α satisfies $x^4 - x^3 + 2x^2 + x + 1$, a $\mathbf{Q}$-irreducible polynomial of discriminant $(15)^2$. Then $2Q = P$, and thus for every $\sigma \in \text{Gal}(\bar{\mathbf{Q}}_2/\mathbf{Q}_2)$ we have $\kappa_2(P)(\sigma) = \sigma(Q) - Q$. The field $\mathbf{Q}(Q)$ $(= \mathbf{Q}(\alpha) = \mathbf{Q}(\sqrt{-3}, \sqrt{5}))$ has degree 4 over $\mathbf{Q}$, with quadratic subfield $\mathbf{Q}(\sqrt{-15})$, while the field $\mathbf{Q}_2(Q)$ $(= \mathbf{Q}_2(\alpha) = \mathbf{Q}_2(\sqrt{-3}) = \mathbf{Q}_2(\sqrt{5}))$ is an unramified quadratic extension of $\mathbf{Q}_2$ $(= \mathbf{Q}_2(\sqrt{-15}))$. Since $M_\mathfrak{p}$ is a totally ramified extension of $\mathbf{Q}_2$, it does not contain $\mathbf{Q}_2(Q)$. Therefore, $\kappa_2(P) \notin \text{Hom}(\text{Gal}(M_\mathfrak{p}/\mathbf{Q}_2), E[2])$, and thus $\kappa_2(P) \notin \text{res}_2(S_2)$. We have

$$(\mathbf{Z}/2\mathbf{Z})^{\text{rank}+2} \hookrightarrow S_2 \hookrightarrow \text{res}_2(S_2) \subsetneq \kappa_2(E(\mathbf{Q}_2)/2E(\mathbf{Q}_2)) \cong (\mathbf{Z}/2\mathbf{Z})^3.$$

Therefore, $\text{rank}(E(\mathbf{Q})) + 2 \leq 2$, so $\text{rank}(E(\mathbf{Q})) = 0$ (proving what Fermat knew),

$$S_2 \cong (\mathbf{Z}/2\mathbf{Z})^2 \cong E(\mathbf{Q})/2E(\mathbf{Q}), \quad \text{III}[2] = 0, \quad \text{and} \quad R = 1.$$

The curve E has CM by $K = \mathbf{Q}(i)$. We have $w(K) = 4$, and Rubin [**128**] showed that

$$\#\text{III} = \frac{L(E,1)}{\Omega}(\text{power of 2}) = \frac{1}{8}(\text{power of 2}).$$

Therefore, Ш = Ш[2] = 0. Recalling that by Example 4.9.1 we have $\prod c_p = 2$, therefore
$$\frac{\#\text{Ш} R \prod c_p}{\#(E(\mathbf{Q})_{\text{tors}})^2} = \frac{1}{8} = \frac{L(E,1)}{\Omega}.$$

Thus, BSD I and II are true for $y^2 = x^3 - x$.

Example 4.9.4. Let E be the elliptic curve $y^2 = x^3 - 25x$. This is a minimal equation, with discriminant $\Delta(E) = 2^6 5^6$ and conductor $N_E = 2^5 5^2$. We have $c_2 = 2$, $c_5 = 4$, $\Omega_5 = \frac{1}{\sqrt{5}}\Omega_1 \approx 2.3452\ldots$. "By night" we compute that $P = (-4,6)$ has infinite order (one can apply Theorem 1.5.1 above), so $\text{rank}(E(\mathbf{Q})) \geq 1$. By (11), we have $w_{E(5)} = -1$, so $L(E,1) = 0$. Computing $\frac{L(E,1+t)}{\varepsilon}$ in PARI for $\varepsilon \approx 10^{-10}$ shows that $L'(E,1) \approx 2.227\ldots$, and so the analytic rank is 1. Again, $E[2] \subset E(\mathbf{Q})$. Here, $M = \mathbf{Q}(\sqrt{2}, i, \sqrt{5})$. There is a unique prime $\mathfrak{p}$ of M above 2, but it is no longer totally ramified. As above, we now have

$$\begin{array}{ccccc}
(\mathbf{Z}/2\mathbf{Z})^{\text{rank}+2} & \xrightarrow{\kappa} & S_2 & \subset & \text{Hom}(\text{Gal}(M_{\mathfrak{p}}/\mathbf{Q}_2), E[2]) \\
& & & & \Big\uparrow \\
(\mathbf{Z}/2\mathbf{Z})^3 & \xrightarrow{\kappa_2} & & & \text{Hom}(\text{Gal}(\bar{\mathbf{Q}}_2/\mathbf{Q}_2), E[2])
\end{array}$$

We have
$$(\mathbf{Z}/2\mathbf{Z})^{\text{rank}+2} \subseteq S_2 \subseteq \kappa_2(E(\mathbf{Q}_2)/2E(\mathbf{Q}_2)) \cong (\mathbf{Z}/2\mathbf{Z})^3,$$
so $\text{rank}(E(\mathbf{Q})) \leq 1$. Thus, $\text{rank}(E(\mathbf{Q})) = 1$, and
$$S_2 \cong (\mathbf{Z}/2\mathbf{Z})^3 \cong E(\mathbf{Q}_2)/2E(\mathbf{Q}_2), \quad \text{Ш}[2] = 0.$$

By Kolyvagin [74], Ш$[p] = 0$ for all odd primes p (Kolyvagin rules out the CM case, but his proof also works in the case we are in). Thus, Ш = 0. One computes that $\hat{h}(P) \approx 0.94974\ldots$. As an exercise, show that P and the torsion points $(5,0)$ and $(0,0)$ together generate $E(\mathbf{Q})$. (Hint: Use Exercise 9.12 in [156], which implies that if an integral point on an elliptic curve (in generalized Weierstrass form with integer coefficients) is a multiple of another point Q, then Q is also integral.) Therefore,
$$R = 2\hat{h}(P) \approx 1.89948\ldots,$$
and $\frac{L'(E,1)}{\Omega R} \approx \frac{1}{2}$. By Gross-Zagier [56], $\frac{L'(E,1)}{\Omega R} = \frac{1}{2}$. Now
$$\frac{\#\text{Ш} \prod c_p}{\#(E(\mathbf{Q})_{\text{tors}})^2} = \frac{1}{2},$$
and BSD I and II are true for $y^2 = x^3 - 25x$.

Example 4.9.5. Let E be the elliptic curve $y^2 = x^3 - (17)^2 x$. This is a minimal equation, with discriminant $\Delta(E) = 2^6 (17)^6$ and conductor $N_E = 2^5 (17)^2$. We have $c_2 = 2$, $c_{17} = 4$, $\Omega_{17} = \frac{1}{\sqrt{17}}\Omega_1 \approx 1.27188\ldots$, $L(E,1) \approx 2.54377\ldots \neq 0$. Therefore, the analytic rank is 0. We have $\frac{L(E,1)}{\Omega_{17}} = 2$. Now, $M = \mathbf{Q}(\sqrt{2}, i, \sqrt{17})$. Taking $\mathfrak{p}$ to be a prime ideal of M above 2, then $M_{\mathfrak{p}}$ ($= \mathbf{Q}_2(i, \sqrt{2})$) is a totally ramified degree 4 extension of $\mathbf{Q}_2$ ($= \mathbf{Q}_2(\sqrt{17})$). As above, we have:

$$
\begin{array}{ccc}
(\mathbf{Z}/2\mathbf{Z})^{\mathrm{rank}+2} \xrightarrow{\kappa} & S_2 \subset & \mathrm{Hom}(\mathrm{Gal}(M/\mathbf{Q}), E[2]) \cong (\mathbf{Z}/2\mathbf{Z})^6 \\
& \searrow^{\mathrm{res}_2} & \downarrow \\
& & \mathrm{Hom}(\mathrm{Gal}(M_{\mathfrak{p}}/\mathbf{Q}_2), E[2]) \cong (\mathbf{Z}/2\mathbf{Z})^4 \\
& & \uparrow \\
(\mathbf{Z}/2\mathbf{Z})^3 \cong E(\mathbf{Q}_2)/2E(\mathbf{Q}_2) \xrightarrow{\kappa_2} & & \mathrm{Hom}(\mathrm{Gal}(\bar{\mathbf{Q}}_2/\mathbf{Q}_2), E[2]) \cong (\mathbf{Z}/2\mathbf{Z})^6
\end{array}
$$

Let
$$P = \left(\frac{1}{4}, \frac{\sqrt{-4623}}{8}\right) \in E(\mathbf{Q}_2).$$
Then for every $\sigma \in \mathrm{Gal}(\bar{\mathbf{Q}}_2/\mathbf{Q}_2)$ we have $\kappa_2(P)(\sigma) = \sigma(Q) - Q$ with $Q = (\alpha, \beta)$ and $2Q = P$, where α and β are roots of irreducible polynomials of degree 4 over $\mathbf{Q}$. The field $\mathbf{Q}_2(Q)$ is an unramified extension of $\mathbf{Q}_2$ of discriminant $(4623)^2$. Since $M_{\mathfrak{p}}$ is a totally ramified extension of $\mathbf{Q}_2$, it does not contain $\mathbf{Q}_2(Q)$. Therefore,
$$\kappa_2(P) \notin \mathrm{res}_2(S_2) \subsetneq \mathrm{im}(\kappa_2) \cong (\mathbf{Z}/2\mathbf{Z})^3.$$
Therefore, $\#\mathrm{res}_2(S_2) \leq 4$. Since $\#\ker(\mathrm{res}_2) \leq 4$, we have $\#S_2 \leq 16$. Since
$$(\mathbf{Z}/2\mathbf{Z})^2 \subseteq S_2 \subseteq (\mathbf{Z}/2\mathbf{Z})^4,$$
we have $\mathrm{III}[2] \subseteq (\mathbf{Z}/2\mathbf{Z})^2$. It is possible to show that $S_2 \cong (\mathbf{Z}/2\mathbf{Z})^4$ (we have
$$S_2 \subset \mathrm{res}_2^{-1}(\mathrm{im}\kappa_2) \cap \mathrm{Hom}(\mathrm{Gal}(M/\mathbf{Q}), E[2]),$$
and one can show equality by checking the right side for the local conditions at 17 and ∞) and/or that $\mathrm{III}[2] \cong (\mathbf{Z}/2\mathbf{Z})^2$ (by constructing elements). Therefore, $\#\mathrm{III} \geq 4$. Rubin [**128**] showed that $\mathrm{III}[p] = 0$ for all odd prime numbers p, and III is finite. One could do a "second descent" to compute that $S_4 \cong (\mathbf{Z}/2\mathbf{Z})^4 \cong S_2$, and therefore that $\mathrm{rank}(E(\mathbf{Q})) = 0$, the 2-primary part of III is $(\mathbf{Z}/2\mathbf{Z})^2$ (since there are no elements of order 4), and $\mathrm{III} \cong (\mathbf{Z}/2\mathbf{Z})^2$. Alternatively, the theorem of Coates-Wiles [**32**] implies that the rank is 0 (since the analytic rank is 0). It follows that BSD I and II are true for $y^2 = x^3 - (17)^2 x$.

CHAPTER 5
ABC and related conjectures

The ABC Conjecture and its variants, although not explicitly stated in terms of elliptic curves, are closely connected with conjectures about elliptic curves, as we shall see below. This chapter relies heavily on [183].

When we write $X \ll_\varepsilon Y$ we mean that there exists a positive constant C_ε, depending only on ε, such that $X \leq C_\varepsilon Y$. Further, $\prod_{p|N}$ means that the product is taken over all prime divisors p of N.

5.1. ABC Conjectures

The ABC Conjecture was conjectured (independently) by Masser and Oesterlé.

ABC Conjecture. *Suppose A, B, and C are positive integers, suppose*
$$\gcd(A, B, C) = 1,$$
and suppose
$$A + B = C.$$
Then
$$C \ll_\varepsilon \left(\prod_{p|ABC} p \right)^{1+\varepsilon}.$$

ABC Conjecture Variant 1. *Suppose $A, B, C \in \mathbf{Z}$, $\gcd(A, B, C) = 1$, and $A + B = C$. Then*
$$\max(|A|, |B|, |C|) \ll_\varepsilon \left(\prod_{p|ABC} p \right)^{1+\varepsilon}.$$

Exercise 5.1.1. *Show that the ABC Conjecture is equivalent to ABC Conjecture Variant 1.*

ABC Conjecture Variant 2.
$$\limsup \frac{\log C}{\log \prod_{p|ABC} p} = 1,$$

where the lim sup *is taken over all positive integers A, B, and C with $A + B = C$ and $\gcd(A, B, C) = 1$.*

Browkin et al. [**16**] proved that the set of limit points is equal to $\left[\frac{1}{3}, 1\right]$ if one assumes the ABC Conjecture. Greaves and Nitaj [**54**] proved that the set of limit points contains $\left[\frac{1}{3}, \frac{36}{37}\right]$.

Exercise 5.1.2. *Show that the ABC Conjecture is equivalent to ABC Conjecture Variant 2. Also, show that*

$$\limsup \frac{\log C}{\log \prod_{p | ABC} p} \geq 1.$$

Hint: Show that letting $A = 1$ and $B = 9^k - 1$ gives infinitely many triples (A, B, C) such that

$$\frac{\log C}{\log \prod_{p | ABC} p} > 1.$$

Weak ABC Conjecture. *Suppose A, B, and C are integers, suppose*

$$\gcd(A, B, C) = 1,$$

and suppose $A + B = C$. Then

$$|ABC|^{1/3} \ll_\varepsilon \left(\prod_{p | ABC} p \right)^{1+\varepsilon}.$$

Clearly, the Weak ABC Conjecture follows from the ABC Conjecture.

Exercise 5.1.3. *Show that the Weak ABC Conjecture implies that Fermat's Last Theorem is true for all sufficiently large exponents.*

Conjecture 5.1.4 (Generalized ABC or n-conjecture). *Fix $n \geq 3$. Suppose*

$$a_1, \ldots, a_n \in \mathbf{Z}, \quad a_1 + \cdots + a_n = 0,$$

and no proper subsum is 0. Then

$$\limsup \frac{\log \max |a_i|}{\log \prod_{p | a_1 \cdots a_n} p} = 2n - 5.$$

Browkin and and Brzezinski [**15**] showed that

$$\limsup \frac{\log \max |a_i|}{\log \prod_{p | a_1 \cdots a_n} p} \geq 2n - 5.$$

A conjecture of Vojta (in analogy with Nevanlinna Theory), applied to the hyperplane $x_1 + \cdots x_n = 0$ in $\mathbf{P}^{n-1}$, implies the following conjecture.

Conjecture 5.1.5 (Vojta's n-conjecture, see [**181**]). *Fix $n \geq 3$. Suppose*

$$a_1, \ldots, a_n \in \mathbf{Z} - 0, \quad \gcd(a_1, \ldots, a_n) = 1, \quad \text{and} \quad a_1 + \cdots + a_n = 0.$$

Then

$$\limsup \frac{\log \max |a_i|}{\log \prod_{p | a_1 \cdots a_n} p} = 1$$

for all such $(a_1, \ldots, a_n)$ outside a proper Zariski-closed subset.

Exercise 5.1.6 ([181]). *Show that one cannot remove the phrase "outside a proper Zariski-closed subset" in Vojta's n-conjecture. (Hint: Either see [181], or use[1] $a_1 = 2^a$, $a_2 = -2^a$, $a_3 = 3^b$, $a_4 = -3^b$.)*

Theorem 5.1.7 (Elkies [42]). *An effective version of the ABC Conjecture over a number field K implies an effective version of the Mordell Conjecture (Faltings' Theorem) over K.*

See [165], [167], [114], and [113] for some additional results on the ABC Conjecture.

5.2. ABC Conjectures for function fields

Even before the ABC Conjecture was stated for number fields, it had been proved for function fields.

Theorem 5.2.1 (Mason [93], see also [79]). *Suppose K is an integral domain of characteristic 0. For $f \in K[t]$, let $\mathrm{rad}(f)$ denote the product of the distinct irreducible factors of f. Suppose f, g, $h \in K[t]$, suppose $\gcd(f,g,h) = 1$, and suppose $f + g = h$. Then*

$$\limsup \frac{\max(\deg(f), \deg(g), \deg(h))}{\deg \mathrm{rad}(fgh)} = 1.$$

Conjecture 5.2.2 (Generalized ABC or n-conjecture: function fields). *Fix $n \geq 3$. If*

$$f_1, \ldots, f_n \in K[t], \quad \gcd(f_1, \ldots, f_n) = 1, \quad \text{and} \quad f_1 + \cdots + f_{n-1} = f_n,$$

then

$$\limsup \frac{\max \deg f_i}{\deg(\mathrm{rad}(f_1 \cdots f_n))} = 2n - 5.$$

This conjecture has been proved for $n = 3$ by Mason's Theorem above, and for $n = 4$ by Browkin and Brzezinski [15]. Browkin showed that when $n = 3$ the set of limit points is $\left[\frac{1}{3}, 1\right]$. Davies [38] proved that for $n \geq 3$ the set of limit points contains $\left[\frac{1}{n}, 2n - 5\right]$.

5.3. Szpiro Conjecture

Let E be an elliptic curve over $\mathbf{Q}$. Let N_E denote the conductor of E and let $\Delta_{\min}(E)$ denote the minimal discriminant (i.e., the discriminant of a minimal Weierstrass equation for E).

Szpiro Conjecture. *If E is an elliptic curve over $\mathbf{Q}$, then*

$$|\Delta_{\min}(E)| \ll_\varepsilon N_E^{6+\varepsilon}.$$

Exercise 5.3.1. *Show that Szpiro's Conjecture implies the Weak ABC Conjecture. Hint: Use that, under the hypotheses of the Weak ABC Conjecture, the conductor of the elliptic curve*

$$y^2 = x(x - A)(x + B)$$

is $\left(\prod_{p \mid ABC} p\right)$ (up to a bounded power of 2), and the minimal discriminant is $(ABC)^2$ (up to a bounded power of 2).

---

[1] I thank Noam Elkies for pointing out this example to me.

The following theorem was conjectured in [**18**].

Theorem 5.3.2 (see [**136**]; see also [**101**], [**46**], [**98**]). *Suppose E is an elliptic curve over $\mathbf{Q}$ with prime conductor p. Then either $|\Delta_{\min}(E)| = p$ or p^2, or else $p = 11$ and $|\Delta_{\min}(E)| = 11^5$, or $p = 17$ and $|\Delta_{\min}(E)| = 17^4$, or $p = 19$ and $|\Delta_{\min}(E)| = 19^3$, or $p = 37$ and $|\Delta_{\min}(E)| = 37^3$. (Therefore $\Delta_{\min}(E)$ divides p^5.)*

This result implies that Szpiro's Conjecture would be true if one restricted to elliptic curves of prime conductor.

In [**115**], Nitaj gives an example (of a curve of conductor $2526810 = 2 \cdot 3 \cdot 5 \cdot 11 \cdot 13 \cdot 19 \cdot 31$) with "large Szpiro ratio":
$$\frac{\log |\Delta_{\min}(E)|}{\log N_E} \approx 8.812.$$

As a corollary of Theorem 4.6.4 above, de Weger proves:

Theorem 5.3.3 (see Corollary on p. 113 of [**183**]). *If Szpiro's Conjecture is true, then there exists an absolute constant C so that*
$$\prod_{p|\Delta(E)} c_p \ll N_E^{C/\log\log(N_E)}.$$

According to de Weger, an interesting open question is to improve his bounds on the product of the "fudge factors" c_p.

Generalized Szpiro Conjecture. *If E is an elliptic curve over $\mathbf{Q}$, then*
$$\max(|\Delta_{\min}(E)|, |g_2^3|) \ll_\varepsilon N_E^{6+\varepsilon},$$
where $|g_2|$ is minimal with the property that E has a model of the form $y^2 = 4x^3 - g_2 x - g_3$ with $g_2, g_3 \in \mathbf{Z}$.

The ABC Conjecture is equivalent to the Generalized Szpiro Conjecture.

A function field analogue of Szpiro's Conjecture (with $\varepsilon = 0$ and an explicit constant) is true (and was proved by Kodaira; see Exercise 3.36 on p. 287 of [**158**]).

For some additional results relating to Szpiro's Conjecture, see [**45**].

5.4. Goldfeld-Szpiro Conjecture

If E is an elliptic curve over $\mathbf{Q}$, let III_E denote the Tate-Shafarevich group $\text{III}(E/\mathbf{Q})$.

Goldfeld-Szpiro Conjecture [**52**]. *If E is an elliptic curve over $\mathbf{Q}$, then*
$$\#\text{III}_E \ll_\varepsilon N_E^{1/2+\varepsilon}.$$

Modulo the Conjecture of Birch and Swinnerton-Dyer (and the Modularity Conjecture) for E, the Goldfeld-Szpiro Conjecture is equivalent to the Szpiro Conjecture (see [**52**]).

Theorem 5.4.1 (Cassels [**27**], see also [**13**], [**77**], [**87**]). *$\#\text{III}_E$ is unbounded, as E varies over elliptic curves over $\mathbf{Q}$.*

Cassels showed the 3-rank of III_E is unbounded (i.e., III_E contains $(\mathbf{Z}/3\mathbf{Z})^k$ with k arbitrarily large) in the family of cubic twists $x^3 + y^3 = d$. Bölling [**13**] showed that the 2-rank of the Tate-Shafarevich group is unbounded for elliptic curves (over a fixed number field) with a fixed j-invariant.

Definition 5.4.2. An elliptic curve E over $\mathbf{Q}$ is *semistable* if its conductor N_E is square-free.

Conjecture 5.4.3 (Mazur, [98]). *There exist semistable elliptic curves over $\mathbf{Q}$ of arbitrarily high conductor N_E with $\text{Ш}_E = 0$.*

In particular, Mazur predicts that
$$\liminf \frac{\#\text{Ш}_E}{\sqrt{N_E}} = 0.$$

5.5. de Weger Conjectures

It was conjectured by de Weger that the bound in the Goldfeld-Szpiro Conjecture is best possible. In other words:

de Weger N_E Conjecture. *For every $\varepsilon > 0$ there exist infinitely many elliptic curves E over $\mathbf{Q}$ with*
$$\#\text{Ш}_E \geq N_E^{1/2-\varepsilon}.$$

Exercise 5.5.1. *Show this is equivalent to:*
For every $\varepsilon > 0$ and every $C > 0$, there exist infinitely many elliptic curves E over $\mathbf{Q}$ with
$$\#\text{Ш}_E > CN_E^{1/2-\varepsilon}.$$
and to:
For every $\varepsilon > 0$ there exist infinitely many elliptic curves E over $\mathbf{Q}$ with
$$\#\text{Ш}_E \gg_\varepsilon N_E^{1/2-\varepsilon}.$$

de Weger $\Delta_{\min}(E)$ Conjecture. *For every $\varepsilon > 0$ there exist infinitely many elliptic curves E over $\mathbf{Q}$ with*
$$\#\text{Ш}_E \geq \Delta_{\min}(E)^{1/12-\varepsilon}.$$

In de Weger's paper [183] he says that the methods of Cassels, Bölling, and Kramer would at best show that there are infinitely many elliptic curves E over $\mathbf{Q}$ with
$$\#\text{Ш}_E \gg N_E^{C/\log\log N}$$
for some $C > 0$. Cassels and Bölling consider twists, adding one prime p at a time. This multiplies the conductor by p^2, but at best multiplies $\#\text{Ш}_E$ by a constant. The prime number theorem would then give the desired result. Mai and Murty [87] show that if the Conjecture of Birch and Swinnerton-Dyer is true, then there are infinitely many elliptic curves E over $\mathbf{Q}$ (of rank and analytic rank 1, all quadratic twists of each other) with
$$\#\text{Ш}_E \geq N_E^{1/4-\varepsilon}.$$

Theorem 5.5.2 (de Weger [183]). *The de Weger N_E Conjecture follows by combining BSD II in the rank 0 case, the Szpiro Conjecture, and a generalization of the Riemann Hypothesis (for the Rankin-Selberg zeta-function associated to the weight $3/2$ modular form associated to E by the Shintani-Shimura lift).*

Theorem 5.5.3 (de Weger [183]). *The de Weger $\Delta_{\min}(E)$ Conjecture follows from BSD II in the rank 0 case.*

The largest value I know of that has been found so far for (the conjectured value of) $\#Ш_E/\sqrt{N_E}$ was found by Nitaj [**116**] and is approximately 42.265. In that example

$$N_E = 6305720190 = 2 \cdot 3 \cdot 5 \cdot 7 \cdot 11 \cdot 19 \cdot 37 \cdot 353$$

and the conjectured order of $Ш_E$ is $(2^3 \cdot 229)^2$.

5.6. Additional remarks

Frey [**46**] has conjectured bounds on the degrees of the modular parametrizations of elliptic curves (in terms of the conductor), which implies a conjecture he has on the heights of elliptic curves, which in turn implies the ABC Conjecture.

See also [**117**], [**79**], [**80**], [**114**], and pp. 84–88 of [**180**] for surveys of some of these conjectures, and for relationships between these conjectures and others.

CHAPTER 6
Some other conjectures

We close with some additional comments and conjectures.

6.1. Integral points on elliptic curves

The starting point for any work on integral points on elliptic curves is the following theorem of Siegel.

Theorem 6.1.1 (Siegel). *Every affine model of an elliptic curve over $\mathbf{Q}$ has only finitely many points over $\mathbf{Z}$.*

For more recent results on integral points on elliptic curves, see for example [170], [171], [48], [163], [178], [169], [47], [161], and [187].

I learned from [159] the statement of the following question:

Question 6.1.2. *Is there is a uniform bound on the number of integral points on minimal (affine) Weierstrass equations for elliptic curves over $\mathbf{Q}$?*

The following conjecture of Lang (see Conjecture IX.3.5 of [156] and p. 140 of [78]) predicts a bound on the number of integral points in terms of the rank of the Mordell-Weil group.

Conjecture 6.1.3 (Lang). *There is an absolute constant C such that if E is an elliptic curve over $\mathbf{Q}$ given by a minimal (affine) Weierstrass equation, then the number of integral points on E is at most $C^{1+\mathrm{rank}(E(\mathbf{Q}))}$.*

Silverman [157] proved Lang's Conjecture for elliptic curves over $\mathbf{Q}$ with everywhere potentially good reduction (i.e., with integral j-invariant).

See [137] for a good discussion of integral points on curves.

6.2. Bounds on numbers of rational points on curves

Faltings [43] proved Mordell's Conjecture that if C is a smooth curve of genus $g \geq 2$ defined over a number field K, then $C(K)$ is finite. The following conjectures predict bounds on the number of points.

Caporaso-Harris-Mazur Conjecture ([**22**]). *If K is a number field and $2 \leq g \in \mathbf{Z}$, then there exists a constant $B(K,g)$ such that no smooth curve of genus g defined over K has more than $B(K,g)$ K-rational points.*

Strong Caporaso-Harris-Mazur Conjecture ([**22**]). *If $2 \leq g \in \mathbf{Z}$, then there exists a constant $N(g)$ such that for every number field K there are only finitely many smooth curves of genus g defined over K with more than $N(g)$ K-rational points.*

See also [**23**], where they show that a conjecture of Lang implies the Caporaso-Harris-Mazur Conjecture.

Theorem 6.2.1 (Brumer and Mestre). *For all g,*
$$N(g) \geq 16(g+1).$$

Theorem 6.2.2 (Shioda [**152**]). *For all g,*
$$B(\mathbf{Q}, g) \geq 8g + 16.$$

Theorem 6.2.3 (Keller-Kulesz [**70**]).
$$B(\mathbf{Q}, 2) \geq 588, \qquad B(\mathbf{Q}, 3) \geq 176.$$

Keller and Kulesz showed that the genus 2 curve
$$y^2 = 278,271,081 x^2 (x^2 - 9)^2 - 229,833,600 (x^2 - 1)^2$$
has at least 588 rational points, and the genus 3 curve
$$y^2 = 7,920,000 (x^2 + 1)^4 - 136,782,591 x^2 (x^2 - 1)^2$$
has at least 176 rational points.

6.3. Conductors of elliptic curves

Let $M(N)$ denote the number of ($\mathbf{Q}$-isomorphism classes of) elliptic curves over $\mathbf{Q}$ of conductor N.

Conjecture 6.3.1 (Brumer-Silverman [**20**]). *With p prime, $M(p)$ is unbounded.*

Brumer and Silverman point out that their conjecture implies the unboundedness of both ranks of elliptic curves over $\mathbf{Q}$ (using Silverman's result bounding the number of integral points in terms of the rank, for elliptic curves with integral j-invariant, and Serre's Theorem 5.3.2 above that if N_E is prime then $\Delta_{\min}(E)$ divides p^5) and 3-ranks of ideal class groups of quadratic fields. Conjecture 6.3.1 would contradict an affirmative answer to Question 6.1.2.

6.4. Abelian varieties, Shimura varieties, and open questions

Abelian varieties are parametrized by Shimura varieties, the way elliptic curves are parametrized by modular curves. As can be seen from the work of Faltings (the Mordell Conjecture) and Wiles (Fermat's Last Theorem), one often cannot remain within the realm of elliptic curves and modular curves, and must use and understand higher dimensional abelian varieties and Shimura varieties. Many results that are known for elliptic curves are still interesting conjectures for abelian varieties. This leaves open wide areas of research for the years to come.

The Mumford-Tate Conjecture (see [**138**], [**39**]) deals with the images of ℓ-adic representations associated to varieties, and is related to the Tate Conjectures [**172**] (which in turn are related to the conjectures of Birch and Swinnerton-Dyer).

As mentioned earlier, see for example [**124**], [**111**], and [**12**] for the conjectures of Beilinson and of Bloch-Kato, which generalize the conjectures of Birch and Swinnerton-Dyer.

There are Modularity Conjectures for elliptic curves over number fields other than **Q**, and for abelian varieties with real multiplication (or more generally, for abelian varieties of GL_2-type). See for example [**125**], [**58**], and [**59**] (see [**50**] and [**49**] for analogues over function fields). See also Ken Ribet's lectures on Serre's Conjectures [**136**], in this volume.

BIBLIOGRAPHY

1. D. Abramovich, *Formal finiteness and the torsion conjecture on elliptic curves. A footnote to a paper "Rational torsion of prime order in elliptic curves over number fields"*, Columbia University Number Theory Seminar (New York, 1992), Astérisque **228** (1995), 3, 5–17.
2. N. Aoki, *Torsion points on abelian varieties with complex multiplication*, in Algebraic cycles and related topics (Kitasakado, 1994), 1–22, World Sci. Publishing, River Edge, NJ, 1995.
3. M. Artin, H. P. F. Swinnerton-Dyer, *The Shafarevich-Tate conjecture for pencils of elliptic curves on K3 surfaces*, Invent. Math. **20** (1973), 249–266.
4. C. Batut, K. Belabas, D. Bernardi, H. Cohen, M. Olivier, *PARI-GP*, ftp://megrez.math.u-bordeaux.fr/pub/pari/ .
5. W. Bauer, *On the conjecture of Birch and Swinnerton-Dyer for abelian varieties over function fields in characteristic $p > 0$*, Invent. Math. **108** (1992), no. 2, 263–287.
6. G. Billing, *Vom Range kubischer Kurven vom Geschlecht Eins in algebraischen Rationalitätsbereichen* (Helsingfors, 23.–26. VIII. 1938), 146–150, IX Congress of Scandinavian Mathematicians, 1939.
7. B. J. Birch, *Conjectures concerning elliptic curves*, in Proc. Sympos. Pure Math. **VIII**, 106–112, Amer. Math. Soc., Providence, R.I., 1965.
8. B. J. Birch, *Diophantine analysis and modular functions*, in Algebraic Geometry (Internat. Colloq., Tata Inst. Fund. Res., Bombay, 1968), 35–42, Oxford Univ. Press, London, 1969.
9. B. J. Birch, *Elliptic curves and modular functions*, in Symposia Mathematica, Vol. IV (INDAM, Rome, 1968/69), 27–32 Academic Press, London, 1970.
10. B. Birch, H. P. F. Swinnerton-Dyer, *Notes on elliptic curves. I*, J. Reine Angew. Math. **212** (1963), 7–25.
11. B. Birch, H. P. F. Swinnerton-Dyer, *Notes on elliptic curves. II*, J. Reine Angew. Math. **218** (1965), 79–108.
12. S. Bloch, K. Kato, *L-functions and Tamagawa numbers of motives*, in The Grothendieck Festschrift, Vol. I, 333–400, Progr. Math. **86**, Birkhäuser Boston, Boston, MA, 1990.
13. R. Bölling, *Die Ordnung der Schafarewitsch-Tate Gruppe kann beliebig gross werden*, Math. Nachr. **67** (1975), 157–179.

14. S. Bosch, W. Lütkebohmert, M. Raynaud, Néron models, Springer, Berlin-Heidelberg-New York, 1990.
15. J. Browkin, J. Brzezinski, *Some remarks on the abc-conjecture*, Math. Comp. **62** (1994), 931–939.
16. J. Browkin, M. Filaseta, G. Greaves, A. Schinzel, *Squarefree values of polynomials and the abc-conjecture*, in Sieve methods, exponential sums, and their applications in number theory (Cardiff, 1995), 65–85, London Math. Soc. Lecture Note Ser., **237**, Cambridge Univ. Press, Cambridge, 1997.
17. A. Brumer, *The average rank of elliptic curves. I*, Invent. Math. **109** (1992), no. 3, 445–472.
18. A. Brumer, K. Kramer, *The rank of elliptic curves*, Duke Math. J. **44** (1977), no. 4, 715–743.
19. A. Brumer, O. McGuinness, *The behavior of the Mordell-Weil group of elliptic curves*, Bull. Amer. Math. Soc. **23** (1990), no. 2, 375–382.
20. A. Brumer, J. H. Silverman, *The number of elliptic curves over* **Q** *with conductor* N, Manuscripta Math. **91** (1996), no. 1, 95–102.
21. D. Bump, S. Friedberg, J. Hoffstein, *Nonvanishing theorems for L-functions of modular forms and their derivatives*, Invent. Math. **102** (1990), no. 3, 543–618.
22. L. Caporaso, J. Harris, B. Mazur, *How many rational points can a curve have?*, in The moduli space of curves (Texel Island, 1994), 13–31, Progr. Math., **129**, Birkhäuser Boston, Boston, MA, 1995.
23. L. Caporaso, J. Harris, B. Mazur, *Uniformity of rational points*, J. Amer. Math. Soc. **10** (1997), no. 1, 1–35.
24. H. Carayol, *Sur les représentations ℓ-adiques associées aux formes modulaires de Hilbert*, Ann. Sci. École Norm. Sup. (4) **19** (1986), no. 3, 409–468.
25. J. W. S. Cassels, *Arithmetic on an elliptic curve*, in Proc. Internat. Congr. Mathematicians (Stockholm, 1962), 234–246, Inst. Mittag-Leffler, Djursholm, 1963.
26. J. W. S. Cassels, *Arithmetic on curves of genus 1 (IV). Proof of the Hauptvermutung*, J. reine angew. Math. **211** (1962), 95–112.
27. J. W. S. Cassels, *Arithmetic on curves of genus 1 (VI). The Tate-Safarevic group can be arbitrarily large*, J. reine angew. Math. **214/215** (1964), 65–70.
28. J. W. S. Cassels, *Arithmetic on curves of genus 1 (VIII). On conjectures of Birch and Swinnerton-Dyer*, J. Reine Angew. Math. **217** (1965), 180–199.
29. J. W. S. Cassels, *Diophantine equations with special reference to elliptic curves*, J. London Math. Soc. **41** (1966), 193–291.
30. J. Chahal, M. Meijer, J. Top, *Sections on certain $j = 0$ elliptic surfaces*, http://www.math.uiuc.edu/Algebraic-Number-Theory/0214/index.html .
31. K. Chandrasekharan, R. Narasimhan, *Functional equations with multiple Gamma factors and the average order of arithmetical functions*, Ann. of Math. **76** (1962), 93–136.
32. J. Coates, A. Wiles, *On the conjecture of Birch and Swinnerton-Dyer*, Invent. Math. **39** (1977), no. 3, 223–251.
33. H. Cohen, A course in computational algebraic number theory, Graduate Texts in Mathematics **138**, Springer-Verlag, Berlin, 1993.
34. Ian Connell, *Apecs*, ftp://ftp.math.mcgill.ca/pub/apecs/ .
35. J. E. Cremona, Algorithms for modular elliptic curves, Second edition, Cambridge Univ. Press, Cambridge, 1997.

36. J. E. Cremona, *mwrank and related programs*, http://www.maths.nott.ac.uk/personal/jec/packages.html .
37. R. M. Damerell, *L-functions of elliptic curves with complex multiplication. I, II*, Acta Arith. **17** (1970), 287–301; **19** (1971), 311–317.
38. D. Davies, *A note on the limit points associated with the generalized abc-conjecture for* **Z**[t], Colloq. Math. **71** (1996), no. 2, 329–333.
39. P. Deligne, J. S. Milne, A. Ogus, K-y. Shih, Hodge cycles, motives, and Shimura varieties, Lecture Notes in Math. **900**, Springer-Verlag, Berlin-New York, 1982.
40. B. Edixhoven, *Rational torsion points on elliptic curves over number fields (after Kamienny and Mazur)*, Sém. Bourbaki, Exp. No. 782, Astérisque **227** (1995), 4, 209–227.
41. M. Eichler, *Quaternäre quadratische Formen und die Riemannsche Vermutung für die Kongruenzzetafunktion*, Arch. Math. **5** (1954), 355–366.
42. N. Elkies, *ABC implies Mordell*, Internat. Math. Res. Notices **7** (1991), 99–109.
43. G. Faltings, *Endlichkeitssätze für abelsche Varietäten über Zahlkörpern*, Invent. Math. **73** (1983), 349–366.
44. E. V. Flynn, *Sequences of rational torsions on abelian varieties*, Invent. math. **106** (1991), 433–442.
45. E. Fouvry, M. Nair, G. Tenenbaum, *L'ensemble exceptionnel dans la conjecture de Szpiro*, Bull. Soc. Math. France **120** (1992), no. 4, 485–506.
46. G. Frey, *Links between solutions of $A - B = C$ and elliptic curves*, in Number theory (Ulm, 1987), 31–62, Lecture Notes in Math. **1380**, Springer, New York-Berlin, 1989.
47. J. Gebel, A. Pethö, H. G. Zimmer, *Computing integral points on elliptic curves*, Acta Arith. **68** (1994), no. 2, 171–192.
48. J. Gebel, A. Pethö, H. G. Zimmer, *On Mordell's equation*, Compositio Math. **110** (1998), no. 3, 335–367.
49. E-U. Gekeler, *Jacquet-Langlands theory over K and relations with elliptic curves*, in Drinfeld modules, modular schemes and applications (Alden-Biesen, 1996), 224–257, World Sci. Publishing, River Edge, NJ, 1997.
50. E-U. Gekeler, M. Reversat, *Jacobians of Drinfeld modular curves*, J. Reine Angew. Math. **476** (1996), 27–93.
51. D. Goldfeld, *Conjectures on elliptic curves over quadratic fields*, in Number theory, Carbondale 1979 (Proc. Southern Illinois Conf., Southern Illinois Univ., Carbondale, Ill., 1979), 108–118, Lecture Notes in Math. **751**, Springer-Verlag, Berlin, 1979.
52. D. Goldfeld, L. Szpiro, *Bounds for the order of the Tate-Shafarevich group*, Compositio Math. **97** (1995), 71–87.
53. F. Gouvêa, B. Mazur, *The square-free sieve and the rank of elliptic curves*, J. Amer. Math. Soc. **4** (1991), 1–23.
54. G. Greaves, A. Nitaj, *Some polynomial identities related to the abc-conjecture*, in Number theory in progress, Vol. 1 (Zakopane-Kościelisko, 1997), 229–236, de Gruyter, Berlin, 1999.
55. R. Greenberg, *On the Birch and Swinnerton-Dyer conjecture*, Invent. Math. **72** (1983), 241–265.
56. B. H. Gross, D. B. Zagier, *Heegner points and derivatives of L-series*, Invent. Math. **84** (1986), no. 2, 225–320.

57. M. Harris, *Systematic growth of Mordell-Weil groups of abelian varieties in towers of number fields*, Invent. Math. **51** (1979), no. 2, 123–141.
58. Y. Hasegawa, **Q**-*curves over quadratic fields*, Manuscripta Math. **94** (1997), no. 3, 347–364.
59. Y. Hasegawa, K. Hashimoto, F. Momose, *Modularity conjecture for* **Q**-*curves and QM-curves*, preprint.
60. D. R. Heath-Brown, *The size of Selmer groups for the congruent number problem. II*, Invent. Math. **118** (1994), no. 2, 331–370.
61. K. Heegner, *Diophantische Analysis und Modulfunktionen*, Math. Z. **56** (1952), 227–253.
62. M. Hindry, J. Silverman, *The canonical height and integral points on elliptic curves*, Invent. math. **93** (1988), 419–450.
63. T. Honda, *Isogenies, rational points and section points of group varieties*, Japan. J. Math. **30** (1960), 84–101.
64. Dale Husemöller, Elliptic curves, Graduate Texts in Mathematics **111**, Springer-Verlag, New York-Berlin, 1987.
65. S. Kamienny, *Torsion points on elliptic curves and q-coefficients of modular forms*, Invent. Math. **109** (1992), 221–229.
66. S. Kamienny, *Torsion points on elliptic curves over fields of higher degree*, Internat. Math. Res. Notices (1992), 129–133.
67. S. Kamienny, B. Mazur, *Rational torsion of prime order in elliptic curves over number fields*, Columbia University Number Theory Seminar (New York, 1992), Astérisque **228** (1995), 3, 81–100.
68. N. Katz, P. Sarnak, *Zeroes of zeta functions and symmetry*, Bull. Amer. Math. Soc. **36** (1999), 1–26.
69. N. Katz, P. Sarnak, Random matrices, Frobenius eigenvalues and monodromy, AMS Colloquium Publications **45**, Amer. Math. Soc., Providence, RI, 1999.
70. W. Keller, L. Kulesz, *Courbes algébriques de genre* 2 *et* 3 *possédant de nombreux points rationnels*, C. R. Acad. Sci. Paris Sér. I Math. **321** (1995), no. 11, 1469–1472.
71. M. Kenku, F. Momose, *Torsion points on elliptic curves defined over quadratic fields*, Nagoya Math. J. **109** (1988), 125–149.
72. S. Kihara, *On an infinite family of elliptic curves with rank* ≥ 14 *over* **Q**, Proc. Japan Acad. Ser. A Math. Sci. **73** (1997), no. 2, 32.
73. Neal Koblitz, Introduction to elliptic curves and modular forms, Graduate Texts in Mathematics **97**, Springer-Verlag, New York, 1993.
74. V. A. Kolyvagin, *Finiteness of* $E(\mathbf{Q})$ *and* $\mathrm{III}(E, \mathbf{Q})$ *for a subclass of Weil curves*, Izv. Akad. Nauk SSSR Ser. Mat. **52** (1988), no. 3, 522–540, 670–671 (= Math. USSR – Izvestija **32** (1989), no. 3, 523–541).
75. V. A. Kolyvagin, *The Mordell-Weil and Shafarevich-Tate groups for Weil elliptic curves*, Izv. Akad. Nauk SSSR Ser. Mat. **52** (1988), no. 6, 1154–1180, 1327; (= Math. USSR – Izvestija **33** (1989), no. 3, 473–499).
76. G. Kramarz, *All congruent numbers less than* 2000, Math. Ann. **273** (1986), 337–340.
77. K. Kramer, *A family of semistable elliptic curves with large Tate-Shafarevitch groups*, Proc. Amer. Math. Soc. **89** (1983), 379–386.
78. S. Lang, Elliptic curves: Diophantine analysis, Grundlehren der Mathematischen Wissenschaften **231**, Springer-Verlag, Berlin-New York, 1978.

79. S. Lang, *Old and new conjectured Diophantine inequalities*, Bull. Amer. Math. Soc. **23** (1990), 37–75.
80. S. Lang, *Die abc-Vermutung*, Elem. Math. **48** (1993), no. 3, 89–99.
81. A. I. Lapin, *Subfields of hyperelliptic fields. I*, Izv. Akad. Nauk SSSR Ser. Mat. **28** (1964), 953–988 (= Amer. Math. Soc. Transl. (2) **69** (1968), 204–240).
82. A. I. Lapin, *On the rational points of an elliptic curve*, Izv. Akad. Nauk SSSR Ser. Mat. **29** (1965), 701–716 (= Amer. Math. Soc. Transl. (2) **66** (1968), 231–245).
83. M. Levin, *On the group of rational points on elliptic curves over function fields*, Amer. J. Math. **90** (1968), 456–462.
84. E. Lutz, *Sur l'equation $y^2 = x^3 - Ax - B$ dans les corps p-adic*, J. Reine Angew. Math. **177** (1937), 238–247.
85. L. Mai, *The average analytic rank of a family of elliptic curves*, J. Number Theory **45** (1993), no. 1, 45–60.
86. L. Mai, *The analytic rank of a family of elliptic curves*, Canad. J. Math. **45** (1993), no. 4, 847–862.
87. L. Mai, M. R. Murty, *A note on quadratic twists of an elliptic curve*, in Elliptic curves and related topics, 121–124, CRM Proc. Lecture Notes **4**, Amer. Math. Soc., Providence, RI, 1994.
88. Y. Manin, *The p-torsion of elliptic curves is uniformly bounded*, Izv. Akad. Nauk SSSR Ser. Mat. **33** (1969), 459–465 (= Math. USSR – Izvestija **3** (1969), 433–438).
89. Y. Manin, *Cyclotomic fields and modular curves*, Uspehi Mat. Nauk **26** (1971), no. 6 (162), 7–71.
90. Y. Manin, *Parabolic points and ζ-functions of modular curves*, Izv. Akad. Nauk SSSR Ser. Mat. **6** (1972), 19–66 (= Math. USSR – Izvestija **6** (1972), no. 1, 19–64).
91. R. Martin, W. McMillen, posting to Number Theory server, March 16, 1998.
92. R. Martin, W. McMillen, posting to Number Theory server, May 2, 2000.
93. R. C. Mason, Diophantine equations over function fields, London Math. Soc. Lecture Note Ser. **96**, Cambridge Univ. Press, Cambridge-New York, 1984.
94. B. Mazur, *Rational points of abelian varieties with values in towers of number fields*, Invent. Math. **18** (1972), 183–266.
95. B. Mazur, *Rational points on modular curves*, in Modular functions of one variable, V (Proc. Second Internat. Conf., Univ. Bonn, Bonn, 1976), 107–148, Lecture Notes in Math. **601**, Springer, Berlin, 1977.
96. B. Mazur, *Modular curves and the Eisenstein ideal*, Publ. math. IHES **47** (1977), 33–186.
97. B. Mazur, *Number theory as Gadfly*, Amer. Math. Monthly **98** (1991), no. 7, 593–610.
98. B. Mazur, *Three lectures about the arithmetic of elliptic curves*, lecture notes for the Arizona Winter School Workshop on Diophantine Geometry Related to the ABC Conjecture, 1998.
99. L. Merel, *Bornes pour la torsion des courbes elliptiques sur les corps de nombres*, Invent. math. **124** (1996), 437–449.
100. J-F. Mestre, *Courbes elliptiques de rang ≥ 11 sur $\mathbf{Q}(t)$*, C. R. Acad. Sci. Paris, **313** (1991), no. 3, 139–142.
101. J-F. Mestre, J. Oesterlé, *Courbes de Weil semi-stables de discriminant une puissance m-iéme*, J. Reine Angew. Math. **400** (1989), 173–184.

102. J. S. Milne, *The Tate-Safarevič group of a constant abelian variety*, Invent. Math. **6** (1968), 91–105.
103. J. S. Milne, *On a conjecture of Artin and Tate*, Ann. of Math. (2) **102** (1975), no. 3, 517–533.
104. J. S. Milne, *Comparison of the Brauer group with the Tate-Safarevič group*, J. Fac. Sci. Univ. Tokyo Sect. IA Math. **28** (1982), no. 3, 735–743.
105. P. Monsky, *Mock Heegner points and congruent numbers*, Math. Z. **204** (1990), no. 1, 45–67.
106. P. Monsky, *Generalizing the Birch-Stephens theorem. I. Modular curves*, Math. Z. **221** (1996), no. 3, 415–420.
107. M. R. Murty, V. K. Murty, *Mean values of derivatives of modular L-series*, Ann. of Math. (2) **133** (1991), no. 3, 447–475.
108. V. K. Murty, Modular elliptic curves, in Seminar on Fermat's Last Theorem (Toronto, ON, 1993–1994), 1–38, CMS Conf. Proc. **17**, Amer. Math. Soc., Providence, RI, 1995.
109. K. Nagao, *An example of elliptic curve over $\mathbf{Q}(T)$ with rank ≥ 13*, Proc. Japan Acad. Ser. A Math. Sci. **70** (1994), no. 5, 152–153.
110. T. Nagell, *Solution de quelque problèmes dans la théorie arithmétique des cubiques planes du premier genre*, Wid. Akad. Skrifter Oslo I (1935), No. 1, 1–25.
111. J. Nekovář, *Beĭlinson's conjectures*, in Motives (Seattle, WA, 1991), 377–400, Proc. Sympos. Pure Math. **55**, Part 1, Amer. Math. Soc., Providence, RI, 1994.
112. A. Néron, *Problèmes arithmétiques et géométriques rattachés à la notion de rang d'une courbe algébrique dans un corps*, Bull. Soc. Math. France **80**, (1952), 101–166.
113. A. Nitaj, *Aspects expérimentaux de la conjecture abc*, in Number theory (Paris, 1993–1994), 145–156, London Math. Soc. Lecture Note Ser. **235**, Cambridge Univ. Press, Cambridge, 1996.
114. A. Nitaj, *La conjecture abc*, Enseign. Math. (2) **42** (1996), no. 1-2, 3–24.
115. A. Nitaj, *Détermination de courbes elliptiques pour la conjecture de Szpiro*, Acta Arith. **85** (1998), no. 4, 351–376.
116. A. Nitaj, *Invariants des courbes de Frey-Hellegouarch et grandes groupes de Tate-Shafarevich*, to appear in Acta Arith.,
    http://www.math.uiuc.edu/Algebraic-Number-Theory/0190/index.html .
117. J. Oesterlé, *Nouvelles approches du "théorème" de Fermat*, Séminaire Bourbaki, Vol. 1987/88, Exp. No. 694, Astérisque **161-162** (1988), 4, 165–186.
118. A. P. Ogg, *Cohomology of abelian varieties over function fields*, Ann. of Math. **76** (1962), 185–212.
119. K. Ono, C. Skinner, *Non-vanishing of quadratic twists of modular L-functions*, Invent. math. **134** (1998), no. 3, 651–660.
120. T. Ooe, J. Top, *On the Mordell-Weil rank of an abelian variety over a number field*, J. Pure Appl. Algebra **58** (1989), no. 3, 261–265.
121. P. Parent, *Bornes effectives pour la torsion des courbes elliptiques sur les corps de nombres*, J. Reine Angew. Math. **506** (1999), 85–116.
122. P. Parent, *Torsion des courbes elliptiques sur les corps cubiques*, preprint, 28 janvier, 1999.
123. B. Poonen, M. Stoll, *The Cassels-Tate pairing on polarized abelian varieties*, Ann. of Math. **150** (1999), no. 3, 1109–1149.

124. M. Rapoport, N. Schappacher, P. Schneider (eds.), *Beilinson's conjectures on special values of L-functions*, Perspectives in Mathematics **4**, Academic Press, Inc., Boston, MA, 1988.

125. K. Ribet, *Abelian varieties over* **Q** *and modular forms*, in Algebra and topology 1992 (Taejon), 53–79, Korea Adv. Inst. Sci. Tech., Taejon, 1992.

126. N. F. Rogers, *Rank computations for the congruent number elliptic curves*, to appear in Experiment. Math.

127. K. Rubin, *Local units, elliptic units, Heegner points and elliptic curves*, Invent. Math. **88** (1987), no. 2, 405–422.

128. K. Rubin, *Tate-Shafarevich groups and L-functions of elliptic curves with complex multiplication*, Invent. Math. **89** (1987), no. 3, 527–559.

129. K. Rubin, *The "main conjectures" of Iwasawa theory for imaginary quadratic fields*, Invent. Math. **103** (1991), 25–68.

130. K. Rubin, A. Silverberg, *Ranks of elliptic curves in families of quadratic twists*, to appear in Experiment. Math.

131. N. Schappacher, R. Schoof, *Beppo Levi and the arithmetic of elliptic curves*, Math. Intelligencer **18** (1996), no. 1, 57–69.

132. P. Schneider, *Zur Vermutung von Birch und Swinnerton-Dyer über globalen Funktionenkörpern*, Math. Ann. **260** (1982), no. 4, 495–510.

133. U. Schneiders, H. G. Zimmer, *The rank of elliptic curves upon quadratic extension*, in Computational number theory (Debrecen, 1989), 239–260, de Gruyter, Berlin, 1991.

134. E. S. Selmer, *A conjecture concerning rational points on cubic curves*, Math. Scand. **2** (1954), 49–54.

135. P. Serf, *Congruent numbers and elliptic curves*, in Computational number theory (Debrecen, 1989), 227–238, de Gruyter, Berlin, 1991.

136. J-P. Serre, *Sur les représentations modulaires de degré 2 de* $\operatorname{Gal}(\overline{\mathbf{Q}}/\mathbf{Q})$, Duke Math. J. **54** (1987), no. 1, 179–230.

137. J-P. Serre, *Lectures on the Mordell-Weil Theorem*, Friedr. Vieweg & Sohn, Braunschweig, 1989.

138. J-P. Serre, *Propriétés conjecturales des groupes de Galois motiviques et des reprsentations ℓ-adiques*, in Motives (Scattle, WA, 1991), 377–400, Proc. Sympos. Pure Math. **55**, Part 1, Amer. Math. Soc., Providence, RI, 1994.

139. J-P. Serre, J. Tate, *Good reduction of abelian varieties*, Ann. of Math. **88** (1968), 492–517.

140. I. R. Šafarevič, *Principal homogeneous spaces defined over a function field*, Trudy Mat. Inst. Steklov. **64** (1961), 316–346 (= Amer. Math. Soc. Transl. (2) **37** (1964), 85–114).

141. G. Shimura, *Correspondances modulaires et les fonctions ζ de courbes algébriques*, J. Math. Soc. Japan **10** (1958), 1–28.

142. G. Shimura, *Sur les intégrales attachées aux formes automorphes*, J. Math. Soc. Japan **11** (1959), 291–311.

143. G. Shimura, *On elliptic curves with complex multiplication as factors of the Jacobians of modular function fields*, Nagoya Math. J. **43** (1971), 199–208.

144. G. Shimura, *On the factors of the jacobian variety of a modular function field*, J. Math. Soc. Japan **25** (1973), 523–544.

145. G. Shimura, *The special values of the zeta functions associated with cusp forms*, Comm. Pure Appl. Math. **29** (1976), no. 6, 783–804.

146. G. Shimura, *On the periods of modular forms*, Math. Ann. **229** (1977), no. 3, 211–221.
147. G. Shimura, *Yutaka Taniyama and his time. Very personal recollections*, Bull. London Math. Soc. **21** (1989), no. 2, 186–196.
148. G. Shimura, Introduction to the arithmetic theory of automorphic functions, Reprint of the 1971 original, Publications of the Math. Soc. Japan **11**, Kanô Memorial Lectures **1**, Princeton Univ. Press, Princeton, NJ, 1994.
149. G. Shimura, Abelian varieties with complex multiplication and modular functions, Princeton Mathematical Series **46**, Princeton Univ. Press, Princeton, NJ, 1998.
150. T. Shioda, *An explicit algorithm for computing the Picard number of certain algebraic surfaces*, Amer. J. Math. **108** (1986), no. 2, 415–432.
151. T. Shioda, *Some remarks on elliptic curves over function fields*, Journées Arithmétiques, 1991 (Geneva), Astérisque **209** (1992), 99–114.
152. T. Shioda, *Genus two curves over $\mathbf{Q}(t)$ with high rank*, Comment. Math. Univ. St. Paul. **46** (1997), no. 1, 15–21.
153. A. Silverberg, *Torsion points on abelian varieties of CM-type*, Compositio Math. **68** (1988), no. 3, 241–249.
154. A. Silverberg, *Points of finite order on abelian varieties*, in p-adic methods in number theory and algebraic geometry, 175–193, Contemp. Math. **133**, Amer. Math. Soc., Providence, RI, 1992.
155. A. Silverberg, *Galois representations attached to points on Shimura varieties*, in Séminaire de Théorie des Nombres, Paris 1990–91, 221–240, Progress in Mathematics **108**, Birkhäuser, Boston, 1993.
156. Joseph H. Silverman, The arithmetic of elliptic curves, Graduate Texts in Mathematics **106**, Springer-Verlag, New York, 1986.
157. J. H. Silverman, *A quantitative version of Siegel's theorem: integral points on elliptic curves and Catalan curves*, J. Reine Angew. Math. **378** (1987), 60–100.
158. J. H. Silverman, Advanced Topics in the Arithmetic of Elliptic Curves, Springer-Verlag, New York, 1994.
159. J. Silverman, *On the rank of elliptic curves*, notes from a talk (see his web page: http://www.math.brown.edu/~jhs/).
160. Joseph H. Silverman, John Tate, Rational points on elliptic curves, Undergraduate Texts in Mathematics, Springer-Verlag, New York, 1992.
161. N. P. Smart, *S-integral points on elliptic curves*, Math. Proc. Cambridge Philos. Soc. **116** (1994), no. 3, 391–399.
162. N. P. Smart, The algorithmic resolution of Diophantine equations, LMS Student Texts **41**, Cambridge Univ. Press, Cambridge, 1998.
163. N. P. Smart, N. M. Stephens, *Integral points on elliptic curves over number fields*, Math. Proc. Cambridge Philos. Soc. **122** (1997), no. 1, 9–16.
164. N. M. Stephens, *The diophantine equation $X^3 + Y^3 = DZ^3$ and the conjectures of Birch and Swinnerton-Dyer*, J. Reine Angew. Math. **231** (1968), 121–162.
165. C. L. Stewart, R. Tijdeman, *On the Oesterlé-Masser conjecture*, Monatsh. Math. **102** (1986), no. 3, 251–257.
166. C. L. Stewart, J. Top, *On ranks of elliptic curves and power-free values of binary forms*, J. Amer. Math. Soc. **8** (1995), 943–973.

167. C. L. Stewart, K. R. Yu, *On the abc conjecture*, Math. Ann. **291** (1991), no. 2, 225–230.
168. P. F. Stiller, *The Picard numbers of elliptic surfaces with many symmetries*, Pacific J. Math. **128** (1987), no. 1, 157–189.
169. R. J. Stroeker, N. Tzanakis, *Solving elliptic Diophantine equations by estimating linear forms in elliptic logarithms*, Acta Arith. **67** (1994), no. 2, 177–196.
170. R. J. Stroeker, N. Tzanakis, *On the elliptic logarithm method for elliptic Diophantine equations: reflections and an improvement*, Experiment. Math. **8** (1999), no. 2, 135–149.
171. R. J. Stroeker, B. M. M. de Weger, *Solving elliptic Diophantine equations: the general cubic case*, Acta Arith. **87** (1999), no. 4, 339–365.
172. J. Tate, *Algebraic cycles and poles of zeta functions*, in Arithmetical Algebraic Geometry (Proc. Conf. Purdue Univ., 1963), 93–110, Harper & Row, New York, 1965.
173. J. Tate, *On the conjectures of Birch and Swinnerton-Dyer and a geometric analog*, Séminaire Bourbaki, Vol. 1965/1966, Exp. No. 306 (1966), 26 pp.
174. J. T. Tate, *The arithmetic of elliptic curves*, Invent. Math. **23** (1974), 179–206.
175. J. T. Tate, I. R. Šafarevič, *The rank of elliptic curves*, Dokl. Akad. Nauk SSSR **175** (1967), no. 4, 770–773 (= Soviet Math. Dokl. **8** (1967), no. 4, 917–920).
176. R. Taylor, A. Wiles, *Ring-theoretic properties of certain Hecke algebras*, Ann. of Math. **141** (1995), no. 3, 553–572.
177. J. B. Tunnell, *A classical Diophantine problem and modular forms of weight 3/2*, Invent. Math. **72** (1983), no. 2, 323–334.
178. N. Tzanakis, *Solving elliptic Diophantine equations by estimating linear forms in elliptic logarithms. The case of quartic equations*, Acta Arith. **75** (1996), no. 2, 165–190.
179. P. Van Mulbregt, *Torsion-points on low-dimensional abelian varieties with complex multiplication*, in p-adic methods in number theory and algebraic geometry, 205–210, Contemp. Math. **133**, Amer. Math. Soc., Providence, RI, 1992.
180. P. Vojta, Diophantine approximations and value distribution theory, Lecture Notes in Math. **1239**, Springer-Verlag, Berlin, 1987.
181. P. Vojta, *A more general ABC conjecture*, lecture notes for the Arizona Winter School Workshop on Diophantine Geometry Related to the ABC Conjecture, 1998.
182. J-L. Waldspurger, *Correspondances de Shimura et quaternions*, Forum Math. **3** (1991), no. 3, 219–307.
183. B. M. M. de Weger, $A + B = C$ *and big* Ш*'s*, Quart. J. Math. Oxford **49** (1998), 105–128.
184. A. Weil, *Über die Bestimmung Dirichletscher Reihen durch Funktionalgleichungen*, Math. Ann. **168** (1967), 149–156.
185. A. Weil, Œuvres scientifiques, Vol. II (1951–1964), Springer-Verlag, New York-Heidelberg, 1979.
186. A. Wiles, *Modular elliptic curves and Fermat's last theorem*, Ann. of Math. **141** (1995), no. 3, 443–551.

187. D. Zagier, *Large integral points on elliptic curves*, Math. Comp. **48** (1987), no. 177, 425–436.
188. D. Zagier, G. Kramarz, *Numerical investigations related to the L-series of certain elliptic curves*, J. Indian Math. Soc. **52** (1987), 51–69.

Lectures on Serre's Conjectures

Kenneth A. Ribet
William A. Stein

IAS/Park City Mathematics Series
Volume 9, 2001

Lectures on Serre's Conjectures

Kenneth A. Ribet[1]
William A. Stein[2]

Preface

We shall begin by discussing some examples of mod ℓ representations of the group $\mathrm{Gal}(\overline{\mathbf{Q}}/\mathbf{Q})$ We'll try to motivate Serre's conjectures by referring first to the case of representations that are unramified outside ℓ; these should come from cusp forms on the full modular group $\mathrm{SL}(2, \mathbf{Z})$. In another direction, one might think about representations coming from ℓ-division points on elliptic curves, or more generally from ℓ-division points on abelian varieties of "GL_2-type." Amazingly, Serre's conjectures imply that all odd irreducible two-dimensional mod ℓ representations of $\mathrm{Gal}(\overline{\mathbf{Q}}/\mathbf{Q})$ may be realized in spaces of ℓ-division points on such abelian varieties. (Richard Taylor has recently obtained a strong result in this direction; see Theorem F of [**118**].) The weak Serre conjecture states that all such representations come from modular forms, and then it takes only a bit of technique to show that one can take the modular forms to have weight two (if one allows powers of ℓ in the level).

Since little work has been done toward proving the weak Serre conjecture, these notes will focus on the bridge between the weak and the strong conjectures. The latter states that each ρ as above comes from the space of cusp forms of a specific weight and level, with these invariants between determined by the local behavior of ρ at ℓ and at primes other than ℓ (respectively). To motivate the strong conjecture, and to begin constructing the bridge, we discuss the local behavior of those ρ that do come from modular forms. For the most part, we look only at forms of weight $k \geq 2$ whose levels N arc prime to ℓ. For these forms, the behavior of ρ at ℓ is described in detail in [**32**], where theorems of P. Deligne and J.-M. Fontaine are recalled. (In [**32**, §6], B. Edixhoven presents a proof of Fontaine's theorem.) Further, the behavior of ρ at primes $p \neq \ell$ may be deduced from H. Carayol's theorems [**11, 12**], which relate the behavior at p of the ℓ-adic representations attached to f with the

---

[1]Math Department, MC 3840, Berkeley, CA 94720-3840
E-mail address: ribet@math.berkeley.edu
[2]Department of Mathematics, Harvard University, Cambridge, MA 02138
E-mail address: was@math.harvard.edu
1991 *Mathematics Subject Classification.* 11.
Key words and phrases. Serre's conjectures, modular forms, Galois representations.
The second author was supported by a Cal@SiliconValley fellowship.
Chapter 5. Appendix: The Shimur Construction in Weight 2. Copyright by Brian Conrad.

©2001 American Mathematical Society

p-adic component of the automorphic representation of GL(2) that one associates with f. (The behavior of ρ at ℓ in the case where ℓ divides N is analyzed in [**89**].)

In [**102**], Serre associates to each ρ a level $N(\rho)$ and a weight $k(\rho)$. These invariants are defined so that $N(\rho)$ is prime to ℓ and so that $k(\rho)$ is an integer greater than 1. As Serre anticipated, if ρ arises from a modular form of weight k and level N, and if k is at least 2 and N is prime to ℓ, then one has $k(\rho) \leq k$ and $N(\rho) \mid N$. To find an f for which $N = N(\rho)$ and $k = k(\rho)$ is to "optimize" the level and weight of a form giving ρ. As Edixhoven explains in his article [**32**], weight optimization follows in a somewhat straightforward manner from the theorems of Deligne and Fontaine alluded to above, Tate's theory of θ-cycles, and Gross's theorem on companion forms [**46**] (see also [**17**]). Moreover, it is largely the case that weight and level optimization can be performed independently.

In [**12**], Carayol analyzes the level optimization problem. He shows, in particular, that the problem breaks down into a series of sub-problems, all but one of which he treats by appealing to a single lemma, the lemma of [**12**, §3]. The remaining sub-problem is the one that intervenes in establishing the implication "Shimura-Taniyama $\Longrightarrow$ Fermat." This problem has been discussed repeatedly [**83, 84, 86, 87**]. In Section 3.3.10, we will explain the principle of [**86**].

The case $\ell = 2$ is the only remaining case for which the level optimization problem has not been resolved. The proof in [**26, 87**] of level optimization for $\ell \geq 3$ does not fully exploit multiplicity one results, but appears to completely break down when $\ell = 2$. In the recent paper [**9**], Kevin Buzzard observed that many new cases of multiplicity one are known and that this can be used to obtain new level optimization results when $\ell = 2$.

In view of these remarks it might be appropriate for us to summarize in a few sentence what is known about the implication "weak Serre conjecture $\Longrightarrow$ strong Serre conjecture." As explained in [**26**], for $\ell \geq 5$ the weak conjecture of Serre implies the strong conjecture about the optimal weight, level, and character. For $\ell = 3$, the weak conjecture implies the strong conjecture, except in a few well-understood situations, where the order of the character must be divisible by ℓ when the level is optimal. The difficulty disappears if one works instead with Katz's definition of a mod ℓ modular form, where the character is naturally defined only mod ℓ. The situation is less complete when $\ell = 2$, but quite favorable. When $\ell = 2$ the situation concerning the weight is explained by Edixhoven in [**32**]: the results of [**17**] do not apply and those of [**46**] rely on unchecked compatibilities.

A certain amount of work has been done on the Hilbert modular case, i.e., the case where **Q** is replaced by a totally real number field F. For this work, the reader may consult articles of Frazer Jarvis [**52, 53, 54**], Kazuhiro Fujiwara [**45**], and Ali Rajaei [**79**]. The authors are especially grateful to Fujiwara for sending them a preliminary version of his manuscript, "Level optimization in the totally real case." However, these notes will treat only the classical case $F = \mathbf{Q}$.

This paper emerged out of a series of lectures that were delivered by the first author at the 1999 IAS/Park City Mathematics Institute. The second author created the text based on the lectures and added examples, diagrams, an exercise section, and the index. Brian Conrad contributed the first appendix, which describes a construction of Shimura. The second appendix, by Kevin Buzzard, contains a proof of some new multiplicity one results when $\ell = 2$

For other expository accounts of Serre's conjectures, the reader may consult the articles of Edixhoven [**33, 34, 35**], H. Darmon [**22**], and R. Coleman [**15**].

The authors would like to thank K. Buzzard and Serre for many useful comments on various drafts of this paper, Conrad and Buzzard for providing the appendices, M. Emerton for his enlightening lecture on Katz's definition of modular forms, N. Jochnowitz for suggestions that improved the exposition in Section 2.2.2, and L. Kilford for help in finding examples of mod 2 representations in Section 3.3.7.

Kenneth A. Ribet
William A. Stein

CHAPTER 1
Introduction to Serre's conjecture

1.1.1. Introduction

Let's start with an elliptic curve $E/\mathbf{Q}$. Nowadays, it's a familiar activity to consider the Galois representations defined by groups of division points of E. Namely, let n be a positive integer, and let $E[n]$ be the kernel of multiplication by n on $E(\overline{\mathbf{Q}})$. The group $E[n]$ is free of rank two over $\mathbf{Z}/n\mathbf{Z}$. After a choice of basis, the action of $\mathrm{Gal}(\overline{\mathbf{Q}}/\mathbf{Q})$ on $E[n]$ is given by a homomorphism

$$\rho_n : \mathrm{Gal}(\overline{\mathbf{Q}}/\mathbf{Q}) \to \mathrm{Aut}(E[n]) \approx \mathrm{GL}(2, \mathbf{Z}/n\mathbf{Z}).$$

This homomorphism is unramified at each prime p that is prime to the product of n with the conductor of E (see Exercise 15). For each such p, the element $\rho_n(\mathrm{Frob}_p)$ is a 2×2 matrix that is well defined up to conjugation. Its determinant is p mod n; its trace is a_p mod n, where a_p is the usual "trace of Frobenius" attached to E and p, i.e., the quantity $1 + p - \#E(\mathbf{F}_p)$. In his 1966 article [**107**], G. Shimura studied these representations and the number fields that they cut out for the curve $E = J_0(11)$. (This curve was also studied by Serre [**91**, pg. 254].) He noticed that for prime values $n = \ell$, the representations ρ_n tended to have large images. In [**93**], Serre proved that for any fixed elliptic curve E, not having complex multiplication, the indices

$$[\mathrm{GL}(2, \mathbf{Z}/n\mathbf{Z}) : \rho_n(\mathrm{Gal}(\overline{\mathbf{Q}}/\mathbf{Q}))]$$

are bounded independently of n. In Shimura's example, Serre proved that

$$[\mathrm{GL}(2, \mathbf{Z}/\ell\mathbf{Z}) : \rho_\ell(\mathrm{Gal}(\overline{\mathbf{Q}}/\mathbf{Q}))] = 1$$

for all $\ell \neq 5$ (see [**93**, §5.5.1]).

In this article, we will be concerned mainly with two-dimensional representations over finite fields. To that end, we restrict attention to the case where $n = \ell$ is prime. The representation ρ_ℓ is "modular" in the familiar sense that it's a representation of a group over a field in positive characteristic. The theme of this course is that it's modular in a different and deeper sense: it comes from a modular form! Indeed, according to a recent preprint of Breuil, Conrad, Diamond and Taylor (see [**7, 19, 114, 117**]), the Shimura-Taniyama conjecture is now a theorem—all elliptic curves over **Q** are modular!! Thus if N is the conductor of E, there is a weight-two newform $f = \sum_{n=1}^{\infty} c_n q^n$ ($q = e^{2\pi i z}$) on $\Gamma_0(N)$ with the property that $a_p = c_p$ for all p prime to N. Accordingly, ρ_ℓ is connected up with modular forms via the relation $\text{tr}(\rho_\ell(\text{Frob}_p)) \equiv c_p \pmod{\ell}$, valid for all but finitely many primes p.

The Shimura-Taniyama conjecture asserts that for each positive integer N there is a bijection between isogeny classes of elliptic curves A over **Q** of conductor N and rational newforms f on $\Gamma_0(N)$ of weight two. Given A, the Shimura-Taniyama conjecture produces a modular form $f = \sum_{n=1}^{\infty} c_n q^n$ whose Dirichlet series is equal to the L-series of A:

$$\sum_{n=1}^{\infty} \frac{c_n}{n^s} = L(f,s) \;=\!=\!=\; L(A,s) = \sum_{n=1}^{\infty} \frac{a_n}{n^s}.$$

The integers a_n encode information about the number of points on A over various finite fields $\mathbf{F}_p$. If p is a prime not dividing N, then $a_p = p + 1 - \#A(\mathbf{F}_p)$; if $p \mid N$,

$$a_p = \begin{cases} -1 & \text{if } A \text{ has non-split multiplicative reduction at } p \\ 1 & \text{if } A \text{ has split multiplicative reduction at } p \\ 0 & \text{if } A \text{ has additive reduction at } p. \end{cases}$$

The integers a_n are obtained recursively from the a_p as follows:

- $a_{p^r} = \begin{cases} a_{p^{r-1}} a_p - p a_{p^{r-2}} & \text{if } p \nmid N \\ a_p^r & \text{if } p \mid N \end{cases}$
- $a_{nm} = a_n a_m,$ if $(n,m) = 1$.

The conjectures made by Serre in [**102**], which are the subject of this paper, concern representations $\rho : \text{Gal}(\overline{\mathbf{Q}}/\mathbf{Q}) \to \text{GL}(2, \overline{\mathbf{F}}_\ell)$. We always require (usually tacitly) that our representations are continuous. The continuity condition just means that the kernel of ρ is open, so that it corresponds to a finite Galois extension K of $\mathbf{Q}$. The representation ρ then embeds $\text{Gal}(K/\mathbf{Q})$ into $\text{GL}(2, \overline{\mathbf{F}}_\ell)$. Since K is a finite extension of $\mathbf{Q}$, the image of ρ is contained in $\text{GL}(2, \mathbf{F})$ for some finite subfield $\mathbf{F}$ of $\overline{\mathbf{F}}_\ell$.

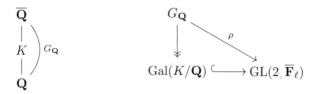

For various technical reasons, the original conjectures of Serre insist that ρ be irreducible. It is nevertheless fruitful to consider the reducible case as well (see [**111**]).

The conjectures state (in particular) that each continuous irreducible ρ that satisfies a necessary parity condition "arises from" (or is associated with) a suitable modular form mod ℓ. To explain what's going on, let's start with

$$\Delta := \sum_{n=1}^{\infty} \tau(n) q^n = q \prod_{i=1}^{\infty} (1 - q^i)^{24},$$

the unique (normalized) cusp form of weight 12 on $\text{SL}(2, \mathbf{Z})$. In [**92**], Serre conjectured the existence of a "strictly compatible" family of ℓ-adic representations of $\mathrm{Gal}(\overline{\mathbf{Q}}/\mathbf{Q})$ whose L-function is the L-function of Δ, namely

$$L(\Delta, s) = \sum_{n=1}^{\infty} \tau(n) n^{-s} = \prod_p (1 - \tau(p) p^{-s} + p^{11-2s})^{-1},$$

where the product is taken over all prime numbers p. The conjectured ℓ-adic representations were constructed soon after by Deligne [24]. Specifically, Deligne constructed, for each prime ℓ, a representation

$$\rho_{\ell^{\infty}} : \mathrm{Gal}(\overline{\mathbf{Q}}/\mathbf{Q}) \to \mathrm{GL}(2, \mathbf{Z}_{\ell}),$$

unramified outside ℓ, such that for all primes $p \neq \ell$,

$$\mathrm{tr}(\rho_{\ell^{\infty}}(\mathrm{Frob}_p)) = \tau(p), \qquad \det(\rho_{\ell^{\infty}}(\mathrm{Frob}_p)) = p^{11}.$$

On reducing $\rho_{\ell^{\infty}}$ mod ℓ, we obtain a representation

$$\rho_{\ell} : \mathrm{Gal}(\overline{\mathbf{Q}}/\mathbf{Q}) \to \mathrm{GL}(2, \mathbf{F}_{\ell})$$

with analogous properties. (Equalities are replaced by congruences mod ℓ.) In other words, the ρ_{ℓ} for Δ are just like the ρ_{ℓ} for an elliptic curve E, except that the integers a_p are replaced by the corresponding values of the τ-function. The determinant of ρ_{ℓ} is the 11th power of the mod ℓ cyclotomic character $\chi : \mathrm{Gal}(\overline{\mathbf{Q}}/\mathbf{Q}) \to \mathbf{F}_{\ell}^*$, i.e., the character giving the action of $\mathrm{Gal}(\overline{\mathbf{Q}}/\mathbf{Q})$ on the group of ℓth roots of unity in $\overline{\mathbf{Q}}$ (see Section 1.1.5).

More generally, take a weight $k \geq 12$ and suppose that $f = \sum_n c_n q^n$ is a nonzero weight-k cusp form for $\mathrm{SL}(2, \mathbf{Z})$ that satisfies $f|T_n = c_n f$ for all $n \geq 1$, T_n being the nth Hecke operator on the space of cusp forms of weight k for $\mathrm{SL}(2, \mathbf{Z})$ (see Section 1.1.5). Then the complex numbers c_n ($n \geq 1$) are algebraic integers. Moreover, the field $E := \mathbf{Q}(\ldots c_n \ldots)$ generated by the c_n is a totally real number field (of finite degree over $\mathbf{Q}$). Thus the c_n all lie in the integer ring $\mathcal{O}_E$ of E. For each ring homomorphism $\varphi : \mathcal{O}_E \to \overline{\mathbf{F}}_{\ell}$, one finds a representation

$$\rho = \rho_{\varphi} : \mathrm{Gal}(\overline{\mathbf{Q}}/\mathbf{Q}) \to \mathrm{GL}(2, \overline{\mathbf{F}}_{\ell}),$$

unramified outside ℓ, such that

$$\mathrm{tr}(\rho(\mathrm{Frob}_p)) = \varphi(c_p), \qquad \det(\rho(\mathrm{Frob}_p)) = p^{k-1}$$

for all $p \neq \ell$. We have $\det \rho = \chi^{k-1}$. Of course, there is no guarantee that ρ is irreducible. We can (and do) suppose that ρ is semisimple by replacing it by its semisimplification. Then ρ is determined up to isomorphism by the displayed trace and determinant conditions; this follows from the Cebotarev density theorem and the Brauer-Nesbitt theorem [21], which states that semisimple representations are determined by their characteristic polynomials.

It is important to note that k is necessarily an even integer; otherwise the space $S_k(\mathrm{SL}(2, \mathbf{Z}))$ of weight-k cusp forms on $\mathrm{SL}(2, \mathbf{Z})$ is easily seen to be 0. Thus the determinant χ^{k-1} of ρ is an odd power of χ. In particular, $\det \rho : \mathrm{Gal}(\overline{\mathbf{Q}}/\mathbf{Q}) \to \mathbf{F}_{\ell}^*$ is unramified outside ℓ and takes the value -1 on complex conjugations $c \in \mathrm{Gal}(\overline{\mathbf{Q}}/\mathbf{Q})$. It's a nice exercise to check that, conversely, all continuous homomorphisms with these properties are odd powers of χ (see Exercise 1).

In the early 1970s, Serre conjectured that all homomorphisms that are "like ρ" come from cusp forms of some weight on $\mathrm{SL}(2, \mathbf{Z})$. Namely, let

$$\rho : \mathrm{Gal}(\overline{\mathbf{Q}}/\mathbf{Q}) \to \mathrm{GL}(2, \overline{\mathbf{F}}_{\ell})$$

be a continuous, irreducible representation that is (1) unramified outside ℓ and (2) of odd determinant, in the sense that $\det \rho(c) = -1 \in \overline{\mathbf{F}}_\ell$ for complex conjugations $c \in \operatorname{Gal}(\overline{\mathbf{Q}}/\mathbf{Q})$. In a May, 1973 letter to Tate, Serre conjectured that ρ is of the form ρ_φ. This means that there is a weight $k \geq 12$, an eigenform $f \in S_k(\operatorname{SL}(2, \mathbf{Z}))$, and a homomorphism $\varphi : \mathcal{O}_E \to \overline{\mathbf{F}}_\ell$ (where $\mathcal{O}_E$ is the ring of integers of the field generated by the coefficients of f) so that ρ_φ and ρ are isomorphic.

To investigate Serre's conjecture, it is fruitful to consider the operation $\rho \mapsto \rho \otimes \chi$ on representations. This "twisting" operation preserves the set of representations that come from modular forms. Indeed, let $\theta = q\frac{d}{dq}$ be the classical differential operator $\sum a_n q^n \mapsto \sum n a_n q^n$. According to Serre and Swinnerton-Dyer [**61, 94, 112**], if f is a mod ℓ form of weight k, then θf is a mod ℓ form of weight $k + \ell + 1$. Then if ρ is associated to f, $\rho \otimes \chi$ is associated with θf. According to a result of Atkin, Serre and Tate (see [**97**, Th. 3] and Section 2.2.1), if ρ comes from an eigenform in some space $S_k(\operatorname{SL}(2, \mathbf{Z}))$, then a suitable twist $\rho \otimes \chi^i$ of f comes from a form of weight $\leq \ell + 1$.

Serre's conjecture thus has the following consequence: each two-dimensional irreducible odd representation of $\operatorname{Gal}(\overline{\mathbf{Q}}/\mathbf{Q})$ over $\overline{\mathbf{F}}_\ell$ that is unramified outside ℓ has a twist (by a power of χ) coming from an eigenform on $\operatorname{SL}(2, \mathbf{Z})$ of weight at most $\ell + 1$. In particular, suppose that $\ell < 11$. Then the spaces $S_k(\operatorname{SL}(2, \mathbf{Z}))$ with $k \leq \ell + 1$ are all 0; as a result, they contain no nonzero eigenforms! The conjecture that all ρ are modular (of level 1) thus predicts that there are *no* representations of the type contemplated if ℓ is 2, 3, 5 or 7. In support of the conjecture, the non-existence statement was proved for $\ell = 2$ by J. Tate in a July, 1973 letter to Serre [**113**]. Soon after, Serre treated the case $\ell = 3$ by methods similar to those of Tate. (See [**113**, p. 155] for a discussion and a reference to a note in Serre's Œuvres.) Quite recently, Sharon Brueggeman considered the case $\ell = 5$; she proved that the conjectured result follows from the Generalized Riemann Hypothesis (see [**8**]). In another direction, Hyunsuk Moon generalized Tate's result and proved that there are only finitely many isomorphism classes of continuous semisimple Galois representations $\rho : G_\mathbf{Q} \to \operatorname{GL}_4(\overline{\mathbf{F}}_2)$ unramified outside 2 such that field $K/\mathbf{Q}$ corresponding to the kernel of ρ is totally real (see [**76**]). Similar work in this direction has been carried out by Joshi [**58**], under additional local hypothesis.

Serre discussed his conjecture with Deligne, who pointed out a number of surprising consequences. In particular, suppose that one takes a ρ coming from an eigenform f' of some weight and of level $N > 1$. On general grounds, ρ has the right to be ramified at primes p dividing N as well as at the prime ℓ. Suppose that, by accident as it were, ρ turned out to be unramified at all primes $p \mid N$. Then the conjecture would predict the existence of a level-1 form f' (presumably of the same weight as f) whose mod ℓ representation was isomorphic to ρ. For example, if $N = \ell^\alpha$ is a power of ℓ, then the conjecture predicts that ρ arises from a form f' of level 1. How could one manufacture the f'?

The passage $f \rightsquigarrow f'$ comes under the rubric of "level optimization". When you take a representation ρ that comes from high level N, and it seems as though that representation comes from a lower level N', then to "optimize the level" is to cough up a form of level N' that gives ρ.

Deligne pointed out also that Serre's conjecture implies that representations ρ over $\overline{\mathbf{F}}_\ell$ are required to "lift" to λ-adic representations of $\operatorname{Gal}(\overline{\mathbf{Q}}/\mathbf{Q})$. In the recent articles [**80**] and [**81**], R. Ramakrishna used purely Galois cohomological techniques to prove results in this direction.

1.1.2. The weak conjecture of Serre

In the mid 1980s, Gerhard Frey began lecturing on a link between Fermat's Last Theorem and elliptic curves (see [**42, 43**]). (Earlier, Hellegouarch had also considered links between Fermat's Last Theorem and elliptic curves; see the MathSciNet review and Appendix of [**48**].) As is now well known, Frey proposed that if $a^\ell + b^\ell$ was a perfect ℓth power, then the elliptic curve $y^2 = x(x - a^\ell)(x + b^\ell)$ could be proved to be non-modular. Soon after, Serre pointed out that the non-modularity contemplated by Frey would follow from suitable level-optimization results concerning modular forms [**101**]. To formulate such optimization results, Serre went back to the tentative conjecture that he had made 15 years earlier and decided to study representations $\rho: \text{Gal}(\overline{\mathbf{Q}}/\mathbf{Q}) \to \text{GL}(2, \overline{\mathbf{F}}_\ell)$ that are not necessarily unramified at ℓ. The results, of course, were the conjectures of [**102**].

An important consequence of these conjectures is the so-called "weak conjecture of Serre." As background, we recall that Hecke eigenforms on congruence subgroups of $\text{SL}(2, \mathbf{Z})$ give rise to two-dimensional representations of $\text{Gal}(\overline{\mathbf{Q}}/\mathbf{Q})$. If we set things up correctly, we get representations over $\overline{\mathbf{F}}_\ell$. More specifically, take integers $k \geq 2$ and $N \geq 1$; these are the weight and level, respectively. Let $f = \sum a_n q^n$ be a normalized Hecke eigenform in the space $S_k(\Gamma_1(N))$ of complex weight-k cusp forms on the subgroup $\Gamma_1(N)$ of $\text{SL}(2, \mathbf{Z})$. Thus f is nonzero and it satisfies $f|T_n = a_n f$ for all $n \geq 1$. Further, there is a character $\varepsilon: (\mathbf{Z}/N\mathbf{Z})^* \to \mathbf{C}^*$ so that $f|\langle d \rangle = \varepsilon(d) f$ for all $d \in (\mathbf{Z}/N\mathbf{Z})^*$, where $\langle d \rangle$ is the diamond-bracket operator. Again, let $\mathcal{O}$ be the ring of integers of the field $\mathbf{Q}(\ldots a_n \ldots)$ generated by the a_n; this field is a number field that is either totally real or a CM field. Consider a ring homomorphism $\varphi: \mathcal{O} \to \overline{\mathbf{F}}_\ell$ as before. Associated to this set-up is a representation $\rho: \text{Gal}(\overline{\mathbf{Q}}/\mathbf{Q}) \to \text{GL}(2, \overline{\mathbf{F}}_\ell)$ with properties that connect it up with f (and φ). First, the representation is unramified at all p not dividing ℓN. Next, for all such p, we have

$$\text{tr}(\rho(\text{Frob}_p)) = a_p, \qquad \det(\rho(\text{Frob}_p)) = p^{k-1}\varepsilon(p);$$

the numbers a_p and $p^{k-1}\varepsilon(p)$, literally in $\mathcal{O}$, are mapped tacitly into $\overline{\mathbf{F}}_\ell$ by φ. The representation ρ is determined up to isomorphism by the trace and determinant identities that are displayed, plus the supplemental requirement that it be semisimple. We are interested mainly in the (generic) case in which ρ is irreducible; in that case, it is of course semisimple.

The construction of ρ from f, k and φ was described in [**24**]. In this article, Deligne sketches a method that manufactures for each non-archimedean prime λ of E a representation $\tilde{\rho}_\lambda: \text{Gal}(\overline{\mathbf{Q}}/\mathbf{Q}) \to \text{GL}(2, E_\lambda)$, where E_λ denotes the completion of E at λ. Given φ, we let $\lambda = \ker \varphi$ and find a model of $\tilde{\rho}_\lambda$ over the ring of integers $\mathcal{O}_\lambda$ of E_λ. Reducing $\tilde{\rho}_\lambda$ modulo λ, we obtain a representation over the finite field $\mathcal{O}_\lambda/\lambda\mathcal{O}_\lambda$, and φ embeds this field into $\overline{\mathbf{F}}_\ell$.

In fact, as Shimura has pointed out, the machinery of [**24**] can be avoided if one seeks only the mod λ representation attached to f (as opposed to the full λ-adic representation $\tilde{\rho}_\lambda$). As the first author pointed out in [**87**], one can use congruences among modular forms to find a form of weight two and level $N\ell^2$ that gives rise to ρ. Accordingly, one can find ρ concretely by looking within the group of ℓ-division points of a suitable abelian variety over $\mathbf{Q}$: the variety $J_1(\ell^2 N)$, which is defined in Section 2.2.3 and in Conrad's Appendix.

Which representations of $\text{Gal}(\overline{\mathbf{Q}}/\mathbf{Q})$ arise in this way (as k, N, f and φ all vary)? As in the case $N = 1$ (i.e., that where $\Gamma_1(N) = \text{SL}(2, \mathbf{Z})$), any ρ that comes from modular forms is an odd representation: we have $\det(\rho(c)) = -1$ when $c \in \text{Gal}(\overline{\mathbf{Q}}/\mathbf{Q})$ is a complex conjugation. To see this, we begin with the fact that $\varepsilon(-1) = (-1)^k$, which generalizes (1.4); this follows from the functional equation that relates $f(\frac{az+b}{cz+d})$ to $f(z)$ when $\begin{pmatrix} a & b \\ c & d \end{pmatrix}$ is an element of $\Gamma_0(N)$ (see Exercise 7). On the other hand, using the Cebotarev density theorem, we find that $\det \rho = \chi^{k-1} \varepsilon$, where χ is again the mod ℓ cyclotomic character and where ε is regarded now as a map $\text{Gal}(\overline{\mathbf{Q}}/\mathbf{Q}) \to \overline{\mathbf{F}}_\ell^*$ in the obvious way, namely by composing $\varepsilon : (\mathbf{Z}/N\mathbf{Z})^* \to \overline{\mathbf{F}}_\ell^*$ with the mod N cyclotomic character. The value on c of the latter incarnation of ε is the number $\varepsilon(-1) = (-1)^k$. Since $\chi(c) = -1$, we deduce that $(\det \rho)(c) = -1$, as was claimed.

Serre's weak conjecture states that, conversely, every irreducible odd representation $\rho : \text{Gal}(\overline{\mathbf{Q}}/\mathbf{Q}) \to \text{GL}(2, \overline{\mathbf{F}}_\ell)$ is modular in the sense that it arises from some f and φ.

A concrete consequence of the conjecture is that all odd irreducible 2-dimensional Galois representations ρ come from abelian varieties over $\mathbf{Q}$. Given ρ, one should be able to find a totally real or CM number field E, an abelian variety A over $\mathbf{Q}$ of dimension $[E : \mathbf{Q}]$ that comes equipped with an action of the ring of integers $\mathcal{O}$ of E, and a ring homomorphism $\varphi : \mathcal{O} \to \overline{\mathbf{F}}_\ell$ with the following property: Let $\lambda = \ker \varphi$. *Then the representation $A[\lambda] \otimes_{\mathcal{O}/\lambda} \overline{\mathbf{F}}_\ell$ is isomorphic to ρ.* (In comparing $A[\lambda]$ and ρ, we use $\varphi : \mathcal{O}/\lambda \hookrightarrow \overline{\mathbf{F}}_\ell$ to promote the 2-dimensional $A[\lambda]$ into a representation over $\overline{\mathbf{F}}_\ell$.)

Much of the evidence for the weak conjecture concerns representations taking values in $\text{GL}(2, \mathbf{F}_q)$ where the finite field $\mathbf{F}_q$ has small cardinality. In his original article [**102**, §5], Serre's discusses a large number of examples of such representations. Serre uses theorems of Langlands [**68**] and Tunnell [**115**] to establish his weak conjecture for odd irreducible representations with values in $\text{GL}(2, \mathbf{F}_2)$ and $\text{GL}(2, \mathbf{F}_3)$. Further, he reports on numerical computations of J.-F. Mestre that pertain to representations over $\mathbf{F}_4$ (and trivial determinant). Additionally, Serre remarks [**102**, p. 219] that the weak conjecture is true for those representations with values in $\text{GL}(2, \overline{\mathbf{F}}_p)$ that are dihedral in the sense that their projective images (in $\text{PGL}(2, \overline{\mathbf{F}}_p)$) are dihedral groups. (See also [**29**, §5] for a related argument.) This section of Serre's paper concludes with examples over $\mathbf{F}_9$ and $\mathbf{F}_7$.

More recently, representations over the fields $\mathbf{F}_4$ and $\mathbf{F}_5$ were treated, under somewhat mild hypotheses, by Shepherd-Barron and Taylor [**105**]. For example, Shepherd-Barron and Taylor show that $\rho : \text{Gal}(\overline{\mathbf{Q}}/\mathbf{Q}) \to \text{GL}(2, \mathbf{F}_5)$ is isomorphic to the 5-torsion representation of an elliptic curve over $\mathbf{Q}$ provided that $\det \rho$ is the mod 5 cyclotomic character. Because elliptic curves over $\mathbf{Q}$ are modular, it follows that ρ is modular.

1.1.3. The strong conjecture

Fix an odd irreducible Galois representation

$$\rho : G_\mathbf{Q} \to \text{GL}(2, \overline{\mathbf{F}}_\ell).$$

As discussed above, the weak conjecture asserts that ρ is modular, in the sense that there exists integers N and k such that ρ comes from some $f \in S_k(\Gamma_1(N))$. The *strong conjecture* goes further and gives a recipe for integers $N(\rho)$ and $k(\rho)$, then asserts that ρ comes from $S_{k(\rho)}(\Gamma_1(N(\rho)))$. In any particular instance, the strong conjecture is, a priori, easier to verify or disprove than the weak conjecture because $S_{k(\rho)}(\Gamma_1(N(\rho)))$ is a finite-dimensional vector space that can be computed (using, e.g., the algorithm in [**73**]). The relation between the weak and strong conjectures is analogous to the relation between the assertion that an elliptic curve is modular of some level and the assertion that an elliptic curve A is modular of a specific level, the conductor of A.

For each prime p, let $I_p \subset G_\mathbf{Q}$ denote an inertia group at p. The optimal level is a product
$$N(\rho) = \prod_{p \neq \ell} p^{n(p)},$$
where $n(p)$ depends only on $\rho|_{I_p}$. The optimal weight $k(\rho)$ depends only on $\rho|_{I_\ell}$. The integer $n(p)$ is a conductor in additive notation. In particular, $n(p) = 0$ if and only if ρ is unramified at p.

View ρ as a homomorphism $G_\mathbf{Q} \to \mathrm{Aut}(V)$, where V is a two-dimensional vector space over $\overline{\mathbf{F}}_\ell$. It is natural to consider the subspace of inertia invariants:
$$V^{I_p} := \{v \in V : \rho(\sigma)v = v, \text{ all } \sigma \in I_p\}.$$
For example, $V^{I_p} = V$ if and only if ρ is unramified at p. Define
$$n(p) := \dim(V/V^{I_p}) + \mathrm{Swan}(V),$$
where the wild term $\mathrm{Swan}(V)$ is the Swan conductor
$$\mathrm{Swan}(V) := \sum_{i=1}^\infty \frac{1}{[G_0 : G_i]} \dim(V/V^{G_i}) \geq 0.$$
Here $G_0 = I_p$ and the $G_i \subset G_0$ are the higher ramification groups.

Suppose that ρ arises from a newform $f \in S_k(\Gamma_1(N))$. A theorem of Carayol [**12**], which was proved independently by Livné [**70**, Prop. 0.1], implies that $N(\rho) \mid N$. It is productive to study the quotient $N/N(\rho)$. Let $\mathcal{O}$ be the ring of integers of the field generated by the Fourier coefficients of f and let $\varphi : \mathcal{O} \to \overline{\mathbf{F}}_\ell$ be the map such that $\varphi(a_p) = \mathrm{tr}(\rho(\mathrm{Frob}_p))$. Let λ be a prime of $\mathcal{O}$ lying over ℓ and E_λ be the completion of $\mathrm{Frac}(\mathcal{O})$ at λ. Deligne [**24**] attached to the pair f, λ a representation
$$\rho_\lambda : G_\mathbf{Q} \to \mathrm{GL}(2, E_\lambda) = \mathrm{Aut}(\tilde{V})$$
where $\tilde{V}$ is a two-dimensional vector space over E_λ. The representation ρ_λ can be conjugated so that its images lies inside $\mathrm{GL}(2, \mathcal{O}_\lambda)$; the reduction of ρ_λ modulo λ is then ρ. The following diagram summarizes the set up:

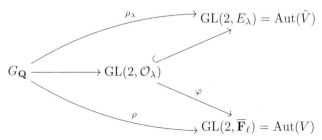

Let $m(p)$ be the power of p dividing the conductor of ρ_λ. In [**12**], Carayol proves that $m(p) = \mathrm{ord}_p N$. As above, $m(p) = \dim(\tilde{V}/\tilde{V}^{I_p}) + $ (wild term), and the wild term is the same as for ρ. The power of p dividing $N/N(\rho)$ is $\dim(\tilde{V}/\tilde{V}^{I_p}) - \dim(V/V^{I_p}) = \dim V^{I_p} - \dim \tilde{V}^{I_p}$. Though $\tilde{V}$ and V are vector spaces over different fields, we can compare the dimensions of their inertia invariant subspaces. The formula

$$\mathrm{ord}_p(N) = n(p) + (\dim V^{I_p} - \dim \tilde{V}^{I_p}) \tag{1.1}$$

indicates how this difference is the deviation of N from the optimal level locally at p. This is the description of $n(p)$ that is used in proving that if ρ is modular at all, then it is possible to refine N and k to eventually discover that ρ arises from a newform in $S_{k(\rho)}(\Gamma_1(N(\rho)))$. After much work (see [**26, 87**]) it has been shown that *for $\ell > 2$ the weak and strong conjectures are equivalent.* See [**9**] for equivalence in many cases when $\ell = 2$.

Rearranging (1.1) into

$$n(p) = \mathrm{ord}_p(N) - (\dim V^{I_p} - \dim \tilde{V}^{I_p})$$

provides us with a way to read off $N(\rho)$ from $\mathrm{ord}_p(N)$, $\dim V^{I_p}$, and $\dim \tilde{V}^{I_p}$. If $f \in S_k(\Gamma_1(N))$ gives rise to ρ and $\ell \nmid N$, then $k(\rho) \leq k$. In contrast, if we allow powers of ℓ in the level then the weight k can always be made equal to 2.

1.1.4. Representations arising from an elliptic curve

Equations for elliptic curves can be found in the Antwerp tables [**4**] and the tables of Cremona [**20**]. For example, consider the elliptic curve B given by the equation $y^2 + y = x^3 + x^2 - 12x + 2$. This is the curve labeled **141A1** in [**20**]; it has conductor $N = 3 \cdot 47$ and discriminant $\Delta = 3^7 \cdot 47$. There is a newform $f \in S_2(\Gamma_0(141))$ attached to B. Because N is square free, the elliptic curve B is *semistable*, in the sense that B has multiplicative reduction at each prime.

The curve B is isolated in its isogeny class; equivalently, for every ℓ the representation

$$\rho_\ell : G_\mathbf{Q} \to \mathrm{Aut}(B[\ell]) \approx \mathrm{GL}(2, \mathbf{F}_\ell)$$

is irreducible (see Exercise 4 and Exercise 5). We will frequently consider the representations ρ_ℓ attached to B. The following proposition shows that because B is semistable, each ρ_ℓ is surjective [**93**].

Proposition 1.1. *If A is a semistable elliptic curve over $\mathbf{Q}$ and ℓ is a prime such that ρ_ℓ is irreducible, then ρ_ℓ is surjective.*

Proof. Serre proved this when ℓ is odd; see [**93**, Prop. 21], [**103**, §3.1]. If ρ_2 isn't surjective, then by [**93**, Prop. 21(b)] and Theorem 2.10 it's unramified outside 2. This contradicts [**113**]. □

To give a flavor of Serre's invariants, we describe $N(\rho_\ell)$ and $k(\rho_\ell)$ for the representations ρ_ℓ attached to B. (Note that we still have not defined $k(\rho)$.) At primes p of bad reduction, after a possible unramified quadratic extension of $\mathbf{Q}_p$, the elliptic curve B is a Tate curve. This implies that for $p \neq \ell$, the representation ρ_ℓ is unramified at p if and only if $\mathrm{ord}_p(\Delta) \equiv 0 \pmod{\ell}$; for more details, see Section 2.2.4.

The optimal level $N(\rho_\ell)$ is a divisor of $3 \cdot 47$; it is divisible only by primes for which ρ_ℓ is ramified, and is not divisible by ℓ. The representation ρ_ℓ is unramified at 3 if and only if $\ell \mid \mathrm{ord}_3(\Delta) = 7$, i.e., when $\ell = 7$. Furthermore, ρ_ℓ is always ramified at 47. First suppose $\ell \notin \{3, 47\}$. If in addition $\ell \neq 7$ then $N(\rho_\ell) = 3 \cdot 47$, and $k(\rho_\ell) = 2$ since $\ell \nmid 3 \cdot 47$. If $\ell = 7$ then $N(\rho_\ell) = 47$, and again $k(\rho_\ell) = 2$. The remaining cases are $\ell = 3$ and $\ell = 47$. If $\ell = 47$ then $N(\rho_\ell) = 3$, and because $\ell - 1$ is the order of the cyclotomic character, $k(\rho_\ell) \equiv 2 \pmod{47-1}$; Serre's recipe then gives $k(\rho_\ell) = 2 + (47 - 1) = 48$. Similarly, when $\ell = 3$, we have $N(\rho_\ell) = 47$ and $k(\rho_\ell) = 2 + (3 - 1) = 4$. The following table summarizes the Serre invariants:

Table 1.1.4. The Serre invariants of ρ_ℓ

ℓ	$N(\rho_\ell)$	$k(\rho_\ell)$
3	47	4
7	47	2
47	3	48
all other ℓ	141	2

To verify the strong conjecture of Serre for $\ell = 3, 47$, we use a standard trade-off of level and weight, which relates eigenforms in $S_2(\Gamma_0(141); \mathbf{F}_\ell)$ to eigenforms in $S_{2+\ell-1}(\Gamma_0(141/\ell); \mathbf{F}_\ell)$ (see Section 3.3.1). The only exceptional prime is $\ell = 7$, for which the minimal weight $k(\rho)$ is 2. The strong conjecture of Serre predicts the existence of an eigenform $g \in S_2(\Gamma_0(47))$ that gives rise to ρ_ℓ. Our initial instinct is to look for an elliptic curve A of conductor 47 such that $A[\ell] \cong B[\ell]$, as $G_{\mathbf{Q}}$-modules. In fact, there are no elliptic curves of conductor 47. This is because $S_2(\Gamma_0(47))$ is four dimensional, having basis the Galois conjugates of a single eigenform $g = \sum c_n q^n$. The Fourier coefficients c_n of g generate the full ring of integers in the field K obtained from $\mathbf{Q}$ by adjoining a root of $h = x^4 - x^3 - 5x^2 + 5x - 1$. The discriminant $1957 = 19 \cdot 103$ of K equals the discriminant of h, so a root of h generates the full ring of integers. The eigenvalue c_2 satisfies $h(c_2) = 0$. Since $h \equiv (x+2)(x^3+4x^2+x+3) \pmod 7$, there is a prime λ lying over 7 such that $\mathcal{O}/\lambda \cong \mathbf{F}_7$; the isomorphism is given by $c_2 \mapsto -2 \bmod 7$. As a check, note that $\#B(\mathbf{F}_2) = 5$ so $a_2 = 3 - 5 = -2 = \varphi(c_2)$. More generally, for $p \nmid 7 \cdot 141$, we have $\varphi(c_p) \equiv a_p \bmod 7$. This equality of traces implies that the representation $\rho_{g,\lambda}$ is isomorphic to $\rho = \rho_{A,7}$, so ρ is modular of level 47.

1.1.5. Background material

In this section, we review the cyclotomic character, Frobenius elements, modular forms, and Tate curves. We frequently write $G_{\mathbf{Q}}$ for $\mathrm{Gal}(\overline{\mathbf{Q}}/\mathbf{Q})$. Many of these basic facts are also summarized in [23].

1.1.5.1. The cyclotomic character

The mod ℓ *cyclotomic character* is defined by considering the group $\boldsymbol{\mu}_\ell$ of ℓth roots of unity in $\overline{\mathbf{Q}}$; the action of the Galois group $G_{\mathbf{Q}}$ on the cyclic group $\boldsymbol{\mu}_\ell$ gives rise to a continuous homomorphism

(1.2) $$\chi_\ell : G_{\mathbf{Q}} \to \mathrm{Aut}(\boldsymbol{\mu}_\ell).$$

Since $\boldsymbol{\mu}_\ell$ is a cyclic group of order ℓ, its group of automorphisms is canonically the group $(\mathbf{Z}/\ell\mathbf{Z})^* = \mathbf{F}_\ell^*$. We emerge with a map $G_{\mathbf{Q}} \to \mathbf{F}_\ell^*$, which is the character in question.

Let A be an elliptic curve and ℓ be a prime number. The Weil pairing e_ℓ (see [**109**, III.8]) sets up an isomorphism of $G_{\mathbf{Q}}$-modules

$$(1.3) \qquad e_\ell : \bigwedge^2 A[\ell] \xrightarrow{\cong} \boldsymbol{\mu}_\ell.$$

The determinant of the representation $\rho_{A,\ell}$ is the mod ℓ cyclotomic character χ_ℓ.

Suppose now that $c \in G_{\mathbf{Q}}$ is the automorphism "complex conjugation." Then the determinant of $\rho_{A,\ell}(c)$ is $\chi_\ell(c)$. Now c operates on roots of unity by the map $\zeta \mapsto \zeta^{-1}$, since roots of unity have absolute value 1. Accordingly,

$$(1.4) \qquad \det \rho_{A,\ell}(c) = -1;$$

one says that $\rho_{A,\ell}$ is *odd*. If $\ell \neq 2$, then $\rho_{A,\ell}(c)$ is conjugate over $\overline{\mathbf{F}}_\ell$ to $\begin{pmatrix} 1 & 0 \\ 0 & -1 \end{pmatrix}$ (Exercise 7). If $\ell = 2$ then the characteristic polynomial of $\rho_{A,\ell}(c)$ is $(x+1)^2$ so $\rho_{A,\ell}(c)$ is conjugate over $\overline{\mathbf{F}}_\ell$ to either the identity matrix or $\begin{pmatrix} 1 & 1 \\ 0 & 1 \end{pmatrix}$.

1.1.5.2. Frobenius elements

Let K be a number field. The Galois group $\mathrm{Gal}(K/\mathbf{Q})$ leaves the ring $\mathcal{O}_K$ of integers of K invariant, so that one obtains an induced action on the ideals of $\mathcal{O}_K$. The set of prime ideals $\mathfrak{p}$ of $\mathcal{O}_K$ lying over p (i.e., that contain p) is permuted under this action. For each $\mathfrak{p}$, the subgroup $D_{\mathfrak{p}}$ of $\mathrm{Gal}(K/\mathbf{Q})$ fixing $\mathfrak{p}$ is called the *decomposition group* of $\mathfrak{p}$. Meanwhile, $\mathbf{F}_{\mathfrak{p}} := \mathcal{O}_K/\mathfrak{p}$ is a finite extension of $\mathbf{F}_p$. The extension $\mathbf{F}_{\mathfrak{p}}/\mathbf{F}_p$ is necessarily Galois; its Galois group is cyclic, generated by the Frobenius automorphism $\varphi_p : x \mapsto x^p$ of $\mathbf{F}_{\mathfrak{p}}$. There is a natural surjective map $D_{\mathfrak{p}} \to \mathrm{Gal}(\mathbf{F}_{\mathfrak{p}}/\mathbf{F}_p)$; its injectivity is equivalent to the assertion that p is unramified in $K/\mathbf{Q}$. Therefore, whenever this assertion is true, there is a unique $\sigma_{\mathfrak{p}} \in D_{\mathfrak{p}}$ whose image in $\mathrm{Gal}(\mathbf{F}_{\mathfrak{p}}/\mathbf{F}_p)$ is φ_p. The automorphism $\sigma_{\mathfrak{p}}$ is then a well-defined element of $\mathrm{Gal}(K/\mathbf{Q})$, the Frobenius automorphism for $\mathfrak{p}$. The various $\mathfrak{p}$ are all conjugate under $\mathrm{Gal}(K/\mathbf{Q})$ and the Frobenius automorphism for the conjugate of $\mathfrak{p}$ by g is $g\sigma_{\mathfrak{p}}g^{-1}$. In particular, the various $\sigma_{\mathfrak{p}}$ are all conjugate; this justifies the practice of writing σ_p for any one of them and stating that σ_p is well defined up to conjugation.

We next introduce the concept of Frobenius elements in $G_{\mathbf{Q}} = \mathrm{Gal}(\overline{\mathbf{Q}}/\mathbf{Q})$. Let p again be a prime and let $\mathfrak{p}$ now be a prime of the ring of integers of $\overline{\mathbf{Q}}$ lying over p. To $\mathfrak{p}$ we associate: (1) its residue field $\mathbf{F}_{\mathfrak{p}}$, which is an algebraic closure of $\mathbf{F}_p$, and (2) a decomposition subgroup $D_{\mathfrak{p}}$ of $G_{\mathbf{Q}}$. There is again a surjective map $D_{\mathfrak{p}} \to \mathrm{Gal}(\mathbf{F}_{\mathfrak{p}}/\mathbf{F}_p)$. The Frobenius automorphism φ_p topologically generates the target group. We shall use the symbol Frob_p to denote any preimage of φ_p in any $D_{\mathfrak{p}}$ corresponding to a prime lying over p, and refer to Frob_p as a Frobenius element for p in $G_{\mathbf{Q}}$. This element is doubly ill defined. The ambiguity in Frob_p results from the circumstance that $\mathfrak{p}$ needs to be chosen and from the fact that $D_{\mathfrak{p}} \to \mathrm{Gal}(\mathbf{F}_{\mathfrak{p}}/\mathbf{F}_p)$ has a large kernel, the inertia subgroup $I_{\mathfrak{p}}$ of $D_{\mathfrak{p}}$. The usefulness of $\mathrm{Frob}_{\mathfrak{p}}$ stems from the fact that the various $\mathfrak{p}$ are all conjugate, so that likewise the subgroups $D_{\mathfrak{p}}$ and $I_{\mathfrak{p}}$ are conjugate. Thus if ρ is a homomorphism mapping $G_{\mathbf{Q}}$ to some other group, the kernel of ρ contains one $I_{\mathfrak{p}}$ if and only if it contains all $I_{\mathfrak{p}}$. In case $I_{\mathfrak{p}} = 0$, one says that ρ is *unramified* at p; the image of Frob_p is then an element of the target that is well defined up to conjugation.

Consider an elliptic curve A over $\mathbf{Q}$ and let ℓ be a prime number. The fixed field of $\rho_{A,\ell}$ is a finite Galois extension $K_\ell/\mathbf{Q}$ whose Galois group G_ℓ is a subgroup of $\mathrm{GL}(2, \mathbf{F}_\ell)$. A key piece of information about the extension $K_\ell/\mathbf{Q}$ is that its discriminant is divisible at most by ℓ and primes dividing the conductor of A. In other words, if $p \neq \ell$ is a prime number at which A has good reduction, then $K_\ell/\mathbf{Q}$ is unramified at ℓ (see Exercise 15); one says that the representation $\rho_{A,\ell}$ is unramified at p. Whenever this occurs, the construction described above produces a Frobenius element σ_p in G_ℓ that is well defined up to conjugation.

Fix again an elliptic curve A and a prime number ℓ, and let $\rho_{A,\ell} : G_\mathbf{Q} \to \mathrm{GL}(2, \mathbf{F}_\ell)$ be the associated representation. For each prime p not dividing ℓ at which A has good reduction the Frobenius $\sigma_p = \rho_{A,\ell}(\mathrm{Frob}_p)$ is well defined only up to conjugation. Nevertheless, the trace and determinant of σ_p are well defined. The determinant of $\rho_{A,\ell}$ is the mod ℓ cyclotomic character χ, so $\sigma_p = \chi(\mathrm{Frob}_p) = p \in \mathbf{F}_\ell$. On the other hand, one has the striking congruence

$$\mathrm{tr}(\rho_{A,\ell}(\mathrm{Frob}_p)) = p + 1 - \#\tilde{A}(\mathbf{F}_p) \pmod{\ell}.$$

1.1.5.3. Modular forms

We now summarize some background material concerning modular forms. Serre's book [**96**] is an excellent introduction (it treats only $N = 1$). One might also read the survey article [**27**] or consult any of the standard references [**65, 66, 75, 108**].

The *modular group* $\mathrm{SL}(2, \mathbf{Z})$ is the group of 2×2 invertible integer matrices. For each positive integer N, consider the subgroup

$$\Gamma_1(N) := \left\{ \begin{pmatrix} a & b \\ c & d \end{pmatrix} \in \mathrm{SL}(2, \mathbf{Z}) : N \mid c \text{ and } a \equiv d \equiv 1 \pmod{N} \right\}.$$

Let $\mathfrak{h}$ be the complex upper half plane. A *cusp form* of integer weight $k \geq 1$ and level N is a holomorphic function $f(z)$ on $\mathfrak{h}$ such that

$$(1.5) \qquad f\left(\frac{az+b}{cz+d}\right) = (cz+d)^k f(z) \quad \text{for all } \begin{pmatrix} a & b \\ c & d \end{pmatrix} \in \Gamma_1(N);$$

we also require that $f(z)$ vanishes at the cusps (see [**108**, §2.1]). We denote by $S_k(\Gamma_1(N))$ the space of weight-k cusp forms of level N. It is a finite dimensional complex vector space. When $k \geq 2$ a formula for the dimension can be found in [**108**, §2.6].

Modular forms are usually presented as convergent Fourier series

$$f(z) = \sum_{n=1}^{\infty} a_n q^n$$

where $q := e^{2\pi i z}$. This is possible because the matrices $\begin{pmatrix} 1 & b \\ 0 & 1 \end{pmatrix}$ lie in $\Gamma_1(N)$ so that $f(z+b) = f(z)$ for all integers b. For the forms that most interest us, the complex numbers a_n are algebraic integers.

The space $S_k(\Gamma_1(N))$ is equipped with an action of $(\mathbf{Z}/N\mathbf{Z})^*$; this action is given by

$$f(z) \mapsto f|\langle \overline{d} \rangle(z) := (cz+d)^{-k} f\left(\frac{az+b}{cz+d}\right)$$

where $\begin{pmatrix} a & b \\ c & d \end{pmatrix} \in \mathrm{SL}(2, \mathbf{Z})$ is any matrix such that $d \equiv \overline{d} \pmod{N}$. The operator $\langle d \rangle = \langle \overline{d} \rangle$ is referred to as a "diamond-bracket" operator.

For each integer $n \geq 1$, the nth *Hecke operator* on $S_k(\Gamma_1(N))$ is an endomorphism T_n of $S_k(\Gamma_1(N))$. The action is generally written on the right: $f \mapsto f|T_n$.

The various T_n commute with each other and are interrelated by identities that express a given T_n in terms of the Hecke operators indexed by the prime factors of n. If $p \nmid N$ is a prime define the operator T_p on $S_k(\Gamma_1(N))$ by

$$f|T_p(z) = \sum_{n=1}^{\infty} a_{np} q^n + p^{k-1} \sum_{n=1}^{\infty} a_n(f|\langle p \rangle) q^{np}.$$

For $p \mid N$ prime, define T_p by

$$f|T_p(z) = \sum_{n=1}^{\infty} a_{np} q^n.$$

The *Hecke algebra* associated to cusp forms of weight k on $\Gamma_1(N)$ is the subring

$$\mathbf{T} := \mathbf{Z}[\ldots T_n \ldots \langle d \rangle \ldots] \subset \operatorname{End}(S_k(\Gamma_1(N)))$$

generated by all of the T_n and $\langle d \rangle$. It is finite as a module over $\mathbf{Z}$ (see Exercise 20). The diamond-bracket operators are really Hecke operators, in the sense that they lie in the ring generated by the T_n; thus $\mathbf{T} = \mathbf{Z}[\ldots T_n \ldots]$.

An *eigenform* is a nonzero element $f \in S_k(\Gamma_1(N))$ that is a simultaneous eigenvector for every element of the Hecke algebra $\mathbf{T}$. Writing $f = \sum a_n q^n$ we find that $a_n = a_1 c_n$ where c_n is the eigenvalue of T_n on f. Since f is nonzero, a_1 is also nonzero, so it is possible to multiply f by a_1^{-1}. The resulting *normalized eigenform* wears its eigenvalues on its sleeve: $f = \sum c_n q^n$. Because f is an eigenform, the action of the diamond bracket operators defines a character $\varepsilon : (\mathbf{Z}/N\mathbf{Z})^* \to \mathbf{C}^*$; we call ε the *character* of f.

Associated to an eigenform $f \in S_k(\Gamma_1(N))$ we have a system $(\ldots a_p \ldots)$, $p \nmid N$, of eigenvalues. We say that f is a *newform* if this system of eigenvalues is not the system of eigenvalues associated to an eigenform $g \in S_k(\Gamma_1(M))$ for some level $M \mid N$ with $M \neq N$. Newforms have been extensively studied (see [**2, 13, 69, 75**]); the idea is to understand where systems of eigenvalues first arise, and then reconstruct the full space $S_k(\Gamma_1(N))$ from newforms of various levels.

1.1.5.4. Tate curves

The Tate curve is a p-adic analogue of the exponentiation of the representation $\mathbf{C}/\Lambda$ of the group of an elliptic curve over $\mathbf{C}$. In this section we recall a few facts about Tate curves; for further details, see [**110**, V.3].

Let K be a finite extension of $\mathbf{Q}_p$; consider an elliptic curve E/K with *split multiplicative* reduction, and let j denote the j-invariant of E. By formally inverting the well-known relation

$$j(q(z)) = \frac{1}{q(z)} + 744 + 196884 q(z) + \cdots$$

between the complex functions $q(z) = e^{2\pi i z}$ and $j(z)$, we find an element $q \in K^*$ with $j = j(q)$ and $|q| < 1$. There is a $\operatorname{Gal}(\overline{\mathbf{Q}}_p/K)$-equivariant isomorphism $E(\overline{\mathbf{Q}}_p) \cong \overline{\mathbf{Q}}_p^*/q^{\mathbf{Z}}$. The Tate curve, which we suggestively denote by $\mathbf{G}_m/q^{\mathbf{Z}}$, is a scheme whose $\overline{\mathbf{Q}}_p$ points equal $\overline{\mathbf{Q}}_p^*/q^{\mathbf{Z}}$.

As a consequence, the group of n-torsion points on the Tate curve is identified with the $\operatorname{Gal}(\overline{\mathbf{Q}}_p/K)$-module $\{\zeta_n^a (q^{1/n})^b : 0 \leq a, b \leq n-1\}$; here ζ_n is a primitive nth root of unity and $q^{1/n}$ is a fixed nth root of q in $\overline{\mathbf{Q}}_p$. In particular, the subgroup generated by ζ_n is invariant under $\operatorname{Gal}(\overline{\mathbf{Q}}_p/K)$, so the local Galois representation on $E[n]$ is reducible. It is also known that the group of connected components of the reduction of the Néron model of E over $\overline{\mathbf{F}}_p$ is a cyclic group whose order is $\mathrm{ord}_p(q)$. The situation is summarized by the following table (taken from [88]):

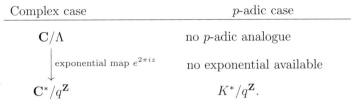

Complex case	p-adic case
$\mathbf{C}/\Lambda$	no p-adic analogue
$\downarrow$ exponential map $e^{2\pi i z}$	no exponential available
$\mathbf{C}^*/q^{\mathbf{Z}}$	$K^*/q^{\mathbf{Z}}$.

Remark 1.2. When E has non-split multiplicative reduction over K, there is an unramified extension L over which E aquires split multiplicative reduction.

1.1.5.5. Mod ℓ modular forms

There are several excellent papers to consult when learning about mod ℓ modular forms. The papers of Serre [95] and Swinnerton-Dyer [112] approach the subject from the point of view of Galois representations. Katz's paper [59] is very geometric. Edixhoven's paper [32] contains a clear description of the basic facts. See also Jochnowitz's paper [56].

CHAPTER 2
Optimizing the weight

In [**102**, §2] Serre associated to an odd irreducible Galois representation
$$\rho : G_{\mathbf{Q}} \to \mathrm{GL}(2, \overline{\mathbf{F}}_\ell)$$
two integers $N(\rho)$ and $k(\rho)$, which are meant to be the minimal level and weight of a form giving rise to ρ.

Conjecture 2.1 (Strong conjecture of Serre). Let $\rho : G_{\mathbf{Q}} \to \mathrm{GL}(2, \overline{\mathbf{F}}_\ell)$ be an odd irreducible Galois representation arising from a modular form. Then ρ arises from a modular form of level $N(\rho)$ and weight $k(\rho)$.

In this chapter, we are concerned with $k(\rho)$. We consider a mod ℓ representation ρ that arises from an eigenform of level N not divisible by ℓ. Using results of Fontaine and Deligne, we motivate Serre's recipe for $k(\rho)$. In [**32**], Edixhoven also defines an "optimal" weight, which sometimes differs from Serre's $k(\rho)$. Our definition is an "average" of the two; for example, we introduce a tiny modification of $k(\rho)$ when $\ell = 2$. We appologize for any confusion this may cause the reader.

Using various arguments involving the Eichler-Shimura correspondence and Tate's θ-cycles, Edixhoven showed in [**31**] that there must exist another form of weight at most $k(\rho)$, also of level N, which gives rise to ρ. Some of Edixhoven's result rely on unchecked compatibilities that are assumed in [**46**]; however, when $\ell \neq 2$ these results were obtained unconditionally by Coleman and Voloch in [**17**]. We sketch some of Edixhoven's arguments to convey the flavor of the subject.

Remark 2.2 (Notation). We pause to describe a notational shorthand which we will employ extensively in this chapter. If $\rho : G \to \mathrm{Aut}(V)$ is a two-dimensional representation over a field $\mathbf{F}$, we will frequently write
$$\rho \sim \begin{pmatrix} \alpha & \beta \\ \gamma & \delta \end{pmatrix}$$
to mean that there is a basis for V with respect to which
$$\rho(x) = \begin{pmatrix} \alpha(x) & \beta(x) \\ \gamma(x) & \delta(x) \end{pmatrix} \in \mathrm{GL}_2(\mathbf{F})$$
for all $x \in G$. If we do not wish to specify one of the entries we will simply write $*$. Thus "$\rho \sim \begin{pmatrix} * & * \\ 0 & 1 \end{pmatrix}$" means that ρ possesses a one-dimensional invariant subspace, and the action on the quotient is trivial.

2.2.1. Representations arising from forms of low weight

We first consider irreducible Galois representations associated to newforms of low weight. Fix a prime ℓ and suppose $f = \sum a_n q^n$ is a newform of weight k and level N, such that $\ell \nmid N$ and $2 \leq k \leq \ell + 1$. Let $\varepsilon : (\mathbf{Z}/N\mathbf{Z})^* \to \mathbf{C}^*$ denote the character of f. Fix a homomorphism φ from the ring of integers $\mathcal{O}$ of $\mathbf{Q}(\ldots a_n \ldots)$ to $\overline{\mathbf{F}}_\ell$. To abbreviate, we often write a_n for $\varphi(a_n)$, thereby thinking of a_n as an element of $\overline{\mathbf{F}}_\ell$. Let $\rho = \rho_{f,\varphi} : G_\mathbf{Q} \to \mathrm{GL}(2, \overline{\mathbf{F}}_\ell)$ be the two-dimensional semisimple odd Galois representation attached to f and φ, and assume that ρ is irreducible.

The recipe for $N(\rho)$ depends on the local behavior of ρ at primes p other than ℓ; the recipe for $k(\rho)$ depends on the restriction $\rho|_{I_\ell}$ of ρ to the inertia group at ℓ. Motivated by questions of Serre, Fontaine and Deligne described $\rho|_{I_\ell}$ in many situations. We distinguish two cases: the ordinary case and the non-ordinary case, which we call the *"supersingular case."*

2.2.1.1. The ordinary case

Deligne (see [**46**, Prop. 12.1]) considered the *ordinary case*, in which ρ arises from a weight-k newform f with $a_\ell(f) \neq 0 \in \overline{\mathbf{F}}_\ell$. He showed that ρ has a one-dimensional unramified quotient β, so $\rho|_{D_\ell} \sim \begin{pmatrix} \alpha & * \\ 0 & \beta \end{pmatrix}$ with $\beta(I_\ell) = 1$ and $\alpha\beta = \chi^{k-1}\varepsilon$. The mod N character ε is also unramified at ℓ because $\ell \nmid N$. Since the mod ℓ cyclotomic character χ has order $\ell - 1$ and $\rho|_{I_\ell} \sim \begin{pmatrix} \chi^{k-1} & * \\ 0 & 1 \end{pmatrix}$, the value of k modulo $\ell - 1$ is determined by $\rho|_{I_\ell}$. In the case when k is not congruent to 2 modulo $\ell - 1$, the restriction $\rho|_{I_\ell}$ determines the minimal weight $k(\rho)$. We will discuss the remaining case in Section 2.2.2.

2.2.1.2. The supersingular case and fundamental characters

Fontaine (see [**32**, §6]) investigated the supersingular case, in which ρ arises from a newform f with $a_\ell(f) = 0 \in \overline{\mathbf{F}}_\ell$. We call such a newform f *supersingular*. To describe the restriction $\rho|_{I_\ell}$ of ρ to the inertia group at ℓ, we introduce the fundamental characters of the tame inertia group. Fix an algebraic closure $\overline{\mathbf{Q}}_\ell$ of the field $\mathbf{Q}_\ell$ of ℓ-adic numbers; let $\mathbf{Q}_\ell^{\mathrm{nr}} \subset \overline{\mathbf{Q}}_\ell$ denote the maximal unramified extension of $\mathbf{Q}_\ell$, and $\mathbf{Q}_\ell^{\mathrm{tm}} \subset \overline{\mathbf{Q}}_\ell$ the maximal tamely ramified extension of $\mathbf{Q}_\ell^{\mathrm{nr}}$. The extension $\mathbf{Q}_\ell^{\mathrm{tm}}$ is the compositum of all finite extensions of $\mathbf{Q}_\ell^{\mathrm{nr}}$ in $\overline{\mathbf{Q}}_\ell$ of degree prime to ℓ. Letting D_ℓ denote the decomposition group, I_ℓ the inertia group, I_t the tame inertia group, and I_w the wild inertia group, we have the following diagram:

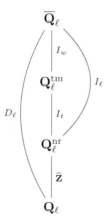

It is a standard fact (see, e.g., [**44**, §8]) that the extensions $\mathbf{Q}_\ell^{\mathrm{nr}}(\sqrt[n]{\ell})$, for all n not divisible by ℓ, generate $\mathbf{Q}_\ell^{\mathrm{tm}}$. For n not divisible by ℓ, the nth roots of unity $\boldsymbol{\mu}_n$ are contained in $\mathbf{Q}_\ell^{\mathrm{nr}}$. Kummer theory (see [**3**]) gives, for each n, a canonical isomorphism

$$\mathrm{Gal}(\mathbf{Q}_\ell^{\mathrm{nr}}(\sqrt[n]{\ell})/\mathbf{Q}_\ell^{\mathrm{nr}}) \xrightarrow{\sim} \boldsymbol{\mu}_n, \qquad \sigma \mapsto \frac{\sigma(\sqrt[n]{\ell})}{\sqrt[n]{\ell}}.$$

Each isomorphism lifts to a map $I_\ell \to \boldsymbol{\mu}_n$ that factors through the tame quotient I_t of I_ℓ. The groups $\boldsymbol{\mu}_n = \boldsymbol{\mu}_n(\overline{\mathbf{Q}}_\ell)$ lie in the ring of integers $\overline{\mathbf{Z}}_\ell$ of $\overline{\mathbf{Q}}_\ell$. Composing any of the maps $I_t \to \boldsymbol{\mu}_n$ with reduction modulo the maximal ideal of $\overline{\mathbf{Z}}_\ell$ gives a mod ℓ character $I_t \to \overline{\mathbf{F}}_\ell^*$, as illustrated:

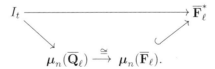

Let $n = \ell^\nu - 1$ with $\nu > 0$. The map $I_t \to \boldsymbol{\mu}_n$ defines a character $\varepsilon : I_\ell \to \mathbf{F}_{\ell^\nu}^*$. Composing with each of the ν field embeddings $\mathbf{F}_{\ell^\nu} \to \overline{\mathbf{F}}_\ell$ gives the ν *fundamental characters* of level ν:

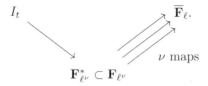

For example, the unique fundamental character of level 1 is the mod ℓ cyclotomic character (see Exercise 16). When $\nu = 2$, there are two fundamental characters, denoted Ψ and Ψ'; these satisfy $\Psi^\ell = \Psi'$ and $(\Psi')^\ell = \Psi$.

Let A be an elliptic curve over $\mathbf{Q}_\ell$ with good supersingular reduction. In [**93**], Serre proved that the representation

$$I_t \to \mathrm{Aut}(A[\ell]) \subset \mathrm{GL}(2, \overline{\mathbf{F}}_\ell)$$

is the direct sum of the two fundamental characters Ψ and Ψ'. One of the characters is

$$I_t \to \mathbf{F}_{\ell^2}^* \subset \mathrm{GL}(2, \mathbf{F}_\ell)$$

where $\mathbf{F}_{\ell^2}^*$ is contained in $\mathrm{GL}(2, \mathbf{F}_\ell)$ as a non-split Cartan subgroup of $\mathrm{GL}(2, \mathbf{F}_\ell)$. More precisely, $\mathbf{F}_{\ell^2}^*$ is embedded in $\mathrm{GL}(2, \mathbf{F}_\ell)$ via the action of the multiplicative group of a field on itself after a choice of basis. More generally, in unpublished joint work, Fontaine and Serre proved in 1979 that if f is a supersingular eigenform of weight $k \leq \ell$, then $\rho|_{I_\ell} : I_\ell \to \mathrm{GL}(2, \overline{\mathbf{F}}_\ell)$ factors through I_t and is a direct sum of the two character Ψ^{k-1} and $(\Psi')^{k-1}$. Note that k is determined by this representation, because it is determined modulo $\ell^2 - 1$.

2.2.2. Representations of high weight

Let D_ℓ be a decomposition group at ℓ and consider a representation $\rho : D_\ell \to \mathrm{GL}(2, \overline{\mathbf{F}}_\ell)$ that arises from a newform f of possibly large weight k. Let ρ^{ss} denote the *semisimplification* of ρ; so $\rho^{\mathrm{ss}} = \rho$ if ρ is irreducible, otherwise ρ^{ss} is a direct sum of two characters α and β. The following lemma of Serre (see [**93**, Prop. 4]) asserts that ρ^{ss} is tamely ramified.

Lemma 2.3. *Any semisimple representation ρ is tame, in the sense that $\rho(I_w) = 0$.*

Proof. Since the direct sum of tame representations is tame, we may assume that ρ is simple.

The wild inertia group I_w is the profinite Sylow ℓ-subgroup of I_ℓ: it is a Sylow ℓ-subgroup because each finite Galois extension of $\mathbf{Q}_\ell^{\mathrm{tm}}$ has degree a power of ℓ, and the order of I_t is prime to ℓ; it is unique, because it is the kernel of $\mathrm{Gal}(\overline{\mathbf{Q}}_\ell/\mathbf{Q}_\ell) \to \mathrm{Gal}(\overline{\mathbf{Q}}_\ell/\mathbf{Q}_\ell^{\mathrm{tm}})$, hence normal.

Because ρ is continuous, the image of D_ℓ is finite and we view ρ as a representation on a vector space W over a finite extension of $\mathbf{F}_\ell$. The subspace

$$W^{I_w} = \{w \in W : \sigma(\tau)w = w \text{ for all } \tau \in I_w\}$$

is invariant under D_ℓ. It is nonzero, as can be seen by writing the finite set W as a disjoint union of its orbits under I_w: since I_w is a pro-ℓ-group, each orbit has size either 1 or a positive power of ℓ. The orbit $\{0\}$ has size 1, and $\#W$ is a power of ℓ, so there must be at least $\ell - 1$ other singleton orbits $\{w\}$; for each of these, $w \in W^{I_w}$.

Since ρ is simple and W^{I_w} is a nonzero D_ℓ-submodule, it follows that $W^{I_w} = W$, as claimed. $\square$

The restriction $\rho^{\mathrm{ss}}|_{I_\ell}$ is abelian and semisimple, so it is given by a pair of characters $\alpha, \beta : I_\ell \to \overline{\mathbf{F}}_\ell^*$. Let n be an integer not divisible by ℓ, and consider the

tower of fields

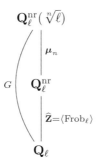

in which $G = \mathrm{Gal}(\mathbf{Q}_\ell^{\mathrm{nr}}(\sqrt[n]{\ell})/\mathbf{Q}_\ell)$, $\boldsymbol{\mu}_n \cong \mathrm{Gal}(\mathbf{Q}_\ell^{\mathrm{nr}}(\sqrt[n]{\ell})/\mathbf{Q}_\ell^{\mathrm{nr}})$, and $\mathrm{Gal}(\mathbf{Q}_\ell^{\mathrm{nr}}/\mathbf{Q}_\ell)$ is topologically generated by a Frobenius element at ℓ. Choose a lift $g \in G$ of Frob_ℓ, and consider an element $h \in \boldsymbol{\mu}_n$ corresponding to an element $\sigma \in \mathrm{Gal}(\mathbf{Q}_\ell^{\mathrm{nr}}(\sqrt[n]{\ell})/\mathbf{Q}_\ell^{\mathrm{nr}})$. Then since g acts as the ℓth powering map on roots of unity,

$$\frac{g\sigma g^{-1}(\sqrt[n]{\ell})}{\sqrt[n]{\ell}} = \frac{g\sigma(\zeta_{g^{-1}}\sqrt[n]{\ell})}{\sqrt[n]{\ell}} = \frac{g(\zeta_{g^{-1}}h\sqrt[n]{\ell})}{\sqrt[n]{\ell}} = \frac{g(h)\sqrt[n]{\ell}}{\sqrt[n]{\ell}} = h^\ell.$$

Applying the conjugation formula $ghg^{-1} = h^\ell$ to ρ^{ss} gives $\rho^{\mathrm{ss}}(ghg^{-1}) = \rho^{\mathrm{ss}}(h^\ell) = \rho^{\mathrm{ss}}(h)^\ell$. The two representations $h \mapsto \rho^{\mathrm{ss}}(h)^\ell$ and $h \mapsto \rho^{\mathrm{ss}}(h)$ of I_t are thus equivalent via conjugation by $\rho^{\mathrm{ss}}(g)$; we have $\rho^{\mathrm{ss}}(g)\rho^{\mathrm{ss}}(h)\rho^{\mathrm{ss}}(g^{-1}) = \rho^{\mathrm{ss}}(ghg^{-1}) = \rho^{\mathrm{ss}}(h)^\ell$. Consequently, the pair of characters $\{\alpha, \beta\}$ is stable under the ℓth power map, so as a set $\{\alpha, \beta\} = \{\alpha^\ell, \beta^\ell\}$. There are two possibilities:

— The *ordinary case*: $\alpha^\ell = \alpha$ and $\beta^\ell = \beta$.
— The *supersingular case*: $\alpha^\ell = \beta \neq \alpha$ and $\beta^\ell = \alpha \neq \beta$.

In the first case α and β take values in $\mathbf{F}_\ell^*$ and in the second case they take values in $\mathbf{F}_{\ell^2}^*$ but not in $\mathbf{F}_\ell^*$. By the results discussed in Section 2.2.1, this terminology is consistent with the terminology introduced above.

We first consider the supersingular case. Let Ψ denote one of the fundamental characters of level 2, and write $\alpha = \Psi^n$, $\beta = \Psi^{n\ell}$, with n an integer modulo $\ell^2 - 1$. Next write the smallest non-negative representative for n in base ℓ: $n = a + \ell b$ with $0 \leq a, b \leq \ell - 1$. Then $\ell n \equiv b + \ell a \pmod{\ell^2 - 1}$. Switching α and β permutes a and b so, relabeling if necessary, we may assume that $a \leq b$. If $a = b$, then $\alpha = \Psi^a(\Psi')^a = \chi^a$, so α takes values in $\mathbf{F}_\ell^*$, which is not the supersingular case; thus we may assume that $0 \leq a < b \leq \ell - 1$. We now factor out by a power of the cyclotomic character:

$$\alpha = \Psi^n = \Psi^a(\Psi')^b = \Psi^a(\Psi')^a(\Psi')^{b-a} = \chi^a(\Psi')^{b-a}$$
$$\beta = \chi^a \Psi^{b-a}.$$

Put another way,

$$\rho^{\mathrm{ss}} \sim \chi^a \otimes \begin{pmatrix} \Psi^{b-a} & 0 \\ 0 & (\Psi')^{b-a} \end{pmatrix}.$$

The untwisted representation is $\begin{pmatrix} \Psi^{k-1} & 0 \\ 0 & (\Psi')^{k-1} \end{pmatrix}$, where $k = 1 + b - a$. Since $2 \leq 1 + b - a \leq \ell - 1$, the weight of the untwisted representation is in the range considered above. Thus we are in good shape to define $k(\rho)$.

Before giving $k(\rho)$ it is necessary to understand how the weight changes upon twisting by a power of the cyclotomic character χ. This problem is addressed by the theory of mod ℓ modular forms, first developed by Serre [**95**] and Swinnerton-Dyer [**112**], then generalized by Katz [**59**]. A brief review of the geometric theory, which gives an excellent definition of mod ℓ modular forms, can be found in [**32**, §2], [**35**, §1], or [**46**, §2].

In [**61**], Katz defined spaces of mod ℓ modular forms, and a q-expansion map

$$\alpha : \bigoplus_{k \geq 0} M_k(\Gamma_1(N); \mathbf{F}_\ell) \to \mathbf{F}_\ell[[q]].$$

This map is not injective, because both the Hasse invariant of weight $\ell - 1$ and the constant 1 have the same q-expansion.

Definition 2.4. The *minimal weight filtration* $w(f) \in \mathbf{Z}$ of an element f of the *ring* of mod ℓ modular forms is the smallest integer k such that the q-expansion of f comes from a modular form of weight k; if no such k exists, do not define $w(f)$.

Definition 2.5. Define the operator $\theta = q\frac{d}{dq}$ on q-expansions by $\theta(\sum a_n q^n) = \sum n a_n q^n$.

For example, if f is an eigenform of weight k, then there is a mod ℓ eigenform θf of weight $k + \ell + 1$, still of level N, whose q-expansion is $\theta(\sum a_n q^n)$.

Theorem 2.6. *Let f be a mod ℓ modular form. Then $w(\theta f) = w(f) + \ell + 1$ if and only if $\ell \nmid w(f)$. In addition, if $\ell \mid w(f)$ then $w(\theta f) < w(f) + \ell + 1$.*

2.2.2.1. The supersingular case

We now give Serre's recipe for $k(\rho)$ in the supersingular case. The minimal weight before twisting is $1 + b - a$, which is a positive integer that is not divisible by ℓ. Each twist by χ adds $\ell + 1$ to the weight, so in the supersingular case we are motivated to define

$$k(\rho) := (1 + b - a) + a(\ell + 1) = 1 + \ell a + b.$$

We have to check that *at each step* the weight is prime to ℓ, so the minimal weight does not drop during any of the a twists by χ. Since $1 < 1 + b - a < \ell$ and

$$(1 + b - a) + a(\ell + 1) \leq (\ell - 1) + (\ell - 2)(\ell + 1) < \ell^2,$$

the weight can only drop if there exists c with $1 \leq c < a$ such that

$$(1 + b - a) + c(\ell + 1) \equiv 0 \pmod{\ell}.$$

If this occurred, then $c \equiv a - b - 1 \pmod{\ell}$. But $1 \leq c < a \leq \ell - 2$, so either $c = a - b - 1$, which implies $c \leq 0$ since $a < b$, or $c = \ell + a - b - 1 = a + \ell - 1 - b \geq a$, which would be a contradiction.

Assume that $\rho : G_\mathbf{Q} \to \mathrm{GL}(2, \overline{\mathbf{F}}_\ell)$ arises from an eigenform f such that $a_\ell(f) = 0 \in \overline{\mathbf{F}}_\ell$. Now we sketch Edixhoven's proof that ρ arises from a mod ℓ eigenform of weight $k(\rho)$.

Let ρ^{ss} denote the semisimplification of the restriction of ρ to a decomposition group at ℓ. The restriction of ρ^{ss} to the inertia group at ℓ is

$$\rho^{\mathrm{ss}}|_{I_\ell} \sim \begin{pmatrix} \Psi^n & 0 \\ 0 & (\Psi')^n \end{pmatrix},$$

where Ψ and $\Psi' = \Psi^\ell$ are the two fundamental characters of level 2. If necessary, reorder Ψ and Ψ' so that $n = a + b\ell$ with $0 \leq a < b \leq \ell - 1$. Then

$$\Psi^n = \Psi^{a+b\ell} = \Psi^a (\Psi')^b = \Psi^a (\Psi')^a (\Psi')^{b-a} = \chi^a (\Psi')^{b-a},$$

and

$$\rho^{ss}|_{I_\ell} \sim \chi^a \otimes \begin{pmatrix} (\Psi')^{b-a} & 0 \\ 0 & \Psi^{b-a} \end{pmatrix}.$$

Recall that, motivated by Fontaine's theorem on Galois representations arising from supersingular eigenforms, we defined

$$k(\rho) = a(\ell+1) + (b-a+1) = 1 + \ell a + b.$$

The first step in showing that ρ arises from a form of weight $k(\rho)$ is to recall the well known result that, up to twist, all systems of mod ℓ eigenvalues occur in weight at most $\ell + 1$. This is the subject of the next section.

2.2.2.2. Systems of mod ℓ eigenvalues

Theorem 2.7. *Suppose ρ is modular of level N and some weight k, and that $\ell \nmid N$. Then some twist $\rho \otimes \chi^{-\alpha}$ is modular of weight $\leq \ell + 1$ and level N.*

This is a general theorem, applying to both the ordinary and supersingular cases. See Serre [**97**, Th. 3] when $N = 1$; significant further work was carried out by Jochnowitz [**55**] and Ash-Stevens [**1**, Thm. 3.5] when $\ell \geq 5$. Two proofs are given in [**32**, Thm. 3.4 and §7]. The original method of Serre, Tate, and Koike for treating questions like this is to use the Eichler-Selberg trace formula. As Serre has pointed out to us, the weight appears in that formula simply as an exponent; this makes more or less clear that a congruence modulo $\ell^2 - 1$ gives information on modular forms mod ℓ.

As a digression, we pause to single out some of the tools involved in one possible proof of Theorem 2.7. Note that by twisting we may assume without loss of generality that $k \geq 2$. The group $\Gamma_1(N)$ acts by matrix multiplication on the real vector space $\mathbf{R}^2$. The Eichler-Shimura correspondence (see [**108**, §8.2]) is an isomorphism of real vector spaces

$$S_k(\Gamma_1(N)) \xrightarrow{\cong} H^1_P(\Gamma_1(N), \operatorname{Sym}^{k-2}(\mathbf{R}^2)).$$

The *parabolic* (or cuspidal) cohomology group H^1_P is the intersection, over all cusps $\alpha \in \mathbf{P}^1(\mathbf{Q})$, of the kernels of the restriction maps

$$\operatorname{res}_\alpha : H^1(\Gamma_1(N), \operatorname{Sym}^{k-2}(\mathbf{R}^2)) \to H^1(\Gamma_\alpha, \operatorname{Sym}^{k-2}(\mathbf{R}^2)),$$

where Γ_α denotes the stabilizer of α. For fixed z_0 in the upper half plane, the Eichler-Shimura isomorphism sends a cusp form f to the class of the cocycle $c : \Gamma_1(N) \to \operatorname{Sym}^{k-2}(\mathbf{R}^2)$ induced by

$$\gamma \mapsto \int_{z_0}^{\gamma(z_0)} \operatorname{Re}\left(f(z) \begin{pmatrix} z \\ 1 \end{pmatrix}^{k-2} dz \right),$$

where $\begin{pmatrix} z \\ 1 \end{pmatrix}^{k-2}$ denotes the image of $\begin{pmatrix} z \\ 1 \end{pmatrix} \otimes \cdots \otimes \begin{pmatrix} z \\ 1 \end{pmatrix} \in \mathrm{Sym}^{k-2}(\mathbf{C}^2)$, and integration is coordinate wise. There is an action of the Hecke algebra $\mathbf{T}$ on
$$H^1_P(\Gamma_1(N), \mathrm{Sym}^{k-2}(\mathbf{R}^2)),$$
such that the Eichler-Shimura correspondence is an isomorphism of $\mathbf{T}$-modules.

The forms whose periods are integral form a lattice $H^1_P(\Gamma_1(N), \mathrm{Sym}^{k-2}(\mathbf{Z}^2))$ inside $H^1_P(\Gamma_1(N), \mathrm{Sym}^{k-2}(\mathbf{R}^2))$. Reducing this lattice modulo ℓ suggests that there is a relationship between mod ℓ modular forms and the cohomology group
$$H^1_P(\tilde{\Gamma}_1(N), \mathrm{Sym}^{k-2}(\mathbf{F}_\ell^2)),$$
where $\tilde{\Gamma}_1(N)$ is the image of $\Gamma_1(N)$ in $\mathrm{SL}(2, \mathbf{F}_\ell)$. Serre and Hida observed that for $k-2 \geq \ell$ the $\tilde{\Gamma}_1(N)$ representations $\mathrm{Sym}^{k-2}(\mathbf{F}_\ell^2)$ are sums of representations arising in $\mathrm{Sym}^{k'-2}(\mathbf{F}_\ell^2)$ for $k' < k$. This essential idea is used in proving that all systems of eigenvalues occur in weight at most $\ell + 1$.

2.2.2.3. The supersingular case revisited

Let ρ be a supersingular mod ℓ representation that arises from some modular form. By Theorem 2.7 there is a form f of weight $k \leq \ell + 1$ such that $\chi^{-\alpha} \otimes \rho \sim \rho_f$. In fact, we may assume that $2 \leq k \leq \ell$; when $k = \ell + 1$ a theorem of Mazur (see [**32**, Thm. 2.8]) implies that there is a form of weight 2 giving rise to ρ_f, and when $k = 1$ we multiply f by the weight $\ell - 1$ Hasse invariant. To show that $w(\theta^\alpha f) = k(\rho)$ we investigate how application of the θ-operator changes the minimal weight. We have $(\rho_f \otimes \chi^\alpha)|_{I_\ell} \sim \begin{pmatrix} \Psi^n & 0 \\ 0 & (\Psi')^n \end{pmatrix}$ with $n = a + b\ell$ and $a < b$. Fontaine's theory (see Section 2.2.1) identifies the characters corresponding to $\rho_f|_{I_\ell}$ as powers Ψ^{k-1} and $(\Psi')^{k-1}$ of the fundamental characters. This gives an equality of unordered sets
$$\{\Psi^{k-1}\chi^\alpha, (\Psi')^{k-1}\chi^\alpha\} = \{\Psi^n, (\Psi')^n\}.$$

It is now possible to compute $w(\theta^\alpha f)$ by considering two cases, corresponding to the ways in which equality of unordered pairs can occur.

Case 1. Suppose that $\Psi^{k-1}\chi^\alpha = (\Psi')^n$. Since $\chi = \Psi^{\ell+1}$, we have
$$\Psi^{k-1+\alpha(\ell+1)} = \Psi^{k-1}\chi^\alpha = (\Psi')^n = (\Psi')^{a+b\ell} = \Psi^{b+a\ell}.$$

Comparing exponents of Ψ gives

(2.1) $$k - 1 + \alpha(\ell+1) \equiv b + a\ell \pmod{\ell^2 - 1},$$

which reduces modulo $\ell + 1$ to $k - 1 \equiv b - a \pmod{\ell + 1}$; because $2 \leq k \leq \ell$, this implies that $k = 1 + b - a$. Reducing (2.1) modulo $\ell - 1$ and substituting $k = 1 + b - a$ gives $b - a + 2\alpha \equiv b + a \pmod{\ell - 1}$; we find the possible solutions $\alpha = a + m(\ell-1)/2$ with m an integer. No solution $\alpha = a + m(\ell-1)/2$, with m odd, satisfies (2.1), so $\alpha = a$ as an integer mod $\ell - 1$. Finally, we apply Theorem 2.6 and argue as in the end of Section 2.2.2, to show that
$$w(\theta^a f) = w(f) + a(\ell + 1) = 1 + b - a + a\ell + a = 1 + b + a\ell = k(\rho).$$

Case 2. Suppose that $\Psi^{k-1}\chi^\alpha = \Psi^n$. Then
$$\Psi^{k-1+\alpha(\ell+1)} = \Psi^{k-1}\chi^\alpha = \Psi^n = \Psi^{a+b\ell}.$$

Comparing powers of Ψ, we obtain

(2.2) $\qquad k - 1 + \alpha(\ell + 1) \equiv a + b\ell \pmod{\ell^2 - 1},$

which reduces modulo $\ell + 1$ to $k - 1 \equiv a - b \pmod{\ell + 1}$; thus $k = \ell + 2 - (b - a)$. The difference $b - a$ must be greater than 1; otherwise $k = \ell + 1$, contrary to our assumption that $2 \leq k \leq \ell$. Reducing (2.2) modulo $\ell - 1$ gives

$$k - 1 + 2\alpha \equiv a + b \pmod{\ell - 1};$$

substituting $k = \ell + 2 - (b - a)$ we find that $\alpha = b - 1 + m(\ell - 1)/2$ with m an integer. If m is odd, then α does not satisfy (2.2), so $\alpha = b - 1$ as an integer modulo $\ell - 1$. It remains to verify the equality $w(\theta^{b-1}f) = w(\rho)$. Unfortunately, $k = \ell + 2 - (b - a)$ is not especially telling. The argument of Case 1 does not apply to compute $w(\theta^\alpha f)$; instead we use θ-cycles.

Because f is supersingular, Fermat's Little Theorem implies that $\theta^{\ell-1}f = f$. We use Tate's theory of θ-cycles (see [**32**, §7] and [**55**]) to compute $w(\theta^{b-1}f)$. The θ-cycle associated to f is the sequence of integers

$$w(f), w(\theta f), w(\theta^2 f), \ldots, w(\theta^{\ell-2}f), w(f).$$

The θ-cycle for any supersingular eigenform must behave as follows (see Theorem 2.6):

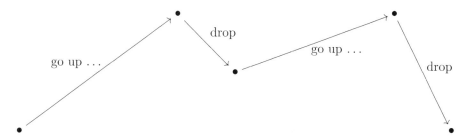

go up ..., drop once, go up ..., drop to original weight

Knowing this, we can deduce the exact θ-cycle. List ℓ numbers starting and ending with k:

$$k, k + (\ell + 1), k + 2(\ell + 1), \ldots, k + (\ell - k)(\ell + 1),$$
$$\ell + 3 - k, (\ell + 3 - k) + (\ell + 1), \ldots, (\ell + 3 - k) + (k - 3)(\ell + 1),$$
$$k$$

The first and second lines contain $\ell + 1 - k$ and $k - 2$ numbers, respectively. All told, ℓ numbers are listed; this must be the θ-cycle.

It is now possible to compute $w(\theta^{b-1}f)$. If

$$b - 1 \leq \ell - k = \ell - (\ell + 2 - b + a) = -2 + b - a,$$

then $a \leq -1$, a contradiction; thus $b - 1 > \ell - k$. It follows that

$$w(\theta^{b-1}f) = \ell + 3 - k + (\ell + 1)(b - 2 - (\ell - k)) = 1 + b + a\ell = k(\rho),$$

verifying Serre's conjecture in this case.

2.2.2.4. The ordinary case

We next turn to the ordinary case, in which

$$\rho|_{I_\ell} \sim \begin{pmatrix} \alpha & * \\ 0 & \beta \end{pmatrix}$$

with $\alpha, \beta : I_\ell \to \mathbf{F}_\ell^*$ powers of the cyclotomic character. View $\rho|_{I_\ell}$ as the twist of a representation in which the lower right entry is 1:

$$\begin{pmatrix} \alpha & * \\ 0 & \beta \end{pmatrix} \sim \beta \otimes \begin{pmatrix} \alpha\beta^{-1} & * \\ 0 & 1 \end{pmatrix}.$$

To determine the minimal weight of a form giving rise to $\rho|_{I_\ell}$, it is necessary to develop an ordinary version of θ-cycles. In general this is complicated, so we make the simplifying assumption that $\beta = 1$; then $\rho|_{I_\ell} \sim \begin{pmatrix} \chi^i & * \\ 0 & 1 \end{pmatrix}$ with $1 \leq i \leq \ell - 1$. Deligne showed that if f is of weight k and $\beta = 1$, then the associated representation is $\begin{pmatrix} \chi^{k-1} & * \\ 0 & 1 \end{pmatrix}$ with $2 \leq k \leq \ell + 1$. Motivated by this, our first reaction is to define $k(\rho)$ to be $i+1$. This definition does not distinguish between the extreme weights 2 and $\ell+1$ because they are congruent modulo $\ell-1$. Given a representation ρ arising from a form of weight either 2 or $\ell + 1$, we cannot, in general, set $k(\rho) = 2$. For example, suppose $f = \Delta$ is the level 1 cusp form of weight 12 and ρ is the associated mod 11 representation. It would be wrong to set $k(\rho) = 2$, because there is no cusp form of weight 2 and level 1.

Warning: When $\ell = 2$ and our $k(\rho)$ is 3, Serre replaced $k(\rho)$ by 4 because there are no weight-3 modular forms whose character is of degree coprime to $\ell = 2$.

2.2.3. Distinguishing between weights 2 and $\ell + 1$

We continue to motivate the definition of $k(\rho)$. Consider a representation $\rho : G_\mathbf{Q} \to \mathrm{GL}(2, \overline{\mathbf{F}}_\ell)$ that arises from a newform f of the optimal level $N = N(\rho)$ and weight k satisfying $2 \leq k \leq \ell+1$. Assume that f is ordinary in the sense that $a_\ell(f) \neq 0 \in \overline{\mathbf{F}}_\ell$. Then, as discussed in Section 2.2.1,

$$\rho|_{I_\ell} \sim \begin{pmatrix} \chi^{k-1} & * \\ 0 & 1 \end{pmatrix},$$

so $\rho|_{I_\ell}$ determines k modulo $\ell - 1$. This suggests a way to define $k(\rho)$ purely in terms of the Galois representation ρ, at least when $k \notin \{2, \ell + 1\}$.

The key to defining $k(\rho)$ when $k = 2$ or $k = \ell + 1$ is good reduction. To understand why this is so, we briefly summarize Shimura's geometric construction of Galois representations associated to newforms of weight 2.

2.2.3.1. Geometric construction of Galois representations

Shimura attached mod ℓ representations to a weight-2 newform $f = \sum a_n q^n$ of level N. Let E be the totally real or CM field $\mathbf{Q}(\ldots a_n \ldots)$. In [108, Thm. 7.14], Shimura described how to associate to f an abelian variety $A = A_f$ over $\mathbf{Q}$ of dimension $[E : \mathbf{Q}]$ furnished with an embedding $E \hookrightarrow \mathrm{End}_\mathbf{Q} A$ (see also Conrad's appendix). The mod ℓ representations attached to f are then found in the ℓ-torsion of A.

Over the complex numbers, the abelian variety A is found as a quotient of the Jacobian of the Riemann surface
$$X_1(N) := \overline{\Gamma_1(N)\backslash \mathfrak{h}} = \Gamma_1(N)\backslash \mathfrak{h} \cup \{\text{cusps}\}.$$
The Riemann surface $X_1(N)$ has a structure of algebraic curve over $\mathbf{Q}$; it is called the *modular curve* of level N. Its Jacobian $J_1(N)$ is an abelian variety over $\mathbf{Q}$ which, by work of Igusa, has good reduction at all primes $\ell \nmid N$. The dimension of $J_1(N)$ equals the genus of $X_1(N)$; for example, when $N = 1$, the curve $X_1(1)$ is isomorphic over $\mathbf{Q}$ to the projective line and $J_1(1) = 0$. There are (at least) two functorial actions of the Hecke algebra $\mathbf{T}$ on $J_1(N)$, and (at least) two definitions of $J_1(N)$. In the next section we will fix choices, and then construct A as the quotient of $J_1(N)$ by the image of the annihilator in $\mathbf{T}$ of f.

2.2.3.1.1. *Hecke operators on $J_1(N)$.* We pause to formulate a careful definition of $X_1(N)$ and of our preferred functorial action of the Hecke operators T_p on $J_1(N)$. For simplicity, we assume that $N > 4$ and $p \nmid N$. Following [**46**, Prop. 2.1] there is a smooth, proper, geometrically connected algebraic curve $X_1(N)$ over $\mathbf{Z}[1/N]$ that represents the functor assigning to each $\mathbf{Z}[1/N]$-scheme S the set of isomorphism classes of pairs (E, α), where E is a generalized elliptic curve over S and $\alpha : (\boldsymbol{\mu}_N)_S \hookrightarrow E^{\mathrm{sm}}[N]$ an embedding of group schemes over S whose image meets every irreducible component in each geometric fiber. Let $X_1(N, p)$ be the fine moduli scheme over $\mathbf{Z}[1/N]$ that represents the functor assigning to each $\mathbf{Z}[1/N]$-scheme S the set of isomorphism classes of triples (E, α, C), where E is a generalized elliptic curve over S, $\alpha : (\boldsymbol{\mu}_N)_S \hookrightarrow E^{\mathrm{sm}}[N]$ an embedding of group schemes over S, and C a locally free subgroup scheme of rank p in $E^{\mathrm{sm}}[p]$, such that $\mathrm{im}(\alpha) \times C$ meets every irreducible component in each geometric fiber of E. Let $\pi_1, \pi_2 : X_1(N, p) \to X_1(N)$ over $\mathbf{Z}[1/N]$ be the two standard degeneracy maps, which are defined on genuine elliptic curves by $\pi_1(E, \alpha, C) = (E, \alpha)$ and $\pi_2(E, \alpha, C) = (E', \alpha' = \varphi \alpha)$, where $E' = E/C$ and $\varphi : E \to E'$ is the associated p-isogeny. The Hecke operator $T_p = (T_p)^*$ acts on divisors D on $X_1(N)_{/\mathbf{Q}}$ by
$$T_p(D) = (\pi_1)_* \circ \pi_2^* D.$$
For example, if (E, α) is a non-cuspidal $\overline{\mathbf{Q}}$-point, then
$$T_p(E, \alpha) = \sum (E', \varphi \circ \alpha \circ [p]^{-1}),$$
The Hecke operator where the sum is over all isogenies $\varphi : E \to E'$ of degree p, and $[p]^{-1}$ is the inverse of pth powering on $\boldsymbol{\mu}_N$. This map on divisors defines an endomorphism T_p of the Jacobian $J_1(N)$ associated to $X_1(N)$ via Picard functoriality.

For each prime p there is an involution $\langle p \rangle$ of $X_1(N)$ called a *diamond bracket operator*, defined functorially by
$$\langle p \rangle(E, \alpha) = (E, \alpha \circ [p]).$$
The diamond bracket operator defines a correspondence, such that the induced map $(\langle p \rangle)^*$ on $J_1(N)$ is
$$(\langle p \rangle)^*(E, \alpha) = (E, \alpha \circ [p^{-1}]).$$
If $(T_p)_*$ denotes the pth Hecke operator as defined in [**46**, §3], then
$$(T_p)_* = T_p \circ (\langle p^{-1} \rangle)^*,$$
Thus our T_p differs from Gross's $(T_p)_*$. Furthermore, upon embedding $X_1(N)$ into $J_1(N)$ and identifying weight-2 cusp forms with differentials on $J_1(N)$, Gross's

$(T_p)_*$ induces, via Albanese functoriality, the usual Hecke action on cusp forms, whereas ours does not. In addition, we could have defined $X_1(N)$ by replacing the group scheme $\boldsymbol{\mu}_N$ by $(\mathbf{Z}/N\mathbf{Z})$. In this connection, see the discussion at the end of Section 5 of [**26**] and [**35**, §2.1].

2.2.3.1.2. *The representations attached to a newform.* Again let $\mathcal{O}$ be the ring of integers of $E = \mathbf{Q}(\ldots a_n \ldots)$, where $f = \sum a_n q^n$ is a weight-2 modular form on $\Gamma_1(N)$. Recall that $A = A_f$ is the quotient of $J_1(N)$ by the image of the annihilator in $\mathbf{T}$ of f. In general, $\mathcal{O}$ need not be contained in $\operatorname{End} A$. However, by replacing A by an abelian variety $\mathbf{Q}$-isogenous to A, we may assume that $\mathcal{O}$ is contained in $\operatorname{End} A$ (see [**108**, pg. 199]). Let λ be a maximal ideal of $\mathcal{O}$ and set

$$A[\lambda] := \{P \in A(\overline{\mathbf{Q}}) : xP = 0 \text{ all } x \in \lambda\}.$$

By [**108**, Prop. 7.20, pg. 190], $\dim_{\mathcal{O}/\lambda} A[\lambda] = 2$, so $A[\lambda]$ affords a 2-dimensional Galois representation, which is well-defined up to semisimplification. Let $\rho_{f,\lambda} : G_\mathbf{Q} \to A[\lambda]^{\mathrm{ss}}$ be the semisimplification of $A[\lambda]$.

2.2.3.1.3. *Good reduction.*

Definition 2.8. A finite group scheme G over $\mathbf{Q}_\ell^{\mathrm{nr}}$ is said to have *good reduction*, or to be *finite flat*, if it extends to a finite flat group scheme over the ring of integers $\mathcal{O}_{\mathbf{Q}_\ell^{\mathrm{nr}}}$ of $\mathbf{Q}_\ell^{\mathrm{nr}}$.

Proposition 2.9. *The representation $\rho_{f,\lambda}$ is finite flat at each prime $p \nmid N$.*

Proof. The finite flat group scheme extending $A[\lambda]$ is the scheme theoretic closure of $A[\lambda]$ in a good model $\mathcal{A}/\mathcal{O}_{\mathbf{Q}_\ell^{\mathrm{nr}}}$ of A. Such a model exists because A has good reduction at p. □

Consider again a Galois representation ρ as in the beginning of Section 2.2.3 such that $\rho|_{I_\ell} \sim \begin{pmatrix} \chi^{k-1} & * \\ 0 & 1 \end{pmatrix}$. If $k \not\equiv 2 \pmod{\ell - 1}$ then $k(\rho)$ is defined to equal k. If $k \equiv 2 \pmod{\ell - 1}$, then

$$k(\rho) := \begin{cases} 2 & \text{if } \rho \text{ is finite flat,} \\ \ell + 1 & \text{otherwise.} \end{cases}$$

2.2.4. Representations arising from elliptic curves

Theorem 2.10. *Suppose $A/\mathbf{Q}$ is a semistable elliptic curve and that $\rho_{A,\ell}$ is irreducible. Let Δ_A denote the minimal discriminant of A. The representation $\rho_{A,\ell}$ is finite flat at ℓ if and only if $\ell \mid \operatorname{ord}_\ell \Delta_A$. If $p \neq \ell$, then $\rho_{A,\ell}$ is unramified at p if and only if $\ell \mid \operatorname{ord}_p \Delta_A$.*

Proof. The first statement is Proposition 5 of [**102**].

When A has good reduction at p, the second statement holds (see Exercise 15). Suppose A has multiplicative reduction at p. There is an unramified extension K of $\mathbf{Q}_p$ such that A has split multiplicative reduction at p. Consider the Tate curve $\mathbf{G}_m/q^\mathbf{Z}$ over K associated to A. Thus $\overline{\mathbf{Q}}_p^*/q^\mathbf{Z} \cong A(\overline{\mathbf{Q}}_p)$ as $\operatorname{Gal}(\overline{\mathbf{Q}}_p/K)$-modules. The ℓ-torsion points $A[\ell]$ correspond to the points $\{\zeta_\ell^a (q^{1/\ell})^b : 0 \leq a,b < \ell\}$ in the Tate curve. The extension $K(\zeta_\ell, q^{1/\ell})$ of K is unramified because $\ell \neq p$ and $\operatorname{ord}_p(q) = \operatorname{ord}_p(\Delta_A)$ is divisible by ℓ. Since an unramified extension of an unramified extension is unramified, the extension $K(\zeta_\ell, q^{1/\ell})$ of $\mathbf{Q}_p$ is unramified, which proves the second part of the theorem. □

2.2.4.1. Frey curves

Using Theorem 2.10 we see that the Shimura-Taniyama conjecture together with Serre's conjecture implies Fermat's Last Theorem. Suppose (a, b, c) is a solution to the Fermat equation $a^\ell + b^\ell = c^\ell$ with $\ell \geq 11$ and $abc \neq 0$. Consider the Frey curve A given by the equation $y^2 = x(x - a^\ell)(x + b^\ell)$; it is an elliptic curve with discriminant $\Delta_A = \frac{((abc)^2)^\ell}{2^8}$. By [**93**, §4.1, Prop. 6] the representation $A[\ell]$ is irreducible. Theorem 2.10 implies that $\rho_{A,\ell}$ is unramified, except possibly at 2 and ℓ. Thus $N(\rho) \mid 2$, and $k(\rho) = 2$ since $\ell \mid \text{ord}_\ell(\Delta_A)$. But there are no cusp forms of level 2 and weight 2. The modularity of A (proved in [**114**, **117**]), together with the weak conjecture of Serre (enough of which is proved in [**84**]), leads to a contradiction.

2.2.4.2. Examples

Using Theorem 2.10 we can frequently determine the Serre invariants $N(\rho)$ and $k(\rho)$ of a representation ρ attached to an elliptic curve. When $N(\rho) < N$, it is illustrative to verify directly that there is a newform of level $N(\rho)$ that also gives rise to ρ. For example, there is a unique weight-2 normalized newform

$$f = q + q^2 - q^3 - q^4 - 2q^5 - q^6 + 4q^7 - 3q^8 + q^9 + \cdots$$

on $\Gamma_0(33)$. One of the elliptic curves associated to f is the curve A given by the equation

$$y^2 + xy = x^3 + x^2 - 11x.$$

The discriminant of A is $\Delta = 3^6 \cdot 11^2$ and the conductor is $N = 3 \cdot 11$. Because A is semistable and there are no elliptic curves 3-isogenous to A, the associated mod 3 representation $\rho = \rho_{A,3} : G_\mathbf{Q} \to \text{Aut}(A[3])$ is surjective (see Section 1.1.4). Since $3 \mid \text{ord}_3 \Delta_A$, the Serre weight and level are $k(\rho) = 2$ and $N(\rho) = 11$. As predicted by Serre's conjecture, there is a weight-2 newform on $\Gamma_0(11)$ such that if B is one of the three elliptic curves of conductor 11 (it does not matter which), then $B[3] \approx A[3]$ as representations of $G_\mathbf{Q}$. Placing the eigenforms corresponding to A and B next to each other, we observe that their Fourier coefficients are congruent modulo 3:

$$\begin{array}{lllllllllll} f_A = & q & +q^2 & -q^3 & -q^4 & -2q^5 & -q^6 & +4q^7 & -3q^8 & +q^9 & + \cdots \\ f_B = & q & -2q^2 & -q^3 & +2q^4 & +q^5 & +2q^6 & -2q^7 & & -2q^9 & + \cdots. \end{array}$$

Next consider the elliptic curve A cut out by the equation

$$y^2 + y = x^3 + x^2 - 12x + 2.$$

It has conductor $N = 141 = 3 \cdot 47$ and discriminant $\Delta = 3^7 \cdot 47$. Since $\text{ord}_3(\Delta)$ is divisible by 7, the mod 7 representation $\rho_{A,7}$ has Serre invariants $k(\rho_{A,7}) = 2$ and $N(\rho_{A,7}) = 47$. In confirmation of Serre's conjecture, we find a form $f \in S_2(\Gamma_0(47))$ that gives rise to $\rho_{A,7}$. The Fourier coefficients of f generate a quartic field.

Next consider $\rho_{A,3}$, whose Serre invariants are $N(\rho_{A,3}) = 47$ and, since 3 does not divide $\text{ord}_3(\Delta)$, $k(\rho_{A,3}) = \ell + 1 = 4$. In $S_4(\Gamma_0(47))$ there are two conjugacy classes of eigenforms, which are defined over fields of degree 3 and 8, respectively. The one that gives rise to $\rho_{A,3}$ is

$$g = q + aq^2 + (-1/2a^2 - 5/2a - 1)q^3 + (a^2 - 8)q^4 + (a^2 + a - 10)q^5 + \cdots,$$

where $a^3 + 5a^2 - 2a - 12 = 0$.

2.2.5. Companion forms

Suppose f is a newform of weight k with $2 \leq k \leq \ell+1$. Let ℓ be an ordinary prime, so $a_\ell(f)$ is not congruent to 0 modulo a prime λ lying over ℓ and

$$\rho_{f,\lambda}|_{I_\ell} \sim \begin{pmatrix} \chi^{k-1} & * \\ 0 & 1 \end{pmatrix}.$$

Is this representation split or not? Put another way, can $*$ be taken equal to 0, after an appropriate choice of basis? For how many ℓ do these representations split? We suspect that the ordinary split primes ℓ are in the minority, among all primes. How can we quantify the number of split primes?

If $* = 0$, then

$$\rho|_{I_\ell} \sim \begin{pmatrix} 1 & 0 \\ 0 & \chi^{k-1} \end{pmatrix},$$

so

$$\rho|_{I_\ell} \otimes \chi^{\ell-k} \sim \begin{pmatrix} \chi^{\ell-k} & 0 \\ 0 & 1 \end{pmatrix}.$$

Assume that $2 \leq 1+\ell-k \leq \ell+1$, so $k(\rho \otimes \chi^{\ell-k}) = 1+\ell-k$. Using the θ-operator we see that $\rho \otimes \chi^{\ell-k}$ is modular, of *some* weight and level. To say that it is modular of Serre's conjectured weight $k(\rho)$ is to make a much strong statement. If $\rho \otimes \chi^{\ell-k}$ is indeed modular of weight $1+\ell-k$, then by definition there exists an eigenform g of weight $1+\ell-k$ with $\rho_g \sim \rho_f \otimes \chi^{\ell-k}$. Such an eigenform g, if it exists, is called a *companion* of f. The existence of g is far from obvious.

We can extend the notion of companion form to the case when $k(\rho) = \ell$. In this case the companion has weight 1. If ρ is unramified at ℓ, then we expect ρ to also arise from a weight-1 eigenform.

The existence of a companion form was proved (assuming unchecked compatibilities) in most cases in which $k < \ell$ by Gross in [**46**] and in a few cases when $k = \ell$. Using new methods, Coleman and Voloch [**17**] proved all cases except $k = \ell = 2$. The arguments of Coleman and Voloch do not require verification of Gross's unchecked compatibilities.

CHAPTER 3
Optimizing the level

Consider an irreducible Galois representation $\rho : G_{\mathbf{Q}} \to \mathrm{GL}(2, \overline{\mathbf{F}}_\ell)$ that arises from a newform of weight k and level N. Serre defined integers $k(\rho)$ and $N(\rho)$, and conjectured that ρ arises from a newform of weight $k(\rho)$ and level $N(\rho)$. In Chapter 2 we sketched Edixhoven's proof that if $\ell \nmid N$ then ρ arises from a newform of weight $k(\rho)$ and level N. In this chapter, we introduce some of the techniques used in proving that ρ arises from a newform of level $N(\rho)$. For more details, see [**84, 87**].

In [**102**, §1.2] Serre defined the *optimal level* $N(\rho)$ as the prime-to-ℓ part of the Artin conductor of ρ. Recall that $N(\rho)$ is a product $\prod p^{n(p)}$ over prime numbers $p \neq \ell$. The integer $n(p)$ is defined by restricting ρ to a decomposition group D_p at p. Consider the sequence of ramification groups $G_0 \supset G_1 \supset \cdots \supset G_i \supset \cdots$ where G_0 is the inertia subgroup I_p of D_p. Let V be a vector space over $\overline{\mathbf{F}}_\ell$ affording the representation ρ, and for each $i \geq 0$ let V_i be the subspace of V consisting of those $v \in V$ that are fixed by G_i. Then

$$n(p) := \sum_{i=0}^{\infty} \frac{1}{(G_0 : G_i)} \dim V/V_i.$$

3.3.1. Reduction to weight 2

The optimal level $N(\rho)$ is not divisible by ℓ. The first step in level optimization is to strip the power of ℓ from N. When ℓ is odd, this is done explicitly in [**87**, §2]; for the case $\ell = 2$ see [**9**, §1]. Many of the arguments and key ideas are due to Serre [**94**]. This proof that ℓ can be stripped from the level uses concrete techniques of Serre [**95**, §3], [**98**, Thm. 5.4], and Queen [**78**, §3]; it involves multiplying f by suitable Eisenstein series and taking traces. Katz's theory of ℓ-adic modular forms suggests an alternative method. A classical form of weight 2 and level $M\ell^m$ is an ℓ-adic form of level M; the mod ℓ reduction of this form is classical of level M and some weight, and is congruent to f. See the appendices of [**60**] and the discussions in [**49**, §1] and [**50**, §1].

The next step is to replace f by a newform of weight between 2 and $\ell + 1$ that gives rise to a twist of ρ. Twisting ρ by the mod ℓ cyclotomic character χ preserves N; this is because $\rho \otimes \chi$ arises from $\theta(f) = q\frac{d}{dq}(f)$, which also has level N. Theorem 2.7 asserts that some twist $\rho \otimes \chi^i$ of ρ arises from a form g of weight between 2 and $\ell+1$. If $\rho \otimes \chi^i$ arises from a newform of level N, then ρ also arises from a newform of the same level, so we can replace f by g and k by the weight of g. By results discussed in Chapter 2, we may assume that $k = k(\rho \otimes \chi^i)$. For the case $\ell = 2$ see [**9**, Prop. 1.3(a)].

We have reduced to considering a representation ρ that arises from a newform f of weight $k(\rho)$ and level N not divisible by ℓ. The weight satisfies $2 \leq k(\rho) \leq \ell+1$, but N need not equal $N(\rho)$. That N is a multiple of $N(\rho)$ is a theorem proved by both Carayol [**12**] and Livné [**70**, Prop. 0.1].

In order to lower N it is convenient to work systematically with form of weight 2. Paradoxically, even though we have just taken all powers of ℓ out of N, we are now going to allow one power of ℓ back into N. This allows us to reduce to weight 2 and realize ρ as a group of torsion points on an abelian variety. An alternative approach (see [**41, 57**]) is to avoid this crutch and work directly with representations coming from arbitrary weights between 2 and $\ell+1$; these are realized in étale cohomology groups. This later approach has the advantage that $X_0(N)$ has good reduction at ℓ.

Reduction to weight 2 is accomplished using a general relationship that originates with ideas of Koike and Shimura. In characteristic ℓ, eigenforms of level N whose weights satisfy $2 < k \leq \ell + 1$ correspond to eigenforms of weight 2 and level ℓN (see [**87**, Thm. 2.2]):

$$\left\{ 2 < k \leq \ell+1, \text{ level } N \right\} \longleftrightarrow \left\{ k = 2, \text{ level } \ell N \right\}.$$

Thus we can and do work with weight 2 and level

$$N^* := \begin{cases} N & \text{if } k = 2, \\ N\ell & \text{if } k > 2. \end{cases}$$

3.3.2. Geometric realization of Galois representations

To understand representations arising from modular forms, it is helpful to realize these representations inside of geometric objects such as $J := J_1(N^*)$. These representations are constructed geometrically with the help of the Hecke algebra

$$\mathbf{T} := \mathbf{Z}[\ldots T_n \ldots],$$

which was defined in Section 2.2.3. Recall that $\mathbf{T}$ is a commutative subring of $\mathrm{End}_{\mathbf{Q}} J$ that is free as a module over $\mathbf{Z}$, and that its rank is equal to the dimension of J. When N is cube free, $\mathbf{T}$ is an order in a product of integer rings of number fields; this is a result of Coleman and Edixhoven (see [**16**, Thm. 4.1]). In contrast, the Hecke operators T_p, for $p^3 \mid N$, are usually not semisimple (see Exercise 3).

It is fruitful to view a newform f as a homomorphism

$$\mathbf{T} \to \mathcal{O} = \mathbf{Z}[\ldots a_n \ldots], \qquad T_n \mapsto a_n.$$

Letting $\varphi : \mathcal{O} \to \overline{\mathbf{F}}_\ell$ be the map sending a_p to $\mathrm{tr}(\rho(\mathrm{Frob}_p)) \in \overline{\mathbf{F}}_\ell$, we obtain an exact sequence $0 \to \mathfrak{m} \to \mathbf{T} \to \overline{\mathbf{F}}_\ell$ with $\mathfrak{m}$ a maximal ideal.

Let $\rho : G_{\mathbf{Q}} \to \mathrm{GL}(2, \overline{\mathbf{F}}_\ell)$ be an irreducible Galois representation that arises from a weight-2 newform f. The next step, after having attached a maximal ideal $\mathfrak{m}$ to f and φ, is to find a $\mathbf{T}/\mathfrak{m}$-vector space affording ρ inside of the group of ℓ-torsion points of J. Following [**71**, §II.7], we consider the $\mathbf{T}/\mathfrak{m}$-vector space

$$J[\mathfrak{m}] := \{ P \in J(\overline{\mathbf{Q}}) : tP = 0 \text{ all } t \in \mathfrak{m} \} \subset J(\overline{\mathbf{Q}})[\ell] \approx (\mathbf{Z}/\ell\mathbf{Z})^{2g}.$$

Since the endomorphisms in **T** are **Q**-rational, $J[\mathfrak{m}]$ comes equipped with a linear action of $G_{\mathbf{Q}}$.

That $\mathrm{tr}(\rho(\mathrm{Frob}_p))$ and $\det(\rho(\mathrm{Frob}_p))$ both lie in the subfield $\mathbf{T}/\mathfrak{m}$ of $\overline{\mathbf{F}}_\ell$ suggests that ρ has a model over $\mathbf{T}/\mathfrak{m}$, in the sense that ρ is equivalent to a representation taking values in $\mathrm{GL}(2, \mathbf{T}/\mathfrak{m}) \subset \mathrm{GL}(2, \overline{\mathbf{F}}_\ell)$.

Lemma 3.1. *The representation ρ has a model $\rho_\mathfrak{m}$ over the finite field $\mathbf{T}/\mathfrak{m}$.*

Proof. This is a classical result of I. Schur. Brauer groups of finite fields are trivial (see e.g., [**100**, X.7, Ex. a]), so the argument of [**99**, §12.2] proves the lemma.

Alternatively, when the residue characteristic ℓ of $\mathbf{T}/\mathfrak{m}$ is odd, the following more direct proof can be used. Complex conjugation acts through ρ as a matrix with distinct $\mathbf{F}_\ell$-rational eigenvalues; another well known theorem of Schur [**90**, IX a] (cf. [**116**, Lemme I.1]) then implies that ρ can be conjugated into a representation with values in $\mathrm{GL}(2, \mathbf{T}/\mathfrak{m})$. $\square$

3.3.3. Multiplicity one

Let $V_\mathfrak{m}$ be a vector space affording $\rho_\mathfrak{m}$. Under the assumption that $\rho_\mathfrak{m}$ is absolutely irreducible, Boston, Lenstra, and Ribet (see [**6**]) proved that $J[\mathfrak{m}]$ is isomorphic as a $G_{\mathbf{Q}}$-module to a sum of copies of $V_\mathfrak{m}$:

$$J[\mathfrak{m}] \approx \bigoplus_{i=1}^{t} V_\mathfrak{m}.$$

The number of copies of $V_\mathfrak{m}$ is called the *multiplicity* of $\mathfrak{m}$. When ℓ is odd, the hypothesis of irreducibility of $\rho_\mathfrak{m}$ is equivalent to absolute irreducibility (see Exercise 3).

Proposition 3.2. *The multiplicity t is at least 1.*

Proof. Let $\mathbf{T} \subset \mathrm{End}(J)$ be the Hecke algebra associated to J. Because $\mathbf{T} \otimes \mathbf{Z}_\ell$ is an algebra of finite rank over the local ring $\mathbf{Z}_\ell$, we have a decomposition

$$\mathbf{T} \otimes \mathbf{Z}_\ell = \bigoplus_{\lambda | \ell} \mathbf{T}_\lambda,$$

where λ runs through the maximal ideals of $\mathbf{T}$ lying over ℓ, and $\mathbf{T}_\lambda$ denotes the completion of $\mathbf{T}$ at λ (see, e.g., [**37**, Cor. 7.6]). The Tate module

$$\mathrm{Tate}_\ell\, J := \mathrm{Hom}(\mathbf{Q}_\ell/\mathbf{Z}_\ell, \cup_{n \geq 1} J[\ell^n]) \cong \varprojlim J[\ell^n]$$

is a free $\mathbf{Z}_\ell$-module of rank equal to twice the dimension of J. For each maximal ideal λ of $\mathbf{T}$ lying over ℓ, let $e_\lambda \in \mathbf{T} \otimes \mathbf{Z}_\ell$ denote the corresponding idempotent; thus $e_\lambda^2 = e_\lambda$ and $\sum_{\lambda | \ell} e_\lambda = 1$. The map $x \mapsto \sum_\lambda e_\lambda x$ gives a decomposition

$$\mathrm{Tate}_\ell\, J \xrightarrow{\cong} \bigoplus_{\lambda | \ell} e_\lambda\, \mathrm{Tate}_\ell\, J.$$

The ring $\mathrm{End}(J) \otimes \mathbf{Z}_\ell$ operates faithfully on $\mathrm{Tate}_\ell\, J$ (see, e.g., [**74**, Lem. 12.2]), so each summand $e_\lambda\, \mathrm{Tate}_\ell\, J$ is nonzero. Set

$$\mathrm{Tate}_\lambda\, J := \mathrm{Hom}(\mathbf{Q}_\ell/\mathbf{Z}_\ell, \cup_{n \geq 1} J[\lambda^n]).$$

We claim that $\mathrm{Tate}_\lambda\, J$ is identified with $e_\lambda\, \mathrm{Tate}_\ell\, J$ under the natural inclusion $\mathrm{Tate}_\lambda\, J \subset \mathrm{Tate}_\ell\, J$. Denote by $\tilde{\lambda}$ the maximal ideal in $\mathbf{T} \otimes \mathbf{Z}_\ell$ generated by λ. Let n be a positive integer, and let I be the ideal in $\mathbf{T}_\lambda$ generated by ℓ^n. Because $\mathbf{T}_\lambda$ is a local ring with maximal ideal $\tilde{\lambda}$, there is an integer m such that $\tilde{\lambda}^m \subset I$. Since I is principal and generated by ℓ^n, and $\mathbf{T}$ acts on $e_\lambda J[\ell^n]$ through $\mathbf{T}_\lambda$, we have

$$e_\lambda J[\ell^n] = (e_\lambda J[\ell^n])[I] \subset (e_\lambda J[\ell^n])[\tilde{\lambda}^m] \subset (e_\lambda J[\ell^n])[\lambda^m] \subset J[\lambda^m].$$

This shows that $e_\lambda \operatorname{Tate}_\ell J \subset \operatorname{Tate}_\lambda J$. Next suppose $\lambda' \neq \lambda$ and let n be a positive integer. Since $\mathbf{T}_\lambda$ acts on $J[\lambda^n]$ through $\mathbf{T}/\lambda^n = \mathbf{T}_\lambda/\tilde{\lambda}^n$, we have $e_{\lambda'} J[\lambda^n] = 0$, so

$$J[\lambda^n] = \sum_{\text{all } \lambda'} e_{\lambda'} J[\lambda^n] = e_\lambda J[\lambda^n].$$

The other inclusion $\operatorname{Tate}_\lambda J = e_\lambda \operatorname{Tate}_\lambda J \subset e_\lambda \operatorname{Tate}_\ell J$, which we need to prove equality, then follows.

We apply the above conclusion with $\lambda = \mathfrak{m}$. Since $\operatorname{Tate}_\mathfrak{m} J \neq 0$, some $J[\mathfrak{m}^r]$ is nonzero; let r be the smallest such integer. Following [**71**, p. 112], observe that for each generating set of elements $a_1, \ldots, a_t$ of the $\mathbf{T}/\mathfrak{m}$-vector space $\mathfrak{m}^{r-1}/\mathfrak{m}^r$, the map $x \mapsto a_1 x \oplus \cdots \oplus a_t x$ is an injection of the module $J[\mathfrak{m}^r]/J[\mathfrak{m}^{r-1}]$ into the direct sum of t copies of $J[\mathfrak{m}]$. Thus $J[\mathfrak{m}]$ is nonzero. $\square$

The special case $t = 1$, in which the multiplicity is one, plays a central role in the development of the theory. A detailed summary of multiplicity one results can be found in [**32**, §9], and some supplementary results are contained in [**117**, Thm. 2.1]. In general, the multiplicity can be greater than one (see [**72**, §13] and [**63**]).

3.3.3.1. Multiplicity one representations

Let $\rho : G_\mathbf{Q} \to \operatorname{GL}(2, \overline{\mathbf{F}}_\ell)$ be an irreducible modular Galois representation such that

$$2 \leq k(\rho) \leq \ell + 1.$$

Consider pairs (N, α) where $N \geq 1$ is an integer with the property that $\ell \nmid N$ if $k(\rho) = 2$ and $\ell \parallel N$ if $k(\rho) > 2$, together with maps $\alpha : \mathbf{T}_N \to \overline{\mathbf{F}}_\ell$, such that $\alpha(T_p) = \operatorname{tr}(\rho(\operatorname{Frob}_p))$ and $\alpha(p\langle p \rangle) = \det(\rho(\operatorname{Frob}_p))$ for almost all p. Here $\mathbf{T}_N$ is the Hecke algebra associated to $S_2(\Gamma_1(N))$. Note that if (N, α) is such a pair and $\mathfrak{m} = \ker(\alpha)$, then

$$\rho \approx \rho_\mathfrak{m} \otimes_{\mathbf{T}/\mathfrak{m}} \overline{\mathbf{F}}_\ell,$$

where $\alpha : \mathbf{T}/\mathfrak{m} \hookrightarrow \overline{\mathbf{F}}_\ell$ and $\rho_\mathfrak{m}$ is the unique (up to isomorphism) semisimple representation over $\overline{\mathbf{F}}_\ell$ such that

$$\operatorname{tr}(\rho_\mathfrak{m}(\operatorname{Frob}_p)) = \alpha(T_p) \qquad \det(\rho_\mathfrak{m}(\operatorname{Frob}_p)) = \alpha(p\langle p \rangle)$$

for almost all p.

Definition 3.3. ρ is a *multiplicity one representation* if $J_1(N)[\ker \alpha]$ has dimension 2 for all pairs (N, α) as above.

Remark 3.4. 1. If $J_1(N)[\ker \alpha]$ has dimension 2 then $\rho_\mathfrak{m} = J_1(N)[\ker \alpha]$ by Eichler-Shimura, see [**6**].
  2. The definition extends to arbitrary modular Galois representations ρ as follows. As explained in Section 2.2.2, every ρ has a twist $\rho \otimes \chi^i$ by some power of the cyclotomic character such that $k(\rho \otimes \chi^i) \leq \ell + 1$. We say that ρ is a *multiplicity one representation* if $\rho \otimes \chi^i$ is a multiplicity one representation.

3.3.3.2. Multiplicity one theorems

Techniques for proving multiplicity one results were pioneered by Mazur in [**71**] who considered $J_0(p)$ with p prime. Let f be an eigenform and fix a nonzero prime λ of the ring generated by the Fourier coefficients of f such that $\rho_{f,\lambda}$ is absolutely irreducible. View the Hecke algebra $\mathbf{T}$ as a subring of $\operatorname{End}(J_0(p))$, and let $\mathfrak{m}$ be the maximal ideal associated to f and λ. Let $V_\mathfrak{m}$ again be a two-dimensional $\mathbf{T}/\mathfrak{m}$-vector space that affords $\rho_\mathfrak{m} : G_\mathbf{Q} \to \operatorname{GL}(2, \mathbf{T}/\mathfrak{m})$. Mazur proved (see Prop. 14.2, ibid.) that $J[\mathfrak{m}] \approx V_\mathfrak{m}$, except perhaps when $\mathfrak{m}$ is ordinary of residue characteristic $\ell = 2$. The missing ordinary case can be treated under suitable hypothesis. If $\rho_\mathfrak{m}$ restricted to a decomposition group at 2 is not contained in the scalar matrices, then $J[\mathfrak{m}] \approx V_\mathfrak{m}$ (see, e.g., [**9**, Prop. 2.4]). The results of Mazur are extended in [**72**] and [**84**, §5].

Theorem 3.5. *An irreducible modular Galois representation* $\rho : G_\mathbf{Q} \to \operatorname{GL}_2(\overline{\mathbf{F}}_\ell)$ *is a multiplicity one representation, except perhaps when all of the following hypothesis on ρ are simultaneously satisfied:*

— $k(\rho) = \ell$;
— ρ *is unramified at* ℓ;
— ρ *is ordinary at* ℓ;
— $\rho|_{D_\ell} \sim \begin{pmatrix} \alpha & * \\ 0 & \beta \end{pmatrix}$ *with* $\alpha = \beta$.

Proof. See [**32**, §9], [**117**, Thm. 2.1], and [**9**, Prop. 2.4] for the case $\ell = 2$. □

In [**46**, §12] Gross proves multiplicity one when $\alpha \neq \beta$, $k(\rho) \leq \ell$, and ρ is ordinary; he uses this result in his proof of the existence of companion forms. In contrast, Coleman and Voloch [**17**] prove the existence of companion forms when $\alpha = \beta$ and $\ell > 2$ using a method that avoids the need for multiplicity one.

Remark 3.6. L. Kilford of London, England has recently discovered an example at prime level 503 in which multiplicity one fails. Let E_1, E_2, and E_3 be the three elliptic curves of conductor 503, and for each $i = 1, 2, 3$, let $\mathfrak{m}_i$ be the maximal ideal of $\mathbf{T} \subset \operatorname{End}(J_0(503))$ generated by 2 and all $T_p - a_p(E_i)$, with p prime. Each of the Galois representations $E_i[2]$ is irreducible, and one can check that $\mathfrak{m}_1 = \mathfrak{m}_2 = \mathfrak{m}_3$. If multiplicity one holds, then $E_1[2] = E_2[2] = E_3[2]$ inside of $J_0(503)$. However, this is not the case, as a modular symbols computation in the integral homology $H_1(X_0(N), \mathbf{Z})$ reveals that $E_1 \cap E_2 = \{0\}$.

3.3.3.3. Multiplicity one for mod 2 representations

For future reference, we now wish to consider multiplicity one in the following rather extreme situation. Suppose that $\ell = 2$, and let ρ be a mod ℓ representation arising from a form of weight either 2 or 3. If the weight is 3 then ρ is not finite at 2; this can be used to deduce multiplicity one by adapting the arguments of [**72**] (see the proof of [**9**, Prop. 2.4]). When the weight is 2, we have the following proposition.

Proposition 3.7. *Let* $\rho : G_\mathbf{Q} \to \operatorname{GL}_2(\overline{\mathbf{F}}_2)$ *be an irreducible Galois representation that arises from a weight-2 form* $f = \sum a_n q^n$ *on* $\Gamma = \Gamma_1(N) \cap \Gamma_0(2)$ *with N odd, and let ε be the character of f. If* $\overline{a}_2^2 \not\equiv \overline{\varepsilon}(2) \in \overline{\mathbf{F}}_2$, *then ρ is a multiplicity one representation.*

Proof. Let $\mathfrak{m}$ be the maximal ideal associated to f in the Hecke algebra $\mathbf{T}$ attached to Γ. Because the weight of f is 2, the representation ρ is finite at 2. If ρ is supersingular then the inertia group I_2 operates through the two fundamental characters of level 2. These both have order $\ell^2 - 1 = 3 \neq 1$, so ρ is ramified and this can be used to deduce multiplicity one. If ρ is ordinary then $\rho|_{D_2} \sim \begin{pmatrix} \alpha & * \\ 0 & \beta \end{pmatrix}$ with β unramified and $\beta(\mathrm{Frob}_2) \equiv T_2 \bmod \mathfrak{m}$. The determinant $\alpha\beta$ of $\rho|_{D_2}$ is $\chi \cdot \varepsilon$ where χ is the mod 2 cyclotomic character and ε is unramified at 2. Since χ, ε, and β are unramified, α is also unramified. Since $\chi(\mathrm{Frob}_2) = 1$ and $\alpha\beta = \chi\varepsilon$, we have $\alpha(\mathrm{Frob}_2) = \beta^{-1}(\mathrm{Frob}_2)\varepsilon(2) = a_2^{-1}\varepsilon(2) \pmod{\mathfrak{m}}$. The further condition, under which we might not know multiplicity one, is $\alpha|_{D_2} = \beta|_{D_2}$; expressed in terms of the image of Frobenius, this becomes $a_2^{-1}\varepsilon(2) \equiv a_2 \pmod{\mathfrak{m}}$, or equivalently, $a_2^2 \equiv \varepsilon(2) \pmod{\mathfrak{m}}$. By hypothesis, this latter condition does not hold. $\square$

3.3.4. The key case

We have set our problem up so that level optimization pertains to weight-2 forms of appropriate level, and takes place on Jacobians of modular curves. This level optimization problem was described, and partially treated, in a paper of Carayol [**12**]. In this paper, Carayol reduced the problem to the following key case.

Key case: Let $\rho : G_{\mathbf{Q}} \to \mathrm{GL}(2, \overline{\mathbf{F}}_\ell)$ be a Galois representation that arises from a weight-2 newform f of level pM, with $p \nmid \ell M$, and character $\varepsilon : (\mathbf{Z}/pM\mathbf{Z})^* \to \mathbf{C}^*$. Assume that ρ is unramified at p, and that ε factors through the natural map $(\mathbf{Z}/pM\mathbf{Z})^* \to (\mathbf{Z}/M\mathbf{Z})^*$. Show that ρ arises from a form of level M.

In the key case, the character ε of f is unramified at p. Thus f, a priori on $\Gamma_1(pM)$, is also on the bigger group $\Gamma_1(M) \cap \Gamma_0(p)$; that is, f lies in $S_2(\Gamma_1(M) \cap \Gamma_0(p))$.

Example 3.8. Consider the representation ρ arising from the 7-division points of the modular elliptic curve A of conductor $N_A = 3 \cdot 47$ and minimal discriminant $\Delta_A = 3^7 \cdot 47$. (The curve A is labeled **141A** in Cremona's notation [**20**].) The newform f corresponding to A is on $\Gamma_0(3 \cdot 47)$. As in Section 1.1.4, since $\mathrm{ord}_3(\Delta_A) = 7$, the representation ρ is unramified at 3 and $N(\rho) = 47$. To optimize the level means to find a form g on $\Gamma_0(47)$ that gives rise to ρ.

Example 3.9 (Frey curves). The elliptic curves that Frey associated in [**42**] to hypothetical solutions of the Fermat equation $x^\ell + y^\ell = z^\ell$ give rise to mod ℓ Galois representations. According to Wiles's theorem [**117**], there is a weight-2 form f of level $2L$, with L big and square free, that gives rise to ρ. At the same time, $N(\rho) = 2$. Taking p to be any odd prime dividing L, we are put in the key case. If we can optimize the level, then we eventually reach a contradiction and thus deduce Fermat's Last Theorem.

The key case divides into two subcases; the more difficult one occurs when the following conditions are both satisfied:

— $p \equiv 1 \pmod{\ell}$;
— $\rho(\mathrm{Frob}_p)$ is a scalar matrix.

The second condition makes sense because $p \nmid N(\rho)$; since $\det(\rho(\mathrm{Frob}_p)) = \chi^{k-1}\varepsilon$, we know the scalar up to ± 1. The complementary case is easier; it can be treated using "Mazur's principle" (see Section 3.3.9). Though Example 3.8 falls into the easier case because $3 \not\equiv 1 \pmod 7$, the proof of Fermat's Last Theorem requires level optimization in both cases.

Consider a modular representation $\rho : G_{\mathbf{Q}} \to \mathrm{GL}(2, \overline{\mathbf{F}}_\ell)$ that arises from a newform of level N and weight $k = k(\rho)$, and assume that $\ell \nmid N$. The goal of level optimization is to show that there is a newform of Serre's optimal level $N(\rho)$ that gives rise to ρ.

As discussed in Section 3.3.1, ρ arises from a newform $f = \sum a_n q^n$ on $\Gamma_1(N^*)$ of weight 2 and some character ε. Thus there is a homomorphism φ from $\mathcal{O} = \mathbf{Z}[\ldots a_n \ldots]$ to $\overline{\mathbf{F}}_\ell$ such that $\varphi(a_p) = \mathrm{tr}(\rho(\mathrm{Frob}_p))$ for all $p \nmid \ell N^*$. Let $\mathbf{T}$ be the Hecke algebra associated to $S_2(\Gamma_1(N^*))$. The maximal ideal $\mathfrak{m}$ of $\mathbf{T}$ associated to ρ is the kernel of the map sending T_n to $\varphi(a_n)$. As was discussed in the previous chapter, the representation ρ is realized geometrically inside the subspace $J[\mathfrak{m}] \subset J[\ell]$ of the ℓ-torsion of the Jacobian J of $X_1(N^*)$.

Problem. Fix a divisor p of $N^*/N(\rho)$. Find a newform whose level is a divisor of N^*/p that also gives rise to ρ.

Lemma 3.10. *Let ρ be as above, and suppose p is a prime such that $p \mid N^*$ but $p \nmid \ell N(\rho)$, so ρ is unramified at p. Let ε_p denote the p part of ε. Then either $\varepsilon_p = 1$ or $p \equiv 1 \pmod \ell$.*

Proof. The character ε is initially defined as a homomorphism $(\mathbf{Z}/N^*\mathbf{Z})^* \to \mathcal{O}^*$; the reduction $\overline{\varepsilon}$ is obtained by composing ε with $\varphi : \mathcal{O} \to \overline{\mathbf{F}}_\ell$. Since ρ is unramified at p, the determinant $\det(\rho) = \chi_\ell^{k-1} \overline{\varepsilon} = \chi_\ell \overline{\varepsilon}$ is also unramified at p. Because χ_ℓ is ramified only at ℓ, the character $\overline{\varepsilon}$ is unramified at p. Let $M = N^*/p^r$ where $r = \mathrm{ord}_p(N^*)$, and write $(\mathbf{Z}/N^*\mathbf{Z})^* \cong (\mathbf{Z}/p^r\mathbf{Z})^* \times (\mathbf{Z}/M\mathbf{Z})^*$. By restricting ε to each factor, we write ε as a product of two characters: $\varepsilon = \varepsilon_p \cdot \varepsilon^{(p)}$ where ε_p is a character of $(\mathbf{Z}/p^r\mathbf{Z})^*$ and $\varepsilon^{(p)}$ is a character of $(\mathbf{Z}/M\mathbf{Z})^*$. The character $\varepsilon^{(p)}$ has conductor dividing M, so it is unramified at p. By class field theory, ε_p is totally ramified at p, so the reduction $\overline{\varepsilon}$ is unramified at p precisely when $\overline{\varepsilon}_p = 1$; equivalently, $\overline{\varepsilon}$ is unramified at p exactly when ε_p has order a power of ℓ. If ε_p is non-trivial, then, since the order of ε_p divides the order $p^{r-1}(p-1)$ of a generator of $(\mathbf{Z}/p^r\mathbf{Z})^*$, a power of ℓ divides $p^{r-1}(p-1)$, so $p \equiv 1 \pmod \ell$ since $\ell \neq p$. □

In addition to his conjectures about the optimal weight and level, Serre also made a conjecture about the optimal character of a form giving rise to ρ. Let p be a prime not dividing $\ell N(\rho)$. Serre's optimal character conjecture implies that ρ, which we know to arise from a form on $\Gamma_1(M) \cap \Gamma_1(p^r)$, arises from a form on $\Gamma_1(M) \cap \Gamma_0(p^r)$, and this has been proved in most cases.

3.3.5. Approaches to level optimization in the key case

As discussed in Section 3.3.4, results of Carayol and Livné (see [12, 70]) reduce the level optimization problem to the following key case. The weight-2 newform f, a priori on $\Gamma_1(N^*)$, is in fact on the bigger group $\Gamma_1(M) \cap \Gamma_0(p)$, where $Mp = N^*$, $p \nmid M$, and ρ is unramified at p. The goal is to show that ρ arises from a newform on $\Gamma_1(M)$. This has been achieved when ℓ is odd, and in many cases when $\ell = 2$, using several level optimization techniques.

I. **Mazur's principle**

If either $\rho(\mathrm{Frob}_p)$ is not a scalar matrix or $p \not\equiv 1 \pmod{\ell}$, then an argument of Mazur, explained in Section 3.3.9, can be used to optimize the level.

II. **Multiplicity one**

It is possible to optimize the level if ρ is a multiplicity one representation, as explained in [84, 9] and Section 3.3.11. The cases in which multiplicity one is known were reviewed in Section 3.3.3. In particular, we do not know multiplicity one in some cases when $k(\rho) = \ell$ and the eigenvalues of Frob_p are not distinct.

III. **Using a pivot**

Suppose that M can be written as a product $M = qK$ with q a prime not dividing pK, that ρ arises from a form on $\Gamma_1(K) \cap \Gamma_0(pq)$, and that ρ is ramified at q and unramified at p. Then q can be used as a "pivot" to remove p from the level. This approach grew out of [83], and was introduced in the short paper [86]. In Section 3.3.10 we describe the approach and discuss the terminology.

IV. **Without multiplicity one**

When ℓ is odd and $\varepsilon = 1$, the level optimization theorem was proved in [87] using an argument that does not require ρ to have multiplicity one. The hypothesis $\ell \neq 2$ is used in the proof of Proposition 7.8 of [87] to force splitting of a short exact sequence. In [26], Diamond extended the results of [87] to cover the case of arbitrary character, still under the assumption that ℓ is odd. One encounters seemingly insurmountable difficulties in trying to push this argument through when $\ell = 2$.

3.3.6. Some commutative algebra

In this section we set up some of the commutative algebra that is required in order to lower levels. There are two injective maps

$$S_2(\Gamma_1(M)) \hookrightarrow S_2(\Gamma_1(M) \cap \Gamma_0(p)).$$

One is the inclusion $f(q) \mapsto f(q)$ and the other is $f(q) \mapsto f(q^p)$ (see Exercise 18). The *p-new subspace* $S_2(\Gamma_1(M) \cap \Gamma_0(p))^{p\text{-new}}$ is the complement, with respect to the Petersson inner product, of the subspace $\mathcal{S}$ generated by the two images of $S_2(\Gamma_1(M))$. The p-new subspace can also be defined algebraically as the kernel of the natural map from $S_2(\Gamma_1(M) \cap \Gamma_0(p))$ to the direct sum of two copies of $S_2(\Gamma_1(M))$.

Let $\mathbf{T}$ denote the Hecke algebra acting on $S_2(\Gamma_1(M) \cap \Gamma_0(p))$. If $p \nmid M$, then T_p acts on $\mathcal{S}$ as a direct sum of two copies of its action on $S_2(\Gamma_1(M))$; otherwise, T_p usually does not act diagonally (see Exercise 19). The image of $\mathbf{T}$ in $\mathrm{End}(\mathcal{S})$ is a quotient $\overline{\mathbf{T}}$ called the *p-new quotient*. A representation ρ associated to a maximal ideal $\mathfrak{m}$ of $\mathbf{T}$ arises from level M if and only if $\mathfrak{m}$ arises by pullback from a maximal ideal of $\overline{\mathbf{T}}$. Because the map $\mathbf{T} \to \overline{\mathbf{T}}$ is surjective, $\mathfrak{m}$ arises from level M if and only if the image of $\mathfrak{m}$ in $\overline{\mathbf{T}}$ is not the unit ideal (see Exercise 21).

3.3.7. Aside: Examples in characteristic two

Sections 3.3.7 and 3.3.8 can be safely skipped on a first reading.

To orient the reader, we focus for the moment on mod 2 representations that arise from elliptic curves. We give examples in which one of the level optimization methods applies but the others do not. We do not consider method **IV** because it is not applicable to mod 2 representations. The hypothesis of the "multiplicity one" method **II** when $\ell = 2$ are discussed after the statement of Theorem 3.19 in Section 3.3.11. We were unable to find an example in which none of the level optimization theorems applies.

We will repeatedly refer to the following theorem, which first appeared in [**85**].

Theorem 3.11. *Suppose ρ arises from a newform in $S_2(\Gamma_0(N))$. Let $p \nmid \ell N$ be a prime satisfying one or both of the identities*

$$\operatorname{tr} \rho(\operatorname{Frob}_p) = \pm(p+1) \pmod{\ell}.$$

Then ρ arises from a newform of level pN.

3.3.7.1. III applies but I and II do not

In this section we give a mod 2 representations in which the pivot hypothesis of **III** is satisfied, but the hypotheses of **I** and **II** are not. Our example is obtained by applying Theorem 3.11 to the mod 2 representation attached to a well-chosen elliptic curve.

We will find an elliptic curve E of conductor $M = qR$ such that $\rho = E[2]$ is absolutely irreducible, ramified at q, unramified at 2, and $\rho(\operatorname{Frob}_2) = \left(\begin{smallmatrix} 1 & 0 \\ 0 & 1 \end{smallmatrix}\right)$. Because of the last condition, [**9**, Prop. 2.4] does not imply that ρ is a multiplicity one representation, so **II** does not apply. Likewise, **I** does not apply because $\rho(\operatorname{Frob}_2)$ is a scalar and the p we will chose will satisfy $p \equiv 1 \pmod{2}$. Next we choose a prime $p \nmid 2qR$ such that $\rho_{E,2}(\operatorname{Frob}_p) = \left(\begin{smallmatrix} 1 & 0 \\ 0 & 1 \end{smallmatrix}\right)$. Let f be the newform associated to E. By Theorem 3.11 there is a newform g of level pqR such that

$$\rho_{g,\lambda} \approx \rho_{E,2}.$$

In particular,

$$\rho_{g,\lambda}(\operatorname{Frob}_p) = \rho_{E,2}(\operatorname{Frob}_p) = \left(\begin{smallmatrix} 1 & 0 \\ 0 & 1 \end{smallmatrix}\right)$$

is scalar and $p \equiv 1 \pmod{2}$, so **I** does not apply. However, method **III** does apply with q used as a pivot.

For example, consider the elliptic curve E defined by the equation

$$y^2 + xy = x^3 - x^2 + 19x - 32.$$

The conductor of E is $N = 19 \cdot 109$, and the discriminant of the field $K = \mathbf{Q}(E[2])$ is $-19^3 \cdot 109^3$. We select $q = 19$ as our pivot. The prime $p = 73$ splits completely in K, so

$$\rho_{E,2}(\operatorname{Frob}_p) = \begin{pmatrix} 1 & 0 \\ 0 & 1 \end{pmatrix}.$$

By Theorem 3.11 there is a form g of level $109 \cdot 19 \cdot 73$ that is congruent to the newform f attached to E modulo a prime lying over 2. Method **III** can be used to optimize the level, but neither method **I** nor **II** applies.

3.3.7.2. II applies but I and III do not

We exhibit a mod 2 representation for which method **II** can be used to optimize the level, but neither method **I** nor **III** applies. Let K be the $\operatorname{GL}_2(\mathbf{F}_2)$-extension

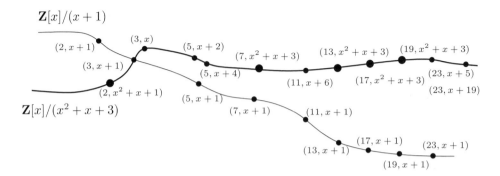

Figure 1. The spectrum of $\mathbf{T} \subset \mathrm{End}(S_2(\Gamma_0(33)))$, with $x = T_3$

of $\mathbf{Q}$ obtained by adjoining all cube roots of 2. Then $K = \mathbf{Q}(E[2])$, where E is the elliptic curve $X_0(27)$ given by the equation $y^2 + y = x^3 - 7$. The prime $p = 31$ splits completely in K, so by Theorem 3.11 there is a newform f of level $31 \cdot 27$ and a maximal ideal λ of the appropriate Hecke algebra such that $\rho_{f,\lambda} \approx E[2]$. Neither method **I** nor **III** can be used to optimize the level of $\rho_{f,\lambda}$. Method **I** doesn't apply because 31 is odd and $\rho_{f,\lambda}(\mathrm{Frob}_{31}) = \left(\begin{smallmatrix} 1 & 0 \\ 0 & 1 \end{smallmatrix}\right)$; method **III** doesn't apply because the only odd prime that is ramified in K is 3, which does not exactly divide $31 \cdot 27$. If D_2 is a decomposition group at 2 then D_2 has image in $\mathrm{GL}_2(\mathbf{F}_2)$ of order 2, so it is not contained in the scalar matrices and **II** can be used to optimize the level of $\rho_{f,\lambda}$.

3.3.8. Aside: Sketching the spectrum of the Hecke algebra

It is helpful to understand the Hecke algebra geometrically using the language of schemes (see, e.g., [38]). The topological space underlying the scheme $\mathrm{Spec}(\mathbf{T})$ is the set of prime ideals of $\mathbf{T}$ endowed with the Zariski topology, in which the closed sets are the set of prime ideals containing a fixed ideal.

We can draw $\mathrm{Spec}(\mathbf{T})$ by sketching a diagram whose irreducible components correspond to the Galois conjugacy classes of eigenforms, and whose intersections correspond to congruences between eigenforms. When the level is not cube free, $\mathbf{T}$ can contain nilpotent elements, and then one might wish to include additional information. If $\sum a_n q^n$ is an eigenform, then the failure of $\mathbf{Z}[\ldots a_n \ldots]$ to be integrally closed can be illustrated by drawing singular points on the corresponding irreducible component; however, we do not do this below.

Example 3.12. The spectrum of the Hecke algebra associated to $\Gamma_0(33)$ is illustrated in Figure 1. The Hecke algebra $\mathbf{T} \subset S_2(\Gamma_0(33))$ has discriminant -99, as does the characteristic polynomial of T_3, so

$$\mathbf{T} = \mathbf{Z}[T_3]/((T_3+1)(T_3^2 + T_3 + 3)) \cong \mathbf{Z}[x]/((x+1)(x^2+x+3)).$$

We sketch a curve corresponding to each of the two irreducible components. Some of the closed points (maximal) ideals are represented as dots. One component corresponds to the unique newform on $\Gamma_0(33)$, and the other corresponds to the two images of the newform on $\Gamma_0(11)$.

Example 3.13. Figure 2 is a diagram of the Hecke algebra associated to $S_2(\Gamma_0(3 \cdot 47))$. We have labeled fewer closed points than in Figure 1. The components are

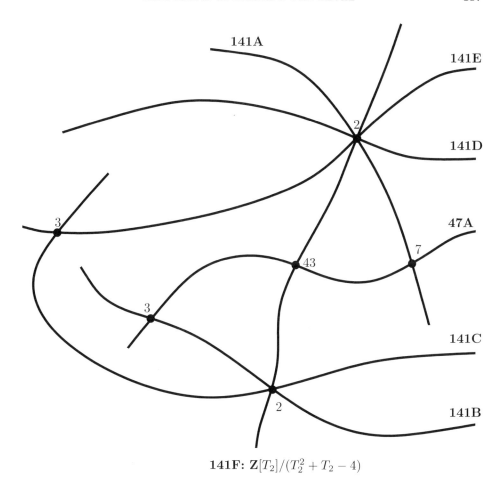

Figure 2. The spectrum of $\mathbf{T} \subset \mathrm{End}(S_2(\Gamma_0(141)))$

labeled by their isogeny class and the level at which they are new (the notation extends that of [**20**]). The component labeled **141F** corresponds to an eigenform whose Fourier coefficients generate a quadratic extension of $\mathbf{Q}$.

The newform corresponding to the elliptic curve A from Example 3.8 is labeled **141A**. Geometrically, the assertion that the level of $\rho_{A,7}$ can be optimized is represented by the characteristic-7 intersection between the component labeled **141A** and the old component **47A** coming from the unique Galois conjugacy class of newforms on $\Gamma_0(47)$.

3.3.9. Mazur's principle

A principle due to Mazur can be used to optimize the level in the key case, provided that a mild hypothesis is satisfied. The principle applies whenever $p \not\equiv 1 \pmod{\ell}$ and also in the case when $p \equiv 1 \pmod{\ell}$ but $\rho(\mathrm{Frob}_p)$ is not a scalar. This principle first appeared in [**84**, §6], then in [**26**, §4], and most recently when $\ell = 2$ in [**9**, pg. 7].

Theorem 3.14 (Mazur's Principle). *Suppose that $\rho : G_{\mathbf{Q}} \to \mathrm{GL}(2, \overline{\mathbf{F}}_\ell)$ arises from a newform f of weight 2 and level Mp, with $p \nmid M$, and character ε of conductor dividing M. Assume that ρ is unramified at p and that either $\rho(\mathrm{Frob}_p)$ is not a scalar matrix or $p \not\equiv 1 \pmod{\ell}$. Then ρ arises from a modular of level dividing M.*

We will require the following basic fact later in the proof.

Lemma 3.15 (Li). *Let $f = \sum a_n q^n$ be a newform on $\Gamma_1(M) \cap \Gamma_0(p)$ of weight k. Then $a_p^2 = \varepsilon(p) p^{k-2}$.*

Proof. Li's proof is an easy application of her generalization to Γ_1 of the Atkin-Lehner theory of newforms [**69**, Thm. 3(iii)]. The newform f is an eigenvector for the operator W_p which is defined on $S_k(\Gamma_1(M) \cap \Gamma_0(p))$ by

$$W_p(f) = p^{k/2} f\left(\frac{apz + b}{Mpz + p}\right),$$

where a and b are integers such that $ap^2 - bMp = p$. By [**69**, Lem. 3],

$$g := T_p(f) + p^{k/2-1} W_p(f)$$

lies in $S_k(\Gamma_1(M))$. For all primes $q \nmid Mp$, the eigenvalue of T_q on the oldform g is the same as the eigenvalue of T_q on the newform f, so $g = 0$. By [**69**, Lem. 2] $W_p^2(f) = \varepsilon(p) f$, so $a_p^2 = \varepsilon(p) p^{k-2}$. □

Remark 3.16. The case of Lemma 3.15 that we will need can also be understood in terms of the local representation $\rho|_{G_p}$, which resembles the mod ℓ representation attached to a Tate curve, in the sense that $\rho|_{G_p} \sim \left(\begin{smallmatrix}\alpha\chi & * \\ 0 & \alpha\end{smallmatrix}\right)$. Our hypothesis include the assumption that ρ is unramified at p, so the two characters $\alpha\chi$ and α are unramified at p. Thus $\alpha(\mathrm{Frob}_p)$ makes sense; we have $\alpha(\mathrm{Frob}_p) = \overline{a}_p(f)$ and $\alpha\chi(\mathrm{Frob}_p) = \overline{a}_p(f) p$. Since $\det(\rho|_{G_p}) = \alpha^2 \chi = \overline{\varepsilon}\chi$, we see that

$$\overline{a}_p^2 = \overline{\varepsilon}(p).$$

This local analysis of ρ was vastly generalized by Langlands in [**67**], which extends the analysis to include many ℓ-adic representations of possibly higher weight. See also [**13**].

Let $\mathbf{T}$ be the Hecke algebra associated to $\Gamma_1(M) \cap \Gamma_0(p)$, and let $\mathfrak{m}$ be the kernel of the following map $\mathbf{T} \to \overline{\mathbf{F}}_\ell$:

$$0 \longrightarrow \mathfrak{m} \longrightarrow \mathbf{T} \xrightarrow{T_n \mapsto \overline{a}_n, \langle d \rangle \mapsto \overline{\varepsilon}(d)} \overline{\mathbf{F}}_\ell.$$

As in Lemma 3.1, the determinants and traces of elements in the image of $\rho = \rho_\mathfrak{m}$ lie in $\mathbf{T}/\mathfrak{m} \subset \overline{\mathbf{F}}_\ell$, so there is a vector space $V \approx (\mathbf{T}/\mathfrak{m})^{\oplus 2}$ that affords $\rho_\mathfrak{m}$.

Next we realize $\rho_\mathfrak{m}$ as a group of division points in a Jacobian. The curve $X_1(Mp)$ corresponding to $\Gamma_1(Mp)$ covers the curve $X_1(M, p)$ corresponding to $\Gamma_1(M) \cap \Gamma_0(p)$. The induced map $J = \mathrm{Jac}(X_1(M, p)) \to J_1(Mp) = \mathrm{Jac}(X_1(Mp))$ has a finite kernel on which the Galois action is abelian.

Just as in Section 2.2.3.1.1, the Hecke algebra associated to $\Gamma_1(M) \cap \Gamma_0(p)$, can be constructed as a ring of correspondences on $X_1(M, p)$, then viewed as a subring $\mathbf{T} \subset \mathrm{End}_{\mathbf{Q}}(J)$. Inside of J we find the nonzero $G_{\mathbf{Q}}$-module $J[\mathfrak{m}] \approx \oplus_{i=1}^t V$. For the purposes of this discussion, we do not need to know that $J[\mathfrak{m}]$ is a direct sum of copies of V. The following weaker assertion, known long ago to Mazur [**71**, §14, pg. 112], will suffice: *$J[\mathfrak{m}]$ is a successive extension of copies of V*. In particular, $V \subset J[\mathfrak{m}]$. A weaker conclusion, true since $\ell \in \mathfrak{m}$, is that $V \subset J[\ell]$,

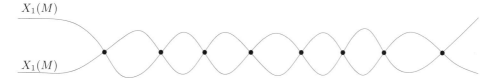

Figure 3. The reduction mod p of the Deligne-Rapoport model of $X_1(M, p)$

Our hypothesis that ρ is unramified at p translates into the inclusion $V \subset J[\ell]^{I_p}$, where I_p is an inertia group at p. By [**104**, Lem. 2], if A is an abelian variety over $\mathbf{Q}$ with good reduction at p, then $A[\ell]^{I_p} \cong A_{\mathbf{F}_p}[\ell]$. However, the modular curve $X_1(M, p)$ has bad reduction at p, so J is likely to have bad reduction at p—in this case it does. We are led to consider the Néron model $\mathcal{J}$ of J (see, e.g., [**5**]), which is a smooth commutative group scheme over $\mathbf{Z}$ satisfying the following property: the restriction map $\mathrm{Hom}_{\mathbf{Z}}(\mathcal{S}, \mathcal{J}) \longrightarrow \mathrm{Hom}_{\mathbf{Q}}(\mathcal{S}_{\mathbf{Q}}, J)$ is bijective for all smooth schemes $\mathcal{S}$ over $\mathbf{Z}$. Passing to the scheme-theoretic closure, we have, inside of $\mathcal{J}$, a two-dimensional $\mathbf{T}/\mathfrak{m}$-vector space scheme $\mathcal{V}$.

In Section 2.2.3.1.1 we only defined $X_1(M, p)$ as a scheme over $\mathbf{Z}[1/Mp]$. Deligne and Rapoport [**25**] extended $X_1(M, p)$ to a scheme over $\mathbf{Z}[1/M]$ and computed the reduction modulo p. The introduction to [**62**] contains a beautiful historical discussion of the difficulties involved in extending modular curves over $\mathbf{Z}$.

We know a great deal about the reduction of $X_1(M, p)$ at p, which is frequently illustrated by the squiggly diagram in Figure 3. This reduction is the union of 2 copies of $X_1(M)_{\mathbf{F}_p}$ intersecting transversely at the supersingular points.

The subspace $S_2(\Gamma_1(M)) \oplus S_2(\Gamma_1(M))$ of $S_2(\Gamma_1(M) \cap \Gamma_0(p))$ is stable under the Hecke algebra $\mathbf{T}$, so there is a map $\mathbf{T} \to \mathrm{End}(S_2(\Gamma_1(M)) \oplus S_2(\Gamma_1(M)))$. The *p-old quotient* of $\mathbf{T}$ is the image $\overline{\mathbf{T}}$. Since the map $\mathbf{T} \to \overline{\mathbf{T}}$ is surjective, the image of $\mathfrak{m}$ in $\overline{\mathbf{T}}$ is an ideal $\overline{\mathfrak{m}}$. To optimize the level in the key case amounts to showing that $\overline{\mathfrak{m}}$ is not the unit ideal.

As is well known (cf. [**71**, Appendix, Prop 1.4]), the results of M. Raynaud [**82**] and Deligne-Rapoport [**25**] combine to produce an exact sequence

$$(3.1) \qquad 0 \longrightarrow \mathcal{T} \longrightarrow \mathcal{J}^0_{\mathbf{F}_p} \longrightarrow J_1(M)_{\mathbf{F}_p} \times J_1(M)_{\mathbf{F}_p} \longrightarrow 0,$$

where $\mathcal{T}$ is a torus, i.e., $\mathcal{T}_{\overline{\mathbf{F}}_p} \approx \mathbf{G}_m \times \cdots \times \mathbf{G}_m$, and $\mathcal{J}^0_{\mathbf{F}_p}$ is the identity component of $\mathcal{J}_{\mathbf{F}_p}$. There is a concrete description of $\mathcal{T}$ and of the maps in the exact sequence. Each object in the sequence is equipped with a functorial action of the Hecke algebra $\mathbf{T}$, and the sequence is $\mathbf{T}$-invariant. The p-old quotient $\overline{\mathbf{T}}$ can be viewed as coming from the action of $\mathbf{T}$ on $J_1(M)_{\mathbf{F}_p} \times J_1(M)_{\mathbf{F}_p}$.

By a generalization of [**104**, Lem. 2], the reduction map $J(\overline{\mathbf{Q}}_p)[\ell]^{I_p} \to \mathcal{J}_{\mathbf{F}_p}(\overline{\mathbf{F}}_p)$ is injective. Thus $V = \mathcal{V}_{\mathbf{F}_p}(\overline{\mathbf{F}}_p) \subset \mathcal{J}_{\mathbf{F}_p}(\overline{\mathbf{F}}_p)$. The component group $\Phi = \mathcal{J}_{\mathbf{F}_p}/\mathcal{J}^0_{\mathbf{F}_p}$ is *Eisenstein*, in the sense that it does not contain irreducible representations arising from eigenforms. Since V is irreducible, as a Galois module Φ does not contain an isomorphic copy of V, so $\mathcal{V}_{\mathbf{F}_p} \subset \mathcal{J}^0_{\mathbf{F}_p}$ and we have the following diagram:

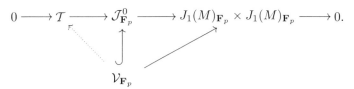

Since $\mathfrak{m}$ acts as 0 on V, the image $\overline{\mathfrak{m}}$ of $\mathfrak{m}$ acts as 0 on the image of V in $J_1(M)_{\mathbf{F}_p} \times J_1(M)_{\mathbf{F}_p}$. If $\overline{\mathfrak{m}} \neq (1)$ then we can optimize the level, so assume $\overline{\mathfrak{m}} = (1)$. Then the image of V in $J_1(M)_{\mathbf{F}_p} \times J_1(M)_{\mathbf{F}_p}$ is 0, so $\mathcal{V}_{\mathbf{F}_p} \hookrightarrow \mathcal{T}$.

Let $\mathcal{X}_p(J) := \mathrm{Hom}(\mathcal{T}, \mathbf{G}_m)$ be the *character group* of $\mathcal{T}$. The action of $\mathbf{T}$ on $\mathcal{T}$ induces an action of $\mathbf{T}$ on $\mathcal{X}_p(J)$. Furthermore, $\mathcal{X}_p(J)$ supports an action of $\mathrm{Gal}(\overline{\mathbf{F}}_p/\mathbf{F}_p)$ which, because tori split over a quadratic extension, factors through the Galois group of $\mathbf{F}_{p^2}$. View the Galois action as an action of $\mathrm{Frob}_p \in D_p = \mathrm{Gal}(\overline{\mathbf{Q}}_p/\mathbf{Q}_p)$. With our conventions, the action of Frobenius on the torus is as follows (cf. [**26**, pg. 31]).

Lemma 3.17. *The Frobenius* Frob_p *acts as* pT_p *on* $\mathcal{T}(\overline{\mathbf{F}}_p)$.

Make the identification $\mathcal{T} \cong \mathrm{Hom}_{\mathbf{Z}}(\mathcal{X}_p(J), \mathbf{G}_m)$, so that
$$V \subset \mathcal{T}(\overline{\mathbf{F}}_p)[\ell] = \mathrm{Hom}_{\mathbf{Z}}(\mathcal{X}_p(J), \boldsymbol{\mu}_\ell).$$
By Lemma 3.17, Frob_p acts on $V \subset \mathcal{T}(\overline{\mathbf{F}}_p)$ as $pa_p \in \mathbf{T}/\mathfrak{m}$, i.e., as a *scalar*. The determinant of ρ is $\chi\varepsilon$, so we have simultaneously
$$\det(\rho(\mathrm{Frob}_p)) = \begin{cases} p\varepsilon(p) & \text{and} \\ (pa_p)^2. \end{cases}$$
By Lemma 3.15, $a_p^2 = \varepsilon(p)$, so $p^2 \equiv p \pmod{\ell}$. Since $p \neq \ell$, this can only happen if $p \equiv 1 \pmod{\ell}$, which completes the proof.

3.3.10. Level optimization using a pivot

In this section we discuss an approach to level optimization that does not rely on multiplicity one results. In this approach, we eliminate a prime p from the level by making use of the rational quaternion algebra that is ramified precisely at p and at a second prime q. The latter prime is, in the simplest case, an appropriate prime number at which ρ ramifies; in more complicated cases, it is an "auxiliary" prime at which ρ is unramified. The central role of q in the argument, and the fact that q stays fixed in the level while p is removed, leads us to refer to q as a "pivot."

The following theorem first appeared in [**86**].

Theorem 3.18. *Let* $\rho : G_{\mathbf{Q}} \to \mathrm{GL}(2, \overline{\mathbf{F}}_\ell)$ *be an irreducible continuous representation that arises from an eigenform f on $\Gamma_1(K) \cap \Gamma_0(pq)$ with p and q distinct primes that do not divide ℓK. Make the **key assumption** that the representation ρ is ramified at q and unramified at p. Then ρ arises from a weight-2 eigenform on $\Gamma_1(K) \cap \Gamma_0(q)$.*

The case $\ell = 2$ is not excluded from consideration.

Before sketching the proof, we describe a famous application. Edixhoven suggested to the first author that such an approach might be possible in the context of

Fermat's Last Theorem. We associate to a (hypothetical) solution $a^\ell + b^\ell + c^\ell = 0$ of the Fermat equation with $\ell > 3$ a Galois representation $E[\ell]$ attached to an elliptic curve E. A theorem of Mazur implies that this representation is irreducible; a theorem of Wiles implies that it arises from a modular form. Using Tate's algorithm, we finds that the discriminant of E is $\Delta_E = \frac{(abc)^{2\ell}}{2^8}$, which is a perfect ℓth power away from 2, and that the conductor of E is $N_E = \mathrm{rad}(abc) = \prod_{p \mid abc} p$. Let $q = 2$; then $E[\ell]$ is ramified at q because $\ell \nmid \mathrm{ord}_2(\Delta_E) = -8$ (see Theorem 2.10), but $E[\ell]$ is unramified at all other primes p, again by Theorem 2.10. To complete the proof of Fermat Last Theorem, we use $q = 2$ as a pivot and inductively remove each odd factor from N. One complication that may arise (the second case of Fermat Last Theorem) is that $\ell \mid N$. Upon removing ℓ from the level (using Section 3.3.1), the weight may initially go up to $\ell + 1$. If this occurs, since $k(\rho) = 2$ we can use [**32**] to optimize the weight back to 2.

As demonstrated by the application to Fermat, in problems of genuine interest the setup of Theorem 3.18 occurs. There are, however, situations in which it does not apply such as the recent applications of level optimization as a key ingredient to a proof of Artin's conjecture for certain icosahedral Galois representations (see [**10**]).

3.3.10.1. Shimura curves

We cannot avoid considering Shimura curves. Denote by $X(K, pq)$ the modular curve associated to $\Gamma_1(K) \cap \Gamma_0(pq)$ and let $J := \mathrm{Jac}(X(K, pq))$ be its Jacobian. Likewise, denote by $X^{pq}(K)$ the Shimura curve associated to the quaternion algebra of discriminant pq. The curve $X^{pq}(K)$ is constructed as follows. Let B be an indefinite quaternion algebra over $\mathbf{Q}$ of discriminant pq. (Up to isomorphism, B is unique.) Let $\mathcal{O}$ be an Eichler order (i.e., intersection of two maximal orders) of level K (i.e., reduced discriminant Kpq) in B. Let Γ_∞ be the group of elements of $\mathcal{O}$ with (reduced) norm 1. After fixing an embedding $B \to M(2, \mathbf{R})$ (an embedding exists because B is indefinite), we obtain in particular an embedding $\Gamma_\infty \hookrightarrow \mathrm{SL}(2, \mathbf{R})$ and therefore an action of Γ_∞ on the upper half-plane $\mathfrak{h}$. Let $X^{pq}(K)$ be the standard canonical model, over $\mathbf{Q}$, of the compact Riemann surface $\Gamma_\infty \backslash \mathfrak{h}$, and let $J' = \mathrm{Jac}(X^{pq}(K))$ denote its Jacobian. The curve $X^{pq}(K)$ is furnished with Hecke correspondences T_n for $n \geq 1$. We write T_n for the endomorphism of J induced by the T_n on $X^{pq}(K)$ via Pic functoriality.

Set $J' := \mathrm{Jac}(X^{pq}(K))$ and $J := \mathrm{Jac}(X(K, pq))$. Work of Eichler, Jacquet-Langlands, and Shimura (see [**36**, **51**, **106**]) has uncovered a deep correspondence between certain automorphic forms and certain cusp forms. Combining their work with the isogeny theorem of Faltings [**40**], we find (noncanonically!) a map $J' \to J$ with finite kernel.

The pq-new part of J is $J_{pq\text{-new}} := \ker(J(K, pq) \longrightarrow J(K, p)^2 \oplus J(K, q)^2)$ where the map is induced by Albanese functoriality from the four maps

$$X(K, pq) \rightrightarrows X(K, p) \qquad \text{and} \qquad X(K, pq) \rightrightarrows X(K, q).$$

The image of $J' \to J$ is the pq-new part of J.

3.3.10.2. Character groups

Amazingly, there seems to be no canonical map $J' \to J$ between the Shimura and classical Jacobians described in the previous section. Surprisingly, there is a canonical relationship between the character groups of J' and J. The Čerednik-Drinfeld theory gives a description of $X^{pq}(K)$ in characteristic p (see [**14**, **30**]). Using this we find a canonical **T**-equivariant exact sequence

$$(3.2) \qquad 0 \to \mathcal{X}_p(J') \to \mathcal{X}_q(J) \to \mathcal{X}_q(J'') \to 0$$

where $J'' = \mathrm{Jac}(X(K,q))^2$. This exact sequence relates a character group "in characteristic p" to two character groups "in characteristic q". We are now prepared to prove the theorem.

3.3.10.3. Proof

Proof of Theorem 3.18. By our key assumption, the representation ρ is ramified at q, so $\mathfrak{m} \subset \mathbf{T}$ is not q-old. We may as well suppose we are in a situation where we can not optimize the level, so we assume that $\mathfrak{m}$ is not p-old either and hope for a contradiction.

Localization is an exact functor, so the localization

$$(3.3) \qquad 0 \longrightarrow \mathcal{X}_p(J')_\mathfrak{m} \longrightarrow \mathcal{X}_q(J)_\mathfrak{m} \longrightarrow \mathcal{X}_q(J'')_\mathfrak{m} \longrightarrow 0$$

of (3.2) is also exact. The Hecke algebra **T** acts on $\mathcal{X}_q(J'')$ through a quotient $\overline{\mathbf{T}}$. Since $\mathfrak{m}$ is not q-old, the image of $\mathfrak{m}$ in $\overline{\mathbf{T}}$ generates the unit ideal. Therefore $\mathcal{X}_q(J'')_\mathfrak{m} = 0$ and we obtain an isomorphism $\mathcal{X}_p(J')_\mathfrak{m} \approx \mathcal{X}_q(J)_\mathfrak{m}$. If R is a **T**-module then $R/\mathfrak{m}R = R_\mathfrak{m}/\mathfrak{m}R_\mathfrak{m}$ so

$$(3.4) \qquad \mathcal{X}_q(J)/\mathfrak{m}\mathcal{X}_q(J) \approx \mathcal{X}_p(J')/\mathfrak{m}\mathcal{X}_p(J').$$

Switching p and q and applying the same argument shows that

$$(3.5) \qquad \mathcal{X}_p(J)/\mathfrak{m}\mathcal{X}_p(J) \approx \mathcal{X}_q(J')/\mathfrak{m}\mathcal{X}_q(J').$$

Both (3.4) and (3.5) are isomorphisms of $\mathbf{T}/\mathfrak{m}$-vector spaces.

By [**6**] we have an isomorphism $J[\mathfrak{m}] \approx \bigoplus_{i=1}^\lambda V$, with $\lambda > 0$ and $J'[\mathfrak{m}] \approx \bigoplus_{i=1}^\nu V$. (It follows from [**51**] that $\nu > 0$, but we will not use this here.) Our hypothesis that V is unramified automatically propagates to all of $J[\mathfrak{m}] \approx \bigoplus_{i=1}^\lambda V$. Since V is irreducible and we are assuming that $\mathfrak{m}$ is not p-old, the same argument as in Section 3.3.9 shows that $J[\mathfrak{m}] \subset \mathcal{T}[\mathfrak{m}]$ where $\mathcal{T}$ is the toric part of $\mathcal{J}_{\mathbf{F}_p}$. This means that $\dim(\mathcal{X}_p(J)/\mathfrak{m}\mathcal{X}_p(J)) \geq 2\lambda$. Using the same argument with J replaced by J' gives that $\dim(\mathcal{X}_p(J')/\mathfrak{m}\mathcal{X}_p(J')) \geq 2\mu$.

As an I_q-module V is an extension of two copies of the trivial character. This follows from results of Langlands [**67**], since ρ is a mod ℓ representation of $G_\mathbf{Q}$ associated to some newform f whose level divides pqK and is divisible by q. (The admissible representation of $\mathrm{GL}(2, \mathbf{Q}_q)$ which is associated to f is a special representation.) Because V is ramified at q and there is an unramified line, we see that $\dim(V^{I_q}) = 1$. Thus $\dim J[\mathfrak{m}]^{I_q} = \lambda$; since $q \neq \ell$ and the action of inertia on character groups is trivial, we see that

$$\mathrm{Hom}(\mathcal{X}_q(J)/\mathfrak{m}\mathcal{X}_q(J), \boldsymbol{\mu}_\ell) \subset J[\mathfrak{m}]^{I_q},$$

so $\dim \mathcal{X}_q(J)/\mathfrak{m}\mathcal{X}_q(J) \leq \lambda$. A similar argument bounds $\dim \mathcal{X}_q(J')/\mathfrak{m}\mathcal{X}_q(J')$. We obtain the following quadruple of inequalities:

$$\dim \mathcal{X}_q(J)/\mathfrak{m}\mathcal{X}_q(J) \leq \lambda,$$
$$\dim \mathcal{X}_q(J')/\mathfrak{m}\mathcal{X}_q(J') \leq \mu,$$
$$\dim \mathcal{X}_p(J)/\mathfrak{m}\mathcal{X}_p(J) \geq 2\lambda,$$
$$\dim \mathcal{X}_p(J')/\mathfrak{m}\mathcal{X}_p(J') \geq 2\mu.$$

Combining these with (3.4, 3.5), we find that

$$2\lambda \leq \dim \mathcal{X}_p(J)/\mathfrak{m}\mathcal{X}_p(J)$$
$$= \dim \mathcal{X}_q(J')/\mathfrak{m}\mathcal{X}_q(J')$$
$$\leq \mu$$

and simulatenously that $2\mu \leq \lambda$. Together these imply that $4\lambda \leq \lambda$ so $\lambda = 0$. But Proposition 3.2 implies that the multiplicity of ρ in $J[\mathfrak{m}]$ is strictly positive. This contradiction implies that our assumption that $\mathfrak{m}$ is not p-old is false, hence $\mathfrak{m}$ is p-old and ρ arises from an eigenform on $\Gamma_1(K) \cap \Gamma_0(q)$. □

3.3.11. Level optimization with multiplicity one

Theorem 3.19. *Suppose $\rho : G_{\mathbf{Q}} \to \mathrm{GL}_2(\overline{\mathbf{F}}_\ell)$ is an irreducible multiplicity one representation that arises from a weight-2 newform f on $\Gamma_1(M) \cap \Gamma_0(p)$ and that p is unramified. Then there is a newform on $\Gamma_1(M)$ that also gives rise to ρ.*

We sketch a proof, under the assumption that $\ell > 2$. Buzzard [**9**] has given a proof when $\ell = 2$; his result has been combined with the results of [**28**] to prove a Wiles-like lifting theorem valid for many representations when $\ell = 2$, and hence (thanks to Taylor) to establish new examples of Artin's conjecture (see [**10**]).

The following diagram illustrates the multiplicity one argument:

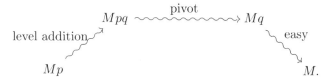

The pivot step is potentially the hardest; though it resembles the pivot step of Section 3.3.10, but the symmetry is broken. In Section 3.3.10 we knew that q could not be removed from the level, but here q can be.

We manufacture q as follows. Pick q to be one of the (infinitely many) primes not dividing $Mp\ell$ such that the following conditions both hold:

1. $\begin{cases} \rho(\mathrm{Frob}_q) \text{ is not a scalar, } or \\ q \not\equiv 1 \pmod{\ell}. \end{cases}$

2. $\begin{cases} \text{the ratio of the eigenvalues of} \\ \rho(\mathrm{Frob}_q) \text{ is either } q \text{ or } 1/q. \end{cases}$

The second condition means that the characteristic polynomial of $\rho(\mathrm{Frob}_q)$ is of the form $(x - a)(x - qa)$ for some $a \in \overline{\mathbf{F}}_\ell^*$.

Lemma 3.20. *There are infinitely many primes q that simultaneously satisfy both of the two conditions listed above.*

Proof. First assume that $\ell > 2$. Using the Cebotarev density theorem, find infinitely many primes q such $\rho(\text{Frob}_q) = \rho(c)$ where c denotes complex conjugation. The eigenvalues of $\rho(c)$ are 1 and -1 (Exercise 8), so their ratio is -1. This ratio is equal to q because
$$-1 = \chi(c) = \det(\rho(\text{Frob}_q)) = \chi(\text{Frob}_q) \equiv q \pmod{\ell},$$
and $q \not\equiv 1 \pmod{\ell}$ because ℓ is odd.

Next assume that $\ell = 2$. Because ρ is irreducible, the image $\rho(G_{\mathbf{Q}}) \subset \text{GL}(2, \overline{\mathbf{F}}_2)$ has even order. After a possible change of basis we find $\left(\begin{smallmatrix} 1 & 1 \\ 0 & 1 \end{smallmatrix}\right) \in \rho(G_{\mathbf{Q}})$. Using Cebotarev density, we find infinitely many q with $\rho(\text{Frob}_q)$ conjugate to $\left(\begin{smallmatrix} 1 & 1 \\ 0 & 1 \end{smallmatrix}\right)$. For such q, condition (1) is satisfied. Condition (2) is also satisfied because the ratio of the eigenvalues is 1 which, because q is an odd prime, is congruent to q modulo $\ell = 2$. $\square$

Sketch of proof of Theorem 3.19. Choose q as in Lemma 3.20. With q thus chosen, we can raise the level. More precisely, there exists a pq-new form on $\Gamma_1(M) \cap \Gamma_0(pq)$. We illustrate this as follows.

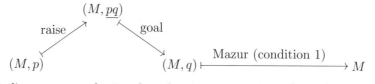

We underline pq to emphasize that the situation at level (M, pq) is symmetrical in p and q.

Let $J = J(M, pq)$; there is a maximal ideal $\mathfrak{m}$ in $\mathbf{T} = \mathbf{Z}[\ldots T_n \ldots] \subset \text{End } J$ attached to the pq-newform f that gives rise to ρ. Applying the multiplicity one hypothesis at level Mpq, we have $J[\mathfrak{m}] = V$ where V is a $\mathbf{T}/\mathfrak{m}$-vector space that supports ρ. In everything so far, M can be divisible by 2; the distinction between whether or not 2 divides M arises mainly in verifying the multiplicity one hypothesis.

Let $J' = J^{pq}(M)$ be the Shimura curve analogue of $J_1(M)$. As described in Section 3.3.10, J' is constructed in a similar manner as $J_1(M)$, but with $M_2(\mathbf{Q})$ replaced by a quaternion algebra. Of primary importance is that $J'[\mathfrak{m}] \approx \bigoplus_{i=1}^{\nu} V$, for some $\nu \geq 1$. This follows *morally* because ρ arises from a pq-new form, though the actual argument is quite involved.

Assume that we cannot optimize the level. We have an exact sequence of character groups
$$0 \longrightarrow \mathcal{X}_p(J') \longrightarrow \mathcal{X}_q(J) \longrightarrow \mathcal{X}_q(J(M,q)^2) \longrightarrow 0.$$
After localizing at $\mathfrak{m}$ as in (3.3), we discover that

(3.6) $\qquad \dim_{\mathbf{T}/\mathfrak{m}} \mathcal{X}_p(J')/\mathfrak{m}\mathcal{X}_p(J') = \dim_{\mathbf{T}/\mathfrak{m}} \mathcal{X}_q(J)/\mathfrak{m}\mathcal{X}_q(J).$

Furthermore, since the component group of J' at p is a quotient of $\mathcal{X}_q(J(M,q)^2)$, we find that $V \hookrightarrow (J'[\mathfrak{m}]^{I_p})^{\text{toric}}$. Thus $\dim \mathcal{X}_p(J')/\mathfrak{m}\mathcal{X}_p(J') \geq 2$, so (3.6) implies that $\dim \mathcal{X}_q(J)/\mathfrak{m}\mathcal{X}_q(J) \geq 2$. The endomorphism Frob_q acts as a scalar (cf. Lemma 3.17) on
$$J[\mathfrak{m}]^{\text{toric}} = \text{Hom}(\mathcal{X}_q(J)/\mathfrak{m}\mathcal{X}_q(J), \mu_\ell).$$

Furthermore, $J[\mathfrak{m}]^{\text{toric}} \subset J[\mathfrak{m}]$ and both $J[\mathfrak{m}]^{\text{toric}}$ and $J[\mathfrak{m}]$ have dimension 2, so Frob_q acts as a scalar on $J[\mathfrak{m}]$. If $q \not\equiv 1 \pmod{\ell}$ then we could use Mazur's principle to optimize the level, so by condition 1 we may assume that $\rho(\text{Frob}_q)$ is not a scalar. This contradiction completes the sketch of the proof. $\square$

CHAPTER 4
Exercises

The following exercises were used in the Park City problem sessions. D. Savitt, K. Kedlaya, and B. Conrad contributed some of the problems. In Section 4.4.2, we provide several solutions, many of which were suggested by students in the problem sessions. The solution of some of the problems in this section requires facts beyond those stated explicitly in this paper.

4.4.1. Exercises

Exercise 1. Suppose $\rho : \text{Gal}(\overline{\mathbf{Q}}/\mathbf{Q}) \to \mathbf{F}_\ell^*$ is a one-dimensional continuous odd Galois representation.

1. Give an example to show that ρ need not be a power of the mod ℓ cyclotomic character.
2. Assume that ρ is unramified outside ℓ. Deduce that ρ is a power of the mod ℓ cyclotomic character.

Exercise 2. The principal congruence subgroup $\Gamma(N)$ of level N is the kernel of the reduction map $\text{SL}(2, \mathbf{Z}) \to \text{SL}(2, \mathbf{Z}/N\mathbf{Z})$. The subgroup $\Gamma_1(N)$ consists of matrices of the form $\begin{pmatrix} 1 & * \\ 0 & 1 \end{pmatrix}$ modulo N. Let $\Gamma \subset \text{SL}(2, \mathbf{Z})$ be a subgroup that contains $\Gamma(N)$ for some N. Show that there exists $g \in \text{GL}(2, \mathbf{Q})$ such that the conjugate $g^{-1}\Gamma g$, which is a subgroup of $\text{GL}(2, \mathbf{Q})$, contains $\Gamma_1(N^2)$.

Exercise 3. Let k be a finite field of characteristic greater than 2, and consider an odd representation $\rho : G_\mathbf{Q} \to \text{GL}(2, k)$. Prove that ρ is irreducible if and only if ρ is absolutely irreducible. (A representation is absolutely irreducible if it remains irreducible after composing with the embedding $\text{GL}(2, \mathbf{F}_\ell) \hookrightarrow \text{GL}(2, \overline{\mathbf{F}}_\ell)$.) Give an example to show that this assertion is false when k has characteristic 2.

Exercise 4. Let $A/\mathbf{Q}$ be an elliptic curve. Show that the group of $\mathbf{Q}$-rational endomorphisms $\text{End}(A)$ of A is equal to $\mathbf{Z}$; that is, integer multiplications are the only $\mathbf{Q}$-rational endomorphisms of A. Assume further that A is isolated in its isogeny class, in the sense that if B is an elliptic curve that is isogenous to A over $\mathbf{Q}$, then A and B are isomorphic over $\mathbf{Q}$. Show that, for every prime number ℓ, the representation
$$\rho_\ell : \text{Gal}(\overline{\mathbf{Q}}/\mathbf{Q}) \to \text{Aut}(A[\ell]) \approx \text{GL}(2, \mathbf{F}_\ell)$$

is irreducible. Must ρ_ℓ be absolutely irreducible?

Exercise 5. Let $A/\mathbf{Q}$ be an elliptic curve and assume that for all ℓ the representation $\rho : \mathrm{Gal}(\overline{\mathbf{Q}}/\mathbf{Q}) \to \mathrm{Aut}(A[\ell])$ is irreducible. Deduce that A is isolated in its isogeny class. This is the converse of Exercise 4.

Exercise 6. Suppose $\rho : \mathrm{Gal}(\overline{\mathbf{Q}}/\mathbf{Q}) \to \mathrm{GL}(2, \mathbf{F}_\ell)$ arises from the ℓ-torsion of an elliptic curve. Verify, using standard properties of the Weil pairing, that $\det(\rho)$ is the mod ℓ cyclotomic character.

Exercise 7. Let $f \in S_k(\Gamma_1(N))$ be a modular form that is an eigenform for all the Hecke operators T_p and for the diamond bracket operators $\langle d \rangle$. Let
$$\varepsilon : (\mathbf{Z}/N\mathbf{Z})^* \to \mathbf{C}^*$$
be the character of f, so $\langle d \rangle f = \varepsilon(d) f$ for all $d \in (\mathbf{Z}/N\mathbf{Z})^*$.

1. Show that f satisfies the following equation: for any $\begin{pmatrix} a & b \\ c & d \end{pmatrix} \in \Gamma_0(N)$,
$$f(z) = \varepsilon(d)(cz+d)^{-k} f\left(\frac{az+b}{cz+d}\right).$$

2. Conclude that $\varepsilon(-1) = (-1)^k$.

3. Choose a prime ℓ and let ρ be one of the mod ℓ Galois representations associated to f. We have $\det(\rho) = \varepsilon \cdot \chi^{k-1}$ where χ is the mod ℓ cyclotomic character. Deduce that ρ is odd, in the sense that $\det(\rho(c)) = -1$ for c complex conjugation.

Exercise 8. Let $\rho : G_\mathbf{Q} \to \mathrm{GL}(2, \overline{\mathbf{F}}_\ell)$ be an odd Galois representation, and let $c \in G_\mathbf{Q}$ denote complex conjugation.

1. Prove that if $\ell \neq 2$ then $\rho(c)$ is conjugate over $\overline{\mathbf{F}}_\ell$ to the matrix $\begin{pmatrix} -1 & 0 \\ 0 & 1 \end{pmatrix}$.
2. Give an example to show that when $\ell = 2$, the matrix $\rho(c)$ need not be conjugate to $\begin{pmatrix} -1 & 0 \\ 0 & 1 \end{pmatrix}$.

Exercise 9. Show that there exists a *non-continuous* homomorphism
$$\rho : \mathrm{Gal}(\overline{\mathbf{Q}}/\mathbf{Q}) \to \{\pm 1\}$$
where $\{\pm 1\}$ has the discrete topology; equivalently, that there is a non-closed subgroup of index two in $\mathrm{Gal}(\overline{\mathbf{Q}}/\mathbf{Q})$. To accomplish this, you must produce a map $\rho : \mathrm{Gal}(\overline{\mathbf{Q}}/\mathbf{Q}) \to \{\pm 1\}$ such that

1. ρ is a homomorphism, and
2. ρ does not factor through $\mathrm{Gal}(K/\mathbf{Q})$ for any *finite* Galois extension $K/\mathbf{Q}$.

Exercise 10. A potential difficulty is that a representation ρ arising from a modular form sometimes takes values in a slightly smaller field than $\mathcal{O}/\lambda$. For example, let f be one of the two conjugate normalized eigenforms in $S_2(\Gamma_0(23))$. Then
$$f = q + \alpha q^2 + (-2\alpha - 1)q^3 + (-\alpha - 1)q^4 + 2\alpha q^5 + \cdots$$
with $\alpha^2 + \alpha - 1 = 0$. The coefficients of f lie in $\mathcal{O} = \mathbf{Z}[\alpha] = \mathbf{Z}[\frac{1+\sqrt{5}}{2}]$. Take λ to be the unique prime of $\mathcal{O}$ lying over 2; then $\mathcal{O}/\lambda \cong \mathbf{F}_4$, so $\overline{\rho}_{f,\lambda}$ is a homomorphism into $\mathrm{GL}(2, \mathbf{F}_4)$. Show that if $p \neq 2$ then $a_p \in \mathbf{Z}[\sqrt{5}]$, so that $\overline{\rho}_{f,\lambda}$ possesses a model over $\mathrm{GL}(2, \mathbf{F}_2)$.

Exercise 11. Let $A/\mathbf{Q}$ be an elliptic curve and $\ell \neq 2$ be a prime.

1. Prove that the field $\mathbf{Q}(A[\ell])$ generated by the coordinates of the points in $A[\ell]$ is strictly larger than $\mathbf{Q}$.
2. Given an example of an elliptic curve A such that $\mathbf{Q}(A[2]) = \mathbf{Q}$.

Exercise 12. Let A be an elliptic curve over $\mathbf{Q}$ defined by a Weierstrass equation $y^2 = x^3 + ax + b$ with $a, b \in \mathbf{Q}$.

1. Describe the Galois representation
$$\rho = \rho_{A,2} : G_\mathbf{Q} \to \mathrm{GL}(2, \mathbf{F}_2).$$
2. Give necessary and sufficient conditions for ρ to be reducible.
3. Choose a prime p, and give an example in which ρ is ramified only at p.

Exercise 13. Let ε and ρ be a pair of continuous homomorphisms from $G_\mathbf{Q}$ to $\mathbf{F}_\ell^*$. Suppose that for all primes p at which both ε and ρ are unramified we have
$$\rho(\mathrm{Frob}_p) = \varepsilon(\mathrm{Frob}_p) p^i \in \mathbf{F}_\ell.$$
Deduce that $\rho = \varepsilon \cdot \chi^i$ where χ is the mod ℓ cyclotomic character.

Exercise 14. Let $A/\mathbf{Q}$ be an elliptic curve of conductor N, and let p be a prime number not dividing N. Denote by $\tilde{A}$ the mod p reduction of A. The Frobenius endomorphism $\Phi = \Phi_p : \tilde{A} \to \tilde{A}$ sends an affine point (x, y) to (x^p, y^p) and fixes ∞. The characteristic polynomial of the endomorphism induced by Φ on the Tate module of $\tilde{A}$ at some (any) prime $\ell \neq p$ is $X^2 - \mathrm{tr}(\Phi) X + \deg(\Phi)$.

1. Show that $\deg(\Phi) = p$.
2. Show that $\mathrm{tr}(\Phi) = p + 1 - \#A(\mathbf{F}_p)$, that is, "$\mathrm{tr}(\Phi) = a_p$."
3. Choose a prime $\ell \nmid pN$. Then $\tilde{A}[\ell]$ is a vector space of dimension two over $\mathbf{F}_\ell$, and Φ induces a map $\tilde{A}[\ell] \to \tilde{A}[\ell]$. Show that this is the same as the map induced by some choice of $\mathrm{Frob}_p \in \mathrm{Gal}(\overline{\mathbf{Q}}/\mathbf{Q})$.
4. Conclude that
$$\mathrm{tr}(\rho_{A,\ell}(\mathrm{Frob}_p)) = p + 1 - \#A(F_p) \pmod{\ell}.$$

Exercise 15. Let $A/\mathbf{Q}$ be an elliptic curve of conductor N, and let ℓ be a prime. Show that any prime p not dividing ℓN is unramified in $\mathbf{Q}(A[\ell])$. You may use the following fact which is proved using formal groups (see, e.g., [**109**, Prop. 3.1]):
Fact: The map $A[\ell] \to \tilde{A}(\overline{\mathbf{F}}_p)$ is injective, where $\tilde{A}$ is the reduction of A modulo p.

Exercise 16. Show that the fundamental character of level 1 is the cyclotomic character $\chi|_{I_t}$. (Hint: This is trickier than it first appears, and requires Wilson's theorem from elementary number theory.)

Exercise 17. For each of the following semistable elliptic curves A, and each ℓ at which $\rho_{A,\ell}$ is *irreducible*, use Theorem 2.10 to compute Serre's minimal weight $k(\rho_{A,\ell})$ and level $N(\rho_{A,\ell})$.

N	$|\Delta|$	reducible ℓ	A
30	$2^4 \cdot 3^5 \cdot 5$	2, 3	$y^2 + xy + y = x^3 + x + 2$
210	$2^{12} \cdot 3^3 \cdot 5 \cdot 7$	2, 3	$y^2 + xy = x^3 - 41x - 39$
330	$2^4 \cdot 3^2 \cdot 5^4 \cdot 11^2$	2	$y^2 + xy = x^3 + x^2 - 102x + 324$
455	$5^3 \cdot 7^4 \cdot 13$	2	$y^2 + xy = x^3 - x^2 - 50x + 111$
2926	$2^8 \cdot 7^3 \cdot 11^4 \cdot 19^2$	2	$y^2 + xy + y = x^3 - x^2 + 1934x - 1935$

Attempt to verify Serre's conjecture directly in some of these cases.

Exercise 18. Let M be a positive integer and let p be a prime. Show that there is an injective linear map
$$S_2(\Gamma_1(M)) \hookrightarrow S_2(\Gamma_1(pM))$$
sending $f(q)$ to $f(q^p)$.

Exercise 19. Let M be an integer such that $S_2(\Gamma_1(M))$ has positive dimension, and let p be a prime (thus $M = 11$ or $M \geq 13$).
1. Let $f \in S_2(\Gamma_1(M))$ be an eigenvector for T_p with eigenvalue λ. Show that T_p acting on $S_2(\Gamma_1(Mp))$ preserves the two-dimensional subspace generated by f and $f(pz)$ (see Section 1.1.5 for the definition of T_p when p divides the level). Show furthermore that if $\lambda^2 \neq 4p$ then T_p is diagonalizable on this 2-dimensional space. What are the eigenvalues of T_p on this space? In fact, one never has $\lambda^2 = 4p$; see [**16**] for more details.
2. Show that for any $r > 2$, the Hecke operator T_p on $S_2(\Gamma_1(Mp^r))$ is not diagonalizable.
3. Deduce that for $r > 2$ the Hecke algebra **T** associated to $S_2(\Gamma_1(Mp^r))$ has nilpotent elements, so it is not an order in a product of rings of integers of number fields.

Exercise 20. Let N be a positive integer. Show that the Hecke algebra $\mathbf{T} = \mathbf{Z}[\ldots T_n \ldots] \subset \mathrm{End}(J_1(N))$ is of finite rank as a **Z**-module.

Exercise 21. Suppose $N = pM$ with $(p, M) = 1$. There is an injection
$$S_2(\Gamma_1(M)) \oplus S_2(\Gamma_1(M)) \hookrightarrow S_2(\Gamma_1(M) \cap \Gamma_0(p))$$
given by $(f, g) \mapsto f(q) + g(q^p)$. The Hecke algebra $\mathbf{T} = \mathbf{T}_N$ acts through a quotient $\overline{\mathbf{T}}$ on the image of $S_2(\Gamma_1(M)) \oplus S_2(\Gamma_1(M))$. Suppose $\mathfrak{m} \subset \mathbf{T}$ is a maximal ideal that arises by pullback from a maximal ideal in $\overline{\mathbf{T}}$. Show that $\rho_{\mathfrak{m}}$ arises from a modular form of level M.

4.4.2. Solutions

Solution 1. 1. Let p be a prime different from ℓ and let
$$\rho : \mathrm{Gal}(\overline{\mathbf{Q}}/\mathbf{Q}) \to \mathrm{Gal}(\mathbf{Q}(\sqrt{p})/\mathbf{Q}) \approx \{\pm 1\} \hookrightarrow \mathbf{F}_\ell^*.$$

2. Let $K = \overline{\mathbf{Q}}^{\mathrm{ker}(\rho)}$. Then $K/\mathbf{Q}$ is abelian and ramified only at ℓ, so $K \subset \mathbf{Q}(\zeta_{\ell^\infty})$. But $[K : \mathbf{Q}] \mid \ell - 1$ so $K \subset \mathbf{Q}(\zeta_\ell)$.

Solution 2. Conjugate using $g = \begin{pmatrix} N & 0 \\ 0 & 1 \end{pmatrix}$.

Solution 3. If ρ is absolutely irreducible then it is irreducible, so assume that ρ is irreducible. If ρ is reducible over the algebraic closure $\overline{k}$ of k, then there is a vector $v \in \overline{k}^{\oplus 2}$ that generates a one-dimensional subspace stable under ρ. In particular, v is stable under complex conjugation, which has characteristic polynomial $x^2 - 1 = (x-1)(x+1)$. Since $-1 \neq 1$, this means that v must lie in one of the two 1-dimensional eigenspaces of complex conjugation, so v is a scalar multiple of an element w of $k^{\oplus 2}$. Then ρ leaves the subspace of $k^{\oplus 2}$ spanned by w invariant, so ρ is reducible, which contradicts our assumption.

Let $\rho : G_{\mathbf{Q}} \to \mathrm{GL}(2, \mathbf{F}_2)$ be any continuous representation whose image is the subgroup generated by $\begin{pmatrix} 0 & 1 \\ 1 & 1 \end{pmatrix}$. Then ρ is irreducible because it has no one-dimensional invariant subspaces over $\mathbf{F}_2$. However, the matrix $\begin{pmatrix} 0 & 1 \\ 1 & 1 \end{pmatrix}$ is diagonalizable over $\mathbf{F}_4$.

Solution 4. Suppose $\varphi \in \mathrm{End}(E)$ is a nonzero endomorphism. The induced map $d\varphi$ on the differentials $H^0(A, \Omega) \approx \mathbf{Q}$ is multiplication by an integer n, so $d(\varphi - n) = 0$ which implies that $\varphi = n$.

Suppose that ρ_ℓ is reducible, so that there is a one-dimensional Galois stable subspace $V \subset A[\ell]$. The quotient $B = A/V$ is then an elliptic curve over $\mathbf{Q}$ and there is an isogeny $\pi : A \to B$ of degree ℓ. Because A is isolated in its isogeny class we have that $B = A$, so there is an endomorphism of A of degree ℓ. But all $\mathbf{Q}$-rational endomorphisms are multiplication by an integer, and multiplication by an integer has degree a perfect square.

The elliptic curve E given by the equation $y^2 = x^3 - 7x - 7$ has the property that $E[2]$ is irreducible but not absolutely irreducible. To see this, note that the splitting field of $x^3 - 7x - 7$ has Galois group cyclic of order 3.

Solution 5. Suppose all $\rho_{A,\ell}$ are irreducible, yet there exists an isogeny $\varphi : A \to B$ with $B \not\approx A$. Choose φ to have minimal possible degree and let $d = \deg(\varphi) > 1$. Let ℓ be the smallest prime divisor of d and choose a point $x \in \ker(\varphi)$ of exact order ℓ. If the order-ℓ cyclic subgroup generated by x is Galois stable, then $\rho_{A,\ell}$ is reducible, which is contrary to our assumption. Thus $\ker(\varphi)$ contains the full ℓ-torsion subgroup $A[\ell]$ of A. In particular, φ factors as illustrated below:

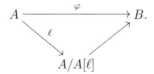

Since $A/A[\ell] \cong A$, there is an isogeny from A to B of degree equal to d/ℓ^2, which contradicts our assumption that d is minimal.

Solution 6. The Weil pairing $(,) : A[\ell] \times A[\ell] \to \boldsymbol{\mu}_\ell$ can be viewed as a map

$$\bigwedge^2 A[\ell] \xrightarrow{\cong} \boldsymbol{\mu}_\ell$$

sending $P \wedge Q$ to (P, Q). For any $\sigma \in \mathrm{Gal}(\overline{\mathbf{Q}}/\mathbf{Q})$, we have $(P^\sigma, Q^\sigma) = (P, Q)^\sigma$. With the action $(P \wedge Q)^\sigma = P^\sigma \wedge Q^\sigma$, the map $\bigwedge^2 A[\ell] \to \boldsymbol{\mu}_\ell$ is a map of Galois modules. To compute $\det(\rho(\sigma))$ observe that if e_1, e_2 is a basis for $A[\ell]$, and $\rho(\sigma) = \begin{pmatrix} a & b \\ c & d \end{pmatrix}$, then

$$\sigma(e_1 \wedge e_2) = (ae_1 + ce_2) \wedge (be_1 + de_2)$$
$$= (ad - bc)e_1 \wedge e_2 = \det(\rho(\sigma))e_1 \wedge e_2$$

Thus $\bigwedge^2 A[\ell]$ gives the one-dimensional representation $\det(\rho)$. Since $\bigwedge^2 A[\ell]$ is isomorphic to $\boldsymbol{\mu}_\ell$ it follows that $\det(\rho) = \chi$.

Solution 7. The definition of $\langle d \rangle$ is as follows: choose any matrix $\sigma_d \in \Gamma_0(N)$ such that $\sigma_d \equiv \begin{pmatrix} d & 0 \\ 0 & d^{-1} \end{pmatrix} \pmod{N}$; then $\langle d \rangle f = f|_{\sigma_d}$. Observe that $\Gamma_1(N)$ is a normal subgroup of $\Gamma_0(N)$ and the matrices σ_d with $(d, N) = 1$ and $d < N$ are a system of coset representatives. Thus any $\begin{pmatrix} a & b \\ c & d \end{pmatrix} \in \Gamma_0(N)$ can be written in the form $\sigma_d \cdot g$ for some $g \in \Gamma_1(N)$. We have

$$f = f|_{\sigma_d g} = (f|_{\sigma_d})|_g = (\varepsilon(d)f)|_g = \varepsilon(d)f|_g = \varepsilon(d)(cz+d)^{-k}f\left(\frac{az+b}{cz+d}\right).$$

Solution 8.
1. Since $c^2 = 1$, the minimal polynomial f of $\rho(c)$ divides $x^2 - 1$. Thus f is either $x + 1$, $x - 1$, or $x^2 - 1$. If $f = x + 1$ then $\rho(c) = -1 = \begin{pmatrix} -1 & 0 \\ 0 & -1 \end{pmatrix}$. This implies that $\det(\rho(c)) = (-1)^2 = 1$, which is a contradiction since $\det(\rho(c)) = -1$ and the characteristic of the base field is odd. If $f = x - 1$, then $\rho(c) = 1$; again a contradiction. Thus the minimal polynomial of $\rho(c)$ is $x^2 - 1 = (x-1)(x+1)$. Since $-1 \neq 1$ there is a basis of eigenvectors for $\rho(c)$ such that the matrix of $\rho(c)$ with respect to this basis is $\begin{pmatrix} -1 & 0 \\ 0 & 1 \end{pmatrix}$.
2. The following example shows that when $\ell = 2$ the matrix of $\rho_{A,\ell}$ need not be conjugate to $\begin{pmatrix} 1 & 0 \\ 0 & -1 \end{pmatrix}$. Let A be the elliptic curve over $\mathbf{Q}$ defined by $y^2 = x(x^2 - a)$ with $a \in \mathbf{Q}$ not square. Then
$$A[2] = \{\infty, (0,0), (\sqrt{a}, 0), (-\sqrt{a}, 0)\}.$$
The action of c on the basis $(0,0), (-\sqrt{a}, 0)$ is represented by the matrix $\begin{pmatrix} 1 & 1 \\ 0 & 1 \end{pmatrix}$, since $c(-\sqrt{a}, 0) = (\sqrt{a}, 0) = (0,0) + (-\sqrt{a}, 0)$.

Solution 9. The extension $\mathbf{Q}(\sqrt{d}, d \in \mathbf{Q}^*/(\mathbf{Q}^*)^2)$ is an extension of $\mathbf{Q}$ with Galois group $X \approx \prod \mathbf{F}_2$. The index-two open subgroups of X correspond to the quadratic extensions of $\mathbf{Q}$. However, Zorn's lemma implies that X contains many more index-two subgroups, which can be seen more precisely as follows.

1. Choose a sequence $p_1, p_2, p_3, \ldots$ of distinct prime numbers. Define $\rho_1 : G_{\mathbf{Q}} \to \prod \mathbf{F}_2$ by
$$\rho_1(\sigma)_i = \begin{cases} 0 & \text{if } \sigma \text{ acts trivially on } \mathbf{Q}(\sqrt{p_i}), \\ 1 & \text{otherwise} \end{cases}$$
Thus ρ_1 is just
$$G_{\mathbf{Q}} \to \mathrm{Gal}(\mathbf{Q}(\sqrt{p_1}, \sqrt{p_2}, \ldots)/\mathbf{Q}) \approx \prod \mathbf{F}_2.$$

2. Let $\oplus \mathbf{F}_2 \subset \prod \mathbf{F}_2$ be the subgroup of elements having only finitely many nonzero coordinates. Then $\prod \mathbf{F}_2 / \oplus \mathbf{F}_2$ is a vector space over $\mathbf{F}_2$ of dimension > 0. By Zorn's lemma, there is a basis $\mathcal{B}$ of $\prod \mathbf{F}_2 / \oplus \mathbf{F}_2$. Let $b \in \mathcal{B}$ and let W be the subspace spanned by $\mathcal{B} - \{b\}$. Then $V = (\prod \mathbf{F}_2 / \oplus \mathbf{F}_2)/W$ is an $\mathbf{F}_2$-vector space of dimensional 1.

3. Let ρ be the composite map

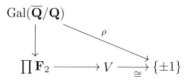

4. Let $H = \ker(\rho) \subset \mathrm{Gal}(\overline{\mathbf{Q}}/\mathbf{Q})$. If $\sigma(\sqrt{p_i}) = -\sqrt{p_i}$ and $\sigma(\sqrt{p_j}) = \sqrt{p_j}$ for $i \neq j$, then $\sigma \in H$. Thus H does not fix any $\mathbf{Q}(\sqrt{p_i})$, so the fixed field of H equals $\mathbf{Q}$. The largest finite Galois group quotient through which ρ

Solution 10. We have $f = f_1 + \alpha f_2$ with $f_1, f_2 \in S_2(\Gamma_0(23))$ and

$$f_1 = q - q^3 - q^4 + \cdots$$
$$f_2 = q^2 - 2q^3 - q^4 + 2q^5 + \cdots.$$

Because $S_2(\Gamma_0(23))$ has dimension 2, it is spanned by f_1 and f_2. Let $\eta(q) = q^{\frac{1}{24}} \prod_{n \geq 1}(1-q^n)$. Then $g = (\eta(q)\eta(q^{23}))^2 \in S_2(\Gamma_0(23))$. Expanding we find that $g = q^2 - 2q^3 + \cdots$, so $g = f_2$. Next observe that g is a power series in q^2 modulo 2:

$$g = q^2 \prod (1-q^n)^2(1-q^{23n})^2$$
$$\equiv q^2 \prod (1-q^{2n})(1-q^{46n}) \pmod{2}$$
$$\equiv q^2 \prod (1+q^{2n}+q^{46n}+q^{48n}) \pmod{2}$$

Thus the coefficient in f_2 of q^p with $p \neq 2$ prime is even, and the proposition follows.

Solution 11.

1. Let $\zeta \in \boldsymbol{\mu}_\ell$ be a primitive ℓth root of unity. Since $\bigwedge^2 A[\ell] \cong \boldsymbol{\mu}_\ell$, there exists $P, Q \in A[\ell]$ such that $P \wedge Q = \zeta$. Since $\ell > 2$ there exists σ such that $\zeta^\sigma \neq \zeta$, hence $P^\sigma \wedge Q^\sigma \neq P \wedge Q$. This is impossible if all ℓ-torsion is rational, since then $P^\sigma = P$ and $Q^\sigma = Q$.
2. Consider the elliptic curve defined by $y^2 = (x-a)(x-b)(x-c)$ where a, b, c are distinct rational numbers.

Solution 12.

1. Let K be the splitting field of $x^3 + ax + b$. Then ρ embeds $\mathrm{Gal}(K/\mathbf{Q})$ in $\mathrm{GL}(2, \mathbf{F}_2)$:

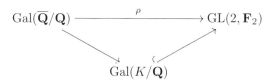

2. The representation ρ is reducible exactly when the polynomial $x^3 + ax + b$ has a rational root.
3. Examples: $y^2 = x(x^2 - 23)$, $y^2 = x^3 + x - 1$.

Solution 13. Consider the character $\tau = \varepsilon\chi/\rho$. By assumption, $\tau(\mathrm{Frob}_p) = 1$ for all unramified p. Let K be an extension of $\mathbf{Q}$ such that τ factors through $\mathrm{Gal}(K/\mathbf{Q})$. For any $\sigma \in \mathrm{Gal}(K/\mathbf{Q})$, the Cebotarev density theorem implies that there are infinitely many primes p such that $\mathrm{Frob}_p = \sigma$. Thus for any σ, $\tau(\sigma) = \tau(\mathrm{Frob}_p) = 1$, so $\tau = 1$ and hence $\rho = \varepsilon\chi$.

Solution 14.

1. See, e.g., [**109**, 2.11].

2. By [**109**, 5.5], $\Phi - 1$ is separable, so $\#A(\mathbf{F}_p) = \deg(\Phi - 1)$. Since Φ has degree p, there exists an isogeny $\overline{\Phi}$ (the dual isogeny, see [**109**, III.6]), such that $\Phi\overline{\Phi} = p$. Letting bars denote the dual isogeny, we have
$$\#A(\mathbf{F}_p) = \deg(\Phi - 1) = (\Phi - 1)\overline{(\Phi - 1)}$$
$$= \Phi\overline{\Phi} - \Phi - \overline{\Phi} + 1$$
$$= p - \operatorname{tr}(\Phi) + 1$$

3. Both maps are pth powering on coordinates.

Solution 15. Since $\ell \neq p$ and A has good reduction at p, the natural map $A[\ell] \to \tilde{A}[\ell]$ is an isomorphism. We have the following commutative diagram

$$\begin{array}{ccc}
\operatorname{Gal}(\mathbf{Q}_p(A[\ell])/\mathbf{Q}_p) & \hookrightarrow & \operatorname{Aut}(A[\ell]) \\
\downarrow & & \downarrow \cong \\
\operatorname{Gal}((\mathcal{O}/\lambda)_{\mathbf{F}_p}) & \longrightarrow & \operatorname{Aut}(\tilde{A}[\ell])
\end{array}$$

It follows that the first vertical map must be injective, which is the same as $\mathbf{Q}_p(A[\ell])$ being unramified over $\mathbf{Q}_p$.

Solution 17. We write $N = N(\rho)$ and $k = k(\rho)$ to save space. The essential tool is Theorem 2.10.
1. $\ell = 5$: $N = 6$, $k = 6$, $\ell > 5$, $N = 30$, $k = 2$.
2. $\ell = 5$: $N = 2 \cdot 3 \cdot 7$, $k = 6$; $\ell = 7$: $N = 2 \cdot 3 \cdot 5$, $k = 8$; $\ell > 7$: $N = 2 \cdot 3 \cdot 5 \cdot 7$, $k = 2$.
3. $\ell = 3$: $N = 2 \cdot 5 \cdot 11$, $k = 4$; $\ell = 5$: $N = 2 \cdot 3 \cdot 11$, $k = 6$; $\ell = 7$: $N = 2 \cdot 3 \cdot 5 \cdot 11$, $k = 2$; $\ell = 11$: $N = 2 \cdot 3 \cdot 5$, $k = 12$; $\ell > 11$: $N = 2 \cdot 3 \cdot 5 \cdot 11$, $k = 2$.
4. $\ell = 3$: $N = 7 \cdot 13$, $k = 2$; $\ell = 5$: $N = 7 \cdot 13$, $k = 6$; $\ell = 7$: $N = 5 \cdot 13$, $k = 8$; $\ell = 13$: $N = 5 \cdot 7$, $k = 14$; $\ell = 11, \ell > 13$: $N = 5 \cdot 7 \cdot 13$, $k = 2$.
5. $\ell = 3$: $N = 2 \cdot 11 \cdot 19$, $k = 2$; $\ell = 7$: $N = 2 \cdot 11 \cdot 19$, $k = 8$; $\ell = 11$: $N = 2 \cdot 7 \cdot 19$, $k = 12$; $\ell = 19$: $N = 2 \cdot 7 \cdot 11$, $k = 20$; ℓ other: $N = 2 \cdot 7 \cdot 11 \cdot 19$, $k = 2$.

Solution 20. One approach is to view $J_1(N)$ as a complex torus, and note that the endomorphism ring is the set of automorphism of a complex vector space that fix a lattice. Another approach is to use the deeper finiteness theorems that are valid in arbitrary characteristic, see, e.g., [**74**, Thm. 12.5].

CHAPTER 5
Appendix by Brian Conrad: The Shimura Construction in Weight 2

The purpose of this appendix[1] is to explain the ideas of Eichler-Shimura for constructing the two-dimensional ℓ-adic representations attached to classical weight-2 Hecke eigenforms. We assume familiarity with the theory of schemes and the theory of newforms, but the essential arithmetic ideas are due to Eichler and Shimura. We warn the reader that a complete proof along the lines indicated below requires the verification of a number of compatibilities between algebraic geometry, algebraic topology, and the classical theory of modular forms. As the aim of this appendix is to explain the key arithmetic ideas of the proof, we must pass over in silence the verification of many such compatibilities. However, we at least make explicit what compatibilities we need. To prove them all here would require a serious digression from our expository goal; see [18, Ch. 3] for details. It is also worth noting that the form of the arguments we present is *exactly* the weight-2 version of Deligne's more general proof of related results in weight > 1, up to the canonical isomorphism

$$\mathbf{Q}_\ell \otimes_{\mathbf{Z}_\ell} \varprojlim \mathrm{Pic}^0_{X/k}[\ell^n](k) \cong H^1_{\text{ét}}(X, \mathbf{Q}_\ell(1)) \cong H^1_{\text{ét,c}}(Y, \mathbf{Q}_\ell(1))$$

for a proper smooth connected curve X over a separably closed field k of characteristic prime to ℓ, and Y a dense open in X. Using ℓ-adic Tate modules allows us to bypass the general theory of étale cohomology which arises in the case of higher weight.

5.5.1. Analytic preparations

Fix $i = \sqrt{-1} \in \mathbf{C}$ for all time. Fix an integer $N \geq 5$ and let $X_1(N)^{\text{an}}$ denote the classical analytic modular curve, the "canonical" compactification of $Y_1(N)^{\text{an}} = \Gamma_1(N) \backslash \mathfrak{h}$, where $\mathfrak{h} = \{z \in \mathbf{C} : \operatorname{Im} z > 0\}$ and $\Gamma_1(N) \subset \mathrm{SL}_2(\mathbf{Z})$ acts on the left via linear fractional transformations. The classical theory identifies the $\mathbf{C}$-vector space $H^0(X_1(N)^{\text{an}}, \Omega^1_{X_1(N)^{\text{an}}})$ with $S_2(\Gamma_1(N), \mathbf{C})$, the space of weight-2 cusp forms. Note that the classical Riemann surface $X_1(N)^{\text{an}}$ has genus 0 if we consider $N < 5$, while

---
[1] Copyright 2001 by Brian Conrad

$S_2(\Gamma_1(N), \mathbf{C}) = 0$ if $N < 5$. Thus, assuming $N \geq 5$ is harmless for what we will do.

The Hodge decomposition for the compact Riemann surface $X_1(N)^{\mathrm{an}}$ supplies us with an isomorphism of $\mathbf{C}$-vector spaces

$$S_2(\Gamma_1(N), \mathbf{C}) \oplus \overline{S_2(\Gamma_1(N), \mathbf{C})}$$
$$\cong H^0(X_1(N)^{\mathrm{an}}, \Omega^1_{X_1(N)^{\mathrm{an}}}) \oplus H^0(X_1(N)^{\mathrm{an}}, \overline{\Omega}^1_{X_1(N)^{\mathrm{an}}})$$
$$\xrightarrow{\sim} H^1(X_1(N)^{\mathrm{an}}, \underline{\mathbf{C}})$$
$$\cong H^1(X_1(N)^{\mathrm{an}}, \underline{\mathbf{Z}}) \otimes_{\mathbf{Z}} \mathbf{C}$$

(where $\underline{A}$ denotes the constant sheaf attached to an abelian group A). This will be called the (weight-2) *Shimura isomorphism*. We want to define "geometric" operations on $H^1(X_1(N)^{\mathrm{an}}, \underline{\mathbf{Z}})$ which recover the classical Hecke operators on $S_2(\Gamma_1(N), \mathbf{C})$ via the above isomorphism.

The "geometric" (or rather, cohomological) operations we wish to define can be described in two ways. First, we can use explicit matrices and explicit "upper-half plane" models of modular curves. This has the advantage of being concrete, but it provides little conceptual insight and encourages messy matrix calculations. The other point of view is to identify the classical modular curves as the base of certain universal analytic families of (generalized) elliptic curves with level structure. A proper discussion of this latter point of view would take us too far afield, so we will have to settle for only some brief indications along these two lines (though this is how to best verify compatibility with the algebraic theory via schemes).

Choose a matrix $\gamma_n \in \mathrm{SL}_2(\mathbf{Z})$ with $\gamma_n \equiv \begin{pmatrix} n^{-1} & * \\ 0 & n \end{pmatrix} \pmod{N}$, for $n \in (\mathbf{Z}/N\mathbf{Z})^*$. The action of γ_n on $\mathfrak{h}$ induces an action on $Y_1(N)^{\mathrm{an}}$ and even on $X_1(N)^{\mathrm{an}}$. Associating to each $z \in \mathfrak{h}$ the data of the elliptic curve $\mathbf{C}/[1, z] = \mathbf{C}/(\mathbf{Z} + \mathbf{Z}z)$ and the point $1/N$ of exact order N, we may identify $Y_1(N)^{\mathrm{an}}$ as a *set* with the set of isomorphism classes of pairs (E, P) consisting of an elliptic curve E over $\mathbf{C}$ and a point $P \in E$ of exact order N. The map $Y_1(N)^{\mathrm{an}} \to Y_1(N)^{\mathrm{an}}$ induced by γ_n can then described on the underlying set by $(E, P) \mapsto (E, nP)$, so it is "intrinsic", depending only on $n \in (\mathbf{Z}/N\mathbf{Z})^*$. We denote by $I_n : X_1(N)^{\mathrm{an}} \to X_1(N)^{\mathrm{an}}$ the induced map on $X_1(N)^{\mathrm{an}}$. Once this data (E, P) is formulated in a relative context over an analytic base, we could define the analytic map I_n conceptually, without using the matrix γ_n. We ignore this point here.

The map $z \mapsto \frac{-1}{Nz}$ on $\mathfrak{h}$ induces a map $Y_1(N)^{\mathrm{an}} \to Y_1(N)^{\mathrm{an}}$ which extends to $w_N : X_1(N)^{\mathrm{an}} \to X_1(N)^{\mathrm{an}}$. More conceptually and more generally, if $\zeta \in \boldsymbol{\mu}_N(\mathbf{C})$ is a primitive Nth root of unity, consider the rule w_ζ that sends $(E, P) \in Y_1(N)^{\mathrm{an}}$ to $(E/P, P' \bmod P)$, where $P' \in E$ has exact order N and $\langle P, P' \rangle_N = \zeta$, with $\langle \, , \, \rangle_N$ the Weil pairing on N-torsion points (following the sign conventions of [**62, 77**]; opposite the convention of [**109**]). More specifically, on $\mathbf{C}/[1, z]$ we have $\langle \frac{1}{N}, \frac{z}{N} \rangle_N = e^{2\pi i/N}$. The map w_ζ extends to an analytic map $X_1(N)^{\mathrm{an}} \to X_1(N)^{\mathrm{an}}$. When $\zeta = e^{2\pi i/N}$, we have $w_\zeta = w_N$ due to the above sign convention.

We have induced pullback maps

$$w_\zeta^*, I_n^* : H^1(X_1(N)^{\mathrm{an}}, \underline{\mathbf{Z}}) \to H^1(X_1(N)^{\mathrm{an}}, \underline{\mathbf{Z}}).$$

We write $\langle n \rangle^*$ rather than I_n^*.

Finally, choose a prime p. Define $\Gamma_1(N, p) \subset \mathrm{SL}_2(\mathbf{Z})$ to be $\Gamma_1(N, p) = \Gamma_1(N) \cap \Gamma_0(p)$ when $p \nmid N$ and $\Gamma_1(N, p) = \Gamma_1(N) \cap \Gamma_0(p)^t$ when $p \mid N$, where the group

$\Gamma_0(p)^t$ is the transpose of $\Gamma_0(p)$. Define $Y_1(N,p)^{\mathrm{an}} = \Gamma_1(N,p)\backslash\mathfrak{h}$ and let $X_1(N,p)^{\mathrm{an}}$ be its "canonical" compactification. Using the assignment

$$z \mapsto (\mathbf{C}/[1,z], \frac{1}{N}, \langle\frac{1}{p}\rangle)$$

when $p \nmid N$ and

$$z \mapsto (\mathbf{C}/[1,z], \frac{1}{N}, \langle\frac{z}{p}\rangle)$$

when $p \mid N$, we may identify the *set* $Y_1(N,p)^{\mathrm{an}}$ with the set of isomorphism classes of triples (E,P,C) where $P \in E$ has exact order N and $C \subset E$ is a cyclic subgroup of order p, meeting $\langle P \rangle$ trivially (a constraint if $p \mid N$). Here and below, we denote by $\langle P \rangle$ the (cyclic) subgroup generated by P.

There are unique analytic maps

$$\pi_1^{(p)}, \pi_2^{(p)} : X_1(N,p)^{\mathrm{an}} \to X_1(N)^{\mathrm{an}}$$

determined on $Y_1(N,p)^{\mathrm{an}}$ by

$$\pi_1^{(p)}(E,P,C) = (E,P)$$

and

$$\pi_2^{(p)}(E,P,C) = (E/C, P \bmod C).$$

For example, $\pi_1^{(p)}$ is induced by $z \mapsto z$ on $\mathfrak{h}$, in terms of the above upper half plane uniformization of $Y_1(N)^{\mathrm{an}}$ and $Y_1(N,p)^{\mathrm{an}}$.

We define

$$T_p^* = (\pi_1^{(p)})_* \circ (\pi_2^{(p)})^* : H^1(X_1(N)^{\mathrm{an}}, \mathbf{Z}) \to H^1(X_1(N)^{\mathrm{an}}, \mathbf{Z})$$

where $(\pi_1^{(p)})_* : H^1(X_1(N,p)^{\mathrm{an}}, \mathbf{Z}) \to H^1(X_1(N)^{\mathrm{an}}, \mathbf{Z})$ is the canonical trace map associated to the finite map $\pi_1^{(p)}$ of compact Riemann surfaces. More specifically, we have a canonical isomorphism

$$H^1(X_1(N,p)^{\mathrm{an}}, \mathbf{Z}) \cong H^1(X_1(N)^{\mathrm{an}}, (\pi_1^{(p)})_*\underline{\mathbf{Z}})$$

since $(\pi_1^{(p)})_*$ is exact on abelian sheaves, and there is a unique trace map of sheaves $(\pi_1^{(p)})_*\underline{\mathbf{Z}} \to \underline{\mathbf{Z}}$ determined on stalks at $x \in X_1(N)^{\mathrm{an}}$ by

(5.1)
$$\prod_{\pi_1^{(p)}(y)=x} \mathbf{Z} \to \mathbf{Z}$$
$$(a_y) \mapsto \Sigma_y e_y a_y$$

where e_y is the ramification degree of y over x via $\pi_1^{(p)}$.

A fundamental compatibility, whose proof we omit for reasons of space, is:

Theorem 5.1. *The weight-2 Shimura isomorphism*

$$\mathrm{Sh}_{\Gamma_1(N)} : S_2(\Gamma_1(N), \mathbf{C}) \oplus \overline{S_2(\Gamma_1(N), \mathbf{C})} \cong H^1(X_1(N)^{\mathrm{an}}, \mathbf{Z}) \otimes_{\mathbf{Z}} \mathbf{C}$$

from (5.5.1) identifies $\langle n \rangle \oplus \overline{\langle n \rangle}$ *with* $\langle n \rangle^* \otimes 1$, $T_p \oplus \overline{T}_p$ *with* $T_p^* \otimes 1$, *and* $w_N \oplus \overline{w}_N$ *with* $w^*_{e^{2\pi i/N}} \otimes 1$.

Let $\mathbf{T}_1(N) \subset \operatorname{End}_{\mathbf{Z}}(H^1(X_1(N)^{\mathrm{an}}, \mathbf{Z}))$ be the subring generated by the T_p^*'s and $\langle n \rangle^*$'s. By Theorem 5.1, this is identified via the Shimura isomorphism with the classical (weight-2) Hecke ring at level N. In particular, this ring is commutative (which can be seen directly via cohomological considerations as well). It is clearly a finite flat $\mathbf{Z}$-algebra.

The natural map

(5.2) $$\mathbf{T}_1(N) \otimes_{\mathbf{Z}} \mathbf{C} \hookrightarrow \operatorname{End}_{\mathbf{C}}(H^1(X_1(N)^{\mathrm{an}}, \mathbf{Z}) \otimes_{\mathbf{Z}} \mathbf{C})$$

induces an *injection* $\mathbf{T}_1(N) \otimes \mathbf{C} \hookrightarrow \operatorname{End}_{\mathbf{C}}(S_2(\Gamma_1(N), \mathbf{C}))$, by Theorem 5.1. This is the classical realization of Hecke operators in weight 2.

Another compatibility we need is between the cup product on $H^1(X_1(N)^{\mathrm{an}}, \mathbf{Z})$ and the (non-normalized) Petersson product on $S_2(\Gamma_1(N), \mathbf{C})$. To be precise, we define an isomorphism $H^2(X_1(N)^{\mathrm{an}}, \mathbf{Z}) \cong \mathbf{Z}$ using the i-orientation of the complex manifold $X_1(N)^{\mathrm{an}}$ (i.e., the "$i dz \wedge d\bar{z}$" orientation), so we get via cup product a (perfect) pairing

$$(,)_{\Gamma_1(N)} : H^1(X_1(N)^{\mathrm{an}}, \mathbf{Z}) \otimes_{\mathbf{Z}} H^1(X_1(N)^{\mathrm{an}}, \mathbf{Z}) \to H^2(X_1(N)^{\mathrm{an}}, \mathbf{Z}) \cong \mathbf{Z}.$$

This induces an analogous pairing after applying $\otimes_{\mathbf{Z}} \mathbf{C}$. For $f, g \in S_2(\Gamma_1(N), \mathbf{C})$ we define

$$\langle f, g \rangle_{\Gamma_1(N)} = \int_{\Gamma_1(N) \backslash \mathfrak{h}} f(z) \overline{g}(z) dx dy$$

where this integral is absolutely convergent since f and g have exponential decay near the cusps. This is a perfect Hermitian pairing.

Theorem 5.2. *Under the weight-2 Shimura isomorphism* $\operatorname{Sh}_{\Gamma_1(N)}$,

$$\left(\operatorname{Sh}_{\Gamma_1(N)}(f_1 + \overline{g}_1), \operatorname{Sh}_{\Gamma_1(N)}(f_2 + \overline{g}_2) \right)_{\Gamma_1(N)} = 4\pi \cdot \left(\langle f_1, g_2 \rangle_{\Gamma_1(N)} - \langle f_2, g_1 \rangle_{\Gamma_1(N)} \right).$$

Note that *both* sides are antilinear in g_1, g_2 and alternating with respect to interchanging the pair (f_1, g_1) and (f_2, g_2). The extra factor of 4π is harmless for our purposes since it does not affect formation of adjoints. What is important is that in the classical theory, conjugation by the involution w_N takes each $T \in \mathbf{T}_1(N)$ to its adjoint with respect to the Petersson product. The most subtle case of this is $T = T_p^*$ for $p \mid N$. For $p \nmid N$ the adjoint of T_p^* is $\langle p^{-1} \rangle^* T_p^*$ and the adjoint of $\langle n \rangle^*$ is $\langle n^{-1} \rangle^*$. These classical facts (especially for T_p^* with $p \mid N$) yield the following important corollary of Theorem 5.2.

Corollary 5.3. *With respect to the pairing* $[x, y]_{\Gamma_1(N)} = (x, w_\zeta^* y)_{\Gamma_1(N)}$ *with* $\zeta = e^{2\pi i/N}$, *the action of* $\mathbf{T}_1(N)$ *on* $H^1(X_1(N)^{\mathrm{an}}, \mathbf{Z})$ *is equivariant. That is,*

$$[x, Ty]_{\Gamma_1(N)} = [Tx, y]_{\Gamma_1(N)}$$

for all $T \in \mathbf{T}_1(N)$. *With respect to* $(,)_{\Gamma_1(N)}$, *the adjoint of* T_p^* *for* $p \nmid N$ *is* $\langle p^{-1} \rangle^* T_p^*$ *and the adjoint of* $\langle n \rangle^*$ *is* $\langle n^{-1} \rangle^*$ *for* $n \in (\mathbf{Z}/N\mathbf{Z})^*$.

Looking back at the "conceptual" definition of w_ζ^* for an arbitrary primitive Nth root of unity $\zeta \in \boldsymbol{\mu}_N(\mathbf{C})$, which gives an analytic involution of $X_1(N)^{\mathrm{an}}$, one can check that $w_{\zeta^j}^* \circ w_\zeta^* = \langle j \rangle^*$ for $j \in (\mathbf{Z}/N\mathbf{Z})^*$. Since $\langle j \rangle^*$ is a unit in $\mathbf{T}_1(N)$ and $\mathbf{T}_1(N)$ is *commutative*, we conclude that Corollary 5.3 is true with $\zeta \in \boldsymbol{\mu}_N(\mathbf{C})$ any primitive Nth root of unity (by reduction to the case $\zeta = e^{2\pi i/N}$).

Our final step on the analytic side is to reformulate everything above in terms of Jacobians. For any compact Riemann surface X, there is an isomorphism of complex Lie groups

$$\text{Pic}_X^0 \cong H^1(X, \mathcal{O}_X)/H^1(X, \underline{\mathbf{Z}}) \tag{5.3}$$

via the exponential sequence

$$0 \to \underline{\mathbf{Z}} \to \mathcal{O}_X \xrightarrow{e^{2\pi i(\cdot)}} \mathcal{O}_X^* \to 1$$

and the identification of the underlying group of Pic_X^0 with

$$H^1(X, \mathcal{O}_X^*) \cong \check{H}^1(X, \mathcal{O}_X^*),$$

where the line bundle $\mathcal{L}$ with trivializations $\varphi_i : \mathcal{O}_{U_i} \cong \mathcal{L}|U_i$ corresponds to the class of the Čech 1-cocycle

$$\{\varphi_j^{-1} \circ \varphi_i : \mathcal{O}_{U_i \cap U_j} \cong \mathcal{O}_{U_i \cap U_j}\} \in \prod_{i<j} H^0(U_i \cap U_j, \mathcal{O}_X^*)$$

for an ordered open cover $\{U_i\}$. Beware that the tangent space isomorphism

$$T_0(\text{Pic}_X^0) \cong H^1(X, \mathcal{O}_X)$$

coming from (5.3) is $-2\pi i$ times the "algebraic" isomorphism arising from

$$0 \to \mathcal{O}_X \to \mathcal{O}_{X[\varepsilon]}^* \to \mathcal{O}_X^* \to 1,$$

where $X[\varepsilon] = (X, \mathcal{O}_X[\varepsilon]/\varepsilon^2)$ is the non-reduced space of "dual numbers over X". This extra factor of $-2\pi i$ will not cause problems. We will use (5.3) to "compute" with Jacobians.

Let $f : X \to Y$ be a finite map between compact Riemann surfaces. Since f is finite flat, there is a natural trace map $f_*\mathcal{O}_X \to \mathcal{O}_Y$, and it is not difficult to check that this is compatible with the trace map $f_*\underline{\mathbf{Z}} \to \underline{\mathbf{Z}}$ as defined in (5.1). In particular, we have a trace map

$$f_* : H^1(X, \mathcal{O}_X) \cong H^1(Y, f_*\mathcal{O}_X) \to H^1(Y, \mathcal{O}_Y).$$

Likewise, we have compatible pullback maps $f^*\mathcal{O}_Y \cong \mathcal{O}_X$ and $f^*\underline{\mathbf{Z}} \cong \underline{\mathbf{Z}}$.

Thus, any such f gives rise to *commutative* diagrams

$$\begin{array}{ccc} H^1(Y, \mathcal{O}_Y) \xrightarrow{f^*} H^1(X, \mathcal{O}_X) & \quad & H^1(X, \mathcal{O}_X) \xrightarrow{f_*} H^1(Y, \mathcal{O}_Y) \\ \uparrow \qquad \qquad \uparrow & & \uparrow \qquad \qquad \uparrow \\ H^1(Y, \underline{\mathbf{Z}}) \xrightarrow{f^*} H^1(X, \underline{\mathbf{Z}}) & & H^1(X, \underline{\mathbf{Z}}) \xrightarrow{f_*} H^1(Y, \underline{\mathbf{Z}}), \end{array}$$

where the columns are induced by the canonical maps $\underline{\mathbf{Z}} \to \mathcal{O}_Y$ and $\underline{\mathbf{Z}} \to \mathcal{O}_X$. Passing to quotients on the columns therefore gives rise to maps

$$f^* : \text{Pic}_Y^0 \to \text{Pic}_X^0, \qquad f_* : \text{Pic}_X^0 \to \text{Pic}_Y^0$$

of analytic Lie groups. These maps are "computed" by

Lemma 5.4. *In the above situation, $f^* = \text{Pic}^0(f)$ is the map induced by Pic^0 functoriality and $f_* = \text{Alb}(f)$ is the map induced by Albanese functoriality. These are dual with respect to the canonical autodualities of Pic_X^0, Pic_Y^0.*

The significance of the theory of Jacobians is that by (5.3) we have a canonical isomorphism

(5.4)
$$T_\ell(\mathrm{Pic}^0_{X_1(N)^{\mathrm{an}}}) \cong H^1(X_1(N)^{\mathrm{an}}, \mathbf{Z}_\ell)$$
$$\cong H^1(X_1(N)^{\mathrm{an}}, \mathbf{Z}) \otimes_{\mathbf{Z}} \mathbf{Z}_\ell,$$

connecting the ℓ-adic Tate module of $\mathrm{Pic}^0_{X_1(N)}$ with the $\mathbf{Z}$-module $H^1(X_1(N)^{\mathrm{an}}, \mathbf{Z})$ that "encodes" $S_2(\Gamma_1(N), \mathbf{C})$ via the Shimura isomorphism. Note that this isomorphism is defined in terms of the analytic construction (5.3) which depends upon the choice of i. The intrinsic isomorphism (compatible with étale cohomology) has $\mathbf{Z}$ above replaced by $2\pi i \mathbf{Z} = -2\pi i \mathbf{Z}$.

Definition 5.5. We define endomorphisms of $\mathrm{Pic}^0_{X_1(N)^{\mathrm{an}}}$ via

$$T_p^* = \mathrm{Alb}(\pi_1^{(p)}) \circ \mathrm{Pic}^0(\pi_2^{(p)}), \quad \langle n \rangle^* = \mathrm{Pic}^0(I_n), \quad w_\zeta^* = \mathrm{Pic}^0(w_\zeta).$$

By Lemma 5.4, it follows that the above isomorphism (5.4) carries the operators on $T_\ell(\mathrm{Pic}^0_{X_1(N)^{\mathrm{an}}})$ over to the ones *previously defined* on $H^1(X_1(N)^{\mathrm{an}}, \mathbf{Z})$ (which are, in turn, compatible with the classical operations via the Shimura isomorphism). By the faithfulness of the "Tate module" functor on complex tori, we conclude that $\mathbf{T}_1(N)$ acts on $\mathrm{Pic}^0_{X_1(N)^{\mathrm{an}}}$ in a unique manner compatible with the above definition, and (5.4) is an isomorphism of $\mathbf{T}_1(N) \otimes_{\mathbf{Z}} \mathbf{Z}_\ell$-modules. We call this the $(\ )^*$-*action* of $\mathbf{T}_1(N)$ on $\mathrm{Pic}^0_{X_1(N)^{\mathrm{an}}}$.

We must warn the reader that under the canonical isomorphism of $\mathbf{C}$-vector spaces

$$S_2(\Gamma_1(N), \mathbf{C}) \cong H^0(X_1(N)^{\mathrm{an}}, \Omega^1_{X_1(N)^{\mathrm{an}}})$$
$$\cong H^0(\mathrm{Pic}^0_{X_1(N)^{\mathrm{an}}}, \Omega^1_{\mathrm{Pic}^0_{X_1(N)^{\mathrm{an}}}})$$
$$\cong \mathrm{Cot}_0(\mathrm{Pic}^0_{X_1(N)^{\mathrm{an}}}),$$

the $(\ )^*$-action of $T \in \mathbf{T}_1(N)$ on $\mathrm{Pic}^0_{X_1(N)^{\mathrm{an}}}$ does *not* go over to the classical action of T on $S_2(\Gamma_1(N), \mathbf{C})$, but rather the adjoint of T with respect to the Petersson pairing. To clear up this matter, we make the following definition:

Definition 5.6.

$$(T_p)_* = \mathrm{Alb}(\pi_2^{(p)}) \circ \mathrm{Pic}^0(\pi_1^{(p)}), \quad \langle n \rangle_* = \mathrm{Alb}(I_n), \quad (w_\zeta)_* = \mathrm{Alb}(w_\zeta).$$

Since $I_n^{-1} = I_{n^{-1}}$ and $w_\zeta^{-1} = w_\zeta$ on $X_1(N)^{\mathrm{an}}$, we have $(w_\zeta)_* = w_\zeta^*$ and $\langle n \rangle_* = \langle n^{-1} \rangle^*$. We claim that the above $(\ )_*$ operators are the *dual* morphisms (with respect to the canonical principal polarization of $\mathrm{Pic}^0_{X_1(N)^{\mathrm{an}}}$) of the $(\ )^*$ operators and induce exactly the *classical* action of T_p and $\langle n \rangle$ on $S_2(\Gamma_1(N), \mathbf{C})$, so we also have a well-defined $(\ )_*$-action of $\mathbf{T}_1(N)$ on $\mathrm{Pic}^0_{X_1(N)^{\mathrm{an}}}$, dual to the $(\ )^*$-action. By Theorem 5.2, Corollary 5.3, and Lemma 5.4, this follows from the following general fact about compact Riemann surfaces. The proof is non-trivial.

Lemma 5.7. *Let X be a compact Riemann surface, and use the i-orientation to define $H^2(X, \underline{\mathbf{Z}}) \cong \mathbf{Z}$. Use $1 \mapsto e^{2\pi i/\ell^n}$ to define $\mathbf{Z}/\ell^n \cong \mu_{\ell^n}(\mathbf{C})$ for all n. The* *diagram*

$$\begin{array}{ccc} H^1(X, \underline{\mathbf{Z}}_\ell) \otimes_{\mathbf{Z}_\ell} H^1(X, \underline{\mathbf{Z}}_\ell) & \xrightarrow{\cup} & \mathbf{Z}_\ell \\ \downarrow \cong & & \downarrow \cong \\ T_\ell(\mathrm{Pic}_X^0) \otimes_{\mathbf{Z}_\ell} T_\ell(\mathrm{Pic}_X^0) & \longrightarrow & \varprojlim \mu_{\ell^n}(\mathbf{C}) \end{array}$$

anticommutes (i.e., going around from upper left to lower right in the two possible ways gives results that are negatives of each other), where the bottom row is the ℓ-adic Weil pairing (with respect to the canonical principal polarization $\mathrm{Pic}_X^0 \cong \widehat{\mathrm{Pic}_X^0}$ for the "second" Pic_X^0 in the lower left.)

Note that the sign doesn't affect formation of adjoints. It ultimately comes from the sign on the bottom of [**77**, pg. 237] since our Weil pairing sign convention agrees with [**77**].

We now summarize our findings in terms of $V_\ell(N) = \mathbf{Q}_\ell \otimes_{\mathbf{Z}_\ell} T_\ell(\mathrm{Pic}_{X_1(N)^{\mathrm{an}}}^0)$, which has a perfect alternating Weil pairing

$$(\, , \,)_\ell : V_\ell(N) \otimes V_\ell(N) \to \mathbf{Q}_\ell(1)$$

and has two $\mathbf{Q}_\ell \otimes \mathbf{T}_1(N)$-actions, via the $(\,)^*$-actions and the $(\,)_*$-actions. Since $(w_\zeta)_* = w_\zeta^*$, we simply write w_ζ for this operator on $V_\ell(N)$.

Theorem 5.8. *Let $\mathbf{T}_1(N)$ act on $V_\ell(N)$ with respect to the $(\,)^*$-action or with respect to the $(\,)_*$-action. With respect to $(\, , \,)_\ell$, the adjoint of T_p for $p \nmid N$ is $\langle p \rangle^{-1} T_p$ and the adjoint of $\langle n \rangle$ is $\langle n \rangle^{-1}$ for $n \in (\mathbf{Z}/N\mathbf{Z})^*$. With respect to*

$$[x, y]_\ell = (x, w_\zeta(y))_\ell$$

for $\zeta \in \boldsymbol{\mu}_N(\mathbf{C})$ a primitive Nth root of unity, the action of $\mathbf{T}_1(N)$ on $V_\ell(N)$ is self-adjoint. In general, adjointness with respect to $(\, , \,)_\ell$ interchanges the $(\,)_*$-action and $(\,)^*$-action.

It should be noted that when making the translation to étale cohomology, the $(\,)^*$-action plays a more prominent role (since this is what makes (5.4) a $\mathbf{T}_1(N)$-equivariant map). However, when working directly with Tate modules and arithmetic Frobenius elements, it is the $(\,)_*$-action which gives the cleaner formulation of Shimura's results.

An important consequence of Theorem 5.8 is

Corollary 5.9. *The $\mathbf{Q}_\ell \otimes_{\mathbf{Z}} \mathbf{T}_1(N)$-module $V_\ell(N)$ is free of rank 2 for either action, and $\mathrm{Hom}_{\mathbf{Q}}(\mathbf{Q} \otimes \mathbf{T}_1(N), \mathbf{Q})$ is free of rank 1 over $\mathbf{Q} \otimes \mathbf{T}_1(N)$ (hence likewise with $\mathbf{Q}$ replaced by any field of characteristic 0).*

Remark 5.10. The assertion about $\mathrm{Hom}_{\mathbf{Q}}(\mathbf{Q} \otimes \mathbf{T}_1(N), \mathbf{Q})$ is equivalent to the intrinsic condition that $\mathbf{Q} \otimes \mathbf{T}_1(N)$ is *Gorenstein*. Also, this freeness clearly makes the two assertions about $V_\ell(N)$ for the $(\,)_*$- and $(\,)^*$-actions equivalent. *For the proof*, the $(\,)^*$-action is what we use. But in what follows, it is the case of the $(\,)_*$-action that we need!

Proof. Using (5.4) and the choice of $(\,)^*$-action on $V_\ell(N)$, it suffices to prove
- $H^1(X_1(N)^{\mathrm{an}}, \mathbf{Q})$ is free of rank 2 over $\mathbf{Q} \otimes \mathbf{T}_1(N)$,
- $\mathrm{Hom}_{\mathbf{Q}}(\mathbf{Q} \otimes \mathbf{T}_1(N), \mathbf{Q})$ is free of rank 1 over $\mathbf{Q} \otimes \mathbf{T}_1(N)$.

Using $[\,,\,]_{\Gamma_1(N)}$, we have

(5.5) $$H^1(X_1(N)^{\mathrm{an}}, \underline{\mathbf{Q}}) \cong \mathrm{Hom}_{\mathbf{Q}}(H^1(X_1(N)^{\mathrm{an}}, \underline{\mathbf{Q}}), \mathbf{Q})$$

as $\mathbf{Q} \otimes \mathbf{T}_1(N)$-modules, so we may study this $\mathbf{Q}$-dual instead. Since $\mathbf{Q} \otimes \mathbf{T}_1(N)$ is semilocal, a finite module over this ring is locally free of constant rank if and only if it is *free* of that rank. But local freeness of constant rank can be checked after faithfully flat base change. Applying this with the base change $\mathbf{Q} \to \mathbf{C}$, and noting that $\mathbf{C} \otimes \mathbf{T}_1(N)$ is semilocal, it suffices to replace $\mathbf{Q}$ by $\mathbf{C}$ above.

Note that *if* the right hand side of (5.5) is free of rank 2, so is the left side, so choosing a basis of the left side and feeding it into the right hand side shows that $\mathrm{Hom}_{\mathbf{Q}}(\mathbf{Q} \otimes \mathbf{T}_1(N)^{\oplus 2}, \mathbf{Q})$ is free of rank 2. In particular, the direct summand $\mathrm{Hom}_{\mathbf{Q}}(\mathbf{Q} \otimes \mathbf{T}_1(N), \mathbf{Q})$ is flat over $\mathbf{Q} \otimes \mathbf{T}_1(N)$ with full support over $\mathrm{Spec}(\mathbf{Q} \otimes \mathbf{T}_1(N))$, so it must be locally free with local rank at least 1 at all points of $\mathrm{Spec}(\mathbf{Q} \otimes \mathbf{T}_1(N))$. Consideration of $\mathbf{Q}$-dimensions then forces $\mathrm{Hom}_{\mathbf{Q}}(\mathbf{Q} \otimes \mathbf{T}_1(N), \mathbf{Q})$ to be locally free of rank 1, hence free of rank 1. In other words, it suffices to show that $\mathrm{Hom}_{\mathbf{Q}}(H^1(X_1(N)^{\mathrm{an}}, \underline{\mathbf{Q}}), \mathbf{Q})$ is free of rank 2 over $\mathbf{T}_1(N) \otimes \mathbf{Q}$, or equivalently that $\mathrm{Hom}_{\mathbf{C}}(H^1(X_1(N)^{\mathrm{an}}, \underline{\mathbf{C}}), \mathbf{C})$ is free of rank 2 over $\mathbf{T}_1(N) \otimes \mathbf{C}$.

Via the Shimura isomorphism (in weight 2), which is compatible with the Hecke actions, we are reduced to showing that $\mathrm{Hom}(S_2(\Gamma_1(N), \mathbf{C}), \mathbf{C})$ is free of rank 1 over $\mathbf{C} \otimes \mathbf{T}_1(N)$. For this purpose, we will study the $\mathbf{C} \otimes \mathbf{T}_1(N)$-equivariant $\mathbf{C}$-bilinear pairing

$$S_2(\Gamma_1(N), \mathbf{C}) \otimes_{\mathbf{C}} (\mathbf{C} \otimes \mathbf{T}_1(N)) \to \mathbf{C}$$
$$(f, T) \mapsto a_1(Tf)$$

were $a_1(\cdot)$ is the "Fourier coefficient of q". This is $\mathbf{C} \otimes \mathbf{T}_1(N)$-equivariant, since $\mathbf{T}_1(N)$ is commutative. It suffices to check that there's no nonzero kernel on either side of this pairing. Since

$$\mathbf{C} \otimes \mathbf{T}_1(N) \to \mathrm{End}_{\mathbf{C}}(S_2(\Gamma_1(N), \mathbf{C}))$$

is *injective* (as noted in (5.2)) and $a_1(TT_n f) = a_n(Tf)$ for $T \in \mathbf{T}_1(N)$, the kernel on the right is trivial. Since $a_1(T_n f) = a_n(f)$, the kernel on the left is also trivial. □

5.5.2. Algebraic preliminaries

Let S be a scheme. An *elliptic curve* $E \to S$ is a proper smooth group scheme with geometrically connected fibers of dimension 1 (necessarily of genus 1). It follows from [**62**, Ch.2] that the group structure is commutative and uniquely determined by the identity section. Fix $N \geq 1$ and assume $N \in H^0(S, \mathcal{O}_S^*)$ (i.e., S is a $\mathbf{Z}[\frac{1}{N}]$-scheme). Thus, the map $N : E \to E$ is finite étale of degree N^2 as can be checked on geometric fibers. A *point of exact order N* on E is a section $P : S \to E$ which is killed by N (i.e., factors through the finite étale group scheme $E[N]$) and induces a point of exact order N on geometric fibers.

It follows from the stack-theoretic methods in [**25**] or the more explicit descent arguments in [**62**] that for $N \geq 5$ there is a proper smooth $\mathbf{Z}[\frac{1}{N}]$-scheme $X_1(N)$ equipped with a finite flat map to $\mathbf{P}^1_{\mathbf{Z}[\frac{1}{N}]}$, such that the open subscheme $Y_1(N)$ lying over $\mathbf{P}^1_{\mathbf{Z}[\frac{1}{N}]} - \{\infty\} = \mathbf{A}^1_{\mathbf{Z}[\frac{1}{N}]}$ is the base of a universal object $(E_1(N), P) \to Y_1(N)$ for elliptic curves with a point of exact order N over variable $\mathbf{Z}[\frac{1}{N}]$-schemes.

Moreover, the fibers of $X_1(N) \to \mathrm{Spec}\,\mathbf{Z}[\frac{1}{N}]$ are *geometrically connected*, as this can be checked on a single geometric fiber and by choosing the complex fiber we may appeal to the fact (whose proof requires some care) that there is an isomorphism $(X_1(N) \times_{\mathbf{Z}[\frac{1}{N}]} \mathbf{C})^{\mathrm{an}} \cong X_1(N)^{\mathrm{an}}$ identifying the "algebraic" data $(\mathbf{C}/[1,z], \frac{1}{N})$ in $Y_1(N)(\mathbf{C}) \subset X_1(N)(\mathbf{C})$ with the class of $z \in \mathfrak{h}$ in $\Gamma_1(N)\backslash\mathfrak{h} = Y_1(N)^{\mathrm{an}} \subset X_1(N)^{\mathrm{an}}$ (and $X_1(N)^{\mathrm{an}}$ *is* connected, as $\mathfrak{h}$ is). These kinds of compatibilities are somewhat painful to check unless one develops a full-blown relative theory of elliptic curves in the analytic world (in which case the verifications become quite mechanical and natural).

Again fixing $N \geq 5$, but now also a prime p, we want an algebraic analogue of $X_1(N,p)^{\mathrm{an}}$ over $\mathbf{Z}[\frac{1}{Np}]$. Let $(E,P) \to S$ be an elliptic curve with a point of exact order N over a $\mathbf{Z}[\frac{1}{Np}]$-scheme S. We're interested in studying triples $(E,P,C) \to S$ where $C \subset E$ is an order-p finite locally free S-subgroup-scheme which is not contained in the subgroup generated by P on geometric fibers (if $p \mid N$). Methods in [25] and [62] ensure the existence of a universal such object $(E_1(N,p), P, C) \to Y_1(N,p)$ for a smooth affine $\mathbf{Z}[\frac{1}{Np}]$-scheme which naturally sits as the complement of a relative Cartier divisor in a proper smooth $\mathbf{Z}[\frac{1}{Np}]$-scheme $X_1(N,p)$ which is finite flat over $\mathbf{P}^1_{\mathbf{Z}[\frac{1}{Np}]}$ (with $Y_1(N,p)$ the preimage of $\mathbf{A}^1_{\mathbf{Z}[\frac{1}{Np}]}$). Base change to $\mathbf{C}$ and analytification recovers $X_1(N,p)^{\mathrm{an}}$ as before, so $X_1(N,p) \to \mathrm{Spec}\,\mathbf{Z}[\frac{1}{Np}]$ has geometrically connected fibers.

There are maps of $\mathbf{Z}[\frac{1}{Np}]$-schemes (respectively, $\mathbf{Z}[\frac{1}{N}]$-schemes)

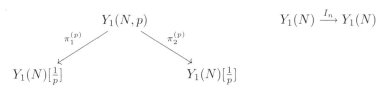

determined by $(E,P,C) \xrightarrow{\pi_1^{(p)}} (E,P)$ and $(E,P,C) \xrightarrow{\pi_2^{(p)}} (E/C, P)$ (which makes sense in $Y_1(N)$ if $p \mid N$ by the "disjointness" condition on C and P) and $I_n(E,P) = (E, nP)$. Although $\pi_2^{(p)}$ is *not* a map over $\mathbf{A}^1_{\mathbf{Z}[\frac{1}{Np}]}$, it can be shown that these all uniquely extend to (necessarily finite *flat*) maps, again denoted $\pi_1^{(p)}, \pi_2^{(p)}, I_n$ between $X_1(N,p)$, $X_1(N)[\frac{1}{p}]$, $X_1(N)$. A proof of this fact requires the theory of minimal regular proper models of curves over a Dedekind base; the analogous fact over $\mathbf{Q}$ is an immediate consequence of basic facts about proper smooth curves over a field, but in order to most easily do some later calculations in characteristic $p \nmid N$ it is convenient to know that we have the map I_p defined on $X_1(N)$ over $\mathbf{Z}[1/N]$ (though this could be bypassed by using liftings to characteristic 0 in a manner similar to our later calculations of T_p in characteristic p).

Likewise, over $\mathbf{Z}[\frac{1}{N}, \zeta_N]$ we can define, for any primitive Nth root of unity $\zeta = \zeta_N^i$ ($i \in (\mathbf{Z}/N\mathbf{Z})^*$), an operator $w_\zeta : Y_1(N)_{/\mathbf{Z}[\frac{1}{N}, \zeta_N]} \to Y_1(N)_{/\mathbf{Z}[\frac{1}{N}, \zeta_N]}$ via $w_\zeta(E,P) = (E/\langle P \rangle, P')$ where $\langle P \rangle$ is the order-N étale subgroup-scheme generated by P and $P' \in (E[N]/\langle P \rangle)(S)$ is uniquely determined by the relative Weil pairing condition $\langle P, P'\rangle_N = \zeta$ (with $P' \in E[N](S)$ here). This really does extend to $X_1(N)_{/\mathbf{Z}[\frac{1}{N}, \zeta_N]}$, and one checks that $w_{\zeta^j} w_\zeta = I_j$ for $j \in (\mathbf{Z}/N\mathbf{Z})^*$. In particular, $w_\zeta^2 = 1$.

Since $X_1(N) \to \mathrm{Spec}\,\mathbf{Z}[\frac{1}{N}]$ is a proper smooth scheme with geometrically connected fibers of dimension 1, $\mathrm{Pic}^0_{X_1(N)_{/\mathbf{Z}[\frac{1}{N}]}}$ is an abelian scheme over $\mathbf{Z}[\frac{1}{N}]$ and hence is the Néron model of its generic fiber. We have scheme-theoretic Albanese and Pic^0 functoriality for finite (flat) maps between proper smooth curves (with geometrically connected fibers) over any base at all, and analytification of such a situation over $\mathbf{C}$ recovers the classical theory of Pic^0 as used in Section 5.5.1.

For example, we have endomorphisms

$$\langle n \rangle^* = \mathrm{Pic}^0(I_n), \quad \langle n \rangle_* = \mathrm{Alb}(I_n)$$

on $\mathrm{Pic}^0_{X_1(N)/\mathbf{Z}[\frac{1}{N}]}$,

$$w_\zeta^* = \mathrm{Pic}^0(w_\zeta) = \mathrm{Alb}(w_\zeta) = (w_\zeta)_*$$

on $\mathrm{Pic}^0_{X_1(N)/\mathbf{Z}[\frac{1}{N},\zeta_N]}$, and

$$T_p^* = \mathrm{Alb}(\pi_1^{(p)}) \circ \mathrm{Pic}^0(\pi_2^{(p)})$$
$$(T_p)_* = \mathrm{Alb}(\pi_2^{(p)}) \circ \mathrm{Pic}^0(\pi_1^{(p)})$$

on $\mathrm{Pic}^0_{X_1(N)/\mathbf{Z}[\frac{1}{Np}]}$. A key point is that by the *Néronian property*, T_p^* and $(T_p)_*$ uniquely extend to endomorphisms of $\mathrm{Pic}^0_{X_1(N)/\mathbf{Z}[\frac{1}{N}]}$, even though the $\pi_i^{(p)}$ do *not* make sense over $\mathbf{Z}[\frac{1}{N}]$ from what has gone before. In particular, it makes sense to study T_p^* and $(T_p)_*$ on the abelian variety $\mathrm{Pic}^0_{X_1(N)/\mathbf{F}_p}$ over $\mathbf{F}_p$ for $p \nmid N$. This will be rather crucial later, but note it requires the Néronian property in the definition.

Passing to the analytifications, the above constructions recover the operators defined on $\mathrm{Pic}^0_{X_1(N)^{\mathrm{an}}}$ in Section 5.5.1. The resulting subring of

$$\mathrm{End}(\mathrm{Pic}^0_{X_1(N)/\mathbf{Z}[\frac{1}{N}]}) \subset \mathrm{End}(\mathrm{Pic}^0_{X_1(N)^{\mathrm{an}}})$$

generated by T_p^*, $\langle n \rangle^*$ (respectively, by $(T_p)_*$, $\langle n \rangle_*$) is identified with $\mathbf{T}_1(N)$ via its $(\ )^*$-action (respectively, via its $(\ )_*$-action) and using

(5.6) $$\varprojlim \mathrm{Pic}^0_{X_1(N)/\mathbf{Z}[\frac{1}{N}]}[\ell^n](\overline{\mathbf{Q}}) \cong T_\ell(\mathrm{Pic}^0_{X_1(N)^{\mathrm{an}}})$$

(using $\overline{\mathbf{Q}} \subset \mathbf{C}$) endows our "analytic" $V_\ell(N)$ with a canonical *continuous* action of $G_\mathbf{Q} = \mathrm{Gal}(\overline{\mathbf{Q}}/\mathbf{Q})$ unramified at all $p \nmid N\ell$ (via Néron-Ogg-Shafarevich) and *commuting* with the action of $\mathbf{T}_1(N)$ (via either the $(\ )^*$-action or the $(\ )_*$-action). We also have an endomorphism $w_\zeta = w_\zeta^* = (w_\zeta)_*$ on $\mathrm{Pic}^0_{X_1(N)/\mathbf{Z}[\frac{1}{N},\zeta_N]}$ and it is easy to see that

$$(g^{-1})^* w_{g(\zeta)} g^* = w_\zeta$$

on $\overline{\mathbf{Q}}$-points, where $g \in \mathrm{Gal}(\overline{\mathbf{Q}}/\mathbf{Q})$ and g^* denotes the natural action of g on $\overline{\mathbf{Q}}$-points (corresponding to base change of degree 0 line bundles on $X_1(N)_{/\overline{\mathbf{Q}}}$). Since $w_\zeta = w_{\zeta^{-1}}$ (as $(E,P) \cong (E,-P)$ via -1), we see that w_ζ is defined over the real subfield $\mathbf{Q}(\zeta_N)^+$. By étale descent, the operator w_ζ is defined over $\mathbf{Z}[\frac{1}{N},\zeta_N]^+$.

In any case, w_ζ acts on $V_\ell(N)$, recovering the operator in Section 5.5.1, and so this conjugates the $(\ )^*$-action to the $(\ )_*$-action, taking each $T \in \mathbf{T}_1(N)$ (for either action on $V_\ell(N)$) to its Weil pairing adjoint, via the canonical principal polarization of the abelian scheme $\mathrm{Pic}^0_{X_1(N)/\mathbf{Z}[\frac{1}{N}]}$. Using Corollary 5.3 and (5.6) we obtain

Lemma 5.11. *Let $\mathbf{T}_1(N)$ act on $V_\ell(N)$ through either the $(\ )^*$-action or the $(\ )_*$-action. Then $\rho_{N,\ell} : G_{\mathbf{Q}} \to \mathrm{Aut}(V_\ell(N)) \cong \mathrm{GL}(2, \mathbf{Q}_\ell \otimes \mathbf{T}_1(N))$ is a continuous representation, unramified at $p \nmid N\ell$.*

The main result we are after is

Theorem 5.12. *Let $\mathbf{T}_1(N)$ act on $\mathrm{Pic}^0_{X_1(N)/\mathbf{Z}[\frac{1}{N}]}$ via the $(\ )_*$-action. For any $p \nmid N\ell$, the characteristic polynomial of $\rho_{N,\ell}(\mathrm{Frob}_p)$ is*

$$X^2 - (T_p)_* X + p\langle p\rangle_*$$

relative to the $\mathbf{Q}_\ell \otimes \mathbf{T}_1(N)$-module structure on $V_\ell(N)$, where Frob_p denotes an arithmetic Frobenius element at p.

The proof of Theorem 5.12 will make essential use of the w_ζ operator. For the remainder of this section, we admit Theorem 5.12 and deduce its consequences. Let $f \in S_2(\Gamma_1(N), \mathbf{C})$ be a *newform* of level N. Let $K_f \subset \mathbf{C}$ be the number field generated by $a_p(f)$ for all $p \nmid N$, where $f = \sum a_n(f) q^n$, so by weak multiplicity one $a_n(f) \in K_f$ for all $n \geq 1$ and the Nebentypus character χ_f has values in K_f. Let $\mathfrak{p}_f \subset \mathbf{T}_1(N)$ be the minimal prime corresponding to f (i.e., the kernel of the map $\mathbf{T}_1(N) \to K_f$ sending each $T \in \mathbf{T}_1(N)$ to its eigenvalue on f).

We now require $\mathbf{T}_1(N)$ to act on $\mathrm{Pic}^0_{X_1(N)/\mathbf{Z}[\frac{1}{N}]}$ via its $(\ )_*$-action.

Definition 5.13. A_f *is the quotient of $\mathrm{Pic}^0_{X_1(N)/\mathbf{Q}}$ by $\mathfrak{p}_f \subset \mathbf{T}_1(N)$.*

By construction, A_f has good reduction over $\mathbf{Z}[\frac{1}{N}]$ and the action of $\mathbf{T}_1(N)$ on $\mathrm{Pic}^0_{X_1(N)/\mathbf{Q}}$ induces an action of $\mathbf{T}_1(N)/\mathfrak{p}$ on A_f, hence an action of $K_f \cong (\mathbf{T}_1(N)/\mathfrak{p}) \otimes_{\mathbf{Z}} \mathbf{Q}$ on A_f in the "up-to-isogeny" category.

Theorem 5.14 (Shimura). *We have $\dim A_f = [K_f : \mathbf{Q}]$ and $V_\ell(A_f)$ is free of rank 2 over $\mathbf{Q}_\ell \otimes_{\mathbf{Q}} K_f$, with Frob_p having characteristic polynomial*

$$X^2 - (1 \otimes a_p(f))X + 1 \otimes p\chi_f(p)$$

for all $p \nmid N\ell$.

Proof. By Lemma 5.11 and Theorem 5.12, we just have to check that the $\mathbf{Q}_\ell \otimes \mathbf{T}_1(N)$-linear map

$$V_\ell(\mathrm{Pic}^0_{X_1(N)/\mathbf{Q}}) \to V_\ell(A_f)$$

identifies the right hand side with the quotient of the left hand side by $\mathfrak{p}_f$. More generally, for any exact sequence

$$B' \to B \to A \to 0$$

of abelian varieties over a field of characteristic prime to ℓ, we claim

$$V_\ell(B') \to V_\ell(B) \to V_\ell(A) \to 0$$

is exact. We may assume the base field is algebraically closed, and then may appeal to Poincaré reducibility (see [**77**, pg. 173]). $\square$

Choosing a place λ of K_f over ℓ and using the natural realization of $K_{f,\lambda}$ as a factor of $\mathbf{Q}_\ell \otimes K_f$, we deduce from Theorem 5.14:

Corollary 5.15. *Let $f \in S_2(\Gamma_1(N), \mathbf{C})$ be a newform and λ a place of K_f over ℓ. There exists a continuous representation*
$$\rho_{f,\lambda} : G_\mathbf{Q} \to GL(2, K_{f,\lambda})$$
unramified at all $p \nmid N\ell$, with Frob_p having characteristic polynomial
$$X^2 - a_p(f)X + p\chi_f(p) \in K_{f,\lambda}[X].$$

5.5.3. Proof of Theorem 5.12

Fix $p \nmid N$ and let
$$J_p = \mathrm{Pic}^0_{X_1(N)/\mathbf{F}_p} \cong \mathrm{Pic}^0_{X_1(N)/\mathbf{Z}[\frac{1}{N}]} \times_{\mathbf{Z}[\frac{1}{N}]} \mathbf{F}_p$$
with $\mathbf{T}_1(N)$ acting through the $(\ )_*$-action. Fix a choice of Frob_p, or more specifically fix a choice of place in $\overline{\mathbf{Q}}$ over p. Note that this determines a preferred algebraic closure $\overline{\mathbf{F}}_p$ as a quotient of the ring of algebraic integers, and in particular a map $\mathbf{Z}[1/N, \zeta_N] \to \overline{\mathbf{F}}_p$. Thus, we may view w_ζ as inducing an endomorphism of the abelian variety $J_p \times_{\mathbf{F}_p} \overline{\mathbf{F}}_p$ over $\overline{\mathbf{F}}_p$ (whereas the elements in $\mathbf{T}_1(N)$ induce endomorphisms of J_p over $\mathbf{F}_p$). The canonical isomorphism
$$V_\ell(\mathrm{Pic}^0_{X_1(N)/\mathbf{Q}}) \cong V_\ell(\mathrm{Pic}^0_{X_1(N)/\mathbf{Z}[\frac{1}{N}]}) \cong V_\ell(J_p)$$
identifies the Frob_p-action on $\overline{\mathbf{Q}}$-points on the left hand side with the (arithmetic) Frobenius action on $\overline{\mathbf{F}}_p$-points on the right hand side. Obviously $V_\ell(J_p)$ is a module over the ring $\mathbf{Q}_\ell \otimes \mathbf{T}_1(N)$ and is free of rank 2 as such. For *any* $\mathbf{F}_p$-schemes Z, Z' and any $\mathbf{F}_p$-map $f : Z \to Z'$ the diagram

(5.7)
$$\begin{array}{ccc} Z & \xrightarrow{f} & Z' \\ F_Z \downarrow & & \downarrow F_{Z'} \\ Z & \xrightarrow{f} & Z' \end{array}$$

commutes, where columns are absolute Frobenius. Taking $Z = \mathrm{Spec}\,\overline{\mathbf{F}}_p$, $Z' = J_p$, we see that the Frob_p action of $V_\ell(J_p)$ through $\overline{\mathbf{F}}_p$-points is *identical* to the action induced by the intrinsic absolute Frobenius morphism $F : J_p \to J_p$ over $\mathbf{F}_p$. Here is the essential input, to be proven later.

Theorem 5.16 (Eichler-Shimura). *In $\mathrm{End}_{\overline{\mathbf{F}}_p}(J_p)$,*
$$(T_p)_* = F + \langle p \rangle_* F^\vee, \qquad w_\zeta^{-1} F w_\zeta = \langle p \rangle_*^{-1} F$$
where $F^\vee$ denotes the dual morphism.

The extra relation involving w_ζ is crucial. The interested reader should compare this with [**108**, Cor. 7.10].

Let us admit Theorem 5.16 and use it to prove Theorem 5.12. We will then prove Theorem 5.16. Using an $\mathbf{F}_p$-rational base point P (e.g., the cusp 0), we get a commutative diagram

$$\begin{array}{ccc} X_1(N)_{/\mathbf{F}_p} & \hookrightarrow & J_p \\ F_{X_1(N)} \downarrow & & \downarrow F \\ X_1(N)_{/\mathbf{F}_p} & \hookrightarrow & J_p \end{array}$$

where $F_{X_1(N)}$ denotes the absolute Frobenius morphism of $X_1(N)_{/\mathbf{F}_p}$, so by Albanese functoriality $F = \mathrm{Alb}(F_{X_1(N)})$. Thus

$$FF^\vee = \mathrm{Alb}(F_{X_1(N)}) \circ \mathrm{Pic}^0(F_{X_1(N)})$$
$$= \deg(F_{X_1(N)}) = p$$

as $X_1(N)_{/\mathbf{F}_p}$ is a smooth *curve*. We conclude from $(T_p)_* = F + \langle p \rangle_* F^\vee$ that

$$F^2 - (T_p)_* F + p\langle p \rangle_* = 0$$

on J_p, hence in $V_\ell(J_p)$. Thus, $\rho_{N,\ell}(\mathrm{Frob}_p)$ satisfies the expected quadratic polynomial

$$X^2 - (T_p)_* X + p\langle p \rangle_* = 0.$$

Let $X^2 - aX + b$ be the *true* characteristic polynomial, which $\rho_{N,\ell}(\mathrm{Frob}_p)$ must also satisfy, by Cayley-Hamilton. We must *prove* that $a = (T_p)_*$, and then $b = p\langle p \rangle_*$ is forced. It is this matter which requires the second relation.

We want $\mathrm{tr}_{\mathbf{Q}_\ell \otimes \mathbf{T}_1(N)}(\rho_{N,\ell}(\mathrm{Frob}_p)) = (T_p)_*$ or equivalently

$$\mathrm{tr}_{\mathbf{Q}_\ell \otimes \mathbf{T}_1(N)}(V_\ell(F)) = (T_p)_*.$$

Using the modified Weil pairing

$$[x, y]_\ell = (x, w_\zeta y)_\ell$$

and using the fact that $V_\ell(J_p) \cong V_\ell(\mathrm{Pic}^0_{X_1(N)/\mathbf{Q}})$ respects Weil pairings (by invoking the relativization of this concept, here over $\mathbf{Z}[\frac{1}{N}]$) we may identify (via Theorem 5.8 and a choice $\mathbf{Q}_\ell(1) \cong \mathbf{Q}_\ell$ as $\mathbf{Q}_\ell$-vector spaces)

$$V_\ell(J_p) \cong \mathrm{Hom}_{\mathbf{Q}_\ell}(V_\ell(J_p), \mathbf{Q}_\ell) := V_\ell(J_p)^*$$

as $\mathbf{Q}_\ell \otimes \mathbf{T}_1(N)$-modules, but taking the F-action over to the $\langle p \rangle_* F^\vee$-action, since adjoints with respect to Weil pairings are dual morphisms and $w_\zeta^{-1} F^\vee w_\zeta$ is dual to $w_\zeta^{-1} F w_\zeta = \langle p \rangle_*^{-1} F = F \langle p \rangle_*^{-1}$ (absolute Frobenius commutes with all morphisms of $\mathbf{F}_p$-schemes!)

Since $V_\ell(J_p)$ is free of rank 2 over $\mathbf{Q}_\ell \otimes \mathbf{T}_1(N)$ and $\mathrm{Hom}_{\mathbf{Q}_\ell}(\mathbf{Q}_\ell \otimes \mathbf{T}_1(N), \mathbf{Q}_\ell)$ is free of rank 1 over $\mathbf{Q}_\ell \otimes \mathbf{T}_1(N)$, by Corollary 5.9, we conclude

$$\mathrm{tr}_{\mathbf{Q}_\ell \otimes \mathbf{T}_1(N)}(F | V_\ell(J_p)) = \mathrm{tr}_{\mathbf{Q}_\ell \otimes \mathbf{T}_1(N)}(\langle p \rangle_* F^\vee | V_\ell(J_p)^*).$$

We wish to invoke the following applied to the $\mathbf{Q}_\ell$-algebra $\mathbf{Q}_\ell \otimes \mathbf{T}_1(N)$ and the $\mathbf{Q}_\ell \otimes \mathbf{T}_1(N)$-module $V_\ell(J_p)$:

Lemma 5.17. *Let $\mathcal{O}$ be a commutative ring, A a finite locally free $\mathcal{O}$-algebra with $\mathrm{Hom}_{\mathcal{O}}(A, \mathcal{O})$ a locally free A-module (necessarily of rank 1). Let M be a finite locally free A-module, $M^* = \mathrm{Hom}_{\mathcal{O}}(M, \mathcal{O})$, so M^* is finite and locally free over A with the same rank as M. For any A-linear map $f : M \to M$ with $\mathcal{O}$-dual $f^* : M^* \to M^*$, automatically A-linear,*

$$\mathrm{char}(f) = \mathrm{char}(f^*)$$

in $A[T]$ (these are the characteristic polynomials).

Proof. Without loss of generality $\mathcal{O}$ is local, so A is semilocal. Making faithfully flat base change to the henselization of $\mathcal{O}$ (or the completion if $\mathcal{O}$ is noetherian or if we first reduce to the noetherian case), we may assume that A is a product of local rings. Without loss of generality, A is then local, so

$$M = \oplus A e_i$$

if free, and $\mathrm{Hom}_\mathcal{O}(A,\mathcal{O})$ is free of rank 1 over A. Choose an isomorphism
$$h: A \cong \mathrm{Hom}_\mathcal{O}(A,\mathcal{O})$$
as A-modules, so the projections
$$\pi_i: M \to Ae_i \cong A$$
satisfy $e_i^* = h(i) \circ \pi_i$ in M^*. These e_i^* are an A-basis of M^* and we compute matrices over A:
$$\mathrm{Mat}_{\{e_i\}}(f) = \mathrm{Mat}_{\{e_i^*\}}(f^*)^t.$$
$\square$

We conclude that
$$\mathrm{tr}_{\mathbf{Q}_\ell \otimes \mathbf{T}_1(N)}(F|V_\ell(J_p)) = \mathrm{tr}_{\mathbf{Q}_\ell \otimes \mathbf{T}_1(N)}(\langle p\rangle_* f^\vee|V_\ell(J_p)).$$
By Theorem 5.16, we have
$$\begin{aligned}2(T_p)_* &= \mathrm{tr}((T_p)_*|V_\ell(J_p)) \\ &= \mathrm{tr}(F + \langle p\rangle_* F^\vee|V_\ell(J_p)) \\ &= 2\,\mathrm{tr}(F|V_\ell(J_p)).\end{aligned}$$
This proves that $\mathrm{tr}(F|V_\ell(J_p)) = (T_p)_*$, so indeed $X^2 - (T_p)_* X + p\langle p\rangle_*$ *is the characteristic polynomial.* Finally, there remains

Proof of Theorem 5.16. It suffices to check the maps coincide on a Zariski dense subset of $J_p(\overline{\mathbf{F}}_p) = \mathrm{Pic}^0(X_1(N)_{/\overline{\mathbf{F}}_p})$. If g is the genus of $X_1(N)_{/\mathbf{Z}[\frac{1}{N}]}$ and we fix an $\overline{\mathbf{F}}_p$-rational base point, we get an induced surjective map
$$X_1(N)^g_{/\overline{\mathbf{F}}_p} \to J_{p/\overline{\mathbf{F}}_p},$$
so for any dense open $U \subset X_1(N)_{\overline{\mathbf{F}}_p}$, $U^g \to (J_p)_{/\overline{\mathbf{F}}_p}$ hits a Zariski dense subset of $\overline{\mathbf{F}}_p$-points. Taking U to be the ordinary locus of $Y_1(N)_{/\overline{\mathbf{F}}_p}$, it suffices to study what happens to a difference $(x) - (x')$ for $x, x' \in Y_1(N)(\overline{\mathbf{F}}_p)$ corresponding to (E, P), (E', P') over $\overline{\mathbf{F}}_p$ with E and E' *ordinary* elliptic curves.

By the commutative diagram (5.7), the map
$$J_p(\overline{\mathbf{F}}_p) \to J_p(\overline{\mathbf{F}}_p)$$
induced by F is the same as the map induced by the pth power map in $\overline{\mathbf{F}}_p$. By *definition* of Pic^0 functoriality, this corresponds to base change of an invertible sheaf on $X_1(N)_{/\overline{\mathbf{F}}_p}$ by the absolute Frobenius on $\overline{\mathbf{F}}_p$. By *definition* of $Y_1(N)_{/\overline{\mathbf{F}}_p}$ as a universal object, such base change induces on $Y_1(N)(\overline{\mathbf{F}}_p)$ *exactly* "base change by absolute Frobenius" on elliptic curves with a point of exact order N over $\overline{\mathbf{F}}_p$. We conclude
$$F((x) - (x')) = (E^{(p)}, P^{(p)}) - ((E')^{(p)}, P^{(p)})$$
where $(\ )^{(p)}$ denotes base change by absolute Frobenius on $\overline{\mathbf{F}}_p$.

Since $p = FF^\vee = F^\vee F$ and F is bijective on $\overline{\mathbf{F}}_p$-points, we have
$$\begin{aligned}F^\vee((x) - (x')) &= pF^{-1}((x) - (x')) \\ &= p((E^{(p^{-1})}, P^{(p^{-1})}) - ((E')^{(p^{-1})}, (P')^{(p^{-1})})).\end{aligned}$$
Thus,
$$\langle p\rangle_* F^\vee((x) - (x')) = p(E^{(p^{-1})}, pP^{(p^{-1})}) - p((E')^{(p^{-1})}, p(P')^{(p^{-1})})$$

so
$$(F + \langle p\rangle_* F^\vee)((x) - (x')) = (E^{(p)}, P^{(p)}) + p(E^{(p^{-1})}, pP^{(p^{-1})})$$
$$- ((E')^{(p)}, (P')^{(p)}) + p((E')^{(p^{-1})}, p(P')^{(p^{-1})}).$$

Computing $(T_p)_*$ on $J_p = \text{Pic}^0_{X_1(N)/\mathbf{Z}[\frac{1}{N}]} \times_{\mathbf{Z}[\frac{1}{N}]} \mathbf{F}_p$ is more subtle because $(T_p)_*$ was defined over $\mathbf{Z}[\frac{1}{Np}]$ (or over $\mathbf{Q}$) as $(\pi_2)_* \pi_1^*$ and was *extended* over $\mathbf{Z}[\frac{1}{N}]$ by the Néronian property. That is, we do *not* have a direct definition of $(T_p)_*$ in characteristic p, so we will need to lift to characteristic 0 to compute. It is *here* that the ordinariness assumption is crucial, for we shall see that, in some sense,
$$(T_p)_*((x) - (x')) = (F + \langle p\rangle_* F^\vee)((x) - (x'))$$
as *divisors* for ordinary points x, x'. This is, of course, much stronger than the mere linear equivalence that we need to prove.

Before we dive into the somewhat subtle calculation of $(T_p)_*((x) - (x'))$, let's quickly take care of the relation $w_\zeta^{-1} F w_\zeta = \langle p\rangle_*^{-1} F$, or equivalently,
$$F w_\zeta = w_\zeta \langle p^{-1}\rangle_* F.$$
All maps here are induced by maps on $X_1(N)_{/\overline{\mathbf{F}}_p}$, with $F = \text{Alb}(F_{X_1(N)})$, $w_\zeta = \text{Alb}(w_{\zeta|_{X_1(N)}})$, $\langle p^{-1}\rangle_* = \text{Alb}(I_{p^{-1}})$. Thus, it suffices to show
$$F_{X_1(N)} \circ w_\zeta = w_\zeta I_{p^{-1}} F_{X_1(N)}$$
on $X_1(N)_{/\overline{\mathbf{F}}_p}$, and we can check by studying $x = (E, P) \in Y_1(N)(\overline{\mathbf{F}}_p)$:
$$F_{X_1(N)} w_\zeta(x) = F_{X_1(N)}(E/P, P') = (E^{(p)}/P^{(p)}, (P')^{(p)})$$
where $\langle P, P'\rangle_N = \zeta$, so $\langle P^{(p)}, (P')^{(p)}\rangle_N = \zeta^p$ by compatibility of the (relative) Weil pairing with respect to base change. Meanwhile,
$$w_\zeta I_{p^{-1}} F_{X_1(N)}(x) = w_\zeta(E^{(p)}, p^{-1} P^{(p)}) = (E^{(p)}/(p^{-1} P^{(p)}), Q)$$
where $\langle p^{-1} P^{(p)}, Q\rangle_N = \zeta$, or equivalently $\langle P^{(p)}, Q\rangle = \zeta^p$. Since $Q = (P')^{(p)}$ is such a point, this second relation is established.

Now we turn to the problem of computing
$$(T_p)_*((x) - (x'))$$
for "ordinary points" $x = (E, P)$, $x' = (E', P')$ as above. Let $R = \mathbf{Z}_p^{\text{un}}$, $W(\overline{\mathbf{F}}_p)$, or more generally any henselian (e.g., complete) discrete valuation ring with residue field $\overline{\mathbf{F}}_p$ and fraction field K of characteristic 0. Since $p \nmid N$, R is a $\mathbf{Z}[\frac{1}{N}]$-algebra. Since $Y_1(N)$ is *smooth* over $\mathbf{Z}[\frac{1}{N}]$, we conclude from the (strict) henselian property that $Y_1(N)(R) \to Y_1(N)(\overline{\mathbf{F}}_p)$ is surjective. Of course, this can be seen "by hand": if (E, P) is given over $\overline{\mathbf{F}}_p$, choose a Weierstrass model $\mathcal{E} \hookrightarrow \mathbf{P}_R^2$ lifting E (this is canonically an elliptic curve, by [**62**, Ch 2]). The finite *étale* group scheme $\mathcal{E}[N]$ is *constant* since R is strictly henselian. Thus there exists a unique closed immersion of group schemes $\mathbf{Z}/N\mathbf{Z} \hookrightarrow \mathcal{E}[N]$ lifting $P: \mathbf{Z}/N\mathbf{Z} \hookrightarrow E[N]$.

Let $(\mathcal{E}, \mathcal{P})$, $(\mathcal{E}', \mathcal{P}')$ over R lift x, x' respectively. We view these sections to $X_1(N)_{/R} \to \text{Spec } R$ as relative effective Cartier divisors of degree 1. Using the reduction map
$$\text{Pic}^0_{X_1(N)/\mathbf{Z}[\frac{1}{N}]}(R) \to J_p(\overline{\mathbf{F}}_p)$$

and the *definition* of $(T_p)_*$, we see that $(T_p)_*((x)-(x'))$ is the image of $(T_p)_*((\mathcal{E},\mathcal{P})-(\mathcal{E}',\mathcal{P}'))$. Now R is *NOT* a $\mathbf{Z}[\frac{1}{Np}]$-algebra but K *is*, and we have an injection (even bijection)

$$\mathrm{Pic}^0_{X_1(N)_{/\mathbf{Z}[\frac{1}{N}]}}(R) \hookrightarrow \mathrm{Pic}^0_{X_1(N)_{/\mathbf{Z}[\frac{1}{N}]}}(K),$$

as $\mathrm{Pic}^0_{X_1(N)_{/\mathbf{Z}[\frac{1}{N}]}} \to \mathrm{Spec}\,\mathbf{Z}[\frac{1}{N}]$ is separated (even proper).

Thus, we will first compute $(T_p)_*((x)-(x'))$ by working with $\overline{K}$-points, where $\overline{K}$ is an algebraic closure of K. Since $p \nmid N$, we have

$$(\pi_2)_* \pi_1^*((\mathcal{E},\mathcal{P})_{/\overline{K}}) = \sum_C (\mathcal{E}_{\overline{K}}/C, \mathcal{P}_{\overline{K}} \bmod C)$$

where C runs through all $p+1$ order-p subgroups of $\mathcal{E}_{/\overline{K}}$. Since $\mathcal{E} \to \mathrm{Spec}\,R$ has *ordinary* reduction, and R is strictly henselian, the connected-étale sequence of $\mathcal{E}[p]$ is the short exact sequence of finite flat R-group schemes

$$0 \to \mu_p \to \mathcal{E}[p] \to \mathbf{Z}/p\mathbf{Z} \to 0.$$

Enlarging R to a finite extension does not change the residue field $\overline{\mathbf{F}}_p$, so we may assume that

$$\mathcal{E}[p]_{/K} \cong \mathbf{Z}/p\mathbf{Z} \times \mathbf{Z}/p\mathbf{Z}.$$

Taking the scheme-theoretic closure in $\mathcal{E}[p]$ of the $p+1$ distinct subgroups of $\mathcal{E}[p]_{/K}$ gives $p+1$ *distinct* finite flat subgroup schemes $\mathcal{C} \subset \mathcal{E}$ realizing the $p+1$ distinct C's over $\overline{K}$.

Exactly one of these $\mathcal{C}$'s is killed by $\mathcal{E}[p] \to \mathbf{Z}/p\mathbf{Z}$ over R, as this can be checked on the generic fiber, so it must be $\mu_p \hookrightarrow \mathcal{E}[p]$. For the remaining $\mathcal{C}$'s, the map $\mathcal{C} \to \mathbf{Z}/p\mathbf{Z}$ is an isomorphism on the generic fiber. We claim these maps

$$\mathcal{C} \to \mathbf{Z}/p\mathbf{Z}$$

over R are isomorphisms. Indeed, if $\mathcal{C}$ is *étale* this is clear, yet $\mathcal{C} \hookrightarrow \mathcal{E}[p]$ is a finite flat closed subgroup-scheme of order p, so a consideration of the closed fiber shows that if $\mathcal{C}$ is *not* étale then it is multiplicative. But $\mathcal{E}[p]$ has a *unique* multiplicative subgroup-scheme since

$$\mathcal{E}[p]^\vee \cong \mathcal{E}[p]$$

by Cartier-Nishi duality and $\mathcal{E}[p]$ has a *unique* order-p *étale* quotient (as any such quotient must kill the μ_p we have inside $\mathcal{E}[p]$.)

Thus,

$$(\pi_2)_*\pi_1^*((\mathcal{E},\mathcal{P})_{/\overline{K}}) = \sum_{\mathcal{C}}(\mathcal{E}/\mathcal{C},\mathcal{P} \bmod \mathcal{C}) - \sum_{\mathcal{C}'}(\mathcal{E}'/\mathcal{C}',\mathcal{P}' \bmod \mathcal{C}')$$
$$\in \mathrm{Pic}^0_{X_1(N)_{/\mathbf{Z}[\frac{1}{N}]}}(R)$$

coincides with $(T_p)_*((\mathcal{E},\mathcal{P}) - (\mathcal{E}',\mathcal{P}'))$ as both induce the same $\overline{K}$-point. Passing to closed fibers,

$$(T_p)_*((x)-(x')) = (E/\mu_p, P \bmod \mu_p) + p(E/\mathbf{Z}/p\mathbf{Z}', P \bmod \mathbf{Z}/p\mathbf{Z})$$
$$- (E'/\mu_p, P' \bmod \mu_p) + p(E'/\mathbf{Z}/p\mathbf{Z}, P' \bmod \mathbf{Z}/p\mathbf{Z})$$

where $E[p] \cong \mu_p \times \mathbf{Z}/p\mathbf{Z}$ and $E'[p] \cong \mu_p \times \mathbf{Z}/p\mathbf{Z}$ are the *canonical* splittings of the connected-étale sequence over the perfect field $\overline{\mathbf{F}}_p$.

Now consider the relative Frobenius morphism
$$F_{E/\overline{\mathbf{F}}_p} : E \to E^{(p)},$$
which sends O to O (and P to $P^{(p)}$) and so is a map of *elliptic curves* over $\overline{\mathbf{F}}_p$. Recall that in characteristic p, for any map of schemes $X \to S$ we define the relative Frobenius map $F_{X/S} : X \to X^{(p)}$ to be the unique S-map fitting into the diagram

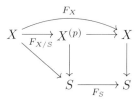

where F_S, F_X are the absolute Frobenius maps. Since $E \to \operatorname{Spec}\overline{\mathbf{F}}_p$ is smooth of pure relative dimension 1, $F_{E/\overline{\mathbf{F}}_p}$ is finite flat of degree $p^1 = p$. It is bijective on points, so $\ker(F_{E/\overline{\mathbf{F}}_p})$ must be connected of order p.

The *only* such subgroup-scheme of E is $\mu_p \hookrightarrow E[p]$ by the *ordinariness*. Thus
$$E/\mu_p \cong E^{(p)}$$
is easily seen to take P mod μ_p to $P^{(p)}$.

Similarly, we have
$$E \xrightarrow{F_{E/\overline{\mathbf{F}}_p}} E^{(p)} \xrightarrow{F^\vee_{E/\overline{\mathbf{F}}_p}} E$$
with composite p, so $F^\vee_{E/\overline{\mathbf{F}}_p}$ is étale of degree p and base extension by $\operatorname{Frob}^{-1} : \overline{\mathbf{F}}_p \to \overline{\mathbf{F}}_p$ gives
$$E^{(p^{-1})} \longrightarrow E \longrightarrow E^{(p^{-1})}$$
with composite p,
$$P^{(p^{-1})} \longmapsto P \longmapsto p \cdot P^{(p^{-1})}.$$

As the second map in this composite is étale of degree p, we conclude
$$(E_{/\mathbf{Z}/p\mathbf{Z}}, P \bmod \mathbf{Z}/p\mathbf{Z}) \cong (E^{(p^{-1})}, pP^{(p^{-1})}).$$
Thus, in $\operatorname{Pic}^0_{X_1(N)}(\overline{\mathbf{F}}_p)$,
$$(T_p)_*((x) - (x')) = (E^{(p)}, P^{(p)}) + p \cdot (E^{(p^{-1})}, p \cdot P^{(p^{-1})})$$
$$- ((E')^{(p)}, (P')^{(p)}) - p \cdot ((E')^{(p^{-1})}, p \cdot (P')^{(p^{-1})})$$
which we have seen is equal to $(F + \langle p \rangle_* F^\vee)((x) - (x'))$.

□

CHAPTER 6
Appendix by Kevin Buzzard: A mod ℓ Multiplicity One Result

In this appendix, we explain how the ideas of [46] can be used to prove the following mild strengthening of the multiplicity one results in §9 of [32].

The setup is as follows. Let f be a normalised cuspidal eigenform of level N, and weight k, defined over $\overline{\mathbf{F}}_\ell$, with $\ell \nmid N$ and $2 \leq k \leq \ell + 1$. Let N^* denote N if $k = 2$, and $N\ell$ if $k > 2$. Let $J_\mathbf{Q}$ be the Jacobian of the curve $X_1(N^*)_\mathbf{Q}$, and let H denote the Hecke algebra in $\mathrm{End}(J_\mathbf{Q})$ generated over $\mathbf{Z}$ by T_p for all primes p, and all the Diamond operators of level N^*. It is well-known (for example by Proposition 9.3 of [46]) that there is a characteristic 0 normalised eigenform F in $S_2(\Gamma_1(N^*))$ lifting f. Let $\mathfrak{m}$ denote the maximal ideal of H associated to F (note that $\mathfrak{m}$ depends only on f and not on the choice of F), and let $\mathbf{F} = H/\mathfrak{m}$, which embeds naturally into $\overline{\mathbf{F}}_\ell$. Suppose the representation $\rho_f : G_\mathbf{Q} \to \mathrm{GL}_2(\overline{\mathbf{F}}_\ell)$ associated to f is absolutely irreducible, and furthermore assume that if $k = \ell + 1$ then ρ_f is not isomorphic to a representation coming from a form of weight 2 and level N.

Theorem 6.1. *If ρ_f is ramified at ℓ, or if ρ_f is unramified at ℓ but $\rho_f(\mathrm{Frob}_\ell)$ is not a scalar matrix, then $J_\mathbf{Q}(\overline{\mathbf{Q}})[\mathfrak{m}]$ has $H/\mathfrak{m}$-dimension two, and hence is a model for (precisely one copy of) ρ_f.*

The motivation for putting ourselves in the setup above is that every absolutely irreducible modular mod ℓ representation has a twist coming from a modular form of level prime to ℓ and weight at most $\ell + 1$. In particular, every modular mod ℓ representation has a twist coming from a form satisfying the conditions of our setup. Furthermore, if f is as in our setup, then Theorems 2.5 and 2.6 of [32] tell us the precise structure of the restriction of ρ_f to D_ℓ, a decomposition group at ℓ. These results are explained in Section 2.2. Using them, it is easy to deduce

Corollary 6.2. *Let ρ be an absolutely irreducible modular mod ℓ representation, such that $\rho_f(D_\ell)$ is not contained within the scalars. Then some twist of ρ comes from a modular form satisfying the conditions of the theorem, and hence ρ is a multiplicity one representation in the sense of Remark 3.4.2.*

The theorem, commonly referred to as a "multiplicity one theorem", is a mild extension of results of Mazur, Ribet, Gross and Edixhoven. It was announced for $\ell = 2$ as Proposition 2.4 of [9] but the proof given there is not quite complete—in fact, the last line of the proof there is a little optimistic. I would hence like to thank Ribet and Stein for the opportunity to correct this oversight in [9].

Proof of Theorem. Firstly we observe that the only case not dealt with by Theorem 9.2 of [32] is the case when $k = \ell$ and ρ_f is unramified at ℓ, with $\rho_f(\text{Frob}_\ell)$ a non-scalar matrix whose eigenvalues are equal. Moreover, using Theorems 2.5 and 2.6 of [32] we see that in this case f must be ordinary at ℓ. We are hence in a position to use the detailed construction of ρ_f given in §§11–12 of [46]. We will follow the conventions set up in the present paper for normalisations of Hecke operators, and so in particular the formulae below differ from the ones in [46] by a twist.

The maximal ideal $\mathfrak{m}$ of H associated to f gives rise as in (12.5) of [46] to an idempotent $e \in H_\ell := H \otimes_{\mathbf{Z}} \mathbf{Z}_l$, such that the completion $H_{\mathfrak{m}}$ of H at $\mathfrak{m}$ is just eH_ℓ. Let G denote $e(J_{\mathbf{Q}_\ell}[\ell^\infty])$, the part of the ℓ-divisible group of J which is associated to $\mathfrak{m}$. Then $H_{\mathfrak{m}}$ acts on G, and it is proved in Propositions 12.8 and 12.9 of [46] that there is an exact sequence of ℓ-divisible groups

$$0 \to G^0 \to G \to G^e \to 0$$

over $\mathbf{Q}_\ell$, which is $H_{\mathfrak{m}}$-stable. Let

$$0 \to T^0 \to T \to T^e \to 0$$

be the exact sequence of Tate modules of these groups. We now explain explicitly, following [46], how the group D_ℓ acts on these Tate modules.

If $k > 2$ then there is a Hecke operator U_ℓ in $H_{\mathfrak{m}}$, and we define $u = U_\ell$. If $k = 2$ then there is a Hecke operator T_ℓ in $H_{\mathfrak{m}}$ and because we are in the ordinary case we know that T_ℓ is a unit in $H_{\mathfrak{m}}$. We define u to be the unique root of the polynomial $X^2 - T_\ell X + \ell \langle \ell \rangle$ in $H_{\mathfrak{m}}$ which is a unit (u exists by an appropriate analogue of Hensel's lemma).

The calculations of Propositions 12.8 and 12.9 of [46] show that, under our conventions, the absolute Galois group D_ℓ of $\mathbf{Q}_\ell$ acts on T^e as $\lambda(u)$, where $\lambda(x)$ denotes the unramified character taking Frob_ℓ to x. Moreover, these propositions also tell us that D_ℓ acts on T^0 via the character $\chi_\ell \lambda(u^{-1}\langle \ell \rangle_N) \chi^{\ell-2}$, where χ_ℓ is the cyclotomic character and χ is the Teichmüller character. The key point is that this character takes values in $H^\times$.

The next key observation is that a standard argument on differentials, again contained in the proof of Propositions 12.8 and 12.9 of [46], shows that $G^e[\mathfrak{m}] = \mathfrak{m}^{-1}\ell T^e / \ell T^e$ has $H_{\mathfrak{m}}/\mathfrak{m}$-dimension 1 and that $G^0[\mathfrak{m}]$ has dimension $d^0 \geq 1$. (Note that the fact that $G^e[\mathfrak{m}]$ has dimension 1 implies, via some simple linear algebra, that the sequence $0 \to G^0[\mathfrak{m}] \to G[\mathfrak{m}] \to G^e[\mathfrak{m}] \to 0$ is exact, as asserted by Gross.) Furthermore, because we can identify $G^0[\mathfrak{m}]$ with $\mathfrak{m}^{-1}\ell T^0 / \ell T^0$, we see that the action of D_ℓ on $G^0[\mathfrak{m}]$ is via a character which takes values in $(H/\mathfrak{m})^\times$. In particular, D_ℓ acts as scalars on $G^0[\mathfrak{m}]$.

Let us now assume that ρ_f is unramified at ℓ, and that $\rho_f(\text{Frob}_\ell)$ is a non-diagonalisable matrix with eigenvalue $\alpha \in H/\mathfrak{m}$. Choose a model ρ for ρ_f defined over $\text{GL}_2(H/\mathfrak{m})$. By the theorem of Boston, Lenstra and Ribet, we know that $G[\mathfrak{m}]$ is isomorphic to a direct sum of d copies of ρ, or more precisely, d copies of the restriction of ρ to D_ℓ. Here d is an integer satisfying $2d = d^0 + d^e$. Hence, if $G[\mathfrak{m}]^\alpha$

denotes the subspace of $G[\mathfrak{m}]$ where Frob_ℓ acts as α, then the $H/\mathfrak{m}$-dimension of $G[\mathfrak{m}]^\alpha$ is at most d. On the other hand, Frob_ℓ acts on $G[\mathfrak{m}]^0$ as a scalar, and hence this scalar must be α, and so we see $G[\mathfrak{m}]^0 \subseteq G[\mathfrak{m}]^\alpha$. Hence $d^0 \leq d = (d^0 + 1)/2$. We deduce that $d^0 \leq 1$ and hence $d^0 = d = 1$ and the theorem is proved. □

We remark that L. Kilford has found examples of mod 2 forms f of weight 2, such that ρ_f is unramified at 2 and $\rho_f(\text{Frob}_2)$ is the identity, and where $J_\mathbf{Q}(\overline{\mathbf{Q}})[\mathfrak{m}]$ has $H/\mathfrak{m}$-dimension 4, and so one cannot hope to extend the theorem to this case. See Remark 3.6 for more details, or [**64**]. A detailed analysis of what is happening in this case, at least in the analogous setting of forms of weight 2 on $J_0(p)$, with p prime, has been undertaken by Emerton in [**39**]. In particular, Emerton proves that multiplicity one fails if and only if the analogue of the exact sequence $0 \to T^0 \to T \to T^e \to 0$ fails to split as a sequence of $H_\mathfrak{m}$-modules.

BIBLIOGRAPHY

1. A. Ash and G. Stevens, *Cohomology of arithmetic groups and congruences between systems of Hecke eigenvalues*, J. Reine Angew. Math. **365** (1986), 192–220.
2. A. O. L. Atkin and J. Lehner, *Hecke operators on $\Gamma_0(m)$*, Math. Ann. **185** (1970), 134–160.
3. B. J. Birch, *Cyclotomic fields and Kummer extensions*, Algebraic Number Theory (Proc. Instructional Conf., Brighton, 1965), Thompson, Washington, D.C., 1967, pp. 85–93.
4. B. J. Birch and W. Kuyk (eds.), *Modular functions of one variable. IV*, Springer-Verlag, Berlin, 1975, Lecture Notes in Mathematics, Vol. 476.
5. S. Bosch, W. Lütkebohmert, and M. Raynaud, *Néron models*, Springer-Verlag, Berlin, 1990.
6. N. Boston, H. W. Lenstra, Jr., and K. A. Ribet, *Quotients of group rings arising from two-dimensional representations*, C. R. Acad. Sci. Paris Sér. I Math. **312** (1991), no. 4, 323–328.
7. C. Breuil, B. Conrad, F. Diamond, and R. Taylor, *On the modularity of elliptic curves over $\mathbf{Q}$, or Wild 3-adic exercises*,
http://www.math.harvard.edu/HTML/Individuals/Richard_Taylor.html
8. S. Brueggeman, *The non-existence of certain Galois extensions unramified outside 5*, Journal of Number Theory **75** (1999), 47–52.
9. K. Buzzard, *On level-lowering for mod 2 representations*, to appear in Mathematics Research Letters.
10. K. Buzzard, M. Dickinson, N. Shepherd-Barron, and R. Taylor, *On icosahedral Artin representations*, in preparation.
11. H. Carayol, *Sur les représentations ℓ-adiques associées aux formes modulaires de Hilbert*, Ann. scient. Éc. Norm. Sup., 4^{eb} série **19** (1986), 409–468.
12. _____, *Sur les représentations galoisiennes modulo ℓ attachées aux formes modulaires*, Duke Math. J. **59** (1989), 785–801.
13. W. Casselman, *On representations of GL_2 and the arithmetic of modular curves*, Modular functions of one variable, II (Proc. Internat. Summer School, Univ. Antwerp, Antwerp, 1972) (Berlin), Springer, 1973, pp. 107–141. Lecture Notes in Math., Vol. 349.
14. I. V. Čerednik, *Uniformization of algebraic curves by discrete arithmetic subgroups of $\text{PGL}_2(k_w)$ with compact quotient spaces*, Mat. Sb. (N.S.) **100(142)**

(1976), no. 1, 59–88, 165.

15. R. F. Coleman, *Serre's conjecture: The Jugentraum of the 20th century*, Mat. Contemp. **6** (1994), 13–18, XII School of Algebra, Part I (Portuguese) (Diamantina, 1992).

16. R. F. Coleman and B. Edixhoven, *On the semi-simplicity of the U_p-operator on modular forms*, Math. Ann. **310** (1998), no. 1, 119–127.

17. R. F. Coleman and J. F. Voloch, *Companion forms and Kodaira-Spencer theory*, Invent. Math. **110** (1992), no. 2, 263–281.

18. B. Conrad, *Modular forms, cohomology, and the Ramanujan conjecture*, in preparation.

19. B. Conrad, F. Diamond, and R. Taylor, *Modularity of certain potentially Barsotti-Tate Galois representations*, J. Amer. Math. Soc. **12** (1999), no. 2, 521–567.

20. J. E. Cremona, *Algorithms for modular elliptic curves*, second ed., Cambridge University Press, Cambridge, 1997.

21. C. W. Curtis and I. Reiner, *Representation theory of finite groups and associative algebras*, Interscience Publishers, a division of John Wiley & Sons, New York-London, 1962, Pure and Applied Mathematics, Vol. XI.

22. H. Darmon, *Serre's conjectures*, Seminar on Fermat's Last Theorem (Toronto, ON, 1993–1994), Amer. Math. Soc., Providence, RI, 1995, pp. 135–153.

23. H. Darmon, F. Diamond, and R. Taylor, *Fermat's last theorem*, Current developments in mathematics, 1995 (Cambridge, MA), Internat. Press, Cambridge, MA, 1994, pp. 1–154.

24. P. Deligne, *Formes modulaires et représentations ℓ-adiques.*, Sém. Bourbaki no. 355, 1968/69 (Berlin and New York), Springer-Verlag, 1971, Lecture Notes in Mathematics, Vol. 179, pp. 139–172.

25. P. Deligne and M. Rapoport, *Les schémas de modules de courbes elliptiques*, Modular functions of one variable, II (Proc. Internat. Summer School, Univ. Antwerp, Antwerp, 1972) (Berlin), Springer, 1973, pp. 143–316. Lecture Notes in Math., Vol. 349.

26. F. Diamond, *The refined conjecture of Serre*, Elliptic curves, modular forms, & Fermat's last theorem (Hong Kong, 1993) (Cambridge, MA), Internat. Press, 1995, pp. 22–37.

27. F. Diamond and J. Im, *Modular forms and modular curves*, Seminar on Fermat's Last Theorem (Providence, RI), Amer. Math. Soc., 1995, pp. 39–133.

28. M. Dickinson, *On the modularity of certain 2-adic galois representations*, Harvard Ph.D. thesis (2000).

29. D. Doud, *S_4 and $\tilde{S}_4$ extensions of $\mathbf{Q}$ ramified at only one prime*, J. Number Theory **75** (1999), no. 2, 185–197.

30. V. G. Drinfeld, *Coverings of p-adic symmetric domains*, Funkcional. Anal. i Prilov zen. **10** (1976), no. 2, 29–40.

31. B. Edixhoven, *L'action de l'algèbre de Hecke sur les groupes de composantes des jacobiennes des courbes modulaires est "Eisenstein"*, Astérisque (1991), no. 196–197, 7–8, 159–170 (1992), Courbes modulaires et courbes de Shimura (Orsay, 1987/1988).

32. ———, *The weight in Serre's conjectures on modular forms*, Invent. Math. **109** (1992), no. 3, 563–594.

33. B. Edixhoven, *Le rôle de la conjecture de Serre dans la démonstration du théorème de Fermat*, Gaz. Math. (1995), no. 66, 25–41.

34. _____, *Erratum and addendum: "The role of Serre's conjecture in the proof of Fermat's theorem"*, Gaz. Math. (1996), no. 67, 19.
35. _____, *Serre's conjecture*, Modular forms and Fermat's last theorem (Boston, MA, 1995) (New York), Springer, 1997, pp. 209–242.
36. M. Eichler, *Quadratische Formen und Modulfunktionen*, Acta Arith. **4** (1958), 217–239.
37. D. Eisenbud, *Commutative algebra with a view toward algebraic geometry*, Springer-Verlag, New York, 1995.
38. D. Eisenbud and J. Harris, *Schemes, The language of modern algebraic geometry*, Springer-Verlag, Berlin, Graduate Texts in Mathematics, Vol. 197.
39. M. Emerton, *Supersingular elliptic curves, theta series and weight two modular forms*, preprint.
40. G. Faltings, *Endlichkeitssätze für abelsche Varietäten über Zahlkörpern*, Invent. Math. **73** (1983), no. 3, 349–366.
41. G. Faltings and B. W. Jordan, *Crystalline cohomology and* $GL(2, \mathbf{Q})$, Israel J. Math. **90** (1995), no. 1-3, 1–66.
42. G. Frey, *Links between stable elliptic curves and certain Diophantine equations*, Ann. Univ. Sarav. Ser. Math. **1** (1986), no. 1, iv+40.
43. _____, *Links between solutions of* $A - B = C$ *and elliptic curves*, Number theory (Ulm, 1987), Springer, New York, 1989, pp. 31–62.
44. A. Fröhlich, *Local fields*, Algebraic Number Theory (Proc. Instructional Conf., Brighton, 1965), Thompson, Washington, D.C., 1967, pp. 1–41.
45. K. Fujiwara, *Level optimization in the totally real case*, in preparation (1999).
46. B. H. Gross, *A tameness criterion for Galois representations associated to modular forms (mod p)*, Duke Math. J. **61** (1990), no. 2, 445–517.
47. R. Hartshorne, *Algebraic geometry*, Springer-Verlag, New York, 1977, Graduate Texts in Mathematics, No. 52.
48. Y. Hellegouarch, *Invitation aux mathématiques de Fermat-Wiles*, Masson, Paris, 1997.
49. H. Hida, *Galois representations into* $GL_2(\mathbf{Z}_p[[X]])$ *attached to ordinary cusp forms*, Invent. Math. **85** (1986), no. 3, 545–613.
50. _____, *Iwasawa modules attached to congruences of cusp forms*, Ann. Sci. École Norm. Sup. (4) **19** (1986), no. 2, 231–273.
51. H. Jacquet and R. P. Langlands, *Automorphic forms on* $GL(2)$, Springer-Verlag, Berlin, 1970, Lecture Notes in Mathematics, Vol. 114.
52. F. Jarvis, *On Galois representations associated to Hilbert modular forms*, J. Reine Angew. Math. **491** (1997), 199–216.
53. _____, *Level lowering for modular mod* ℓ *representations over totally real fields*, Math. Ann. **313** (1999), no. 1, 141–160.
54. _____, *Mazur's principle for totally real fields of odd degree*, Compositio Math. **116** (1999), no. 1, 39–79.
55. N. Jochnowitz, *A study of the local components of the Hecke algebra mod* ℓ, Trans. Amer. Math. Soc. **270** (1982), no. 1, 253–267.
56. _____, *The index of the Hecke ring,* T_k, *in the ring of integers of* $T_k \otimes \mathbf{Q}$, Duke Math. J. **46** (1979), no. 4, 861–869.
57. B. W. Jordan and R. Livné, *Conjecture "epsilon" for weight* $k > 2$, Bull. Amer. Math. Soc. (N.S.) **21** (1989), no. 1, 51–56.
58. K. Joshi, *Remarks on methods of Fontaine and Faltings*, Internat. Math. Res. Notices **1999**, no. 22, 1199–1209.

59. N. M. Katz, *p-adic properties of modular schemes and modular forms*, Modular functions of one variable, III (Proc. Internat. Summer School, Univ. Antwerp, Antwerp, 1972) (Berlin), Springer, 1973, pp. 69–190. Lecture Notes in Mathematics, Vol. 350.
60. _____, *Higher congruences between modular forms*, Ann. of Math. (2) **101** (1975), 332–367.
61. _____, *A result on modular forms in characteristic p*, Modular functions of one variable, V (Proc. Second Internat. Conf., Univ. Bonn, Bonn, 1976) (Berlin), Springer, 1977, pp. 53–61. Lecture Notes in Math., Vol. 601.
62. N. M. Katz and B. Mazur, *Arithmetic moduli of elliptic curves*, Princeton University Press, Princeton, N.J., 1985.
63. C. Khare, *Multiplicities of mod p Galois representations*, Manuscripta Math. **95** (1998), no. 2, 181–188.
64. L. J. P. Kilford, *Some examples of non-Gorenstein Hecke algebras associated to modular forms*, in preparation.
65. A. W. Knapp, *Elliptic curves*, Princeton University Press, Princeton, NJ, 1992.
66. S. Lang, *Introduction to modular forms*, Springer-Verlag, Berlin, 1995, With appendixes by D. Zagier and Walter Feit, Corrected reprint of the 1976 original.
67. R. P. Langlands, *Modular forms and ℓ-adic representations*, Proceedings of the International Summer School, University of Antwerp, RUCA, July 17–August 3, 1972 (Berlin) (P. Deligne and W. Kuyk, eds.), Springer, 1973, pp. 361–500. Lecture Notes in Math., Vol. 349.
68. _____, *Base change for* GL(2), Princeton University Press, Princeton, N.J., 1980.
69. W-C. Li, *Newforms and functional equations*, Math. Ann. **212** (1975), 285–315.
70. R. Livné, *On the conductors of mod ℓ Galois representations coming from modular forms*, J. Number Theory **31** (1989), no. 2, 133–141.
71. B. Mazur, *Modular curves and the Eisenstein ideal*, Inst. Hautes Études Sci. Publ. Math. (1977), no. 47, 33–186 (1978).
72. B. Mazur and K. A. Ribet, *Two-dimensional representations in the arithmetic of modular curves*, Astérisque (1991), no. 196-197, 6, 215–255 (1992), Courbes modulaires et courbes de Shimura (Orsay, 1987/1988).
73. L. Merel, *Universal Fourier expansions of modular forms*, On Artin's conjecture for odd 2-dimensional representations (Berlin), Springer, 1994, pp. 59–94.
74. J. S. Milne, *Abelian varieties*, Arithmetic geometry (Storrs, Conn., 1984), Springer, New York, 1986, pp. 103–150.
75. T. Miyake, *Modular forms*, Springer-Verlag, Berlin, 1989, Translated from the Japanese by Yoshitaka Maeda.
76. H. Moon, *Finiteness results on certain mod p Galois representations*, to appear in J. Number Theory.
77. D. Mumford, *Abelian varieties*, Published for the Tata Institute of Fundamental Research, Bombay, 1970, Tata Institute of Fundamental Research Studies in Mathematics, No. 5.
78. C. Queen, *The existence of p-adic Abelian L-functions*, Number theory and algebra (New York), Academic Press, 1977, pp. 263–288.

79. A. Raji, *On the levels of modular mod ℓ Galois representations of totally real fields*, Princeton University Ph.D. thesis, 1998.
80. R. Ramakrishna, *Lifting Galois representations*, Invent. Math. **138** (1999), no. 3, 537–562.
81. _____, *Deforming Galois representations and the conjectures of Serre and Fontaine-Mazur*, preprint, ftp://math.cornell.edu/pub/ravi (2000).
82. M. Raynaud, *Spécialisation du foncteur de Picard*, Inst. Hautes Études Sci. Publ. Math. No. **38** (1970), 27–76.
83. K. A. Ribet, *From the Taniyama-Shimura conjecture to Fermat's last theorem*, Ann. Fac. Sci. Toulouse Math. (5) **11** (1990), no. 1, 116–139.
84. _____, *On modular representations of* $\mathrm{Gal}(\overline{\mathbf{Q}}/\mathbf{Q})$ *arising from modular forms*, Invent. Math. **100** (1990), no. 2, 431–476.
85. _____, *Raising the levels of modular representations*, Séminaire de Théorie des Nombres, Paris 1987–88, Birkhäuser Boston, Boston, MA, 1990, pp. 259–271.
86. _____, *Lowering the levels of modular representations without multiplicity one*, International Mathematics Research Notices (1991), 15–19.
87. _____, *Report on mod ℓ representations of* $\mathrm{Gal}(\overline{\mathbf{Q}}/\mathbf{Q})$, Motives (Seattle, WA, 1991), Amer. Math. Soc., Providence, RI, 1994, pp. 639–676.
88. A. Robert, *Elliptic curves*, Springer-Verlag, Berlin, 1973, Notes from postgraduate lectures given in Lausanne 1971/72, Lecture Notes in Mathematics, Vol. 326.
89. T. Saito, *Modular forms and p-adic Hodge theory*, Invent. Math. **129** (1997), 607–620.
90. I. Schur, *Arithmetische Untersuchungen über endliche Gruppen linearer Substitutionen*, Sitz. Pr. Akad. Wiss. (1906), 164–184, Gesam. Abhl., **I**, 177–197, Springer-Verlag, Berlin-Heidelberg-New York-Tokyo, 1973.
91. J-P. Serre, *Groupes de Lie l-adiques attachés aux courbes elliptiques*, Les Tendances Géom. en Algébre et Théorie des Nombres, Éditions du Centre National de la Recherche Scientifique, Paris, 1966, pp. 239–256 (= Collected Papers **70**).
92. _____, *Une interprétation des congruences relatives à la fonction τ de Ramanujan*, Séminaire Delange-Pisot-Poitou n^o **14** (1967–68) (= C.P. **80**).
93. _____, *Propriétés galoisiennes des points d'ordre fini des courbes elliptiques*, Invent. Math. **15** (1972), no. 4, 259–331 (= C.P. **94**).
94. _____, *Congruences et formes modulaires [d'après H. P. F. Swinnerton-Dyer]*, Séminaire Bourbaki, 24e année (1971/1972), Exp. No. 416 (Berlin), Springer, 1973, pp. 319–338. Lecture Notes in Math., Vol. 317 (= C.P. **95**).
95. _____, *Formes modulaires et fonctions zêta p-adiques*, Proceedings of the International Summer School, University of Antwerp, RUCA, July 17–August 3, 1972 (Berlin), Springer, 1973, pp. 191–268. Lecture Notes in Math., Vol. 350 (= C.P. **97**).
96. _____, *A Course in Arithmetic*, Springer-Verlag, New York, 1973, Translated from the French, Graduate Texts in Mathematics, No. 7.
97. _____, *Valeurs propres des opérateurs de Hecke modulo ℓ*, Astérisque **24–25** (1975), 109–117 (= C.P. **104**).
98. _____, *Divisibilité de certaines fonctions arithmétiques*, Enseign. Math. (2) **22** (1976), no. 3-4, 227–260 (= C.P. **108**).

99. _____, *Linear representations of finite groups*, Springer-Verlag, New York, 1977, Translated from the second French edition by Leonard L. Scott, Graduate Texts in Mathematics, Vol. 42.

100. _____, *Local fields*, Springer-Verlag, New York, 1979, Translated from the French by Marvin Jay Greenberg.

101. _____, *Lettre à J.-F. Mestre*, Current trends in arithmetical algebraic geometry (Arcata, Calif., 1985), Amer. Math. Soc., Providence, RI, 1987, pp. 263–268 (= C.P. **142**).

102. _____, *Sur les représentations modulaires de degré 2 de* $\mathrm{Gal}(\overline{\mathbf{Q}}/\mathbf{Q})$, Duke Math. J. **54** (1987), no. 1, 179–230 (= C.P. **143**).

103. _____, *Travaux de Wiles (et Taylor, ...)*, Partie I, Séminaire Bourbaki, **803** (1995) (= C.P. **168**).

104. J-P. Serre and J. T. Tate, *Good reduction of abelian varieties*, Ann. of Math. (2) **88** (1968), 492–517 (= C.P. **79**).

105. N. I. Shepherd-Barron and R. Taylor, *Mod 2 and mod 5 icosahedral representations*, J. Amer. Math. Soc. **10** (1997), no. 2, 283–298.

106. H. Shimizu, *On zeta functions of quaternion algebras*, Ann. of Math. (2) **81** (1965), 166–193.

107. G. Shimura, *A reciprocity law in non-solvable extensions*, J. Reine Angew. Math. **221** (1966), 209–220.

108. _____, *Introduction to the arithmetic theory of automorphic functions*, Princeton University Press, Princeton, NJ, 1994, Reprint of the 1971 original, Kan Memorial Lectures, 1.

109. J. H. Silverman, *The arithmetic of elliptic curves*, Springer-Verlag, New York, 1992, Corrected reprint of the 1986 original.

110. _____, *Advanced topics in the arithmetic of elliptic curves*, Springer-Verlag, New York, 1994.

111. C. M. Skinner and A. J. Wiles, *Ordinary representations and modular forms*, Proc. Nat. Acad. Sci. U.S.A. **94** (1997), no. 20, 10520–10527.

112. H. P. F. Swinnerton-Dyer, *On ℓ-adic representations and congruences for coefficients of modular forms*, Proceedings of the International Summer School, University of Antwerp, RUCA, July 17–August 3, 1972 (Berlin), Springer, 1973, pp. 1–55. Lecture Notes in Math., Vol. 350.

113. J. T. Tate, *The non-existence of certain Galois extensions of* $\mathbf{Q}$ *unramified outside* 2, Contemporary Math. **174** (1994), 153–156.

114. R. Taylor and A. J. Wiles, *Ring-theoretic properties of certain Hecke algebras*, Ann. of Math. (2) **141** (1995), no. 3, 553–572.

115. J. Tunnell, *Artin's conjecture for representations of octahedral type*, Bull. Amer. Math. Soc. (N.S.) **5** (1981), no. 2, 173–175.

116. J.-L. Waldspurger, *Quelques propriétés arithmétiques de certaines formes automorphes sur* $\mathrm{GL}(2)$, Compositio Math. **54** (1985), no. 2, 121–171.

117. A. J. Wiles, *Modular elliptic curves and Fermat's last theorem*, Ann. of Math. (2) **141** (1995), no. 3, 443–551.

118. R. Taylor, *Remarks on a conjecture of Fontaine and Mazur*, preprint (2000).

Deformations of Galois Representations

Fernando Q. Gouvêa

Deformations of Galois Representations

Fernando Q. Gouvêa[1]

Introduction

These notes were prepared for a short graduate course which I was invited to teach at the Park City Mathematics Institute in the summer of 1999. The topic is the theory of deformations of Galois representations, which was created by Barry Mazur and has become, especially after Wiles' fundamental work on the modularity conjecture for elliptic curves, an important number-theoretical tool. This revised version of the notes includes several supplements, including three appendices. The first, by Mark Dickinson, gives a proof of the existence of the universal deformation that works directly from Grothendieck's theorem instead of using the Schlessinger criteria. The second, by Tom Weston, gives a detailed account of how to prove a theorem of Flach mentioned in Lecture 5. The third, by Matthew Emerton, gives an introduction to the theory of p-adic modular forms which is sketched very briefly in Lecture 7. The second and third appendix are write-ups of talks given at PCMI. I am grateful to all three authors for their permission to include their work in these notes.

I have tried to sprinkle problems throughout the write-up. These are of various kinds. Some simply ask the reader to fill in the details of an argument or to supply the proof of a theorem; these are mostly straightforward, but keep in mind that the notion of "straightforward" is highly dependent on each person's background. Other problems ask the reader to work out a specific example; I hope these will be helpful in understanding the material. Some problems are open-ended suggestions that the reader might want to investigate. A few problems ask questions whose answers I do not know, but which seemed natural to me as I was preparing these notes. Some of these are bound to be embarrassingly easy, while others may be quite hard.

---

[1]Department of Mathematics, Colby College, Waterville, Maine 04901 USA
E-mail address: fqgouvea@colby.edu
1991 *Mathematics Subject Classification.* Primary 11F80; Secondary 11F03, 11F33, 11R39, 14D15.
Key words and phrases. Galois representations, deformations, universal deformations.
Project sponsored by the National Security Agency under Grant number MDA904-98-1-0012. The United States Government is authorized to reproduce and distribute reprints.

©2001 American Mathematical Society

The first six lectures form the core material for the course on deformations of Galois representations. I have tried to make them a useful introduction to the subject. The last two lectures have a different character. The seventh is a broad survey of how deformation theory interacts with the theory of modular forms, with special focus on (various forms of) the issue of deciding which deformations are modular. The last lecture is a brief account of the material in [**100**] and [**62**], describing the construction of the "infinite fern" in the deformation space of a modular residual representation. Due to this difference in style, the problems disappear when we get to the last two lectures. (The last two lectures are in fact a sort of meta-problem: fill in the details in this account.)

Several other accounts of the basic theory are available, and all were enormously useful (and at times intimidating!) as I was preparing my own notes. First of all, one should mention Mazur's three excellent (and very different) accounts of the theory: his original paper [**97**], his more elementary account in [**100**] (which puts special emphasis on modular representations), and the expository account [**101**] in the proceedings [**33**] of the Boston University conference on Fermat's Last Theorem. In the same volume one also finds the articles [**36**], which gives an alternative construction of the universal deformation, and [**32**], which looks closely at the theory of flat deformations. Another extended account of the theory is the unpublished book [**42**] by Charles Doran and Siman Wong. Shorter surveys include Mazur's own summary in [**100**], my survey in [**60**], and portions of the various surveys of the proof of Fermat's Last Theorem, such as [**119**], [**34**], and [**114**]. These are all worth a look.

I would like to thank the organizers of the 1999 Park City Mathematics Institute for their invitation to teach this course. Thanks also to the several people who made suggestions for this revised version, including Brian Conrad, Ralph Greenberg, Armand Brumer, and Blair Kelly. Special thanks to Mark Dickinson, my teaching assistant at PCMI, who made extensive suggestions, handled problem sessions, and saw to it that copies were made and distributed.

FERNANDO GOUVÊA
DEPARTMENT OF MATHEMATICS
COLBY COLLEGE
WATERVILLE, ME 04901
fqgouvea@colby.edu

LECTURE 1
Galois Groups and Their Representations

Our main concern throughout these lectures will be to study the representations of the absolute Galois group of a field. More often than not, the field in question will be $\mathbb{Q}$, the field of rational numbers, but we will also need to consider number fields (finite extensions of $\mathbb{Q}$), their various completions, and finite fields. So we'll start by saying some general things about Galois groups of infinite extensions, then quickly specialize to the cases that interest us. We will then try to collect what is known about the groups we want to study. This will give this lecture something of the nature of a survey; we have tried to add details only when they are not easily found in the standard references.

Galois groups of infinite algebraic extensions

In this section we give a very brief sketch of the Galois theory of infinite algebraic extensions. References for this material are [**112**, Chapter 1] and [**109**, Chapter IV].

Let K be a perfect field, and let F be a (finite or infinite) normal extension of K. The Galois group
$$G(F/K) = \mathrm{Gal}(F/K)$$
is defined, as usual, to consist of all automorphisms of F which induce the identity on K. When F/K is infinite, this is an infinite group. In the special case in which $F = \overline{K}$ is an algebraic closure of K, we will call this the *absolute Galois group of* K and we will denote it by G_K.

For finite extensions, the Galois correspondence nicely matches up subgroups of the Galois group with subextensions. The main difficulty with which we have to deal in this section is the fact that the naïve generalization of this correspondence does not work for infinite extensions. The easiest way to see this is to consider an example that will keep coming up: the absolute Galois group of a finite field.

Example. Let p be a prime number, $K = \mathbb{F}_p$ be the finite field with p elements, and let $F = \overline{\mathbb{F}}_p$ be an algebraic closure. The *Frobenius automorphism* $\phi = \phi_p : F \longrightarrow F$ (which we will often simply call "the Frobenius" or "the Frobenius at p") is defined by $\phi(x) = x^p$. Recall that for each n there is only one extension of $\mathbb{F}_p$ inside $\mathbb{F}$

which has degree n; this extension is fixed by ϕ^n. Let $Z \subset G_{\mathbb{F}_p}$ be the subgroup generated by ϕ. It is easy to see that Z is an infinite cyclic group and that its fixed subfield is $\mathbb{F}_p$. Galois theory for extensions of finite degree would then lead us to expect that $Z = G_{\mathbb{F}_p}$, but this is very far from being true in our case. To see this, choose any sequence of integers a_n such that we have $a_n \equiv a_m \pmod{m}$ whenever $m|n$. We define an automorphism ψ of $\mathbb{F}$ by requiring $\psi|_{\mathbb{F}_{p^n}} = \phi^{a_n}$. The conditions on the sequence a_n mean that these definitions are compatible. Since every element of $\mathbb{F}$ belongs to some subfield of finite degree over $\mathbb{F}_p$, we see that this defines an automorphism $\psi \in G_{\mathbb{F}_p}$. But $\psi \in Z$ if and only if the sequence a_n is constant, i.e., if there is some integer a such that $a_n = a$ for all n. Since there are many non-constant sequences of this type, we have shown that $G_{\mathbb{F}_p}$ is in fact much larger than Z.

Problem 1.1. Check the details in this construction. Specifically, show that many non-constant sequences $\{a_n\}$ exist and that the conditions defining ψ are indeed compatible.

Problem 1.2. How big *is* $G_{\mathbb{F}_p}$? For example, is it a countable set?

As usual, the way to fix the problem is to introduce a topology on our infinite Galois group, and then to show that the Galois correspondence will work provided we work only with *closed* subgroups. The definition of the topology is quite natural; basically, we need something that reduces to the discrete topology when F/K is finite and which selects the right subgroups for the correspondence to work.

Here is the formal definition. First of all, we'll say an extension of fields is a *Galois extension* if it is algebraic, normal, and separable. (In what follows, we will always assume that the base field K is perfect, so that we need not worry about separability.)

Definition 1.1. Let F/K be a (finite or infinite) Galois extension. For each *finite* Galois subextension K'/K, consider the Galois group $G(K'/K)$, and whenever we have two finite subextensions $K' \subset K''$ consider the homomorphism

$$G(K''/K) \longrightarrow G(K'/K)$$

given by restriction. This whole package defines an inverse system of groups, and we define the Galois group of F over K to be the inverse limit[1]

$$G(F/K) = \varprojlim_{K'/K} G(K'/K)$$

with its natural profinite topology.

This definition nicely generalizes the example above; after all, an element of the inverse limit is exactly given by an indexed set $\{\sigma_{K'}\}$ such that $\sigma_{K'} \in G(K'/K)$ and such that the various $\sigma_{K'}$ are compatible under restriction, which is very similar to the way we constructed the sequence $\{a_n\}$. The main thing we have added is the topology. It's worth noticing that if F is actually a finite extension of K, then the group is finite and the topology is just the discrete topology.

Problem 1.3. Let F/K be an infinite Galois extension, and let G be the group of automorphisms of F which induce the identity on K. For each *finite* Galois subextension K'/K, let $G(F/K')$ denote the normal subgroup of G consisting of all automorphisms which induce the identity on K'. Define a topology on G by defining a basis of

---

[1] See the complements to lecture 1 for a quick overview of inverse limits.

neighborhoods of each $\sigma \in G$ to be the set of all cosets $\sigma G(F/K')$, where K' runs through all finite Galois extensions of K. Show that this yields the same group and the same topology as in the definition above.

Problem 1.4. Show that $G(F/K)$ is Hausdorff, compact, and totally disconnected.

Problem 1.5. Let G be a topological group. Show that all open subgroups of G are also closed. If G is compact, show that all open subgroups are of finite index in G. Conversely, show that a closed subgroup of finite index in a topological group G is open.

Groups that are inverse limits of a projective system of finite groups are called *profinite*. As this discussion suggests, they appear quite naturally in arithmetical and algebraic contexts. One can often show that profinite groups have properties that are very close to the properties of finite groups. There will be several examples in what follows. A quick overview of profinite groups appears in the complements to lecture 1. For more information, see [**109**, Appendix C], [**139**] (or its English translation [**140**]), and [**143**].)

The immediate result of topologizing our group is that we now get a good Galois correspondence:

Theorem 1.1. *Let F/K be a (finite or infinite) Galois extension. The map*

$$K' \mapsto G(F/K')$$

defines a bijective inclusion-reversing correspondence between subextensions K'/K and closed subgroups of $G(F/K)$. The inverse correspondence is given by

$$H \mapsto F^H,$$

where, as usual, F^H denotes the subfield of F consisting of those elements which are fixed by every element of H.

In particular, the open subgroups (which are also closed and of finite index, see above) correspond to the finite subextensions.

Problem 1.6. Prove the theorem.

Problem 1.7. Consider the field $F \subset \overline{\mathbb{Q}}$ which is the compositum of all quadratic extensions of $\mathbb{Q}$. Describe the Galois group $G(F/\mathbb{Q})$ in as much detail as you can. Show that it has many subgroups of finite index which are not closed. (In fact, most of its subgroups of finite index are not closed!)

Problem 1.8. Let G_1 and G_2 be profinite groups. Show that a continuous injective homomorphism $G_1 \longrightarrow G_2$ is an isomorphism from G_1 onto a closed subgroup of G_2.

Problem 1.9. Show that every profinite group arises as a Galois group for some Galois extension.

Let's reconsider our initial example, and then take a brief look at the other main examples we'll need to consider.

Example. Let $\mathbb{F}$ be an algebraic closure of $\mathbb{F}_p$, and consider subfields $K \subset \mathbb{F}$. Then we know that there is a unique finite subextension $K/\mathbb{F}_p$ of degree n, which we call $\mathbb{F}_{p^n}$, and we have

$$G(\mathbb{F}_{p^n}/\mathbb{F}_p) \cong \mathbb{Z}/n\mathbb{Z},$$

where the isomorphism is obtained by mapping the Frobenius element ϕ to 1. The argument above shows directly that $G_{\mathbb{F}_p} \cong \hat{\mathbb{Z}}$, where $\hat{\mathbb{Z}}$ is the procyclic group

$$\hat{\mathbb{Z}} = \varprojlim_n \mathbb{Z}/n\mathbb{Z},$$

where the homomorphisms $\mathbb{Z}/n\mathbb{Z} \longrightarrow \mathbb{Z}/m\mathbb{Z}$ used to obtain the limit are defined, whenever $m|n$, to be simply reduction modulo m. The set Z which we considered above, consisting of all the integral powers of ϕ, is a dense subset of $G_{\mathbb{F}_p}$. Hence we say that ϕ *topologically generates* $G_{\mathbb{F}_p}$.

Problem 1.10. Show that the natural map $\hat{\mathbb{Z}} \longrightarrow \prod_p \mathbb{Z}_p$ is an isomorphism.

Problem 1.11. Show that $\mathbb{Z}_p$ (thought of as an additive group) is also a profinite group topologically generated by a single element, but is not isomorphic to $\hat{\mathbb{Z}}$.

The example of the absolute Galois group of a finite field is one for which we can get a precise description. This is far from being the case for the other two groups that will be at the center of our attention: the absolute Galois groups of $\mathbb{Q}_p$ and of $\mathbb{Q}$. We will consider both groups more carefully in the next section.

Before we go on to that, let's set up one more bit of the abstract theory.

Definition 1.2. Let G be a topological group. We define a *G-module* to be an abelian topological group M together with a *continuous* map

$$G \times M \longrightarrow M$$
$$(\sigma, m) \mapsto \sigma m$$

which satisfies the following conditions (where we write the operation on M additively):

i. $1m = m$, for all $m \in M$,
ii. $\sigma(m+n) = \sigma m + \sigma n$, for all $\sigma \in G$ and $m, n \in M$, and
iii. $(\sigma\tau)m = \sigma(\tau m)$, for all $\sigma, \tau \in G$ and all $m \in M$.

The most common situation is when G is a profinite group and M has the discrete topology. In this case, the condition of continuity can be translated to a simple group-theoretic condition:

Problem 1.12. Suppose G is profinite and M has the discrete topology. For each subgroup $H \subset G$, write M^H for the set of elements of M which are fixed by every element of H. Show that the map $G \times M \longrightarrow M$ is continuous if and only if we have

$$M = \bigcup_H M^H,$$

where H runs through all the *open* subgroups of G.

Problem 1.13. Let A be a topological ring whose undelying abelian group is profinite. We say that A is a *profinite ring*.

i. Prove that the natural map of topological rings

$$A \longrightarrow \varprojlim_I A/I$$

is an isomorphism, where I runs through the closed *ideals* of finite index in A. (To begin with, you need to show that if $A \neq 0$ such proper closed ideals $I \subset A$ of finite index do exist.)

ii. Let A be a complete noetherian local ring, and give A its "natural" topology, that is, the topology defined by the powers of its maximal ideal. Show that A is profinite if and only if its residue field is finite.

iii. Give an example of a profinite local ring which is not noetherian.

Problem 1.14. Let A be a profinite ring and let M be a finite free A-module with a continuous action of a profinite group G. For each open subgroup $H \subset G$, let M^H be defined as above. Show that the natural map
$$M \longrightarrow \varprojlim_{H} M^H$$
is a topological isomorphism. How does this relate to problem 1.12?

Given a G-module M, we can define the cohomology groups $\mathrm{H}^i(G, M)$ as in [152]. The case of interest for us, of course, will be when G is a Galois group. In addition to [152], see also [156] for a short introduction to Galois Cohomology. The books [135], [139], [140], and [111] contain more extensive treatments, the last being especially complete.

We conclude this section with two problems that deal with an idea that will later be very important for us: the notion of a *pro-p-group*.

Problem 1.15. Let p be a prime. A profinite group G is called a *pro-p-group* if every finite quotient of G is a p-group. (For example, $\mathbb{Z}_p$, as an additive group, is a pro-p-group.) Let $\Gamma_2(\mathbb{Z}_p)$ denote the kernel of the reduction mod p map
$$\mathrm{GL}_2(\mathbb{Z}_p) \longrightarrow \mathrm{GL}_2(\mathbb{F}_p).$$
Show that $\Gamma_2(\mathbb{Z}_p)$ is a pro-p-group.

Problem 1.16. Let G be a profinite group and p be a prime. Define another profinite group $G^{(p)}$ by
$$G^{(p)} = \varprojlim_{H} G/H,$$
where H runs through the open normal subgroups of G whose index in G is a power of p.

i. Show that there is a canonical continuous group homomorphism $\pi : G \longrightarrow G^{(p)}$ and that any continuous group homomorphism from G to a finite discrete p-group factors through π.

ii. Let $G = \hat{\mathbb{Z}}$. What is $G^{(p)}$?

iii. Is $G^{(p)}$ a quotient of G?

iv. Formulate and prove a universal property of $G \longrightarrow G^{(p)}$ in the category of profinite groups.

The Galois group of $\mathbb{Q}$

The Galois group which will mainly concern us is $G_\mathbb{Q} = G(\overline{\mathbb{Q}}/\mathbb{Q})$, the absolute Galois group of $\mathbb{Q}$. In this section we gather together some basic information about this group (and also try to point out that there is much about it that is still quite mysterious). Much of what we do would also apply to the Galois group of a general number field. All of this is to be found in standard references on algebraic number theory; our summary is inspired by the material in [101], [34, Section 2.1] and [112, Chapter 1].

Let's begin with what one might call the "local structure" of $G_\mathbb{Q}$. For each prime number p, there is a canonical inclusion of $\mathbb{Q}$ into its completion $\mathbb{Q}_p$. When we go to algebraic closures, however, there are many different inclusions $\overline{\mathbb{Q}} \longrightarrow \overline{\mathbb{Q}}_p$ (this is equivalent to saying that there are many ways to extend the p-adic valuation on $\mathbb{Q}$ to $\overline{\mathbb{Q}}$). Once we choose such an embedding, we get an inclusion of Galois groups $G_{\mathbb{Q}_p} \hookrightarrow G_\mathbb{Q}$. Changing the embedding changes this inclusion by conjugation. The image of $G_{\mathbb{Q}_p}$ is called a *decomposition group at p*. We will usually want to identify $G_{\mathbb{Q}_p}$ with its image in $G_\mathbb{Q}$, but when we do so we will have to bear in mind that the picture we have in mind is unique only up to conjugation.

We know quite a bit about the structure of algebraic extensions of $\mathbb{Q}_p$. First of all, there is a maximal unramified extension $\mathbb{Q}_p^{\mathrm{ur}}$, and we know that

$$G(\mathbb{Q}_p^{\mathrm{ur}}/\mathbb{Q}_p) \cong G(\overline{\mathbb{F}}_p/\mathbb{F}_p),$$

where $\overline{\mathbb{F}}_p$ is the residue field of the valuation ring of $\mathbb{Q}_p^{\mathrm{ur}}$ (which is an algebraic closure of $\mathbb{F}_p$). The restriction map then gives a surjective homomorphism

$$G(\overline{\mathbb{Q}}_p/\mathbb{Q}_p) \longrightarrow G(\overline{\mathbb{F}}_p/\mathbb{F}_p).$$

The kernel of this homomorphism is called the *inertia group* (at p) and we will denote it by I_p. Recall that $G(\overline{\mathbb{F}}_p/\mathbb{F}_p)$ is topologically generated by the Frobenius automorphism ϕ_p. We will call any lift of ϕ_p to $G_{\mathbb{Q}_p}$ a Frobenius automorphism, and we will confuse things even further by using the same notation ϕ_p for any such element. (To be fair, we usually do this in contexts where we are looking at the image of $G_{\mathbb{Q}_p}$ via a map whose kernel contains I_p. In this case, ϕ_p is any of many elements in a coset of I_p, but the image of ϕ_p is well-defined.)

The structure of the inertia group I_p is somewhat more complicated. It has a large normal Sylow pro-p-subgroup, which we denote by W_p and which is known as the *wild inertia group*. The quotient I_p/W_p is sometimes called the *tame inertia group*, and it is the better understood part of I_p. In fact, there is a (non-canonical) isomorphism

$$I_p/W_p \cong \prod_{\ell \neq p} \mathbb{Z}_\ell,$$

and if ϕ_p is any Frobenius element and $\bar{\sigma} \in I_p/W_p$, we have $\phi_p \bar{\sigma} \phi_p^{-1} = \bar{\sigma}^p$.

Problem 1.17. The group I_p/W_p corresponds to an extension of $\mathbb{Q}_p^{\mathrm{ur}}$. Describe that extension and the map from its Galois group to $\prod_{\ell \neq p} \mathbb{Z}_\ell$. (You will need to make a choice of a compatible sequence of ℓ^n-th roots of unity, which is why the isomorphism is non-canonical.)

The names of those subgroups reflect their origins in the theory of algebraic number fields. In fact, a Galois extension K of $\mathbb{Q}_p$ corresponds to a surjective homomorphism $G_{\mathbb{Q}_p} \longrightarrow G(K/\mathbb{Q}_p)$, and the extension will be called *unramified* if the image of I_p under this map is trivial. Similarly, we'll say the extension is *tamely ramified* if the image of W_p is trivial, and *wildly ramified* if not.

This analysis of the structure of $G_{\mathbb{Q}_p}$ can be continued, producing still smaller subgroups of W_p known as the "higher ramification groups." See, for example, [**135**] for details of all this.

Problem 1.18. Above we worked with a Galois extension $K/\mathbb{Q}_p$. How do we handle an extension which is not Galois? What changes?

Putting together the whole picture of what this says about the full Galois group $G_\mathbb{Q}$, we see that for each prime number p we have a complex package of information:

a set of subgroups

$$W_p \subset I_p \subset G_{\mathbb{Q}_p} \subset G_{\mathbb{Q}}$$

(with both W_p and I_p normal in $G_{\mathbb{Q}_p}$) and a set of Frobenius elements at p. The last inclusion depends on the choice of the embedding of $\overline{\mathbb{Q}}$ into $\overline{\mathbb{Q}}_p$, and hence the whole picture is only determined up to conjugation.

As before, we can translate this group-theoretical picture in terms of algebraic number fields. A Galois extension $K/\mathbb{Q}$ corresponds to a surjective homomorphism $G_{\mathbb{Q}} \to G(K/\mathbb{Q})$, and the extension $K/\mathbb{Q}$ will be unramified at p when the images of all the inertia groups at p are trivial. Note that in this case there is a well-defined (up to conjugation) image of the Frobenius element ϕ_p in $G(K/\mathbb{Q})$. In finite extensions, all but a finite number of primes will be unramified, and if $K \neq \mathbb{Q}$ at least one prime will be ramified:

Theorem 1.2. *If $K/\mathbb{Q}$ is a finite extension, then K is ramified at finitely many primes (to be specific, they are the primes dividing the discriminant of $K/\mathbb{Q}$). Every non-trivial extension of $\mathbb{Q}$ is ramified at at least one prime.*

The first part of this theorem is relatively easy (and true over a general number field). The second is a theorem of Minkowski (and *not* true over a general number field).

Problem 1.19. Above, we described the "local picture" only for non-archimedean primes. We also need to consider the prime at infinity. Make the proper definitions. In particular, show that there is a well-defined conjugacy class of elements of order two in $G_{\mathbb{Q}}$; we call any element of this conjugacy class a "complex conjugation."

Problem 1.20. Describe all this for G_K, where K is a number field. Does anything significant change?

An important class of extensions of $\mathbb{Q}$ are the *cyclotomic* extensions obtained by adjoining roots of unity to $\mathbb{Q}$. Let m be a positive integer and let ζ_m be a primitive m-th root of unity. Then we know that $G(\mathbb{Q}(\zeta_m)/\mathbb{Q})$ is abelian, and in fact isomorphic to the group of units of $\mathbb{Z}/m\mathbb{Z}$. If we take the union K_ℓ of all $\mathbb{Q}(\zeta_m)$ as m ranges over all the powers of a prime ℓ, these isomorphisms compile to give an isomorphism between $G(K_\ell/\mathbb{Q})$ and $\mathbb{Z}_\ell^\times$. Composing this with the surjective map $G_{\mathbb{Q}} \longrightarrow G(K_\ell/\mathbb{Q})$ gives a homomorphism

$$\epsilon_\ell : G_{\mathbb{Q}} \longrightarrow \mathbb{Z}_\ell^\times.$$

This is called the ℓ-adic *cyclotomic character*; it can be described by saying that, for any ℓ-power root of unity ζ and any element $\sigma \in G_{\mathbb{Q}}$ we have

$$\sigma(\zeta) = \zeta^{\epsilon_\ell(\sigma)}.$$

The extension K_ℓ (or, equivalently, the ℓ-adic cyclotomic character) is ramified only at ℓ and infinity. If $p \neq \ell$, then, it makes sense to talk about the image of a Frobenius element at p under ϵ_ℓ.

Problem 1.21. Prove that if $p \neq \ell$ we have $\epsilon_\ell(\phi_p) = p$.

The Galois group of any cyclotomic extension, and hence also of any Galois subextension, is abelian. The *Kronecker-Weber Theorem* asserts that all abelian extensions of $\mathbb{Q}$ are of this kind. In keeping with our main theme, we can restate this in terms of characters of the Galois group. Given any group G, write G^{ab} for its abelianization, i.e., its (unique) maximal continuous[2] abelian quotient, i.e., the quotient of G by the closed subgroup topologically generated by the commutators of G. Then we have:

Theorem 1.3 (Kronecker-Weber). *For each prime p, let ϵ_p denote the p-adic cyclotomic character. The product of all the ϵ_p, which maps $G_{\mathbb{Q}}$ to the product of all $\mathbb{Z}_p^\times$, induces an isomorphism*

$$(\prod \epsilon_p) : G_{\mathbb{Q}}^{ab} \xrightarrow{\cong} \prod_p \mathbb{Z}_p^\times \cong \hat{\mathbb{Z}}^\times.$$

Problem 1.22. Check that this isomorphism is equivalent to the theorem as it is usually stated: all abelian extensions of $\mathbb{Q}$ are contained in some cyclotomic field.

There is also a local Kronecker-Weber theorem, and it too can be restated in terms of characters. Let $\pi : G_{\mathbb{Q}_p} \longrightarrow G_{\mathbb{F}_p}$ be the standard projection and let ϵ_p be the p-adic cyclotomic character. Then we have:

Theorem 1.4 (Local Kronecker-Weber). *The map $\pi \times \epsilon_p$ induces an isomorphism*

$$\pi \times \epsilon_p : G_{\mathbb{Q}_p}^{ab} \xrightarrow{\cong} G_{\mathbb{F}_p} \times \mathbb{Z}_p^\times.$$

The Kronecker-Weber theorem is the first piece of Class Field Theory over $\mathbb{Q}$. In general, Class Field Theory provides a detailed description of the abelian extensions of any number field, and so it can be used to study Galois groups and their representations. The theory is too complex to summarize here, so we refer readers to the literature for more details. There are several accounts of Class Field Theory available; one of the more accessible ones is [112].

There are many other questions to ask about the Galois group of $\mathbb{Q}$. For example, the following is a famous conjecture:

Conjecture. Any finite group can be obtained as a discrete quotient of $G_{\mathbb{Q}}$.

Much work has been done in the direction of this conjecture (for example, see [138]), but the full conjecture remains very much open. One reason for mentioning it here is to point out that it implies that $G_{\mathbb{Q}}$ must be quite complicated!

Various attempts have been made to come up with conjectural descriptions of $G_{\mathbb{Q}}$; one of the most interesting is Grothendieck's theory of "dessins d'enfants" (see [126] for details).

Restricting the ramification

As we saw above, a finite extension $K/\mathbb{Q}$ can be only be ramified at a finite number of primes. This is not true for infinite extensions (consider the example in Problem 1.7, for example), but there are good reasons to expect that the "natural" Galois representations (more precision later) are all finitely ramified. This section considers the Galois theory with bounded ramification.

We'll fix a finite set S of primes, including the prime at infinity. (This need not be done in general, but for our purposes we'll always want to allow ramification at infinity.) We want to consider extensions $K/\mathbb{Q}$ which are ramified only at the primes belonging to S; we describe these as "unramified outside S." Putting all

---

[2]In general, we will consider only quotients that are quotients in the category of profinite groups, i.e., quotients by closed subgroups. When we want to emphasize this, we will speak of "continuous quotients."

such K together gives $\mathbb{Q}_S$, the maximal extension of $\mathbb{Q}$ which is unramified outside S. This is easily checked to be a Galois extension of $\mathbb{Q}$; we want to study the group

$$G_{\mathbb{Q},S} = G(\mathbb{Q}_S/\mathbb{Q}).$$

This is a quotient of the full $G_\mathbb{Q}$, of course, but in many ways it is much easier to understand.

Before we consider the main results, let's point out that we can make the same definition for a general number field K, and a set of primes[3] S of K (again, including all the archimedean primes), yielding a group that we will call $G_{K,S}$. Notice that in both cases we are putting no restrictions on the ramification at infinity.

Problem 1.23. Suppose S is a set of primes in $\mathbb{Q}$ and S_1 is the set of primes in K lying over the primes in S. Is there a simple description of the relation between G_{K,S_1} and $G_{\mathbb{Q},S}$?

Problem 1.24. Show that any open subgroup of $G_{K,S}$ is again of the form G_{K_1,S_1}, for some finite extension K_1/K.

Problem 1.25. What would change if we decided to restrict the ramification at the archimedean primes?

The first important fact about $G_{K,S}$ is the following finiteness result:

Theorem 1.5 (Hermite-Minkowski). *Let K be a finite extension of $\mathbb{Q}$, let S be a finite set of primes and let d be a positive integer. There are only finitely many extensions F/K of degree d which are unramified outside S.*

An important consequence of this theorem is the fact that the set

$$\mathrm{Hom}_{\mathrm{cont}}(G_{K,S}, \mathbb{Z}/p\mathbb{Z})$$

is finite, since each nontrivial continuous homomorphism corresponds to an extension of degree p, unramified outside S. Putting this together with Problem 1.24 gives the following crucial (for us) result:

Theorem 1.6. *Let p be a prime number, K a number field, and S a finite set of (non-archimedean) primes. Let $G \subset G_{K,S}$ be an open subgroup. Then there exist only a finite number of continuous homomorphisms from G to $\mathbb{Z}/p\mathbb{Z}$.*

Mazur calls this the *p-finiteness condition*, and we will use it in an essential way to understand the deformations of a Galois representation.

Problem 1.26. Show that the p-finiteness condition also holds for any of the $G_{\mathbb{Q}_\ell}$. (Is anything special about the case $\ell = p$?)

How big are the groups we are considering? As we saw, the absolute Galois group of a finite field is topologically generated by one element, the Frobenius. For a local field, one can also show a finite generation result:

Theorem 1.7. *If K is a finite extension of $\mathbb{Q}_p$, then G_K is topologically finitely generated.*

---

[3]We actually get a choice here. We could keep S as a set of primes in $\mathbb{Q}$, and say that an extension of K is "unramified outside S" if it is unramified at all primes of K that do not lie above a prime belonging to S. This point of view has its advantages, but of course choosing a set of primes in K is more general.

For this and much more about the Galois group of a local field, see [**76**], [**77**], and [**161**], which together give a detailed description of G_K in this case.

The situation for the $G_{K,S}$ is much more complicated. Shafarevich conjectured[4] that this group is also topologically finitely generated, but so far this remains an open question. (The p-finiteness property, which would follow from finite generation, is a sort of replacement for this still-unknown result.) We do know that $G_{K,S}$ is topologically *countably* generated.

Problem 1.27. Prove this. In other words, show that there exists a countable set of elements that generate a dense subgroup of $G_{K,S}$. (This is actually quite easy.)

Problem 1.28. (A test situation; as far as I know, this is an open problem.) For each elliptic curve E defined over $\mathbb{Q}$, with good reduction outside 2, let $\mathbb{Q}(T_2(E))$ be the extension of $\mathbb{Q}$ obtained by adjoining the coordinates of the 2^n-division points for all n (equivalently, it is the field fixed by the kernel of the 2-adic representation attached to E). Let K be the compositum of the $\mathbb{Q}(T_2(E))$ as E runs through all such elliptic curves. Is $G(K/\mathbb{Q})$ topologically finitely generated?

For every prime p, we can carry the local picture from the previous section over to $G_{\mathbb{Q},S}$ and get homomorphisms $G_{\mathbb{Q}_p} \longrightarrow G_{\mathbb{Q},S}$. When $p \notin S$, the image of the inertia group I_p is trivial, and therefore there is a well-defined Frobenius element ϕ_p in $G_{\mathbb{Q},S}$ which generates the image of $G_{\mathbb{Q}_p}$. It seems natural to conjecture[5] that the image of $G_{\mathbb{Q}_p}$ in $G_{\mathbb{Q},S}$ is as large as possible:

Conjecture. With the notations above, we have:

  i. If $p \in S$, the map $G_{\mathbb{Q}_p} \longrightarrow G_{\mathbb{Q},S}$ is an inclusion.
  ii. If $p \notin S$, the kernel of the map $G_{\mathbb{Q}_p} \longrightarrow G_{\mathbb{Q},S}$ is exactly I_p, so that we get an inclusion $G_{\mathbb{Q}_p}/I_p \hookrightarrow G_{\mathbb{Q},S}$.

This very natural conjecture seems to be quite difficult to prove.

As pointed out above, for each $p \notin S$ we have a well-defined Frobenius element ϕ_p in the image of $G_{\mathbb{Q}_p}$. The whole local picture, however, is defined only up to conjugation, so if we do not want to fix the homomorphisms $G_{\mathbb{Q}_p} \longrightarrow G_{\mathbb{Q},S}$ we should think of the Frobenius element as a conjugacy class of elements of $G_{\mathbb{Q},S}$. One of the most significant results about the set of all these conjugacy classes is the following density result.

Theorem 1.8 (Chebotarev). *Let $K/\mathbb{Q}$ be a Galois extension that is unramified outside a finite set S of primes. Let T be a finite set of primes containing S. For each prime $p \notin T$, there is a well-defined Frobenius conjugacy class $[\phi_p] \subset G(K/\mathbb{Q})$. The union of all these Frobenius conjugacy classes is dense in $G(K/\mathbb{Q})$.*

Problem 1.29. What does this say when K is a *finite* extension of $\mathbb{Q}$? (Easy question, but the fact is worth noting.)

Problem 1.30. Let ζ_m be a primitive m-th root of unity, and let $K = \mathbb{Q}(\zeta_m)$. In this case the Galois group is known completely explicitly, and we also know what the Frobenius elements are. What does the Chebotarev theorem tell us in this situation?

---

[4]The reference is [**142**], but we should note that there Shafarevich simply asks whether it is the case that $G_{K,S}$ is finitely generated for any number field (and whether the number of generators can be bounded in terms of the number of elements of S). His main reason for posing the question is that the analogous statement is true for function fields over $\mathbb{C}$.
[5]Thanks to Ralph Greenberg for mentioning this issue to me.

Problem 1.31. Is the set of (topological) generators given by the Chebotarev theorem countable?

Finally, the abelianization of $G_{\mathbb{Q},S}$ is easily understood by using the Kronecker-Weber theorem.

Problem 1.32. Show that $G_{\mathbb{Q},S}^{\mathrm{ab}}$ is isomorphic to $\prod_{p \in S} \mathbb{Z}_p^\times$.

As before, we can use Class Field Theory to understand the abelianization in the case of a number field.

Galois representations

Why consider the representations of $G_{\mathbb{Q},S}$? One of the reasons is simply that such representations arise naturally, for example from the theory of elliptic curves and modular forms. Another reason, as Mazur has pointed out, is the fact that the whole picture we want to study, which includes not only $G_{\mathbb{Q},S}$ but also all the maps $G_{\mathbb{Q}_p} \longrightarrow G_{\mathbb{Q},S}$, is only defined up to conjugation. Group representations are well-suited to this situation. For example, for $p \notin S$ the Frobenius elements ϕ_p are only defined up to conjugation, but the characteristic polynomial of the image of ϕ_p under a representation is well-defined (and therefore so are the trace and determinant of the image).

Let's make the formal definition:

Definition 1.3. A Galois representation (defined over A, unramified outside S) is a continuous homomorphism
$$\rho : G_{\mathbb{Q},S} \longrightarrow \mathrm{GL}_n(A),$$
where A is some topological ring and n is a positive integer. Two Galois representations ρ_1 and ρ_2 are *equivalent* if there is a matrix $P \in \mathrm{GL}_n(A)$ such that $P^{-1}\rho_1 P = \rho_2$.

Given such a thing, we can consider the free A-module of rank n and give it a continuous action of $G_{\mathbb{Q},S}$ by defining $g \cdot m = \rho(g)m$. Conversely, given a finite free A-module M of rank n with a continuous action of $G_{\mathbb{Q},S}$, we can get a representation ρ as above by choosing a basis for M. Changing the basis changes ρ into an equivalent representation.

If we have a finite free A-module M with a continuous action of a profinite group G such that
$$M = \varprojlim_H M^H$$
as H runs through the open normal subgroups of G, then we can canonically make M into a module over the completed group ring $A[[G]]$, defined as
$$A[[G]] = \varprojlim_H A[G/H],$$
where H runs through the open normal subgroups of G and $A[G/H]$ is the usual group ring of the finite group G/H over A. In problem 1.14, we checked that the condition on M is automatically verified when A is a profinite ring. Hence, giving (up to equivalence) a representation of G defined over a profinite ring A is the same as giving a continuous $A[[G]]$-module M which is finite and free as an A-module. This point of view is also occasionally useful.

There is one final point of view which is occasionally useful. Given a representation

$$\rho : G \longrightarrow \mathrm{GL}_n(A)$$

defined over a profinite ring A, we can extend it by linearity to the completed group ring $A[[G]]$, to get a continuous homomorphism of A-algebras

$$A[[G]] \longrightarrow \mathrm{M}_n(A).$$

Conversely, the restriction to G of any such homomorphism gives a representation in the usual sense.

In what follows, we will mostly stick to the first point of view, but every once in a while, when A is a profinite ring (as it will often be), it will be convenient to switch to the other versions.

The standard choices for the ring A are the following:

i. $A = \mathbb{C}$. These are known as "Artin representations," and are the most classical. Because of the topologies involved, the image of $G_{\mathbb{Q},S}$ in $\mathrm{GL}_n(A)$ must be *finite*.
ii. A is a finite field. These representations arise naturally from elliptic curves and modular forms, and they are the ones that Serre's conjecture tries to describe.
iii. $A = \mathbb{Z}_p$ or $\mathbb{Q}_p$ or finite extensions thereof. These also arise from elliptic curves and modular forms. Since $\mathbb{Z}_p$ (or a finite extension) also carries a profinite topology, this case gives the best "match" in topologies. In particular, the image of $G_{\mathbb{Q},S}$ is not necessarily finite in this case.

Our main interest is in the last two cases, and in the relation between them, so we will choose to work with rings A that are generalizations of those two situations. Specifically, we will assume A is a complete noetherian local ring with finite residue field. Note, by problem 1.13, that such an A is automatically a profinite ring.

In a sense, we are interested in trying to understand all the (finitely ramified) Galois representations into $\mathrm{GL}_n(A)$. For $n = 1$, this is essentially already done, since describing all such representations amounts to describing the abelianizations G_K^{ab}, and this is basically what Class Field Theory does. (If $K \neq \mathbb{Q}$ this is not quite true, since one runs into such difficulties as Leopoldt's Conjecture, but one still has quite good control of the situation.) Hence, we'll focus on $n \geq 2$. In fact, things are already so interesting for $n = 2$ that we'll often restrict ourselves to that case, which is also the case where Serre's conjecture applies and where we get representations from elliptic curves and modular forms.

Studying "all the representations" is far too vague to serve as a guideline for investigation, however, and so we have to come up with a more specific program. The point of view we will take, then, is to start with a given representation into $\mathrm{GL}_n(\mathsf{k})$, where k is a finite field, and then to consider all the representations which "lift" this representation to $\mathrm{GL}_n(A)$, where A runs through all complete noetherian local rings with residue field k. It turns out that we can make this into a well-defined question and (even better!) that the question has an interesting answer.

LECTURE 1. GALOIS GROUPS AND THEIR REPRESENTATIONS

Complements to Lecture 1

The first lecture makes intensive use of both the notion of an inverse limit and the theory of profinite groups. We give brief summaries of each of these, with some references.

Inverse limits

Inverse limits make sense for various kinds of mathematical objects. We could phrase everything in the language of categories (as in the next lecture), but we avoid that for now. Hence, we discuss inverse limits of sets, of groups, and of rings; the reader should note, however, that we do not really use many specific properties of these objects.

We start with a partially ordered set I, which we want to be a *directed set*. This just means that given $i, j \in I$ there exists $k \in I$ such that $i \leq k$ and $j \leq k$. For example, I could be the set of all positive integers with the usual order, or the set of positive integers ordered by divisibility.

To give an *inverse system* we must specify:

- a directed set I,
- for each $i \in I$, a group (or ring, or set) G_i,
- for each pair $i, j \in I$ such that $i \leq j$, a group (or ring, or set) homomorphism
  $$\phi_{ij} : G_j \longrightarrow G_i.$$

(Of course, a "set homomorphism" is just an arbitrary function.) We could add requirements of continuity if our objects carried natural topologies, and so on. We require that this data satisfy the "obvious" conditions:

- ϕ_{ii} is the identity, and
- "all triangles commute," that is, if $i \leq j \leq k$ then $\phi_{ik} = \phi_{ij} \circ \phi_{jk}$.

Here are some famous examples:

Problem 1.33. Let I be the set of positive integers with the usual order, let p be a fixed prime number, and for each $n \in I$ let G_n be the ring $\mathbb{Z}/p^n\mathbb{Z}$. If $n \leq m$, let ϕ_{nm} be "reduction mod p^n." Check that this defines an inverse system.

Problem 1.34. Let I be the set of positive integers ordered by divisibility. For each $n \in I$, get G_n be the ring $\mathbb{Z}/n\mathbb{Z}$, and whenever $n|m$ let ϕ_{nm} be "reduction modulo n." Check that this defines an inverse system.

Problem 1.35. Let F/K be a field extension, and let I be the set of finite Galois subextensions, ordered by inclusion. For each $K' \in I$, let $G_{K'} = G(K'/K)$, and if $K' \subset K''$ let $\phi_{K'K''}$ be the restriction map $G(K''/K) \longrightarrow G(K'/K)$. Check that this defines an inverse system.

Of course, we can also make a trivial inverse system by making all the G_i be the same and taking all the maps to be the identity map.

Now we can define an inverse limit. Given an inverse system of groups (or rings, or sets) as above, we'll say a group (or ring, or set) G is the *inverse limit* of the system if it satisfies two conditions:

- G comes equipped with homomorphisms $\psi_i : G \longrightarrow G_i$ for every $i \in I$ making all triangles commute: if $i \leq j$, then $\psi_i = \phi_{ij} \circ \psi_j$.
- G is "universal" among groups (rings, sets) with this property, i.e., given any other group (ring, set) G' with such a set of homomorphisms, there exists a unique homomorphism $G' \longrightarrow G$ through which they all factor.

In this case, we write
$$G = \varprojlim_i G_i.$$

(This notation is somewhat abusive, since it doesn't sufficiently specify the inverse system. In most cases, however, it's easy to figure out which inverse system is meant.)

Of course, this kind of definition is useless without some kind of concrete construction to show that such things exist. So here is a constructive description of G (that can also serve as proof that inverse limits of groups, rings, and sets exist). Consider the product
$$P = \prod_{i \in I} G_i.$$

Elements of P are sequences indexed by I, which we write as $(g_i)_{i \in I}$, where $g_i \in G_i$. We get the inverse limit G by considering the piece of this product set which contains the "coherent" sequences, i.e.,
$$G = \{(g_i)_{i \in I} | \phi_{ij}(g_j) = g_i \text{ whenever } i \leq j\}.$$

Problem 1.36. Check that this works, that is, that this G is a group (ring, set) and is the inverse limit.

Problem 1.37. Check that if each of the G_i is a topological group and the ϕ_{ij} are all continuous, then G is the inverse limit of the G_i as topological groups.

In most of the situations we work with, the G_i are *finite* groups (rings, sets). If we give them the discrete topology, they are then compact as topological spaces. The product P, with the product topology, is then compact, and it's easy to see that G is a closed subset of P. Hence G carries a natural compact Hausdorff topology.

The inverse limits of the three inverse systems we considered above are the ring $\mathbb{Z}_p$ of p-adic integers, the ring $\hat{\mathbb{Z}}$, and the Galois group $G(F/K)$. The first of these is a good example on which to test the theory, particularly if you think of $\mathbb{Z}_p$ in terms of p-adic expansions. (See [**59**, Sections 1.2 and 3.3] for a very elementary treatment and [**133**, Chapter 2] for a more sophisticated version.)

Finally, two problems to give you a chance to play with these ideas and extend them at the same time:

Problem 1.38. Define an "exact sequence of inverse systems" and decide whether inverse limits preserve exactness. (Not easy!)

Problem 1.39. Define "direct systems" and "direct limits" by turning all the arrows around. As an example, let I be the positive integers ordered by divisibility, and for each n let $G_n = \mathbb{Z}/n\mathbb{Z}$ (thought of as an additive group). Whenever $n|m$, define $\phi_{mn} : \mathbb{Z}/n\mathbb{Z} \longrightarrow \mathbb{Z}/m\mathbb{Z}$ by mapping 1 to m/n. Can you describe the direct limit?

Profinite groups

The definition of a profinite group is a direct application of the ideas we have just discussed:

Definition 1.4. A *profinite group* is a topological group which can be represented as the inverse limit of an inverse system of finite groups (thought of as carrying the discrete topology).

It follows that profinite groups are compact, by the discussion above. Some of the topological properties of profinite groups we discussed (or set as problems) in the lecture. Here are three others: suppose G is a profinite group, so that

$$G = \varprojlim_i G_i,$$

and let $K_i = \text{Ker}(\psi_i : G \longrightarrow G_i)$. Then, since G_i is discrete, K_i is an open subgroup of G.

Problem 1.40. Show that the K_i, $i \in I$, form a basis of open neighborhoods of the identity in G.

Problem 1.41. Show that a subgroup of G is open if and only if it is closed and of finite index.

Problem 1.42. Show that any closed subgroup of a profinite group is the intersection of all open subgroups containing it.

As we pointed out in the lecture, profinite groups are *totally disconnected*, that is, the connected component of any point is the singleton set consisting only of that point. It turns out that this is equivalent to being profinite:

Theorem 1.9. *Let G be a (Hausdorff and) compact topological group. The following are equivalent:*
  i. G is profinite,
  ii. G is totally disconnected,
  iii. G has a set of open normal subgroups which is a full system of neighborhoods of the identity.

See [143] for a proof.

The property of being profinite is preserved under taking closed subgroups, quotients (by closed subgroups), arbitrary direct products, and inverse limits. As the lecture suggests, many of the properties of finite subgroups still make sense for profinite subgroups. For example, it makes sense to talk about pro-p-groups, meaning profinite groups all of whose finite discrete quotients are p-groups.

For more on profinite groups, see [109, Appendix C], [139] (or its English translation [140]), and [143].)

LECTURE 2
Deformations of Representations

The basic situation we want to study is as follows. We are given either a number field K and a finite set of primes S, or a local field F, and we are given a representation of either $G_{K,S}$ or G_F into $\mathrm{GL}_n(\mathsf{k})$, where k is a finite field. We want to try to understand all possible lifts of this representation to $\mathrm{GL}_n(A)$, where A is a complete noetherian local ring with residue field k. It is not exactly clear, of course, what "understand all possible lifts" means, and so the main goal of this lecture is to make our question precise. We begin by discussing some of the historical motivation for the theory, then develop (the simplest form of) the precise deformation problem we want to study.

Why deform Galois representations?

Nowadays, the obvious reason to study deformations of Galois representations is that they played a crucial role in the proof of the modularity conjecture for elliptic curves over $\mathbb{Q}$ (work of Wiles, Taylor, Diamond, Breuil, and Conrad). However, the theory predates that work, and so the original motivation was different.

Historically, the first (p-adic) Galois representations to be carefully studied were those coming from elliptic curves. Every elliptic curve defined over $\mathbb{Q}$ with good reduction outside a set of primes S gives us representations of $G_{\mathbb{Q},S}$ into both $\mathrm{GL}_2(\mathbb{Z}_p)$ and $\mathrm{GL}_2(\mathbb{F}_p)$, and (usually) the representation over $\mathbb{Z}_p$ is a "lift" of the representation over $\mathbb{F}_p$. This already gives us a first example of a residual representation and one of its deformations.

The second classical source of Galois representations are modular forms, and once again one sees the same pattern: one gets a pair of representations, in characteristic zero and in characteristic p, which are (usually[1]) an example of a residual representation and a deformation.

The real push towards a careful study of such deformations, however, seems to have been inspired by Hida's results on the theory of ordinary p-adic modular forms,

---
[1]We say "usually" because of the following problem. We'll often want to work with a *semisimple* residual representation, which means we'll sometimes have to pass from a representation to its semisimplification. This means that the residual representation coming, for example, from a modular form may fail to be the reduction of the representation in characteristic zero..

which yielded representations which were described at the time as "very large." In particular, specializing Hida's large representations in different ways produced a large number of deformations of a residual representation, some of which did not look much like any of the "usual" representations. This seems to have led Mazur, in his seminal paper [**97**], to pose the question of understanding the deformations in general.

To explain this more fully, we give a very loose description of Hida's work on ordinary p-adic modular forms (see [**71**], [**70**] or [**72**], for example), focusing only on the features that are relevant to our theme. Our summary assumes the reader is familiar with the standard theory of modular forms; see [**13**].

Fix a prime $p \geq 5$, an integer N, not divisible by p, and an integer $k \geq 2$. Let $M_k(\mathrm{N}, \mathbb{Z}_p)$ be the $\mathbb{Z}_p$-module of modular forms of weight k on $\Gamma_1(\mathrm{N}) \cap \Gamma_0(p)$ with coefficients in $\mathbb{Z}_p$, and consider the submodule $M_k(\mathrm{N}, \mathbb{Z}_p)^0$ of "ordinary" modular forms, i.e., the submodule spanned by the eigenforms for the $\mathrm{U} = \mathrm{U}_p$ operator whose eigenvalues are p-adic units. (We could consider finite extensions of $\mathbb{Z}_p$ in exactly the same way.) Hida constructed a Hecke algebra $\mathbf{T}^0 = \mathbf{T}^0(\mathrm{N})$ attached to the whole family of spaces $M_k(\mathrm{N}, \mathbb{Z}_p)^0$, for $k \geq 2$. If we let $\Gamma = 1 + p\mathbb{Z}_p$ and $\Lambda = \mathbb{Z}_p[[\Gamma]]$ be the usual Iwasawa algebra, $\mathbf{T}^0$ is a finite flat Λ-algebra, and any Hecke eigenform in any of our spaces corresponds to a homomorphism $\mathbf{T}^0 \longrightarrow \mathbb{Z}_p$.

Suppose we have such an eigenform $f \in M_k(\mathrm{N}, \mathbb{Z}_p)^0$. Write $\bar{f}$ for the reduction of f modulo p, which we can think of as a modular form of weight k over the finite field $\mathbb{F}_p$; the fact that f is ordinary translates, modulo p, into the assertion that $\bar{f}$ is not in the kernel of U. Because it is an eigenform, $\bar{f}$ corresponds to a homomorphism $\mathbf{T}^0 \longrightarrow \mathbb{F}_p$. The kernel of this homomorphism is a maximal ideal $\mathfrak{m} = \mathfrak{m}(f) \subset \mathbf{T}^0$. Let $R(f) = \mathbf{T}^0_{\mathfrak{m}}$ be the completion of $\mathbf{T}^0$ at the ideal $\mathfrak{m}$. Then $R(f)$ is a complete local Λ-algebra, and is finite and flat over Λ.

Now we bring in Galois representations. One of the fundamental facts in the theory is that every time we have a Hecke eigenform, it gives rise to a two-dimensional Galois representation. Since our f is an eigenform, we get a representation

$$\rho_f : G_\mathbb{Q} \longrightarrow \mathrm{GL}_2(\mathbb{Z}_p).$$

Reducing modulo p (and taking the semi-simplification if necessary, see below) gives a representation

$$\bar{\rho}_f : G_\mathbb{Q} \longrightarrow \mathrm{GL}_2(\mathbb{F}_p).$$

Hida's work showed that there exists a "big" representation

$$\rho_H : G_\mathbb{Q} \longrightarrow \mathrm{GL}_2(R(f))$$

which "interpolates" all the representations ρ_g coming from ordinary eigenforms $g \in M_{k'}(\mathrm{N}, \mathbb{Z}_p)^0$ for various weights k' such that the q-expansions of f and g coincide modulo p.

More precisely, suppose we find an eigenform $g \in M_{k'}(\mathrm{N}, \mathbb{Z}_p)^0$ such that $\bar{f} = \bar{g}$, where bars indicate reduction modulo p. Then it turns out that the "residual representations" $\bar{\rho}_f$ and $\bar{\rho}_g$ are the same (up to equivalence). Hida's theorem says that there exists a homomorphism $R(f) \longrightarrow \mathbb{Z}_p$ such that composing ρ_H with this homomorphism gives ρ_g. Thus, our big representation is somehow parametrizing all lifts of the residual representation $\bar{\rho}_f$ which are of a certain type. We can think of ρ_H as an *analytic family* of Galois representations, all of which have the same reduction modulo p. It is natural then, to ask about such families in general, and that question leads at once to the deformation theory.

Furthermore, it quickly became clear that there are other homomorphisms $R(f) \longrightarrow \mathbb{Z}_p$, ones that *do not* correspond to modular forms (or at least not to the usual kind of modular forms). Specializing the representation via one of these homomorphisms can produce rather strange representations (for example, see the final sections of [**106**]). This suggests, once again, that it makes sense to try to study "all the deformations."

The deformation functor

Mazur created the theory of deformations of Galois representations in his paper [**97**], which is one of the fundamental references for this section (and for much of what follows also). See also Mazur's notes from the Boston University conference, [**101**], and the notes [**42**] by Doran and Wong.

What we want to do is imitate the situation in Hida's theory, in maximal generality. So we'll start with a profinite group Π (which will later be a Galois group of some kind) and a representation of Π into matrices over a finite field. The basic question will be: can we describe all lifts of this representation to (appropriate) p-adically complete rings?

Let k be a finite[2] field of characteristic p. For this section, we make no assumptions on p. We will want to start with a representation into $\mathrm{GL}_n(\mathrm{k})$, and consider its lifts.

Let Π be a profinite group. In order for the theory to work, we need to know that Π satisfies the finiteness condition which we considered, in the case of Galois groups, in our first lecture.

Condition Φ_p: *For every open subgroup of finite index $\Pi_0 \subset \Pi$ there exist only a finite number of continuous homomorphisms $\Pi_0 \longrightarrow \mathbb{F}_p$.*

We already know, by Theorem 1.6 and Problem 1.26, that the Condition Φ_p holds for $G_{\mathbb{Q}_\ell}$ and for $G_{\mathbb{Q},S}$, where ℓ is a prime and S is a finite set of primes. It's worth pointing out that the condition can be stated in several equivalent forms.

Let's first set up some notation. First, the *pro-p-completion* of the profinite group Π is
$$\Pi^{(p)} = \varprojlim_{N} \Pi/N,$$
where we take the limit over all closed normal subgroups whose index is (finite and) a power of p. Second, the *p-Frattini quotient* of Π is the maximal continuous abelian quotient of Π which is of exponent p.

Problem 2.1. Show that the p-Frattini quotient exists and that it is the image of a surjective continuous homomorphism from $\Pi^{(p)}$.

The following lemma gives several equivalent ways of stating condition Φ_p.

Lemma 2.1. *Let Π_0 be a profinite group. The following conditions are equivalent:*

  i. *the pro-p-completion of Π_0 is topologically finitely generated,*
  ii. *the abelianization of the pro-p-completion of Π_0 is a $\mathbb{Z}_p$-module of finite rank,*

---
[2]The construction works just as well for something like the algebraic closure of $\mathbb{F}_p$, but the case of a finite field is the most significant for us, and we'll simplify things by restricting ourselves to this case.

iii. the p-Frattini quotient of Π_0 is finite,
iv. the set of continuous homomorphisms from Π_0 to $\mathbb{F}_p$ is finite.

Proof. Clearly a set of topological generators of the pro-p-completion becomes a set of generators over $\mathbb{Z}_p$ in the abelianization, and becomes a basis of the p-Frattini-quotient as a vector space over $\mathbb{F}_p$. Hence, it's clear that (1) $\Rightarrow$ (2) $\Rightarrow$ (3). Since any homomorphism $\Pi_0 \longrightarrow \mathbb{F}_p$ must factor through the p-Frattini-quotient, (3) and (4) are equivalent. To conclude the proof, we use the profinite version of the Burnside Basis Theorem, which says that if the image in the p-Frattini quotient of a set $\{g_1, \ldots, g_r\}$ of elements of the pro-p-group $\Pi_0^{(p)}$ is a basis for the quotient as a vector space over $\mathbb{F}_p$ then $g_1, \ldots, g_r$ topologically generate $\Pi_0^{(p)}$. It follows that (3) $\Rightarrow$ (1), and we're done. $\square$

Problem 2.2. (The pro-p version of the Burnside Basis Theorem) Let G be a pro-p-group, i.e., an inverse limit of finite p-groups, and let $\mathrm{Fr}(G)$ be its pro-p-Frattini quotient. Prove that any lifting to G of a basis of $\mathrm{Fr}(G)$ as a vector space over $\mathbb{F}_p$ is a set of topological generators for G.

As we saw above, (finitely ramified) Galois groups satisfy condition Φ_p. This can be vastly generalized; see [**84**].

Having stated our crucial assumption about the profinite group Π, let's go on to talk about its representations, according to the "program" we have outlined. We want to start, then, with a homomorphism

$$\overline{\rho}: \Pi \longrightarrow \mathrm{GL}_n(\mathsf{k}),$$

and we want to consider lifts of $\overline{\rho}$, that is, homomorphisms

$$\rho: \Pi \longrightarrow \mathrm{GL}_n(R)$$

where R is a ring together with a homomorphism $\pi: R \longrightarrow \mathsf{k}$ such that the image of ρ under the homomorphism $\mathrm{GL}_n(R) \longrightarrow \mathrm{GL}_n(\mathsf{k})$ induced by π is our residual representation $\overline{\rho}$, i.e., the diagram

$$\begin{array}{ccc} & & \mathrm{GL}_n(R) \\ & {}^{\rho}\nearrow & \downarrow \pi \\ \Pi & \xrightarrow{\overline{\rho}} & \mathrm{GL}_n(\mathsf{k}) \end{array}$$

is commutative. If we want to do this correctly, however, we need to be a bit more precise about the rings we will be considering. So we set this up in the language of categories.

Choose and fix a finite field k of characteristic p. Let $\mathcal{C}$ denote the category whose objects are complete noetherian local rings with residue field k and whose morphisms are local homomorphisms $R_1 \longrightarrow R_2$ of complete noetherian local rings which induce the identity on k.[3] In particular, this means that if $\mathfrak{m}$ is the maximal ideal of R, then we are requiring that, first, $R/\mathfrak{m} = \mathsf{k}$, and second,

$$R = \varprojlim_j R/\mathfrak{m}^j.$$

---

[3]To be absolutely precise, we need to make our objects be complete noetherian local rings $(R, \mathfrak{m})$ together with a *fixed* isomorphism $R/\mathfrak{m} \cong ks$, but we'll refrain from being picky about this.

(It is useful to recall that by the Krull intersection theorem the intersection of the $\mathfrak{m}^j$ is 0, which is equivalent to saying that the natural topology on a noetherian local ring is always Hausdorff.)

Sometimes it is also convenient to consider the full subcategory $\mathcal{C}^0$ whose objects are artinian local rings with residue field k. (Notice that the maximal ideal of an artinian local ring is always nilpotent, and hence such rings are automatically complete and noetherian.) Following Mazur, we will call the objects of $\mathcal{C}$ "coefficient rings" and we will call the morphisms "coefficient ring homomorphisms."

Notice that (as is implicit from our use of the word "complete") all coefficient rings carry a natural topology, in which the powers of the maximal ideal are a basis of neighborhoods of 0. Coefficient ring homomorphisms are continuous with respect to this topology.

Problem 2.3. Prove that objects of $\mathcal{C}$ are pro-objects of $\mathcal{C}^0$. Specifically, prove that if R is a complete noetherian local ring with maximal ideal $\mathfrak{m}$, then for every n the quotient $R/\mathfrak{m}^n$ is an object of $\mathcal{C}^0$, and R is the inverse limit of the $R/\mathfrak{m}^n$.

Problem 2.4. With the notations in the previous problem, show that the topology on $R/\mathfrak{m}^n$ is discrete, and that the topology on R is the inverse limit topology.

Problem 2.5. How serious are the restrictions on "coefficient ring homomorphisms?" Find examples of ring homomorphisms between objects of $\mathcal{C}$ which are not "coefficient ring homomorphisms."

The "simplest" example of a (non-artinian) element of $\mathcal{C}$ is the ring $W(\mathsf{k})$ of Witt vectors. Since k is finite, this is simply the (unique) unramified extension of $\mathbb{Z}_p$ whose residue field is k. When $\mathsf{k} = \mathbb{F}_p$, then $W(\mathsf{k})$ is $\mathbb{Z}_p$ itself. See [135] for more information on rings of Witt vectors.

Problem 2.6. Show that any coefficient ring R in $\mathcal{C}$ carries a canonical $W(\mathsf{k})$ algebra structure. (That is, show that every such R has a unique coefficient ring homomorphism $W(\mathsf{k}) \longrightarrow R$.)

Problem 2.7. Show that in fact every coefficient ring is a quotient of a power series ring in several variables with coefficients in $W(\mathsf{k})$.

As the last two problems show, our coefficient rings are automatically $W(\mathsf{k})$-algebras. It often happens, however, that we want to modify this somewhat. For example, suppose that we start the game with a representation that comes from a modular form. Then we have at hand not only a residual representation defined over a finite field k, but also a particular lift to a discrete valuation ring $\mathcal{O}$ that may very well not be $W(\mathsf{k})$. In such a situation, we may decide that we want to restrict the whole game to coefficient rings that are $\mathcal{O}$-algebras. This amounts to working in a slightly different category.

Let Λ be an object of $\mathcal{C}$, that is a complete noetherian local ring with residue field k. We'll define $\mathcal{C}_\Lambda$ to be the category whose objects are complete noetherian local Λ-algebras with residue field k and whose morphisms are coefficient-ring homomorphisms which are also Λ-algebra homomorphisms. As before we let $\mathcal{C}_\Lambda^0$ be the full subcategory of artinian Λ-algebras with residue field k. Of course, $\mathcal{C}$ is the same as $\mathcal{C}_{W(\mathsf{k})}$.

Problem 2.8. Is it true that every element of $\mathcal{C}_\Lambda$ is a quotient of a power series ring in several variables over Λ?

Given a coefficient ring R (i.e. an object of $\mathcal{C}$, or of $\mathcal{C}_\Lambda$ if we have fixed a different base ring), we will write π for the canonical projection $R \longrightarrow \mathsf{k}$ and also,

by abuse of language, for the map it induces from $\mathrm{GL}_n(R)$ to $\mathrm{GL}_n(\mathsf{k})$. Finally, we let
$$\Gamma_n(R) = \mathrm{Ker}\left(\mathrm{GL}_n(R) \xrightarrow{\pi} \mathrm{GL}_n(\mathsf{k})\right).$$

Definition 2.1. Let R be a coefficient ring. We say two homomorphisms
$$\rho_1, \rho_2 : \Pi \longrightarrow \mathrm{GL}_n(R)$$
are *strictly equivalent* if there exists $M \in \Gamma_n(R)$ such that $\rho_1 = M^{-1}\rho_2 M$.

The idea, of course, is that strictly equivalent homomorphisms give the same homomorphism when we compose them with $\pi : R \longrightarrow \mathsf{k}$. This will give the right notion of equivalence for the theory we have in mind.

Suppose now that we start with a representation (i.e., a continuous group homomorphism)
$$\overline{\rho} : \Pi \longrightarrow \mathrm{GL}_n(\mathsf{k}).$$
We will call this a *residual representation*. For the rest of this section we will assume that we have chosen and fixed such a residual representation. We are finally ready to make the crucial definition.[4]

Definition 2.2. Let $\overline{\rho}$ be a residual representation and let R be a coefficient ring. A *deformation* of $\overline{\rho}$ to R is a strict equivalence class of continuous homomorphisms
$$\rho : \Pi \longrightarrow \mathrm{GL}_n(R)$$
which reduce to $\overline{\rho}$ via the projection π, that is, such that $\pi \circ \rho = \overline{\rho}$.

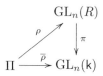

If we want to be precise, we really have to say that we require $\pi \circ \varphi = \overline{\rho}$ for any homomorphism φ in the strict equivalence class of ρ. Of course, this will be true for all φ in the strict equivalence class if and only if it is true for one of them, so this quibble isn't really very serious. We will, in fact, routinely confuse a homomorphism with its strict equivalence class, and deal with the possible confusions this will generate as they arise.

We think of this as defining a functor
$$\mathbf{D} = \mathbf{D}_{\overline{\rho}} : \mathcal{C} \rightsquigarrow \underline{\mathrm{Sets}}$$
where
$$\mathbf{D}_{\overline{\rho}}(R) = \{\text{deformations of } \overline{\rho} \text{ to } R\}.$$

Similarly, we define the functor $\mathbf{D}_{\overline{\rho}, \Lambda}$ by restricting to the subcategory $\mathcal{C}_\Lambda$. We will often drop the $\overline{\rho}$ from the notation, since our residual representation will typically be fixed for the whole discussion.

Lemma 2.2. $\mathbf{D}$ *and* $\mathbf{D}_\Lambda$ *are functors.*

---

[4]People who are familiar with deformation theory in a geometric context should note that we are really talking of "infinitesimal" or "formal" deformations here.

Problem 2.9. Prove the lemma. (The main things here are to recall what it means to be a functor and to remember that deformations are strict equivalence classes of homomorphisms.)

Problem 2.10. Translate this into the language of free R-modules with a continuous action of Π. A residual representation becomes a k-vector space with a continuous action of Π and a deformation must be some kind of free R-module with a continuous action of Π. Make the appropriate definition and compare the resulting functor with the one(s) we have just defined.

Recall that the categories $\mathcal{C}$ and $\mathcal{C}^0$ are related because objects of $\mathcal{C}$ are pro-objects of $\mathcal{C}^0$. Specifically, if R is a complete noetherian local ring with residue field k, then, if $\mathfrak{m}$ is the maximal ideal of R, we have

$$R = \varprojlim_k R/\mathfrak{m}^k.$$

Now, if we have a functor $\mathbf{F}$ on $\mathcal{C}$, the sets $\mathbf{F}(R/\mathfrak{m}^k)$ will form an inverse system, and there will be compatible morphisms $\mathbf{F}(R) \longrightarrow \mathbf{F}(R/\mathfrak{m}^k)$. These compile to give a canonical morphism

$$\mathbf{F}(R) \longrightarrow \varprojlim_k \mathbf{F}(R/\mathfrak{m}^k).$$

Definition 2.3. We say a functor $\mathbf{F}$ on $\mathcal{C}$ is *continuous* when the canonical morphism

$$\mathbf{F}(R) \longrightarrow \varprojlim_k \mathbf{F}(R/\mathfrak{m}^k)$$

is an isomorphism.

Lemma 2.3. $\mathbf{D}$ *and* $\mathbf{D}_\Lambda$ *are continuous functors.*

Proof. We work with $\mathbf{D}$; the proof for $\mathbf{D}_\Lambda$ is the same.

Recall, first, that

$$\mathrm{GL}_n(R) = \varprojlim_k \mathrm{GL}_n(R/\mathfrak{m}^k)$$

and

$$\Gamma_n(R) = \varprojlim_k \Gamma_n(R/\mathfrak{m}^k).$$

Furthermore, note that the maps

$$\mathrm{GL}_n(R/\mathfrak{m}^{k+1}) \longrightarrow \mathrm{GL}_n(R/\mathfrak{m}^k)$$

and

$$\Gamma_n(R/\mathfrak{m}^{k+1}) \longrightarrow \Gamma_n(R/\mathfrak{m}^k)$$

are all surjective.

If deformations were simply homomorphisms, the continuity would now follow at once. However, deformations are strict equivalence classes of homomorphisms, and so we have to be a bit more careful.

The canonical map

$$\mathbf{D}(R) \longrightarrow \varprojlim_k \mathbf{D}(R/\mathfrak{m}^k)$$

maps a deformation $\rho = \rho_R$ of $\bar{\rho}$ to R to the coherent sequence $\{\rho_k\}$, where ρ_k is the deformation to $R/\mathfrak{m}^k$ obtained by reducing (any homomorphism representing) ρ modulo $\mathfrak{m}^k$.

To show that the canonical map is surjective, we need to show that any coherent sequence $\{\rho_k\}$ comes from a deformation ρ to R. For this, it is enough to show that we can choose the homomorphisms representing ρ_k so as to have a coherent sequence *of homomorphisms*. For $k = 1$, we must have $\rho_1 = \bar{\rho}$, so there is no choice at this level. Assume we have chosen homomorphisms $r_1, \ldots, r_k$ representing the deformations $\rho_1, \ldots, \rho_k$ and forming a coherent sequence. Let r' be any homomorphism representing ρ_{k+1}. The assumption that the sequence $\{\rho_k\}$ is coherent means that there exists $M_k \in \Gamma(R/\mathfrak{m}^k)$ such that $M_k^{-1}(r' \pmod{\mathfrak{m}^k}) M_k = r_k$. Choosing a lift M_{k+1} of M_k to $\Gamma(R/\mathfrak{m}^{k+1})$ and setting $r_{k+1} = M_{k+1}^{-1} r' M_{k+1}$ extends the coherent sequence to level $k + 1$. By induction, we get a coherent sequence $\{r_k\}$ of homomorphisms $\Pi \to \mathrm{GL}_n(R/\mathfrak{m}^k)$. Taking the inverse limit of these homomorphisms then gives a deformation $\rho : \Pi \longrightarrow \mathrm{GL}_n(R)$ whose reduction modulo $\mathfrak{m}^k$ is ρ_k. This proves the canonical map is surjective.

To show that the canonical map is injective, we need to show that if ρ and ρ' are homomorphisms $\Pi \longrightarrow \mathrm{GL}_n(R)$ such that $\rho_k = \rho \pmod{\mathfrak{m}^k}$ and $\rho'_k = \rho' \pmod{\mathfrak{m}^k}$ are strictly equivalent for all k, then ρ and ρ' are strictly equivalent. The assumption is that for all k we can find $M_k \in \Gamma(R/\mathfrak{m}^k)$ such that

$$\rho_k = M_k^{-1} \rho'_k M_k.$$

It is clear that we can choose the M_k such that $M_{k+1} \equiv M_k \pmod{\mathfrak{m}^k}$, giving a coherent sequence and therefore and element of $\Gamma(R)$ such that $\rho = M^{-1} \rho' M$, as desired. This proves the canonical map is injective. $\square$

The continuity of our functor is an important technical tool: basically it shows that **D** is completely determined by its values on the full subcategory $\mathcal{C}^0$. We will use this in a crucial way later, when we use the Schlessinger criteria for representability, which apply to functors on artinian rings.

We should note a final variation on the basic idea. Suppose we have a lift ρ_A of $\bar{\rho}$ to a coefficient ring A. Then it makes sense to look only at those deformations which are actually deformations of our fixed lift to A, that is, deformations to coefficient rings R with a map to A such that the induced deformation is ρ_A. This leads to a slightly modified functor again, for which we need to make two changes:

i. First, we work with the category whose objects are coefficient rings (or coefficient Λ-algebras) that come with an "A-augmentation," that is, a coefficient ring (Λ-algebra) homomorphism $R \longrightarrow A$. In [**101**], Mazur calls these "A-augmented coefficient rings (or Λ-algebras)." We call this category $\mathcal{C}(A)$ (or $\mathcal{C}_\Lambda(A)$).

ii. Second, we change the definition of strict equivalence to allow conjugation only by matrices in the kernel of the map $\mathrm{GL}_n(R) \longrightarrow \mathrm{GL}_n(A)$ induced by the augmentation.

As before, one needs to also consider the full subcategory $\mathcal{C}^0_\Lambda(A)$ of A-augmented artinian local Λ-algebras with residue field k. See Mazur's discussion in [**101**] for more about this "relative" version of the theory.

Universal deformations: why representable functors are nice

The question we want to ask about our deformation functor is whether it is *representable*. This means the following. Given any coefficient ring R, we can define a set-valued functor $\mathbf{h}_R$ on $\mathcal{C}$ by setting, for each coefficient ring S,

$$\mathbf{h}_R(S) = \mathrm{Hom}(R, S),$$

where of course Hom indicates coefficient ring homomorphisms, and where the action on homomorphisms $S_1 \longrightarrow S_2$ defined by composition. We say that a functor $\mathbf{F}$ is *representable* if it is isomorphic to the functor $\mathbf{h}_\mathcal{R}$ for some coefficient ring $\mathcal{R}$. Hence, to ask whether $\mathbf{D}$ is representable is to ask whether there exists a coefficient ring (or coefficient Λ-algebra) $\mathcal{R}_{\bar{\rho}}$ such that we have

$$\mathbf{D}_{\bar{\rho}}(R) = \mathrm{Hom}(\mathcal{R}_{\bar{\rho}}, R)$$

for every coefficient ring R and this identification is "functorial," i.e., transforms well under homomorphisms. Let's explore that idea a little in this section.

Assume, then, that the ring $\mathcal{R} = \mathcal{R}_{\bar{\rho}}$ exists. First of all, consider the case $R = \mathcal{R}$. Since the identity is a homomorphism from $\mathcal{R}$ to itself, it corresponds to a deformation

$$\boldsymbol{\rho} : \Pi \longrightarrow \mathrm{GL}_n(\mathcal{R}).$$

This will turn out to deserve to be called the "universal" deformation. To see why, consider any deformation ρ to a coefficient ring R. By our assumption that the functor is represented by $\mathcal{R}$, this deformation must correspond to a (better: exactly one) coefficient ring homomorphism $\varphi : \mathcal{R} \longrightarrow R$, and the morphism mapping $\boldsymbol{\rho}$ to ρ must be "composition with φ." In other words, given any deformation ρ to a coefficient ring R, there is a coefficient ring homomorphism $\varphi : \mathcal{R} \longrightarrow R$ such that $\rho = \varphi \circ \boldsymbol{\rho}$. Thus, the ring $\mathcal{R}$ parametrizes all possible deformations, and the deformation $\boldsymbol{\rho}$ is "universal," because every deformation is derived from it.

The upshot, then, is that if we can show that our functor is representable we will get a large ring, which we will call the *universal deformation ring of* $\bar{\rho}$ and denote by $\mathcal{R}_{\bar{\rho}}$, and a representation

$$\boldsymbol{\rho} : \Pi \longrightarrow \mathrm{GL}_n(\mathcal{R}_{\bar{\rho}}),$$

which we will call the *universal deformation of* $\bar{\rho}$.

Problem 2.11. Show that any representable functor is continuous.

As we noted above, we sometimes want to work not with the functor $\mathbf{D}$ but rather with the functor $\mathbf{D}_\Lambda$ which is the result of restricting our attention to coefficient rings which are also Λ-algebras. As we pointed out above, $\mathbf{D}$ is the same as $\mathbf{D}_{W(k)}$, so it would seem that we should work directly with the more general case. It turns out, however, that the moving from one case to another is quite easy:

Theorem 2.4. *If $\mathbf{D}$ is representable by a coefficient-ring $\mathcal{R}$, then $\mathbf{D}_\Lambda$ is representable by $\mathcal{R}_\Lambda = \mathcal{R} \hat{\otimes}_{W(k)} \Lambda$.*

Proof. This is essentially clear if one understands what a completed tensor product is. First of all, note that $\mathcal{R}_\Lambda$ is a coefficient ring and a Λ-algebra. (This is why we need a *completed* tensor product: the tensor product of two coefficient rings need not be complete, so we must pass to the completion to obtain a coefficient ring again.) Next, there is a canonical coefficient-ring homomorphism $\mathcal{R} \longrightarrow \mathcal{R}_\Lambda$, which induces a deformation $\boldsymbol{\rho}_\Lambda$ of $\bar{\rho}$ to $\mathcal{R}_\Lambda$. We claim that this is the universal deformation to coefficient Λ-algebras. To see this, just note that any map from $\mathcal{R}$ to a coefficient Λ-algebra A factors through the canonical map $\mathcal{R} \longrightarrow \mathcal{R}_\Lambda$. □

Problem 2.12. Show that the completed tensor product (over $W(\mathsf{k})$) of two coefficient rings is a coefficient ring. In fact, show that completed tensor product of R_1 and R_2 is the inverse limit over pairs (i,j) of the tensor products $R_1/\mathfrak{m}_{R_1}^i$ tensor $R_2/\mathfrak{m}_{R_2}^j$.

Problem 2.13. Suppose $\mathcal{R} = W(\mathsf{k})[[X_1, X_2, \ldots, X_k]]/I$, where I is the closed ideal generated by $f_1, f_2, \ldots, f_s$. Describe $\mathcal{R}_\Lambda$.

Problem 2.14. More generally, suppose two coefficient rings R_1 and R_2 are given explicitly as quotients of power series rings in several variables over $W(\mathsf{k})$. Describe $R_1 \hat{\otimes}_{W(\mathsf{k})} R_2$.

Problem 2.15. It is clearly not always possible to recover $\mathcal{R}$ from $\mathcal{R}_\Lambda$ (for example, consider $\Lambda = \mathsf{k}$). Is it ever possible to "descend" from $\mathcal{R}_\Lambda$ to $\mathcal{R}$?

As we learn from algebraic geometry, one can associate to a ring such as $\mathcal{R}$ a geometric object $\operatorname{Spec} \mathcal{R}$, and an "$A$-valued point on $\operatorname{Spec} \mathcal{R}$" is the same as a ring homomorphism $\mathcal{R} \longrightarrow A$. Since coefficient-ring homomorphisms $\mathcal{R} \longrightarrow A$ correspond to deformations, this suggests that we should call $\operatorname{Spec} \mathcal{R}$ the "universal deformation space" of $\overline{\rho}$. There is something to be careful of here, however: the "A-valued points of $\operatorname{Spec} \mathcal{R}$" include *all* ring-homomorphisms $\mathcal{R} \longrightarrow A$, and of course not all such homomorphisms will induce a deformation of $\overline{\rho}$ to A (one will always get a representation $\Pi \longrightarrow \operatorname{GL}_n(A)$, but it need not be continuous nor, even if continuous, need it be a lift of $\overline{\rho}$). A better way to obtain a "deformation space" from the universal deformation ring $\mathcal{R}$ is to consider its formal spectrum $\operatorname{Spf} \mathcal{R}$ as a formal scheme over $\operatorname{Spf} W(\mathsf{k})$ or the associated rigid-analytic space $\operatorname{Spf} \mathcal{R}^{\mathrm{rig}}$ (which, however, is not quasi-caompact; see [**35**] for the properties of the functor $(-)^{\mathrm{rig}}$).

Suppose, for example, that $\mathsf{k} = \mathbb{F}_p$ and $\mathcal{R} = \mathbb{Z}_p[[X_1, X_2, X_3]]$ (as will actually be the case in one of our examples). Then we want to think of the associated space as a three-dimensional space over $\mathbb{Z}_p$, with three parameters corresponding to the three variables. But the dimension of $\operatorname{Spec} \mathcal{R}$ (which is the same as the Krull dimension of the ring) is four, not three. On the other hand, the relative dimension of $\operatorname{Spf} \mathcal{R}$ over $\operatorname{Spf} \mathbb{Z}_p$ is indeed three.

Representable functors and fiber products

Suppose we are working in some category, and we are given objects A, B, and C and maps $\alpha : A \longrightarrow C$ and $\beta : B \longrightarrow C$. Visualize this as the beginning of a commutative diagram

which we want to complete to a commutative "diamond." If a "universal" solution to this problem exists, we call it the *fiber product of A and B over C*, which we denote by $A \times_C B$. This comes with maps $p : A \times_C B \longrightarrow A$ and $q : A \times_C B \longrightarrow B$

such that the diagram

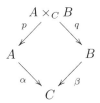

commutes, and is the universal such object in the following sense. Suppose we have another ring D and maps $D \longrightarrow A$ and $D \longrightarrow B$ such that the diagram

is commutative. Then there exists a unique map $D \longrightarrow A \times_C B$ through which both the maps $D \longrightarrow A$ and $D \longrightarrow B$ factor. It is easy to see that if the fiber product exists it is unique up to unique isomorphism.

In the category of sets, the fiber product is given by
$$A \times_C B = \{(a,b) \in A \times B \mid \alpha(a) = \beta(b)\}.$$

Problem 2.16. Check that this set does have the universal property described above.

Now let's consider this in the context of representable functors. We continue to work in a general category. If R and S are objects of our category, we write $\text{Hom}(R, S)$ for the set of morphisms from R to S. Then we can translate the universal property of the fiber product into the statement that for any object D we have
$$\text{Hom}(D, A \times_C B) = \text{Hom}(D, A) \times_{\text{Hom}(D,C)} \text{Hom}(D, B),$$
since the set on the right consists exactly of the pairs of morphisms $D \longrightarrow A$ and $D \longrightarrow B$ that make the diagram commute. If we think of D as the object representing a functor from our category to sets given by $\mathbf{F}(R) = \text{Hom}(D, R)$, we can read this statement as saying that "representable functors commute with fiber products." In other words, if $\mathbf{F}$ is a representable functor from a category where fiber products exist to the category of sets, then we have
$$\mathbf{F}(A \times_C B) = \mathbf{F}(A) \times_{\mathbf{F}(C)} \mathbf{F}(B),$$
where of course the object on the right is a fiber product of sets. Following Mazur, we will call this property the *Mayer-Vietoris* property of representable functors.

For general functors, all we know is that $\mathbf{F}(A \times_C B)$ fits into the diagram

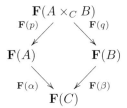

Hence, by the universal property of fiber products, we get a map

$$\mathbf{F}(A \times_C B) \longrightarrow \mathbf{F}(A) \times_{\mathbf{F}(C)} \mathbf{F}(B),$$

but a priori there is no reason to expect this map to have any special properties. On the other hand, when $\mathbf{F}$ is representable, this function will be a bijection.

The upshot of this discussion is the following: when we are working in a category in which fiber products exist, the Mayer-Vietoris property is a *necessary* condition for a functor to be representable. As we will see below, it is very close to being also sufficient.

How can we apply this in our situation? It turns out that we *cannot* apply it to $\mathcal{C}$, the category of all coefficient rings (i.e., complete noetherian local rings with residue field k), but we can use it if we work with the smaller category $\mathcal{C}^0$ of artinian local rings with residue field k; this is the main reason to bring up $\mathcal{C}^0$ in the first place. The reason is this: if A, B, and C are commutative rings and α and β are ring homomorphisms, then $A \times_C B$ has a natural ring structure that makes it the fiber product in the category of rings. The property of being local is preserved, and the property of having residue field k is also preserved. But the fiber product of noetherian rings doesn't need to be noetherian.

Problem 2.17. Let $A = \mathsf{k}[[X,Y]]$, $B = \mathsf{k}$, $C = \mathsf{k}[[X]]$. Let $\alpha : A \longrightarrow C$ be the map that sends Y to 0 and let $\beta : B \longrightarrow C$ be the inclusion of k in $\mathsf{k}[[X]]$. Note that A, B, and C are objects of $\mathcal{C}$ and that α and β are morphisms in $\mathcal{C}$. Show that the ring $A \times_C B$ is not noetherian, and hence is not an object of $\mathcal{C}$.

Problem 2.18. Show that if both α and β are surjective, then $A \times_C B$ is an object of $\mathcal{C}$, i.e., a complete noetherian local ring with residue field k, and is the fiber product of A and B over C in $\mathcal{C}$.

Problem 2.19. Show that if A, B, and C are in $\mathcal{C}^0$, i.e., are artinian rings with residue field k, then $A \times_C B$ is an object of $\mathcal{C}^0$.

Problem 2.20. Show that the same is true in the categories $\mathcal{C}^0_\Lambda$ and $\mathcal{C}^0_\Lambda(A)$.

Problem 2.21. Suppose we work in some subcategory $\mathcal{Z}$ of the category of commutative rings. Suppose we are given objects A, B, and C and morphisms $A \longrightarrow C$ and $B \longrightarrow C$. Let $A \times_C B$ be the ring-theoretical fiber product, i.e., the ring defined by

$$A \times_C B = \{(a,b) \in A \times B \mid \alpha(a) = \beta(b)\}$$

with the natural operations. Is it true that if $A \times_C B$ is an object of $\mathcal{Z}$ then it is the fiber product in $\mathcal{Z}$? Is it true that if $A \times_C B$ is not an object of $\mathcal{Z}$ then there is no fiber product of A and B over C in $\mathcal{Z}$?

Recall that objects of $\mathcal{C}$ are "pro-objects" of $\mathcal{C}^0$, that is, that any object of $\mathcal{C}$ is an inverse limit of objects of $\mathcal{C}^0$. To be specific, if R is a complete noetherian local ring with maximal ideal $\mathfrak{m}$, then $R/\mathfrak{m}^n$ is artinian and we have

$$R = \varprojlim_n R/\mathfrak{m}^n.$$

Suppose that our functor is *continuous*, which, as we explained above, means that

$$\mathbf{F}(R) = \varprojlim_n \mathbf{F}(R/\mathfrak{m}^n).$$

(As we noted above, the deformation functors do have this property.) Then $\mathbf{F}$ is completely determined by its values on the smaller category $\mathcal{C}^0$.

Furthermore, it may happen that **F** is not representable as a functor on $\mathcal{C}^0$, but that there exists an object $\mathcal{R}$ of the larger category $\mathcal{C}$ such that we have

$$\mathbf{F}(A) = \mathrm{Hom}(\mathcal{R}, A)$$

for every artinian coefficient ring A. In this case, we say that the functor **F** on the category $\mathcal{C}^0$ is *pro-representable*.

Problem 2.22. Show that if **F** is continuous, then it is pro-representable as a functor on $\mathcal{C}^0$ if and only if it is representable as a functor on $\mathcal{C}$.

It is easy to see that *pro-representable functors have the Mayer-Vietoris property*. In fact, this is quite close to being a *sufficient* condition for a functor on $\mathcal{C}^0$ to be pro-representable.

To state the exact theorem, let's introduce the coefficient ring of "dual numbers,"

$$\mathsf{k}[\varepsilon] = \mathsf{k}[X]/(X^2)$$

where $\varepsilon = X \pmod{X^2}$. If we are working with coefficient Λ-algebras, we make $\mathsf{k}[\varepsilon]$ into a Λ-algebra via the map

$$\Lambda \longrightarrow \Lambda/\mathfrak{m}_\Lambda = \mathsf{k} \hookrightarrow \mathsf{k}[\varepsilon].$$

Then we have:

Theorem 2.5 (Grothendieck). *Let*

$$\mathbf{F} : \mathcal{C}^0_\Lambda \rightsquigarrow \underline{Sets}$$

be a (covariant) functor such that $\mathbf{F}(\mathsf{k})$ *consists of a single element. Then* **F** *is pro-representable if and only if*

i. **F** *satisfies the Mayer-Vietoris property, and*
ii. $\mathbf{F}(\mathsf{k}[\varepsilon])$ *is a finite set.*

For the proof, see [**67**].

The statement of this theorem is one of the few places where we are *really* using our assumption that k is a finite field, but we are using it only to state this result before having explained why $\mathbf{F}(\mathsf{k}[\varepsilon])$ is a k-vector space. Once we know that, we can replace the finiteness assumption above with finite dimensionality over k.

As Mazur says in [**101**], this result "is easy to prove (it is a good exercise) but ... is difficult to use because its hypothesis is hard to check." The problem of course, is that the Mayer-Vietoris condition involves checking something for all diagrams

That is clearly hard to do in general. Schlessinger's theorem (to be discussed in the next lecture) should be viewed as basically a simplification of this result. On the other hand, see the complements to Lecture 3 for a proof that proceeds directly from Grothendieck's theorem.

The finiteness (or finite-dimensionality) condition is there to guarantee that the representing object is noetherian. If we were willing to work in a larger category (e.g., the category of all Λ-algebras which are inverse limits of objects of $\mathcal{C}_\Lambda^0$), then we could drop this finiteness assumption (which in fact is not in Grothendieck's original theorem).

Problem 2.23. (Some category theory needed.) Show that the first condition in the theorem, together with the condition that $\mathbf{F}(\mathsf{k})$ is a singleton, is equivalent to saying that $\mathbf{F}$ preserves all finite limits. (Grothendieck calls functors which preserve all finite limits *left exact*.)

The tangent space

For this section, we fix a coefficient ring Λ and work in the category $\mathcal{C}_\Lambda$ of coefficient Λ-algebras. We let $\mathfrak{m}_\Lambda$ denote the maximal ideal of Λ. Let R be a coefficient Λ-algebra, and let $\mathfrak{m}_R$ be its maximal ideal. The *Zariski cotangent space* of R is defined to be
$$t_R^* = \mathfrak{m}_R/(\mathfrak{m}_R^2, \mathfrak{m}_\Lambda),$$
where
$$(\mathfrak{m}_R^2, \mathfrak{m}_\Lambda) = \mathfrak{m}_R^2 + (\text{image of } \mathfrak{m}_\Lambda)R$$
is the ideal of R generated by the square $\mathfrak{m}_R^2$ of the maximal ideal of R and the image in R of the maximal ideal of Λ. Notice that t_R^* is a module over $\Lambda/\mathfrak{m}_\Lambda$, that is, it is a k-vector space.

The *Zariski tangent space* of R is, of course, the dual of the cotangent space:
$$t_R = \text{Hom}_\mathsf{k}(\mathfrak{m}_R/(\mathfrak{m}_R^2, \mathfrak{m}_\Lambda), \mathsf{k}).$$
Since R is noetherian, t_R^* is a finite-dimensional vector space, so that there is no problem with the duality here.

Problem 2.24. Let R be a noetherian local ring with residue field k and define the tangent space t_R as above. Prove that t_R is a finite-dimensional vector space over $R/\mathfrak{m}_R$. Is the converse true? (Well, first of all, what would the converse say?)

Problem 2.25. Let $f : B \longrightarrow A$ be a morphism in $\mathcal{C}_\Lambda$. Show that f induces a k-linear transformation $f_* : t_B^* \longrightarrow t_A^*$ of cotangent spaces. Show that f is surjective if and only if f_* is surjective. (This is Lemma 1.1 in [**125**].)

Problem 2.26. Use the duality between the tangent and cotangent spaces to reinterpret the previous problem in terms of tangent spaces.

Suppose we have a functor $\mathbf{F}$ as above which is represented by R. We'd like to reinterpret this construction in terms of the functor. The crucial observation is the following.

Lemma 2.6. *If $\mathbf{F}$ is a functor which is represented by R, there is a natural bijection*
$$\text{Hom}_\mathsf{k}(\mathfrak{m}_R/(\mathfrak{m}_R^2, \mathfrak{m}_\Lambda), \mathsf{k}) \cong \text{Hom}_\Lambda(R, \mathsf{k}[\varepsilon]),$$
where Hom_k *means* k-*vector space homomorphisms and* Hom_Λ *means homomorphisms of coefficient Λ-algebras.*

The basic point is that a homomorphism of coefficient Λ-algebras
$$R \longrightarrow \mathsf{k}[\varepsilon],$$
because it must induce the identity on residue fields, must have the form
$$r \mapsto \bar{r} + \varphi(r)\varepsilon,$$

where $\bar{r} = r \pmod{\mathfrak{m}}$ denotes the image of r in the residue field k, and $\varphi(r) \in$ k. Furthermore, since the map must be a homomorphism of Λ-algebras, φ is completely determined by its values on elements $r \in \mathfrak{m}_R$. Working out what φ must look like yields the Lemma.

Problem 2.27. Fill in the details to give a proof of the Lemma.

We have shown, then, that if a functor **F** is represented by a coefficient Λ-algebra R, then $\mathbf{F}(\mathsf{k}[\varepsilon]) = t_R$, at least as sets. To make this really work, we have to explain how to think of $\mathbf{F}(\mathsf{k}[\varepsilon])$, which a priori is just a set, as a k-vector space. Of course, we want to do that in such a way as to make the bijection in the Lemma be k-linear, and therefore an isomorphism of k-vector spaces.

It turns out that there is a natural vector space structure on $\mathbf{F}(\mathsf{k}[\varepsilon])$ that arises simply from the fact that **F** is a reasonably nice functor. One part of this is easy: an element α of k gives an automorphism of $\mathsf{k}[\varepsilon]$ by

$$a + b\varepsilon \mapsto a + \alpha b\varepsilon,$$

(yes, this is actually a ring homomorphism!) and therefore, by functoriality, gives an automorphism of $\mathbf{F}(\mathsf{k}[\varepsilon])$. This gives a scalar multiplication by k.

The addition is a bit harder. Since we are assuming that **F** is representable, we know it has the Mayer-Vietoris property. We apply it to the diagram

where both arrows are the canonical projection onto the residue field. Since there is only one coefficient Λ-algebra homomorphism $R \longrightarrow$ k, $\mathbf{F}(\mathsf{k})$ consists of only one element, and therefore the fiber product $\mathbf{F}(\mathsf{k}[\varepsilon]) \times_{\mathbf{F}(\mathsf{k})} \mathbf{F}(\mathsf{k}[\varepsilon])$ is just a product. Hence the Mayer-Vietoris property says, in this situation, that

$$\mathbf{F}\left(\mathsf{k}[\varepsilon] \times_{\mathsf{k}} \mathsf{k}[\varepsilon]\right) \cong \mathbf{F}(\mathsf{k}[\varepsilon]) \times \mathbf{F}(\mathsf{k}[\varepsilon]).$$

Now, we have a homomorphism of coefficient Λ-algebras

$$\mathfrak{p} : \mathsf{k}[\varepsilon] \times_{\mathsf{k}} \mathsf{k}[\varepsilon] \longrightarrow \mathsf{k}[\varepsilon]$$

defined by

$$\mathfrak{p}(x + y_1\varepsilon, x + y_2\varepsilon) = x + (y_1 + y_2)\varepsilon.$$

(The notation $\mathfrak{p}$ is meant to recall "plus," or perhaps the abbreviation of "piu" used by the early Italian algebraists.) Then we put all this together to define the addition: the composition

$$\mathbf{F}(\mathsf{k}[\varepsilon]) \times \mathbf{F}(\mathsf{k}[\varepsilon]) \cong \mathbf{F}\left(\mathsf{k}[\varepsilon] \times_{\mathsf{k}} \mathsf{k}[\varepsilon]\right) \xrightarrow{\mathbf{F}(\mathfrak{p})} \mathbf{F}(\mathsf{k}[\varepsilon])$$

gives the vector addition.

Problem 2.28. Check everything! In particular, check that if **F** is represented by R, then

   i. $\mathfrak{p}$ is indeed a homomorphism of coefficient Λ-algebras,
   ii. these two operations do make **F** a vector space over k,

iii. with these definitions the natural bijection in Lemma 2.6 is in fact an isomorphism of k-vector spaces.

The reason to go through this effort of translation is the following: we can now try to define the "tangent space of a functor" by following this template. Of course, we did need to use the Mayer-Vietoris property, but we used it only for the specific diagram above. So we have in fact proved the following:

Proposition 2.7. *Let*
$$\mathbf{F} : \mathcal{C}_\Lambda^0 \rightsquigarrow \underline{Sets}$$
be a (covariant) functor such that $\mathbf{F}(\mathsf{k})$ *consists of a single element. Suppose that the natural map*
$$\mathbf{F}(\mathsf{k}[\varepsilon] \times_\mathsf{k} \mathsf{k}[\varepsilon]) \longrightarrow \mathbf{F}(\mathsf{k}[\varepsilon]) \times \mathbf{F}(\mathsf{k}[\varepsilon])$$
is a bijection. Then $\mathbf{F}(\mathsf{k}[\varepsilon])$ *has a natural vector space structure over* k.

Problem 2.29. In the discussion above, we were assuming that $\mathbf{F}$ was represented by R. So to prove the proposition we need to check that the only properties of $\mathbf{F}$ that we really used are the ones listed in the statement of the proposition. Do that.

We will refer to the assumption that the map
$$\mathbf{F}(\mathsf{k}[\varepsilon] \times_\mathsf{k} \mathsf{k}[\varepsilon]) \longrightarrow \mathbf{F}(\mathsf{k}[\varepsilon]) \times \mathbf{F}(\mathsf{k}[\varepsilon])$$
is a bijection as the *tangent space hypothesis over* k. When it is satisfied, we will write
$$t_\mathbf{F} = \mathbf{F}(\mathsf{k}[\varepsilon])$$
and call this the *tangent space of the functor* $\mathbf{F}$.

In [**101**], Mazur suggests that we say the functor $\mathbf{F}$ is *nearly representable* if it satisfies the tangent space hypothesis and the tangent space $t_\mathbf{F}$ is finite-dimensional over k. See Mazur's article for further discussion, and also for a discussion of how to adapt this to the "relative" case in which the category is $\mathcal{C}_\Lambda(A)$.

Finally, we note in passing another interpretation of the tangent space:

Problem 2.30. Show that t_R is naturally isomorphic to $\mathrm{Der}_\Lambda(R, \mathsf{k})$, the k-vector space of Λ-algebra derivations from R to k. (This is another reason to think of t_R as the tangent space.)

Complements to lecture 2

The language of categories and functors is a particularly convenient way to think about the deformation theory. Basically, category theory tries to make precise the idea that in a mathematical "universe of discourse" there is typically a collection of objects which we study (e.g., sets, groups, rings, topological spaces, complete noetherian local rings with residue field k, etc.), and for each such collection of objects we have a "correct" notion of function between our objects (e.g., for the list above, they would be: functions in general, group homomorphisms, ring homomorphisms, continuous functions, local homomorphisms inducing the identity on residue fields). Such a "universe of discourse" is called a *category*.

Functors connect different categories, transforming the objects of one to objects of the other and doing the same to the functions, while preserving some obvious structure (the identity function and compositions of functions). We could even speak of a "category of categories," in which the appropriate functions would be the functors. The functors we are interested in are set-valued, that is, they associate a set to each coefficient rings. Mathematics is full of functors, but the most famous ones are certainly the various functors that attach algebraic objects to various geometric objects (homology and cohomology, etc.).

Making this impressionistic description precise is the business of category theory, of which we need only a small amount. The basic notions are discussed in most algebra textbooks; for example, see [**90**, I, §11]. For more information, a good reference is MacLane's [**92**], which contains much more material than we have used (or will use).

LECTURE 3
The Universal Deformation: Existence

Our goal for this lecture is to prove that, under suitable hypotheses, the deformation functor is indeed (pro-)representable. Our proof will be very similar to Mazur's original proof (in [97]; see also [8], [10], [9], [115]), which is based on Schlessinger's criteria for pro-representability of a functor on a category of artinian rings first given in [125].

Since the publication of [97], several other approaches to proving the representability of the deformation functor (or, equivalently, the existence of a universal deformation) have been found. One is a "direct" approach that constructs the universal deformation ring by generators and relations from what is known about the Galois group. Constructions in this style have been given by Faltings (see [34]) and by Lenstra and de Smit (see [36]). We will see a little bit of this point of view when we discuss "explicit" deformations.

Another approach is based on the notion of a "pseudo-character," which is basically a function that "looks like" the character of a representation. This has been studied by Nyssen [113] and Rouquier [121], who find conditions for a pseudo-character to be the trace of a representation and use them to construct the universal deformation.

As before, k will denote a finite field of characteristic p and Π will denote a profinite group satisfying hypothesis Φ_p. We will assume we are given a residual representation

$$\overline{\rho} : \Pi \longrightarrow \mathrm{GL}_n(\mathsf{k})$$

whose lifts we want to understand.

We let $\mathcal{C}$ stand for the category whose objects are complete noetherian local rings with residue field k and whose morphisms are local homomorphisms which induce the identity on k. We use the shorthand expression "coefficient ring" for an object of $\mathcal{C}$. We let $\mathcal{C}^0$ be the full subcategory of $\mathcal{C}$ whose objects are artinian coefficient rings. If Λ is an object of $\mathcal{C}$, we write $\mathcal{C}_\Lambda$ for the category whose objects are complete noetherian local Λ-algebras with residue field k and whose morphisms are local Λ-algebra homomorphisms which induce the identity on residue fields. We use the shorthand expression "coefficient Λ-algebra" for an object of $\mathcal{C}_\Lambda$. Finally, we write $\mathcal{C}_\Lambda^0$ for the full subcategory of artinian local Λ-algebras with residue field k.

If R is a coefficient ring, we write $\Gamma_n(R)$ for the kernel of the map $\mathrm{GL}_n(R) \longrightarrow \mathrm{GL}_n(k)$ given by reduction modulo the maximal ideal, so that $\Gamma_n(R) = 1 + \mathrm{M}_n(\mathfrak{m})$, i.e., it consists of matrices whose off-diagonal elements are in $\mathfrak{m}$ and whose diagonal elements belong to $1 + \mathfrak{m}$.

Given the residual representation $\bar{\rho}$, we have defined set-valued covariant functors $\mathbf{D}$ and $\mathbf{D}_\Lambda$ whose value on a coefficient ring (resp, a coefficient Λ-algebra) R is the set of deformations of $\bar{\rho}$ to R. The functors depend on $\bar{\rho}$, of course, but since $\bar{\rho}$ will usually be fixed throughout we omit it from the notation; if it is necessary to emphasize the dependence on $\bar{\rho}$, we will write $\mathbf{D}_{\bar{\rho}}$ and $\mathbf{D}_{\bar{\rho},\Lambda}$, respectively.

Schlessinger's criteria

As we saw above, we can think of the deformation functor $\mathbf{D}$ (or $\mathbf{D}_\Lambda$) as a functor on the category $\mathcal{C}^0$ (or $\mathcal{C}^0_\Lambda$) of artinian coefficient rings (or Λ-algebras). From this point of view, our goal is to show that these functions are pro-representable. As we saw from Grothendieck's theorem, pro-representability is closely related to the Mayer-Vietoris property, that is, to how our functors act on fiber products. In [125], Schlessinger obtained a set of criteria for pro-representability of functors on categories of artinian rings which are much easier to apply. In this section, we recall Schlessinger's criteria in preparation for using them, in the next section, to prove that the deformation functors are pro-representable.

Let $\mathbf{F}$ be a covariant functor

$$\mathbf{F} : \mathcal{C}^0_\Lambda \rightsquigarrow \underline{\mathrm{Sets}},$$

and assume that $\mathbf{F}(k)$ consists of one element. We want to give sufficient conditions for $\mathbf{F}$ to be pro-representable by a ring $\mathcal{R}$ in $\mathcal{C}_\Lambda$.

In general, if R is an artinian coefficient Λ-algebra, we write $\mathfrak{m}_R$ for the maximal ideal in R. If R and S are two coefficient Λ-algebras, we say a homomorphism

$$\phi : R \longrightarrow S$$

is *small* if it is surjective and if $\mathrm{Ker}(\phi)$ is principal and is annihilated by $\mathfrak{m}_R$.

Problem 3.1. Show that any surjective homomorphism in $\mathcal{C}^0_\Lambda$ factors as the composition of small homomorphisms.

The prototypical example of a small homomorphism, which will be of great importance in what follows, is the homomorphism

$$k[\varepsilon] \longrightarrow k,$$

where $k[\varepsilon]$, as above, is the ring of dual numbers.

To set up the Schlessinger criteria, consider rings R_0, R_1, and R_2 in $\mathcal{C}^0_\Lambda$, and suppose we have morphisms

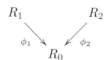

Let

$$R_3 = R_1 \times_{R_0} R_2 = \{(r_1, r_2) \in R_1 \times R_2 \mid \phi_1(r_1) = \phi_2(r_2)\}$$

be the fiber product of R_1 and R_2 over R_0, which is again an artinian coefficient Λ-algebra. Since **F** is a functor, we get a map

(∗) $$\mathbf{F}(R_3) \longrightarrow \mathbf{F}(R_1) \times_{\mathbf{F}(R_0)} \mathbf{F}(R_2)$$

If **F** is representable, we know (this is just the Mayer-Vietoris condition again) that the map (∗) is a bijection. We also know that if (∗) is a bijection in the case where $R_1 = R_2 = \mathsf{k}[\varepsilon]$ and $R_0 = \mathsf{k}$, then $\mathbf{F}(\mathsf{k}[\varepsilon])$ has a natural k-vector space structure.

Now we can state the Schlessinger conditions, which we label as **H1**, **H2**, **H3**, and **H4** (the "H" stands for hull; see below). They are basically a weakened form of the conditions in Grothendieck's theorem. The first two specify that the map (∗) should be nice when the map $R_2 \longrightarrow R_0$ is particularly simple.

H1: If the map $R_2 \longrightarrow R_0$ is small, then (∗) is surjective.

H2: If $R_0 = \mathsf{k}$ and $R_2 = \mathsf{k}[\varepsilon]$, then (∗) is bijective.

If **H2** holds, applying it to the case when $R_1 = R_2 = \mathsf{k}[\varepsilon]$ shows that the tangent space hypothesis over k is satisfied, and hence we can think of $t_\mathbf{F} = \mathbf{F}(\mathsf{k}[\varepsilon])$ as a k-vector space; as before, we call this the tangent space of **F**. Schlessinger's third condition is:

H3: The vector space $t_\mathbf{F} = \mathbf{F}(\mathsf{k}[\varepsilon])$ is finite-dimensional.

The fourth condition is another Mayer-Vietoris variant:

H4: If $R_1 = R_2$, the maps $R_i \longrightarrow R_0$ are the same, and $R_i \longrightarrow R_0$ is small, then (∗) is bijective.

Theorem 3.1 (Schlessinger). *Let **F** be a set-valued covariant functor on $\mathcal{C}_\Lambda^0$ such that **F**(k) has exactly one element. If **F** satisfies conditions **H1** to **H4**, then **F** is pro-representable. In particular, there exists an object $\mathcal{R}$ of $\mathcal{C}_\Lambda$ such that $\mathbf{F}(A) = \mathrm{Hom}(\mathcal{R}, A)$ for every A in $\mathcal{C}_\Lambda^0$.*

Schlessinger's theorem is in fact more general: he shows that if **F** satisfies only conditions **H1** to **H3** then it has a "hull" which satisfies some of the properties one would expect the representing object to have. (See [**125**], [**101**], and the problems at the end of this section for more discussion of what this means.) This is the reason for the otherwise rather peculiar ordering of the four conditions.

It's also worth noting that since representable functors *do* satisfy all four conditions, the theorem gives necessary and sufficient conditions for representability. Similarly, **H1** to **H3** are necessary and sufficient conditions for the existence of a hull.

Let $d = \dim_\mathsf{k} t_\mathbf{F}$. The proof constructs $\mathcal{R}$ as an inverse limit of quotients of $\Lambda[[X_1, X_2, \ldots, X_d]]$. See [**125**] for the details. It is probably worth pointing out that the proof does not give much information about the resulting ring beyond the fact that it is a quotient of $\Lambda[[X_1, X_2, \ldots, X_d]]$.

We'll often want to apply Schlessinger's criteria to a subfunctor of a functor which we already know is representable. It turns out to be quite easy to do this. We say a set-valued functor $\mathbf{F}_1$ on $\mathcal{C}_\Lambda^0$ is a *subfunctor* of **F** if, for every coefficient Λ-algebra R, we have $\mathbf{F}_1(R) \subset \mathbf{F}(R)$. (If we want to be more precise, we'd have to say that there exists a natural transformation $\mathbf{F}_1 \longrightarrow \mathbf{F}$ which induces the inclusions $\mathbf{F}_1(R) \subset \mathbf{F}(R)$ for every R.)

Proposition 3.2. *Let $\mathbf{F}_1$ be a subfunctor of $\mathbf{F}$ such that $\mathbf{F}_1(k) = \mathbf{F}(k)$ is a singleton, and suppose $\mathbf{F}$ is pro-representable, and so satisfies conditions **H1** to **H4**. If $\mathbf{F}_1$ satisfies condition **H1**, then $\mathbf{F}_1$ satisfies the other three conditions, and therefore is also pro-representable.*

Problem 3.2. Prove the proposition. (The main point is that the restriction of an injective homomorphism is automatically injective.)

Problem 3.3. Check that the proposition is also true if we replace "is pro-representable" by "has a hull" (equivalently, if we omit property **H4**) in both hypothesis and conclusion.

Problem 3.4. In the situation of the proposition, let $\mathcal{R}$ be the coefficient Λ-algebra that represents $\mathbf{F}$. Prove that the object representing $\mathbf{F}_1$ is a quotient of $\mathcal{R}$.

An alternative approach is suggested by Mazur in [**101**], which does not depend on knowing that $\mathbf{F}$ is pro-representable. We need one bit of language first: in any category where fiber products exist, let's say that a diagram

is *Cartesian* if the induced map $D \longrightarrow A \times_C B$ is an isomorphism. Now suppose $\mathbf{F}_1$ is a subfunctor of a covariant set-valued functor on $\mathcal{C}_\Lambda^0$, and suppose that $\mathbf{F}_1(k) = \mathbf{F}(k)$. Given a diagram in $\mathcal{C}_\Lambda^0$

we consider the commutative square

$$\begin{array}{ccc} \mathbf{F}_1(A \times_C B) & \longrightarrow & \mathbf{F}_1(A) \times_{\mathbf{F}_1(C)} \mathbf{F}_1(B) \\ \downarrow & & \downarrow \\ \mathbf{F}(A \times_C B) & \longrightarrow & \mathbf{F}(A) \times_{\mathbf{F}(C)} \mathbf{F}(B) \end{array}$$

in which the vertical arrows are inclusions and the horizontal arrows are obtained as in our previous discussion. If every such diagram is Cartesian, Mazur says that $\mathbf{F}_1 \subset \mathbf{F}$ is *relatively representable*.

Problem 3.5. Show that if $\mathbf{F}_1 \subset \mathbf{F}$ is relatively representable, then, for each i, $\mathbf{F}_1$ satisfies **H**i if $\mathbf{F}$ does, and similarly for the tangent space hypothesis.

The point, then, is that Mazur's condition allows one to transfer each property separately, while our previous theorem only applies when we already know that $\mathbf{F}$ is pro-representable. Nevertheless, in most of the cases with which we will be working the deformation functor will indeed be pro-representable, so that Proposition 3.2 is good enough.

Universal Deformations exist

We now apply Schlessinger's theorem to the deformation functor

$$\mathbf{D}_\Lambda : \mathcal{C}_\Lambda \rightsquigarrow \underline{\text{Sets}}$$

given by

$$\mathbf{D}_\Lambda(R) = \{\text{deformations of } \overline{\rho} \text{ to } R\}.$$

As we will see, the first three conditions will always hold, but the fourth will depend on what $\overline{\rho}$ is.

Definition 3.1. Let $\overline{\rho}$ be a residual representation. We let

$$C(\overline{\rho}) = \text{Hom}_\Pi(\mathsf{k}^n, \mathsf{k}^n) = \{P \in \text{M}_n(\mathsf{k}) \mid P\overline{\rho}(g) = \overline{\rho}(g)P \text{ for all } g \in \Pi\}.$$

As the definition suggests, we can think of k^n as a Π-module via $\overline{\rho}$, and then $C(\overline{\rho})$ is its ring of Π-module endomorphisms. More generally, if A is a coefficient Λ-algebra and ρ is a deformation of $\overline{\rho}$ to A, we can use ρ to make A^n a Π-module, and then make the analogous definition:

Definition 3.2. Let $\overline{\rho}$ be a residual representation, and let ρ be a deformation of $\overline{\rho}$ to a coefficient Λ-algebra A. We define

$$C_A(\rho) = \text{Hom}_\Pi(A^n, A^n) = \{P \in \text{M}_n(A) \mid P\rho(g) = \rho(g)P \text{ for all } g \in \Pi\}.$$

In particular $C(\overline{\rho}) = C_\mathsf{k}(\overline{\rho})$. We will be especially interested in the case where $C(\overline{\rho}) = \mathsf{k}$, that is, where the only matrices in $\text{M}_n(\mathsf{k})$ that commute with the image of $\overline{\rho}$ are the scalar matrices.

Theorem 3.3 (Mazur, Ramakrishna). *Suppose Π is a profinite group that satisfies property Φ_p, $\overline{\rho} : \Pi \longrightarrow \text{GL}_n(\mathsf{k})$ is a continuous representation, and Λ is a complete noetherian ring with residue field k. Then the deformation functor $\mathbf{D}_\Lambda$ always satisfies properties **H1**, **H2**, and **H3**. Furthermore, if $C(\overline{\rho}) = \mathsf{k}$, then $\mathbf{D}_\Lambda$ also satisfies property **H4**.*

Mazur essentially proved this theorem in [**97**], except that he showed property **H4** under the assumption that $\overline{\rho}$ is absolutely irreducible. Ramakrishna pointed out in [**115**] that the hypothesis could be weakened as above. Note that by Schur's Lemma (see below) we do know that $C(\overline{\rho}) = \mathsf{k}$ when $\overline{\rho}$ is absolutely irreducible.

We will prove the theorem by a series of lemmas. We fix the following notation throughout.

Let R_0, R_1 and R_2 be artinian coefficient Λ-algebras, and suppose we are given $\phi_1 : R_1 \longrightarrow R_0$ and $\phi_2 : R_2 \longrightarrow R_0$ as above. Let

$$E_i = \text{Hom}_{\overline{\rho}}(\Pi, \text{GL}_n(R_i))$$

be the set of homomorphisms from Π to $\text{GL}_n(R_i)$ which reduce to $\overline{\rho}$ modulo the maximal ideal. Then $\Gamma_n(R_i)$ acts on E_i by conjugation and (because deformations are strict equivalence classes of homomorphisms) we have

$$\mathbf{D}_\Lambda(R_i) = E_i / \Gamma_n(R_i).$$

The only difficulties in the proof arise in passing to the quotient from E_i to $\mathbf{D}_\Lambda(R_i)$.

The map in $(*)$ is

$$b : E_3/\Gamma_n(R_3) \longrightarrow E_1/\Gamma_n(R_1) \times_{E_0/\Gamma_n(R_0)} E_2/\Gamma_n(R_2).$$

If $R_2 \longrightarrow R_0$ is surjective, then $\Gamma_n(R_2) \longrightarrow \Gamma_n(R_0)$ is also surjective (see problem 3.24).

Lemma 3.4. *Property **H1** is true.*

Proof. Suppose $R_2 \longrightarrow R_0$ is small (in fact, all we need to know is that it is surjective). We want to show that if we have a pair (ρ_1, ρ_2) of deformations to R_1 and R_2 which induce the same deformation to R_0, we can paste them together to get a deformation to R_3. This is clear for homomorphisms (i.e., for elements of E_i), so we need to check that we can pick representatives for the strict equivalence classes so that they "match" when projected down to R_0.

To do this, we pick any two representatives ϕ_1 and ϕ_2. The assumption is that their images in R_0 are strictly equivalent, so that there is an element $\overline{M}$ of $\Gamma_n(R_0)$ such that conjugating the image of ϕ_2 by $\overline{M}$ gives the image of ϕ_1. Since $R_2 \longrightarrow R_0$ is surjective, so is $\Gamma_n(R_2) \longrightarrow \Gamma_n(R_0)$; hence, we can lift $\overline{M}$ to $M \in \Gamma_n(R_2)$. Then ϕ_1 and $M^{-1}\phi_2 M$ are group homomorphisms that have the same image in $\mathrm{GL}_n(R_0)$, and hence they define a homomorphism $\phi_3 \in E_3$. The strict equivalence class of ϕ_3 maps to (ρ_1, ρ_2), and so we have proved that $(*)$ is surjective. □

This settles **H1**, but we still need to consider when it is that b is injective. For this, we call on two lemmas. Let $\phi_2 \in E_2$ and let $\phi_0 \in E_0$ be its image. Set

$$G_i(\phi_i) = \{g \in \Gamma_n(R_i) \mid g \text{ commutes with the image of } \phi_i \text{ in } \mathrm{GL}_n(R_i)\}.$$

(Note that this is similar but not identical to $C_{R_i}(\phi_i)$ defined above. In particular, this is a subgroup of $\Gamma_n(R_i)$ while the other is a ring.) The first result is:

Lemma 3.5. *If for all $\phi_2 \in E_2$ the map*

$$G_2(\phi_2) \longrightarrow G_0(\phi_0)$$

is surjective, then the map b is injective.

Proof. Suppose ϕ and ψ are elements of E_3 that induce elements ϕ_i and ψ_i in E_i for each $i = 0, 1, 2$. Saying that ϕ and ψ have the same image under $(*)$ means that for each $i = 1, 2$ there is an $M_i \in \Gamma_n(R_i)$ such that $\phi_i = M_i^{-1} \psi_i M_i$. Mapping down to E_0 we see that

$$\phi_0 = \overline{M}_1^{-1} \psi_0 \overline{M}_1 = \overline{M}_2^{-1} \psi_0 \overline{M}_2,$$

and so that $\overline{M}_2 \overline{M}_1^{-1}$ commutes with the image of ϕ_0, i.e., $\overline{M}_2 \overline{M}_1^{-1} \in G_0(\phi_0)$.

Now use the surjectivity assumption to find $N \in G_2(\phi_2)$ which maps to $\overline{M}_2 \overline{M}_1^{-1}$. Let $N_2 = N^{-1} M_2$. Then we have

$$N_2^{-1} \psi_2 N_2 = M_2^{-1} N \psi_2 N^{-1} M_2 = M_2^{-1} \psi_2 M_2 = \phi_2.$$

On the other hand, the image of N_2 in $\Gamma_n(R_0)$ is

$$\overline{N}_2 = (\overline{M}_2 \overline{M}_1^{-1})^{-1} \overline{M}_2 = \overline{M}_1.$$

Since M_1 and N_2 have the same image in $\Gamma_n(R_0)$, the pair (M_1, N_2) defines an element $M \in \Gamma_n(R_3)$ and we have $M^{-1} \psi M = \phi$. Thus, ϕ and ψ are strictly equivalent, and we are done. □

Lemma 3.6. *Property **H2** is true.*

Proof. If $R_0 = \mathsf{k}$ and $R_2 = \mathsf{k}[\varepsilon]$, then we already know $(*)$ is surjective by **H1**. Injectivity will follow if we know that the map $G_2(\phi_2) \longrightarrow G_0(\phi_0)$ is always surjective. But when $R_0 = \mathsf{k}$, $G_0 = \Gamma_n(R_0)$ consists only of the identity matrix, and $G_0(\phi_0)$, which is a subgroup, is again just the identity. So the surjectivity holds (trivially) and we are done. □

Lemma 3.7. *Property **H3** is true.*

Proof. Let $\Pi_0 = \operatorname{Ker}\bar{\rho}$ and let ρ be a lift of $\bar{\rho}$ to $\mathsf{k}[\varepsilon]$. If $x \in \Pi_0$, we have $\bar{\rho}(x) = 1$, and hence $\rho(x) \in \Gamma_n(\mathsf{k}[\varepsilon])$. Hence, ρ determines a map from $\Pi_0 = \operatorname{Ker}\bar{\rho}$ to $\Gamma_n(\mathsf{k}[\varepsilon])$. Two lifts that determine the same map must be identical. Now, Π_0 is an open subgroup of Π and by problem 3.23, we know $\Gamma_n(\mathsf{k}[\varepsilon])$ is a (finite) p-elementary abelian group. By property Φ_p, there are finitely many maps $\Pi_0 = \operatorname{Ker}\bar{\rho}$ to $\Gamma_n(\mathsf{k}[\varepsilon])$. Hence, $\mathbf{D}_\Lambda(\mathsf{k}[\varepsilon])$ is a finite set, and we're done. □

Problem 3.6. This proof relies, once again, on the fact that k is a *finite* field. Is it possible to modify it to cover the case of an infinite field of characteristic p?

Lemma 3.8. *If $C(\bar{\rho}) = \mathsf{k}$, then for any i the group $G_i(\phi_i) \subset R_i$, i.e., $G_i(\phi_i)$ consists of the scalar matrices in $\Gamma_n(R_i)$.*

Proof. We prove the stronger assertion that for any deformation ρ of $\bar{\rho}$ to an artinian coefficient ring A we have $C_A(\rho) = A$. Our argument follows the one given in [42].

Since the map $A \longrightarrow \mathsf{k}$ is surjective, it factors as a sequence of small extensions. Since we know that $C_\mathsf{k}(\bar{\rho}) = \mathsf{k}$, the lemma will follow, by induction, from the claim that if $C_B(\rho_B) = B$ and $A \longrightarrow B$ is small, then $C_A(\rho_A) = A$.

To prove this, take $c \in C_A(\rho_A)$. By our assumption, the image of c in $\mathrm{M}_n(B)$ is a scalar matrix. Suppose $c \mapsto \bar{r}$, where the scalar $\bar{r} \in B$ is the image of $r \in A$. Then we can write $c = r + tM$ where t is a generator of the kernel of $A \longrightarrow B$ and $M \in \mathrm{M}_n(A)$.

Now, c commutes with the image of ρ_A, so that we must have, for every $g \in \Pi$,
$$(r + tM)\rho_A(g) = \rho_A(g)(r + tM),$$
which, since scalars commute with everything, boils down to
$$M\rho_A(g) = \rho_A(g)M.$$
Reducing modulo the maximal ideal $\mathfrak{m}_A$ and using the fact that $C(\bar{\rho}) = \mathsf{k}$, we see that M must be of the form $M = s + M_1$, where $s \in A$ is a scalar and all the entries of M_1 belong to $\mathfrak{m}_A$. But, since $A \longrightarrow B$ is small, we have $t\mathfrak{m}_A = 0$, from which it follows that $M = r + ts$ is a scalar. □

Problem 3.7. Show that if $C(\bar{\rho}) = \mathsf{k}$ then two lifts ρ and ρ' of $\bar{\rho}$ to a coefficient Λ-algebra A are equivalent if and only if they are strictly equivalent.

Lemma 3.9. *Suppose $C(\bar{\rho}) = \mathsf{k}$. Then property **H4** is true.*

Proof. From the previous lemma, $G_i(\phi_i)$ consists only of scalars (of the form $1 + \mathfrak{m}_{R_i}$, in fact), and the lemma follows. □

The upshot is:

Theorem 3.10 (Mazur, Ramakrishna). *Suppose Π is a profinite group satisfying condition Φ_p and*
$$\bar{\rho} : \Pi \longrightarrow \mathrm{GL}_n(\mathsf{k})$$
is a continuous representation such that $C(\bar{\rho}) = \mathsf{k}$. Then there exists a ring $\mathcal{R} = \mathcal{R}(\Pi, \mathsf{k}, \bar{\rho})$ in $\mathcal{C}_\Lambda$ and a deformation $\boldsymbol{\rho}$ of $\bar{\rho}$ to $\mathcal{R}$,
$$\boldsymbol{\rho} : \Pi \longrightarrow \mathrm{GL}_n(\mathcal{R})$$

such that any deformation of $\bar{\rho}$ to a coefficient Λ-algebra A is obtained from $\boldsymbol{\rho}$ via a unique morphism $\mathcal{R} \longrightarrow A$.

We call $\mathcal{R}$ the *universal deformation ring* and $\boldsymbol{\rho}$ the *universal deformation* of $\bar{\rho}$. The ring $\mathcal{R}(\Pi, \mathsf{k}, \bar{\rho})$ is unique in the following strong sense.

Theorem 3.11 (Mazur). *Suppose*
$$\bar{\rho} : \Pi \longrightarrow \mathrm{GL}_n(\mathsf{k})$$
is a continuous representation such that $C(\bar{\rho}) = \mathsf{k}$. If $\bar{\rho}'$ is a representation equivalent to $\bar{\rho} \otimes \chi$, where χ is a representation of dimension one, then there is a canonical isomorphism
$$r(\bar{\rho}', \bar{\rho}) : \mathcal{R}(\Pi, \mathsf{k}, \bar{\rho}) \longrightarrow \mathcal{R}(\Pi, \mathsf{k}, \bar{\rho}')$$
mapping the universal deformation of $\bar{\rho}$ to the universal deformation of $\bar{\rho}'$. This system of canonical isomorphisms satisfies the natural compatibility conditions.

Proof. This is basically immediate from the definition. See the next lecture for some hints about what is involved, and [**97**] for the details. □

Absolutely irreducible representations

Given the important role of the hypothesis that $C(\bar{\rho}) = \mathsf{k}$ in the theorem, it's important to ask which representations have this property. The most important part of the answer is *Schur's Lemma*, which says absolutely irreducible representations satisfy the condition $C(\bar{\rho}) = \mathsf{k}$.

Definition 3.3. A representation $\bar{\rho} : \Pi \longrightarrow \mathrm{GL}_n(\mathsf{k})$ is called *reducible* if the representation space k^n (with the Π-action given by $\bar{\rho}$) has a proper subspace that is invariant under the action of Π. It is called *irreducible* if no such subspace exists. Finally, we say that $\bar{\rho}$ is *absolutely irreducible* if there is no extension k'/k such that $\bar{\rho} \otimes_{\mathsf{k}} \mathsf{k}'$ is reducible.

The idea of "absolute" irreducibility is just this: it's perfectly possible for a representation to have no invariant subspaces as given, but for the subspaces to appear once we move to a larger field. For example, suppose we send a cyclic group of order 4 to $\mathrm{GL}_2(\mathbb{R})$, representing the generator by a ninety degree rotation. Then this representation has no fixed lines (i.e., no real eigenvalues). But if we base-change to $\mathbb{C}$, two fixed lines will appear. Hence our representation was irreducible but not absolutely irreducible.

Problem 3.8. Let $\bar{\mathsf{k}}$ be an algebraic closure of k. Show that $\bar{\rho}$ is absolutely irreducible if and only if $\bar{\rho} \otimes_{\mathsf{k}} \bar{\mathsf{k}}$ is irreducible.

The main reason we like absolutely irreducible representations is the following result:

Theorem 3.12 (Schur's Lemma). *If $\bar{\rho} : \Pi \longrightarrow \mathrm{GL}_n(\mathsf{k})$ is absolutely irreducible, then $C(\bar{\rho}) = \mathsf{k}$.*

The proof can be found in any standard text on group representation theory. For example, see page 7 of [**55**].

Hence, absolutely irreducible representations have universal deformations. Of course, there are other representations that satisfy $C(\bar{\rho}) = \mathsf{k}$. The following problems ring some changes on these ideas.

Problem 3.9. Find an example of a reducible representation which nevertheless satisfies the condition $C(\bar{\rho}) = k$.

Problem 3.10. Show that if $\bar{\rho} : \Pi \longrightarrow \mathrm{GL}_2(k)$ is irreducible and its image contains an element of exact order 2 and determinant -1, then $C(\bar{\rho}) = k$. (The case of characteristic 2 has to be considered separately.)

Problem 3.11. Show that $\bar{\rho} : \Pi \longrightarrow \mathrm{GL}_2(k)$ is absolutely irreducible if and only if it is irreducible and $C(\bar{\rho}) = k$. (The same is true for representations to $\mathrm{GL}_n(k)$, but it's slightly harder to prove.)

Problem 3.12. We can extend a representation $\bar{\rho} : \Pi \longrightarrow \mathrm{GL}_n(k)$ to a continuous homomorphism of k-algebras

$$f_{\bar{\rho}} : k[[\Pi]] \longrightarrow \mathrm{M}_n(k).$$

Show that $\bar{\rho}$ is absolutely irreducible if and only if $f_{\bar{\rho}}$ is onto. (Note that Schur's Lemma follows from this result.)

Example: the case $n = 1$

As we will see in the next section, it can often be quite difficult to get concrete information about the universal deformation ring. In this section, we look at one example in which it is possible to get a complete description.

Let's consider, then, what happens when $n = 1$. If we consider the uniqueness statement above, we see that the deformation ring in the one-dimensional case does not depend (up to canonical isomorphism) on the representation at all (of course, the universal deformation *does* depend on the representation). We will confirm that by constructing the ring directly.

So let's start with a one-dimensional representation, that is, a character

$$\bar{\rho} = \bar{\chi} : \Pi \longrightarrow k^\times = \mathrm{GL}_1(k),$$

and try to describe the deformation ring explicitly. The basic idea is as follows: there is a canonical ("Teichmüller") lift of $\bar{\chi}$ to the ring of Witt vectors $W(k)$, and thence to Λ (which is a $W(k)$-algebra). Once that lift is given, the "rest" of any lift must differ from this one by a homomorphism with values in an abelian pro-p-group, and these we can describe in a somewhat explicit manner.

So let $\Gamma = \Pi^{\mathrm{ab},(p)}$ be the abelianization of the pro-p-completion (see Lecture 2 for the definition) of Π, and let $\gamma : \Pi \longrightarrow \Gamma$ be the canonical projection. Note that any map from Π to an abelian pro-p-group must factor through Γ.

Problem 3.13. Prove that any homomorphism from Π to an abelian pro-p-group must factor through the abelianization of its pro-p-completion. (Easy, but helps you remember the definitions.)

Problem 3.14. Is there a difference between the abelianization of the pro-p-completion and the pro-p-completion of the abelianization?

Problem 3.15. Let $\Pi = G_{\mathbb{Q},S}$. Describe Γ. (Use the description of the abelianization of Π in Lecture 1. You'll need to distinguish the cases $p \in S$ and $p \notin S$.)

Problem 3.16. Let $\Pi = G_{\mathbb{Q}_\ell}$. Describe Γ. (Use the description of the abelianization of Π in Lecture 1. You'll need to distinguish the cases $\ell = p$ and $\ell \neq p$.)

As usual, we will let Λ be a coefficient ring and work in the category $\mathcal{C}_\Lambda$ of coefficient Λ-algebras. (If we want to work with all coefficient rings we just take

$\Lambda = W(\mathsf{k})$.) Let $\Lambda[[\Gamma]]$ be the completed group ring over Λ, that is,
$$\Lambda[[\Gamma]] = \varprojlim_H \Lambda[\Gamma/H]$$
where H ranges through the open normal subgroups of Γ and $\Lambda[\Gamma/H]$ is the usual group ring of the (finite) group Γ/H over Λ. If $u \in \Gamma$, we write $[u]$ for the corresponding element in the group ring.

Problem 3.17. Show that $\Lambda[[\Gamma]]$ is a coefficient Λ-algebra, i.e., an object of $\mathcal{C}_\Lambda$.

Since Λ is a complete noetherian ring, and hence is henselian, the units of Λ split (canonically) into a product:
$$\Lambda^\times \cong \mathsf{k}^\times \times (1 + \mathfrak{m}_\Lambda).$$
Using this splitting we get a canonical lift of $\overline{\chi}$ to Λ, which we will call $\chi_0 : \Pi \longrightarrow \Lambda$.

Now we can state the theorem. Recall that $\gamma : \Pi \longrightarrow \Gamma$ is the canonical projection. Then we have:

Proposition 3.13. *The universal deformation ring for a character $\overline{\chi} : \Pi \longrightarrow \mathsf{k}^\times$ is $\mathcal{R}(\Pi, \mathsf{k}, \overline{\chi}) = \Lambda[[\Gamma]]$ and the universal deformation is given by*
$$\boldsymbol{\chi}(x) = \chi_0(x)[\gamma(x)].$$

Proof. First of all, we know, by hypothesis Φ_p, that Γ is finitely generated as a $\mathbb{Z}_p$-module. If r is the number of generators, then we know that $\Lambda[[\Gamma]]$ is a quotient of the power series ring $\Lambda[[X_1, X_2, \ldots, X_r]]$, and hence is a coefficient Λ-algebra. It's also clear that $\boldsymbol{\chi}$ is a character. It remains to show, then, that this is indeed the universal deformation.

Consider, then, a lift $\chi : \Pi \longrightarrow A^\times$ to some coefficient Λ-algebra A. Let $\psi = \chi_0^{-1}\chi$. Then it's clear that ψ is a character of Π taking values in $1 + \mathfrak{m}_A$. Since $1 + \mathfrak{m}_A$ is an abelian pro-p-group, ψ must factor through the homomorphism $\gamma : \Pi \longrightarrow \Gamma$; this defines a map $f_\chi : \Gamma \longrightarrow 1 + \mathfrak{m}_A$ which extends to a homomorphism of Λ-algebras $f_\chi : \Lambda[[\Gamma]] \longrightarrow A$. We then have $\chi = f_\chi \circ \boldsymbol{\chi}$. Thus, $\Lambda[[\Gamma]]$ is the universal deformation ring and $\boldsymbol{\chi}$ is the universal deformation of χ. $\square$

Note that, as we pointed out above, $\Lambda[[\Gamma]]$ is independent of χ.

Problem 3.18. Check that $1 + \mathfrak{m}_A$ is indeed a pro-p-group.

Problem 3.19. Given two characters $\overline{\chi}_1, \overline{\chi}_2 : \Pi \longrightarrow \mathsf{k}^\times$, Theorem 3.11 says that there must be an isomorphism
$$r(\overline{\chi}_1, \overline{\chi}_2) : \Lambda[[\Gamma]] \longrightarrow \Lambda[[\Gamma]].$$
Can you describe this isomorphism?

Here's an interesting consequence of this calculation. Suppose we have a residual representation
$$\overline{\rho} : \Pi \longrightarrow \mathrm{GL}_n(\mathsf{k}).$$
Then we can look at its determinant, $\det \overline{\rho}$, which is a one-dimensional representation. If ρ is a deformation of $\overline{\rho}$ to a ring R, then clearly $\det \rho$ is a deformation of $\det \overline{\rho}$ to R. In particular, it follows that $\det \boldsymbol{\rho}$ is a deformation of $\det \overline{\rho}$ to the universal ring $\mathcal{R}(\Pi, \mathsf{k}, \overline{\rho})$. By universality, it follows that there is a map
$$\Lambda[[\Gamma]] \longrightarrow \mathcal{R} = \mathcal{R}(\Pi, \mathsf{k}, \overline{\rho}).$$
This homomorphism, which we will call the determinant homomorphism, makes $\mathcal{R}$ a $\Lambda[[\Gamma]]$-algebra. This extra structure is sometimes important.

Problem 3.20. Let $K = \mathbb{Q}$, p be an odd prime, $S = \{p\}$, $\Lambda = \mathbb{Z}_p$, and $\Pi = G_{\mathbb{Q},S}$. Show that $\Gamma \cong 1 + p\mathbb{Z}_p$ and therefore that the deformation ring $\mathbb{Z}_p[[\Gamma]]$ is the usual "Iwasawa algebra."

Extra Problems

This section collects a few extra problems related to the material in this lecture. First, a case where the deformation theory is easy to compute.

Problem 3.21. Suppose Π is a *finite* group of order not divisible by p, and suppose that $\bar{\rho} : \Pi \longrightarrow \mathrm{GL}_n(\mathsf{k})$ is an inclusion. Show that there exists a lift of $\bar{\rho}$ to $\mathrm{GL}_n(W(\mathsf{k}))$ (and hence, via the canonical map $W(\mathsf{k}) \longrightarrow \Lambda$, to $\mathrm{GL}_n(\Lambda)$). Show that every deformation of $\bar{\rho}$ factors through this lift, and hence that the universal deformation ring is $\mathcal{R}_{\bar{\rho}} = \Lambda$.

For other subgroups of $\mathrm{GL}_n(\mathsf{k})$, the problem is much harder. For example:

Problem 3.22. Suppose $\Pi = \mathrm{SL}_n(\mathsf{k}) \subset \mathrm{GL}_n(\mathsf{k})$ and $\bar{\rho}$ is the inclusion. Can you determine the universal deformation ring?

The next two problems have to do with the group $\Gamma_n(R)$, i.e., the kernel of the reduction map from $\mathrm{GL}_n(R)$ to $\mathrm{GL}_n(\mathsf{k})$.

Problem 3.23. Show that $\Gamma_n(\mathsf{k}[\varepsilon])$ is a finite p-elementary abelian group.

Problem 3.24. Show that if $R \longrightarrow S$ is a surjective homomorphism in $\mathcal{C}_\Lambda^0$ then the induced homomorphism of groups

$$\Gamma_n(R) \longrightarrow \Gamma_n(S)$$

is also surjective.

Finally, a few problems related to the notion of a "pro-representable hull" (the definition appears below).

Problem 3.25. Suppose $\mathbf{F}$ and $\mathbf{G}$ are set-valued functors on $\mathcal{C}_\Lambda^0$ such that both $\mathbf{F}(\mathsf{k})$ and $\mathbf{G}(\mathsf{k})$ are singletons. We say that a morphism of functors $\xi : \mathbf{F} \longrightarrow \mathbf{G}$ is *smooth* if, given any surjective homomorphism $B \longrightarrow A$ of artinian coefficient Λ-algebras, any element $f \in \mathbf{F}(A)$ and any lifting of $\xi(f) \in \mathbf{G}(A)$ to an element $g \in \mathbf{G}(B)$, there exists a lifting $f' \in \mathbf{F}(B)$ of f such that $\xi(f') = g$.

  i. Show that this condition is equivalent to saying that for any surjective homomorphism $B \longrightarrow A$ of artinian Λ-algebras the natural mapping
  $$\mathbf{F}(B) \longrightarrow \mathbf{F}(A) \times_{\mathbf{G}(A)} \mathbf{G}(B)$$
  is surjective.
  ii. Show that we can replace $\mathcal{C}_\Lambda^0$ by $\mathcal{C}_\Lambda$ in the definition.
  iii. Show that if $\mathbf{F} \longrightarrow \mathbf{G}$ is smooth, the map $\mathbf{F}(B) \longrightarrow \mathbf{G}(B)$ is surjective for every coefficient Λ-algebra B.
  iv. Suppose $\mathbf{F}$ is represented by a coefficient Λ-algebra R and $\mathbf{G}$ is represented by a coefficient Λ-algebra S. Then a morphism $\mathbf{F} \longrightarrow \mathbf{G}$ corresponds to a homomorphism $S \longrightarrow R$. If $\mathbf{F} \longrightarrow \mathbf{G}$ is smooth, what does that tell you about the corresponding ring homomorphism?

If R is an object of $\mathcal{C}_\Lambda$, let $\mathbf{h}_R$ denote the functor $\mathrm{Hom}(R, \cdot)$ that maps each coefficient Λ-algebra A to the set $\mathrm{Hom}(R, A)$.

Problem 3.26. Suppose $\mathbf{F}$ is a functor on $\mathcal{C}_\Lambda^0$ such that $\mathbf{F}(\mathsf{k})$ is a singleton. Show that any element $\rho \in \mathbf{F}(R)$ induces a morphism of functors $\mathbf{h}_R \longrightarrow \mathbf{F}$.

Problem 3.27. Let R_1 and R_2 be coefficient Λ-algebras and let $\varphi : R_1 \longrightarrow R_2$ be a homomorphism. Show that the following statements are equivalent:
- The homomorphism $\varphi : R_1 \longrightarrow R_2$ is surjective.
- The corresponding morphism of functors $\mathbf{h}_{R_2} \longrightarrow \mathbf{h}_{R_2}$ is injective.

Problem 3.28. Suppose $\mathbf{F}$ is a functor on $\mathcal{C}^0_\Lambda$ such that $\mathbf{F}(k)$ is a singleton, R is a coefficient Λ-algebra, and $\rho \in \mathbf{F}(R)$. We say the pair (R, ρ) is a *pro-representable hull* for $\mathbf{F}$ if
- the map $\mathbf{h}_R \longrightarrow \mathbf{F}$ induced by ρ is smooth, and
- the induced map $t_R \longrightarrow t_\mathbf{F}$ of tangent spaces is an isomorphism.

Prove that any two pro-representable hulls for $\mathbf{F}$ are isomorphic.

Problem 3.29. In the situation of the previous problem, prove that if $\mathbf{F}$ is representable by $\mathcal{R}$, then $\mathcal{R}$ (together with the universal element $\rho \in \mathbf{F}(\mathcal{R})$ corresponding to the identity $\mathcal{R} \longrightarrow \mathcal{R}$) is a hull of $\mathbf{F}$. Are there other hulls of $\mathbf{F}$?

Problem 3.30. We mentioned above that since the deformation functor $\mathbf{D}_{\overline{\rho}}$ always satisfies conditions **H1** to **H3**, it has a hull $(\mathcal{R}, \rho)$. The representation ρ is sometimes called a *versal deformation* of $\overline{\rho}$. Translate the definition of a hull into deformation-theoretical terms: what properties does a versal deformation have?

Finally a meta-problem: in this and the following lectures, how much of the theory survives if all we have is a hull (or a "versal deformation")?

Complements to lecture 3

Mark Dickinson has given an alternative proof of the main theorem in this lecture that sidesteps the Schlessinger conditions, working instead with Grothendieck's theorem from the previous chapter. See Appendix 1 for Dickinson's proof.

LECTURE 4
The Universal Deformation: Properties

Now that we have proved that (under certain conditions) universal deformations exist, we want to find out more about them and about the universal deformation rings. This turns out, of course, to be quite difficult. In this lecture we will look at what can be said in general about the deformation ring and its properties. We will continue to use the notations we established above. In particular, Π is a profinite group satisfying the hypothesis Φ_p. More and more, however, we want to think of Π as being either $G_{K,S}$ for some number field K and some set of primes S including the archimedean primes or the absolute Galois group of a local field.

Functorial properties

The simplest properties of the universal deformation rings might be described as "functorial" properties. Basically, they are derived from various constructions involving group representations, together with the universality properties of the universal deformation. These properties are worked out in detail in [**97**], and we won't spend too much time on them. For this whole section, assume that $\bar{\rho}$ satisfies the condition $C(\bar{\rho}) = \mathsf{k}$, so that the deformation functor $\mathbf{D}_\Lambda$ is representable.

The kind of properties we want to consider here are those which arise simply from the fact that the functor is representable, together with the fact that GL_n itself has functorial properties (specifically, it is an affine group scheme of finite type over $\mathbb{Z}$). One example of this kind of property already appeared in the previous lecture, when we considered the determinant function

$$\det : \mathrm{GL}_n \longrightarrow \mathrm{GL}_1 \,.$$

Since the determinant is a homomorphism of affine group schemes over Λ, it sends deformations of $\bar{\rho}$ into deformations of $\det \bar{\rho}$. By the universal property, this gives a homomorphism of coefficient Λ-algebras from the completed group ring $\Lambda[[\Gamma]]$ (where $\Gamma = \Pi^{ab,(p)}$, as defined in the previous lecture) to the universal deformation ring $\mathcal{R}(\bar{\rho})$.

Another example of the same thing is as follows. Suppose we have two residual representations $\bar{\rho}$ and $\bar{\rho}'$ which are equivalent. Then there is a matrix $x \in$

$\mathrm{GL}_n(W(\mathsf{k}))$ (which we can think of as in $\mathrm{GL}_n(\Lambda)$) such that $\overline{\rho} = \overline{x}^{-1}\overline{\rho}'\overline{x}$. Conjugation by x is an isomorphism of group schemes over Λ,
$$\delta_x : \mathrm{GL}_n \longrightarrow \mathrm{GL}_n,$$
and it transforms deformations of $\overline{\rho}'$ into deformations of $\overline{\rho}$. Therefore, it determines an isomorphism of deformation rings
$$r(\delta_x) : \mathcal{R}(\overline{\rho}) \longrightarrow \mathcal{R}(\overline{\rho}').$$

Problem 4.1. Show that the isomorphism $r(\delta_x)$ is determined by the representations $\overline{\rho}$ and $\overline{\rho}'$. (In other words, it does not depend on our choice of the particular matrix $x \in \mathrm{GL}_n(W(\mathsf{k}))$.)

Since $r(\delta_x)$ depends only on the residual representations $\overline{\rho}$ and $\overline{\rho}'$, we can denote it by $r(\overline{\rho}', \overline{\rho})$. This proves part of the uniqueness assertion in Theorem 3.11.

Problem 4.2. Play the same game with the "transpose-inverse" automorphism of GL_n to show that the universal deformation ring of $\overline{\rho}$ and of its contragredient (i.e., the representation $g \mapsto (\overline{\rho}(g)^{-1})^t$) are canonically isomorphic.

Problem 4.3. Suppose $C(\overline{\rho}_1) = \mathsf{k}$, $C(\overline{\rho}_2) = \mathsf{k}$, and $C(\overline{\rho}_1 \otimes \overline{\rho}_2) = \mathsf{k}$, so that all three representations have universal deformation rings. Given a deformation ρ_1 of $\overline{\rho}_1$ to a ring A_1 and a deformation ρ_2 of $\overline{\rho}_2$ to a ring A_2, show that the tensor product $\rho_1 \otimes \rho_2$ is a deformation of $\overline{\rho}_1 \otimes \overline{\rho}_2$ to the ring $A_1 \hat{\otimes} A_2$, and deduce a natural homomorphism
$$\mathcal{R}(\overline{\rho}_1 \otimes \overline{\rho}_2) \longrightarrow \mathcal{R}(\overline{\rho}_1) \hat{\otimes}_\Lambda \mathcal{R}(\overline{\rho}_2).$$

Problem 4.4. Still in the situation of the previous problem, suppose we pick a lift ρ_1 of $\overline{\rho}_1$ to $\mathrm{GL}_n(\Lambda)$. By the universal property, this corresponds to a map $h_1 : \mathcal{R}(\overline{\rho}_1) \longrightarrow \Lambda$. Use this and the map defined in the previous problem to deduce that there exists a homomorphism
$$\mathcal{R}(\overline{\rho}_1 \otimes \overline{\rho}_2) \longrightarrow \mathcal{R}(\overline{\rho}_2).$$
This is called "contraction with the lift ρ_1."

Problem 4.5. In the situation of the previous problem, suppose $\overline{\rho}_1$ is one-dimensional, i.e., is a character. In this case we sometimes refer to the homomorphism obtained in that problem as the "twisting homomorphism" corresponding to ρ_1. Show that the twisting homomorphism is in fact an isomorphism and that it satisfies the obvious "homomorphic" property with respect to ρ_1. (This problem, together with what was done above, completes the proof of Theorem 3.11.)

We can continue in this vein to consider, for example, what happens when we change the group Π (e.g., by restricting to a subgroup) and what happens when we change the base field k. See [97] for a careful discussion of all this.

Tangent spaces and cohomology groups

Fix a residual representation $\overline{\rho} : \Pi \longrightarrow \mathrm{GL}_n(\mathsf{k})$ such that $C(\overline{\rho}) = \mathsf{k}$ and a coefficient ring Λ. For this section, let $\mathbf{D} = \mathbf{D}_{\overline{\rho},\Lambda}$, in order to make the notation lighter. We know, then, that the functor $\mathbf{D}$ is representable; denote the representing coefficient Λ-algebra by $\mathcal{R}$. We have already introduced the tangent space $t_\mathbf{D} = \mathbf{D}(\mathsf{k}[\varepsilon])$ of the functor $\mathbf{D}$. Since $\mathbf{D}$ is represented by $\mathcal{R}$, we know that
$$t_\mathbf{D} = \mathbf{D}(\mathsf{k}[\varepsilon]) = \mathrm{Hom}_\Lambda(\mathcal{R}, \mathsf{k}[\varepsilon]) = \mathrm{Hom}_\mathsf{k}(\mathfrak{m}_\mathcal{R}/(\mathfrak{m}_\mathcal{R}^2, \mathfrak{m}_\Lambda), \mathsf{k}).$$

This is true for any representable functor; for the deformation functor, we can say a bit more. Suppose we know $\overline{\rho}(g) = a$, where a is some matrix in $\mathrm{GL}_n(\mathsf{k})$, and suppose that ρ_1 is a deformation of $\bar{\rho}$ to $k[\varepsilon]$. Then we must have $\rho_1(g) = (1+b_g\varepsilon)a$ for some matrix $b_g \in M_n(k)$. In other words, ρ_1 determines (and is determined by) a map $b : \Pi \longrightarrow M_n(k)$ mapping g to the matrix b_g. (Equivalently, the point is that $GL_n(k[\varepsilon])$ is the semi-direct product of $1 + \varepsilon M_n(k)$ and $GL_n(k)$, and the first group is isomorphic to the additive group $M_n(k)$.)

Imposing the condition that ρ_1 be a homomorphism boils down to saying that the map

$$g \mapsto b_g$$

should be a cocycle with values in $M_n(k)$, where we make Π act on $M_n(k)$ via conjugation:

$$g \cdot b = \bar{\rho}(g) b \bar{\rho}(g)^{-1}.$$

The k-vector space $M_n(k)$ with this action of Π is usually called the *adjoint representation* of $\bar{\rho}$, and denoted $\mathrm{Ad}(\bar{\rho})$. One checks, then, that this association gives an isomorphism

$$t_\mathbf{D} \cong H^1(\Pi, \mathrm{Ad}(\bar{\rho})).$$

This gives a connection between the deformation theory and the cohomology of $\mathrm{Ad}(\bar{\rho})$ which is quite important and which we will continue to examine in the next section.

Problem 4.6. Check that the map $g \mapsto b_g$ is a cocycle and that the cocycles corresponding to strictly equivalent lifts differ by a coboundary.

Problem 4.7. Check that the map $t_\mathbf{D} \longrightarrow H^1(\Pi, \mathrm{Ad}(\bar{\rho}))$ defined above is indeed an isomorphism of k-vector spaces.

Notice that when Π is a Galois group, this puts us in the realm of Galois cohomology, which means it puts at our disposal an enormous array of techniques and theorems. We already have a simple numerical consequence:

Corollary 4.1. *Retain the assumptions and notations of this section, so that, in particular, a universal deformation of $\bar{\rho}$ exists and the universal deformation ring is $\mathcal{R}$. Let $d_1 = \dim_k H^1(\Pi, \mathrm{Ad}(\bar{\rho}))$. Then $\mathcal{R}$ is a quotient of a power series ring in d_1 variables over Λ.*

In other words, $\mathcal{R}$ fits into an exact sequence

$$0 \longrightarrow I \longrightarrow \Lambda[[X_1, X_2, \ldots, X_{d_1}]] \longrightarrow \mathcal{R} \longrightarrow 0.$$

One possible approach to understanding $\mathcal{R}$, then, is to try to determine the dimension d_1 and the ideal I. In many situations, in fact, the value of d_1 is the crucial piece of information.

Tangent spaces and extensions of modules

We have already given several different interpretations of the tangent space to the deformation functor. One often uses still another one, relating the tangent space to extensions of Π-modules. We'll work out the basic correspondence, and leave the details to the reader.

Let, then

$$\bar{\rho} : \Pi \longrightarrow GL_n(k)$$

be a residual representation, $\mathbf{D}$ be the deformation functor, and $t_{\mathbf{D}} = \mathbf{D}(\mathsf{k}[\varepsilon])$ be its tangent space. We want to establish a correspondence between elements of $t_{\mathbf{D}}$ and extensions of $\bar{\rho}$ by $\bar{\rho}$, by which we mean k-vector spaces E with an action of Π such that there is an exact sequence in the category of $\mathsf{k}[[\Pi]]$-modules

$$0 \longrightarrow V_{\bar{\rho}} \longrightarrow E \longrightarrow V_{\bar{\rho}} \longrightarrow 0,$$

where by $V_{\bar{\rho}}$ we mean[1] k^n with the action of Π given by the residual representation $\bar{\rho}$.

Suppose, first, that we are given an element of $t_{\mathbf{D}}$, that is, a deformation ρ of $\bar{\rho}$ to $\mathsf{k}[\varepsilon]$. Let M be $\mathsf{k}[\varepsilon]^n$ with the action of Π given by ρ. Clearly M is of dimension $2n$ as a vector space over k. To see that M fits into an exact sequence as above, consider the submodule εM and the quotient $M/\varepsilon M$. These are both clearly n-dimensional, and in fact they are both isomorphic to $V_{\bar{\rho}}$ as Π-modules.

Problem 4.8. Check that both εM and $M/\varepsilon M$ are free k-modules of dimension n on which Π acts via $\bar{\rho}$. (If $\{e_1, e_2, \ldots, e_n\}$ is a basis of M over $\mathsf{k}[\varepsilon]$, check that

$$\{\varepsilon e_1, \varepsilon e_2, \ldots, \varepsilon e_n\}$$

is a basis of εM over k and that

$$\{(e_1 \bmod \varepsilon M), (e_2 \bmod \varepsilon M), \ldots, (e_n \bmod \varepsilon M)\}$$

is a basis of $M/\varepsilon M$ over k. Then check that the action of Π is correct.)

Thus, if we identify both εM and $M/\varepsilon M$ with $V_{\bar{\rho}}$, we have the exact sequence we wanted. This shows that every element of $t_{\mathbf{D}}$ determines an extension. For the converse, suppose we are given a $2n$-dimensional k-vector space E which fits into an exact sequence

$$0 \longrightarrow V_{\bar{\rho}} \xrightarrow{\alpha} E \xrightarrow{\beta} V_{\bar{\rho}} \longrightarrow 0.$$

We then make E into a $\mathsf{k}[\varepsilon]$-module by defining multiplication by ε to be

$$\alpha \circ \beta : E \xrightarrow{\beta} V_{\bar{\rho}} \xrightarrow{\alpha} E$$

(this reverse composition makes sense, since the image of β is the same as the domain of α). It's easy to see that $(\alpha \circ \beta)^2 = 0$, since $\beta \circ \alpha = 0$ by the exactness of our sequence. In addition, since both α and β are homomorphisms of Π-modules, this $\mathsf{k}[\varepsilon]$-module structure commutes with the action of Π. We can now check directly that this makes E into a free $\mathsf{k}[\varepsilon]$-module of rank n with an action of Π, and therefore defines a representation $\Pi \longrightarrow \mathrm{GL}_n(\mathsf{k}[\varepsilon])$, which is clearly a deformation of $\bar{\rho}$.

Problem 4.9. Check the details. (For example, why is E a free $\mathsf{k}[\varepsilon]$-module? Why is the representation defined by E a lifting of $\bar{\rho}$? Do isomorphic E's give strictly equivalent deformations?)

Problem 4.10. Check that strictly equivalent deformations correspond to isomorphic extensions and conversely.

---

[1] This is somewhat nonstandard notation, but the usual notation, which is to call this space $\bar{\rho}$ (thereby identifying the homomorphism with the Π-module it induces), is a bit too confusing to use here.

There is a standard k-vector space structure on the set of isomorphism classes of extensions of $V_{\bar\rho}$ by $V_{\bar\rho}$. This vector space is denoted $\mathrm{Ext}^1_{k[[\Pi]]}(V_{\bar\rho}, V_{\bar\rho})$, and what we have shown so far is that there is a bijection

$$\mathbf{D}(k[\varepsilon]) \longrightarrow \mathrm{Ext}^1_{k[[\Pi]]}(V_{\bar\rho}, V_{\bar\rho}).$$

Problem 4.11. (For those who know the vector space structure on Ext.) Show that this bijection is in fact an isomorphism of k-vector spaces.

Another point of view here is to think in terms of matrices. If E is a $2n$-dimensional k-vector space on which Π acts, the existence of an exact sequence

$$0 \longrightarrow V_{\bar\rho} \xrightarrow{\alpha} E \xrightarrow{\beta} V_{\bar\rho} \longrightarrow 0$$

amounts to saying that the representation

$$\rho_E : \Pi \longrightarrow \mathrm{GL}_{2n}(k)$$

corresponding to E can be put into the block form

$$\rho_E(g) = \begin{pmatrix} \bar\rho(g) & A_g \\ 0 & \bar\rho(g) \end{pmatrix}$$

with $A_g \in \mathrm{M}_n(k)$.

Problem 4.12. Show that the correspondence $g \mapsto A_g \bar\rho(g)^{-1}$ is a 1-cocycle and therefore determines an element of $\mathrm{H}^1(\Pi, \mathrm{Ad}(\bar\rho))$. Show that the resulting map

$$\mathrm{Ext}^1_{k[[\Pi]]}(V_{\bar\rho}, V_{\bar\rho}) \longrightarrow \mathrm{H}^1(\Pi, \mathrm{Ad}(\bar\rho))$$

is an isomorphism.

Thus, we have established canonical isomorphisms

$$t_{\mathbf{D}} = \mathbf{D}(k[\varepsilon]) \cong \mathrm{H}^1(\Pi, \mathrm{Ad}(\bar\rho)) \cong \mathrm{Ext}^1_{k[[\Pi]]}(V_{\bar\rho}, V_{\bar\rho}).$$

This gives us one more way to get at the tangent space (and especially its dimension).

Obstructed and unobstructed deformation problems

The use of the notation d_1 for the dimension of the tangent space probably signals that there is a d_2 about to show up. This is indeed the case. We deepen the connection between deformation theory and cohomology by using a standard idea in deformation theory: we try to compute the obstruction to lifting a homomorphism.

Keep the notations and assumptions as above. Suppose we have rings R_1 and R_0 in $\mathcal{C}_\Lambda$, and a surjective coefficient Λ-algebra homomorphism $R_1 \longrightarrow R_0$ with kernel I satisfying $I \cdot \mathfrak{m}_{R_1} = 0$ (in particular, we could be working with a small homomorphism). Because of the last assumption, we can (and do) view I as a k-vector space. Suppose we are given a homomorphism $\rho : \Pi \longrightarrow \mathrm{GL}_n(R_0)$. What keeps us from finding a deformation to R_1?

Well, we can certainly find a set-theoretic lift, i.e., a function $\gamma : \Pi \longrightarrow \mathrm{GL}_n(R_1)$ that lifts ρ. To test whether this is a homomorphism, we would have to compute

$$c(g_1, g_2) = \gamma(g_1 g_2) \gamma(g_2)^{-1} \gamma(g_1)^{-1}$$

for every $g_1, g_2 \in \Pi$. If γ were a homomorphism, we would have $c(g_1, g_2) = 1$. Since it *is* a homomorphism modulo I, we do know that

$$c(g_1, g_2) = 1 + d(g_1, g_2)$$

with $d(g_1, g_2) \in M_n(I) \cong \mathrm{Ad}(\overline{\rho}) \otimes_k I$. It now isn't too hard to check that $d(g_1, g_2)$ is a 2-cocycle with values in $\mathrm{Ad}(\overline{\rho}) \otimes_k I$, and that replacing γ by a different lift changes this cocycle by a coboundary.

Problem 4.13. Check this!

Therefore, the cocycle $d(g_1, g_2)$ gives an element $\mathcal{O}(\rho_0)$ in the cohomology $\mathrm{H}^2(\Pi, \mathrm{Ad}(\overline{\rho}) \otimes_k I) \cong \mathrm{H}^2(\Pi, \mathrm{Ad}(\overline{\rho})) \otimes_k I$, and this element is trivial if and only if there exists a homomorphism $\Pi \longrightarrow \mathrm{GL}_n(R_1)$ lifting ρ_0. We might call $\mathcal{O}(\rho_0)$ the obstruction class of ρ_0 relative to $R_1 \longrightarrow R_0$.

In general, one can't readily compute obstruction classes. However, the fact that liftings exist exactly when $\mathcal{O}(\rho_0) = 0$ means that if $\mathrm{H}^2(\Pi, \mathrm{Ad}(\overline{\rho})) = 0$ the deformation problem should be especially simple. And this is indeed the case, as Mazur showed.

Theorem 4.2. *Suppose $C(\overline{\rho}) = \mathsf{k}$ and let $\mathcal{R} = \mathcal{R}(\Pi, \mathsf{k}, \overline{\rho})$ be the universal deformation ring representing the deformation functor $\mathbf{D}_\Lambda$. Let*

$$d_1 = \dim \mathrm{H}^1(\Pi, \mathrm{Ad}(\overline{\rho})) \quad \text{and} \quad d_2 = \dim \mathrm{H}^2(\Pi, \mathrm{Ad}(\overline{\rho})).$$

Then we have

$$(**) \qquad \mathrm{Krull}\dim(\mathcal{R}/\mathfrak{m}_\Lambda \mathcal{R}) \geq d_1 - d_2.$$

Furthermore, if $d_2 = 0$ we have equality in (**), and in fact

$$\mathcal{R} \cong \Lambda[[X_1, X_2, \ldots, X_{d_1}]].$$

Proof. We already know that there is a surjective homomorphism of coefficient Λ-algebras

$$\Lambda[[T_1, T_2, \ldots, T_{d_1}]] \longrightarrow \mathcal{R}$$

which induces an isomorphism on tangent spaces. Reducing modulo the maximal ideal gives a homomorphism $\mathsf{k}[[T_1, T_2, \ldots, T_{d_1}]] \longrightarrow \mathcal{R}/\mathfrak{m}_\Lambda \mathcal{R}$ which still induces an isomorphism on tangent spaces, and therefore is still surjective. Let J be the kernel of this surjection. To save on notation, write $F = \mathsf{k}[[T_1, T_2, \ldots, T_{d_1}]]$ and let $\mathfrak{m}_F$ be its maximal ideal. We have an exact sequence

$$0 \longrightarrow J \longrightarrow F \longrightarrow \mathcal{R}/\mathfrak{m}_\Lambda \mathcal{R} \longrightarrow 0$$

in which the surjective homomorphism $F \longrightarrow \mathcal{R}/\mathfrak{m}_\Lambda \mathcal{R}$ induces an isomorphism of tangent spaces. What we need to prove is that the minimal number of generators for J is at most d_2.

Since $\mathfrak{m}_F J \subset J$, the sequence of k-vector spaces

$$0 \longrightarrow J/\mathfrak{m}_F J \longrightarrow F/\mathfrak{m}_F J \longrightarrow \mathcal{R}/\mathfrak{m}_\Lambda \mathcal{R} \longrightarrow 0$$

is still exact and the map on the right still induces an isomorphism of tangent spaces. Hence, the Krull dimension of $\mathcal{R}/\mathfrak{m}_\Lambda \mathcal{R}$ is at least $d_1 - \dim_\mathsf{k}(J/\mathfrak{m}_F J)$. (Equivalently, the minimal number of generators for J is at most $\dim_\mathsf{k}(J/\mathfrak{m}_F J)$.)

Let $\boldsymbol{\rho}_p$ be image of the universal deformation $\boldsymbol{\rho}$ under the quotient map $\mathcal{R} \longrightarrow \mathcal{R}/\mathfrak{m}_\Lambda \mathcal{R}$. It is clear that $\boldsymbol{\rho}_p$ is universal among deformations of $\overline{\rho}$ to Λ-algebras killed by $\mathfrak{m}_\Lambda$ (equivalently, to k-algebras). The construction above gives a cohomology class

$$\mathcal{O}(\boldsymbol{\rho}_p) \in \mathrm{H}^2(\Pi, \mathrm{Ad}(\overline{\rho})) \otimes J/\mathfrak{m}_F J$$

which is the obstruction to lifting $\boldsymbol{\rho}_p$ to $F/\mathfrak{m}_F J$.

LECTURE 4. THE UNIVERSAL DEFORMATION: PROPERTIES

Consider the k-linear map

$$\mathrm{Hom}_\mathsf{k}(J/\mathfrak{m}_F J, \mathsf{k}) \xrightarrow{\alpha} \mathrm{H}^2(\Pi, \mathrm{Ad}(\bar{\rho}))$$

given by

$$f \mapsto (1 \otimes f)(\mathcal{O}(\boldsymbol{\rho}_p)).$$

If we can show α is injective, then we will have $\dim_\mathsf{k}(J/\mathfrak{m}_F J) \leq d_2$, whence we can conclude that the Krull dimension of $\mathcal{R}/\mathfrak{m}_\Lambda \mathcal{R}$ is at least $d_1 - d_2$, as claimed.

To prove the injectivity of α, let f be a nonzero element in the kernel, let A be the quotient of $F/\mathfrak{m}_F J$ by the kernel of f and let I be the image of $J/\mathfrak{m}_F J$ in the quotient, so that $I = (J/\mathfrak{m}_F J)/\mathrm{Ker}(f) = \mathrm{Im}(f) = \mathsf{k}$. Then we get an exact sequence

$$0 \longrightarrow I \longrightarrow A \longrightarrow \mathcal{R}/\mathfrak{m}_\Lambda \mathcal{R} \longrightarrow 0$$

where I is isomorphic to k and which still induces an isomorphism on tangent spaces (check!). But now the obstruction to lifting $\boldsymbol{\rho}_p$ to A vanishes. Thus we get a deformation of $\bar{\rho}$ to A lifting $\boldsymbol{\rho}_p$. But A is a k-algebra and $\boldsymbol{\rho}_p$ is universal among lifts to such rings, so this lift must be induced by a homomorphism $\mathcal{R}/\mathfrak{m}_\Lambda \mathcal{R} \longrightarrow A$. This means that the sequence splits, but this and the fact that $I \neq 0$ contradict the fact that $A \longrightarrow \mathcal{R}/\mathfrak{m}_\Lambda \mathcal{R}$ induces an isomorphism of tangent spaces. Thus, there cannot be a nonzero element f in the kernel of α, that is, α is injective as claimed. This proves the inequality.

The last assertion follows at once. If $d_2 = 0$, then the kernel of the surjective homomorphism of coefficient Λ-algebras

$$\Lambda[[T_1, T_2, \ldots, T_{d_1}]] \longrightarrow \mathcal{R}$$

has at most 0 generators. Hence, $\mathcal{R} \cong \Lambda[[T_1, T_2, \ldots, T_{d_1}]]$ in this case; in particular, equality holds in $(**)$. $\square$

When we are in the situation where $d_2 = 0$, we say the lifting problem is *unobstructed*.

In the examples where the deformation ring of an absolutely irreducible $\bar{\rho}$ has been explicitly computed (see the next lecture for how one might do such a thing), the Krull dimension has always turned out to be equal to $d_1 - d_2$. Hence, one might conjecture that one always has the equality.

Conjecture. If $\bar{\rho} : \Pi \longrightarrow \mathrm{GL}_n(\mathsf{k})$ is an absolutely irreducible residual representation, and $\mathcal{R}$ is the universal deformation ring, we have

$$\mathrm{Krull}\dim(\mathcal{R}/\mathfrak{m}_\Lambda \mathcal{R}) = d_1 - d_2.$$

We refer to this as the "Dimension Conjecture." Böckle has shown in [6] that this conjecture does not hold in some cases where $\bar{\rho}$ is reducible but still has a universal deformation (because it satisfies $C(\bar{\rho}) = \mathsf{k}$). He has also been able to show that it is true in many cases.

Mazur points out in [97] that one should think of this conjecture as a generalization of Leopoldt's Conjecture. To see why will require computing d_1 and d_2 when Π is a global Galois group, and we will do this in the next section.

Galois representations

Let K be a number field, S a finite set of primes in K. We will assume S includes all the primes above p and also (as we have assumed from the beginning) the primes at infinity. Let $S_\infty \subset S$ be the set of primes at infinity. Finally, let $\Pi = G_{K,S}$ and let

$$\overline{\rho} : \Pi \longrightarrow \mathrm{GL}_n(\mathsf{k})$$

be a residual representation such that $C(\overline{\rho}) = \mathsf{k}$, and let $\mathcal{R}$ be its universal deformation ring. From our work in the previous section, we know a lower bound for the dimension of $\mathcal{R}$, expressed in terms of the dimensions of two cohomology groups. The goal of this section is to compute that bound somewhat more explicitly using known results about Galois cohomology. It turns out that we will be able to obtain a better formula for $d_1 - d_2$, but not for d_1 and d_2 separately.

The result we need from Galois cohomology is Tate's global Euler characteristic formula. Here's what it says. We take an extension $K/\mathbb{Q}$ of degree d, S a finite set of primes in K including all the infinite primes, M a *finite* $G_{K,S}$-module such that S contains all the primes that divide the order of M. (In our application, the order of M will be a power of p.) For each prime v of K, let K_v be the completion at v. In particular, if $v \in S_\infty$, K_v is either $\mathbb{R}$ or $\mathbb{C}$. Then the global Euler characteristic formula (see [156], [68], or [107]) says that

$$\frac{\#\mathrm{H}^0(G_{K,S}, M) \cdot \#\mathrm{H}^2(G_{K,S}, M)}{\#\mathrm{H}^1(G_{K,S}, M)} = \frac{1}{(\#M)^d} \prod_{v \in S_\infty} \#\mathrm{H}^0(G_{K_v}, M).$$

In our situation M will be $\mathrm{Ad}(\overline{\rho})$, which is a k-vector space and hence will have order a power of p and S will include all primes above p. Since the cohomology groups in this case will also be k-vector spaces, all of the groups in the formula have order a power of p, and we can translate the formula into a statement about dimensions:

$$\dim \mathrm{H}^0(G_{K,S}, M) - \dim \mathrm{H}^1(G_{K,S}, M) + \dim \mathrm{H}^2(G_{K,S}, M) =$$
$$= \sum_{v \in S_\infty} \dim \mathrm{H}^0(G_{K_v}, M) - d \dim M,$$

where all of the dimensions are dimensions over k.

Now let $M = \mathrm{Ad}(\overline{\rho})$, and write $d_i = \dim \mathrm{H}^i(G_{K,S}, \mathrm{Ad}(\overline{\rho}))$ as before. Then the formula becomes

$$d_0 - d_1 + d_2 = \sum_{v \in S_\infty} \dim \mathrm{H}^0(G_{K_v}, \mathrm{Ad}(\overline{\rho})) - dn^2,$$

and hence

$$d_1 - d_2 = d_0 + dn^2 - \sum_{v \in S_\infty} \dim \mathrm{H}^0(G_{K_v}, \mathrm{Ad}(\overline{\rho})).$$

But d_0 we can compute:

$$\mathrm{H}^0(G_{K,S}, \mathrm{Ad}(\overline{\rho})) = (\mathrm{Ad}(\overline{\rho}))^{G_{K,S}}$$

is the set of matrices in $\mathrm{M}_n(\mathsf{k})$ fixed by the conjugation action of $G_{K,S}$, i.e., it is $C(\overline{\rho}) = \mathsf{k}$. So $d_0 = 1$. The upshot, then, is the following.

Proposition 4.3. *Let K be a number field of degree d over $\mathbb{Q}$, let $\bar{\rho}: G_{K,S} \longrightarrow \mathrm{GL}_n(\mathsf{k})$ be a residual representation such that $C(\bar{\rho}) = \mathsf{k}$, and let $\mathcal{R}$ be its universal deformation ring. Then*

$$\mathrm{Krull\ dim}\, \mathcal{R}/\mathfrak{m}_\Lambda \mathcal{R} \geq 1 + dn^2 - \sum_{v \in S_\infty} \dim \mathrm{H}^0(G_{K_v}, \mathrm{Ad}(\bar{\rho})).$$

The advantages of this formula are two. First, it only refers to the 0-th cohomology groups, which are just the fixed points under the Galois action and so are relatively easy to compute. Second, the groups acting are the G_{K_v} for v an archimedean prime, so that G_{K_v} has order two if v is real and order one if v is complex.

Let's work out what the formula gives in the two most interesting cases. First, let K be a number field and $\bar{\rho}$ be a character (i.e., a one-dimensional representation). As we saw above, the universal deformation does not depend on $\bar{\rho}$ in this case; in fact, we showed that

$$\mathcal{R} = \Lambda[[G_{K,S}^{ab,(p)}]].$$

Now notice that

$$\mathcal{R}/\mathfrak{m}_\Lambda \mathcal{R} = \mathsf{k}[[G_{K,S}^{ab,(p)}]],$$

so the Krull dimension of this ring is equal to the rank of $G_{K,S}^{ab,(p)}$ as a $\mathbb{Z}_p$-module, or, equivalently, to the rank of $\mathrm{Hom}(G_{K,S}, \mathbb{Z}_p)$ as a $\mathbb{Z}_p$-module.

On the other hand, the formula above says that the Krull dimension is at least $1 + r_2$, where r_2 is the number of complex primes of K. So we have shown that

$$\mathrm{rank}_{\mathbb{Z}_p} \mathrm{Hom}(G_{K,S}, \mathbb{Z}_p) \geq 1 + r_2,$$

which is in fact a well-known result. The assertion that these two numbers are equal is equivalent to the *Leopoldt Conjecture* for the field K. This is why we said that the general dimension conjecture should be viewed as a vastly generalized Leopoldt Conjecture.

The Leopoldt Conjecture is known to be true for abelian extensions of $\mathbb{Q}$, and for abelian extensions of quadratic imaginary fields, so in that case the deformation ring has the expected dimension. The general case seems very elusive.

Problem 4.14. Check these computations.

Problem 4.15. Look up the classical statement of the Leopoldt Conjecture and explain why it is equivalent to

$$\mathrm{rank}_{\mathbb{Z}_p} \mathrm{Hom}(G_{K,S}, \mathbb{Z}_p) = 1 + r_2.$$

The next case in which we want to go through the computation is the case that is related to modular forms and elliptic curves: $n = 2$, p an odd prime, $K = \mathbb{Q}$, S containing p and ∞. In this situation there is only one infinite prime, and G_∞ is a group of order two generated by the complex conjugation σ. Since $\sigma^2 = 1$ and p is odd, $\bar{\rho}(\sigma)$ is a matrix of order 2 in $\mathrm{GL}_2(\mathsf{k})$, and hence we must have

$$\bar{\rho}(\sigma) \sim \pm \begin{pmatrix} 1 & 0 \\ 0 & 1 \end{pmatrix} \quad \text{or} \quad \bar{\rho}(\sigma) \sim \begin{pmatrix} 1 & 0 \\ 0 & -1 \end{pmatrix}.$$

In the first case, $\det \bar{\rho}(\sigma) = 1$, and we call $\bar{\rho}$ an *even* representation. In the second, $\det \bar{\rho}(\sigma) = -1$ and we say $\bar{\rho}$ is *odd*.

Now it's easy to compute the dimension d_0 of $\mathrm{H}^0(G_\infty, \mathrm{Ad}(\bar{\rho}))$. If $\bar{\rho}$ is even, then $\bar{\rho}(\sigma)$ is a scalar matrix, and hence the action of G_∞ on $\mathrm{Ad}(\bar{\rho})$ is trivial, so $d_0 = 4$.

If $\bar\rho$ is odd, then an easy computation shows that $d_0 = 2$. Plugging all this into the formula gives:

Proposition 4.4. *Let p be an odd prime, let S be a set of rational primes including p and ∞, let $\bar\rho : G_{\mathbb{Q},S} \longrightarrow \mathrm{GL}_2(\mathsf{k})$ be a residual representation satisfying $C(\bar\rho) = \mathsf{k}$, and let $\mathcal{R}$ be the universal deformation ring of $\bar\rho$. Then:*
- *if $\bar\rho$ is even, then Krull $\dim \mathcal{R}/\mathfrak{m}_\Lambda \mathcal{R} \geq 1$, and*
- *if $\bar\rho$ is odd, then Krull $\dim \mathcal{R}/\mathfrak{m}_\Lambda \mathcal{R} \geq 3$.*

One conjectures, at least when $\bar\rho$ is absolutely irreducible, that both of these inequalities are in fact equalities, i.e., that the cohomological constraints on the dimension are the only constraints. This is indeed the case in many cases that have been computed explicitly.

It is worth noting that the representations coming from elliptic curves and from modular forms are always odd.

Problem 4.16. What happens if $p = 2$? In this case, the distinction between $\det \bar\rho(\sigma) = 1$ and $\det \bar\rho(\sigma) = -1$ vanishes; can the computation still be done?

Problem 4.17. (Hard) Work out $d_1 - d_2$ when $\Pi = G_{\mathbb{Q}_p}$.

LECTURE 5
Explicit Deformations

So far, we have developed an elaborate theory about the universal deformation, but (except for the case of GL_1) we have not been able to get our hands on one in any concrete way. In this lecture, we want to discuss a point of view, going back to the work of Boston in [8], [9], and [10], which allows us, in many cases, to get a rather explicit description of the universal deformation ring of a Galois representation. We will try to describe the basic idea, and then give some sample theorems that have been proved using this method.

The basic setup

The first thing to do is to "see" our residual representation $\bar{\rho}$ in terms of field extensions. Take, as before, S to be a finite set of primes in $\mathbb{Q}$ including p and ∞, $\mathbb{Q}_S$ the maximal extension of $\mathbb{Q}$ unramified outside S, $\Pi = G_{\mathbb{Q},S} = G(\mathbb{Q}_S/\mathbb{Q})$, and suppose $\bar{\rho} : \Pi \longrightarrow GL_n(k)$ is absolutely irreducible or, more generally, assume that $C(\bar{\rho}) = k$. (This is mostly for convenience; the method also allows us to compute "versal deformations" in cases where the deformation functor is not representable.) We want to understand the deformation theory of $\bar{\rho}$.

Let $\Pi_0 = \text{Ker}(\bar{\rho})$, and let K be the fixed field of Π_0, so that we have a tower of fields $\mathbb{Q} \subset K \subset \mathbb{Q}_S$, with Galois groups $H = G(K/\mathbb{Q}) \cong \text{Im}(\bar{\rho})$ and $G(\mathbb{Q}_S/K) \cong \Pi_0$. Let S_1 be the set of primes of K which lie above S.

$$\begin{array}{l} \mathbb{Q}_S \\ | \quad G(\mathbb{Q}_S/K)\cong\Pi_0=\text{Ker}(\bar{\rho}) \\ K \\ | \quad G(K/\mathbb{Q})=H\cong\text{Im}(\bar{\rho}) \\ \mathbb{Q} \end{array}$$

Let $\rho : \Pi \longrightarrow GL_n(\mathcal{R})$ be the universal deformation of $\bar{\rho}$. As above, write $\Gamma_n(\mathcal{R})$ for the kernel of the natural projection $GL_n(\mathcal{R}) \longrightarrow GL_n(k)$. If $\gamma \in \Pi_0$, then $\bar{\rho}(\gamma) = 1$, and therefore $\rho(\gamma) \in \Gamma_n(\mathcal{R})$. Thus, the restriction of ρ to Π_0 gives a homomorphism $\Pi_0 \longrightarrow \Gamma_n(\mathcal{R})$. We note, then, that

Lemma 5.1. *For any ring R in $\mathcal{C}$, $\Gamma_n(R)$ is a pro-p-group.*

Proof. This appeared as a problem in a previous lecture. The basic idea of the proof is this: one writes R as the inverse limit of the quotients $R/\mathfrak{m}^k$, where $\mathfrak{m}$ is

the maximal ideal in R. Then
$$\Gamma_n(R) = \varprojlim_k \Gamma_n(R/\mathfrak{m}^k).$$

To prove the lemma, note first that $\Gamma_n(R/\mathfrak{m}) = \Gamma_n(\mathsf{k}) = \{1\}$ is a p-group, and then consider each of the transition homomorphisms
$$\Gamma_n(R/\mathfrak{m}^k) \longrightarrow \Gamma_n(R/\mathfrak{m}^{k-1}).$$

By induction, we know the image is a p-group. The kernel consists of those matrices whose off-diagonal entries are in the ideal $\mathfrak{m}^{k-1}/\mathfrak{m}^k$ and whose diagonal entries are in $1 + \mathfrak{m}^{k-1}/\mathfrak{m}^k$, i.e., it is
$$1 + \mathrm{M}_n(\mathfrak{m}^{k-1}/\mathfrak{m}^k).$$

One checks easily that this multiplicative group is isomorphic to the *additive* group $\mathrm{M}_n(\mathfrak{m}^{k-1}/\mathfrak{m}^k)$, which is easily seen to be a p-group. □

Problem 5.1. To complete the proof of the Lemma, check that the additive group $\mathrm{M}_n(\mathfrak{m}^{k-1}/\mathfrak{m}^k)$ is a p-group.

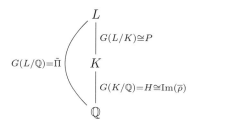

So here's what we have: the universal deformation ρ induces a homomorphism $\Pi_0 \longrightarrow \Gamma_n(\mathcal{R})$, and $\Gamma_n(\mathcal{R})$ is a pro-p-group. Therefore the homomorphism $\Pi_0 \longrightarrow \Gamma_n(\mathcal{R})$ must factor through some pro-p quotient of Π_0. Any such quotient will be the Galois group of a pro-p-extension of K, so let L be the maximal pro-p-extension of K unramified outside S_1. Then $P = G(L/K)$ is a pro-p-group (in fact, it is the maximal continuous pro-p-quotient of Π_0), and we see that ρ must factor through $\tilde{\Pi} = G(L/\mathbb{Q})$. (Boston and Mazur call $\tilde{\Pi}$ the *p-completion of Π relative to $\overline{\rho}$.*)

We've shown, then, that the universal deformation ρ must factor through the quotient $\tilde{\Pi}$ of $G_{\mathbb{Q},S}$. It follows that all the deformations must factor through $\tilde{\Pi}$. Hence, the upshot of this discussion is that we can replace Π with $\tilde{\Pi}$ when studying the deformation theory of $\overline{\rho}$. The crucial feature of $\tilde{\Pi}$ is that it has a big normal subgroup P which is a pro-p-group, and the quotient $\tilde{\Pi}/P$ is isomorphic to the image of $\overline{\rho}$, so that the sequence
$$1 \longrightarrow P \longrightarrow \tilde{\Pi} \longrightarrow \mathrm{Im}(\overline{\rho}) \longrightarrow 1$$
is exact. The basic idea is now the following: to understand deformations of $\overline{\rho}$ to a coefficient Λ-algebra R, we need to understand all maps from P to $\Gamma_n(R)$ and then to consider how they may be extended to all of $\tilde{\Pi}$ in such a way as to be a deformation of $\overline{\rho}$. It turns out that we know enough about pro-p-extensions of number fields that in many cases this program can be pushed through to give a good description of $\mathcal{R}$ and often also of the universal deformation ρ.

Rather than do this in the most general case, we will focus on the simpler case of "tame representations."

Definition 5.1. We say a residual representation $\overline{\rho}$ is *tame* if the order of $\mathrm{Im}(\overline{\rho})$ is not divisible by p.

Notice that in the tame case, the sequence
$$1 \longrightarrow P \longrightarrow \tilde{\Pi} \longrightarrow \mathrm{Im}(\overline{\rho}) \longrightarrow 1$$
tells us that $\tilde{\Pi}$ is a profinite group with a normal pro-p-Sylow subgroup, which allows us to get at the structure of $\tilde{\Pi}$ in a pretty explicit way.

Group theory

In this section, we summarize some group-theoretical results that will help us understand the universal deformation of a tame representation. The relevance of these results was first pointed out by Boston, whose exposition in [**9**] we follow.

As we saw above, when the residual representation $\overline{\rho}$ is tame, any deformation factors through a profinite group $\tilde{\Pi}$ which has a normal pro-p-Sylow subgroup P such that
$$\tilde{\Pi}/P \cong \mathrm{Im}(\overline{\rho}).$$
In this situation, we can apply the following theorem:

Theorem 5.2 (Schur-Zassenhaus)**.** *Let G be a profinite group with normal pro-p-Sylow subgroup P of finite index in G. Let $\pi : G \longrightarrow G/P$ be the projection on the quotient. Then G contains a subgroup A such that π induces an isomorphism $A \xrightarrow{\cong} G/P$. Furthermore, any two subgroups with this property are conjugate by an element of P.*

(See [**120**, p. 246].)

As a consequence, G is the semi-direct product of P and A. We will exploit this later to define a homomorphism on G by defining it on A and on P in a compatible way.

Problem 5.2. Suppose G is as above, G' is a topological group, and we are given continuous homomorphisms $\alpha : A \longrightarrow G'$ and $\beta : P \longrightarrow G'$. Under what conditions do α and β together define a homomorphism $G \longrightarrow G'$?

The group P is a pro-p-group. If P is topologically finitely generated, it is a quotient of a free pro-p-group on finitely many generators. The minimal number of generators for P is called the *generator rank* of P and sometimes denoted $d(P)$. The kernel of the map from the free pro-p-group on $d(P)$ generators to P is itself finitely generated; the minimal number of generators for the kernel is called the *relation rank* of P, and sometimes denoted $r(P)$.

The Burnside Basis Theorem (Problem 2.2) provides a way to compute the generator rank. To make the notation simpler, let $\overline{P} = \mathrm{Fr}(P)$ be the Frattini quotient of P, i.e., the maximal p-elementary abelian (continuous) quotient of P. The Burnside Basis Theorem says that if $x_1, x_2, \ldots, x_d$ are elements of P such that their image in $\overline{P}$ generates $\overline{P}$, then $x_1, x_2, \ldots, x_d$ topologically generate P.

The following strengthening of the theorem is due to Boston (see [**9**]).

Theorem 5.3. *Let G be a profinite group with normal pro-p-Sylow subgroup P of finite index in G, and let A be a subgroup of G mapping isomorphically to G/P. Let A act on P and on $\overline{P}$ by conjugation. If $\overline{V}$ is an $\mathbb{F}_p[A]$-submodule of $\overline{P}$, then there exists an A-invariant subgroup V of P with $\dim_{\mathbb{F}_p} \overline{V}$ generators which maps onto $\overline{V}$ under π.*

For example, consider the case when $\overline{V}$ is one-dimensional. Then the assumption is that we have $\overline{x} \in \overline{P}$ such that $a^{-1}\overline{x}a = \overline{x}^{\phi(a)}$ for some character $\phi : A \longrightarrow \mathbb{F}_p^\times$. In this case, the theorem says that $\overline{x}$ can be lifted to an element $x \in P$ such that $a^{-1}xa = x^{\psi(a)}$, where $\psi : A \longrightarrow \mathbb{Z}_p^\times$ is a character lifting ϕ. (Since A is finite, this must be a character of finite order, and hence must be the Teichmüller lift of ϕ.)

One final bit of general group representation theory: suppose $H \subset \mathrm{GL}_n(\mathsf{k})$ is a subgroup whose order is prime to p, and let M be the adjoint representation of H, that is $M = \mathrm{M}_n(\mathsf{k})$ with H acting by conjugation. Suppose V is some other finite-dimensional k-vector space with an action of H; as usual, we can think of M and V as modules over the group ring $\mathsf{k}[H]$. Since the order of H is not divisible by p, Maschke's Theorem says that every $\mathsf{k}[H]$-module can be written as a direct sum of irreducible $\mathsf{k}[H]$-modules.

Definition 5.2. We will say V is *prime to adjoint* if V and M have no irreducible sub-representations in common.

The "prime-to-adjoint" condition plays a big role in understanding certain deformation problems. See [97, section 1.12], [9, section 2], [42, section 3.6], and [7] for instances where this condition plays a significant role.

Problem 5.3. Show that the subspace of M consisting of matrices whose trace is zero is an H-invariant subspace. What is the complementary subspace? (You may want to assume the p does not divide n.)

Problem 5.4. Let R be a coefficient ring, and let
$$K_r = \mathrm{Ker}(\Gamma_n(R) \longrightarrow \Gamma_n(R/\mathfrak{m}^r)).$$
Suppose H is a finite subgroup of $\mathrm{GL}_n(\mathsf{k})$ of order prime to p. Show that K_{r-1}/K_r has a natural $\mathsf{k}[H]$-module structure, and that it is isomorphic to a multiple of the adjoint representation of H (that is, a direct sum of several copies of the adjoint representation).

Problem 5.5. Let R be a coefficient ring, let X be a topologically finitely generated closed subgroup of $\Gamma_n(R)$, and let H be a finite subgroup of $\mathrm{GL}_n(R)$ whose order is not divisible by p and which normalizes X. Suppose that the Frattini quotient $\bar{X}$, viewed as a $\mathsf{k}[H]$-module, is prime-to-adjoint. Show that X must be trivial. (Hint: Using the notation of the previous problem, let r be minimal such that X is not contained in K_r, and consider the image of X in K_{r-1}/K_r.)

The last two problems are taken from [9, section 2]. The last one is particularly significant: it highlights the importance of the adjoint representation and explains why the prime-to-adjoint condition is so significant.

Pro-p-extensions

In the picture above, we have a number field K, a set of primes S_1 including all primes above p and all archimedean primes, and an extension L/K which is the maximal pro-p-extension of K unramified outside S_1. Our goal in this section is to collect some of the known facts about the Galois group $P = G(L/K)$. We again follow Boston's exposition in [9].

The main result we want to quote gives the generator and relation ranks of P in terms of the arithmetic of K. (For a first hint about why this is possible, notice that the Burnside Basis Theorem allows one to reduce the question about the generator rank to class field theory, since $\overline{P}$ is abelian.)

Let r_2 denote the number of complex primes of K. As usual, if v is a (finite or infinite) prime of K, we write K_v for the completion of K at v. For any field F, let $\delta(F) = 1$ if F contains any primitive p-th roots of unity, and $\delta(F) = 0$ otherwise. Let $H = G(K/\mathbb{Q}) \cong \operatorname{Im}(\bar\rho)$.

Let Z_S be the set of nonzero elements $x \in K$ such that the fractional ideal (x) generated by x is the p-th power of some ideal and such that x is a p-th power in each completion K_v for $v \in S_1$. Of course, if x is already a p-th power in K, then $x \in Z_S$. Notice that both $(K^\times)^p$ and Z_S are stable under the Galois action of H. Let B_S denote the $\mathbb{F}_p[H]$-module $Z_S/(K^\times)^p$.

We can now (finally) state the theorem about P.

Theorem 5.4. *Let $d(P)$ and $r(P)$ denote, as above, the generator and relation ranks of P. Then*
$$r(P) = \left(\sum_{v \in S_1} \delta(K_v)\right) - \delta(K) + \dim_{\mathbb{F}_p} B_S,$$
and
$$d(P) = r_2 + 1 + r(P).$$
In particular, P is topologically finitely generated.

For what follows, we will need to know something about $\overline{P}$ as an H-module. To set up the theorem, let $\overline{E}$ denote the units of K modulo p-th powers, and let $\overline{E}_v$ denote the units of K_v modulo p-th powers. If the class number of K is prime to p, then we can deduce from global class field theory an exact sequence of $\mathbb{F}_p[H]$-modules
$$0 \longrightarrow B_S \longrightarrow \overline{E} \longrightarrow \bigoplus_{v \in S_1} \overline{E}_v \longrightarrow \overline{P} \longrightarrow 0.$$

For each rational prime ℓ, let H_ℓ be a decomposition subgroup of H at ℓ, and let H_∞ be the subgroup of H generated by a complex conjugation. Let $\mu_p(K)$ be the group of p-th roots of unity in K, which we think of as a module over H and over the H_ℓ. Note that it's perfectly possible for $\mu_p(K)$ to be trivial.

Theorem 5.5 (Boston-Mazur). *Under the hypotheses above, if H has order prime to p, then we have the following isomorphisms of $\mathbb{F}_p[H]$-modules:*
$$\bigoplus_{v \in S_1} \overline{E}_v \cong \mathbb{F}_p[H] \oplus \left(\bigoplus_{\ell \in S} \operatorname{Ind}_{H_\ell}^H \mu_p\right)$$
$$\overline{E} \oplus \mathbb{F}_p \cong \mu_p \oplus \operatorname{Ind}_{H_\infty}^H \mathbb{F}_p$$

See [**10**] for the proof. The point is this: using this theorem and the exact sequence above, we can determine the decomposition of $\overline{P}$ into irreducible subrepresentations of H. Later, we'll want to look for homomorphisms from P to $\Gamma_n(R)$, where R is a coefficient ring. We will often be able to reduce this to looking for H-homomorphisms from $\overline{P}$ to appropriate quotients of submodules of $\Gamma_n(R)$. By the results in the previous section, the latter only involve irreducible subrepresentations contained in the adjoint representation of H. So the subrepresentations of $\overline{P}$ which are *not* contained in the adjoint representation will have to have trivial image.

Tame representations

Keep notations as above, and assume now that $\bar{\rho}$ is tame, that is, that the order of $H = \text{Im}(\bar{\rho})$ is not divisible by p. Recall that in this case P is a normal pro-p-Sylow subgroup of $\tilde{\Pi}$. By the Schur-Zassenhaus theorem, $\tilde{\Pi}$ is the semidirect product of P and a subgroup $A \cong G/P = H$. Because $\Gamma_n(W(\mathsf{k}))$ is pro-p, we can use Schur-Zassenhaus again to find a subgroup H_1 of $\text{GL}_n(W(\mathsf{k}))$ which is isomorphic to H and therefore we can find a lift

$$\rho_1 : \tilde{\Pi} \longrightarrow \text{GL}_n(W(\mathsf{k}))$$

inducing an isomorphism from A to H_1. We get, then, an induced inclusion $\sigma : A \longrightarrow \text{GL}_n(W(\mathsf{k}))$, which we will fix from now on.

Problem 5.6. Show that the lift ρ_1 is unique up to strict equivalence, and conclude that any two choices of the inclusion σ are conjugate by an element of $\Gamma_n(W(\mathsf{k}))$.

For any coefficient ring R we have a canonical homomorphism $W(\mathsf{k}) \longrightarrow R$ and hence a homomorphism $\sigma_R : A \longrightarrow \text{GL}_n(R)$. We let A act on $\Gamma_n(R)$ by conjugation via this homomorphism.

Given all this setup, recall that any deformation of $\bar{\rho}$ induces a homomorphism from $P = \text{Ker}\,\bar{\rho}$ to $\Gamma_n(R)$. We can make this into a precise correspondence by taking into account the A-actions on both sides.

Define a set-valued covariant functor $\mathbf{E}_{\bar{\rho}}$ on $\mathcal{C}$ by defining, for each coefficient ring R,

$$\mathbf{E}_{\bar{\rho}}(R) = \text{Hom}_A(P, \Gamma_n(R)),$$

where Hom_A denotes the set of continuous homomorphisms from P to $\Gamma_n(R)$ which commute with the A action. We want to compare this functor to the deformation functor $\mathbf{D}_{\bar{\rho}}$.

Notice that since $\tilde{\Pi}$ is the semidirect product of P and A (and we have been careful to take the A-action into account), any element $\phi \in \mathbf{E}_{\bar{\rho}}(R)$, together with the inclusion σ_R, defines a deformation of $\bar{\rho}$ to R. Hence, there is a natural morphism of functors $\mathbf{E}_{\bar{\rho}} \longrightarrow \mathbf{D}_{\bar{\rho}}$.

Theorem 5.6 (Boston). *The functor $\mathbf{E}_{\bar{\rho}}$ is always representable. Furthermore,*
  i. *If $C(\bar{\rho}) = \mathsf{k}$, the natural morphism of functors $\mathbf{E}_{\bar{\rho}} \longrightarrow \mathbf{D}_{\bar{\rho}}$ is an isomorphism.*
  ii. *Otherwise, the morphism is smooth and induces an isomorphism on tangent spaces.*

Proof. Let's first prove that we have an isomorphism when $C(\bar{\rho}) = \mathsf{k}$. To lighten the notation, write $\mathbf{E} = \mathbf{E}_{\bar{\rho}}$ and $\mathbf{D} = \mathbf{D}_{\bar{\rho}}$. Given a coefficient ring R, we claim that the induced map $\mathbf{E}(R) \longrightarrow \mathbf{D}(R)$ is a bijection.

To see that it is surjective, suppose ρ is a deformation of $\bar{\rho}$ to R. Then ρ induces a lift $A \longrightarrow \text{GL}_n(R)$. Since all such lifts are conjugate by elements of $\Gamma_n(R)$ (see the problem above), we can choose a homomorphism ψ in the strict equivalence class of ρ such that $\psi|_A = \sigma_R$. Then $\psi|_P$ is an element of $E(R)$ which maps to the strict equivalence class of ψ, that is, to ρ.

To see that the map is injective, suppose ϕ_1 and ϕ_2 produce strictly equivalent lifts ψ_1 and ψ_2 of $\bar{\rho}$. Since both ψ_1 and ψ_2 induce σ on A, the matrix realizing the strict equivalence must be an element of $\Gamma_n(R)$ acting trivially on A by conjugation, i.e., commuting with the image of A. However, under our assumption that $C(\bar{\rho}) = \mathsf{k}$, the only elements of $\Gamma_n(R)$ commuting with the image of A are the scalars. Hence

LECTURE 5. EXPLICIT DEFORMATIONS

ψ_1 and ψ_2 differ by conjugation by a scalar, i.e., don't differ at all. In particular, their restrictions to P are the same, and hence $\phi_1 = \phi_2$.

We leave the case when $C(\bar{\rho}) \neq \mathsf{k}$ to the reader, and proceed to prove that $\mathbf{E}$ is representable. Choose generators $x_1, x_2, \ldots, x_d$ of P. The image of x_r in $\Gamma_n(R)$ is a matrix

$$\begin{pmatrix} 1 + m_{11}^{(r)} & m_{12}^{(r)} & \ldots & m_{1n}^{(r)} \\ m_{21}^{(r)} & 1 + m_{22}^{(r)} & \ldots & m_{2n}^{(r)} \\ \ldots & & & \\ m_{n1}^{(r)} & m_{n2}^{(r)} & \ldots & 1 + m_{nn}^{(r)} \end{pmatrix},$$

where the $m_{ij}^{(r)}$ are in the maximal ideal of R.

We will construct the ring representing $\mathbf{E}$ as a quotient of the power series ring $W(\mathsf{k})[[T_{11}^{(1)}, \ldots, T_{nn}^{(d)}]]$ in dn^2 variables. Let $\mathcal{F}$ be the free pro-p-group on $x_1, x_2, \ldots, x_d$, so that we get an exact sequence of groups

$$1 \longrightarrow N \longrightarrow \mathcal{F} \longrightarrow P \longrightarrow 1.$$

A homomorphism from P to $\Gamma_n(R)$ is exactly the same as a homomorphism from $\mathcal{F}$ to $\Gamma_n(R)$ such that N is in the kernel. We begin by defining a homomorphism from the free pro-p-group $\mathcal{F}$ to $\Gamma_n(W(\mathsf{k})[[T_{ij}^{(r)}]])$ such that the image of x_r is the matrix

$$\begin{pmatrix} 1 + T_{11}^{(r)} & T_{12}^{(r)} & \ldots & T_{1n}^{(r)} \\ T_{21}^{(r)} & 1 + T_{22}^{(r)} & \ldots & T_{2n}^{(r)} \\ \ldots & & & \\ T_{n1}^{(r)} & T_{n2}^{(r)} & \ldots & 1 + T_{nn}^{(r)} \end{pmatrix}.$$

Requiring that N be in the kernel amounts to requiring that certain equations involving the $T_{ij}^{(r)}$ hold. Requiring that the A-actions commute with the homomorphism imposes further equations. Let I be the ideal of $W(\mathsf{k})[[T_{ij}^{(r)}]]$ generated by all these equations. If we let $\mathcal{R} = W(\mathsf{k})[[T_{ij}^{(r)}]]/I$, we have produced a homomorphism $\phi : P \longrightarrow \Gamma_n(\mathcal{R})$. It is clear that this is the universal such homomorphism, and hence that $\mathcal{R}$ represents the functor $\mathbf{E}$. $\square$

Problem 5.7. Check the remaining assertions in the theorem.

It's worth noting that this proof sheds some light on the issue of what it means for the functor $\mathbf{D}$ is representable, at least in the case when $\bar{\rho}$ is tame. As usual, the question of representability turns out to be connected to whether there are "extra automorphisms." In our situation, the question turns out to be whether the lift $A \longrightarrow \mathrm{GL}_n(R)$ is unique.

This result has been generalized by Böckle; see [**5**], for example.

Since the two functors $\mathbf{E}$ and $\mathbf{D}$ are isomorphic, so are their tangent spaces. Hence

$$\dim_{\mathsf{k}} t_{\mathbf{D}} = \dim_{\mathsf{k}} \mathrm{Hom}_A(P, \Gamma_n(\mathsf{k}[\varepsilon])).$$

Note, now that $\Gamma_n(\mathsf{k}[\varepsilon])$ is isomorphic to $\mathrm{Ad}(\bar{\rho})$ as an A-module; in particular, it is a p-elementary abelian group, so every homomorphism from P to $\Gamma_n(\mathsf{k}[\varepsilon])$ must factor through $\bar{P}$. Hence

$$\dim_{\mathsf{k}} t_{\mathbf{D}} = \dim_{\mathsf{k}} \mathrm{Hom}_A(\bar{P}, \mathrm{Ad}(\bar{\rho})).$$

Now, since the order of A is not divisible by p, both $\overline{P}$ and $\mathrm{Ad}(\overline{\rho})$ can be decomposed as a sum or irreducible A-modules. The dimension on the right, then, can be computed in terms of the number of irreducible A-modules which appear in the decomposition of both $\overline{P}$ and $\mathrm{Ad}(\overline{\rho})$. We are back to the situation described above: if we can determine which sub-representations occur in $\overline{P}$, we can compute the dimension of the tangent space.

Suppose $C(\overline{\rho}) = \mathsf{k}$ and let $\mathcal{R}$ be the universal deformation ring. Note that if we can compute the dimension d of the tangent space, we know that there is a surjective homomorphism of coefficient rings

$$W(\mathsf{k})[[T_1, \ldots, T_d]] \longrightarrow \mathcal{R}$$

inducing the identity on tangent spaces. Thus, when $\overline{\rho}$ is tame we can read the dimension off from the structure of $\overline{P}$ as an $\mathbb{F}_p[H]$-module. This crucial idea, together with the fact that what matters are the sub-representations which occur in the adjoint representation, has been dubbed the "prime-to-adjoint" principle by Böckle.

Here's a sample theorem:

Theorem 5.7 (Boston). *Let p be an odd prime. Suppose that $\overline{\rho} : G_{\mathbb{Q},S} \longrightarrow \mathrm{GL}_2(\mathbb{F}_p)$ is odd and absolutely irreducible. Let $H = \mathrm{Im}(\overline{\rho})$, and suppose that p does not divide the order of H, so that $\overline{\rho}$ is tame. Let K be the field fixed by the kernel of $\overline{\rho}$, and let S_1 be the set of primes of K which lie above the primes in S. Let*

$$V = \mathrm{coker}\left(\mu_p(K) \longrightarrow \bigoplus_{v \in S_1} \mu_p(K_v)\right),$$

and let $B = B_S$ defined as above. Both V and B are $\mathbb{F}_p[H]$-modules. Suppose that the class number of K is not divisible by p and that both V and B are relatively prime to $\mathrm{Ad}(\overline{\rho})$ as $\mathbb{F}_p[H]$-modules. Then

$$\mathcal{R}(\overline{\rho}) \cong \mathbb{Z}_p[[T_1, T_2, T_3]],$$

and one can give an explicit description of $\boldsymbol{\rho}$ on (well-chosen) generators for $\tilde{\Pi} = G(L/\mathbb{Q})$.

Proof. (Just a sketch.) From Theorem 5.5, one can see that $\overline{P}$ is generated by

  i. an element $\overline{x}$ which is fixed under H,
  ii. an element $\overline{y}$ such that $\overline{y}^c = (\overline{y})^{-1}$, where c is a chosen complex conjugation in G,
  iii. other elements which generate a prime-to-adjoint $\mathbb{F}_p[H]$-module.

Using theorem 5.3, we see that P is generated by x (fixed under H), y such that $y^c = y^{-1}$, and other generators which are prime-to-adjoint. From this we see that the dimension of the tangent space is three and we can define a deformation to $\mathbb{Z}_p[[T_1, T_2, T_3]]$ by

$$x \mapsto \begin{pmatrix} 1+T_1 & 0 \\ 0 & 1+T_1 \end{pmatrix}$$

$$y \mapsto \begin{pmatrix} (1+T_2T_3)^{1/2} & T_2 \\ T_3 & (1+T_2T_3)^{1/2} \end{pmatrix}$$

and mapping all the other generators (which are prime-to-adjoint) to the identity. This gives the universal deformation. □

While this result covers a case where the deformation problem is unobstructed, Boston's work also includes various results for obstructed cases. See, for example, [**9**].

Flach, Boston, and Ullom have similar results for deformations of residual representations $\bar\rho$ which come from the p-division points of an elliptic curve. For example, the following is a result of Flach. Suppose E is an elliptic curve over $\mathbb{Q}$, $p \geq 5$, and assume that E has good reduction at p. Take
$$S = \{\text{primes of bad reduction}\} \cup \{p, \infty\},$$
let $E[p]$ be the points of order p on E, and let
$$\bar\rho = \bar\rho_E : G_{\mathbb{Q},S} \longrightarrow \mathrm{GL}_2(\mathbb{F}_p)$$
be the representation given by the action of $G_{\mathbb{Q},S}$ on $E[p]$.

Theorem 5.8 (Flach). *Suppose that*
  i. $\bar\rho : G_{\mathbb{Q},S} \longrightarrow \mathrm{GL}_2(\mathbb{F}_p)$ *is surjective,*
  ii. for all $r \in S$, $\mathrm{H}^0(\mathbb{Q}_r, E[p] \otimes E[p]) = 0$,
  iii. p *does not divide* $\Omega^{-1} L(M, 2)$, *where* $M = \mathrm{Sym}^2(E)$ *and* Ω *is a transcendental period.*
Then $\mathcal{R}(\bar\rho) \cong \mathbb{Z}_p[[T_1, T_2, T_3]]$.

The point is to show that the deformation problem is unobstructed by exploiting the fact that $\mathrm{Sym}^2(E)$ is closely related to $\mathrm{Ad}(\bar\rho)$. See Appendix 2 for an expository account of the details and [**50**] for the original publication.

Boston and Ullom have obtained results of this type for the case of an elliptic curve with complex multiplication in [**11**]. Their result also covers some cases where the deformation problem is obstructed.

The most far-reaching results based on this method have been obtained by Böckle. For example, in [**6**], he shows under mild extra hypotheses that if $\bar\rho$ is tame and absolutely irreducible then the rigid space $\mathrm{Spf}\,\mathcal{R}(\bar\rho)^{\mathrm{rig}}$ has the expected Krull dimension. In [**5**] he proves a generalization of a result of Mazur (which we will mention in a later lecture) that implies, in many cases (and without a tameness assumption), that $\mathcal{R}(\bar\rho)$ has the predicted dimension. Finally (for example, in [**7**]), he has also studied carefully the case in which $\bar\rho$ is Borel (therefore reducible) but still satisfies $C(\bar\rho) = \mathsf{k}$.

LECTURE 6
Deformations With Prescribed Properties

Suppose we have a residual representation $\bar{\rho}$ which satisfies $C(\bar{\rho}) = k$. Then we have a universal deformation ring. As we pointed out before, we can think of this ring as defining a universal deformation *space* whose points correspond to actual deformations of $\bar{\rho}$. For example, in the previous lecture we saw cases where the universal deformation ring is $W(k)[[T_1, T_2, T_3]]$; the corresponding space is three-dimensional (over $W(k)$), and the T_i can be thought of as giving "coordinates" for our space: for each triple (m_1, m_2, m_3) in the maximal ideal of some $W(k)$-algebra, we get a representation by mapping T_i to m_i. We would like to understand what these "coordinates" mean in terms of the representations themselves. One strategy for doing this is to consider interesting subspaces (equivalently, quotient rings of the universal deformation ring). The natural way to do this is to consider subspaces of the universal deformation space that correspond to deformations that have certain interesting or desirable properties. This idea was first considered in Mazur's original paper [**97**], where he discusses ordinary deformations and also looks at several other possible conditions.

Even more important is the observation that in many circumstances we do not want to consider all deformations, but rather only those satisfying certain conditions. The best known example of this is when we try to prove modularity of certain deformations. In that situation, it is natural to require our deformations to have those properties that we know modular representations will have. We will discuss modular representations more carefully in the next lecture.

The basic idea for today, then, is this: suppose that the residual representation has a certain property. Then one can ask which deformations retain that property. For well-chosen properties, this allows us to define a representable sub-functor of the general deformation functor, and therefore to obtain a "universal deformation with the given property", which will correspond to a quotient of the universal deformation ring (or, from the geometric point of view, will define a subspace of the full deformation space.)

Deformation Conditions

Let's begin with a general account of "deformation conditions," that is, conditions that give rise to "good" subfunctors of the deformation functor. We follow the discussion in [**101**].

What do we expect of a "deformation condition?" Well, the first thing we need is that the association

$$R \rightsquigarrow \{\text{deformations of } \bar\rho \text{ satisfying our condition}\}$$

be a subfunctor of $\mathbf{D}_{\bar\rho}$, so that we have a deformation problem to work with. In addition, we would like this functor to be "relatively representable," that is, we would like our subfunctor to be sufficiently well behaved so that it is representable whenever $\mathbf{D}_{\bar\rho}$ is.

Recall that we can interpret a representation

$$\rho : \Pi \longrightarrow \mathrm{GL}_n(A)$$

by saying that it gives us a free A-module of rank n with a continuous A-linear action of the profinite group Π. In practice, all of the deformation conditions that have been useful have amounted to specifying properties that this Π-module should have. For technical reasons, it makes sense to specify these properties for representations where A is an artinian coefficient Λ-algebra.

Before we state the definition, let's introduce some terminology. Let A and A_1 be artinian coefficient Λ-algebras. If we are given a representation $\rho : \Pi \longrightarrow \mathrm{GL}_n(A)$ and a homomorphism of artinian coefficient Λ-algebras $\alpha : A \longrightarrow A_1$, then we get a representation $\rho_1 : \Pi \longrightarrow \mathrm{GL}_n(A_1)$ by composing ρ with the homomorphism $\mathrm{GL}_n(A) \longrightarrow \mathrm{GL}_n(A_1)$ induced by α. We will call ρ_1 the *push-forward of ρ by α*, and sometimes denote it by $\alpha_*\rho$. Of course, this works just as well for deformations of a residual representation $\bar\rho$; in fact, the push-forward operation is what makes $\mathbf{D}_{\bar\rho}$ a functor (we could have denoted the push-forward map by $\mathbf{D}_{\bar\rho}(\alpha)$ instead of α_*).

It's perhaps worth remarking that if we interpret a representation to $\mathrm{GL}_n(A)$ as giving a Π-module structure to the free A-module of rank n, then the push-forward operation is just the tensor product: if M is the free A module of rank n with a continuous linear Π-action, then the push-forward is $M \otimes_A A_1$, where the map $\alpha : A \longrightarrow A_1$ gives the A-module structure on A_1.

Now we are ready to define a (good) deformation condition. Informally, we want this to be a condition on deformations that is satisfied by the residual representation $\bar\rho$ (otherwise there's no point) and that is functorial (that is, preserved by push-forwards). Finally, we want the resulting functor to be relatively representable, so we require that our condition behave well with respect to fiber products and subrings.

Definition 6.1. Let $\bar\rho$ be a residual representation of dimension n. A *deformation condition* on deformations of $\bar\rho$ is a property $\mathfrak{Q}$ of n-dimensional representations of Π defined over artinian coefficient Λ-algebras (equivalently, of A-modules which are free of rank n over A and have a continuous Π-action) which satisfies the following conditions.

  i. The residual representation $\bar\rho$ has property $\mathfrak{Q}$.

ii. Given a deformation $\rho : \Pi \longrightarrow \mathrm{GL}_n(A)$ of $\bar\rho$ and a homomorphism of coefficient Λ-algebras $\alpha : A \longrightarrow A_1$, if ρ has property $\mathcal{Q}$ then so does the push-forward $\alpha_*\rho$.

iii. Let

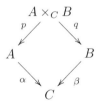

be a fiber product diagram in $\mathcal{C}^0_\Lambda$, and let

$$\rho : \Pi \longrightarrow \mathrm{GL}_n(A \times_C B)$$

be a deformation of $\bar\rho$. Then ρ has property $\mathcal{Q}$ if and only if both $p_*\rho$ and $q_*\rho$ have property $\mathcal{Q}$.

iv. Let $\alpha : A \longrightarrow A_1$ be an injective homomorphism of coefficient Λ-algebras and let $\rho : \Pi \longrightarrow \mathrm{GL}_n(A)$ be a deformation of $\bar\rho$. If $\alpha_*\rho$ has property $\mathcal{Q}$ then so does ρ.

It's probably worthwhile to comment a bit on the role of these four conditions. The first is clearly necessary for our subfunctor not to be trivial (and to make sure its value on k is a singleton). The second makes sure that we are going to get a functor. The third is clearly related to the Schlessinger criteria, and its role will be to make sure that the subfunctor is relatively representable. As Mark Dickinson explained to me, the fourth in fact follows from (ii) and (iii)—see the complements to this lecture for a proof.

Given a deformation condition, we can define a subfunctor of the deformation functor $\mathbf{D}_{\bar\rho}$:

Definition 6.2. Let $\mathcal{Q}$ be a deformation condition for $\bar\rho$. We define a functor

$$\mathbf{D}_\mathcal{Q} : \mathcal{C}^0_\Lambda \rightsquigarrow \underline{\mathrm{Sets}}$$

by setting, for each artinian coefficient Λ-algebra A,

$$\mathbf{D}_\mathcal{Q}(A) = \{\text{deformations of } \bar\rho \text{ to } A \text{ which have property } \mathcal{Q}\}.$$

We can then extend $\mathbf{D}_\mathcal{Q}$ to all of $\mathcal{C}_\Lambda$ by continuity: if R is a coefficient Λ-algebra,

$$\mathbf{D}_\mathcal{Q}(R) = \varprojlim_k \mathbf{D}_\mathcal{Q}(R/\mathfrak{m}^k).$$

In other words, we are saying that a deformation of $\bar\rho$ to a coefficient Λ-algebra R has property $\mathcal{Q}$ if and only if its reductions modulo $\mathfrak{m}^k$ have property $\mathcal{Q}$ for all $k \geq 1$.

Problem 6.1. Check that $\mathbf{D}_\mathcal{Q}$ is a subfunctor of $\mathbf{D}_{\bar\rho}$. (The main thing to check is that it does the right thing when we have a homomorphism of artinian coefficient Λ-algebras.)

Theorem 6.1. *If $\mathcal{Q}$ is a deformation condition for $\bar\rho$, then $\mathbf{D}_\mathcal{Q}$ satisfies conditions **H1**, **H2**, and **H3** in Schlessinger's theorem. If $C(\bar\rho) = \mathrm{k}$, then $\mathbf{D}_\mathcal{Q}$ also satisfies property **H4**, and therefore is representable by a ring $\mathcal{R}_\mathcal{Q}$ which is a quotient of the universal deformation ring $\mathcal{R}(\bar\rho)$.*

Proof. This is pretty much immediate from conditions (ii) and (iii) in the definition of a deformation problem. □

Problem 6.2. Check the details!

Suppose we have a deformation condition $\mathcal{Q}$. Then we can consider the tangent spaces of both $\mathbf{D}_\mathcal{Q}$ and $\mathbf{D}_{\overline{\rho}}$:

$$\mathbf{D}_\mathcal{Q}(\mathsf{k}[\varepsilon]) \subset \mathbf{D}_{\overline{\rho}}(\mathsf{k}[\varepsilon]).$$

Recall that we have a cohomological interpretation of the larger space:

$$\mathbf{D}_{\overline{\rho}}(\mathsf{k}[\varepsilon]) \cong \mathrm{H}^1(\Pi, \mathrm{Ad}(\overline{\rho})).$$

Definition 6.3. We define $\mathrm{H}^1_\mathcal{Q}(\Pi, \mathrm{Ad}(\overline{\rho}))$ to be the subspace of $\mathrm{H}^1(\Pi, \mathrm{Ad}(\overline{\rho}))$ corresponding, under this isomorphism, to the subspace $\mathbf{D}_\mathcal{Q}(\mathsf{k}[\varepsilon])$.

It's important to emphasize that this definition, as Mazur says, is really only a "promissory note." What it does is hint at the fact that for most of the interesting deformation conditions $\mathcal{Q}$ it will be possible to describe this cohomology group in an intrinsic way.

Problem 6.3. Use a similar dodge to define the Ext group $\mathrm{Ext}^1_{\mathsf{k}[[\Pi]],\mathcal{Q}}(V_{\overline{\rho}}, V_{\overline{\rho}})$. Can you describe this intrinsically in terms of extensions of Π-modules?

We go on to consider several different possible choices for the deformation condition $\mathcal{Q}$.

Deformations with fixed determinant

Probably the most natural restriction we can put on deformations is to fix their determinant. To see why this is something we might want to do, remember that representations attached to elliptic curves have determinant equal to the cyclotomic character, and that representations attached to modular forms have determinant equal to a character of finite order times a power (related to the weight of the modular form) of the cyclotomic character. Thus, for example, when we are trying to show that an elliptic curve is modular by studying its Galois representation it usually suffices to look only at the deformations that have determinant equal to the cyclotomic character.

In order for it to make sense to say that "all deformations have determinant δ," we need to take δ to be a character with values in $\Lambda^\times$, which we can then view as taking values in any coefficient Λ-algebra via the structural map.

Definition 6.4. Let δ be a continuous homomorphism

$$\delta : \Pi \longrightarrow \Lambda^\times,$$

and for every coefficient Λ-algebra R let δ_R be the composition

$$\delta_R : \Pi \xrightarrow{\delta} \Lambda^\times \longrightarrow R^\times.$$

We say a deformation ρ of $\overline{\rho}$ to R *has determinant δ* if $\det \rho = \delta_R$.

Notice that this abuses the language somewhat, since after all the determinant of a deformation "with determinant δ" is actually not δ! In practice, this does not lead to any trouble, so we won't worry too much about it.

Now suppose $\overline{\rho}$ itself has determinant δ. Let "det $= \delta$" be shorthand for the property of having determinant δ.

Lemma 6.2. *Suppose $\bar{\rho}$ has determinant δ. Then "det $= \delta$" is a deformation condition.*

Problem 6.4. Prove the lemma. (This is quite straightforward.)

Suppose, now, that $C(\bar{\rho}) = \mathsf{k}$. Then the lemma implies that there exists a quotient $\mathcal{R}_{\det=\delta}$ of the universal deformation ring $\mathcal{R}$ corresponding to those deformations whose determinant is δ. It's tangent space is not too hard to pin down: let $\mathrm{Ad}^0(\bar{\rho})$ denote the subspace of $\mathrm{Ad}(\bar{\rho})$ consisting of those matrices whose trace is zero. It's clear that $\mathrm{Ad}^0(\bar{\rho})$ is stable under the conjugation action of Π. Then we have

Lemma 6.3. *If $p \nmid n$, the tangent space of the functor $\mathbf{D}_{\det=\delta}$ is given cohomologically by*

$$\mathbf{D}_{\det=\delta}(\mathsf{k}[\varepsilon]) = \mathrm{H}^1(\Pi, \mathrm{Ad}^0(\bar{\rho})) \subset \mathrm{H}^1(\Pi, \mathrm{Ad}(\bar{\rho})).$$

If $p|n$, then $\mathrm{H}^1(\Pi, \mathrm{Ad}^0(\bar{\rho}))$ is in general no longer a subset of $\mathrm{H}^1(\Pi, \mathrm{Ad}(\bar{\rho}))$, but the inclusion $\mathrm{Ad}^0(\bar{\rho}) \hookrightarrow \mathrm{Ad}(\bar{\rho})$ still induces a map from one to the other; in this case

$$\mathbf{D}_{\det=\delta}(\mathsf{k}[\varepsilon]) = \mathrm{Im}\left(\mathrm{H}^1(\Pi, \mathrm{Ad}^0(\bar{\rho})) \longrightarrow \mathrm{H}^1(\Pi, \mathrm{Ad}(\bar{\rho}))\right).$$

Proof. The proof is the same in both cases. Essentially, we just repeat the argument, given in lecture 4, connecting lifts to $\mathsf{k}[\varepsilon]$ with elements in the cohomology, and then note that the diagram

$$\begin{array}{ccc} 1+\varepsilon \mathrm{M}_n(\mathsf{k}) & \longrightarrow & \mathrm{M}_n(\mathsf{k}) \\ \downarrow{\scriptstyle\det} & & \downarrow{\scriptstyle\mathrm{Tr}} \\ 1+\varepsilon\mathsf{k} & \longrightarrow & \mathsf{k} \end{array}$$

is commutative, where the horizontal maps are the isomorphism $1 + \varepsilon b \mapsto b$. The requirement of fixed determinant forces the $1 + \varepsilon \mathrm{M}_n(\mathsf{k})$ part of the lift to have determinant 1, which translates to trace 0 when we go to $\mathrm{M}_n(\mathsf{k})$. Thus, we get elements in $\mathrm{H}^1(\Pi, \mathrm{Ad}(\bar{\rho}))$ which are represented by cocycles taking values in $\mathrm{Ad}^0(\bar{\rho})$, and hence belong to the image of $\mathrm{H}^1(\Pi, \mathrm{Ad}^0(\bar{\rho}))$ in $\mathrm{H}^1(\Pi, \mathrm{Ad}(\bar{\rho}))$. When $p \nmid n$, this image is just $\mathrm{H}^1(\Pi, \mathrm{Ad}^0(\bar{\rho}))$ itself. $\square$

Problem 6.5. Suppose $\bar{\rho}$ satisfies $C(\bar{\rho}) = \mathsf{k}$ and has determinant δ. Let $\mathcal{R}$ be the universal deformation ring of $\bar{\rho}$, let ρ be the universal deformation, and let $\mathcal{R}_{\det=\delta}$ be the universal "deformation with determinant δ" ring. Let $\delta_{\mathcal{R}}$ be

$$\delta_{\mathcal{R}} : \Pi \xrightarrow{\delta} \Lambda^\times \longrightarrow \mathcal{R}^\times,$$

the composition of δ with the structure homomorphism of $\mathcal{R}$ as a Λ-algebra. Show that $\mathcal{R}_{\det=\delta}$ is the quotient of $\mathcal{R}$ by the closed ideal generated by the elements

$$\delta_{\mathcal{R}}(g) - \det \rho(g),$$

where g runs through a set of topological generators of Π.

Problem 6.6. (From [6].) Assume p does not divide n. Suppose $\bar{\rho}$ satisfies $C(\bar{\rho}) = \mathsf{k}$ and has determinant δ. Consider the three universal objects

　i. $\mathcal{R}$ is the universal deformation ring and ρ the universal deformation,
　ii. $\mathcal{R}_{\det=\delta}$ is the universal ring for deformations of determinant δ and ρ_δ is the universal deformation of determinant δ,

iii. $\Lambda[[\Gamma]]$ is the universal deformation ring of the trivial character and ϵ is the universal deformation.

Show that
$$\mathcal{R} = \mathcal{R}_{\det=\delta}\hat{\otimes}_\Lambda \Lambda[[\Gamma]]$$
and that
$$\rho = \rho_\delta \otimes \epsilon.$$

The result in this last problem is used by Böckle in [**6**] to reduce the dimension conjecture for $\mathcal{R}$ to a question about the dimension of $\mathcal{R}_{\det=\delta}$. In general, it is not too hard to go back and forth between deformations in general (perhaps even subject to other deformation conditions) and deformations (of the same type) with fixed determinant.

Categorical deformation conditions

One interesting class of deformation conditions was introduced by Ramakrishna in [**115**]. The idea is to require that our deformations (at least at the artinian level) define Π-modules that belong to a particularly nice subcategory of the category of Λ-modules of finite length with a continuous Λ-linear action of Π.

To set this up, let $\mathcal{P}$ be a full subcategory of the category of Λ-modules of finite length with a continuous Λ-linear action of Π, and assume $\mathcal{P}$ is closed under passage to sub-objects, quotients, and finite direct sums. Given $\bar{\rho}$, we say that a deformation $\rho : \Pi \longrightarrow \mathrm{GL}_n(R)$ is of type $\mathcal{P}$ if the Π-modules defined by all of its quotients $\rho_k : \Pi \longrightarrow \mathrm{GL}_n(R/\mathfrak{m}^k)$, viewed as Λ-modules with a continuous linear action of Π, are in the category $\mathcal{P}$.

Theorem 6.4 (Ramakrishna). *Suppose $\bar{\rho}$ is of type $\mathcal{P}$. The condition of "being of type $\mathcal{P}$" is a deformation condition.*

Proof. We need to prove that the property of "being of type $\mathcal{P}$" is preserved by push-forwards and that it works well with fiber products. Suppose first that A and A_1 are artinian coefficient rings, that we have a coefficient ring homomorphism $\alpha : A \longrightarrow A_1$, and that ρ is a deformation of $\bar{\rho}$ to A which has property $\mathcal{P}$. Let $M = A^n$ and let $M_1 = A_1^n$, both of which are endowed with continuous linear actions of Π via ρ (and its push-forward to A_1). We think of both M and M_1 as Λ-modules of finite length, and we know that M is in the subcategory $\mathcal{P}$.

Since both A and A_1 are of finite length, there exists an artinian coefficient ring B such that

- $\alpha : A \longrightarrow A_1$ factors through B, that is, there exist coefficient ring homomorphisms $\alpha_1 : A \longrightarrow B$ and $\alpha_2 : B \longrightarrow A_1$ such that $\alpha = \alpha_2 \circ \alpha_1$,
- B is free of finite rank as an A-module, and
- $\alpha_2 : B \longrightarrow A_1$ is surjective.

(For example, we can take B to be a quotient of a power series ring over A by a power of its maximal ideal.) Now the pushforward of M via α_1 is simply a direct sum M^r of copies of M, and therefore is an object of $\mathcal{P}$, and M_1, which is the pushforward of M^r under α_2, is a quotient of M^r. Since $\mathcal{P}$ is closed under finite direct sums and under quotients, it follows that M_1 is an object of $\mathcal{P}$. This proves condition (ii) in the definition of a deformation condition.

Property (iii) is easier. Suppose we have homomorphisms of artinian coefficient rings $A \longrightarrow C$ and $B \longrightarrow C$. Let $R = A \times_C B$, and suppose we have a deformation ρ of $\bar{\rho}$ to R. We already know that if ρ has property $\mathcal{P}$ then so do the push-forwards to A and B. For the converse, let ρ_A and ρ_B be the push-forwards, and suppose both have property $\mathcal{P}$. Notice that R is a subring of $A \oplus B$, and therefore R^n is a submodule of $A^n \oplus B^n$. Since we know both A^n and B^n are in $\mathcal{P}$ and $\mathcal{P}$ is closed under direct sums and sub-objects, it follows that R^n is in $\mathcal{P}$, and we are done. $\square$

The most important example of a property of this type is the one Ramakrishna considered in his original paper and which was later used by Wiles. Suppose $\Pi = G_{\mathbb{Q}_\ell}$ is the absolute Galois group of $\mathbb{Q}_\ell$, and we let $\mathcal{P} = \mathcal{P}_{fl}$ be the category of all $G_{\mathbb{Q}_\ell}$ representations ρ (over artinian rings A) such that the deformation space of ρ is isomorphic to the $G_{\mathbb{Q}_\ell}$-module obtained from the generic fiber of a finite flat group scheme over $\mathrm{Spec}(\mathbb{Z}_\ell)$. The category $\mathcal{P}_{fl}$ is closed under passage to sub-objects, quotients, and direct sums, and therefore defines a deformation condition. We call deformations satisfying this condition "flat deformations."

One can also define the analogous condition for representations of $G_{\mathbb{Q},S}$ if $\ell \in S$. In this case we have a homomorphism $G_{\mathbb{Q}_\ell} \longrightarrow G_{\mathbb{Q},S}$; we say a deformation $\rho : G_{\mathbb{Q},S} \longrightarrow \mathrm{GL}_n(A)$ is "flat at ℓ" if its composition with this homomorphism defines a representation of $G_{\mathbb{Q}_\ell}$ which is of type $\mathcal{P}_{fl}$.

Problem 6.7. Work out a usable description of the tangent space of the functor $\mathbf{D}_{fl}$ of flat deformations of representations of $G_{\mathbb{Q}_\ell}$. (Hint: what we would like to have is an identification of the tangent space with an Ext^1 group in the category of finite flat group schemes. To make this work, you will need to use Raynaud's result in [116] on uniqueness of models; this requires $e < p - 1$, and hence you'll want to assume $p > 2$.)

Ordinary deformations

We now restrict ourselves to the case $n = 2$ to talk about the condition of being "ordinary." This actually appears with two different meanings in the literature. We follow Mazur's definition (which I think is a minority opinion); to other authors what we are defining here are "co-ordinary" deformations, i.e., deformations whose dual is ordinary. In many situations, this makes no difference (e.g., because the universal deformation rings of $\bar{\rho}$ and of its contragredient are canonically isomorphic).

Ordinary deformations were first considered by Mazur in connection with Hida's theory of ordinary p-adic modular forms. Let $\mathcal{O}$ be a discrete valuation ring which is finite over $\mathbb{Z}_p$, and suppose we have a Hecke eigenform f of weight k and level N and defined over $\mathcal{O}$ such that $\mathrm{U}_p(f) = \lambda f$ with $\lambda \in \mathcal{O}^\times$ a p-adic unit. Then the attached Galois representation

$$\rho_f : G_{\mathbb{Q},S} \longrightarrow \mathrm{GL}_2(\mathcal{O})$$

has the following property. If we set

$$M = \mathcal{O} \times \mathcal{O}$$

endowed with the $G_{\mathbb{Q},S}$-action defined by ρ_f, and if I_p is an inertia subgroup at p, then the submodule M^{I_p} of vectors fixed under I_p is $\mathcal{O}$-free of rank one, and is a direct summand.[1]

---

[1] Caveat: for this to be true, one must define the representation attached to a modular form f in terms of the geometric Frobenius transformation, which induces $x \mapsto x^{p^{-1}}$ on the residue field. As

One can show, in fact, that all of the deformations which Hida's theory gives us have this property. Thus, it makes sense to try to work out which piece of the full deformation space is cut out by this condition.

Definition 6.5. Fix Π and k as above, let R be a ring in $\mathcal{C}$, and choose a closed subgroup $I \subset \Pi$. Let

$$\rho : \Pi \longrightarrow \mathrm{GL}_2(R)$$

be a representation, and let $M = R \times R$ with the Π-module structure determined by ρ. We say ρ is I-ordinary if the sub-R-module $M^I \subset M$ is free of rank 1 over R and a direct summand of M.

Notice that by this definition a representation satisfying $M^I = M$ (we might call it I-unramified) does *not* qualify as being I-ordinary.

Problem 6.8. Suppose $\bar\rho$ is I-ordinary and that ρ is a representation lifting $\bar\rho$. Check that if ρ is strictly equivalent to σ and ρ is ordinary, then σ is ordinary. In other words, the property of being ordinary is invariant under strict equivalence.

Theorem 6.5. *Suppose $\bar\rho$ is I-ordinary. Then the condition of being I-ordinary is a deformation condition for $\bar\rho$.*

Proof. This is pretty much straightforward. The first condition we need to check is that if a deformation $\rho : \Pi \longrightarrow \mathrm{GL}_2(A)$ is I-ordinary and $\pi : A \longrightarrow A'$ is a homomorphism of coefficient Λ-algebras, then the deformation $\pi_*\rho : \Pi \longrightarrow \mathrm{GL}_2(A')$ obtained from π is also ordinary. Changing basis if necessary, we can assume that the image of any $x \in I$ under ρ is of the form

$$\rho(x) = \begin{pmatrix} 1 & * \\ 0 & * \end{pmatrix}.$$

But then it's clear that the image of this matrix under π is a matrix of the same form, which shows that $\pi_*\rho$ is I-ordinary.

The second condition asks us to look at a fiber product situation, So suppose we have rings R_0, R_1, and R_2 in $\mathcal{C}_0$, and morphisms

$$\phi_1 : R_1 \longrightarrow R_0 \quad \text{and} \quad \phi_2 : R_2 \longrightarrow R_0.$$

Let

$$R_3 = R_1 \times_{R_0} R_2.$$

We need to show that if we have a deformation to R_3 such that the induced deformations to R_1 and R_2 are both I-ordinary, then so is R_3. The basic strategy is similar to what we have done before: we must find homomorphisms in the strict equivalence classes of the deformations to R_1 and R_2 such that the rank-one subspaces fixed by I map to the same subspace of $R_0 \times R_0$, and so on. We leave the details to the reader. □

Problem 6.9. Finish the proof of the theorem.

---

a result, our "representation attached to a modular form f" is the contragredient of the usual one constructed in étale cohomology. With the arithmetic Frobenius, one has an invariant quotient instead of an invariant subspace.

Given an I-ordinary residual representation
$$\bar{\rho}: \Pi \longrightarrow \mathrm{GL}_2(\mathsf{k}),$$
it now makes sense to consider the subfunctor $\mathbf{D}_I$ of $\mathbf{D}$ such that, for any ring R in $\mathcal{C}_\Lambda$,
$$\mathbf{D}_I(R) = \{I\text{-ordinary deformations of } \bar{\rho} \text{ to } R\}.$$
Since being I-ordinary is a deformation condition, this functor is representable whenever $\mathbf{D}$ is.

Corollary 6.6. *Suppose $\bar{\rho}$ is I-ordinary and that $C(\bar{\rho}) = \mathsf{k}$. Then there exists a universal I-ordinary deformation of $\bar{\rho}$. Specifically, there exists a ring $\mathcal{R}_I = \mathcal{R}_I(\Pi, \mathsf{k}, \bar{\rho}, I)$ and an I-ordinary deformation*
$$\boldsymbol{\rho}_I: \Pi \longrightarrow \mathrm{GL}_2(\mathcal{R}_I)$$
such that any I-ordinary deformation of $\bar{\rho}$ to a ring A is $\mathcal{C}_\Lambda$ is obtained from $\boldsymbol{\rho}_I$ via a unique homomorphism $\mathcal{R}_I \longrightarrow A$.

It is easy to see that one can extend this result to show that if we consider a set of closed subgroups $I_1, I_2, \ldots, I_n$ and the residual representation $\bar{\rho}$ is ordinary for each of these subgroups, then there exists a universal deformation of that type.

Problem 6.10. A representation $\rho: \Pi \longrightarrow \mathrm{GL}_2(A)$ is called I-co-ordinary if its representation space M has a submodule M_1 which is free of rank 1 as an A-module, and is a direct summand of M, and such that M/M_1 is invariant under I. Show that being I-co-ordinary is a deformation condition.

As discussed above, the tangent space to the I-ordinary deformation subfunctor corresponds to a subspace
$$\mathrm{H}^1_I(\Pi, \mathrm{Ad}(\bar{\rho})) \subset \mathrm{H}^1(\Pi, \mathrm{Ad}).$$

Problem 6.11. Let $V_{\bar{\rho}}$ be the representation space for $\bar{\rho}$, and let V^I be the subspace fixed by inertia. Assume $\bar{\rho}$ is ordinary, so that V^I is one-dimensional. Let $\mathrm{Ad}_I(\bar{\rho})$ denote the subspace of $\mathrm{Ad}(\bar{\rho})$ consisting of those matrices which correspond to endomorphisms of V which factor through V/V^I. Show that
$$\mathrm{II}^1_I(\Pi, \mathrm{Ad}(\bar{\rho})) = \mathrm{H}^1(\Pi, \mathrm{Ad}_I(\bar{\rho})) \subset \mathrm{H}^1(\Pi, \mathrm{Ad}).$$

Deformation conditions for global Galois representations

Now let's go back to the situation of most interest for us, when $\Pi = G_{\mathbb{Q},S}$ is the Galois group of the maximal extension of $\mathbb{Q}$ unramified outside a finite set of primes S. As always, we will assume that S contains both p and the prime at infinity. The first thing to consider is why one would want to impose deformation conditions to begin with, and which conditions they might be.

Suppose, for example, that our residual representation $\bar{\rho}$ is the Galois representation arising from the p-division points of an elliptic curve E, and let's assume $\bar{\rho}$ is absolutely irreducible. Then we can take S to consist of the primes of bad reduction of E, plus p and ∞, and we have at hand at least one lift to characteristic zero, the representation
$$\rho_{E,p}: G_{\mathbb{Q},S} \longrightarrow \mathrm{GL}_2(\mathbb{Z}_p)$$
arising from the Galois action on the Tate module of E. In this situation, the thing to note is that we know quite a bit about these representations. For example, we

know that the determinant of ρ is the p-adic cyclotomic character, and we know that (the prime-to-p part of) the Artin conductor of $\rho_{E,p}$ is equal to (the prime-to-p part of) the conductor of E. So, for each of the primes in S, we have some information on how the representation behaves when restricted to the decomposition group at that prime. If we are trying to make our deformation problem "as tight as possible" (for example, so that $\rho_{E,p}$ is a deformation captured by our problem but at the same time so that it is possible that every representation captured by our problem is modular), we need to impose deformation conditions that reflect the properties of these deformations. These conditions are of two types. First, there is the condition on the determinant. As the discussion above suggests, this is not too serious an issue. Second, there are local conditions at primes $\ell \in S$. These are deformation conditions that are posed in terms of the restriction of ρ to the decomposition groups $G_{\mathbb{Q}_\ell}$ (as we did above for the example of the "finite flat" condition).

Let $\bar\rho : G_{\mathbb{Q},S} \longrightarrow \mathrm{GL}_n(k)$ be a residual representation. Formally, a *global Galois deformation problem* $\mathcal{Q}$ is the problem of representing a subfunctor of $\mathbf{D}_{\bar\rho}$ defined by giving, for each non-archimedean prime $\ell \in S$, a deformation condition $\mathcal{Q}_\ell$ for the local residual representation $\bar\rho|_{G_{\mathbb{Q}_\ell}}$. A *global Galois deformation problem with fixed determinant* is the same, with an added "$\det = \delta$" condition.

The following is easy to check:

Lemma 6.7. *A global Galois deformation problem is a deformation condition for representations of $G_{\mathbb{Q},S}$.*

Proof. Easy, given that we know that each of the local "pieces" is a deformation condition. □

As always, we want to understand the tangent space to the subfunctor $\mathbf{D}_\mathcal{Q}$ associated to a global Galois deformation problem. The main thing we need to be careful about is that the local conditions $\mathcal{Q}_\ell$ are defined only in terms of the restriction of the representations to $G_{\mathbb{Q}_\ell}$. As before, let $\mathrm{H}^1_\mathcal{Q}(G_{\mathbb{Q},S}, \mathrm{Ad}(\bar\rho))$ be the subspace of $\mathrm{H}^1(G_{\mathbb{Q},S}, \mathrm{Ad}(\bar\rho))$ corresponding to the tangent space of $\mathbf{D}_\mathcal{Q}$, and let $\mathrm{H}^1_{\mathcal{Q}_\ell}(G_{\mathbb{Q}_\ell}, \mathrm{Ad}(\bar\rho))$ be the subspace of $\mathrm{H}^1(G_{\mathbb{Q}_\ell}, \mathrm{Ad}(\bar\rho))$ corresponding to the tangent space of the subfunctor of the local deformation functor defined by the local condition $\mathcal{Q}_\ell$.

Theorem 6.8. *The diagram*

$$\begin{array}{ccc} \mathrm{H}^1_\mathcal{Q}(G_{\mathbb{Q},S}, \mathrm{Ad}(\bar\rho)) & \longrightarrow & \mathrm{H}^1(G_{\mathbb{Q},S}, \mathrm{Ad}(\bar\rho)) \\ \downarrow & & \downarrow \\ \bigoplus_{\ell \in S} \mathrm{H}^1_{\mathcal{Q}_\ell}(G_{\mathbb{Q}_\ell}, \mathrm{Ad}(\bar\rho)) & \longrightarrow & \bigoplus_{\ell \in S} \mathrm{H}^1(G_{\mathbb{Q}_\ell}, \mathrm{Ad}(\bar\rho)) \end{array}$$

where the horizontal arrows are inclusions and the vertical arrow on the right is the restriction map on cohomology, is Cartesian, that is, it identifies $\mathrm{H}^1_\mathcal{Q}(G_{\mathbb{Q},S}, \mathrm{Ad}(\bar\rho))$ with the set of elements of $\mathrm{H}^1_\mathcal{Q}(G_{\mathbb{Q},S}, \mathrm{Ad}(\bar\rho))$ which, for each $\ell \in S$, map under restriction to the image of $\mathrm{H}^1_{\mathcal{Q}_\ell}(G_{\mathbb{Q}_\ell}, \mathrm{Ad}(\bar\rho))$.

Proof. Straightforward from the definitions. □

This shows that the tangent space to the functor $\mathbf{D}_\mathcal{Q}$ is a kind of "Selmer group," i.e., a part of the global cohomology group defined by local conditions for each $\ell \in S$.

Representations that are ordinary at p

An interesting application of the ideas in this lecture is when $\bar{\rho}$ is a two-dimensional global Galois representation, so that $\Pi = G_{\mathbb{Q},S}$ for a finite set of primes S, and we choose only one local deformation condition, requiring that our deformations be I-ordinary, when $I = I_p$ is an inertia subgroup at p. We say that such representations are *ordinary at p*.

So suppose $\bar{\rho}$ is ordinary at p, and consider the functor $\mathbf{D}^0$ which is defined by

$$\mathbf{D}^0(R) = \{\text{deformations of } \bar{\rho} \text{ to } R \text{ which are ordinary at } p\}.$$

As before, this is representable, and we call the representing ring $\mathcal{R}^0(\bar{\rho})$ the *universal ordinary deformation ring*. This is a quotient of the full universal deformation ring and parametrizes deformations of $\bar{\rho}$ which are ordinary at p.

There are two reasons to give special attention to the universal ordinary deformation ring. The first is that, in many cases, it follows from Wiles' work that if $\bar{\rho}$ is modular, then any ordinary deformation of $\bar{\rho}$ is also modular (perhaps in an extended sense). Another way of saying this is that we can prove, in many cases, that the ring $\mathcal{R}^0(\bar{\rho})$ can be identified with a certain p-adic Hecke algebra (a localization of the Hida algebra, to be discussed further in the next lecture).

The other reason the universal ordinary deformation ring is interesting is that, in certain cases, the homomorphism $\mathcal{R}(\bar{\rho}) \longrightarrow \mathcal{R}^0(\bar{\rho})$ is well understood. For example, one can prove the following result.

Proposition 6.9 (Mazur, Martin). *Let $S = \{p, \infty\}$, $\mathsf{k} = \mathbb{F}_p$, and let*

$$\bar{\rho}: G_{\mathbb{Q},S} \longrightarrow \mathrm{GL}_2(\mathbb{F}_p)$$

be ordinary (at p). Let ω denote the Teichmüller character $G_{\mathbb{Q},S} \longrightarrow \mathbb{F}_p^\times$. Suppose either that

 i. $\det \bar{\rho} \neq 1, \omega, \omega^{-1}, \omega^{\frac{p-1}{2}}$, *or that*

 ii. $\bar{\rho}$ *is tamely ramified.*

Then the kernel of the canonical homomorphism $\mathcal{R}(\bar{\rho}) \longrightarrow \mathcal{R}^0(\bar{\rho})$ is generated by two elements.

See [98] and [93]. Böckle has now improved considerably on this; see [5]. The reason such results are interesting is that they give us a way from lifting information about $\mathcal{R}^0$ to information about $\mathcal{R}$.

Corollary 6.10. *Under the assumptions of the Proposition, we have*

$$\mathrm{Krull\ dim}\left(\mathcal{R}^0(\bar{\rho})/p\mathcal{R}^0(\bar{\rho})\right) \geq 1.$$

If in addition we know that $\dim t_{\mathcal{R}^0(\bar{\rho})} \leq 1$, then we have $\mathcal{R}^0(\bar{\rho}) \cong \mathbb{Z}_p[[T]]$ and $\mathcal{R}(\bar{\rho})$ is a power series ring in two variables over $\mathcal{R}^0(\bar{\rho})$.

One might describe this result as saying that the ordinary deformation ring "controls" the full deformation ring. In practical terms, this means that studying ordinary deformations (when they exist, that is, when $\bar{\rho}$ is ordinary) can lead to results about all deformations.

Complements to lecture six

As we pointed out above, condition (iv) in the definition of a "deformation condition" follows from conditions (i) to (iii). The argument we give here is due to Mark Dickinson.

Suppose, then, that we have a deformation condition $\mathcal{Q}$ satisfying conditions (i) to (iii) in the definition. Then (by property (ii)) we have a subfunctor $\mathbf{D}_\mathcal{Q}$ of the deformation functor $\mathbf{D}$, and (by property (i)) we know that $\mathbf{D}_\mathcal{Q}(\mathsf{k}) = \mathbf{D}(\mathsf{k})$ are both sets with exactly one element. The result we want to prove is essentially the content of the following proposition:

Proposition 6.11 (Dickinson)**.** *Let $\mathbf{F}$ be a set-valued functor on $\mathcal{C}^0$, and let $\mathbf{G}$ be a subfunctor of $\mathbf{F}$. Suppose that the following condition holds:*

> **Property (*):** *If $A \longrightarrow C$ and $B \longrightarrow C$ are homomorphisms of artinian coefficient rings, and the push-forwards to $\mathbf{F}(A)$ and $\mathbf{F}(B)$ of an element x in $\mathbf{F}(A \times_C B)$ are in $\mathbf{G}(A)$ and $\mathbf{G}(B)$, respectively, then x is in $\mathbf{G}(A \times_C B)$.*

Suppose $j : A \longrightarrow B$ is an inclusion of artinian coefficient rings and x is an element of $\mathbf{F}(A)$ whose push-forward by j lies in $\mathbf{G}(B)$. Then x is in $\mathbf{G}(A)$.

Proof. The main ingredient in the proof is the notion of an *equalizer* in the category of artinian coefficient rings, Suppose we have two homomorphisms of artinian coefficient rings $f, g : B \longrightarrow R$; we say a homomorphism $j : A \longrightarrow B$ is the equalizer of f and g if two conditions are satisfied:

  i. The composite maps are equal: $f \circ j = g \circ j$.
  ii. The map j is "universal" in the sense that given any other homomorphism $h : A' \longrightarrow B$ such that $f \circ h = g \circ h$ there exists a homomorphism $h' : A' \longrightarrow A$ such that $h = j \circ h'$.

In other words, any other homomorphism which "equalizes" f and g factors through the equalizer map j. It is easy to see that equalizers are always injective.

The strategy of the proof is, first, to show that the result is true when $j : A \longrightarrow B$ is the equalizer of some pair of homomorphisms, and second, to show that we can deduce the general case from this one.

Step One: *If $j : A \longrightarrow B$ is an equalizer in the category of artinian coefficient rings, then the result holds.* Suppose there are homomorphisms $f, g : B \longrightarrow R$ such that j is the equalizer of f and g. Then we have a Cartesian diagram[2]

$$\begin{array}{ccc} A & \xrightarrow{j} & B \\ {\scriptstyle f \circ j = g \circ j} \downarrow & & \downarrow {\scriptstyle f \times_\mathsf{k} g} \\ R & \xrightarrow{\text{diag}} & R \times_\mathsf{k} R \end{array}$$

Now $j_* x \in \mathbf{G}(B)$ and, by functoriality, $(f \circ j)_* x \in \mathbf{G}(R)$, and Property (*) tells us that $x \in \mathbf{G}(A)$.

Step Two: *If the inclusion j is not a surjection, then there exists a proper subring C of B which contains the image of j and for which the inclusion $C \longrightarrow B$ is an equalizer.* If j is not surjective, then neither is the induced map on cotangent spaces, and so the dual map

$$\operatorname{Hom}(B, \mathsf{k}[\varepsilon]) \longrightarrow \operatorname{Hom}(A, \mathsf{k}[\varepsilon])$$

is not injective. Thus, there exist distinct maps $f, g : B \longrightarrow \mathsf{k}[\varepsilon]$ which agree on A. We take C to be the equalizer of these two maps.

Wrap-up: The result now follows by induction on the length over $W(\mathsf{k})$ of B/A. If j is surjective (the base case), there is nothing to prove. If not, we use

---
[2]i.e., a fiber product diagram, see page 274.

step two to factor $j : A \longrightarrow B$ as $A \longrightarrow C \longrightarrow B$ where the second map is an equalizer. By step one, the image of x in $\mathbf{F}(C)$ lies in $\mathbf{G}(C)$. Since the length of C/A is smaller than the length of B/A, x is in $\mathbf{G}(A)$ by the induction hypothesis, and we are done. $\square$

LECTURE 7
Modular Deformations

The idea that "naturally occurring" Galois representations should "come from" modular forms has become an important number-theoretical principle, and the success of Wiles, Taylor, Diamond, Conrad, and Breuil in proving that every elliptic curve over $\mathbb{Q}$ is modular strengthens this expectation. Our theory has produced for us a plethora of representations; is there any sense in which they all come from modular forms? Alternatively, can we make sense of the idea that "most" of them do?

First of all, in a deformation theory setting the question makes the most sense when we start from an absolutely irreducible odd two-dimensional Galois representation $\bar{\rho} : G_{\mathbb{Q},S} \longrightarrow \mathrm{GL}_2(\mathsf{k})$ *which is attached to a modular form*. The Serre Conjecture claims that all such residual representations are indeed attached to a modular form, and we know that if this is true then (with some caveats if $p=2$ or $p=3$) we can require this modular form to have "optimal weight and level" in the sense of [**117**]. If we know that the residual representation is modular, then we can ask whether the deformations are modular too.

The first difficulty one faces is making precise what is meant by "modular form." The most obvious thing is to consider the classical modular forms (we'll recall the theory below). It's clear, however, that unless we add some very stringent deformation conditions there is no chance that *all* deformations will be modular, though it makes sense to ask whether the modular deformations are "dense" (in various senses of the word) in the space of all deformations.

The other option is to generalize the notion of "modular." The correct theory here would be some sort of p-adic theory of modular forms, which should be the "p-adic completion" of the usual theory. Such a theory is in fact available, and produces enough "modular deformations" so that there is some chance that these are in fact all the deformations. We will only mention some of the bare facts of this theory, and then explain (in the next lecture) how in at least one case one can prove that "all deformations are (in this extended sense) modular." This still leaves us, of course, with the problem of "locating" the deformations which are modular in the "classical" sense among the others.

An introduction to modular forms can be found in [**13**]; for more details, see the references therein. For the p-adic theory, see the expository account by Matthew Emerton in Appendix 3, which includes a survey of the literature.

Classical modular forms and their representations

We begin from the "classical" theory of modular forms. Let N be an integer relatively prime to p. Given non-negative integers k and ν, we will write $S_k(\Gamma_1(Np^\nu), \mathbb{Z}_p)$ for the space of cuspidal modular forms of weight k on $\Gamma_1(Np^\nu)$ defined over $\mathbb{Z}_p$ (i.e., whose q-expansions have coefficients in $\mathbb{Z}_p$). What this means is the following. The classical space $S_k(\Gamma_1(Np^\nu), \mathbb{C})$ of complex modular forms of weight k on $\Gamma_1(Np^\nu)$ has a basis consisting of modular forms whose q-expansions at infinity are such that all the coefficients are integers. By a "modular form defined over $\mathbb{Z}_p$" we just mean a $\mathbb{Z}_p$-linear combination of this integral basis. Equivalently, if we write $S_k(\Gamma_1(Np^\nu), \mathbb{Z})$ for the subspace of $S_k(\Gamma_1(Np^\nu), \mathbb{C})$ consisting of modular forms whose q-expansion coefficients at infinity are integral, then $S_k(\Gamma_1(Np^\nu), \mathbb{Z})$ is a finite $\mathbb{Z}$-module and it is known that

$$S_k(\Gamma_1(Np^\nu), \mathbb{C}) = S_k(\Gamma_1(Np^\nu), \mathbb{Z}) \otimes_\mathbb{Z} \mathbb{C},$$

and our definition amounts to defining the $\mathbb{Z}_p$-space analogously:

$$S_k(\Gamma_1(Np^\nu), \mathbb{Z}_p) = S_k(\Gamma_1(Np^\nu), \mathbb{Z}) \otimes_\mathbb{Z} \mathbb{Z}_p.$$

The choice of $\mathbb{Z}_p$ as the ring over which our forms are defined is made for simplicity; any discrete valuation ring $\mathcal{O}$ finite over $\mathbb{Z}_p$ would work just as well, and the definition would be the same.

This definition has a somewhat arbitrary "feel" to it, of course, not least because it privileges the cusp at infinity over the other cusps (the only reason to do this is not to have to bother with adjoining roots of unity to our base ring). In fact, one can do much better by using a geometric definition. Suppose $N \geq 5$ (and, as we agreed above, not divisible by p), and let $\mathbb{Z}_{(p)}$ be the localization of $\mathbb{Z}$ at p (so that $\mathbb{Z}_{(p)} = \mathbb{Q} \cap \mathbb{Z}_p$). Then there is an algebraic curve $X_1(Np^\nu)$ defined over $\mathbb{Z}_{(p)}$ which parametrizes (generalized) elliptic curves with an appropriately-defined Np^ν-level structure (basically, the choice of a point of order Np^ν, but this can't be taken literally when $\nu > 0$, so one must find a more sophisticated description that will work in our situation). There is a canonical invertible sheaf ω on $X_1(Np^\nu)$ constructed in terms of the universal (generalized) elliptic curve over $X_1(Np^\nu)$, and we can define $S_k(\Gamma_1(Np^\nu), \mathbb{Z}_{(p)})$ as the global sections of ω^k. If we extend scalars from $\mathbb{Z}_{(p)}$ to $\mathbb{C}$ this space of global sections is exactly the classical space $S_k(\Gamma_1(Np^\nu), \mathbb{C})$, and thus it makes sense to define $S_k(\Gamma_1(Np^\nu), \mathbb{Z}_p)$ as the global sections of ω^k over $X_1(Np^\nu)_{/\mathbb{Z}_p}$. This gives *almost* the same space as above; in fact, it is contained in the space above with finite index equal to a power of p. The difference is that the geometric definition automatically forces integrality (suitably defined) at *all* the cusps rather than only at the cusp at infinity. The case of $N < 5$ can then be dealt with (at least for $p \neq 2, 3$) by taking fixed points under the appropriate group. For an introduction to this view of things, start with Appendix 3, then see [**79**] for the case $\nu = 0$; for the case $\nu > 0$ it is harder to give good references,[1] but [**39**] and [**85**] are the natural starting points.

---

[1]This is particularly true if one wants to define the Hecke operators geometrically, the trickiest one in our case being the U_p operator.

LECTURE 7. MODULAR DEFORMATIONS

In what follows, one can work with either definition of $S_k(\Gamma_1(Np^\nu), \mathbb{Z}_p)$. In fact, more often than not we will want to work with $S_k(\Gamma_1(Np^\nu), \mathcal{O})$, where $\mathcal{O}$ is a discrete valuation ring which is finite over $\mathbb{Z}_p$. If $\mathcal{O}$ contains enough roots of unity, then it makes sense to require that the q-expansions at all the cusps have coefficients in $\mathcal{O}$, and we recover the same space as the one given by the geometric definition. It's important to note, however, that it is the geometric definition that really allows us to understand the situation.

The space $S_k(\Gamma_1(Np^\nu), \mathbb{Z}_p)$ is a finite free $\mathbb{Z}_p$-module on which act the Hecke operators T_ℓ, $\ell \nmid Np$, defined in the usual way. One can also define[2] operators U_ℓ for $\ell | N$ and, when $\nu > 0$, an operator U_p which acts on q-expansions by

$$U_p\left(\sum a_n q^n\right) = \sum a_{np} q^n.$$

We will be particularly interested in *eigenforms*, that is, forms which are simultaneous eigenfunctions all for the Hecke operators.

There is another family of operators acting on our space, the diamond operators[3] $\langle n \rangle$, $n \in (\mathbb{Z}/N\mathbb{Z})^\times$. If a modular form is an eigenfunction for these operators, then we have $\langle n \rangle f = \varepsilon(n) f$ for some character ε which we call the *tame nebentypus character* of f. We extend this action to an action of $\mathbb{Z}_p^\times \times (\mathbb{Z}/N\mathbb{Z})^\times$, which we call the "double diamond" action $f \mapsto \langle x, y \rangle f$. The action of $x \in \mathbb{Z}_p^\times$ on a form f is determined by a combination of its weight and the p-part of its nebentypus character. In particular, if f is of weight k on $\Gamma_1(Np^\nu)$, and has nebentypus character $\varepsilon = \varepsilon_N \varepsilon_p$, where ε_N is a character on $(\mathbb{Z}/N\mathbb{Z})^\times$ and ε_p is a character on $(\mathbb{Z}/p^\nu\mathbb{Z})^\times$, then

$$\langle x, y \rangle f = \varepsilon_N(y) \varepsilon_p(x) x^k f.$$

(For more details about, and a more natural definition of, this action, see [57].) We will require eigenforms to also be eigenfunctions for the diamond action (which will usually require them to be defined over an extension $\mathcal{O}$ of $\mathbb{Z}_p$, as we mentioned above).

The reason for "twisting" the diamond action by the factor x^k is that this results in operators that act on the *sum* (over k) of the spaces $S_k(\Gamma_1(Np^\nu), \mathbb{Z}_p)$ in a way that preserves integrality of q-expansions. (See [81] for an example of how this can be used to find congruences between modular forms). This will be helpful for the p-adic theory in the next section.

We will say an eigenform is *normalized* if its q-expansion (at infinity) is of the form

$$f(q) = q + a_2 q^2 + a_3 q^3 + \ldots.$$

In this case, it is easy to see that $T_\ell(f) = a_\ell f$ for all $\ell \nmid Np$.

Let $\mathcal{O}$ be the valuation ring in a finite extension K of $\mathbb{Q}_p$. Given a form $f \in S_k(\Gamma_1(Np^\nu), \mathcal{O})$ which is an eigenform for the T_ℓ for all $\ell \nmid Np$ and for the diamond operators, one can construct a Galois representation ρ_f which is attached

---

[2] There seems to be no agreement about whether one should call these operators U_ℓ or simply define T_ℓ differently when $\ell | N$. The advantage of retaining the distinction is that then the action of T_ℓ on q-expansions is always the same as one varies k, N, and ν, and is different from the action of U_ℓ. In the case when $\ell = p$, the distinction between T_p (which acts on our space when $\nu = 0$) and U_p (which acts when $\nu > 0$) does matter for what we want to do.

[3] Notice the notational weirdness here: these are only *some* of the usual diamond operators on $S_k(\Gamma_1(Np^\nu), \mathbb{Z}_p)$. We will add the p-part when we move to the "double diamonds" in a minute.

to f in the following sense. Suppose we have $T_\ell f = a_\ell f$ and $\langle \ell, \ell \rangle f = \lambda(\ell) f$ for all $\ell \nmid Np$. Let

$$S = \{\text{primes dividing } N\} \cup \{p, \infty\},$$

and let $\Phi_\ell \in G_{\mathbb{Q},S}$ denote a geometric Frobenius transformation at ℓ. Then one can construct a Galois representation

$$\rho_f : G_{\mathbb{Q},S} \longrightarrow \mathrm{GL}_2(\mathcal{O})$$

such that $\rho_f \otimes_{\mathcal{O}} K$ is semisimple, and such that f and ρ_f are related by the formulas

$$\det \rho_f(\Phi_\ell) = \frac{1}{\ell}\lambda(\ell) \quad \text{and} \quad \mathrm{Tr}\, \rho_f(\Phi_\ell) = a_\ell$$

for each $\ell \nmid Np$. (Readers familiar with the usual definition should remember that if f is of weight k with nebentypus χ we have $\lambda(\ell) = \chi(\ell)\ell^k$.) It is easy to see (because $\rho_f \otimes_{\mathcal{O}} K$ is semisimple) that ρ_f is completely determined (up to equivalence over K) by f.

How is the representation attached to f obtained? The construction is due to Eichler, Shimura, Deligne, and Serre, and it is quite complicated. When the form f is of weight 2, one finds the representation by considering the Jacobian of the modular curve $X_1(Np^\nu)$. There is an action of the Hecke algebra on the Jacobian, and the eigenform gives a map from the Hecke algebra to R which allows us to "cut out" a piece of the étale cohomology of the Jacobian where the Galois group acts as we want (this is the dual of the usual approach, which finds the representation in the Tate module of the Jacobian). For general weights k, things get a lot more complicated. See [117, Appendix 2] for an account of how this works for $k = 2$, and [144], [37], and [40] for the beginnings of the rest of story.

Given such a ρ_f associated to a form f defined over some discrete valuation ring $\mathcal{O}$ which is a finite extension of $\mathbb{Z}_p$, we can reduce this modulo the maximal ideal $\mathfrak{m} \subset \mathcal{O}$. The resulting representation (which is defined over the residue field of $\mathcal{O}$) may not be semisimple, so we take its semisimplification, call it the reduction modulo the maximal ideal of the representation attached to f, and denote it by $\bar{\rho}_f$. The actual reduction may depend on the homomorphism $G_{\mathbb{Q},S} \longrightarrow \mathrm{GL}_2(\mathcal{O})$ rather than just on its equivalence class; the semisimplification, however, is the same for any homomorphism in the equivalence class. (Notice that if $\bar{\rho}_f$ is absolutely irreducible, we do not need to take the semisimplification step, and $\bar{\rho}_f$ is just the reduction of ρ_f. We will always restrict to this situation in what follows.)

From our point of view, we should note that if we have two normalized eigenforms f and g (perhaps of different weights and with different values of ν) which are congruent modulo $\mathfrak{m}$ (in the sense that their q-expansion coefficients are congruent modulo $\mathfrak{m}$) then their Hecke and diamond eigenvalues are the same modulo p, so that the reductions modulo p of their Galois representations are the same also. In other words, if $\bar{f} = \bar{g}$, then the two representations ρ_f and ρ_g are (different) deformations of the same residual representation $\bar{\rho} = \bar{\rho}_f = \bar{\rho}_g$. In fact, we can weaken this somewhat: all we need is for the eigenvalues for T_ℓ with $\ell \nmid Np$ to be the same, and this will happen if

$$a_n(f) \equiv a_n(g) \pmod{\mathfrak{m}}$$

for all n such that $\gcd(n, Np) = 1$, where $a_n(f)$ (resp. $a_n(g)$) denotes the n-th q-expansion coefficient of f (resp. g).

In order to be able to think about whether all (or many) deformations are modular, we need to collect some information about these representations. The properties of residual modular representations are discussed in detail in [**117**], to which we refer the reader. We record here only a few useful facts, especially having to do with ramification. Given a residual Galois representation

$$\bar{\rho}: G_{\mathbb{Q},S} \longrightarrow \mathrm{GL}_2(\mathsf{k}),$$

one may measure its ramification outside p by its conductor, which we denote by $\mathrm{N}(\bar{\rho})$. Since $\bar{\rho}$ is certainly not ramified at primes that do not divide Np, the conductor is a product of powers of primes dividing N,

$$\mathrm{N}(\bar{\rho}) = \prod_{\ell \in S - \{p\}} \ell^{n(\ell,\bar{\rho})},$$

where the numbers $n(\ell,\bar{\rho})$ are defined as follows: choose a place of $\overline{\mathbb{Q}}$ over ℓ, and let $I = I_\ell$ be the corresponding inertia group; let $\overline{V} = \mathsf{k} \times \mathsf{k}$ with the $G_{\mathbb{Q},S}$-action given by $\bar{\rho}$, and let $\overline{V}_0$ be the subspace of $\overline{V}$ fixed by $\bar{\rho}(I)$; then

$$n(\ell,\bar{\rho}) = 2 - \dim \overline{V}_0 + sw(\bar{\rho}),$$

where $sw(\bar{\rho})$ is the Swan conductor of (the restriction to a decomposition group at ℓ of) $\bar{\rho}$. (For a definition, see, for example, [**137**].) Note that if $\dim \overline{V}_0 = 2$, then $\bar{\rho}$ is unramified at ℓ and $n(\ell,\bar{\rho}) = 0$ (and in particular $sw(\bar{\rho}) = 0$ in this case). We know that $\bar{\rho}$ is tamely ramified at ℓ if and only if $sw(\bar{\rho}) = 0$, and that $\bar{\rho}$ is ℓ-ordinary exactly when $\dim \overline{V}_0 = 1$.

Of course, the conductor of $\bar{\rho}$ can be much smaller than the tame level N of the modular form from which it comes. In fact, that is a major theme of the recent work on Serre's conjecture reported in [**117**].

Now let's consider the situation for a lift of $\bar{\rho}$ to characteristic zero. As before, let K be a finite extension of $\mathbb{Q}_p$, let $\mathcal{O}$ be its valuation ring, and assume the residue field of $\mathcal{O}$ is k. The conductor of a deformation

$$\rho: G_{\mathbb{Q},S} \longrightarrow \mathrm{GL}_2(\mathcal{O})$$

of $\bar{\rho}$ (which may or may not be modular) is defined in an analogous way, as

$$\mathrm{N}(\rho) = \prod_{\ell \in S - \{p\}} \ell^{n(\ell,\rho)},$$

where we take $V = \mathcal{O} \times \mathcal{O}$ with the action of $G_{\mathbb{Q},S}$ given by ρ, V_0 is the submodule of invariants under the action of an inertia group at ℓ, and set

$$n(\ell,\rho) = 2 - \mathrm{rank}_{\mathcal{O}} V_0 + sw(\rho) = 2 - \mathrm{rank}_{\mathcal{O}} V_0 + sw(\bar{\rho}).$$

(The fact that the Swan conductors of ρ and of $\bar{\rho}$ are equal is well known; it is so because the wild inertia group at ℓ is a pro-ℓ-group, while, as noted in Lecture 5, the kernel of the reduction map $\mathrm{GL}_2(A) \longrightarrow \mathrm{GL}_2(\mathsf{k})$ is a pro-p-group, so that all the wild ramification will already occur in the $\bar{\rho}$.)

Note that we have eliminated powers of p from the conductor; this is inevitable, since we are not assuming ρ is a part of a compatible family of ℓ-adic representations. (For a modular deformation, we have an ℓ-adic representation for every prime ℓ, and the exponents in the conductor agree. This allows us to define the p-part of the conductor by looking at an ℓ-adic representation for $\ell \neq p$. If we want to do deformation theory, however, we have to stick to a p-adic setting.) That it is the "tame level" that is detectable from the representation is also pleasantly consistent with the fact that only the prime-to-p part of the level is relevant in the context of p-adic modular forms, as we will see below.

What do we know about these representations (and specifically, their conductors)? In characteristic zero, we know

- If f is a newform of level Np^ν, then the (prime-to-p part of the) Artin conductor of the representation ρ_f is equal to N. (This is an important theorem of Carayol.)
- The local representations $\rho_f|_{G_{\mathbb{Q}_\ell}}$, for $\ell|N$, $\ell \neq p$, are well-understood via the "local Hecke correspondence." (See [15] or the more accessible [14] for the details, which are complicated.)

Modulo p, we know a little more. In particular, the recent work on Serre's conjecture involved a quite detailed understanding of the local representations modulo p. See [117] for more details.

For the deformation theory, it is important to compare the situation in characteristic zero with the situation in characteristic p. For example, the following result captures what can happen to the conductor:

Proposition 7.1. *Let $\overline{\rho}$ be a residual Galois representation*

$$\overline{\rho}: G_{\mathbb{Q},S} \longrightarrow \mathrm{GL}_2(k),$$

and let ρ be any deformation of $\overline{\rho}$ to characteristic zero. Then, for each $\ell \in S$:

  i. *if $\overline{\rho}$ is unramified at ℓ, then $n(\ell, \rho) \leq 2$;*
  ii. *if $\overline{\rho}$ is ℓ-ordinary, then $n(\ell, \rho) \leq n(\ell, \overline{\rho}) + 1$;*
  iii. *if $\overline{\rho}$ and ρ are both ℓ-ordinary, then we have $n(\ell, \rho) = n(\ell, \overline{\rho})$;*
  iv. *if $\overline{\rho}$ is ramified at ℓ but not ℓ-ordinary, then $n(\ell, \rho) = n(\ell, \overline{\rho})$.*

Proof. This all follows immediately from the fact that

$$n(\ell, \rho) - n(\ell, \overline{\rho}) = \dim \overline{V}_0 - \operatorname{rank} V_0,$$

since $\operatorname{rank} V_0 \leq \dim \overline{V}_0 \leq 2$. □

The local representation $\rho_f|_{G_{\mathbb{Q}_p}}$ is somewhat harder to understand. The main tool for studying it is Fontaine's theory of p-adic $G_{\mathbb{Q}_p}$-modules. Using this language, we can describe what we know about the local representation attached to a modular form by saying that it is either *crystalline* (if $\nu = 0$) or, in general, *potentially semistable*. See [52] for definitions, discussion, and a study of the deformation problems attached to such conditions.

We can also use the p-adic Hodge theory of Tate and Sen to study the local representation. We know that the local representation $\rho_f|_{G_{\mathbb{Q}_p}}$ is of Hodge-Tate type, and the Hodge-Tate-Sen weights (as defined in [129] and [130]; see also [97] and [100]) attached to it are $(0, k-1)$. In particular, any representation that comes from a classical modular form must be in the "Sen null subspace" of the deformation space, which consist of those deformations which have one of their Sen weights equal to zero. This is known to be a codimension one analytic subspace of the (rigid-analytic subpsace attached to the) full deformation space. See the discussion in §7 of [100] for more details.

One can "move" from the Sen null subspace by twisting deformations by characters of infinite order which reduce to the trivial character modulo p (sometimes called "wild" characters of infinite order). See [97] and [100] for a more careful discussion of how twisting can be interpreted as the action of a formal group on the deformation space that is "essentially transversal" to the Sen null subspace. (We will get back to this in the next lecture.)

One final bit of information is also quite useful. If the eigenvalue for the p-th Hecke operator (either T_p or U_p, depending on whether $\nu = 0$ or $\nu > 0$) is a p-adic unit, then the local representation $\rho_f|_{G_{\mathbb{Q}_p}}$ is p-ordinary (see [106], for example). For residual representations, the converse is true (see [78]); the converse for representations in characteristic zero is known in some cases (basically, whenever one knows an "all ordinary deformations are modular" theorem, see below).

So let's summarize what we know. In the full deformation space, which we expect to be three-dimensional by the dimension conjecture (see page 289), the representations that come from (classical) modular forms sit in various smaller subspaces: first, they all sit in the Sen null subspace, which (since we know it is of codimension one) we expect to be two-dimensional. Second, a representation coming from a modular of weight k has determinant equal to (a finite character times) the $(k-1)$-th power of the cyclotomic character, and we can consider the subspace of the deformation space corresponding to deformations with that determinant. This again gives a space of codimension one. If the modular form is one whose p-th Hecke eigenvalue is a p-adic unit, the representation is p-ordinary, and we can consider the subspace corresponding to p-ordinary deformations (by contrast with the others, this is usually one-dimensional). If we're even more ambitious, we might try to pin down the subspace corresponding to deformations that are potentially semi-stable (if there is one). Finally, we can try to get the conductor right by controlling the ramification at the primes $\ell \neq p$.

This raises a number of interesting questions about the deformation space. Which of these conditions are "deformation conditions" in the sense of Lecture 6, and hence define algebraic subspaces of the deformation space? How do these various subspaces intersect?

On the other hand, this discussion also points out that modular deformations are very special, and that they at best fill up only a small part of the deformation space. If we hope to have a theorem saying something like "all deformations are modular," we will either have to restrict the meaning of "deformation" (by introducing deformation conditions, as in Lecture 6) or extend the meaning of "modular."

p-adic modular forms

One way of extending the meaning of a "modular deformation" is to work with "p-adic modular forms." In this section we give a very rough description of what this theory looks like. See Matthew Emerton's survey in Appendix 3 for a much more careful description, and [134], [79], [82], [57], [26], [27], and [28] for various accounts of the details.

The theory of p-adic modular forms gives us a very large space $\mathbf{V}(N, \mathbb{Z}_p)$ of "parabolic p-adic modular functions defined over $\mathbb{Z}_p$." We can go from forms over $\mathbb{Z}_p$ to forms over more general p-adically complete and separated rings R simply by defining $\mathbf{V}(N, R) = \mathbf{V}(N, \mathbb{Z}_p) \hat{\otimes} R$. Rather than describe this space and its construction, we note only that for any weight k and any $\nu \geq 0$ there exist inclusions

$$S_k(\Gamma_1(Np^\nu), \mathbb{Z}_p) \hookrightarrow \mathbf{V}(N, \mathbb{Z}_p),$$

and that the union of the images of these inclusions is dense in $\mathbf{V}(N, \mathbb{Z}_p)$ with respect to the p-adic topology derived from the q-expansions. Thus, $\mathbf{V}(N, \mathbb{Z}_p)$

contains every single one of the eigenforms we have considered so far, plus many more that are obtained by some sort of limiting process.[4]

There are naturally-defined Hecke operators T_ℓ (for $\ell \nmid Np$) and diamond operators $\langle x, y \rangle$ on $\mathbf{V}(N, \mathbb{Z}_p)$. The Hecke operators act as expected on q-expansions, and both the Hecke and the diamond operators restrict to the ones we defined above when we apply them to classical modular forms. (The other Hecke operators, and in particular the U_p operator, also extend, and the U_p operator in fact plays an important role in the theory. For now, however, we will stick with a smaller set of operators.)

The easiest way to define the Hecke operators on $\mathbf{V}(N, \mathbb{Z}_p)$ is to use the fact (first proved by Hida) that for any fix ν the union over k of the spaces $S_k(\Gamma_1(Np^\nu), \mathbb{Z}_p)$ is dense in $\mathbf{V}(N, \mathbb{Z}_p)$. Using this, we can define the Hecke algebra $\mathbf{T}$ as the inverse limit over k of the (restricted) Hecke algebras[5] acting on $S_k(\Gamma_1(Np^\nu), \mathbb{Z}_p)$. The action of $\mathbf{T}$ extends to all of $\mathbf{V}(N, \mathbb{Z}_p)$ by continuity. We can check that $\mathbf{T}$ is independent of ν, so that we get an algebra of operators on $\mathbf{V}(N, \mathbb{Z}_p)$, and that the T_ℓ act as expected on q-expansions. Each of the Hecke algebras at finite level has a natural p-adic topology, and we give $\mathbf{T}$ the inverse limit topology. Any p-adic modular function $f \in \mathbf{V}(N, \mathbb{Z}_p)$ determines a continuous map $\mathbf{T} \longrightarrow \mathbb{Z}_p$ by mapping an operator T to $a_1(Tf)$, the first q-expansion coefficient of Tf. This map is a homomorphism if and only if f is a normalized eigenform; in that case, it maps each operator to its eigenvalue. More generally, for any p-adically complete and separated ring R, any normalized eigenform $f \in \mathbf{V}(N, R)$ gives a continuous homomorphism $\mathbf{T} \longrightarrow R$.

(One might suspect that any such homomorphism determines a form f. This would certainly be true if we had included *all* the Hecke operators in $\mathbf{T}$. It is a little more problematic with our setup.)

Now suppose R is a ring in $\mathcal{C}$ and suppose that we have a form $f \in \mathbf{V}(N, R)$ which is an eigenform for T_ℓ (for all $\ell \nmid Np$) and for the diamond operators. Then f determines a homomorphism $\phi_f : \mathbf{T} \longrightarrow R$, and hence, after reduction modulo the maximal ideal, a homomorphism $\mathbf{T} \longrightarrow \mathsf{k}$. Let $\mathfrak{m}$ be the kernel of this homomorphism, and let $\mathcal{R}_m(\bar{f}) = \mathbf{T}_\mathfrak{m}$ be the completion of $\mathbf{T}$ at $\mathfrak{m}$.[6] Given another eigenform $g \in \mathbf{V}(N, R)$, the corresponding homomorphism ϕ_g factors through $\mathcal{R}_m(\bar{f})$ if and only if $\bar{f}$ and $\bar{g}$ have the same Hecke eigenvalues for all T_ℓ with $\ell \nmid Np$ and for all the diamond operators (i.e., if and only if the eigenvalues of f and of g are the same modulo the maximal ideal). In other words, $\mathcal{R}_m(\bar{f})$ is a kind of "universal deformation ring" for the packet of Hecke eigenvalues coming from the residual eigenform $\bar{f}$.

Now let $\bar{\rho}$ be the residual Galois representation attached to $\bar{f}$. If $\bar{\rho}$ is absolutely irreducible, then it has a universal deformation ring $\mathcal{R}(\bar{\rho})$. To get a "universal modular deformation" of $\bar{\rho}$, we need to construct a deformation of $\bar{\rho}$ to the completed Hecke algebra $\mathcal{R}_m(\bar{f})$. By the universal property of $\mathcal{R}(\bar{\rho})$ constructing such a thing

---

[4]Eigenforms in $\mathbf{V}(N, \mathbb{Z}_p)$ are sometimes limits of classical eigenforms, but more commonly they are limits of classical modular forms which are not eigenforms, but which are "eigenforms modulo p^n" for bigger and bigger n.

[5]That is, the $\mathbb{Z}_p$-submodule of the algebra of endomorphisms of $S_k(\Gamma_1(Np^\nu), \mathbb{Z}_p)$ which is generated by the T_ℓ with $\ell \nmid Np$ and by the diamond operators.

[6]The m in the notation $\mathcal{R}_m(\bar{f})$ stands for "modular." The point is that (in many cases) this ring will turn out to give a "universal modular deformation" of the representation $\bar{\rho}$ corresponding to $\bar{f}$.

is the same as constructing a homomorphism $\mathcal{R}(\bar\rho) \longrightarrow \mathcal{R}_m(\bar f)$. This turns out to be possible.

What we get is the following. Fix the "tame level" N, and let
$$S = \{\text{primes dividing } N\} \cup \{p, \infty\}.$$
Suppose we have an eigenform $\bar f \in \mathbf{V}(N, \mathbb{F}_p)$. Since the classical forms are dense in $\mathbf{V}(N, R)$, we know that $\bar f$ is equal to the reduction of a classical eigenform, and in fact we can assume that this classical eigenform is of level N. In particular, there is a Galois representation attached to $\bar f$; let $\bar\rho : G_{\mathbb{Q},S} \longrightarrow \mathrm{GL}_2(k)$ be the attached Galois representation.

Theorem 7.2 (Gouvêa, Hida). *There exists a representation*
$$\boldsymbol{\rho}_m : G_{\mathbb{Q},S} \longrightarrow \mathrm{GL}_2(\mathcal{R}_m(\bar f))$$
such that, if Φ_ℓ is a geometric Frobenius at $\ell \nmid Np$,
$$\det \boldsymbol{\rho}_m(\Phi_\ell) = \frac{1}{\ell}\langle \ell, \ell \rangle \qquad \text{and} \qquad \operatorname{Tr} \boldsymbol{\rho}_m(\Phi_\ell) = \mathrm{T}_\ell.$$

In [**57**], this theorem is proved by constructing a homomorphism from the universal deformation ring of $\bar\rho$ to the completed Hecke algebra. In the original version, there were technical assumptions which have since been removed by work of Carayol (see [**16**] and [**101**, §6]). Hida's proof (in [**73**]; see also [**72**]) gets around these technicalities by using the theory of pseudo-representations (also known as pseudo-characters).

A (perhaps surprising) consequence of this theorem is that *there exists a Galois representation attached to a p-adic eigenform f even when that eigenform is not a classical modular form*, provided that the residual representation attached to a classical eigenform that is congruent to f is absolutely irreducible. These representations are considerably more mysterious than the representations attached to classical eigenforms. For example, they are not necessarily of Hodge-Tate type (but see [**86**] for a positive result along these lines).

If $\bar\rho$ is absolutely irreducible, so that the theorem applies, we call $\boldsymbol{\rho}_m$ the *universal modular deformation* of $\bar\rho$. It parametrizes all the deformations of $\bar\rho$ which come from p-adic modular forms (which of course includes the ones which come from classical modular forms). Since the universal modular deformation is itself a deformation of $\bar\rho$, we get a homomorphism
$$\mathcal{R}(\bar\rho) \longrightarrow \mathcal{R}_m(\bar f).$$
The trace of the Frobenius Φ_ℓ in the universal deformation ring must map to T_ℓ in $\mathcal{R}_m(\bar f)$. Since the Hecke operators topologically generate $\mathcal{R}_m(\bar f)$, the homomorphism is surjective.

Thus, we have constructed a "universal modular deformation" which cuts out the portion of the deformation space which corresponds to deformations attached to p-adic modular forms. This can be viewed as a sort of Zariski closure of the points in the deformation space corresponding to modular deformations in the classical sense. (We do not expect it to be the closure in the p-adic topology, because p-adic eigenforms need not be limits of classical *eigen*forms.) The question, then, is how big a portion of the deformation space we have obtained. It will be the full deformation space exactly when the homomorphism
$$\mathcal{R}(\bar\rho) \longrightarrow \mathcal{R}_m(\bar f)$$

is an isomorphism. As we will see in the next lecture, in at least one case we know that it is in fact the full deformation space, i.e., we know that every deformation is "pro-modular" (i.e., comes from a p-adic modular form). On the other hand, we will see that unless $N = 1$ we cannot escape imposing some sort of deformation condition before we can get a positive result.

The game we just played with the big p-adic Hecke algebra can also be played with classical Hecke algebras. Thus, suppose we have an eigenform

$$f \in S_k(\Gamma_1(N), \mathbb{Z}_p),$$

and let $\bar\rho$ be the residual representation attached to f. (Eigenforms defined over finite extensions of $\mathbb{Z}_p$ work exactly the same way; we work with $\mathbb{Z}_p$ for simplicity.) Let $\mathbf{T}_k(N)$ be the subalgebra of the endomorphisms of $S_k(\Gamma_1(N), \mathbb{Z}_p)$ generated by the Hecke operators T_ℓ for $\ell \nmid Np$ and by the diamond operators. As before, the eigenform f determines a homomorphism $\mathbf{T}_k(N) \longrightarrow \mathbb{Z}_p$; reducing modulo the maximal ideal gives a homomorphism $\mathbf{T}_k(N) \longrightarrow \mathbb{F}_p$ whose kernel is a maximal ideal $\mathfrak{m}$. Let $\mathbf{T}(\bar{f})$ be the completion of $\mathbf{T}_k(N)$ at the maximal ideal $\mathfrak{m}$. Then there is a surjective homomorphism from the universal modular deformation ring $\mathcal{R}_m(\bar{f})$ to $\mathbf{T}(\bar{f})$, and therefore there exists a deformation ρ_f of $\bar\rho$ to $\mathbf{T}(\bar{f})$. (Note that this is generally not the same as the Galois representation attached to f. In fact, they are the same only if $\mathbf{T}(\bar{f}) = \mathbb{Z}_p$.) This deformation parametrizes all deformations of $\bar\rho$ which come from classical modular forms of weight k and level N, i.e., it plays the same role, for weight k and level N, that the big Hecke algebra $\mathcal{R}_m(\bar{f})$ plays for all p-adic modular forms.

If we could determine the representation-theoretic properties of ρ_f with sufficient precision, we could hope to write down deformation conditions that would restrict our deformation problem to "those deformations that look as if they come from modular forms of weight k and level N." One part of this is not difficult: to fix the weight, we need a determinant condition. Fixing the level, as before, boils down to imposing local deformation conditions at the primes dividing N. The subtle part is finding a deformation condition that will restrict us to classical modular forms. This would presumably be a local condition at p. In Wiles' work, for example, this condition was either that the deformation be p-ordinary or that it be flat at p.

If we successfully find the correct set $\mathcal{Q}$ of deformation conditions, then we get a ring $\mathcal{R}_{\mathcal{Q}}(\bar\rho)$ that gives us the universal deformation subject to those conditions. This should give us a surjective homomorphism

$$\mathcal{R}_{\mathcal{Q}}(\bar\rho) \longrightarrow \mathbf{T}(\bar{f}).$$

As before, asking whether this map is in fact an isomorphism amounts to asking whether "all deformations (of this kind) are modular (of weight k and level N)." In comparison to the overall question about whether all deformations are modular, this is both more precise (in particular, a yes answer to the question about the big deformation ring does not imply a yes answer here without some extra work) and perhaps more accessible. One reason for this is that the Hecke algebras $\mathbf{T}_k(N)$ and $\mathbf{T}(\bar{f})$ are relatively well understood, while the big Hecke algebra $\mathcal{R}_m(f)$ is much more mysterious (as are, of course, the universal deformation rings).

This approach to the problem, which works with fixed weight and level, is what appears in the work of Wiles, Taylor, Diamond, Breuil, and Conrad on the Shimura-Taniyama conjecture. For the most part, they work with weight $k = 2$, and consider several different deformation problems attached to $k = 2$ and varying levels. (For the crucial application to the representation attached to an elliptic curve, it's essential *not* to impose the condition that the residual modular form $\bar{f}$ is of minimal level.) The hardest part of this work, as noted above, is to pin down the deformation conditions that will restrict us to classical modular forms. As we will see in the next section, this is easy to do in the ordinary case, because of Hida's "control theorem." In the non-ordinary case, they use (variants of) the flat deformation condition we discussed in Lecture 6 to force the restriction to classical modular forms.

The ordinary case

The case of representations coming from ordinary p-adic modular forms is much better understood than the general case. We say that a p-adic modular function $f \in \mathbf{V}(N, R)$ is *ordinary* if f belongs to the R-submodule of $\mathbf{V}(N, R)$ (topologically) spanned by generalized eigenforms[7] for the U_p operator corresponding to eigenvalues which are p-adic units. As Hida shows, there exists an idempotent e in the endomorphism ring of $\mathbf{V}(N, \mathbb{Z}_p)$ which commutes with the Hecke and diamond operators and "picks out" the ordinary forms, so that $ef = f$ if and only if f is ordinary. On the representation theory side, Mazur and Wiles have shown in [**106**] that if f is ordinary then the associated representation is also I_p-ordinary in the representation-theoretic sense discussed above, that is, the subspace fixed by inertia in the representation space is of rank one and a direct summand. This nice match between a "modular" condition and a representation condition, together with the fact that thanks to Hida we know quite a lot about the Hecke algebra associated to ordinary modular forms, allows us to set things up nicely ... and some of the questions actually have answers.

First, we let $\mathbf{T}^0 = e\mathbf{T}$ and call it the ordinary part of the Hecke algebra (with our definitions, e does not belong to $\mathbf{T}$, so that $\mathbf{T}^0$ is not actually a "part" of $\mathbf{T}$; nevertheless, this wording is instructive). As before, an ordinary eigenform $\bar{f} \in \mathbf{V}(N, \mathbf{k})$ gives a map $\mathbf{T}^0 \longrightarrow \mathbf{k}$ whose kernel is a maximal ideal $\mathfrak{m}$, and we write $\mathcal{R}_m^0(\bar{f})$ for the completion of $\mathbf{T}^0$ at $\mathfrak{m}$. Then Hida proves the following theorem.

Theorem 7.3 (Hida). *If $p \geq 5$, and $\bar{\rho}$ is absolutely irreducible, there exists a representation*

$$\boldsymbol{\rho}_h : G_{\mathbb{Q}, S} \longrightarrow \mathrm{GL}_2(\mathcal{R}_m^0(\bar{f}))$$

such that, if Φ_ℓ is a geometric Frobenius at $\ell \nmid Np$,

$$\det \boldsymbol{\rho}_h(\Phi_\ell) = \frac{1}{\ell}\langle \ell, \ell \rangle \qquad \text{and} \qquad \mathrm{Tr}\, \boldsymbol{\rho}_h(\Phi_\ell) = \mathrm{T}_\ell.$$

We can think of Hida's representation as the "universal modular-ordinary deformation" of $\bar{\rho}$, since it parametrizes all deformations of $\bar{\rho}$ which come from ordinary p-adic modular forms. One of the more important things about $\mathcal{R}_m^0(\bar{f})$ is that one knows what sort of ring it is: Hida showed that it is a finite flat algebra over $\Lambda = \mathbb{Z}_p[[\Gamma]]$. One also knows, by work of Mazur and Wiles, that the representation $\boldsymbol{\rho}_h$ is ordinary in the sense of representation theory.

Suppose $\bar{\rho}$ is absolutely irreducible and attached to an ordinary modular form $\bar{f}$, so that we also know that $\bar{\rho}$ is ordinary in the representation-theoretic sense.

---

[7]An eigenform f for the U_p operator satisfies $(U_p - \lambda)f = 0$ for some λ; we say f is a *generalized eigenform* attached to the eigenvalue λ if it satisfies $(U_p - \lambda)^n f = 0$ for some $n \geq 1$.

Then we have constructed various deformation rings: the universal deformation ring $\mathcal{R}(\bar{\rho})$, the universal modular deformation $\mathcal{R}_m(\bar{f})$, the universal ordinary deformation $\mathcal{R}^0(\bar{\rho})$, and the universal modular-ordinary deformation $\mathcal{R}^0_m(\bar{f})$. These fit together in a diagram:

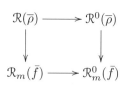

with all of the maps in the diagram surjective. Around 1990, Mazur and I both stated the conjecture that the vertical maps are in fact isomorphisms, that is, that *all deformations of a modular residual representation are (p-adically) modular.* Similarly, we would conjecture that *all ordinary deformations of a residual representation coming from an ordinary modular form themselves come from ordinary p-adic modular forms.* As they stand, both conjectures seem unlikely to be true unless $N=1$, simply because, as the notation indicates, the top row depends on $\bar{\rho}$ and the bottom row depends on $\bar{f}$. This seems innocuous until we realize that in the deformation theory we simply fixed a set of primes at which we allowed ramification, while in the modular theory we fixed the tame level N. If we take a form f of level N and think of it as of level N^2, say, the top row does not change, while the bottom row does. This requires us to make things a little more precise before we make our conjecture.

Imposing deformation conditions

There are several choices as to how to proceed.

  i. We can impose local conditions that force the conductor to remain equal to N. If we assume that our form was chosen with "optimal level," this amounts to making ρ be "as unramified as possible."
  ii. In addition to the local conditions, we can impose conditions that force the representations to look like those that come from classical modular forms. For example, in the weight two case, we could follow Wiles and Taylor (see above) and require that the determinant be the cyclotomic character, impose local conditions at $\ell \neq p$, and at p restrict ourselves to representations that are p-ordinary (if $\bar{\rho}$ is) or finite flat (if $\bar{\rho}$ is).
  iii. We can also *relax* some of the local conditions, provided we understand where to look for modular forms that produce representations satisfying the relaxed conditions.

Here's a sketch of the first approach.

To be able to formulate the right conjecture about modularity, we must understand the relation between the level N of a modular form and the (tame part of the) conductor of the attached Galois representation. Suppose we have an absolutely irreducible representation $\bar{\rho}$ which comes from a modular form $\bar{f} \in S_k(\Gamma_1(N), \mathsf{k})$ (as we mentioned above, there is no loss of generality in assuming $\bar{f}$ is of level exactly N).

For this whole section,[8] let $p \geq 5$. To measure the ramification of $\bar{\rho}$, we use, as discussed above, its conductor. Since $\bar{\rho}$ is attached to a modular form of level N, it follows from the discussion above that the conductor of $\bar{\rho}$ will be a divisor of N, and in fact Proposition 7.1 gives a quite precise description of how the two can differ. This lets us "control the conductor" in the deformation theory by imposing local deformation conditions at the primes dividing N.

We describe the strategy in the "minimal case," in which we assume that we have chosen our modular form $\bar{f}$ so that the level N is optimal. (That this is possible for $p \geq 5$ is one of the main theorems described in [**117**].) Once we have done that, the conductor of $\bar{\rho}$ will be exactly equal to N. We will look for deformations that "look as if they might correspond to forms of level N" by imposing local ramification conditions at some of the primes $\ell | N$.

As usual, let
$$S = \{\text{primes dividing } N\} \cup \{p, \infty\}.$$

Now suppose that we take a discrete valuation ring R which is a finite extension of $\mathbb{Z}_p$ and has residue field k, and suppose we have a classical eigenform $g \in S_{k'}(\Gamma_1(N), R)$ such that $\bar{g} = \bar{f}$. Then it follows from Carayol's main theorem in [**15**] that the conductor of the corresponding representation ρ_g is exactly N.

On the other hand, suppose ρ_1 is unramified outside S and $\bar{\rho}_1 = \bar{\rho}$. Then the conductor of ρ_1 can indeed be bigger than N. Proposition 7.1 shows that this will happen if and only if there exists a prime $\ell \in S$ such that $\bar{\rho}$ is I_ℓ-ordinary and ρ_1 is not I_ℓ-ordinary[9]. This gives us the clue about how to fix the problem.

Suppose, then, that $\bar{\rho}$ is an absolutely irreducible Galois representation arising from a modular form $\bar{f} \in S_k(\Gamma_1(N), \mathsf{k})$ which is the reduction of a classical modular form $f \in S_k(\Gamma_1(N), R)$ (with R as above), and assume that the conductor of $\bar{\rho}$ is exactly N, i.e., that f is of optimal level. Let
$$S = \{\text{primes dividing } N\} \cup \{p, \infty\},$$
and let
$$S_0 = \{\ell | N \text{ such that } \bar{\rho} \text{ is } I_\ell\text{-ordinary}\}.$$

(Let's note in passing that one can determine which primes are in S_0 in strictly "modular" terms—see [**58**] for the details.) By our assumptions on f, the representation ρ_f is also I_ℓ-ordinary for every $\ell \in S_0$. Let $\mathfrak{Q}$ denote the condition that any deformation $\bar{\rho}$ be I_ℓ-ordinary for all $\ell \in S_0$. This is easily checked to be a deformation condition, and hence it defines a deformation ring:
$$\mathcal{R}_N(\bar{\rho}) = \mathcal{R}_\mathfrak{Q}(\bar{\rho})$$

This is the universal deformation ring for deformations unramified outside S and ordinary at each $\ell \in S_0$, and we have a corresponding universal deformation
$$\boldsymbol{\rho}_N : G_{\mathbb{Q},S} \longrightarrow \mathrm{GL}_2(\mathcal{R}_N(\bar{\rho})).$$

We might call these the *universal level N deformation ring* and the *universal level N deformation*, respectively.

---

[8]For $p = 2$ or 3 one needs to be careful with the issue of adjusting the level, and the theory of p-adic forms requires a bit more care, so we prefer not to consider them here.

[9]This is true only because of our assumption that $\bar{f}$ has been chosen of minimal level. Without this assumption, it could also be the case that $\bar{\rho}$ is unramified at ℓ while ρ is ramified. When one considers a non-minimal deformation problem, this case must also be taken into account.

If $\bar{\rho}$ is ordinary at p, we can do the analogous thing with the added deformation condition of being I_p-ordinary, and define $\mathcal{R}_N^0(\bar{\rho})$, the universal ($p$-)ordinary level N deformation ring.

With these definitions, we can show that the homomorphism $\mathcal{R}(\bar{\rho}) \longrightarrow \mathcal{R}_m(\bar{f})$ and (if $\bar{\rho}$ is ordinary at p and f is an ordinary modular form) the homomorphism $\mathcal{R}^0(\bar{\rho}) \longrightarrow \mathcal{R}_m^0(\bar{f})$ factor through the level N deformation rings, giving maps

(I) $$\mathcal{R}_N(\bar{\rho}) \longrightarrow \mathcal{R}_m(\bar{f})$$

and (if $\bar{\rho}$ is ordinary at p and f is an ordinary modular form)

(II) $$\mathcal{R}_N^0(\bar{\rho}) \longrightarrow \mathcal{R}_m^0(\bar{f})$$

Both of these maps are known to be surjective (see above, or [58] for more detail), and it is now reasonable to conjecture that they are in fact isomorphisms.

Conjecture. The maps (I) and (II) above are isomorphisms.

This conjecture is due to Mazur, though it seems that it was first stated in print in [58]. What it says is that any deformation of a modular residual representation is p-adically modular, i.e., attached to a p-adic modular form. The work of Wiles, Taylor-Wiles, et. al. is sufficient to establish in many cases that (II) is an isomorphism. As for (I), it seems much harder to get a handle on it, basically because we do not really know very much about the big Hecke algebra. We will sketch later an argument (involving the "infinite fern" construction) that proves that (I) is true in a particularly simple case (basically, when $N = 1$ and the deformation problem is unobstructed).

Suppose we can prove (I). Then we know that every (appropriately ramified) deformation comes from a p-adic modular form of level N. The question of how to locate the deformations attached to classical modular forms now amounts to asking whether we can locate the classical modular forms within the p-adic modular forms. One important result is Hida's "control theorem." Let k be an integer; we say that a p-adic modular form f is of weight k if the left diamond operators act via k-th powers:

$$\langle x, 1 \rangle f = x^k f \qquad \text{for all } x \in \mathbb{Z}_p^\times.$$

Theorem 7.4 (Hida). *Let f be an ordinary p-adic modular form of weight $k \geq 2$ and level N. Then f is a classical modular modular form of weight k on $\Gamma_1(N) \cap \Gamma_0(p)$.*

Two remarks are in order. First, if $k \geq 3$ the form f is in fact of level N. (This follows, for example, from the discussion of p-old and p-new forms in the next lecture.) Second, Hida in fact shows a much more general result which captures classical forms of level Np^ν with $\nu > 0$; the main change is that one must consider characters of $\mathbb{Z}_p^\times$ that are more complicated than raising to the k-th power.

For the non-ordinary case, things are much less satisfactory. Coleman has proved a generalized control theorem (we will discuss it in the next lecture), but it applies to p-adic modular forms which have a special property (they are "overconvergent"). The problem is that we do not yet know how to distinguish the representations attached to overconvergent forms from representations that come from non-overconvergent forms.

LECTURE 8
p-adic families and infinite ferns

The goal of this final lecture is to explain how Mazur and I showed, in a very special case, that "all deformations are modular." This involves using Coleman's work on families of modular forms and the theory of p-old and p-new forms to produce an intricate structure inside the deformation ring. The existence of this structure, together with the assumption that the deformation problem is unobstructed, then yields the result.

Here's the basic setup. We'll assume we are given a residual representation

$$\bar{\rho} : G_{\mathbb{Q}, \{p, \infty\}} \longrightarrow \mathrm{GL}_2(\mathbb{F}_p)$$

which is absolutely irreducible and comes from some eigenform of weight k and level 1 defined over $\mathbb{Z}_p$. Let $f \in S_k(\Gamma_1(1), \mathbb{Z}_p)$ be the eigenform attached to $\bar{\rho}$.

As before, we can consider the universal deformation ring $\mathcal{R}(\bar{\rho})$ and the universal modular deformation ring $\mathcal{R}_m(\bar{f})$ (which is just a completion of the big p-adic Hecke algebra). Since we are assuming that $N = 1$, we don't have to worry about imposing local deformation conditions. As before, we then have a homomorphism

$$\mathcal{R}(\bar{\rho}) \longrightarrow \mathcal{R}_m(\bar{f}),$$

and we want to prove this is in fact an isomorphism.

The crucial assumption we will make is the following.

Assume that the deformation problem for $\bar{\rho}$ is unobstructed.

In particular, we have $\mathcal{R}(\bar{\rho}) \cong \mathbb{Z}_p[[T_1, T_2, T_3]]$. This allows us to think of the deformation space in a very concrete way: every triple of p-adic integers $(a_1, a_2, a_3) \in p\mathbb{Z}_p \times p\mathbb{Z}_p \times p\mathbb{Z}_p$ defines a homomorphism

$$\mathcal{R}(\bar{\rho}) = \mathbb{Z}_p[[T_1, T_2, T_3]] \longrightarrow \mathbb{Z}_p,$$

and this describes all such homomorphisms. Hence, we can think of the space of deformations of $\bar{\rho}$ to $\mathbb{Z}_p$ as a "cube with side $p\mathbb{Z}_p$," i.e., a kind of affine three-dimensional space. (The same is true over any extension of $\mathbb{Z}_p$ also, of course.)

Our goal is to exploit two very simple ideas (the "slope" of an eigenform and the theory of p-old and p-new forms), together with a powerful theorem of Coleman, to produce a large number of points in our space that are attached to (classical) modular forms. Under a mild technical assumption on the form f, there turn out to be enough such points that one can conclude that the homomorphism

$$\mathcal{R}(\bar{\rho}) \longrightarrow \mathcal{R}_m(\bar{f})$$

must in fact be an isomorphism. Rather than give the full proof, we will set up the ideas that allow us to construct many modular points and use them to construct a very complex object inside the deformation space. The details of how to prove that the existence of this object implies that the homomorphism above must be an isomorphism can be found in [**62**].

The slope of an eigenform

As before, let N be a number not divisible by p. We'll want to consider modular forms on $\Gamma_1(N)$, but for the p-adic theory it's important to work with the U_p operator rather than the more natural T_p. Since the space of forms on $\Gamma_1(N)$ is not stable under U_p, we move to the next largest space that is, i.e., we look at modular forms on the group $\Gamma_1(N) \cap \Gamma_0(p)$. Notice that $\Gamma_1(N) \supset \Gamma_1(N) \cap \Gamma_0(p) \supset \Gamma_1(Np)$.

Suppose, then, that $f \in S_k(\Gamma_1(N) \cap \Gamma_0(p), \mathbb{C}_p)$, where $\mathbb{C}_p$ is the completion of an algebraic closure of $\mathbb{Q}_p$. (We extend the field to $\mathbb{C}_p$ to avoid having to worry about the field of definition of our eigenforms. This way, our definition is as general as possible.) Suppose f is an eigenform under the action of U_p, and that the eigenvalue is λ_p.

Definition 8.1. If $f \in S_k(\Gamma_1(N) \cap \Gamma_0(p), \mathbb{C}_p)$ and $U_p(f) = \lambda_p f$, we define the *slope* of f to be the p-adic valuation of λ_p:

$$\text{slope}(f) = \text{ord}_p(\lambda_p),$$

where the p-adic valuation ord_p is normalized by $\text{ord}_p(p) = 1$.

The reason for the name "slope" is the following. We have an operator U_p acting on a finite-dimensional vector space $S_k(\Gamma_1(N) \cap \Gamma_0(p), \mathbb{C}_p)$, and so we can compute its characteristic polynomial $P_k(t) = \det(1 - tU_p)$. We have $P_k(t) \in \mathbb{Z}_p[t]$, because the U_p operator (like all the other Hecke operators) is in fact rationally defined. We can construct, in the usual way, the *Newton polygon* of this polynomial. (This is the lower convex hull of the points $(i, \text{ord}_p(c_i))$, where c_i is the i-th coefficient of $P_k(t)$; see, for example, [**59**] for more on Newton polygons.) The slopes of the eigenforms in $S_k(\Gamma_1(N) \cap \Gamma_0(p), \mathbb{C}_p)$ are exactly the slopes of the line segments making up the polygon, and the length of (the projection on the x-axis of) the segments gives the number of times each slope occurs among the eigenforms in this space.

We have considered, in the previous lecture, the situation in which $\text{slope}(f) = 0$, i.e., in which the eigenvalue λ_p is a p-adic unit, and we called such eigenforms *ordinary*. As we pointed out, the Galois representation attached to such a form is also (I_p-)ordinary in the representation-theoretic sense, i.e., the representation space contains a one-dimensional direct summand which is fixed under the image of the inertia group at p.

When f has non-zero slope, it is far less clear how the slope may be understood in terms of the representation. In fact, as we'll soon see, there are almost always forms of different slope which produce the same Galois representation.

On the other hand, the slope plays a very important role in the theory of p-adic modular forms, especially in the case of "overconvergent" p-adic modular forms. A first example of this, mentioned above, is Hida's control theorem (Theorem 7.4), which shows that ordinary p-adic modular forms of low weight are automatically classical.

p-old and p-new

To understand a little better the slopes of modular forms in $S_k(\Gamma_1(N) \cap \Gamma_0(p), \mathbb{C}_p)$, we need to introduce the idea of p-old and p-new forms, and then consider what it tells us about slopes.

The starting point is to notice that there are two inclusion maps
$$S_k(\Gamma_1(N), \mathbb{C}_p) \hookrightarrow S_k(\Gamma_1(N) \cap \Gamma_0(p), \mathbb{C}_p).$$
The first is essentially the "forgetful" map: a form that is modular under the action of the larger group $\Gamma_1(N)$ is certainly also modular under the subgroup $\Gamma_1(N) \cap \Gamma_0(p)$. This gives an inclusion
$$i_p : S_k(\Gamma_1(N), \mathbb{C}_p) \hookrightarrow S_k(\Gamma_1(N) \cap \Gamma_0(p), \mathbb{C}_p)$$
which induces the identity map on q-expansions, so that
$$(i_p f)(q) = f(q).$$
The second map is a little bit harder to describe; let's just say that there is an inclusion
$$v_p : S_k(\Gamma_1(N), \mathbb{C}_p) \hookrightarrow S_k(\Gamma_1(N) \cap \Gamma_0(p), \mathbb{C}_p)$$
which acts on q-expansions by replacing q by q^p:
$$(v_p f)(q) = f(q^p).$$
The map v_p is a standard tool in the theory of newforms, which is due to Atkin, Lehner, Miyake, Casselman, and Li; see, for example, [108] for more details.

The subspace of $S_k(\Gamma_1(N) \cap \Gamma_0(p), \mathbb{C}_p)$ spanned by the images of i_p and v_p is called the *p-old subspace*, and the eigenforms in this space are called *p-old eigenforms*. There is a natural inner product on $S_k(\Gamma_1(N) \cap \Gamma_0(p), \mathbb{C}_p)$, and using that inner product we can define the *p-new subspace* as the orthogonal complement of the p-old subspace. Eigenforms that belong to the p-new subspace are called *p-new eigenforms*, or sometimes just *p-newforms*. It is known that every eigenform is either in the p-old space or in the p-new space.

Our goal here is to understand what this structure can tell us about the slopes of eigenforms. For the p-new part, the answer turns out to be very simple:

Theorem 8.1. *The slope of a p-new eigenform of weight k on $\Gamma_1(N) \cap \Gamma_0(p)$ is always $(k-2)/2$. More specifically, if ϵ is the nebentypus character of f and a_p is the eigenvalue for U_p, we have $a_p^2 = \epsilon(p) p^{k-2}$.*

The proof can be found in the standard accounts of the theory of newforms; for example, see [91, Theorem 3]. An important consequence for the application we want to make is the contrapositive: if the slope of f is not $(k-2)/2$, then f must be p-old.

From our point of view, this tells us that all p-newforms have the same slope: $(k-2)/2$. The situation for p-oldforms is very different, and in some ways much more interesting.

To understand what happens, consider an eigenform $f \in S_k(\Gamma_1(N), \mathbb{C}_p)$, and let a_p be the eigenvalue of f under the action of the p-th Hecke operator T_p. As above, f has two images, $i_p f$ and $v_p f$, in the bigger space $S_k(\Gamma_1(N) \cap \Gamma_0(p), \mathbb{C}_p)$. If $\ell \neq p$, both $i_p f$ and $v_p f$ are also eigenforms under the action of the ℓ-th Hecke operator, and the eigenvalue is the same as the one for f. What we need to analyze, then, is the action of U_p. This is easy to work out. As before, let ϵ be the nebentypus character of f (a Dirichlet character modulo N, therefore). Then we have

$$U_p(i_p f) = a_p i_p f - \epsilon(p) p^{k-1} v_p f$$
$$U_p(v_p f) = i_p f$$

In other words, the two-dimensional subspace spanned by $i_p f$ and $v_p f$ is stable under U_p, which acts as the matrix

$$\begin{pmatrix} a_p & 1 \\ -\epsilon(p) p^{k-1} & 0 \end{pmatrix}.$$

The characteristic polynomial of U_p is then $t^2 - a_p t + \epsilon(p) p^{k-1}$.

Suppose that there are two distinct roots of this polynomial, λ_1 and λ_2. Then we can construct two eigenforms

$$f_1 = i_p f - \lambda_2 v_p f$$
$$f_2 = i_p f - \lambda_1 v_p f,$$

And then we will have $U_p f_1 = \lambda_1 f_1$ and $U_p f_2 = \lambda_2 f_2$. Notice that f_1 and f_2 are still eigenforms for all the other Hecke operators, with the same eigenvalues as f. In this situation, we call f_1 and f_2 *twin eigenforms*.

What can we say about the slopes of f_1 and f_2? Well, we know that $\lambda_1 \lambda_2 = \epsilon(p) p^{k-1}$, so we know that

$$\text{slope}(f_1) + \text{slope}(f_2) = k - 1.$$

We can, and do, pick the indices so that $\text{slope}(f_1) \leq \text{slope}(f_2)$. In addition, we know that $\lambda_1 + \lambda_2 = a_p$, from which we can conclude that one of two things happen:

- if $\text{ord}_p(a_p) < (k-1)/2$, then we have $\text{slope}(f_1) = \text{ord}_p(a_p)$ and $\text{slope}(f_2) = k - 1 - \text{ord}_p(a_p)$;
- if $\text{ord}_p(a_p) \geq (k-1)/2$, then we have $\text{slope}(f_1) = \text{slope}(f_2) = (k-1)/2$.

In particular, we have

$$0 \leq \text{slope}(f_1) \leq \text{slope}(f_2) \leq k - 1.$$

All of this depends on assuming that $\lambda_1 \neq \lambda_2$. If there were only one eigenvalue, then U_p would not be diagonalizable on the two-dimensional subspace spanned by $i_p f$ and $v_p f$. Notice that this would imply that $a_p = 2\lambda_1$ and hence $\text{ord}_p(a_p) = (k-1)/2$ (unless $p = 2$, in which case we would have $\text{ord}_p(a_p) = (k+1)/2$). The conjecture is that this cannot happen:

Conjecture (Ulmer). *The action of U_p on $S_k(\Gamma_1(N) \cap \Gamma_0(p), \mathbb{C}_p)$ is semisimple. In particular, we always have $\lambda_1 \neq \lambda_2$.*

This has been proved for the case of forms of weight $k = 2$ on $\Gamma_0(Np)$ by Coleman and Edixhoven in [**30**]. Under the assumption that the Tate Conjecture is true, they show that in fact it is true for all weights $k \geq 2$. Ulmer has shown a different conditional result in [**153**]: the case $k = 3$ follows from the conjecture of Birch and Swinnerton-Dyer for elliptic curves over function fields.

Experimentally, the conjecture certainly seems to hold. In computations for the case $N = 1$, small primes $p \leq 100$ and small weights $k \leq 100$, we always find

that
$$\mathrm{ord}_p(a_p) < \frac{k-1}{2}.$$
In fact, we "almost always" find that a much stronger inequality holds, namely that
$$\mathrm{ord}_p(a_p) < \frac{k-1}{p+1},$$
where by "almost always" we mean that the number of counterexamples seems to be quite small. We will discuss these computations in a forthcoming paper.

Every p-old eigenform must arise in this way from an eigenform in level N. The upshot, then, is that p-oldforms usually (if we assume Ulmer's conjecture, always) come in twin pairs f_1 and f_2 whose slopes add up to $k-1$. These twin forms have the same eigenvalues under all the Hecke operators except for U_p. In particular, since the Galois representation is determined by the eigenvalues of the T_ℓ with $\ell \nmid Np$, the Galois representations attached to f_1 and to f_2 (or, for that matter, to f) are exactly the same. It is this fact, that a single representation can "come from" forms of different slopes, that will fuel the construction of the "infinite fern" in the deformation space.

p-adic families of modular forms

The final ingredient in our witches' brew is Coleman's theorem on p-adic families of modular forms. To state a version of this theorem, suppose that we start with an eigenform $f \in S_{k_0}(\Gamma_1(N) \cap \Gamma_0(p), \mathbb{Z}_p)$ (note that now we are asking for coefficients in $\mathbb{Z}_p$!) and suppose that the slope of f is not equal to $k_0 - 1$. Coleman has proved (see [**26**] and [**28**]) that any such eigenform fits into a one-parameter p-adic analytic family of overconvergent p-adic modular forms with Fourier coefficients in $\mathbb{Z}_p$ which are eigenforms for T_ℓ for all $\ell \neq p$ and for U_p, have constant slope α, are all congruent modulo p, and where the "one parameter" is given by the weight. Furthermore, he has shown in [**26**] that the forms in this family corresponding to weights that are rational integers bigger than $\alpha + 1$ are classical modular forms; this, of course, is Coleman's generalization of Hida's "control theorem" for ordinary modular forms. (For an expository formulation of some of these results of Coleman, see [**100**].)

One can think of this analytic family as a family of q-expansions
$$f_k = q + a_2(k)q^2 + a_3(k)q^3 + \ldots,$$
where each of the $a_n(k)$ is an analytic function of k and where specialization to the original weight k_0 gives the form from which we started. Since each f_k is an (p-adic) eigenform, there is an associated representation ρ_k; since the f_k are all congruent modulo p, these ρ_k are all deformations of the residual representation attached to our original form. Hence, Coleman's theorem gives us a "p-adic analytic curve" in the deformation space, consisting of representations all of which are attached to forms of slope α.

Infinite ferns

We now focus on the special case we want to study more closely. Let $N = 1$ and assume $\bar\rho$ is an absolutely irreducible representation coming from a (classical) modular form f of weight k and level 1 with coefficients in $\mathbb{Z}_p$. As above, we can think of f as an oldform on $\Gamma_0(p)$. Take $S = \{p, \infty\}$, and $\Lambda = \mathbb{Z}_p$. Let $\mathcal{R}(\bar{\rho})$ be the universal deformation ring and (as before) let $\mathcal{R}_m(\bar{f})$ be the completion of the big p-adic Hecke algebra at the maximal ideal corresponding to f. **We assume that the deformation problem is unobstructed,** so that in particular we have $\mathcal{R}(\bar{\rho}) \cong \mathbb{Z}_p[[T_1, T_2, T_3]]$. Let X be the universal deformation space, which is just the "cube" $p\mathbb{Z}_p \times p\mathbb{Z}_p \times p\mathbb{Z}_p$, thought of as a p-adic analytic space.[1] Let X_0 be the "Sen null subspace" which we discussed in the previous lecture, i.e., the subspace of X corresponding to representations one of whose Sen weights is zero. As we pointed out above, X_0 is a analytic subspace of X of dimension two.

Let B^1 be the rigid-analytic closed unit ball over $\mathbb{Q}_p$, so that $B^1(\mathbb{Q}_p) = \mathbb{Z}_p$. Here is the version of Coleman's theorem which we will use:

Theorem 8.2 (Coleman). *Let f be an eigenform of level p and weight κ_0 and let $x \in X(\mathbb{Q}_p)$ be the point such that the representation ρ_x is attached to f. Suppose the p-adic valuation of the eigenvalue of U_p acting on f is not equal to $\kappa_0 - 1$. Then there exists an open neighborhood $D \subset B^1$ of $\kappa_0 \in \mathbb{Z} \subset \mathbb{Z}_p = B^1(\mathbb{Q}_p)$, and a p-adic analytic mapping*

$$z \colon D \longrightarrow X_0 \subset X$$

of D to the subspace X_0 of the p-adic analytic manifold X, and a p-adic analytic function

$$u \colon D \longrightarrow B^1,$$

such that, for an arithmetic progression of (positive, rational) integers $\mathcal{K} \subset D$ which is topologically dense in D, the image of each $\kappa \in \mathcal{K}$ under the mapping z is a point $z(\kappa)$ whose associated representation is the representation attached to a modular eigenform f_κ of level p, weight κ, and U_p-eigenvalue equal to $u(\kappa)$. Finally, $f_{\kappa_0} = f$.

The crucial point is that the family depends not only on the representation we start with, but also on the *slope* of the modular form attached to that representation. As a result, when the Galois representation ρ is attached (in the above sense) to a pair of "twins," it follows that there exist *two distinct* one-parameter families of deformations of ρ. This is what allows us to construct the structures Mazur and I call "infinite ferns" in the deformation space of ρ.

Let $\Gamma \subset \mathbb{Z}_p^\times$ denote the group of 1-units in $\mathbb{Z}_p$, i.e., the multiplicative group of p-adic integers congruent to 1 mod p. Twisting the representation ρ_x corresponding to a point $x \in X(\mathbb{Q}_p)$ by a one-dimensional "wild" character $\psi \colon G_{\mathbb{Q}, \{p\}} \longrightarrow \Gamma$,

$$\rho_x \mapsto \psi \otimes \rho_x,$$

induces a p-adic analytic action of the group of wild characters (i.e., the formal group in one parameter, call it $\Psi = \mathrm{Hom}_{\mathrm{cont}}(G_{\mathbb{Q}, \{p\}}, \Gamma)$) on the p-adic manifold X. For a very brief discussion of this action, see §5 of [**100**]. Let us denote the point of X which corresponds to the representation $\psi \otimes \rho_x$ by $\psi \circ x$.

As noted above, X contains the p-adic analytic "surface" X_0. This space is "essentially transversal" to the action of Ψ on X, in the sense that the natural

---

[1] Well, really $X = \mathrm{Spf}(\mathcal{R}(\bar{\rho}))^{\mathrm{rig}}$ is the open unit rigid-analytic 3-ball over $\mathbb{Q}_p$. The "cube" is actually $X(\mathbb{Q}_p)$, i.e., it consists of the points in X that are defined over $\mathbb{Q}_p$. All the points in the "infinite fern" we are about to construct *are* in fact defined over $\mathbb{Q}_p$, so the mental image of the "cube" is not too misleading.

mapping

$$\pi \colon X_0 \times \Psi \longrightarrow X$$
$$(x_0, \psi) \mapsto \psi \circ x_0$$

has fibers of cardinality ≤ 2 (in fact, π is the restriction to $X_0 \times \Psi$ of a mapping of degree 2 on analytic spaces) and π is unramified off the locus $X_{00} \times \Psi \subset X_0 \times \Psi$, where X_{00} is the analytic subset of X whose $\mathbb{Q}_p$-points x are those for which the Hodge-Tate-Sen weights of the associated representation ρ_x are $\{0,0\}$ (cf. the main proposition of §8 in [**100**]).

We need one further assumption before we can proceed:

We assume the slope of the modular form f is not equal to 0 or to $k-1$.

Note that forms of slope 0 always come in twin pairs with forms of slope $k-1$, so that it makes sense to assume both of those together. We call forms whose slope is 0 or $k-1$ "forms of critical slope;" thus, our assumption is that f is of non-critical slope.[2]

Given that whole setup, we do the following.

- To begin with, we have a modular residual representation whose associated modular form $f = f_0$ is of weight $k = k_0$ and slope α_0 not equal to 0 or $k_0 - 1$.
- Use Coleman's theorem to produce a curve of deformations containing f_0, each attached to a modular form of some weight and slope α_0. (This might be the curve C in the picture.)
- If the initial form is not a newform, then it has a twin, and we can construct a curve corresponding to that twin. It goes through the same initial point, since twin forms give the same representation, but it corresponds to forms of weight $k_0 - 1 - \alpha_0$. (If the first curve is C, this is $C^{(\kappa)}$.)
- "Move" along either of the families to a classical form f_1 of weight k_1 and slope α_0. Make sure $k_1 > \alpha_0 + 1$, $k_1 \neq 2\alpha_0 + 1$, and $k_1 \neq 2\alpha_0 + 2$. Notice that there are infinitely many integers k_1 with these properties. In fact, the set of such integers is dense in $\mathbb{Z}_p$.
- The last inequality means that f_1 is p-old, so it has a twin $\tilde{f}_1$ of weight k_1 and slope $\alpha_1 = k_1 - 1 - \alpha_0$. Notice that $\alpha_1 \neq 0, k_1 - 1, \alpha_0$. (This is where we need to know that the initial form did not have critical slope.)
- Now repeat the process starting from f_1. In fact, repeat this at all points that satisfy the weight-slope constraints.

This produces an amazingly complex structure, which we call an "infinite fern," made up of infinitely many analytic arcs, all contained in X_0. (See figure 1 and the discussion in §18 of [**100**].) In the diagram, each curve segment corresponds to a modular arc in X_0, which is an embedded p-adic analytic image of a disk in $\mathbb{Z}_p$. For a topologically dense set of the points κ on any given modular arc C there is another arc $C^{(\kappa)}$ crossing C at κ. More pictorially, calling any given modular arc a "spine" and calling the modular arcs crossing it "needles," we have that each "spine" has a topologically dense set of "needles."

---

[2]This is actually a minor assumption: if our initial form happens to have slope 0, we can replace it by another form of non-zero slope which produces the same residual representation, at the cost of enlarging the base ring over which we are working, using the trick given on page 111 of [**57**] (but note that this forces us to move from $\mathbb{Z}_p$ to some ramified extension).

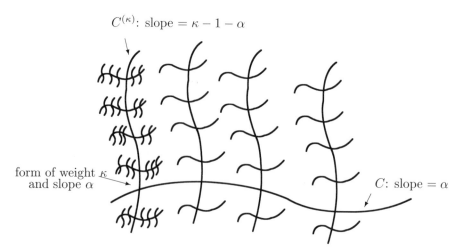

Figure 1. An infinite fern

In fact, not only do we get an infinite fern growing around our initial point, but also we see that every modular form satisfying our constraints which is congruent to f corresponds to a point in X_0 which has a topological neighborhood containing the image of an infinite fern. This structure "fills up" the subvariety X_0, and we can "thicken it" by considering all possible twists of the whole structure. This gives a structure that is sufficiently "big" to prove that the modular points must be Zariski-dense in the full deformation space. Using this, we can prove the following result:

Proposition 8.3. *Suppose $\bar{\rho}$ is absolutely irreducible, unramified outside p and infinity, and attached to a (classical) modular form on $\Gamma_0(p)$ of non-critical slope and with Fourier coefficients in $\mathbb{Z}_p$. Suppose also that the deformation problem associated to $\bar{\rho}$ is unobstructed. Then the map $\mathcal{R}(\bar{\rho}) \longrightarrow \mathcal{R}(\bar{f})$ given by the deformation theory is an isomorphism.*

It is possible that a similar approach will work whenever we know the dimension (perhaps even the local dimension?) of the deformation space.

In the case where the residual representation comes from an ordinary modular form, another approach to theorems of this sort seems plausible: one can use Wiles' methods to show that all ordinary deformations are modular, and then use Böckle's results (in [5]) on the relation between the ordinary deformation space and the full deformation space to compare the full deformation space to the modular deformation space. Böckle has recently announced a result along these lines.

APPENDIX 1
A criterion for existence of a universal deformation ring
by Mark Dickinson

The purpose of this note is to give an elementary proof of the existence of the universal deformation ring, using a representability criterion of Grothendieck. In particular the proof makes no use of Schlessinger's criteria or of noetherian hypotheses. I would like to thank Brian Conrad for suggesting that this be written up and Fernando Gouvêa for allowing me to include it here. I would also like to thank Sam Williams for helpful comments.

Let k be a field and let Λ be a topological ring which is the inverse limit of a system of artinian local rings, each with the discrete topology and with residue field k. (For example let k be finite and Λ the ring of Witt vectors of k.) We define two full subcategories of the category of topological Λ-algebras R. Let $\mathscr{C}_\Lambda^{fl}$ be the full subcategory whose objects are discrete finite-length local Λ-algebras R for which the structure map induces an isomorphism on residue fields, and define $\mathscr{C}_\Lambda$ to be the full subcategory whose objects are those arising as an inverse limit of objects of $\mathscr{C}_\Lambda^{fl}$. (The objects of $\mathscr{C}_\Lambda$ are examples of 'pseudo-compact' rings; for basic properties of these rings see section 0 of [56].)

Now let G be a profinite group, d a positive integer and $\bar{\rho} \colon G \to \mathrm{GL}_d(k)$ a continuous representation of G. Assume that the only matrices in $\mathrm{M}_d(k)$ which commute with every element of the image of $\bar{\rho}$ are the scalar matrices.

Definition 9.1. A *lifting* of $\bar{\rho}$ (to R) is a pair (R, ρ) consisting of an object R of $\mathscr{C}_\Lambda$ together with a continuous representation $\rho \colon G \to \mathrm{GL}_d(R)$ whose pushforward by the natural reduction map $R \to k$ is conjugate to $\bar{\rho}$. Two liftings (R, ρ) and (R, σ) of $\bar{\rho}$ to the same ring are *conjugate* if the representations ρ and σ are conjugate. A *deformation* of $\bar{\rho}$ to R is a conjugacy class of liftings of $\bar{\rho}$ to R; the notation (R, ρ) will also be used for the deformation represented by a lifting (R, ρ).

Note that, if we assume $C(\bar{\rho}) = k$, the conjugating matrix relating two (conjugate) representations (R, ρ) and (R, σ) lifting $\bar{\rho}$ can be chosen to be congruent to the identity modulo the maximal ideal of R.

If (R, ρ) is a deformation of $\bar{\rho}$ and $\phi \colon R \to S$ is a morphism of $\mathscr{C}_\Lambda$ then the pushforward $(S, \phi_* \rho)$ of (R, ρ) by ϕ is a deformation of $\bar{\rho}$ to S. Thus there is a well-defined functor

$$\mathrm{Def}\colon \mathscr{C}_\Lambda \to \mathbf{Sets}$$

which sends an object R of $\mathscr{C}_\Lambda$ to the set of deformations of $\bar\rho$ to R. Suppose that certain deformations of $\bar\rho$ are designated 'of type $\mathscr{P}$', and that the pushforward of a deformation (R, ρ) of type $\mathscr{P}$ by a morphism $\phi\colon R \to S$ is again of type $\mathscr{P}$. Then one can define a subfunctor

$$\mathrm{Def}_\mathscr{P}\colon \mathscr{C}_\Lambda \to \mathbf{Sets}$$

of Def which sends an object R to the set of deformations of $\bar\rho$ to R of type $\mathscr{P}$.

Definition 9.2. We define a *universal deformation* of $\bar\rho$ of type $\mathscr{P}$ to be a deformation $(R_\mathscr{P}^{\mathrm{univ}}, \rho_\mathscr{P}^{\mathrm{univ}})$ of $\bar\rho$ of type $\mathscr{P}$ such that for any given deformation (R, ρ) of $\bar\rho$ of type $\mathscr{P}$ there is a unique morphism $\phi\colon R_\mathscr{P}^{\mathrm{univ}} \to R$ for which the pushforward of $(R_\mathscr{P}^{\mathrm{univ}}, \rho_\mathscr{P}^{\mathrm{univ}})$ by ϕ is equal to (R, ρ).

By the Yoneda Lemma, to give such a universal deformation is equivalent to giving an object $R_\mathscr{P}^{\mathrm{univ}}$ of $\mathscr{C}_\Lambda$ along with an isomorphism

$$\mathrm{Hom}_{\mathscr{C}_\Lambda}(R_\mathscr{P}^{\mathrm{univ}}, -) \cong \mathrm{Def}_\mathscr{P}$$

of functors; thus a universal deformation of $\bar\rho$ of type $\mathscr{P}$ exists if and only if the functor $\mathrm{Def}_\mathscr{P}$ is representable. We call the object $R_\mathscr{P}^{\mathrm{univ}}$ a *universal deformation ring* for deformations of $\bar\rho$ of type $\mathscr{P}$. The following theorem tells us when we can expect a universal deformation of type $\mathscr{P}$ to exist.

Theorem 9.1. *The following three conditions are necessary and sufficient for the existence of a universal deformation of $\bar\rho$ of type $\mathscr{P}$:*

- *the trivial deformation $(k, \bar\rho)$ of $\bar\rho$ is of type $\mathscr{P}$,*
- *given a diagram $R \xrightarrow{\phi} T \xleftarrow{\psi} S$ in $\mathscr{C}_\Lambda^{\mathrm{fl}}$, any deformation of $\bar\rho$ to the fiber product $R \times_T S$ whose pushforwards to R and to S are both of type $\mathscr{P}$ is itself of type $\mathscr{P}$, and*
- *if R in $\mathscr{C}_\Lambda$ is a filtered limit of objects $(R_i)_{i \in \mathscr{I}}$ of $\mathscr{C}_\Lambda^{\mathrm{fl}}$ and the pushforward of a deformation (R, ρ) by the natural reduction map $R \to R_i$ is of type $\mathscr{P}$ for each i, then (R, ρ) is of type $\mathscr{P}$.*

Note especially that in the case where *every* deformation of $\bar\rho$ is of type $\mathscr{P}$ the conditions of the theorem are trivially satisfied and so a universal deformation exists.

To prove the theorem, we first give Grothendieck's criterion for a set-valued covariant functor on $\mathscr{C}_\Lambda$ to be representable. Recall that a functor is called left exact if it is compatible with the formation of finite limits and that being left exact is equivalent to taking terminal objects (resp., fiber products) to terminal objects (resp., fiber products).

Proposition 9.2. *A functor $X\colon \mathscr{C}_\Lambda \to \mathbf{Sets}$ is representable if and only if the restriction of X to $\mathscr{C}_\Lambda^{\mathrm{fl}}$ is left exact and X preserves filtered limits, taken in $\mathscr{C}_\Lambda$, of objects of $\mathscr{C}_\Lambda^{\mathrm{fl}}$.*

Proof. First note that the functor which sends an object R of $\mathscr{C}_\Lambda$ to the system of finite-length discrete quotients of R gives an equivalence of categories between the category $\mathscr{C}_\Lambda$ and the category of pro-objects of $\mathscr{C}_\Lambda^{\mathrm{fl}}$, as defined in section A2 of [67]. By the corollary to Proposition 3.1 of section A of [67] there is an object R

of $\mathscr{C}_\Lambda$ and an isomorphism $\mathrm{Hom}_{\mathscr{C}_\Lambda}(R,-) \cong X$ of functors on $\mathscr{C}_\Lambda^{\mathrm{fl}}$ if and only if the restriction of X to $\mathscr{C}_\Lambda^{\mathrm{fl}}$ is left exact. This isomorphism extends to an isomorphism of functors on $\mathscr{C}_\Lambda$ if and only if X preserves filtered limits of objects of $\mathscr{C}_\Lambda^{\mathrm{fl}}$. $\square$

So in order to prove Theorem 9.1 it is enough to check that the functor $\mathrm{Def}_{\mathscr{P}}$ satisfies the hypotheses of Proposition 9.2. We first prove that the full deformation functor $\mathrm{Def}\colon \mathscr{C}_\Lambda \to \mathbf{Sets}$ satisfies these hypotheses. We begin with the following easy consequence of the fact that the centralizer of the image of $\bar\rho$ is the scalar matrices.

Lemma 9.3. *Let W be a subspace of a finite-dimensional k-vector space V, and B a $d \times d$ matrix with entries in V. If $B\bar\rho(g) - \bar\rho(g)B$ has entries in W for all g in G, then $B = vI + C$ for some element v of V and a matrix C all of whose entries lie in W.*

Proof. Take a basis $\{e_1, \ldots, e_r\}$ of W over k and extend it to a basis $\{e_1, \ldots, e_s\}$ of V. Write B as $\sum_{i=1}^{s} B_i e_i$ where each B_i is an element of $\mathrm{M}_d(k)$. Then for $r < i \leq s$ the matrix B_i commutes with each element of the image of $\bar\rho$ and hence is a scalar matrix. $\square$

Lemma 9.4. *Suppose that R is an object of $\mathscr{C}_\Lambda^{\mathrm{fl}}$ and (R, ρ) is a lifting of $\bar\rho$. Then any element of $\mathrm{M}_d(R)$ which centralizes the image of ρ is a scalar matrix.*

Proof. Since R is a local artinian ring the nth power of the maximal ideal $\mathfrak{m}_R$ of R is trivial for some $n \geq 1$; we prove the result by induction on n. For $n = 1$ the result is immediate, since we assumed that only the scalar matrices in $\mathrm{M}_d(k)$ centralize the image of $\bar\rho$. Suppose that $n > 1$ and that A is an element of $\mathrm{M}_d(R)$ such that $A\rho = \rho A$. By the induction hypothesis the reduction of A modulo $\mathfrak{m}_R^{n-1}$ is a scalar matrix and we can write $A = \lambda I + B$ for some element λ of R and some matrix B with entries in the finite-dimensional k-vector space $\mathfrak{m}_R^{n-1}$. Then $(\lambda I + B)\rho = \rho(\lambda I + B)$ and so $B\rho - \rho B = 0$. Since B has entries killed by $\mathfrak{m}_R$, we can rewrite this equation as $B\bar\rho - \bar\rho B = 0$. So B is a scalar matrix by Lemma 9.3, hence so is $A = \lambda I + B$. $\square$

Lemma 9.5. *The functor*

$$\mathrm{Def}\colon \mathscr{C}_\Lambda \to \mathbf{Sets}$$

which sends an object R of $\mathscr{C}_\Lambda$ to the set of deformations of $\bar\rho$ over R is representable.

Proof. By Proposition 9.2 we need to show that Def preserves filtered limits of objects of $\mathscr{C}_\Lambda^{\mathrm{fl}}$ and that Def restricted to $\mathscr{C}_\Lambda^{\mathrm{fl}}$ is left exact. To show the former requires, not surprisingly, an application of Zorn's Lemma. Let R be the filtered limit of a system $(R_i)_{i \in \mathscr{I}}$ of objects and maps of $\mathscr{C}_\Lambda^{\mathrm{fl}}$ indexed by a set $\mathscr{I}$ and suppose that for each $i \in \mathscr{I}$ we are given a deformation (R_i, ρ_i) of $\bar\rho$ and that for every morphism $i \to j$ of $\mathscr{I}$ the pushforward of (R_i, ρ_i) by the corresponding map $R_i \to R_j$ is equal to (R_j, ρ_j). We must show that there is a unique deformation (R, ρ) whose pushforward by each natural projection map $R \to R_i$ is equal to (R_i, ρ_i). For each object i of $\mathscr{I}$ let S_i be the set of liftings of $\bar\rho$ to R_i which represent (R_i, ρ_i) and which reduce to $\bar\rho$, and consider the system of subsets of S_i consisting of the orbits of subgroups of $\mathrm{GL}_d(R_i)$ of the form $I + \mathrm{M}_d(J)$ for some proper ideal J of R_i. Using these subsets we can apply Theorem 1 of section 7.4 of Chapter 3 of [**12**] to deduce that the inverse limit over i in $\mathscr{I}$ of the sets S_i is non-empty, hence that there is a deformation (R, ρ) as desired. To show that this deformation is unique, suppose that (R, σ) is another such and that for every object i of $\mathscr{I}$ there is a matrix A_i, unique up to scalar multiplication by Lemma 9.4, which conjugates the pushforward to R_i of ρ to the pushforward of σ. If we assume that both ρ and σ reduce to $\bar\rho$ then it follows from the triviality of the centralizer of the image of $\bar\rho$ that each A_i reduces to a scalar matrix. Thus we may assume that the top left entry of A_i is equal to 1 for each i in $\mathscr{I}$; then the A_i form a compatible system of matrices and so give a matrix A with entries in R which conjugates ρ to σ.

To check that Def is left exact on $\mathscr{C}_\Lambda^{\mathrm{fl}}$ it suffices to check that it preserves fiber products and the terminal object. The only deformation of $\bar\rho$ to k is $(k, \bar\rho)$ itself, so $\mathrm{Def}(k)$ is a one point set and Def preserves the terminal object. It remains to show that Def preserves fiber products in $\mathscr{C}_\Lambda^{\mathrm{fl}}$. Suppose that we have a fiber product

$$\begin{array}{ccc} R \times_T S & \longrightarrow & R \\ \downarrow & & \downarrow \phi \\ S & \xrightarrow{\psi} & T \end{array}$$

of objects of $\mathscr{C}_\Lambda^{\mathrm{fl}}$, and suppose that (R, ρ) and (S, σ) are liftings of $\bar\rho$ whose pushforwards $(T, \phi_*\rho)$ and $(T, \psi_*\sigma)$ are conjugate. We will show that it is possible to replace (R, ρ) and (S, σ) with conjugate liftings whose pushforwards to T are identical. Then we obtain a lifting $(R \times_T S, \pi)$ of $\bar\rho$ whose pushforwards to R and S are conjugate to (R, ρ) and (S, σ) respectively; a similar argument to the one above for filtered limits shows that if this lifting exists then it is unique up to conjugation.

We suppose that the nth power of the maximal ideal $\mathfrak{m}_T$ of T is zero and prove the existence of π as above by induction on n. We may assume that ρ and σ each reduce to $\bar\rho$ (rather than just to a conjugate of $\bar\rho$), and in the case $n = 1$ there is nothing more to do; now suppose that $n > 1$ and that

$$\phi_*\rho = \psi_*\sigma \quad \text{modulo } \mathfrak{m}_T^{n-1}.$$

By assumption there is a matrix C in $\mathrm{GL}_d(T)$ such that $C\psi_*\sigma C^{-1} = \phi_*\rho$. The reduction of C modulo $\mathfrak{m}_T^{n-1}$ centralizes the image of the reduction of the representation $\psi_*\sigma$ modulo $\mathfrak{m}_T^{n-1}$ so is a scalar matrix by Lemma 9.4. So without loss of generality we may assume that $C = I + L$ for some matrix L with entries in $\mathfrak{m}_T^{n-1}$. Now

$$(I + L)\psi_*\sigma(I - L) = \phi_*\rho$$

and since L is annihilated by $\mathfrak{m}_T$ we can rewrite this as

$$L\bar\rho - \bar\rho L = \phi_*\rho - \psi_*\sigma.$$

From Lemma 9.3 it follows that $L = \lambda I + \phi(M) - \psi(N)$ for some element λ of $\mathfrak{m}_T^{n-1}$ and for some matrices M and N with entries in R and S respectively; by replacing M and N by $M - P$ and $N + P$ for some P in $\mathrm{M}_d(\Lambda)$ lifting the reduction of M to $\mathrm{M}_d(k)$ we may assume that M and N have entries in the maximal ideals $\mathfrak{m}_R$ and $\mathfrak{m}_S$ of R and S. Then $\phi(M)(\phi(M) - \psi(N)) = 0$ and it follows that $I + L = (1 + \lambda)(I - \phi(M))^{-1}(I - \psi(N))$ and hence that

$$\psi_*((I - N)\sigma(I - N)^{-1}) = \phi_*((I - M)\rho(I - M)^{-1})$$

thus giving conjugates of ρ and σ whose pushforwards to T are identical, as required. □

Proof of Theorem 9.1. The proof of Theorem 9.1 now follows easily: Def is left exact and preserves filtered limits by Lemma 9.5 and Proposition 9.2. Now the first two conditions in the statement of the theorem assert that the subfunctor $\text{Def}_{\mathscr{P}}$ of Def is left exact on $\mathscr{C}_\Lambda^{\text{fl}}$ and the third that $\text{Def}_{\mathscr{P}}$ preserves filtered limits of objects of $\mathscr{C}_\Lambda^{\text{fl}}$. So by Proposition 9.2 again the functor $\text{Def}_{\mathscr{P}}$ is representable (equivalently, a universal deformation of $\bar\rho$ of type $\mathscr{P}$ exists) if and only if the conditions of the theorem are satisfied. □

To end we note that Grothendieck's results also yield a criterion for the universal deformation ring to be noetherian. Let $k[\varepsilon]$ denote the object $k[X]/(X^2)$ of $\mathscr{C}_\Lambda^{\text{fl}}$ (where ε corresponds to the image of X). It is a k-vector space object of the category $\mathscr{C}_\Lambda^{\text{fl}}$, with addition $k[\varepsilon] \times_k k[\varepsilon] \to k[\varepsilon]$ defined by sending $(a\varepsilon, b\varepsilon)$ to $(a+b)\varepsilon$ and scalar multiplication by a defined by sending $b\varepsilon$ to $ab\varepsilon$. For any left exact functor $X\colon \mathscr{C}_\Lambda^{\text{fl}} \to \mathbf{Sets}$, these maps provide the set $X(k[\varepsilon])$ with the structure of a k-vector space, and for any natural transformation of left-exact functors $\alpha\colon X \to Y$ the map $\alpha_{k[\varepsilon]}\colon X(k[\varepsilon]) \to Y(k[\varepsilon])$ is k-linear. In particular if the deformation functor $\text{Def}_{\mathscr{P}}$ is representable then $\text{Def}_{\mathscr{P}}(k[\varepsilon])$ is a k-vector space and is isomorphic to $\text{Hom}_{\mathscr{C}_\Lambda}(R_{\mathscr{P}}^{\text{univ}}, k[\varepsilon])$.

Proposition 9.6. *Suppose that Λ is noetherian and that the functor $\text{Def}_{\mathscr{P}}$ satisfies the conditions of Theorem 9.1 and so is representable. Then the universal deformation ring $R_{\mathscr{P}}^{\text{univ}}$ is noetherian if and only if the k-vector space $\text{Def}_{\mathscr{P}}(k[\varepsilon])$ is finite-dimensional; furthermore, if $\text{Def}_{\mathscr{P}}(k[\varepsilon])$ has dimension d then $R_{\mathscr{P}}^{\text{univ}}$ is a quotient of the power-series ring $\Lambda[[X_1,\ldots,X_d]]$.*

Proof. This follows from Proposition 5.1 of section A of [**67**]. □

Mark Dickinson
Department of Mathematics
Harvard University
Cambridge, MA 02138
dickinso@math.harvard.edu

APPENDIX 2
An overview of a theorem of Flach
by Tom Weston

In recent years, the study of the deformation theory of Galois representations has become of central importance in arithmetic algebraic geometry. The most fundamental question in this field is the explicit determination of universal deformation rings associated to given residual representations. It was observed by Mazur in his first paper [**97**, Section 1.6] on the subject that the solution of this problem is immediate if a certain Galois cohomology group associated to the residual representation vanishes.

The goal of this appendix is to provide an overview of a theorem of Flach which yields the vanishing of this cohomology group for many mod l representations coming from rational elliptic curves. We do not seek to give a complete proof; we hope only to make clear the main ideas. In the process we will touch on many facets of arithmetic algebraic geometry, including Tate's duality theorems in Galois cohomology, generalized Selmer groups, Kolyvagin's theory of Euler systems and the geometry of modular curves.

The work we will describe actually has another, more direct, application: it can be used in many cases to prove the Taylor-Wiles isomorphism between a certain universal deformation ring and a certain Hecke algebra. We will not touch on this aspect; for details, see [**99**] or [**158**].

We have tried to keep the prerequisites to a minimum. The main requirement is a good familiarity with Galois cohomology. The algebraic geometry we use is mostly at the level of [**146**, Chapters 1 and 2], with the exception of Appendix B, which is significantly more advanced. Some familiarity with elliptic curves is helpful, although with the exception of Appendix A we will use little more than the Tate module and the existence of the Weil pairing.

I would like to thank Brian Conrad, Matthew Emerton and Karl Rubin for teaching me much of the material presented here. I would also like to thank Fernando Gouvêa for encouraging the writing of this paper. Above all, I would like to thank Barry Mazur for his constant help and insights; I can only hope that his point of view is visible in the mathematics below.

Unobstructed deformation problems

Let $G_{\mathbb{Q},S}$ be the maximal quotient of the absolute Galois group of $\mathbb{Q}$ unramified away from a finite set of places S. Let l be a prime number and let $\bar\rho: G_{\mathbb{Q},S} \to \mathrm{GL}_2(\mathbb{F}_l)$ be a Galois representation. Under certain additional hypotheses (for example, if $\bar\rho$ is absolutely irreducible) we can associate a *universal deformation ring* $\mathcal{R}(\bar\rho)$ to such a residual representation; see Lecture 3 in these notes, and [**97**] or [**101**] for details.

In general the determination of the structure of the ring $\mathcal{R}(\bar\rho)$ is quite difficult. However, in at least one case the determination is easy: let $\mathrm{ad}(\bar\rho)$ be the $G_{\mathbb{Q},S}$-module of 2×2 matrices over $\mathbb{F}_l$ on which $\gamma \in G_{\mathbb{Q},S}$ acts via conjugation by $\bar\rho(\gamma)$. If

$$\mathrm{H}^2(G_{\mathbb{Q},S}, \mathrm{ad}(\bar\rho)) = 0,$$

then $\mathcal{R}(\bar\rho)$ is isomorphic to a power series ring in $\dim_{\mathbb{F}_l} \mathrm{H}^1(G_{\mathbb{Q},S}, \mathrm{ad}(\bar\rho))$ variables over $\mathbb{Z}_l$; see [**97**, Section 1.6, Proposition 2]. If this is the case, we say that the deformation problem for $\bar\rho$ is *unobstructed*.

Our goal in this paper is to explain the main ideas of the proof of the following theorem of Flach (this is [**50**, Theorem 2]).

Theorem 10.1. *Let E be an elliptic curve over $\mathbb{Q}$, let $l \geq 5$ be a prime and let S be the set of places of $\mathbb{Q}$ at which E has bad reduction, together with l and ∞. Let $\rho: G_{\mathbb{Q},S} \to \mathrm{GL}_2(\mathbb{Z}_l)$ be the representation of $G_{\mathbb{Q},S}$ on the l-adic Tate module of E and let $\bar\rho: G_{\mathbb{Q},S} \to \mathrm{GL}_2(\mathbb{F}_l)$ be the residual representation. Assume further that:*

- *E has good reduction at l;*
- *ρ is surjective;*
- *For all $p \in S - \{\infty\}$, $E[l] \otimes E[l]$ has no $G_{\mathbb{Q}_p}$-invariants;*
- *l does not divide the rational number $L(\mathrm{Sym}^2 T_l E, 0)/\Omega$, where $\mathrm{Sym}^2 T_l E$ is the symmetric square of the l-adic Tate module of E and Ω is a certain period.*

Then the deformation problem for $\bar\rho$ is unobstructed.

In Appendix A we discuss precisely how stringent these hypotheses are; the main result is that for fixed E which does not have complex multiplication, then they are satisfied for a set of primes l of density 1. We will explain the fourth hypothesis in Section 5.

We should note that this theorem uses in a crucial way the fact that E is modular, and thus stating it in the form above relies heavily on the recent proof of the Shimura-Taniyama conjecture.

Galois modules and the calculus of Tate twists

We begin with some formalities on Galois actions and certain commutative algebra operations. Let S be a finite set of places of $\mathbb{Q}$ including the prime l. Let M and N be $\mathbb{Z}_l$-modules with $G_{\mathbb{Q},S}$-actions. We make the tensor product $M \otimes_{\mathbb{Z}_l} N$ a $G_{\mathbb{Q},S}$-module via the diagonal action: $\gamma(m \otimes n) = \gamma m \otimes \gamma n$. We make $\mathrm{Hom}_{\mathbb{Z}_l}(M,N)$ a $G_{\mathbb{Q},S}$-module via the adjoint action: $\gamma f(m) = \gamma \cdot f(\gamma^{-1} m)$ for $f \in \mathrm{Hom}_{\mathbb{Z}_l}(M,N)$. Note that the $G_{\mathbb{Q},S}$-invariants of $\mathrm{Hom}_{\mathbb{Z}_l}(M,N)$ are precisely the $G_{\mathbb{Q},S}$-equivariant homomorphisms $\mathrm{Hom}_{\mathbb{Z}_l[G_{\mathbb{Q},S}]}(M,N)$. Throughout this paper we assume that the base ring for any of these constructions is $\mathbb{Z}_l$; we will usually omit it from the notation.

Now assume that M is free over $\mathbb{Z}_l$ of rank 2. We define the *symmetric square* $\mathrm{Sym}^2 M$ of M to be the submodule of $M \otimes M$ which is invariant under the automorphism of $M \otimes M$ interchanging the two factors. If x, y is a basis for M, then $x \otimes x, x \otimes y + y \otimes x, y \otimes y$ is a basis for $\mathrm{Sym}^2 M$, so that $\mathrm{Sym}^2 M$ is free over $\mathbb{Z}_l$ of rank 3. In fact, if $l \neq 2$, then $\mathrm{Sym}^2 M$ is a direct summand of $M \otimes M$; the complimentary summand is the alternating square $\wedge^2 M$, which has basis $x \otimes y - y \otimes x$:

$$(1) \qquad M \otimes M = \wedge^2 M \oplus \mathrm{Sym}^2 M.$$

One checks easily that $\mathrm{Sym}^2 M$ is stable under the action of $G_{\mathbb{Q},S}$, so that it can also be considered as a $G_{\mathbb{Q},S}$-module. We also have an induced action of $G_{\mathbb{Q},S}$ on $\wedge^2 M$, and with these actions the decomposition (1) is a decomposition of $G_{\mathbb{Q},S}$-modules.

The module of endomorphisms $\mathrm{End}(M)$ admits a similar decomposition if $\ell \neq 2$. (As always we let $G_{\mathbb{Q},S}$ act on $\mathrm{End}(M)$ via the adjoint action.) The scalar matrices in $\mathrm{End}(M)$ are a free $\mathbb{Z}_l$-module of rank 1 with trivial Galois action, since conjugation is trivial on scalars. Thus, for $\ell \neq 2$ we have a canonical decomposition

$$\mathrm{End}(M) = \mathbb{Z}_l \oplus \mathrm{End}^0(M)$$

where $\mathrm{End}^0(M)$ denotes the trace zero matrices in $\mathrm{End}(M)$ and the first summand corresponds to the scalar matrices. (We always take $\mathbb{Z}_l$ itself to have trivial $G_{\mathbb{Q},S}$-action.)

Now let E be a rational elliptic curve (i.e., an elliptic curve defined over $\mathbb{Q}$), and let l be an odd prime. Recall that the *l-adic Tate module* of E is the free $\mathbb{Z}_l$-module of rank 2 defined by

$$T_l E = \varprojlim E[l^n].$$

If S is any set of places of $\mathbb{Q}$ including l and the places where E has bad reduction, then $T_l E$ carries a natural action of $G_{\mathbb{Q},S}$. Upon choosing a basis for $T_l E$ we can view this as a representation

$$\rho : G_{\mathbb{Q}} \to \mathrm{GL}_2(\mathbb{Z}_l).$$

The Galois module which will actually prove most relevant to the proof of Theorem 10.1 is the symmetric square of $T_l E$.

We can perform a similar construction with the l-power roots of unity: this Tate module, $\varprojlim \mu_{l^n}$, is written as $\mathbb{Z}_l(1)$. Thus $\mathbb{Z}_l(1)$ is a free $\mathbb{Z}_l$-module of rank 1 on which $G_{\mathbb{Q},S}$ acts via the ℓ-adic cyclotomic character

$$\varepsilon : G_{\mathbb{Q},S} \to \mathbb{Z}_l^*;$$

here S is any set of places of $\mathbb{Q}$ containing l and ∞. For any $n > 0$, define $\mathbb{Z}_l(n)$ to be the tensor product (over $\mathbb{Z}_l$) of $\mathbb{Z}_l(1)$ with itself n times. Define $\mathbb{Z}_l(-1)$ to be the integral Pontrjagin dual $\mathrm{Hom}(\mathbb{Z}_l(1), \mathbb{Z}_l)$ of $\mathbb{Z}_l(1)$, and define $\mathbb{Z}_l(-n)$ as the tensor product of $\mathbb{Z}_l(-1)$ with itself n times. Thus $\mathbb{Z}_l(n)$ is a free $\mathbb{Z}_l$-module of rank one on which $G_{\mathbb{Q},S}$ acts via ε^n. If M is any $\mathbb{Z}_l$-module with an action of $G_{\mathbb{Q}}$, we define its nth *Tate twist* by

$$M(n) = M \otimes_{\mathbb{Z}_l} \mathbb{Z}_l(n).$$

$M(n)$ is isomorphic to M as a $\mathbb{Z}_l$-module, but they usually have different $G_{\mathbb{Q}}$-actions.

A key property of the Tate module of an elliptic curve is that the Weil pairings

$$E[l^n] \otimes E[l^n] \to \mu_{l^n}$$

compile to yield a perfect, skew-symmetric, Galois equivariant pairing
$$e : T_l E \otimes T_l E \to \mathbb{Z}_l(1).$$
See [**146**, Proposition III.8.3]. Since e is skew-symmetric, this implies that $\wedge^2 T_l E \cong \mathbb{Z}_l(1)$. We record some additional consequences below.

Lemma 10.2. *The Weil pairing induces a Galois equivariant isomorphism*
$$\operatorname{End}(T_l E)(1) \cong T_l E \otimes T_l E.$$
This isomorphism restricts to an isomorphism
$$\operatorname{End}^0(T_l E)(1) \cong \operatorname{Sym}^2 T_l E$$
of direct summands.

Proof. The Weil pairing yields a duality isomorphism
$$T_l E \cong \operatorname{Hom}(T_l E, \mathbb{Z}_l(1)),$$
essentially by the definition of a perfect pairing. Galois equivariance of the Weil pairing implies precisely that this identification respects Galois action, thanks to the definition of the adjoint Galois action. Tensoring with $T_l E$ now yields the first statement of the lemma, since $\operatorname{Hom}(T_l E, T_l E(1))$ is visibly isomorphic to $\operatorname{End}(T_l E)(1)$.

Explicitly, the above isomorphism sends $t \otimes t' \in T_l E \otimes T_l E$ to the function $t' \otimes e(t, \cdot) \in \operatorname{Hom}(T_l E, T_l E \otimes \mathbb{Z}_l(1))$. To check the second statement, we can ignore Galois actions and we simply have to check that symmetric elements of $T_l E \otimes T_l E$ correspond to trace zero matrices in $\operatorname{End}(T_l E)$. This follows immediately from the fact that the Weil pairing is alternating; we leave it as an exercise. $\square$

If M is any finite free $\mathbb{Z}_l$-module with an action of $G_{\mathbb{Q},S}$, we define its *integral Cartier dual* M^* to be the $G_{\mathbb{Q},S}$-module $\operatorname{Hom}(M, \mathbb{Z}_l(1))$. (Often the term *Cartier dual* is used for the module $\operatorname{Hom}(M, \mathbb{Q}_l/\mathbb{Z}_l(1))$.)

Lemma 10.3. *The Weil pairing induces an isomorphism*
$$(T_l E \otimes T_l E)^* \cong T_l E \otimes T_l E(-1).$$
This isomorphism restricts to an isomorphism
$$(\operatorname{Sym}^2 T_l E)^* \cong (\operatorname{Sym}^2 T_l E)(-1).$$

Proof. In general, if A and B are any free $\mathbb{Z}_l$-modules with $G_{\mathbb{Q},S}$-actions, then $(A \otimes B)^* \cong A^* \otimes B^*(-1)$, as one checks easily from the definition. In our case, the Weil pairing shows that $(T_l E)^* \cong T_l E$, and the first statement follows. The second statement is immediate once the first isomorphism has been made explicit; we omit the details. $\square$

First reductions

In this section we will use various global duality theorems of Tate to reduce our calculation of $H^2(G_{\mathbb{Q},S}, \operatorname{ad}(\bar{\rho}))$ to the vanishing of a certain Shafarevich-Tate group. For the remainder of the paper we fix a rational elliptic curve E and a prime l satisfying the hypotheses of Theorem 10.1. Let S be the corresponding set of places of $\mathbb{Q}$. Note that as Galois modules $\operatorname{ad}(\bar{\rho}) \cong \operatorname{End}(E[l])$; we will use the notation $\operatorname{End}(E[l])$ from now on.

We begin with the following small piece of the Poitou-Tate exact sequence (see [**156**, Section 8] for statements and [**107**, Chapter 1, Section 4] for a proof; here we are using the fact that $l \neq 2$ to eliminate the terms at infinity):

$$\prod_{p \in S - \{\infty\}} \mathrm{H}^0(\mathbb{Q}_p, E[l] \otimes E[l]) \to \mathrm{Hom}\bigl(\mathrm{H}^2(G_{\mathbb{Q},S}, (E[l] \otimes E[l])^*), \mathbb{Z}/l\mathbb{Z}\bigr) \to$$

$$\mathrm{H}^1(G_{\mathbb{Q},S}, E[l] \otimes E[l]) \to \prod_{p \in S} \mathrm{H}^1(\mathbb{Q}_p, E[l] \otimes E[l]).$$

Tensoring the first isomorphism of Lemma 10.2 with $\mathbb{Z}/l\mathbb{Z}$ yields an isomorphism

$$E[l] \otimes E[l] \cong \mathrm{End}(E[l])(1).$$

Together with Lemma 10.3, this implies that

$$(E[l] \otimes E[l])^* \cong \mathrm{End}(E[l]).$$

Thus the above exact sequence contains a term which is the dual vector space to $\mathrm{H}^2(G_{\mathbb{Q},S}, \mathrm{End}(E[l]))$; since to prove Theorem 10.1 we need to show that this group vanishes, we see that it will suffice to show that the two groups

$$\prod_{p \in S - \{\infty\}} \mathrm{H}^0(\mathbb{Q}_p, E[l] \otimes E[l])$$

$$\mathrm{III}^1(G_{\mathbb{Q},S}, E[l] \otimes E[l])) = \ker\left(\mathrm{H}^1(G_{\mathbb{Q},S}, E[l] \otimes E[l]) \to \prod_{p \in S} \mathrm{H}^1(\mathbb{Q}_p, E[l] \otimes E[l])\right)$$

both vanish.

The vanishing of the first of these groups is the third hypothesis in the statement of Theorem 10.1. For the second, we first write

$$E[l] \otimes E[l] = \wedge^2 E[l] \oplus \mathrm{Sym}^2 E[l] \cong \mu_l \oplus \mathrm{Sym}^2 E[l].$$

(The isomorphism of $\wedge^2 E[l]$ and μ_l comes from the Weil pairing.) One sees immediately from this that there is a corresponding decomposition

$$\mathrm{III}^1(G_{\mathbb{Q},S}, E[l] \otimes E[l]) \cong \mathrm{III}^1(G_{\mathbb{Q},S}, \mu_l) \oplus \mathrm{III}^1(G_{\mathbb{Q},S}, \mathrm{Sym}^2 E[l]).$$

The first term is easily dealt with. We will need the following results, which will also be useful later when dealing with Selmer groups.

Lemma 10.4. *Let A be a $G_\mathbb{Q}$-module which is unramified away from a finite set of primes S. Then*

$$\mathrm{H}^1(G_{\mathbb{Q},S}, A) \cong \ker\left(\mathrm{H}^1(\mathbb{Q}, A) \to \prod_{p \notin S} \mathrm{H}^1(I_p, A)\right).$$

Here $I_p \subseteq G_\mathbb{Q}$ is the inertia group at p.

Proof. See [**156**, Proposition 6] for a proof. The idea is simply that cohomology classes for $G_{\mathbb{Q},S}$ are automatically unramified away from S and therefore are trivial when restricted to the corresponding inertia groups. □

Lemma 10.5. *Let p be a prime different from l. Then the maximal pro-l quotient of the inertia group I_p is isomorphic to $\mathbb{Z}_l$ as a topological group. If $G_{\mathbb{Q}_p}$ is made to act on I_p by conjugation, then the maximal pro-l quotient of I_p is isomorphic to $\mathbb{Z}_l(1)$ as a $G_{\mathbb{Q}_p}$-module.*

Proof. Recall that $I_p = \text{Gal}(\bar{\mathbb{Q}}_p/\mathbb{Q}_p^{\text{ur}})$. It is shown in [**54**, Section 8, Corollary 3] that the maximal pro-l quotient of this group is $\text{Gal}(\mathbb{Q}_p^{\text{ur}}(p^{1/l^\infty})/\mathbb{Q}_p^{\text{ur}})$. This is seen to be isomorphic to $\mathbb{Z}_l(1)$ using the isomorphisms

$$\text{Gal}(\mathbb{Q}_p^{\text{ur}}(p^{1/l^\infty})/\mathbb{Q}_p^{\text{ur}}) \cong \varprojlim \text{Gal}(\mathbb{Q}_p^{\text{ur}}(p^{1/l^n})/\mathbb{Q}_p^{\text{ur}}) \cong \varprojlim \mu_{\ell^n} \cong \mathbb{Z}_l(1).$$

We leave the verification that the conjugation action of $G_{\mathbb{Q}_p}$ is cyclotomic as an exercise. $\square$

Lemma 10.6. $\text{III}^1(G_{\mathbb{Q},S}, \mu_l) = 0$.

Proof. By definition,

$$\text{III}^1(G_{\mathbb{Q},S}, \mu_l) = \ker\left(\text{H}^1(G_{\mathbb{Q},S}, \mu_l) \to \prod_{p \in S} \text{H}^1(\mathbb{Q}_p, \mu_l)\right).$$

Lemma 10.4 shows that we can rewrite this as

$$\text{III}^1(G_{\mathbb{Q},S}, \mu_l) = \ker\left(\text{H}^1(\mathbb{Q}, \mu_l) \to \prod_{p \notin S} \text{H}^1(I_p, \mu_l) \times \prod_{p \in S} \text{H}^1(\mathbb{Q}_p, \mu_l)\right).$$

We will compute these groups.

We begin by working in some generality. Let K be any perfect field of characteristic different from l, and consider the exact sequence

$$0 \to \mu_l \to \bar{K}^* \xrightarrow{l} \bar{K}^* \to 0$$

of G_K-modules. Hilbert's theorem 90 (see [**140**, Chapter II.1, Proposition 1]) says that $\text{H}^1(K, \bar{K}^*) = 0$, so the long exact sequence in G_K-cohomology coming from the short exact sequence above yields an isomorphism

$$K^* \otimes_{\mathbb{Z}} \mathbb{Z}/l\mathbb{Z} \cong \text{H}^1(K, \mu_l).$$

This applies in particular to the fields $\mathbb{Q}_p$:

$$\mathbb{Q}_p^* \otimes_{\mathbb{Z}} \mathbb{Z}/l\mathbb{Z} \cong \text{H}^1(\mathbb{Q}_p, \mu_l).$$

For $p \notin S$ we also need to compute $\text{H}^1(I_p, \mu_l)$. Since $p \notin S$, we know that $p \neq l$; thus I_p acts trivially on μ_l. Thus $\text{H}^1(I_p, \mu_l) \cong \text{Hom}(I_p, \mu_l)$. Any such homomorphism must factor through the maximal pro-l quotient of I_p, and now Lemma 10.5 shows that this group is just μ_ℓ. It is easily checked that the restriction map

$$\text{H}^1(\mathbb{Q}_p, \mu_l) \to \text{H}^1(I_p, \mu_l) \cong \text{Hom}(\mu_\ell, \mu_\ell) \cong \mathbb{Z}/l\mathbb{Z}$$

is just the natural map

$$\mathbb{Q}_p^\times \otimes_{\mathbb{Z}} \mathbb{Z}/l\mathbb{Z} \cong (\mathbb{Z} \times \mathbb{Z}_p^\times) \otimes_{\mathbb{Z}} \mathbb{Z}/l\mathbb{Z} \to \mathbb{Z}/l\mathbb{Z}$$

which is trivial on $\mathbb{Z}_p^\times$; that is, it is the p-adic valuation map modulo l.

The group $\text{III}^1(G_{\mathbb{Q},S}, \mu_l)$ is therefore the kernel of the map

$$\mathbb{Q}^\times \otimes_{\mathbb{Z}} \mathbb{Z}/l\mathbb{Z} \to \prod_{p \notin S} \mathbb{Z}/l\mathbb{Z} \times \prod_{p \in S-\{\infty\}} \mathbb{Q}_p^\times \otimes_{\mathbb{Z}} \mathbb{Z}/l\mathbb{Z}.$$

(Since $l \neq 2$ the term at ∞ vanishes.) Our calculations above show that this kernel consists only of elements of $\mathbb{Q}^\times \otimes_{\mathbb{Z}} \mathbb{Z}/l\mathbb{Z}$ which have p-adic valuation divisible by l for all primes p. (In fact, at $p \in S$ the conditions are even stronger, but we won't need that.) Unique factorization in $\mathbb{Z}$ implies that any such rational number is an lth-power in $\mathbb{Q}^\times$, and therefore is zero in $\mathbb{Q}^\times \otimes \mathbb{Z}/l\mathbb{Z}$. (Here we also need to use the fact that the units $\mathbb{Z}^\times$ are just ± 1 and disappear on tensoring with $\mathbb{Z}/l\mathbb{Z}$.) Thus $\text{III}^1(G_{\mathbb{Q},S}, \mu_l) = 0$, as claimed. Note that the fact that $\mathbb{Z}$ has unit rank 0 and class number 1 was essential to this argument. $\square$

We have now reduced the proof of Theorem 10.1 to showing that the Shafarevich-Tate group $\text{III}^1(G_{\mathbb{Q},S}, \text{Sym}^2 E[l])$ is trivial. We will first show that it embeds into an a priori larger group. Before we can do this, however, we need to define the Selmer groups of our Galois modules.

Selmer groups

Recall that the Selmer group of an elliptic curve over $\mathbb{Q}$ is defined to be the subgroup of $\text{H}^1(\mathbb{Q}, T_l E \otimes \mathbb{Q}_l/\mathbb{Z}_l)$ of cohomology classes which for all p are locally in the image of $E(\mathbb{Q}_p)$ under the Kummer map. (See [**63**].) We will be working with $\text{Sym}^2 T_l E \otimes \mathbb{Q}_l/\mathbb{Z}_l$, but here we no longer have any natural geometric object on which to base our local conditions in the definition of the Selmer group. The key to the general definition of a Selmer group is the fact that the image of the local Kummer map consists precisely of those cohomology classes which are unramified, in a sense which we shall make precise below.

Let us fix some notation for the remainder of the paper: set $T = \text{Sym}^2 T_l E$ (a free $\mathbb{Z}_l$-module of rank 3), $V = T \otimes_{\mathbb{Z}_l} \mathbb{Q}_l$ (a 3 dimensional $\mathbb{Q}_l$-vector space) and $A = V/T = \text{Sym}^2 T_l E \otimes_{\mathbb{Z}_l} \mathbb{Q}_l/\mathbb{Z}_l$ (which is isomorphic as an abelian group to $(\mathbb{Q}_l/\mathbb{Z}_l)^3$). T, V and A are to be regarded as three different incarnations of the same Galois module. We will also need to consider $A^* = \text{Hom}_{\mathbb{Z}_l}(T, \mu_{l^\infty})$, which is also isomorphic as an abelian group to $(\mathbb{Q}_l/\mathbb{Z}_l)^3$. Lastly, for technical reasons we will later need to consider the finite modules $T_n = T/l^n T$ and $A_n^* = A^*[l^n]$ (both of which are isomorphic to $(\mathbb{Z}/l^n\mathbb{Z})^3$ as abelian groups).

To formalize the notion of a Selmer group, we wish to define "unramified subgroups" (the usual term is *finite subgroups*) $\text{H}^1_f(\mathbb{Q}_p, A)$ of $\text{H}^1(\mathbb{Q}_p, A)$ for every prime p. We will then define the *Selmer group* $\text{H}^1_f(\mathbb{Q}, A)$ of A by

$$\text{H}^1_f(\mathbb{Q}, A) = \ker\left(\text{H}^1(\mathbb{Q}, A) \to \prod_p \text{H}^1(\mathbb{Q}_p, A)/\text{H}^1_f(\mathbb{Q}_p, A)\right)$$
$$= \{c \in \text{H}^1(\mathbb{Q}, A) \mid \text{res}_p c \in \text{H}^1_f(\mathbb{Q}_p, A) \text{ for all } p\}.$$

Here res_p is the restriction map from $\text{H}^1(\mathbb{Q}, A)$ to $\text{H}^1(\mathbb{Q}_p, A)$.

The definition of $\text{H}^1_f(\mathbb{Q}_p, A)$ is fairly straightforward for primes p which do not lie in S. Indeed, the most obvious notion of an unramified cocycle is one which becomes trivial when restricted to inertia; that is, it should lie in the kernel of the restriction map

$$\text{H}^1(\mathbb{Q}_p, A) \to \text{H}^1(I_p, A)$$

where I_p is the inertia subgroup of $G_{\mathbb{Q}_p}$. The inflation-restriction sequence (see [**156**, Proposition 2 and the discussion following]) identifies this kernel with

$$\text{H}^1(G_{\mathbb{Q}_p}/I_p, A) = \text{H}^1(\mathbb{F}_p, A).$$

(Here we are using that I_p acts trivially on A to see that A is a $G_{\mathbb{Q}_p}/I_p$-module.) We take this as our definition of the finite subgroup, at least for $p \notin S$:

$$\text{H}^1_f(\mathbb{Q}_p, A) = \text{H}^1(\mathbb{F}_p, A),$$

or more honestly its image in $\text{H}^1(\mathbb{Q}_p, A)$ under inflation.

Note that the inflation-restriction exact sequence now takes the form
$$0 \to \mathrm{H}^1_f(\mathbb{Q}_p, A) \to \mathrm{H}^1(\mathbb{Q}_p, A) \to \mathrm{H}^1(I_p, A)^{G_{\mathbb{F}_p}} \to \mathrm{H}^2(\mathbb{F}_p, A).$$

In fact, $\mathrm{H}^2(\mathbb{F}_p, A)$ vanishes because $G_{\mathbb{F}_p} \cong \hat{\mathbb{Z}}$; see [**140**, Chapter II.3]. We will call $\mathrm{H}^1(I_p, A)^{G_{\mathbb{F}_p}}$ the *singular quotient* of $\mathrm{H}^1(\mathbb{Q}_p, A)$ and write it as $\mathrm{H}^1_s(\mathbb{Q}_p, A)$, so that we have an exact sequence
$$0 \to \mathrm{H}^1_f(\mathbb{Q}_p, A) \to \mathrm{H}^1(\mathbb{Q}_p, A) \to \mathrm{H}^1_s(\mathbb{Q}_p, A) \to 0.$$

In analogy with this definition, for $p \in S$ it might seem most reasonable to define the finite part of $\mathrm{H}^1(\mathbb{Q}_p, A)$ also as the kernel of $\mathrm{H}^1(\mathbb{Q}_p, A) \to \mathrm{H}^1(I_p, A)$, which equals $\mathrm{H}^1(\mathbb{F}_p, A^{I_p})$. However, for reasons which will not become apparent in this paper this definition turns out to be inadequate. It turns out that the correct definition is as follows: first assume $p \neq l$. Let $\pi : V \to A$ be the natural quotient map. We define
$$\mathrm{H}^1_f(\mathbb{Q}_p, V) = \mathrm{H}^1(\mathbb{F}_p, V^{I_p})$$
and
$$\mathrm{H}^1_f(\mathbb{Q}_p, A) = \pi_* \mathrm{H}^1_f(\mathbb{Q}_p, V).$$
Here $\pi_* : \mathrm{H}^1(\mathbb{Q}_p, V) \to \mathrm{H}^1(\mathbb{Q}_p, A)$ is the induced map on cohomology. Despite appearances, $\pi_* \mathrm{H}^1_f(\mathbb{Q}_p, V)$ need not be the same as $\mathrm{H}^1(\mathbb{F}_p, V^{I_p})$, at least for $p \in S$. For later use, let us also set $\mathrm{H}^1_f(\mathbb{Q}_p, V) = \mathrm{H}^1(\mathbb{F}_p, V)$ for $p \notin S$. One checks easily that in this case $\pi_* \mathrm{H}^1_f(\mathbb{Q}_p, V)$ does agree with our previous definition of $\mathrm{H}^1_f(\mathbb{Q}_p, A)$.

The definition of $\mathrm{H}^1_f(\mathbb{Q}_l, A)$ is much more subtle. The generally accepted definition (which recovers the usual definition in the case of the Tate module of an elliptic curve) is
$$\mathrm{H}^1_f(\mathbb{Q}_p, V) = \ker\bigl(\mathrm{H}^1(\mathbb{Q}_p, V) \to \mathrm{H}^1(\mathbb{Q}_p, V \otimes B_{\mathrm{crys}})\bigr)$$
and $\mathrm{H}^1_f(\mathbb{Q}_p, A) = \pi_* \mathrm{H}^1_f(\mathbb{Q}_p, V)$. Here B_{crys} is one of Fontaine's "big rings". We will not concern ourselves very much with this definition in this paper, although in many ways it is one of the most important topics. It is possible to make this definition much more concrete, but even that does not really make this condition any easier to deal with.

In passing, we should note that $\mathrm{H}^1(\mathbb{R}, A) = 0$ (since $l \neq 2$), so we need not concern ourselves with any definitions at infinity. We have now defined $\mathrm{H}^1_f(\mathbb{Q}_p, A)$ for all primes p, and with it the Selmer group $\mathrm{H}^1_f(\mathbb{Q}, A)$. Of course, we can mimic the identical construction with A^* instead of A, and thus we also have a Selmer group $\mathrm{H}^1_f(\mathbb{Q}, A^*)$.

We can also redo the construction for the Galois module T. Note that T is fundamentally quite different from A, in that it is free over $\mathbb{Z}_l$ rather than isomorphic to several copies of $\mathbb{Q}_l/\mathbb{Z}_l$. The definitions are nevertheless quite analogous. Let $i : T \to V$ be the natural inclusion, and for all p define
$$\mathrm{H}^1_f(\mathbb{Q}_p, T) = i_*^{-1} \mathrm{H}^1_f(\mathbb{Q}_p, V).$$

We also define singular quotients $\mathrm{H}^1_s(\mathbb{Q}_p, T) = \mathrm{H}^1(\mathbb{Q}_p, T)/\mathrm{H}^1_f(\mathbb{Q}_p, T)$. In this case, it will turn out that the singular part of the local cohomology is that which we can most easily work with. One can also define a Selmer group for T, although we will have no need to consider it.

We will also need corresponding subgroups for the finite modules T_n and $A^*[l^n]$. For any prime p, we simply take $\mathrm{H}^1_f(\mathbb{Q}_p, A^*_n)$ to be the inverse image of $\mathrm{H}^1_f(\mathbb{Q}_p, A^*)$

under the natural map $\mathrm{H}^1(\mathbb{Q}_p, A_n^*) \to \mathrm{H}^1(\mathbb{Q}_p, A^*)$. Similarly, we define $\mathrm{H}_f^1(\mathbb{Q}_p, T_n)$ to be the image of $\mathrm{H}_f^1(\mathbb{Q}_p, T)$ under the natural map $\mathrm{H}^1(\mathbb{Q}_p, T) \to \mathrm{H}^1(\mathbb{Q}_p, T_n)$. One can now define singular quotients and Selmer groups in the usual way.

It is worth noting the general philosophy: we took the natural definition of unramified cocycles in $\mathrm{H}^1(\mathbb{Q}_p, V)$ (with the exception of the case $p = l$, which was more complicated) and we then let these choices propagate down to all of the related Galois modules.

Returning to the case of $T_l E$, recall that one defines the Shafarevich-Tate group $\mathrm{III}(E/\mathbb{Q})$ as the cokernel of the Kummer map

$$E(\mathbb{Q}) \otimes_\mathbb{Z} \mathbb{Q}_l/\mathbb{Z}_l \to \mathrm{H}_f^1(\mathbb{Q}, T_l E).$$

As before we have no obvious analogue of $E(\mathbb{Q})$ in our situation. The work of Bloch and Kato suggests that the correct analogue is the following: let

$$\mathrm{H}_f^1(\mathbb{Q}, V) = \ker\left(\mathrm{H}^1(\mathbb{Q}, V) \to \prod_p \mathrm{H}^1(\mathbb{Q}_p, V)/\mathrm{H}_f^1(\mathbb{Q}_p, V)\right)$$

be the Selmer group for V, and define the "rational points of A" to be

$$A(\mathbb{Q}) = \pi_* \mathrm{H}_f^1(\mathbb{Q}, V) \subseteq \mathrm{H}^1(\mathbb{Q}, A).$$

Note that it follows immediately from the definition of the $\mathrm{H}_f^1(\mathbb{Q}_p, A)$ that $A(\mathbb{Q})$ actually lies in $\mathrm{H}_f^1(\mathbb{Q}, A)$, although it could conceivably be smaller. We define the Shafarevich-Tate group $\mathrm{III}(A/\mathbb{Q})$ to be the quotient $\mathrm{H}_f^1(\mathbb{Q}, A)/A(\mathbb{Q})$, so that there is an exact sequence

$$0 \to A(\mathbb{Q}) \to \mathrm{H}_f^1(\mathbb{Q}, A) \to \mathrm{III}(A/\mathbb{Q}) \to 0.$$

$\mathrm{III}(A/\mathbb{Q})$ is to be thought of as elements of the Selmer group which appear over A but not over V. Note also that despite the similar notation, $\mathrm{III}(A/\mathbb{Q})$ is not the same as any of the Shafarevich-Tate groups we considered earlier.

Again, we can make analogous definitions for $A^*(\mathbb{Q})$, yielding an exact sequence

$$0 \to A^*(\mathbb{Q}) \to \mathrm{H}_f^1(\mathbb{Q}, A^*) \to \mathrm{III}(A^*/\mathbb{Q}) \to 0.$$

We are now in a position to finish our reductions of the previous section.

Lemma 10.7. $\mathrm{III}^1(G_{\mathbb{Q},S}, \mathrm{Sym}^2 E[l])$ *injects into* $\mathrm{H}_f^1(\mathbb{Q}, A)$.

Proof. As an abelian group A is isomorphic to $(\mathbb{Q}_l/\mathbb{Z}_l)^3$, so multiplication by l is surjective on A. Furthermore, the kernel of multiplication by l on A,

$$\mathrm{Sym}^2 T_l E \otimes \frac{1}{l}\mathbb{Z}_l/\mathbb{Z}_l,$$

naturally identifies with $\mathrm{Sym}^2 E[l]$, so there is an exact sequence

(2) $$0 \to \mathrm{Sym}^2 E[l] \to A \xrightarrow{l} A \to 0.$$

The fact that $E[l] \otimes E[l]$ is assumed to have no $G_{\mathbb{Q}_p}$-invariants for any $p \in S - \{\infty\}$ insures that the direct summand $\mathrm{Sym}^2 E[l]$ has no $G_{\mathbb{Q},S}$-invariants; indeed, knowing that it had no invariants at any one place would suffice. It follows easily from this that A itself has no $G_{\mathbb{Q},S}$-invariants, as if there were any, then they could be realized in $\mathrm{Sym}^2 E[l]$ by multiplication by an appropriate power of l. Thus the long exact sequence in $G_{\mathbb{Q},S}$-cohomology associated to (2) yields an injection

$$\mathrm{H}^1(G_{\mathbb{Q},S}, \mathrm{Sym}^2 E[l]) \hookrightarrow \mathrm{H}^1(G_{\mathbb{Q},S}, A).$$

Furthermore, under this injection $\text{III}^1(G_{\mathbb{Q},S}, \text{Sym}^2 E[l])$ maps into $\text{III}^1(G_{\mathbb{Q},S}, A)$. Indeed, there is a commutative diagram

(3)
$$\begin{array}{ccc} H^1(G_{\mathbb{Q},S}, \text{Sym}^2 E[l]) & \longrightarrow & H^1(G_{\mathbb{Q},S}, A) \\ \downarrow & & \downarrow \\ \prod_{p \in S} H^1(\mathbb{Q}_p, \text{Sym}^2 E[l]) & \longrightarrow & \prod_{p \in S} H^1(\mathbb{Q}_p, A) \end{array}$$

This shows that any element of $\text{III}^1(G_{\mathbb{Q},S}, \text{Sym}^2 E[l])$, which is by definition trivial in each $H^1(\mathbb{Q}_p, \text{Sym}^2 E[l])$, is automatically trivial in each $H^1(\mathbb{Q}_p, A)$, and thus lies in $\text{III}^1(G_{\mathbb{Q},S}, A)$. In other words, the induced map on the kernels of the vertical maps in (3) is the desired injection

$$\text{III}^1(G_{\mathbb{Q},S}, \text{Sym}^2 E[l]) \hookrightarrow \text{III}^1(G_{\mathbb{Q},S}, A).$$

To prove the lemma it will therefore suffice to show that $\text{III}^1(G_{\mathbb{Q},S}, A)$ injects into $H^1_f(\mathbb{Q}, A)$.

It follows from Lemma 10.4 that there is an injection

$$H^1(G_{\mathbb{Q},S}, A) \hookrightarrow H^1(\mathbb{Q}, A)$$

and that the composite maps

(4) $\quad H^1(G_{\mathbb{Q},S}, A) \to H^1(\mathbb{Q}, A) \to H^1(I_p, A) \cong H^1(\mathbb{Q}_p, A)/H^1_f(\mathbb{Q}_p, A)$

are zero for all $p \notin S$. We must show that the image of $\text{III}^1(G_{\mathbb{Q},S}, A)$ in $H^1(\mathbb{Q}, A)$ lies in $H^1_f(\mathbb{Q}, A)$. By (4), this image is automatically locally unramified for all $p \notin S$. Furthermore, an argument using a diagram analogous to (3) above shows that the maps

$$\text{III}^1(G_{\mathbb{Q},S}, A) \to H^1(\mathbb{Q}, A) \to H^1(\mathbb{Q}_p, A)$$

are zero for $p \in S$. Thus the image of $\text{III}^1(G_{\mathbb{Q},S}, A)$ trivially lands in $H^1_f(\mathbb{Q}_p, A)$ for such p. Thus now $\text{III}^1(G_{\mathbb{Q},S}, A)$ maps to $H^1_f(\mathbb{Q}_p, A)$ for every prime p, so its image in $H^1(\mathbb{Q}, A)$ lies in $H^1_f(\mathbb{Q}, A)$. Thus $\text{III}^1(G_{\mathbb{Q},S}, A)$ injects into $H^1_f(\mathbb{Q}, A)$, which proves the lemma. $\square$

At this point, we have reduced the proof of Theorem 10.1 to the vanishing of the Selmer group $H^1_f(\mathbb{Q}, A)$.

The L-function of $\text{Sym}^2 T_l E$

Before we complete our final reformulation of Theorem 10.1, we should give the long promised explanation of the term $L(\text{Sym}^2 T_l E, 0)/\Omega$. Recall that the L-function of the Tate module of an elliptic curve (also known as the L-function of the elliptic curve) is defined using the characteristic polynomials of Frobenius elements acting on $T_l E$. We use an analogous method to define $L(\text{Sym}^2 T_l E, s)$. Specifically, the action of $G_{\mathbb{Q}_p}$ on T is unramified for every $p \notin S$, so it makes sense to talk about the action of an arithmetic Frobenius element Fr_p on T. Let $P_p(t)$ be the characteristic polynomial of Fr_p^{-1} acting on T:

$$P_p(t) = \det(1 - (\text{Fr}_p|_T)t).$$

For $p \in S - \{l\}$, Fr_p is only well-defined acting on the inertia invariants T^{I_p}, so we define

$$P_p(t) = \det(1 - (\text{Fr}_p|_{T^{I_p}})t).$$

These are all initially polynomials with coefficients in $\mathbb{Z}_l$, but it turns out that $P_p(t)$ actually has coefficients in $\mathbb{Z}$ and the polynomial $P_p(t)$ does not depend on the distinguished prime l, so long as $l \neq p$. (Again, this is all completely analogous to the $T_l E$ case.)

This suggests that to define the factor $P_l(t)$ itself, we should not work directly with T, but rather switch to $\operatorname{Sym}^2 T_p E$ for some $p \neq l$. $\operatorname{Sym}^2 T_p E$ is unramified at l (since E is assumed to have good reduction at l), so Fr_l is well-defined here and we define
$$P_l(t) = \det(1 - \operatorname{Fr}_l |_{\operatorname{Sym}^2 T_p E} t).$$
As before this is independent of the choice of auxiliary $p \neq l$.

We now define
$$L(T, s) = \prod_p P_p(p^{-s})^{-1}.$$
It is shown in [**19**] that $L(T, s)$ is an entire function of s.

Since E is modular, one can use the work of Shimura [**145**] to compute special values of this L-function. Let N be the conductor of E and fix a modular parameterization
$$\phi : X_0(N) \to E$$
of E. We assume that ϕ is minimal in the sense that $\deg \phi$ is as small as possible for our fixed E and N. Let $f(z)$ be the newform corresponding to ϕ; this means that for all p not dividing N, the pth Fourier coefficient of $f(z)$ equals $p + 1 - \#E(\mathbb{F}_p)$. Let ω be the Néron differential on E. (If E is given in the form $y^2 = x^3 + ax + b$, ω is just $dx/2y$.) Since ϕ is defined over $\mathbb{Q}$, one can show that the pullback of ω under ϕ must be a rational multiple of the differential $2\pi i f(z) dz$ on $X_0(N)$. We define the *Manin constant* $c \in \mathbb{Q}^\times$ by the equality
$$\phi^* \omega = c 2\pi i f(z) dz.$$
Work of Mazur shows that c is divisible only by 2 and primes of bad reduction for E; see [**95**, Corollary 4.1].

We also use ω to define the period Ω by
$$\Omega = \pi i \int_{E(\mathbb{C})} \omega \wedge \bar{\omega}.$$
Shimura's formula is

(5) $$\frac{L(T, 0)}{\Omega} = \frac{\deg \phi}{Nc^2} \prod_{p \in S'} P_p(1).$$

Here S' is the subset of S of places where E has potentially good reduction. Note in particular that $L(T, 0)/\Omega$ is rational and non-zero.

We now state the theorem which we will prove in the remainder of this paper and explain how it implies Theorem 10.1.

Theorem 10.8. *Let E be a rational elliptic curve and let $\phi : X_0(N) \to E$ be a modular parameterization. Let l be a prime such that*
- *E has good reduction at l;*
- *$l \geq 5$;*
- *The Tate module representation $\rho : G_{\mathbb{Q}, S} \to \operatorname{GL}_2(\mathbb{Z}_l)$ is surjective.*

Then $\deg \phi \cdot \operatorname{H}_f^1(\mathbb{Q}, A^*) = 0$.

To see that this implies Theorem 10.1, recall that we had already reduced the proof to showing that $\mathrm{H}^1_f(\mathbb{Q}, A)$ vanishes. We had also assumed that l does not divide $L(T,0)/\Omega$. We first must show that l does not divide $\deg\phi$ either. To see this, by (5) we need to show that the rational number

$$\frac{1}{Nc^2}\prod_{p\in S'} P_p(1)$$

has no factors of l in the denominator. But N and c^2 are divisible only by 2 and primes of bad reduction for E, and each $P_p(1)$ is an integer, so this is clear. Thus l does not divide $\deg\phi$.

Now, by Theorem 10.8, $\mathrm{H}^1_f(\mathbb{Q}, A^*)$ is annihilated by the l-adic unit $\deg\phi$. But $\mathrm{H}^1_f(\mathbb{Q}, A^*)$ is an l-power torsion group since A^* is, and it follows that it must be 0. This in turn implies that both $A^*(\mathbb{Q})$ and $\mathrm{III}(A^*/\mathbb{Q})$ vanish, as they are a subgroup and a quotient of $\mathrm{H}^1_f(\mathbb{Q}, A^*)$, respectively. Flach has shown (see [**49**]) that the vanishing of $A^*(\mathbb{Q})$ implies that of $A(\mathbb{Q})$. Furthermore, he constructs (generalizing ideas of Cassels and Tate; see [**48**]) a perfect pairing

$$\mathrm{III}(A/\mathbb{Q}) \otimes \mathrm{III}(A^*/\mathbb{Q}) \to \mathbb{Q}_l/\mathbb{Z}_l;$$

thus the vanishing of $\mathrm{III}(A^*/\mathbb{Q})$ implies that of $\mathrm{III}(A/\mathbb{Q})$. Since both $A(\mathbb{Q})$ and $\mathrm{III}(A/\mathbb{Q})$ vanish, this implies that the Selmer group $\mathrm{H}^1_f(\mathbb{Q}, A)$ vanishes, which completes the proof of Theorem 10.1.

Kolyvagin's theory of Euler systems

Until the mid-eighties the problem of bounding Selmer groups was nearly hopeless; there were no methods which worked in any generality. This changed dramatically with the work of Thaine and Rubin and finally Kolyvagin's theory of Euler systems. Since we only seek to annihilate $\mathrm{H}^1_f(\mathbb{Q}, A^*)$, rather than actually bound its order, we will need only the rudiments of the theory. There are several good sources for more extensive treatments: see [**155**, Chapter 15] for a nice introduction, [**66**] and [**122**] for applications to the arithmetic of elliptic curves, and [**124**] for a general theory. In fact, all of these sources deal with a slightly different type of Euler system than we will use. We will have more to say about this later.

We only sketch the main ideas. For a proof of the result we will need, see [**50**, Proposition 1.1] or [**157**].

Fix a power l^n of l and set $T_n = T/l^n T$, $A_n^* = A^*[l^n]$. We must work with these finite modules for technical reasons; passing from them to the full modules will be easy. The basic idea is the following: recall that since $A_n^* = \mathrm{Hom}(T_n, \mu_{l^n})$ there is a Tate local duality

$$\mathrm{H}^1(\mathbb{Q}_p, T_n) \otimes \mathrm{H}^1(\mathbb{Q}_p, A_n^*) \to \mathbb{Q}_l/\mathbb{Z}_l;$$

see [**156**, Theorem 1]. One can show easily that $\mathrm{H}^1_f(\mathbb{Q}_p, T_n)$ and $\mathrm{H}^1_f(\mathbb{Q}_p, A_n^*)$ are exact annihilators of each other (see [**157**, Lectures 5 and 6]), so that restricting the right-hand factor to $\mathrm{H}^1_f(\mathbb{Q}_p, A_n^*)$ gives a perfect pairing

(6) $$\mathrm{H}^1_s(\mathbb{Q}_p, T_n) \otimes \mathrm{H}^1_f(\mathbb{Q}_p, A_n^*) \to \mathbb{Q}_l/\mathbb{Z}_l.$$

These local pairings sum to a global pairing

(7) $$\left(\bigoplus_p \mathrm{H}^1_s(\mathbb{Q}_p, T_n)\right) \otimes \mathrm{H}^1_f(\mathbb{Q}, A_n^*) \to \mathbb{Q}_l/\mathbb{Z}_l;$$

the pairing of an element $(c_p) \in \oplus \mathrm{H}^1_s(\mathbb{Q}_p, T_n)$ and $d \in \mathrm{H}^1_f(\mathbb{Q}, A_n^*)$ is simply the sum of the local pairings (6) of c_p and $\mathrm{res}_p\, d$; since $c_p = 0$ for almost all p, this is well-defined. This pairing is not perfect, but it does have the following key property, which is a consequence of global class field theory: the image of $\mathrm{H}^1(\mathbb{Q}, T_n)$ under the natural map

$$\mathrm{H}^1(\mathbb{Q}, T_n) \to \prod_p \mathrm{H}^1(\mathbb{Q}_p, T_n) \to \prod_p \mathrm{H}^1_s(\mathbb{Q}_p, T_n)$$

actually lands in $\oplus_p \mathrm{H}^1_s(\mathbb{Q}_p, T_n)$ (see [**157**, Lecture 7, Section 2.1]; this is one place where it is critical that we dropped to a finite quotient of T) and it is orthogonal to all of $\mathrm{H}^1_f(\mathbb{Q}, A_n^*)$ under the global pairing (7); see [**157**, Lecture 8].

The significance of this to our problem is the following: suppose that for lots of primes r we can exhibit elements $c_r \in \mathrm{H}^1(\mathbb{Q}, T_n)$ with the property that they restrict to 0 in $\mathrm{H}^1_s(\mathbb{Q}_p, T_n)$ for all p *except* for r itself, where they restrict to something non-zero. Under the global pairing, the image of c_r in $\oplus \mathrm{H}^1_s(\mathbb{Q}_p, T_n)$ is orthogonal to all of $\mathrm{H}^1_f(\mathbb{Q}, A_n^*)$. But the definition of the global pairing together with the fact that c_r restricts to 0 in $\mathrm{H}^1_s(\mathbb{Q}_p, T_n)$ away from r now shows that $\mathrm{res}_r\, c_r \in \mathrm{H}^1_s(\mathbb{Q}_r, T_n)$ is orthogonal to the image of $\mathrm{H}^1_f(\mathbb{Q}, A_n^*)$ in $\mathrm{H}^1_f(\mathbb{Q}_r, A_n^*)$ under the Tate local pairing at r. If $\mathrm{res}_r\, c_r$ generates a submodule of $\mathrm{H}^1_s(\mathbb{Q}_r, T_n)$ of small index, then this orthogonality and the fact that the Tate local pairing is perfect will force $\mathrm{H}^1_f(\mathbb{Q}, A_n^*)$ to map into a small subgroup of $\mathrm{H}^1_f(\mathbb{Q}_r, A_n^*)$. Since we can do this for lots of r, we obtain conditions on the local behavior of $\mathrm{H}^1_f(\mathbb{Q}, A_n^*)$ at many primes r. Hopefully if we could do this for enough primes r we could somehow show that the local conditions are so stringent that the group $\mathrm{H}^1_f(\mathbb{Q}, A_n^*)$ itself must be small.

Before we state all of this somewhat more formally, we prove the following fundamental lemma. We will call a prime p *good* if it is not in S and if a Frobenius element at p acts on $E[l]$ as complex conjugation. This is equivalent to Fr_p being a complex conjugation element on the splitting field $\mathbb{Q}(E[l])$ of $E[l]$. This field is a finite extension of $\mathbb{Q}$ since $E[l]$ is finite, so by the Chebotarev density theorem there are infinitely many good primes.

Lemma 10.9. *Assume $p \notin S$. Then*

$$\mathrm{H}^1_s(\mathbb{Q}_p, T) \cong T(-1)^{G_{\mathbb{F}_p}}.$$

If p is good, then this group is a free $\mathbb{Z}_l$-module of rank 1. In particular, each $\mathrm{H}^1_s(\mathbb{Q}_p, T_n)$ is a free $\mathbb{Z}/l^n\mathbb{Z}$-module of rank 1.

Proof. For the first isomorphism, recall that

$$\mathrm{H}^1_s(\mathbb{Q}_p, T) \cong \mathrm{H}^0(\mathbb{F}_p, \mathrm{H}^1(I_p, T)).$$

Since I_p acts trivially on T, $\mathrm{H}^1(I_p, T)$ is nothing more than $\mathrm{Hom}(I_p, T)$. Now, T is a pro-l group, so only the pro-l part of I_p can map to it non-trivially. By Lemma 10.5, this quotient is isomorphic to $\mathbb{Z}_l(1)$ as a $G_{\mathbb{F}_p}$-module. We conclude that

$$\mathrm{H}^0(\mathbb{F}_p, \mathrm{H}^1(I_p, T)) \cong \mathrm{H}^0(\mathbb{F}_p, \mathrm{Hom}(\mathbb{Z}_l(1), T)).$$

But $\mathrm{Hom}(\mathbb{Z}_l(1), T)$ is canonically isomorphic to $\mathrm{Hom}(\mathbb{Z}_l, T(-1))$, which in turn is just $T(-1)$. This proves the first statement.

Now assume that p is good. This means that Fr_p acts on $E[l]$ as complex conjugation. In particular, it is a non-scalar involution, which one easily shows implies that it acts diagonally on $E[l]$ with eigenvalues 1 and -1. A Nakayama's lemma argument together with Hensel's lemma and a dimension count allows one to conclude that there is a basis x, y of $T_l E$ over $\mathbb{Z}_l$ with respect to Fr_p acts diagonally; that is, $\mathrm{Fr}_p(x) = ux$ and $\mathrm{Fr}_p(y) = vy$, and we must have

$$u \equiv -v \equiv 1 \pmod{l}.$$

Note that uv is the determinant of Fr_p acting on $T_l E$, which is just $\varepsilon(\mathrm{Fr}_p) = p$ since $T_l E$ has cyclotomic determinant by the Weil pairing. In particular, $p \equiv -1 \pmod{l}$.

A basis for $T = \mathrm{Sym}^2 T_l E$ is given by $x \otimes x, x \otimes y + y \otimes x, y \otimes y$. Fr_p acts on the first by multiplication by u^2; on the second by multiplication by $uv = p$; and on the third by multiplication by v^2. Note that $u^2 \equiv v^2 \equiv 1 \pmod{l}$, which implies that neither u^2 nor v^2 equals p.

Now consider $T(-1)$. Fr_p acts on $\mathbb{Z}_l(1)$ by multiplication by $\varepsilon(\mathrm{Fr}_p) = p$, so it acts on $\mathbb{Z}_l(-1)$ by multiplication by p^{-1}. Thus Fr_p acts on our basis of $T(-1)$ by multiplication by $u^2 p^{-1}$, 1 and $v^2 p^{-1}$ respectively. As we saw above, the first and last terms are different from 1. It follows that only the rank one subspace of multiples of $x \otimes y + y \otimes x$ is invariant under the $G_{\mathbb{F}_p}$-action, so $\mathrm{H}^1_s(\mathbb{Q}_p, T) = T(-1)^{G_{\mathbb{F}_p}}$ is free of rank one over $\mathbb{Z}_l$, as claimed.

The result for T_n follows exactly the same argument. □

We are now in a position to give a precise definition of the sort of set of cohomology classes we seek: let η be an integer. We define a *Flach system of depth η for T_n* to be a collection of cohomology classes $c_r \in \mathrm{H}^1(\mathbb{Q}, T_n)$, one for each good prime r, such that:

- $\mathrm{res}_p c_r$ lies in $\mathrm{H}^1_f(\mathbb{Q}_p, T_n)$ for $p \neq r$;
- $\mathbb{Z}_l \cdot \mathrm{res}_r c_r$ contains $\eta \mathrm{H}^1_s(\mathbb{Q}_r, T_n)$.

The second condition is equivalent to the quotient $\mathrm{H}^1_s(\mathbb{Q}_r, T_n)/\mathbb{Z}_l \cdot \mathrm{res}_r c_r$ being annihilated by η. By Lemma 10.9, $\mathrm{H}^1_s(\mathbb{Q}_r, T_n)$ is a free $\mathbb{Z}/l^n\mathbb{Z}$-module of rank 1, so this condition is reasonable. Note that to check both of the conditions in the definition above, we simply need a good understanding of the singular restriction maps

$$\mathrm{H}^1(\mathbb{Q}, T_n) \to \mathrm{H}^1(\mathbb{Q}_p, T_n) \to \mathrm{H}^1_s(\mathbb{Q}_p, T_n)$$

for all primes p.

Of course, it is trivial and not very useful to write down a Flach system of depth l^n for T_n; to make this a useful notion, we will want η to be independent of n.

Recall that we have assumed that the Tate module representation $\rho : G_{\mathbb{Q}} \to \mathrm{GL}_2(\mathbb{Z}_l)$ is surjective. This implies immediately that $E[l]$ is an absolutely irreducible $G_{\mathbb{Q}}$-representation, which in turn one can check implies that $A^*[l] \cong (\mathrm{Sym}^2 E[l])^*$ is absolutely irreducible. Even though we are assuming all of these hypotheses, we will include the relevant ones in the hypotheses of each result below.

Given what we have said so far, the following lemma is fairly straightforward.

Lemma 10.10. *Assume that T_n admits a Flach system of depth η. Then for every $d \in \mathrm{H}^1_f(\mathbb{Q}, A^*_n)$ and every good prime r, $\mathrm{res}_r d$ lies in $\mathrm{H}^1_f(\mathbb{Q}_p, A^*_n)[\eta]$.*

Proof. See [157, Lecture 15, Section 1.2]. □

More difficult is the next result, which goes from this local annihilation result to a global annihilation result. Let K be the fixed field of the kernel of $G_{\mathbb{Q}, S}$

acting on $E[l^n]$; it is a finite extension of $\mathbb{Q}$ since $E[l^n]^*$ is finite. Note that this field also lies in the kernel of the $G_{\mathbb{Q},S}$ action on $A^*[l^n]$ (since its Galois action is entirely derived from $E[l^n]$ and the cyclotomic character) and that there is a natural inflation injection
$$\mathrm{H}^1(K/\mathbb{Q}, A_n^*) \hookrightarrow \mathrm{H}^1(\mathbb{Q}, A_n^*).$$

Lemma 10.11. *Assume that $l \neq 2$ and that $A^*[l]$ is absolutely irreducible as a $G_{\mathbb{Q},S}$-module. Let $d \in \mathrm{H}^1(\mathbb{Q}, A_n^*)$ be such that $\mathrm{res}_r\, d = 0$ for every good prime r. Then d lies in the image of $\mathrm{H}^1(K/\mathbb{Q}, A_n^*)$.*

Proof. See [157, Lecture 15, Section 1.3]. □

Lemma 10.10 and Lemma 10.11 combine to show that
$$\eta \mathrm{H}_f^1(\mathbb{Q}, A_n^*) \subseteq \mathrm{H}^1(K/\mathbb{Q}, A_n^*).$$
Since the l^n-torsion representation $\rho_n : G_{\mathbb{Q},S} \to \mathrm{GL}_2(\mathbb{Z}/l^n\mathbb{Z})$ is surjective, we will have $\mathrm{Gal}(K/\mathbb{Q}) \cong \mathrm{GL}_2(\mathbb{Z}/l^n\mathbb{Z})$. Furthermore, Lemma 10.2 and Lemma 10.3 show that $A_n^* \cong \mathrm{ad}^0(\rho_n)$. The next result, which is purely a statement about group cohomology, now finishes our proof, at least at the level of l^n-torsion.

Lemma 10.12. *Let $\mathrm{GL}_2(\mathbb{Z}/l^n\mathbb{Z})$ act on $\mathrm{End}^0(\mathbb{Z}/l^n\mathbb{Z})$ via the adjoint action. Then*
$$\mathrm{H}^1(\mathrm{GL}_2(\mathbb{Z}/l^n\mathbb{Z}), \mathrm{End}^0(\mathbb{Z}/l^n\mathbb{Z})) = 0.$$

Proof. See [34, Lemma 2.48] and [50]. □

Combining all of this, we have the following theorem.

Theorem 10.13. *Let E be a rational elliptic curve and let l be a prime. Let $\rho : G_\mathbb{Q} \to \mathrm{GL}_2(\mathbb{Z}_l)$ be the associated Tate module representation. Assume that ρ is surjective. Further assume that for every good prime r there is a class $c_r \in \mathrm{H}^1(\mathbb{Q}, T)$ such that*
- *$\mathrm{res}_p\, c_r$ lies in $\mathrm{H}_f^1(\mathbb{Q}_p, T)$ for $r \neq p$;*
- *$\mathbb{Z}_l \cdot \mathrm{res}_r\, c_r$ contains $\eta \mathrm{H}_s^1(\mathbb{Q}_r, T_n)$.*

Then η annihilates $\mathrm{H}_f^1(\mathbb{Q}, A^*)$.

Proof. The given Flach system for T induces one for each T_n. The results to this point have thus shown that $\eta \mathrm{H}_f^1(\mathbb{Q}, A_n^*) = 0$ for each n. But any class $d \in \mathrm{H}_f^1(\mathbb{Q}, A^*)$ must be annihilated by some power of l, so it lies in the image of some $\mathrm{H}_f^1(\mathbb{Q}, A_n^*)$. (Note that $\mathrm{H}_f^1(\mathbb{Q}, A_n^*)$ maps into $\mathrm{H}_f^1(\mathbb{Q}, A^*)$ since we defined the finite subgroups for A_n^* using those for A^*.) Thus $\eta d = 0$, which completes the proof. □

The proof of Theorem 10.13 is purely a Galois cohomology argument, and therefore there is no actual need to assume that the representation ρ comes from an elliptic curve. For example, in [51] Galois representations coming from more general modular forms are considered.

The machinery we have given is sufficient for annihilation and finiteness results. To actually obtain a bound on the order of $\mathrm{H}_f^1(\mathbb{Q}, A^*)$, one has to exhibit classes not only for prime levels (like the c_r) but also for composite levels. Kolyvagin's derivative construction is then used to turn these classes into better and better annihilators. We should note, however, that with most Euler systems which have been studied the classes c_n are defined over larger and larger fields, depending on n. In our case, the classes would all be defined over $\mathbb{Q}$. Such an Euler system is often called a *geometric Euler system*, and there is not yet a general theory of such objects. For one example, see [**123**]. In fact, no one has succeeded in extending the Flach system above to a full geometric Euler system; this was the "gap" in the original proof of semistable Shimura-Taniyama by Wiles, which was eventually filled in by Taylor-Wiles using different methods.

The Flach map

We continue to let E be an elliptic curve over $\mathbb{Q}$ and $\phi : X_0(N) \to E$ a modular parameterization. It remains to construct a Flach system for T of depth $\deg \phi$. This construction lies at the heart of Flach's proof. These classes will come from certain well-chosen geometric objects on the surface $E \times E$, although in order to actually exhibit them we will need to work on the surface $X_0(N) \times X_0(N)$, which has a much richer intrinsic geometry. These objects are then transformed into classes in $\mathrm{H}^1(\mathbb{Q}, T)$ via a certain Chern class map. The key to Flach's construction is that it is possible to read off local properties of these classes in $\mathrm{H}^1_s(\mathbb{Q}_p, T)$ from corresponding local properties of the associated geometric objects. That is, we will begin with a map $\sigma : \mathcal{C}(E \times E) \to \mathrm{H}^1(\mathbb{Q}, T)$, where $\mathcal{C}(E \times E)$ will be defined in a moment purely geometrically. We cannot describe the image of σ directly (we can't even really describe $\mathrm{H}^1(\mathbb{Q}, T)$ effectively), but there is (for p not lying in S) a commutative diagram

(8)
$$\begin{array}{ccc} \mathcal{C}(E \times E) & \longrightarrow & \cdot \\ \sigma \downarrow & & \downarrow \\ \mathrm{H}^1(\mathbb{Q}, T) & \longrightarrow \mathrm{H}^1(\mathbb{Q}_p, T) \longrightarrow & \mathrm{H}^1_s(\mathbb{Q}_p, T) \end{array}$$

to be filled in later. Since all we care about for the production of our Flach system is the restriction of classes to $\mathrm{H}^1_s(\mathbb{Q}_p, T)$, to check that classes c_r really form a Flach system we will be able to bypass the complicated $\mathrm{H}^1(\mathbb{Q}, T)$ entirely and work instead with much more concrete geometric objects.

We begin by defining $\mathcal{C}(E \times E)$, which will involve working with curves lying in the surface $E \times E$. Let C be any projective geometrically integral algebraic curve over $\mathbb{Q}$; we do *not* assume that C is non-singular. We will be interested in rational functions on C which have trivial Weil divisor. (Recall that the *Weil divisor* of the function f on C is the formal sum of the points at which it has zeros minus the formal sum of the points at which it has poles, all counted with multiplicity. Often Weil divisors are only defined for *nonsingular* curves, but it is possible to define them more generally. The simplest approach is to define Weil divisors on singular curves by considering Weil divisors on their normalizations and then identifying points which become identified on the singular model. We will see an example of this in a moment.) If C is nonsingular, then it is a standard fact that the only such functions are constant. However, if C is singular, it is possible to exhibit non-constant rational functions with trivial Weil divisor.

For an example, consider a curve C with a nodal singularity $P \in C(\mathbb{Q})$. Let C' be its normalization, with P_1 and P_2 the points lying above P. Let f be a rational function on C' with divisor $nP_1 - nP_2$ for some n. (Such a function may or may not exist for a general C'; it will certainly exist if C' has genus 0, for example.) C' and C are birational, so f can also be interpreted as a rational function on C, and both P_1 and P_2 map to P. Thus f has trivial Weil divisor on C, even though it is a non-constant rational function.

Now consider the non-singular projective surface $E \times E$. We define $\mathcal{C}(E \times E)$ as follows: elements are pairs (C, f) of (possibly singular) curves C contained in $E \times E$ together with a rational function f on C with trivial Weil divisor. We also require that both C and f are defined over $\mathbb{Q}$. Flach defines a map

$$\sigma : \mathcal{C}(E \times E) \to \mathrm{H}^1(\mathbb{Q}, T)$$

which is what we will use to generate our Flach system. For the definition of σ, which involves étale cohomology and algebraic K-theory, see Appendix B.

Let us try briefly to explain the underlying philosophy by analogy with algebraic topology. Begin with the genus one complex curve $E(\mathbb{C})$, which we can regard as $\mathbb{C}/\Lambda$ for an appropriate lattice Λ. One way to obtain Λ is as the integral homology group $\mathrm{H}_1(E(\mathbb{C}), \mathbb{Z})$ (just thing about the standard homology generators on a torus), which is a lattice of maximal rank in $\mathrm{H}_1(E(\mathbb{C}), \mathbb{R}) \cong \mathbb{C}$. From this point of view the l^n-torsion on E is $\frac{1}{l^n}\Lambda$, so that the l-adic Tate module of E is $\Lambda \otimes_{\mathbb{Z}} \mathbb{Z}_l$; thus we can regard the l-adic Tate module of E as $\mathrm{H}_1(E(\mathbb{C}), \mathbb{Z}_l)$. Of course, we have lost any trace of Galois actions, but let us not concern ourselves with this at the moment.

Now consider the complex surface $E(\mathbb{C}) \times E(\mathbb{C})$ and its second homology group $\mathrm{H}_2(E(\mathbb{C}) \times E(\mathbb{C}), \mathbb{Z})$. The Künneth theorem shows that this group surjects onto

$$\mathrm{H}_1(E(\mathbb{C}), \mathbb{Z}) \otimes_{\mathbb{Z}} \mathrm{H}_1(E(\mathbb{C}), \mathbb{Z}).$$

Tensoring this with $\mathbb{Z}_l$ yields

$$\mathrm{H}_1(E(\mathbb{C}), \mathbb{Z}_l) \otimes_{\mathbb{Z}_l} \mathrm{H}_1(E(\mathbb{C}), \mathbb{Z}_l)$$

which by the above discussion contains $\mathrm{Sym}^2 T_l E = T$ as a direct summand. Combining all of this we see that we have a map

$$\mathrm{H}_2(E(\mathbb{C}) \times E(\mathbb{C}), \mathbb{Z}_l) \to T.$$

By Poincaré duality, we can also regard this as a map

(9) $$\mathrm{H}^2(E(\mathbb{C}) \times E(\mathbb{C}), \mathbb{Z}_l)^{\vee} \to T$$

where the $\vee$ denotes the Poincaré dual. We will return to this map later.

Now consider a pair (C, f). The curve C has real dimension 2, and therefore determines in a natural way an element of the homology group

$$\mathrm{H}_2(E(\mathbb{C}) \times E(\mathbb{C}), \mathbb{Z}).$$

If f has non-trivial divisor on C, then we could also use this divisor (which has real dimension 0) to determine an element of

$$\mathrm{H}_0(E(\mathbb{C}) \times E(\mathbb{C}), \mathbb{Z}).$$

In our case, however, f has trivial divisor. In this situation, the pair (C, f) does not determine anything of dimension 0, but still somehow contains more information than just C itself. This extra information has the effect of cutting down the relevant dimension by 1, so that (C, f) determines an element of

$$\mathrm{H}_1(E(\mathbb{C}) \times E(\mathbb{C}), \mathbb{Z}).$$

(This is where the algebraic K-theory comes in; K-theory is very good at keeping track of dimensions like this which may not make all that much sense purely geometrically.) Applying Poincaré duality we can regard this as an element of $\mathrm{H}^3(E(\mathbb{C}) \times E(\mathbb{C}), \mathbb{Z})^\vee$. So far, then, we have a map

$$\mathcal{C}(E \times E) \to \mathrm{H}^3(E(\mathbb{C}) \times E(\mathbb{C}), \mathbb{Z})^\vee. \tag{10}$$

(At this point we should confess that this map turns out to just be the zero map. It will nevertheless serve our motivational purposes.)

Étale cohomology is the algebraic analogue of singular cohomology, and the first miracle of étale cohomology is that the preceding construction can be carried out over $\bar{\mathbb{Q}}$ rather than $\mathbb{C}$, so long as we always use l-adic coefficients. Thus the pair (C, f) should give rise to an element of the dual $\mathrm{H}^3_{\text{ét}}(E_{\bar{\mathbb{Q}}} \times E_{\bar{\mathbb{Q}}}, \mathbb{Z}_l)^\vee$ of the étale cohomology group $\mathrm{H}^3_{\text{ét}}(E_{\bar{\mathbb{Q}}} \times E_{\bar{\mathbb{Q}}}, \mathbb{Z}_l)$; this last cohomology group is isomorphic to $\mathrm{H}^3(E(\mathbb{C}) \times E(\mathbb{C}), \mathbb{Z}_l)$ as an abelian group, but has the advantage of having a Galois action.

In fact, since (C, f) is defined over $\mathbb{Q}$, this element should be Galois invariant, yielding a map

$$\mathcal{C}(E \times E) \to \left(\mathrm{H}^3_{\text{ét}}(E_{\bar{\mathbb{Q}}} \times E_{\bar{\mathbb{Q}}}, \mathbb{Z}_l)^\vee\right)^{G_{\mathbb{Q}}}$$

analogous to (10). Unfortunately, $\mathrm{H}^3_{\text{ét}}(E_{\bar{\mathbb{Q}}} \times E_{\bar{\mathbb{Q}}}, \mathbb{Z}_l)^\vee$ can be shown to have no non-zero Galois invariants, so at the moment all of this work has produced 0.

The second miracle of étale cohomology is that we can carry our construction out over $\mathbb{Q}$, rather than $\bar{\mathbb{Q}}$. Thus (C, f) yields an element of $\mathrm{H}^3_{\text{ét}}(E \times E, \mathbb{Z}_l)^\vee$. This group is no longer isomorphic to any singular cohomology group, but rather is a complicated combination of various $\mathrm{Gal}(\bar{\mathbb{Q}}/\mathbb{Q})$-cohomology groups of étale cohomology groups of $E_{\bar{\mathbb{Q}}} \times E_{\bar{\mathbb{Q}}}$. It admits a natural map via a spectral sequence to

$$\mathrm{H}^0\left(\mathbb{Q}, \mathrm{H}^3_{\text{ét}}(E_{\bar{\mathbb{Q}}} \times E_{\bar{\mathbb{Q}}}, \mathbb{Z}_l)^\vee\right).$$

However, as we said above, this group vanishes, and from this one shows that the spectral sequence yields a map

$$\mathrm{H}^3_{\text{ét}}(E \times E, \mathbb{Z}_l)^\vee \to \mathrm{H}^1\left(\mathbb{Q}, \mathrm{H}^2_{\text{ét}}(E_{\bar{\mathbb{Q}}} \times E_{\bar{\mathbb{Q}}}, \mathbb{Z}_l)^\vee\right).$$

Thus we finally have a map

$$\mathcal{C}(E \times E) \to \mathrm{H}^1\left(\mathbb{Q}, \mathrm{H}^2_{\text{ét}}(E_{\bar{\mathbb{Q}}} \times E_{\bar{\mathbb{Q}}}, \mathbb{Z}_l)^\vee\right).$$

Combining this with the étale analogue of (9), we finally obtain our desired Flach map

$$\sigma : \mathcal{C}(E \times E) \to \mathrm{H}^1(\mathbb{Q}, T).$$

We now discuss the local behavior of σ. Let p be a prime not lying in S. We will define a map

$$d_p : \mathcal{C}(E \times E) \to \mathrm{Div}(E_{\mathbb{F}_p} \times E_{\mathbb{F}_p})$$

where $\mathrm{Div}(E_{\mathbb{F}_p} \times E_{\mathbb{F}_p})$ is the group of Weil divisors (defined over $\mathbb{F}_p$) on the non-singular surface $E_{\mathbb{F}_p} \times E_{\mathbb{F}_p}$. (Recall that a Weil divisor on a surface is a formal sum of curves on the surface.) d_p is the map which will go on the top of (8) above. To define $d_p(C, f)$, first consider the reduction of C modulo p. (Technically, by the reduction of C modulo p we mean the base change to $\mathbb{F}_p$ of the scheme-theoretic closure of C in a model of E over $\mathbb{Z}_p$. However, one loses nothing by simply regarding this as considering the equations defining C modulo p.) This may well have several geometric components $C_1, \ldots, C_n$, even if in characteristic 0 it did not. We claim that if the function f has a zero or pole at any point of a component C_i, then it has a zero or pole of the same order along the entire component C_i. (Actually, poles and zeros can combine at the points where the C_i intersect, but this doesn't matter much.) The idea is that if f had a zero or pole at an isolated point of C_i, then we could lift this to a zero or pole of f over $\mathbb{Q}$, which is not possible by the definition of $\mathcal{C}(E \times E)$. Given this, for any component C_i we can let m_i be the order of the zero or pole of f on C_i: of course, we could have $m_i = 0$. We define

$$d_p(C, f) = \sum m_i C_i.$$

The last thing we will need to connect the geometry to the behavior of cohomology classes is a map from $\mathrm{Div}(E_{\mathbb{F}_p} \times E_{\mathbb{F}_p})$ to $\mathrm{H}^1_s(\mathbb{Q}_p, T)$. Recall that by Lemma 10.9 we have $\mathrm{H}^1_s(\mathbb{Q}_p, T) \cong T(-1)^{G_{\mathbb{F}_p}}$. This in turn is isomorphic to $\mathrm{End}_{G_{\mathbb{F}_p}}^0(T_l E)$, by Lemma 10.2. That is, the singular quotient at p corresponds precisely to trace zero $G_{\mathbb{F}_p}$-equivariant maps of the l-adic Tate module of E. The map we seek is a standard one in algebraic geometry, called a *cycle class map*:

$$s : \mathrm{Div}(E_{\mathbb{F}_p} \times E_{\mathbb{F}_p}) \to \mathrm{End}_{G_{\mathbb{F}_p}}(T_l E) \to \mathrm{End}_{G_{\mathbb{F}_p}}^0(T_l E)$$

(the second map is simply projection onto a direct summand). We will not give a general description of the map $\mathrm{Div}(E_{\mathbb{F}_p} \times E_{\mathbb{F}_p}) \to \mathrm{End}_{G_{\mathbb{F}_p}}(T_l E)$, except in the special case we will need. (In fact, if $\mathrm{End}_{G_{\mathbb{F}_p}}(T_l E)$ is given an appropriate cohomological interpretation, then the cycle class map is really just an algebraic version of part of the algebraic topology construction discussed above.) Let $g : E_{\mathbb{F}_p} \to E_{\mathbb{F}_p}$ be some map. Let Γ_g be the graph of g, by which we mean the image of the product map

$$\mathrm{id} \times g : E_{\mathbb{F}_p} \to E_{\mathbb{F}_p} \times E_{\mathbb{F}_p}.$$

Then Γ_g has codimension 1 in $E_{\mathbb{F}_p} \times E_{\mathbb{F}_p}$, and thus is an element of $\mathrm{Div}(E_{\mathbb{F}_p} \times E_{\mathbb{F}_p})$. The image of Γ_g under s is nothing other than the endomorphism of $T_l E$ induced by the map g (or more honestly its projection onto the trace zero direct summand). This endomorphism is $G_{\mathbb{F}_p}$-equivariant since g is defined over $\mathbb{F}_p$.

We are finally in a position to state the fundamental local description of the map σ: for every prime p not in S, there is a commutative diagram

(11)
$$\begin{array}{ccc}
\mathcal{C}(E \times E) & \xrightarrow{d_p} & \mathrm{Div}(E_{\mathbb{F}_p} \times E_{\mathbb{F}_p}) \\
{\scriptstyle \sigma} \downarrow & & \downarrow {\scriptstyle s} \\
\mathrm{H}^1(\mathbb{Q}, T) \longrightarrow \mathrm{H}^1(\mathbb{Q}_p, T) \longrightarrow \mathrm{H}^1_s(\mathbb{Q}_p, T) & \xrightarrow{\cong} & \mathrm{End}_{G_{\mathbb{F}_p}}^0(T_l E)
\end{array}$$

"All" we have to do to generate our Flach system, then, is to exhibit appropriate elements of $\mathcal{C}(E \times E)$ and compute their image in $\mathrm{End}_{G_{\mathbb{F}_p}}^0(T_l E)$ via the clockwise maps. Unfortunately, in general this would be extremely difficult. That is, given an arbitrary surface S, it is a very difficult problem in algebraic geometry to write down many particularly useful curves on S. In order to do this in our case we will have to take full advantage of the fact that E is modular.

The geometry of modular curves

In this section we review the facts we will need about the geometry of modular curves. For a more thorough treatment and references to the standard sources, see [**41**, Part II].

Let N be a positive integer and let $\Gamma_0(N)$ be the usual congruence subgroup of $\mathrm{SL}_2(\mathbb{Z})$. Recall that orbits for the $\Gamma_0(N)$-action on the upper half plane $\mathfrak{H}$

correspond to isomorphism classes of pairs of complex elliptic curves E and cyclic subgroups of $E(\mathbb{C})$ of order N. Furthermore, the orbit space $\Gamma_0(N)\backslash\mathfrak{H}$ can be given the structure of a non-compact Riemann surface. We will write $Y_0(N)^{\text{an}}$ for this Riemann surface. (The "an" is for analytic.) We can also compactify $Y_0(N)^{\text{an}}$ by adding a finite number of points called the *cusps*; we write $X_0(N)^{\text{an}}$ for the resulting compact Riemann surface.

It is a classical fact of algebraic geometry that every compact Riemann surface can be realized as a nonsingular projective complex algebraic curve; that is, there is an algebraic curve $X_0(N)_{\mathbb{C}}$ over the complex numbers such that the $\mathbb{C}$-valued points of $X_0(N)_{\mathbb{C}}$ recover $X_0(N)^{\text{an}}$. A fundamental fact in the arithmetic theory of modular curves is that this curve can actually be canonically defined over the rational numbers $\mathbb{Q}$. That is, the polynomial equations which define $X_0(N)_{\mathbb{C}}$ can be canonically chosen in such a way that all of the coefficients are rational. We will write $X_0(N)_{\mathbb{Q}}$ for this nonsingular projective rational algebraic curve. The subscheme of cusps of $X_0(N)_{\mathbb{Q}}$ is canonically defined over $\mathbb{Q}$, so $Y_0(N)^{\text{an}}$ can also be realized as the complex points of a nonsingular algebraic curve $Y_0(N)_{\mathbb{Q}}$; of course, $Y_0(N)_{\mathbb{Q}}$ is only quasi-projective.

Now that we have a model for our modular curve over $\mathbb{Q}$, we can ask how the equations for $X_0(N)_{\mathbb{Q}}$ reduce modulo primes p. The fundamental result is that $X_0(N)_{\mathbb{Q}}$ reduces to a nonsingular projective algebraic curve over $\mathbb{F}_p$ for every prime p which does not divide N. (Perhaps the most compact way to say this is that $X_0(N)_{\mathbb{Q}}$ is the generic fiber of a canonical smooth proper $\mathbb{Z}[1/N]$-scheme. This description allows one to work with $X_0(N)_{\mathbb{Q}}$ and all of its reductions simultaneously, which is often convenient; nevertheless, we will content ourselves below with working one prime at a time.) From now on we will just write $X_0(N)$ when it does not matter what field (of characteristic 0 or p not dividing N) we are working over; the behavior of the modular curves over the various fields is virtually identical.

If p is a prime which divides N, then $X_0(N)$ will pick up singularities over $\mathbb{F}_p$, but at least in the case that p divides N exactly once it is possible to very explicitly describe $X_0(N)_{\mathbb{F}_p}$: it has two irreducible components, each isomorphic to $X_0(N/p)_{\mathbb{F}_p}$ (which is nonsingular by what we said above), and they intersect transversally at a finite number of points.

The other question one might ask about our models $Y_0(N)$ is whether or not they still classify pairs of elliptic curves and cyclic subgroups of order N. Let us say that a pair of an elliptic curve E and a cyclic subgroup C of order N is *defined over a field K* (of characteristic 0 or p not dividing N) if E is defined over K and if C is mapped to itself under the action of every element of the absolute Galois group of K. (Note that we do not require that C is fixed pointwise by the Galois group, which is equivalent to C actually lying in $E(K)$.) We could hope that the K-rational points of $Y_0(N)$ correspond to the pairs (E, C) as above which are defined over K. If this were the case, we would call these modular curves *fine moduli spaces* for classifying such pairs. Unfortunately, this is simply not true. This is well-known in the case $N = 1$: $Y_0(1)_{\mathbb{Q}}$ is isomorphic to the affine line $\mathbb{A}^1_{\mathbb{Q}}$ via the j-invariant, but the j-invariant is not enough to determine the isomorphism class of elliptic curves over fields which are not algebraically closed. The problem in the general case is similar, although somewhat less severe.

However, these modular curves are at least *coarse moduli spaces*. We will not give the technical definition of this, except to say that it means that the modular curves are as close to fine moduli spaces for classifying pairs as it is possible for them to be. In particular, every pair of an elliptic curve E and a cyclic subgroup C of order N, defined over a field K, does give rise to a point of $Y_0(N)(K)$.

In passing, we should note that it is also possible to give a modular interpretation to the cusps of $X_0(N)$ in terms of *generalized elliptic curves*. We will not give a description of this here; it is, however, extremely useful for computations involving the cusps.

We can use our moduli interpretations to define various maps between modular curves. (Actually, the description we gave above is not enough to actually make rigorous the upcoming definitions. However, we assure the reader that these constructions can be made entirely rigorous with a more thorough understanding of the "moduli interpretation" of $X_0(N)$.) Let K be a field as above, and let r be a prime not dividing N. A pair of an elliptic curve E and a cyclic subgroup C of order Nr, defined over K, gives rise in a natural way to a corresponding pair with respect to N: take the same elliptic curve E and take the unique cyclic subgroup of C of order N; call it C_N. We define
$$j_r : X_0(Nr) \to X_0(N)$$
to be the corresponding map; that is, it sends the point corresponding to the pair (E, C) to the point corresponding to the pair (E, C_N). (As we said above, this requires more work to be a rigorous definition, but it is possible to give it a better interpretation.) The fact that we can make this definition on the level of points for any field K (of the appropriate characteristics) insures that j_r can actually be defined over any of the fields $\mathbb{Q}$ or $\mathbb{F}_p$ for p not dividing N. Note that there is a slight subtlety (which we will ignore) for the field $\mathbb{F}_r$, as there we have not said anything about the moduli interpretation of $X_0(Nr)$.

There is a second way to obtain a map between these modular curves: let C_r be the unique subgroup of C of order r, and now send the pair (E, C) to the pair $(E/C_r, C/C_r)$. This gives rise to a second map
$$j'_r : X_0(Nr) \to X_0(N).$$
We define the rth *Hecke correspondence* T_r on $X_0(N)$ to be the image of the product map
$$j_r \times j'_r : X_0(Nr) \to X_0(N) \times X_0(N).$$

It can be shown that T_r is a singular curve which is birational to $X_0(Nr)$. Furthermore, it is possible to give a very precise description of T_r in characteristic r. Recall that a curve (or more generally any scheme) over $\mathbb{F}_r$ has a Frobenius endomorphism induced by the r-power map on the function field. Define $\Gamma \subseteq X_0(N)_{\mathbb{F}_r} \times X_0(N)_{\mathbb{F}_r}$ as the graph of the Frobenius map $\text{Fr} : X_0(N)_{\mathbb{F}_r} \to X_0(N)_{\mathbb{F}_r}$; that is, Γ is the image of the product map
$$\text{id} \times \text{Fr} : X_0(N)_{\mathbb{F}_r} \to X_0(N)_{\mathbb{F}_r} \times X_0(N)_{\mathbb{F}_r}.$$
We define the *Verschiebung* $\Gamma' \subseteq X_0(N)_{\mathbb{F}_r} \times X_0(N)_{\mathbb{F}_r}$ to be the image of the transpose map
$$\text{Fr} \times \text{id} : X_0(N)_{\mathbb{F}_r} \to X_0(N)_{\mathbb{F}_r} \times X_0(N)_{\mathbb{F}_r}.$$
Note that $T_{r,\mathbb{F}_r}$, Γ and Γ' are all of codimension one in the surface $X_0(N)_{\mathbb{F}_r} \times X_0(N)_{\mathbb{F}_r}$, and thus are divisors. The *Eichler-Shimura relation* states that there is an equality
$$T_{r,\mathbb{F}_r} = \Gamma + \Gamma'.$$

of divisors on $X_0(N)_{\mathbb{F}_r} \times X_0(N)_{\mathbb{F}_r}$. In fact, each of Γ and Γ' is the isomorphic image of one of the irreducible components of $X_0(Nr)_{\mathbb{F}_r}$ (both isomorphic to $X_0(N)_{\mathbb{F}_r}$) which we discussed above. This relation will be the key to our computations below.

Some modular units

Modular curves will be of use to us since the surface $X_0(N) \times_{\mathbb{Q}} X_0(N)$ has all of the fairly explicit divisors T_r. Our basic plan at this point is to find an appropriate rational function f_r on $T_{r,\mathbb{Q}}$ (for each r not dividing N) such that $(T_{r,\mathbb{Q}}, f_r) \in \mathcal{C}(X_0(N)_{\mathbb{Q}} \times X_0(N)_{\mathbb{Q}})$. We will then map the pair $(T_{r,\mathbb{Q}}, f_r)$ via $\phi \times \phi$ into $\mathcal{C}(E_{\mathbb{Q}} \times E_{\mathbb{Q}})$, and then by σ into $\mathrm{H}^1(\mathbb{Q}, T)$. If we can also arrange for f_r to have trivial divisor away from characteristic r, then our geometric description of the local behavior of σ will show that f_r maps to 0 in each $\mathrm{H}^1_s(\mathbb{Q}_p, T)$ for p not dividing rlN (recall that our geometric description of the Flach map broke at the primes in S) and we will be well on our way to constructing the desired Flach system.

Since $T_{r,\mathbb{Q}}$ is birational to $X_0(Nr)_{\mathbb{Q}}$, to exhibit rational functions on $T_{r,\mathbb{Q}}$ we can work instead on $X_0(Nr)_{\mathbb{Q}}$. Recall that rational functions on $X_0(Nr)_{\mathbb{Q}}$ are modular functions of level Nr and weight 0, with coefficients in $\mathbb{Q}$. We will define such a function as the ratio of two modular forms of the same weight.

We want f_r to have trivial divisor over $\mathbb{Q}$, so we should start with modular forms with especially simple divisors. Perhaps the best known is $\Delta(z)$, the unique cusp form of level 1 and weight 12. Δ is initially defined on $X_0(1)_{\mathbb{Q}}$, and has a simple zero at the unique cusp ∞ and no other zeros or poles. (Δ is a pluri-canonical form, not a function, so it can have more zeros than poles.) Pulling back Δ via the natural map $\pi : X_0(N)_{\mathbb{Q}} \to X_0(1)_{\mathbb{Q}}$ yields a form $\pi^*\Delta$ on $X_0(N)_{\mathbb{Q}}$ (this is really nothing more than reinterpreting Δ as having level N) which will have zeros of various orders at the cusps and no other zeros or poles:

$$\mathrm{div}_{X_0(N)_{\mathbb{Q}}} \Delta = \sum_{\text{cusps } c_i} n_i c_i$$

for some integers n_i.

We will pull back Δ to $X_0(Nr)_{\mathbb{Q}}$ via the two maps j_r and j'_r. In order to understand the divisors of these forms, we need to know how the cusps behave under these maps. The basic fact is that the preimage of a cusp c_i of $X_0(N)_{\mathbb{Q}}$ under j_r consists of two cusps $c_{i,1}$ and $c_{i,2}$ of $X_0(Nr)_{\mathbb{Q}}$; j_r is unramified at $c_{i,1}$ and ramified of degree r at $c_{i,2}$. Under j'_r we have the opposite behavior: $c_{i,1}$ and $c_{i,2}$ are again the only two points in the preimage of c_i, but now $c_{i,2}$ is unramified and $c_{i,1}$ is ramified of degree r. Combining all of this, we find that

$$\mathrm{div}_{X_0(Nr)_{\mathbb{Q}}} j_r^* \Delta = \sum n_i c_{i,1} + r n_i c_{i,2}$$
$$\mathrm{div}_{X_0(Nr)_{\mathbb{Q}}} j_r'^* \Delta = \sum r n_i c_{i,1} + n_i c_{i,2}.$$

Both of these forms have weight 12, since Δ does, so their ratio is a rational function f_r on $X_0(Nr)_{\mathbb{Q}}$ with divisor

$$\mathrm{div}_{X_0(Nr)_{\mathbb{Q}}} f_r = \sum (1-r) n_i (c_{i,1} - c_{i,2}).$$

We now think of f_r as a rational function on the singular curve $T_{r,\mathbb{Q}}$, which is birational to $X_0(Nr)_{\mathbb{Q}}$. As we said above, both $c_{i,1}$ and $c_{i,2}$ map to c_i under j_r and

j'_r. Since $T_{r,\mathbb{Q}}$ is the image of $X_0(Nr)_\mathbb{Q}$ under the map $j_r \times j'_r$, the divisor of f_r on $T_{r,\mathbb{Q}}$ is

$$\operatorname{div}_{T_{r,\mathbb{Q}}} f_r = \sum (1-r) n_i \big((j_r(c_{i,1}), j'_r(c_{i,1})) - (j_r(c_{i,2}), j'_r(c_{i,2})) \big) = 0.$$

Thus $(T_{r,\mathbb{Q}}, f_r) \in \mathcal{C}(X_0(N)_\mathbb{Q} \times X_0(N)_\mathbb{Q})$, as desired.

We can now define the Flach classes $c_r \in \mathrm{H}^1(\mathbb{Q}, T)$: we first map $(T_{r,\mathbb{Q}}, f_r)$ to $\mathcal{C}(E_\mathbb{Q} \times E_\mathbb{Q})$ via $(\phi \times \phi)_*$. That is, let $T'_{r,\mathbb{Q}}$ be the image of $T_{r,\mathbb{Q}}$ under $\phi \times \phi$ and let f'_r be the rational function on $T'_{r,\mathbb{Q}}$ induced by f_r. (f'_r is really the norm of f_r in the finite extension of function fields $k(X_0(N)_\mathbb{Q})/k(E_\mathbb{Q})$ induced by ϕ.) One easily checks that

$$\operatorname{div}_{T'_{r,\mathbb{Q}}} f'_r = (\phi \times \phi)_* \operatorname{div}_{T_{r,\mathbb{Q}}} f_r = 0,$$

so $(T'_{r,\mathbb{Q}}, f'_r) \in \mathcal{C}(E_\mathbb{Q} \times E_\mathbb{Q})$. We now map this pair to $\mathrm{H}^1(\mathbb{Q}, T)$ via the Flach map σ. This is the class c_r. Note that we defined these classes for all r not dividing N, even though we only need them for good r. (In fact, one can even define classes for r dividing N with some care.)

Local behavior of the c_r

To complete the construction of our Flach system, we need to analyze the local behavior of the classes c_r in $\mathrm{H}^1_s(\mathbb{Q}_p, T)$ for all p. Specifically, we need to show that they map to 0 for all $p \neq r$ and we need to compute them explicitly for $p = r$. We do this using our description in terms of divisors and cycle classes. We distinguish several cases; for simplicity, we assume that r is good, although this is not critical to these computations.

p does not divide Nlr

This is the easiest case. $d_p(T'_{r,\mathbb{Q}}, f'_r)$ is just the divisor of f'_r on $T'_{r,\mathbb{F}_p}$, and the analysis of the preceding section of the divisor of f_r over $\mathbb{Q}$ goes through in exactly the same way over $\mathbb{F}_p$. Thus

$$\operatorname{div}_{T'_{r,\mathbb{F}_p}} f'_r = (\phi \times \phi)_* \operatorname{div}_{T_{r,\mathbb{F}_p}} f_r = 0,$$

so $d_p(T'_{r,\mathbb{Q}}, f'_r) = 0$. Commutativity of the diagram (11) now shows that c_r maps to 0 in $\mathrm{III}^1_s(\mathbb{Q}_p, T)$, since it is the image of the pair $(T'_{r,\mathbb{Q}}, f'_r)$ which already maps to 0 in $\operatorname{Div}(E_{\mathbb{F}_p} \times E_{\mathbb{F}_p})$. In particular, there is no need to know anything at all about the map s in this case.

p divides N

This is the case of bad reduction of E and the local diagram we used in the first case does not hold in this setting. Flach uses two different arguments to handle this case. If E has potentially multiplicative reduction at p, then one can give a very explicit description of the $G_{\mathbb{Q}_p}$-action on V, and one can compute that $\mathrm{H}^1_f(\mathbb{Q}_p, V) = \mathrm{H}^1(\mathbb{Q}_p, V)$. It follows that $\mathrm{H}^1_s(\mathbb{Q}_p, T) = 0$, so that there is no local condition to check! If E has potentially good reduction at p, Flach mimics the argument above in the case of good reduction, using the Néron model of E; see [**50**, pp. 324–325] and [**51**, Section 5.5.2 and Section 6].

$p = l$

In this case we again do not have the local diagram to fall back on. Flach uses results of Faltings to conclude that $\mathrm{res}_l\, c_r$ lies in $\mathrm{H}^1_f(\mathbb{Q}_l, T)$; see [**50**, pp. 322-324].

$p = r$

This is the key computation. Recall that the Eichler-Shimura relation says that $T_{r,\mathbb{F}_r}$ can be written as a sum $\Gamma + \Gamma'$ of the graph of Frobenius and the Verschiebung. We will work with each piece separately.

We begin with Γ:

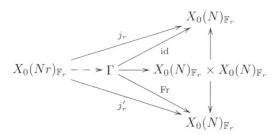

(Only one of the irreducible components of $X_0(Nr)_{\mathbb{F}_r}$ maps to Γ, which is why we have used a dotted line there.) The function on Γ corresponding to $j_r^* \Delta$ is just the pull back of Δ under the identity map; thus $\mathrm{div}_\Gamma\, j_r^* \Delta$ will just be the usual linear combination of points of Γ corresponding to cusps. The function on Γ corresponding to $j_r'^* \Delta$ is the pull back of Δ under Fr. Fr is purely inseparable, and purely inseparable maps are trivial on differentials; see [**146**, Chapter 2, Proposition 4.2]. Thus $\mathrm{Fr}^* \Delta = j_r'^* \Delta$ will pick up a zero on Γ in addition to the usual cuspidal divisor. One can check that this zero has order 6 (essentially because Δ has weight $12 = 2 \cdot 6$). As always, the cuspidal parts of the divisor cancel, and we conclude that

$$\mathrm{div}_\Gamma\, f_r = \mathrm{div}_\Gamma\, j_r^* \Delta - \mathrm{div}_\Gamma\, j_r'^* \Delta = -6\Gamma.$$

The computation for Γ' is virtually identical, except that id and Fr are interchanged. Thus

$$\mathrm{div}_{\Gamma'}\, f_r = 6\Gamma'.$$

We conclude that

$$\mathrm{div}_{T_{r,\mathbb{F}_r}}\, f_r = 6(\Gamma' - \Gamma).$$

The next step is to push our whole construction forward to $E \times E$. If we let Γ_E and Γ'_E be the image of

$$\mathrm{id} \times \mathrm{Fr} : E_{\mathbb{F}_r} \to E_{\mathbb{F}_r} \times E_{\mathbb{F}_r}$$

$$\mathrm{Fr} \times \mathrm{id} : E_{\mathbb{F}_r} \to E_{\mathbb{F}_r} \times E_{\mathbb{F}_r}$$

respectively, then it is clear that $\phi \times \phi$ maps Γ onto Γ_E and Γ' onto Γ'_E. Since each point of Γ_E and Γ'_E is the image of $\deg \phi$ points of Γ and Γ', we have the equalities

$$(\phi \times \phi)_* \Gamma = (\deg \phi) \Gamma_E$$
$$(\phi \times \phi)_* \Gamma' = (\deg \phi) \Gamma'_E$$

as divisors. We conclude finally that

$$d_p(\phi \times \phi)_*(T_{r,\mathbb{Q}}, f_r) = 6(\deg \phi)(\Gamma'_E - \Gamma_E) \in \mathrm{Div}(E_{\mathbb{F}_r} \times E_{\mathbb{F}_r}).$$

We now need to compute the image of this under the cycle class map s. Our description of s shows that Γ_E, being the graph of Frobenius at r maps to precisely the endomorphism of $T_l E$ given by Fr_r. (This is well defined since $T_l E$ is unramified at r.) r is assumed to be good, so the proof of Lemma 10.9 shows that we can choose a basis x, y for $T_l E$ with respect to which Fr_r has matrix

$$\begin{pmatrix} u & 0 \\ 0 & v \end{pmatrix}$$

with $u \equiv -v \equiv 1 \pmod{l}$ and $uv = r$. This matrix is the image of Γ_E in $\mathrm{End}_{G_{\mathbb{F}_p}}(T_l E)$.

To make the corresponding calculation for Γ'_E we will need to reinterpret it as a graph. Since Fr has degree r, there is a map $V : E_{\mathbb{F}_r} \to E_{\mathbb{F}_r}$ with the property that $V \circ \mathrm{Fr} = \mathrm{Fr} \circ V$ is the multiplication by r map on $E_{\mathbb{F}_r}$; see [**146**, Chapter 3, Section 6]. Γ'_E is the image of the map

$$\mathrm{Fr} \times \mathrm{id} : E_{\mathbb{F}_r} \to E_{\mathbb{F}_r} \times E_{\mathbb{F}_r}.$$

If we precompose with the map $V : E_{\mathbb{F}_r} \to E_{\mathbb{F}_r}$, which is a surjective map of degree r, the literal image will not change, but each point will pick up a multiplicity of r. Thus the image of the map $\mathrm{Fr} \circ V \times V = r \times V$ is $r\Gamma'_E$. We claim that we can cancel the two r's, which leaves us with the fact that Γ'_E is the graph of V. The easiest way to do this is to pretend for the moment that multiplication by r has an inverse r^{-1} on E. (Of course, this is absurd, but it is somewhat less absurd when one does the entire computation in the range $\mathrm{End}(T_l E)$, where r is invertible.) Then an argument similar to the one above for precomposing with V shows that the image of $r \times V$ is r^2 times the image of $\mathrm{id} \times V r^{-1}$. This means that $s(r\Gamma'_E) = r s(\Gamma'_E)$ is the same as $r^2 V r^{-1}$, where now V is regarded as an endomorphism of $T_l E$. In other words, $s(\Gamma'_E)$ is just the endomorphism induced by V.

Since $V \circ \mathrm{Fr} = r$, this implies that the cycle class of Γ'_E has matrix

$$r \begin{pmatrix} u & 0 \\ 0 & v \end{pmatrix}^{-1} = \begin{pmatrix} v & 0 \\ 0 & u \end{pmatrix}.$$

We conclude that $6(\deg \phi)(\Gamma'_E - \Gamma_E)$ maps to

$$6(\deg \phi) \begin{pmatrix} (v-u) & 0 \\ 0 & (u-v) \end{pmatrix}$$

in $\mathrm{End}_{G_{\mathbb{F}_r}}(T_l E)$, and even in $\mathrm{End}^0_{G_{\mathbb{F}_r}}(T_l E)$ since this matrix already has trace 0. This, then, is the image of c_r in $\mathrm{H}^1_s(\mathbb{Q}_r, T)$.

Recall that $\mathrm{H}^1_s(\mathbb{Q}_r, T) \cong \mathrm{End}^0_{G_{\mathbb{F}_r}}(T_l E)$ is a free $\mathbb{Z}_l$-module of rank 1. One easily checks that the matrix

$$\begin{pmatrix} 1 & 0 \\ 0 & -1 \end{pmatrix}$$

is a generator of this module. Combined with our computation above, we find that $6(\deg \phi)(v - u)$ annihilates

$$\mathrm{H}^1_s(\mathbb{Q}_r, T) / \mathbb{Z}_l \cdot \mathrm{res}_r c_r.$$

But 6 is an l-adic unit, and $v - u \equiv -2 \pmod{l}$, so $v - u$ is as well. We conclude that $\deg \phi$ annihilates this module, and thus that the c_r form a Flach system of depth $\deg \phi$ for T. This concludes the proof of Theorem 10.13, and with it the proof of Theorem 10.1.

Appendix A: On the local Galois invariants of $E[l] \otimes E[l]$

The purpose of this appendix is to prove the following result.

Theorem 10.14. *Let E be an elliptic curve over $\mathbb{Q}$ without complex multiplication. Let $\phi : X_0(N) \to E$ be a modular parameterization of E and let S_0 be the set of places of $\mathbb{Q}$ at which E has bad reduction. Then the set of primes such that*

- *E has good reduction at l;*
- *The l-adic representation $\rho : G_{\mathbb{Q}, S_0 \cup \{l\}} \to \mathrm{GL}_2(\mathbb{Z}_l)$ is surjective;*
- *For all $p \in S_0$, $E[l] \otimes E[l]$ has no $G_{\mathbb{Q}_p}$-invariants;*
- *$E[l] \otimes E[l]$ has no $G_{\mathbb{Q}_l}$-invariants;*
- *l does not divide the degree of ϕ;*

has density 1 in the set of all primes.

Of course, the first and fifth conditions are obviously satisfied for almost all l. That the second condition is satisfied for almost all l is a result of Serre; see [132]. The new content is in the third and fourth conditions. We will show that the third condition is also satisfied for almost all l, and that the fourth condition is satisfied for a set of primes of density 1.

Recall that by the Weil pairing we can write

$$E[l] \otimes E[l] \cong \mu_l \oplus \mathrm{Sym}^2 E[l].$$

The analysis of $\mathbb{Q}_p$-rational points on the first of these factors is immediate from the fact that (for $p \neq 2$) $\mathbb{Q}_p$ contains precisely the $(p-1)$-th roots of unity: it has $\mathbb{Q}_p$-rational points if and only if $p \equiv 1 \pmod{l}$.

We begin the analysis of $\mathrm{Sym}^2 E[l]$ with a modification of the argument of [20, Lemma 2.3(i)].

Lemma 10.15. *Let E be an elliptic curve over $\mathbb{Q}_p$ and let l be any prime. Then $\mathrm{H}^0(\mathbb{Q}_p, \mathrm{Sym}^2 E[l]) \neq 0$ if and only if $E(K)$ has non-trivial l-torsion for some quadratic extension K of $\mathbb{Q}_p$.*

Proof. We first set some notation. Let $\varepsilon : G_{\mathbb{Q}_p} \to \mathbb{Z}_l^*$ be the cyclotomic character; its image has finite index in $\mathbb{Z}_l^*$. Let $\rho : G_{\mathbb{Q}_p} \to \mathrm{GL}(E[l])$ and $\varphi : G_{\mathbb{Q}_p} \to \mathrm{GL}(\mathrm{Sym}^2 E[l])$ be the Galois representations associated to E. By the Weil pairing we have $\det \rho = \varepsilon$.

If x is a K-rational l-torsion point for some quadratic extension K of $\mathbb{Q}_p$, then one checks immediately that $x \otimes x \in E[l] \otimes E[l]$ is $G_{\mathbb{Q}_p}$-invariant, which proves one direction of the lemma.

Suppose, then, that there exists $t \in \mathrm{Sym}^2 E[l]$ such that $\varphi(\tau)t = t$ for all $\tau \in G_{\mathbb{Q}_p}$. Thus 1 is an eigenvalue of $\varphi(\tau)$ for every $\tau \in G_{\mathbb{Q}_p}$.

Now choose $\sigma_0 \in G_{\mathbb{Q}_p}$ such that $\varepsilon(\sigma_0)$ is not a root of unity; this is certainly possible since the image of ε has finite index. Let λ and μ be the eigenvalues of $\rho(\sigma_0)$. Then the eigenvalues of $\varphi(\sigma_0)$ are λ^2, $\lambda\mu = \varepsilon(\sigma_0)$ and μ^2. Since one of these is 1 and $\varepsilon(\sigma_0)$ is not a root of unity, we can assume without loss of generality that $\lambda^2 = 1$.

Set $\sigma = \sigma_0^2$. The eigenvalues of $\rho(\sigma)$ are $\lambda^2 = 1$ and μ^2. We have $\mu^2 \neq 1$ (since $\lambda^2 \mu^2 = \varepsilon(\sigma_0)^2$ is not a root of unity), so we can choose a basis x, y for $E[l]$ of eigenvectors for $\rho(\sigma)$, with eigenvalues 1 and $\varepsilon(\sigma)$ respectively. $\varepsilon(\sigma)^2$ is still not 1, from which one easily computes (using the basis $x \otimes x$, $x \otimes y + y \otimes x$, $y \otimes y$ of $\mathrm{Sym}^2 E[l]$) that t is a scalar multiple of $x \otimes x$. It follows easily that x is rational over some quadratic extension of $\mathbb{Q}_p$. □

We now state the general analysis of torsion in elliptic curves over local fields, coming from an analysis of the formal group and the component group of the Néron model.

Proposition 10.16. *Let E be an elliptic curve over a finite extension K of $\mathbb{Q}_p$ and assume $p \neq l$. Let k be the residue field of K.*
- *If E has good reduction over K, then $E(K)$ has non-trivial l-torsion only if l divides $\#E(k)$.*
- *If E has non-split multiplicative reduction over K, then $E(K)$ has non-trivial l-torsion only if $p \equiv 1 \pmod{l}$ or $l \leq 3$.*
- *If E has split multiplicative reduction over K, then $E(K)$ has non-trivial l-torsion only if $p \equiv 1 \pmod{l}$ or $l \leq 11$.*
- *If E has additive reduction over K, then $E(K)$ has non-trivial l-torsion only if $l \leq 3$.*

Proof. By [**146**, Proposition VII.2.1] there is an exact sequence
$$0 \to E_1(\mathfrak{m}_K) \to E_0(K) \to \tilde{E}_{ns}(k) \to 0$$
where E_1 is the formal group of E, $\mathfrak{m}_K$ is the maximal ideal of the ring of integers of K, $E_0(K)$ is the set of points of $E(K)$ with non-singular reduction and $\tilde{E}_{ns}(k)$ are the non-singular points of the reduction. By [**146**, Proposition IV.3.2(b)], $E_1(K)$ has no non-trivial l-torsion, so any l-torsion in $E(K)$ must appear in $E(K)/E_0(K)$ or $\tilde{E}_{ns}(k)$. The proposition now follows from the determination of $\tilde{E}_{ns}(k)$ in the various cases (see [**146**, Proposition VII.5.1]) and the analysis of the component group of the Néron model of E (see [**146**, Theorem VII.6.1] and use that the minimal discriminant has valuation at most 11). □

An entirely similar argument yields the following result for the case $p = l$.

Proposition 10.17. *Let E be an elliptic curve over a quadratic extension K of $\mathbb{Q}_l$ and assume $l \geq 5$. Let k be the residue field of K.*
- *If E has good reduction over K, then $E(K)$ has non-trivial l-torsion only if l divides $\#E(k)$.*
- *If E has non-split multiplicative reduction over K, then $E(K)$ has no non-trivial l-torsion.*
- *If E has split multiplicative reduction over K, then $E(K)$ has non-trivial l-torsion only if $l \leq 11$.*

Proof. The only difference with Proposition 10.16 is the possibility of torsion in $E_1(K)$, but this is ruled out by [**146**, Theorem IV.6.1] and the fact that the valuation of l in K is at most 2. We can make no statement about the case of additive reduction since then $\tilde{E}_{ns}(k)$ always has l-torsion. □

The last ingredient of the proof of Theorem 10.14 is some additional analysis of l-torsion in the case of good reduction in characteristic l. Note that if $K/\mathbb{Q}_l$ is a quadratic extension, then the residue field k is either $\mathbb{F}_l$ or $\mathbb{F}_{l^2}$.

Consider first the case that $k = \mathbb{F}_l$. Then by the Riemann hypothesis for elliptic curves over finite fields (see [**146**, Theorem V.1.1]) we know that
$$-2\sqrt{l} \leq \#E(\mathbb{F}_l) - l - 1 \leq 2\sqrt{l}.$$

It follows easily that for $l \geq 7$ the only way to have l divide $\#E(\mathbb{F}_l)$ is to have $\#E(\mathbb{F}_l) = l$.

Now consider the case $k = \mathbb{F}_{l^2}$. This time the Riemann hypothesis shows that the only way to have l divide $\#E(\mathbb{F}_{l^2})$ is to have
$$\#E(\mathbb{F}_{l^2}) \in \{l^2 - l, l^2, l^2 + l, l^2 + 2l\}.$$
Let α, β be the eigenvalues of Frobenius at l acting on the p-adic Tate module of E for some $p \neq l$; we have $\alpha\beta = l$. Then by [**146**, Section V.2],
$$\#E(\mathbb{F}_l) = 1 + l - \alpha - \beta$$
$$\#E(\mathbb{F}_{l^2}) = 1 + l^2 - \alpha^2 - \beta^2.$$
Since $\alpha\beta = l$, we have
$$\alpha^2 + 2\alpha\beta + \beta^2 = \alpha^2 + 2l + \beta^2,$$
and we conclude that
$$|l + 1 - \#E(\mathbb{F}_l)| = \sqrt{1 + 2l + l^2 - \#E(\mathbb{F}_{l^2})}.$$

This equation has several consequences. First, suppose that $\#E(\mathbb{F}_{l^2}) = l^2 - l$. Then $3l + 1$ must be a perfect square, say n^2. Thus $3l = n^2 - 1$, which implies that $l = 5$. Similarly, the case $\#E(\mathbb{F}_{l^2}) = l^2$ can not occur, and if $\#E(\mathbb{F}_{l^2}) = l^2 + l$, then $l = 3$. If $\#E(\mathbb{F}_{l^2}) = l^2 + 2l$, then we find that
$$\#E(\mathbb{F}_l) \in \{l, l + 2\}.$$

We now state and prove a more precise version of the unresolved part of Theorem 10.14.

Theorem 10.18. *Let E be an elliptic curve over $\mathbb{Q}$ and let S_0 be the set of places of $\mathbb{Q}$ at which E has bad reduction. Let l be a prime such that*
- *$l \geq 13$;*
- *l does not divide $p - 1$ for any $p \in S_0$;*
- *l does not divide $\#E(\mathbb{F}_p)$ or $\#E(\mathbb{F}_{p^2})$ for any $p \in S_0$;*
- *E has good reduction at l;*
- *$\#E(\mathbb{F}_l)$ is not l or $l + 2$.*

Then $\mathrm{H}^0(\mathbb{Q}_p, E[l] \otimes E[l]) = 0$ for all $p \in S_0 \cup \{l\}$. In particular, the set of such l has density 1 in the set of all primes.

Proof. The second condition insures that μ_l has no $\mathbb{Q}_p$-rational points for any $p \in S_0$. To show that $\mathrm{Sym}^2 E[l]$ has no $\mathbb{Q}_p$-rational points for $p \in S_0$, we must (by Lemma 10.15) show that $E(K)$ has no non-trivial l-torsion for any quadratic extension of $\mathbb{Q}_r$. This possibility is ruled out by the first three conditions and Proposition 10.16. Note that we do need to consider the case of good reduction here, as even though E has bad reduction over $\mathbb{Q}_p$, it may attain good reduction over K.

To show that $\mathrm{H}^0(\mathbb{Q}_l, E[l] \otimes E[l]) = 0$, note first that μ_l has no $\mathbb{Q}_l$-rational points, so we must only consider $\mathrm{Sym}^2 E[l]$. By Proposition 10.17 and the first and fourth hypotheses, it suffices to show that l does not divide $\#E(\mathbb{F}_l)$ or $\#E(\mathbb{F}_{l^2})$, and this follows from the preceding discussion and the fifth hypothesis.

It remains to show that the set of such l has density 1. It is clear that the first four conditions eliminate only finitely many primes l. It is shown as a very special case of [**136**, Theorem 20] that the fifth condition is satisfied for a set of primes of density 1. This completes the proof. □

Appendix B: The definition of the Flach map

In this section we give the formal definition of the Flach map. For conceptual clarity we will work in a more general setting. Let X be a nonsingular projective variety of dimension n, defined over $\mathbb{Q}$.

Let X^p denote the set of irreducible subschemes of X of codimension p. Quillen has constructed a spectral sequence from the filtration by codimension of support:
$$E_1^{pq} = \bigoplus_{x \in X^p} K_{-p-q} k(x) \Rightarrow K_{-p-q}(X);$$
here $k(x)$ is the function field of the scheme x and the $K_i k(x)$ are Quillen's K-groups. There is an analogous spectral sequence in étale cohomology:
$$(E_1^{pq})'(\mathcal{F}) = \bigoplus_{x \in X^p} \mathrm{H}_{\mathrm{\acute{e}t}}^{q-p}(\mathrm{Spec}\, k(x), \mathcal{F}(-p)) \Rightarrow \mathrm{H}^{p+q}(X, \mathcal{F})$$
where $\mathcal{F}$ is some Tate twist of the constant étale sheaf $\mathbb{Z}_l$. For any integer i, these spectral sequence are connected by Chern class maps
$$E_r^{pq} \to (E_r^{p,q+2i})'(\mathbb{Z}_l(i))$$
constructed by Gillet.

Now fix an integer m between 0 and n and assume
- $\mathrm{H}_{\mathrm{\acute{e}t}}^{2m+1}(X_{\bar{\mathbb{Q}}}, \mathbb{Z}_l(m+1))$ has no $G_{\mathbb{Q}}$-invariants.

This is implied by the Weil conjectures if this cohomology group is torsion free, as is the case when X is a curve or a product of curves. We define the Flach map
$$\sigma_m : E_2^{m,-m-1} \to \mathrm{H}^1(\mathbb{Q}, \mathrm{H}_{\mathrm{\acute{e}t}}^{2m}(X_{\bar{\mathbb{Q}}}, \mathbb{Z}_l(m+1)))$$
as the composition of three maps. The first is the Chern class map above with $p = m$, $q = -m-1$ and $i = m+1$:
$$E_2^{m,-m-1} \to (E_2^{m,m+1})'(\mathbb{Z}_l(m+1)).$$
The second is an edge map in the étale cohomology spectral sequence above:
$$(E_2^{m,m+1})'(\mathbb{Z}_l(m+1)) \to \mathrm{H}_{\mathrm{\acute{e}t}}^{2m+1}(X, \mathbb{Z}_l(m+1)).$$
(To see that there really is an edge map from this term, one uses the fact that terms of this spectral sequence below the diagonal always vanish, as is clear from the expression above.) This last group appears in the Hochschild-Serre spectral sequence
$$\mathrm{H}^p(\mathbb{Q}, \mathrm{H}_{\mathrm{\acute{e}t}}^q(X_{\bar{\mathbb{Q}}}, \mathbb{Z}_l(m+1))) \Rightarrow \mathrm{H}_{\mathrm{\acute{e}t}}^{p+q}(X, \mathbb{Z}_l(m+1)).$$
Our assumption above that $\mathrm{H}^0(\mathbb{Q}, \mathrm{H}_{\mathrm{\acute{e}t}}^{2m+1}(X_{\bar{\mathbb{Q}}}, \mathbb{Z}_l(m+1))) = 0$ insures that there is an edge map
$$\mathrm{H}_{\mathrm{\acute{e}t}}^{2m+1}(X, \mathbb{Z}_l(m+1)) \to \mathrm{H}^1(\mathbb{Q}, \mathrm{H}^{2m}(X_{\bar{\mathbb{Q}}}, \mathbb{Z}_l(m+1)))$$
and gives the last map in the definition of σ_m.

The map considered in the text is a slight variant of this. Take $X = E \times E$ and $m = 1$, so that we have a map
$$\sigma_1 : E_2^{1,-2} \to \mathrm{H}^1(\mathbb{Q}, \mathrm{H}_{\mathrm{\acute{e}t}}^2(E_{\bar{\mathbb{Q}}} \times E_{\bar{\mathbb{Q}}}, \mathbb{Z}_l(2))).$$

We now show how to manipulate these terms to obtain the map

$$\sigma : \mathcal{C}(E \times E) \to \mathrm{H}^1(\mathbb{Q}, \mathrm{Sym}^2 T_l E)$$

of the text. Working from the expression above for the Quillen spectral sequence, we see that $E_2^{1,-2}$ is the cohomology of a sequence

$$K_2 k(E \times E) \to \bigoplus_{x \in (E \times E)^1} k(x)^* \to \bigoplus_{y \in (E \times E)^2} \mathbb{Z}.$$

Quillen computes that the second map is just the divisor map sending a term $f \in k(x)^*$ to $\oplus_{y \in x} m_y$, where m_y is the order of f at y. The kernel of this map is precisely the group $\mathcal{C}(E \times E)$; σ_1 is defined on the quotient of this group by the image of $K_2 k(E \times E)$, so we can also regard it as defined on $\mathcal{C}(E \times E)$ itself. This takes care of the domain.

Next, the Kunneth theorem implies that $\mathrm{H}^2_{\mathrm{\acute{e}t}}(E_{\bar{\mathbb{Q}}} \times E_{\bar{\mathbb{Q}}}, \mathbb{Z}_l(2))$ is torsion free and that there is a projection

$$\mathrm{H}^2_{\mathrm{\acute{e}t}}(E_{\bar{\mathbb{Q}}} \times E_{\bar{\mathbb{Q}}}, \mathbb{Z}_l(2)) \to \mathrm{H}^1_{\mathrm{\acute{e}t}}(E_{\bar{\mathbb{Q}}}, \mathbb{Z}_l(1)) \otimes_{\mathbb{Z}_l} \mathrm{H}^1_{\mathrm{\acute{e}t}}(E_{\bar{\mathbb{Q}}}, \mathbb{Z}_l(1)).$$

The Kummer sequence naturally identifies $\mathrm{H}^1_{\mathrm{\acute{e}t}}(E_{\bar{\mathbb{Q}}}, \mathbb{Z}_l(1))$ with the l-adic Tate module $T_l E$, so projecting onto the symmetric direct summand yields a map

$$\mathrm{H}^2_{\mathrm{\acute{e}t}}(E_{\bar{\mathbb{Q}}} \times E_{\bar{\mathbb{Q}}}, \mathbb{Z}_l(2)) \to \mathrm{Sym}^2 T_l E$$

which is easily used to finish the definition of the map σ.

Returning to the general case, let us now investigate the local behavior of σ_m. Let p be a prime different from l at which X has good reduction (meaning that $X_{\mathbb{Q}_p}$ is the generic fiber of a proper smooth $\mathrm{Spec}\,\mathbb{Z}_p$-scheme $\mathfrak{X}$) and make the additional assumption:

- $\mathrm{H}^{2m+1}_{\mathrm{\acute{e}t}}(X_{\bar{\mathbb{Q}}_p}, \mathbb{Z}_l(m+1))$ has no $G_{\mathbb{Q}_p}$-invariants;

Let $T = \mathrm{H}^{2m}(X_{\bar{\mathbb{Q}}}, \mathbb{Z}_l(m+1))$. Then there is a commutative diagram

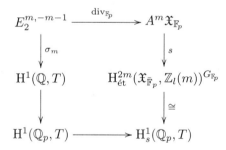

Here $A^m \mathfrak{X}_{\mathbb{F}_p}$ is the codimension m Chow group of $\mathfrak{X}_{\mathbb{F}_p}$, which is just the analogue of the Picard group in higher codimension; $\mathrm{div}_{\mathbb{F}_p}$ sends a pair (x,f) of a cycle and a rational function to its divisor in characteristic p; s is the usual cycle class map in étale cohomology; and the bottom right isomorphism is the natural analogue of the isomorphism of Lemma 10.9, using the smooth base change theorem to identify $\mathrm{H}^{2m}_{\mathrm{\acute{e}t}}(\mathfrak{X}_{\bar{\mathbb{F}}_p}, \mathbb{Z}_l(m))$ with $T(-1)$. The diagram of the text follows immediately from this one.

Flach defines the map σ (in the case $X = E \times E$, $m = 1$) using a related method in [50]. There he proves the commutativity of the above local diagram (in a slightly different form) through explicit computations. In [51] he gives the construction of σ (this time in the case $X = X_0(N) \times X_0(N)$, $m = 1$) we gave above and writes down the local diagram, although his proof of commutativity is somewhat incomplete and does not immediately generalize. The general case, which relies heavily on purity conjectures of Grothendieck (which have been proven in the relevant cases by Raskind and Thomason), is the subject of [**158**, Chapters 6 and 7]. The construction of maps similar to σ also appear in the work of Kato; see [**4**] and [**127**]. Mazur offers an alternative construction of the Flach map in [**99**], without any explicit dependence on K-theory. There he also studies some algebraic properties of the map which are not immediately apparent and which permit some Euler system type conclusions even without the existence of an Euler system.

Tom Weston
Department of Mathematics
Harvard University
Cambridge, MA 02138
weston@math.harvard.edu

APPENDIX 3
An introduction to the p-adic geometry of modular curves
by Matthew Emerton

The aim of this appendix is to give some feeling for the ideas behind the p-adic theory of modular curves and modular forms (at a more down to earth level than that of [**57**] or [**79**], which are the standard references on this topic). It provides a written approximation to an informal lecture given on this topic by the author at the 1999 PCMI meeting. It was in the informal spirit of the lecture to make many assertions and allusions without providing details. That spirit may also permeate this written version, but some attempt has been made to give relevant references to the literature. In fact, although our coverage of the bibliography is (of course and inevitably) incomplete, we do mention many of the major papers in the field, and the final section is devoted to annotating briefly some of those references.

As one might guess, the background and sophistication that is assumed of the reader varies widely from section to section. A large part of the presentation is devoted to treating the case $p = 2$, where many phenomena can be discovered and investigated via explicit calculation; in these parts of the presentation we assume comparatively little background. On the other hand, in order to understand the theory behind the calculations, and the generalization to arbitrary primes p, more sophisticated ideas (such as the consideration of elliptic curves and modular forms defined over rings other than fields, as well as the techniques of formal and rigid analytic geometry) cannot be avoided. Nevertheless, we have attempted to make the discussion as clear and intuitive as possible; we leave it to the reader and the references to fortify the intuition with correct mathematics, while the author will accept responsibility for any lack of clarity.

Acknowledgments: I would like to thank David Ben-Zvi, Robert Coleman, Brian Conrad and Mike Roth for their helpful comments on earlier drafts of the present work, as well as Fernando Gouvêa for all his assistance in the TEXnical preparation of this appendix.

The curve $X_0(2)$

All the essential ideas in the p-adic theory can already be seen in the case $p = 2$, and this case has the merit that one can easily perform explicit computations. We begin by describing some of these, working first over $\mathbb{C}$.

We let $\mathcal{H}^*$ denote the *extended upper half-plane*, that is, the union of the usual complex upper half-plane with the set $\mathbb{Q} \cup \{\infty\}$. The group $\mathrm{SL}_2(\mathbb{Z})$ acts naturally on $\mathcal{H}^*$ by linear fractional transformations; the quotient is the *modular curve of level one*, which (a little idiosyncratically, but for reasons of consistency) we will denote by $X_0(1)$. It is well known that $\mathrm{SL}_2(\mathbb{Z})$ acts transitively on $\mathbb{Q} \cup \{\infty\}$, and we denote the corresponding point of $X_0(1)$ simply by ∞, and refer to it as the *cusp* of $X_0(1)$ (although of course it is not a singular point; the usage of this term is purely for historical reasons). Let us also write $Y_0(1) = X_0(1) \setminus \{\infty\}$. It is well known that the points of $Y_0(1)$ are in a one-to-one correspondence with the isomorphism classes of elliptic curves over $\mathbb{C}$; the orbit of a point τ in the upper half-plane corresponds to the elliptic curve $E_\tau := \mathbb{C}/\Lambda_\tau$, where Λ_τ is the lattice $2\pi i \tau \mathbb{Z} + 2\pi i \mathbb{Z}$. (The $2\pi i$ factors are included just for the purposes of having a good normalization when it comes to comparing the analytic and algebraic theory of elliptic curves.)

From the algebraic theory of elliptic curves, we know that elliptic curves over any field are also classified up to isomorphism by their j-invariant. If we write $j(\tau) := j(E_\tau)$, then the function $j(\tau)$ is an analytic function on the open subset $Y_0(1)$, which induces an analytic isomorphism $Y_0(1) \xrightarrow{\sim} \mathbb{A}^1$. (This is essentially tautological if one has a sufficient understanding of $X_0(1)$ as a moduli space; alternatively, see [**133**] for a down-to-earth account of j as an analytic function on the upper half-plane and $Y_0(1)$). The function $j(\tau)$ has a simple pole at ∞, and thus this isomorphism extends to an isomorphism

$$X_0(1) \xrightarrow{\sim} \mathbb{P}^1,$$

and by virtue of this isomorphism we will also regard j as a coordinate on $X_0(1)$, and refer to $X_0(1)$ as the j-line.

The curve $X_0(2)$ is obtained by taking the quotient of $\mathcal{H}^*$ by the congruence subgroup $\Gamma_0(2)$ of $\mathrm{SL}_2(\mathbb{Z})$. There are two orbits of $\Gamma_0(2)$ on $\mathbb{Q} \cup \{\infty\}$, the orbit of the point 0 and that of the point ∞. We refer to these points as the *cusps* of $X_0(2)$, and we label them simply by 0 and ∞. We also write $Y_0(2) = X_0(2) \setminus \{0, \infty\}$.

Of course the points $Y_0(2)$ also have an interpretation in terms of moduli of elliptic curves. Namely, given a point τ in the upper half-plane, we can form the curves E_τ and $E_{2\tau}$, and observe that there is an isogeny between them:

(I) $$E_\tau \longrightarrow E_{2\tau}$$

via the map $z \bmod \Lambda_\tau \mapsto 2z \bmod \Lambda_{2\tau}$. The kernel of this isogeny has order two, and is generated by the point $\pi i \bmod \Lambda_\tau$. One easily checks that if we modify τ by an element of $\Gamma_0(2)$ then the isomorphism class of this two-isogeny remains unchanged, and that in fact the points of $Y_0(2)$ are in bijection with the isomorphism classes of two-isogenies of elliptic curves over $\mathbb{C}$.

We were able to explicitly describe $X_0(1)$ as an algebraic curve: it is the j-line. Is there a similar description of $X_0(2)$? Here is a first attempt: since the orbit of a point τ under $\Gamma_0(2)$ determines the two-isogeny (I) up to isomorphism, we certainly know its source and target up to isomorphism, and so to the orbit $\tau \bmod \Gamma_0(2)$ in $Y_0(2)$ we can associate the two complex numbers $j(\tau)$ and $j(2\tau)$, which do classify the isomorphism class of E_τ and $E_{2\tau}$. Thus we get a map

(II) $$Y_0(2) \longrightarrow Y_0(1) \times Y_0(1) \xrightarrow{\sim} \mathbb{A}^2.$$

Since the source of this map is one-dimensional over $\mathbb{C}$, its image in $Y_0(1) \times Y_0(1)$ must be a curve, whose equation will express a (somewhat complicated, as it turns out) relation between $j(\tau)$ and $j(2\tau)$. Such relations are classically called *modular equations*. From our point of view this modular equation has two disadvantages: firstly, it is a little hard to compute explicitly; secondly, the map (II) is not an isomorphism onto its image – it turns out that the image has singularities. What this means in modular terms is that for certain special choices of τ, there is one or more two-isogeny between E_τ and $E_{2\tau}$ which is not isomorphic to the isogeny (I).

There is an alternative approach to describing $X_0(2)$ as an algebraic curve, which we now present. We first proceed algebraically: thus we assume given a two isogeny $\psi : E_1 \longrightarrow E_2$ between two elliptic curves over $\mathbb{C}$. Let ω_2 be a regular differential on E_2, and let $\omega_1 := \psi^*\omega_2$ be the pulled-back regular differential on E_1.

Recall the algebraic definition of modular forms (of level one): a modular form f of weight k is a "rule" which to any pair consisting of an elliptic curve E and a non-zero regular differential ω on E attaches a number $f(E,\omega)$ depending only on the isomorphism class of the pair (E,ω), such that for any non-zero scalar λ, $f(E,\lambda\omega) = \lambda^{-k}f(E,\omega)$, and which "behaves well in families". (We won't make precise the meaning of this statement here; see [**38**] or [**79**].)

There is a canonical modular form of weight 12, the discriminant Δ (see [**38**]). Returning to our two isogeny $\psi : E_1 \longrightarrow E_2$, we define

$$j_2(\psi) = 2^{12} \frac{\Delta(E_1,\omega_1)}{\Delta(E_2,\omega_2)}.$$

Note that if we multiply ω_2 by a non-zero scalar λ, both the numerator and denominator of $j_2(\psi)$ are scaled by the same amount λ^{-12}, and so $j_2(\psi)$ remains invariant. Thus it really does depend only on the isogeny ψ, and not on the auxiliary choice of ω_2.

Now let us interpret this in the analytic picture: on $E_\tau = \mathbb{C}/\Lambda_\tau$ there is a canonical differential ω_τ obtained by reducing the differential dz on $\mathbb{C}$ modulo Λ. (This reduction is possible because the differential dz is invariant under translation by elements of the lattice Λ_τ.) The function

$$\Delta(\tau) := \Delta(E_\tau,\omega_\tau)$$

is a modular form (in the classical analytic sense explained in [**133**] for example) of weight twelve and level one – in fact, it is the unique normalized *cuspform* of this weight and level. Its q-expansion is given by the famous formula

$$\Delta(\tau) = q \prod_{n=1}^{\infty} (1-q^n)^{24}.$$

(Recall that in the context of modular forms, q denotes the exponential $q = e^{2\pi i \tau}$. The product formula for the q-expansion of Δ is proved in [**133**], for example.) If we let ψ_τ denote the isogeny (I), then we see that $\psi_\tau^*\omega_{2\tau} = 2\omega_\tau$, and so

(III) $$j_2(\psi_\tau) = 2^{12} \frac{\Delta(E_\tau, 2\omega_\tau)}{\Delta(E_{2\tau},\omega_{2\tau})} = \frac{\Delta(E_\tau,\omega_\tau)}{\Delta(E_{2\tau},\omega_{2\tau})}.$$

We define the function j_2 on $Y_0(2)$ via the formula

$$j_2(\tau) = j_2(\psi_\tau) \stackrel{\text{(III)}}{=} \frac{\Delta(\tau)}{\Delta(2\tau)} = \frac{q \prod_{n=1}^\infty (1-q^n)^{24}}{q^2 \prod_{n=1}^\infty (1-q^{2n})^{24}} = q^{-1} \prod_{n=1}^\infty (1+q^n)^{-24}.$$

This is an analytic function $Y_0(2) \longrightarrow \mathbb{A}^1$, and from its q-expansion (and the easily verified fact that $q = e^{2\pi i \tau}$ is a uniformizer in the neighbourhood of the point ∞ of $X_0(2)$) we see that is has a simple pole at the point ∞ of $X_0(2)$. Thus j_2 *necessarily* extends to an *isomorphism* $X_0(2) \xrightarrow{\sim} \mathbb{P}^1$. Because of this, we will also refer to $X_0(2)$ as the j_2-line.

Note that we conclude incidentally that any two-isogeny ψ is determined up to isomorphism by the invariant $j_2(\psi)$. One could also establish this fact (with more difficulty) by pure algebra, rather than resorting to the consideration of modular curves, as we did above. However, one motivation for introducing the modular curves (or moduli spaces in general) is to simplify the proof of results such as this.

Now that we have our algebraic description of $X_0(2)$ as the j_2-line, it will be interesting to return to and reinterpret slightly the map (II). The first coordinate of this map extends to a map $X_0(2) \longrightarrow X_0(1)$, which is the natural map arising from the inclusion of $\Gamma_0(2)$ in $\mathrm{SL}_2(\mathbb{Z})$. We denote this map by B_1. The second coordinate of (II) also extends to a map $X_0(2) \longrightarrow X_0(1)$, which we denote by B_2. There is also a natural automorphism of $X_0(2)$, given by the construction of dual isogenies: if $\psi: E_1 \longrightarrow E_2$ is a two-isogeny, it has a dual two-isogeny $\check{\psi}: E_2 \longrightarrow E_1$. If ψ is the isogeny ψ_τ of (I) then $\check{\psi}_\tau$ is the isogeny $E_{2\tau} \longrightarrow E_\tau$ arising from the inclusion of lattices $\Lambda_{2\tau} \subset \Lambda_\tau$. This is isomorphic to the two-isogeny $\psi_{-1/2\tau}$. Thus we see that constructing dual isogenies yields an automorphism of $Y_0(2)$ given by the formula

$$\tau \bmod \Gamma_0(2) \mapsto \frac{-1}{2\tau} \bmod \Gamma_0(2).$$

This extends to an automorphism of $X_0(2)$ which interchanges 0 and ∞. We denote this automorphism by w_2; it is an *involution* (has order two), and is often referred to as the Atkin-Lehner involution. The map B_2 is easily described via w_2: since w_2 interchanges the source and target of a two-isogeny, we see that $B_2 = B_1 \circ w_2$.

How do we describe w_2 in terms of the coordinate j_2 on $X_0(2)$? This is an easy computation, using the modularity properties of Δ; one could make it either algebraically or analytically, but the latter is probably simpler:

$$j_2(-1/2\tau) = \frac{\Delta(-1/2\tau)}{\Delta(-1/\tau)} = \frac{2^{12}\tau^{12}\Delta(2\tau)}{\tau^{12}\Delta(\tau)} = \frac{2^{12}}{j_2(\tau)}.$$

Note that from this formula we also find the value of j_2 at the cusp 0: $j_2(0) = j_2(w_2(\infty)) = 2^{12}/j_2(\infty) = 0$, since j_2 has a pole at ∞.

Here is another question: how do we describe the function B_1 in terms of the coordinates j and j_2?

Lemma 11.1. *The map $B_1 : X_0(2) \longrightarrow X_0(1)$ is described by the equation*

(IV) $$j \circ B_1 = (j_2 + 256)^3/j_2^2.$$

It is ramified at the point $512 = j_2(i)$ over the point $1728 = j(i)$ with degree two, at the point $-256 = j_2((-1+\sqrt{-3})/2)$ over the point $0 = j((-1+\sqrt{-3})/2)$ with degree three, and at the point $0 = j_2(0)$ over the point $\infty = j(\infty)$ with degree two.

Proof. The map from the upper half-plane to $Y_0(1)$ is ramified over two points: the point $j = 1728$, corresponding to the $\mathrm{SL}_2(\mathbb{Z})$ orbit of $z = i$, with ramification degree two, and the point $j = 0$, corresponding to the $\mathrm{SL}_2(\mathbb{Z})$ orbit of $z = (-1 + \sqrt{-3})/2$, with ramification degree three.

The map from the upper half-plane to $Y_0(2)$ is ramified over one point, corresponding to the $\Gamma_0(2)$ orbit of $z = (1+i)/2$, with ramification degree two.

Now $\Gamma_0(2)$ has index three in $\mathrm{SL}_2(\mathbb{Z})$, and so the map B_1 has degree three, and from the preceding two paragraphs we see that the resulting degree three map $Y_0(2) \longrightarrow Y_0(1)$ is ramified at the $\Gamma_0(2)$ orbit of $z = i$, with ramification degree two, and is totally ramified over the point $j = 0$.

Of the two cusps ∞ and 0 on $X_0(2)$, the cusp ∞ is unramified over the cusp ∞ on $X_0(1)$ with respect to B_1 (we saw this implicitly above, when we remarked that q is a uniformizer in a neighbourhood of ∞ on $X_0(2)$, since it is also a uniformizer in a neighbourhood of ∞ on $X_0(1)$), and hence the cusp 0 must be ramified of degree two over the cusp ∞ on $X_0(1)$.

Now pull back j via B_1 to a function on $X_0(2)$. Since we know the zeroes and poles of j on $X_0(1)$, as well as the ramification structure of B_1 above each of these points, we see that $j \circ B_1$ has a pole of order one at ∞, a pole of order two at 0, a zero of order three at $j_2((-1+\sqrt{-3})/2)$, and no other zeroes or poles. Thus $j \circ B_1$ is a scalar multiple of $(j_2 - a)^3/j_2^2$, where $a = j_2((-1+\sqrt{-3})/2)$. One computes that
$$(j_2 - a)^3/j_2^2 = 1/q - (24 + 3a) + \cdots,$$
while
$$j = 1/q + 744 + \cdots$$
(see [**133**]). Comparing these expressions we find that $a = -256$ and hence that
$$j \circ B_1 = (j_2 + 256)^3/j_2^2.$$
From this equation we see that $-256 = j_2((-1+\sqrt{-3})/2)$ is the point lying over $0 = j((-1+\sqrt{-3})/2)$, and also that
$$j \circ B_1 - 1728 = (j_2 + 64)(j_2 - 512)^2/j_2^2,$$
and thus, from the above description of the ramification of B_1, that $j_2(i) = 512$ and $j_2((1+i)/2) = -64$. We have now verified the asserted formula for the map B_1 and also the claims about its ramification. $\square$

The equation $B_2 = B_1 \circ w_2$ allows us to compute that

(V) $\quad j \circ B_2 = ((j_2 + 256)^3/j_2^2) \circ w_2 = (2^{12}/j_2 + 256)^3/(2^{12}/j_2)^2 = (j_2 + 16)^3/j_2.$

With sufficient enthusiasm, one can eliminate j_2 from the two equations (IV) and (V) and thus find the equation relating $j \circ B_1$ and $j \circ B_2$; this will be the modular equation which describes the image of the map (II). Classically, one of the roles of higher level modular curves such as $X_0(2)$ (or more precisely, the modular functions such as j_2 that are defined on them) was to simplify the shape of the modular equations. In modern terms, we can see this as being related to the simplification that comes from replacing a singular curve by its normalization.

Canonical subgroups of elliptic curves over $\mathbb{C}_2$

The Lefschetz principle assures us that algebraic geometry is the same when studied over any algebraically closed field of characteristic zero. In this section we will take our ground field to be the completion of the algebraic closure of the field of 2-adic numbers $\mathbb{Q}_2$. This field (which is algebraically closed as well as complete) is denoted $\mathbb{C}_2$; it provides a natural location for conducting 2-adic analysis, just as $\mathbb{C}$ provides a natural location for performing classical analysis. In particular, $\mathbb{C}_2$ is equipped with a non-archimedean absolute value, which we denote by $|\ |$, and which we normalize by the condition that $|2| = 1/2$ (although the normalization won't be important).

By the Lefschetz principle, all the algebro-geometric observations about moduli of elliptic curves over $\mathbb{C}$ which we made in the preceding section apply equally well to elliptic curves over $\mathbb{C}_2$. Thus we see that there is j-line whose non-cuspidal points classify isomorphism classes of elliptic curves over $\mathbb{C}_2$, as well as a j_2-line, whose points (other than ∞ and 0) classify isomorphism classes of two-isogenies between elliptic curves over $\mathbb{C}_2$. There is an involution w_2 of the j_2-line given by passing to the dual isogeny, there are two maps B_1 and B_2 mapping the j_2-line to the j-line, and all the formulas for w_2, B_1 and B_2 that we proved in the preceding section continue to hold true.

Let us consider the formula (IV) for B_1. We can rewrite this in the form

$$\frac{256}{j \circ B_1} = \frac{256/j_2}{(1 + 256/j_2)^2}.$$

Now if $|j_2| > |256|$, then we may expand the right-hand side of this equation in a power series, and so obtain the equation

$$\frac{256}{j \circ B_1} = \sum_{n=1}^{\infty} (-1)^{n-1} n \left(\frac{256}{j_2}\right)^n.$$

Since the leading coefficient of this series is 1 (and so in particular is non-zero) we see by the implicit function theorem (which in this non-archimedean context is just a formal manipulation of power-series) that B_1 establishes an *isomorphism* between the region $|j_2| > |256|$ on the j_2-line and the region $|j| > |256|$ on the j-line. Each of these regions is a disk centred at the point ∞ on the appropriate line; let us denote them by D_2 and D_1 respectively. Thus we see that B_1 induces a 2-adic analytic isomorphism from D_2 to D_1, and so has a 2-adic analytic inverse $B_1^{-1}: D_1 \longrightarrow D_2$. So what?

Well, in modular terms, this means that if E is an elliptic curve over $\mathbb{C}_2$ whose j-invariant satisfies the inequality $|j(E)| > |256|$, then there is a naturally determined two-isogeny whose source is E. Equivalently, thinking of a two-isogeny as being determined by its kernel, we see that there is a naturally determined subgroup of E of order two. Since E has three distinct subgroups of order two, it seems pretty remarkable that there is any way to distinguish one of them as being naturally determined! (Caveat: since the point $B_1^{-1}(j(E))$ in D_2 only determines a two-isogeny up to isomorphism, the subgroup of order two is determined only up to the application of automorphisms of E. Now if E has no automorphism besides ± 1, then since these both fix any subgroup of E, we see that the isomorphism class of the isogeny does determine a well-defined subgroup of order two of E. What if E has extra automorphisms? The point $j = 0$ (whose corresponding elliptic curve has six automorphisms) is not in the disk D_1, so it remains to consider the point $j = 1728$ (whose corresponding elliptic curve has four automorphisms). Now if $j(E) = 1728$, let α denote an automorphism of E of order four. Then $\alpha - 1$ has degree two as an endomorphism of E, and so α fixes exactly one subgroup of order two of E, and interchanges the other two. Since $B_1^{-1}(1728)$ lies in D_2, we see by lemma 1.4 and its proof that $B_1^{-1}(1728) = -64$, and since this is *not* a ramification point of B_1, the kernel of the corresponding two-isogeny must be *fixed* by α. Thus we see that the caveat is no caveat at all: even taking into account possible extra automorphisms of E, we see that the map B_1^{-1} *does* determine an order two subgroup of every E with $j(E)$ lying in D_1.) This subgroup is called the *canonical subgroup* of E.

Now we consider the composite $\phi := B_2 \circ B_1^{-1} : D_1 \longrightarrow X_0(1)$, which is called the *Deligne-Tate map*. The map B_1^{-1} is described by a formula of the form $256/B_1^{-1}(j) = \sum_{n=1}^{\infty} a_n (256/j)^n$, for certain integers a_n such that $a_1 = 1$. The map B_2 is described by (V), which can be rewritten in the form

$$\frac{256}{j} \circ B_2 = \frac{(16/j_2)^2}{(1 + 16/j_2)^3}.$$

Thus we find that

$$\frac{256}{\phi(j)} = \sum_{n=2}^{\infty} b_n \left(\frac{16}{j}\right)^2,$$

for some integers b_n such that $b_1 = 1$. Actually, although ϕ is defined on all of D_1, this power-series formula only converges on the subdisk D_1' defined by the inequality $|j_2| > |16|$; it shows that the restriction of ϕ to D_1' is a degree two map whose image is D_1 and which satisfies the formula

$$|\phi(j)| = |j|^2$$

for any point j of D_1'.

What is the modular interpretation of the Deligne-Tate map? Well, if we start with an elliptic curve E such that $j(E)$ lies in D_1, then $B_1^{-1}(j)$ associates to E its canonical subgroup C. Then B_2 associates to the pair (E, C) the *target* of the isogeny whose kernel is C, which is just E/C. Thus in modular terms ϕ associates to any elliptic curve E with $j(E)$ in D_1 its quotient by its canonical subgroup. Now if $j = j(E)$ in fact lies in D_1', we saw that $\phi(j)$ lies in D_1, and so E/C also has a canonical subgroup, which pulls back to a subgroup C' of E of order four.

Lemma 11.2. *In the above notation, C' is a cyclic subgroup of E*

Proof. Either C' is a cyclic subgroup of E of order four, or it is the full two-torsion subgroup of E. In the latter case, we would see that E/C' is isomorphic to E, and thus that $E \longrightarrow E/C$ and $E/C \longrightarrow E/C'$ are dual isogenies.

On the other hand, both of these isogenies have j_2 lying in D_2 (since they are in the image of B_1^{-1}), and the first even satisfies $|j_2| > |16|$, because $|j(E)| > |16|$, and B_1^{-1} preserves absolute values. If $|j_2| > |16|$, then $|2^{12}/j_2| < |256|$, and so $2^{12}/j_2$ does not lie in D_2. Thus they *cannot* be dual isogenies, and we have proved the lemma. □

More generally, if we let $D_1^{(n)}$ denote the disk in the j-line determined by the inequality $|j| > |2^{2^{3-n}}|$, then we see that $\phi^n : D_1^{(n)} \longrightarrow D_1$, and that as a consequence if E is an elliptic curve over $\mathbb{C}_2$ for which $j(E)$ lies in $D_1^{(n)}$, then E has a canonical subgroup of order 2^{n+1}, which is cyclic (by the easily proved analogue of lemma 2.2).

The parameter q

This section and the next are a digression on some points of theory which we will need to understand the preceding calculations better. The subject of our first digression is an alternative approach to describing elliptic curves, using the parameter q rather than τ. This leads to a more theoretical interpretation of the q-expansion of a modular form, and to a modular interpretation of the cusp ∞ on the j-line.

Let us begin by considering the affine j-line $Y_0(1)$ over the complex numbers. If τ is a point in $\mathcal{H}$ its image in $Y_0(1)$ corresponds to the elliptic curve $E_\tau = \mathbb{C}/(2\pi i \tau \mathbb{Z} + 2\pi i \mathbb{Z})$. We can take this quotient in two steps, by noting that the exponential function induces an isomorphism $\mathbb{C}/2\pi i \mathbb{Z} \xrightarrow{\sim} \mathbb{C}^\times$ (where $\mathbb{C}^\times$ denotes the multiplicative group of non-zero complex numbers.) Under the exponential, the lattice $2\pi i \tau \mathbb{Z} + 2\pi i \mathbb{Z}$ has image equal to the cyclic subgroup $q^\mathbb{Z}$ of $\mathbb{C}^\times$, where $q = e^{2\pi i \tau}$. Thus we can describe E_τ as the quotient $\mathbb{C}^\times/q^\mathbb{Z}$. Thinking of $\mathbb{C}^\times$ as a cylinder of infinite length (which it is topologically), this is just the familiar description of a torus as being obtained by gluing together the ends of a finite length cylinder (where in our case, the finite length cylinder will be a fundamental domain for the action of $q^\mathbb{Z}$ on $\mathbb{C}^\times$), as in the following picture:

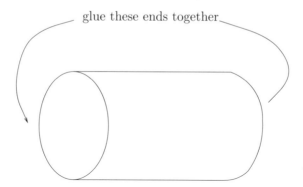

fundamental domain for multiplication by q on $\mathbb{C}^\times$

We will use this description of elliptic curves to obtain a geometric description of the q-expansions of modular forms. We begin with a discussion of differentials. Let z denote the coordinate on $\mathbb{C}$ and t the coordinate on $\mathbb{C}^\times$, so that $t = e^z$. Then the differential dz on $\mathbb{C}$ (which is invariant under the action of $2\pi i \mathbb{Z}$ by translation) descends to the differential dt/t on $\mathbb{C}^\times$, and since this differential is invariant under the action of $q^\mathbb{Z}$ by multiplication, it descends to a differential on $\mathbb{C}^\times/q^\mathbb{Z}$.

Let D^* denote the punctured open unit disk, consisting of those complex numbers q such that $1 > |q| > 0$. As q ranges over all elements of D^*, the elliptic curves $\mathbb{C}^\times/q^\mathbb{Z}$ form a family of elliptic curves lying over D^*, and the differential dt/t on $\mathbb{C}^\times$ descends to a relative differential on this family. Denote this family of elliptic curves by E_{Tate}, and this relative differential by ω_{Tate}. If f is any modular form of some weight k, then we may evaluate f on each member of the family, and thus obtain a function on D^*. More succinctly, we evaluate f on the elliptic curve E_{Tate} over the base D^* to obtain an element $f(E_{\text{Tate}}, \omega)$ of the ring of functions on D^* (necessarily holomorphic – this is a consequence of the condition that f behave well in families). But such a function is just a convergent power-series in q, and this is precisely the q-expansion of the modular form f. Thus the existence of q-expansions of modular forms is related to the existence of the family of elliptic curve E_{Tate} over D^*, equipped with its canonical relative differential ω_{Tate}.

We now turn to an analysis of the cusp ∞ in $X_0(1)$. As τ tends towards the point ∞ in the boundary of $\mathcal{H}^*$, the value of q tends to zero. Thus in the preceding description of E_τ, the circumference of the finite cylinder is remaining fixed, while its length is tending towards zero. What happens when τ reach the point ∞? One way to describe what happens is to remember that the circumferential circle of this cylinder, and the circle obtained by gluing the endpoints of a longitudinal interval, represent two independent generators of the fundamental group of E_τ. What we just observed is that as τ goes to ∞, one of these loops stays a fixed length, while the length of the other tends to zero. This makes it reasonable to think of the point $\tau = \infty$ as corresponding to the curve obtained by completely shrinking one of these loops to a point, as in the following picture:

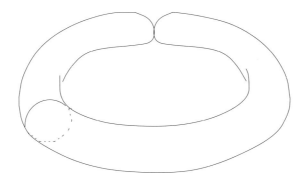

The indicated generator of the fundamental group is pinched off to zero

In short, the point ∞ on $X_0(1)$ corresponds to a rational curve with one node. In fact one can show that there is a canonical way to extend the family of elliptic curves E_{Tate} over the punctured disk D^* to a family of curves over the (unpunctured) open unit disk D whose fibre at $q = 0$ is the above rational nodal curve.

If one is working over a field K different from $\mathbb{C}$, the above analytic construction of the family E_{Tate} is not available. However, one can construct in an analogous way an elliptic curve E_{Tate} over the field $K((q))$ (which one thinks of as the ring of functions on the *formal punctured disk*), equipped with a canonical nowhere-zero differential ω_{Tate}. Furthermore, the j-invariant of E_{Tate} is an element of $K((q))$ given precisely by the usual formula for j as a function of q [**38, 79**]. Letting $\mathbb{G}_m$ denote the multiplicative group over $K((q))$, one can even describe E_{Tate} as a quotient $\mathbb{G}_m/q^{\mathbb{Z}}$, in a suitable geometric sense, and (if t denotes the coordinate on $\mathbb{G}_m$) ω_{Tate} is obtained by descending the differential dt/t on $\mathbb{G}_m$. Finally, the elliptic curve E_{Tate} extends to a curve over the ring $K[[q]]$, whose fibre over $q = 0$ is a nodal curve. (See [**39**] for the rather technical details of the construction of E_{Tate} over $K[[q]]$ and even its construction over $\mathbb{Z}[[q]]$. With regard to this, note that it is quite important that the Tate curve over $K[[q]]$ for any field K is obtained from a given elliptic curve defined over $\mathbb{Z}[[q]]$, and also that, when $K = \mathbb{C}$, the Tate curve over the formal power-series ring $\mathbb{C}[[q]]$ is 'the same' as the analytic Tate curve over the open unit disk D described in the preceding paragraph; that is, the power-series in $\mathbb{Z}[[q]]$ which appear in the Weierstrass equation for the Tate curve converge on the open unit disc D, and so describe an analytic family of elliptic curves over D, which is the analytic Tate curve described above. These two facts allow one to use analytic reasoning over $\mathbb{C}$ to draw conclusions about the Tate curve which can then be applied in other situations, for example in the p-adic analytic context that we will consider below.)

One can unify the treatment of elliptic curves and singular curves of the type that correspond to the cusp $j = \infty$ by introducing the notion of *generalized elliptic curve* [**39**]. Since such a uniform treatment is possible, at various points in the following text the conceptual distinction between elliptic curves and rational nodal curves will become a little blurred.

If f is any modular form over K then one *defines* the q-expansion of f to be the element of $K((q))$ obtained by evaluating f on the pair $(E_{\text{Tate}}, \omega_{\text{Tate}})$ [**38**, **79**]. The q-expansion principle asserts that if f and g are two modular forms of the same weight having the same q-expansion, then they are equal. This is essentially a corollary of the fact that any function on a Zariski open subset of $Y_0(1)$ is determined by its resriction to the formal punctured disk around the cusp $j = \infty$. (More algebraically, and perhaps more precisely, any localization of $K[j]$ injects into the fraction field $K(j)$, and thus into $K((q))$; the point is that $K[j]$ is an integral domain, for any K, or scheme-theoretically speaking, that $Y_0(1)$ is geometrically irreducible and reduced.) (See [**79**] for the details.) The usual requirement of the q-expansion of a modular form, that it lie in $k[[q]]$ rather than just $k((q))$, in the light of this section corresponds to the fact that modular forms extend from functions on elliptic curves to functions on generalized elliptic curves (and so can be evaluated on the fibre of E_{Tate} over the point $q = 0$, for example) [**39**].

The Hasse invariant

In this section we recall briefly the theory of the Hasse invariant of elliptic curves in positive characteristic. We will need to treat elliptic curves defined over rings other than fields. Thus we let p be a prime and R a ring of characteristic p.

Let E be an elliptic curve over R. Recall that there is canonically associated to E another elliptic curve $E^{(p)}$ and a canonical isogeny $\pi : E \longrightarrow E^{(p)}$ of degree p, the *Frobenius*. Furthermore, if ω is a nowhere-zero regular differential on E, then it induces canonically a nowhere-zero regular differential $\omega^{(p)}$ on $E^{(p)}$.

The easiest way to explain these constructions is to note that given E and ω we can find (at least locally over R) a Weierstrass equation

$$y^2 + a_1 xy + a_3 y = x^3 + a_2 x^2 + a_4 x + a_6$$

for E such that $\omega = dx/(2y + a_1 x + a_3)$. Then $E^{(p)}$ is given by the Weierstrass equation

$$y^2 + a_1^p xy + a_3^p y = x^3 + a_2^p x^2 + a_4^p x + a_6^p,$$

the differential $\omega^{(p)}$ is equal to $dx/(2y + a_1^p x + a_3^p)$, and the isogeny π is the map $(x, y) \mapsto (x^p, y^p)$.

Let us suppose that E and ω are given as above, and let $\check{\pi}$ be the dual isogeny to π. Then we can construct *two* differentials on $E^{(p)}$: the nowhere-zero regular differential $\omega^{(p)}$, and also the pulled-back differential $(\check{\pi})^*\omega$. Define $A(E, \omega)$ to be that element of R such that

$$(\check{\pi})^*\omega = A(E, \omega)\omega^{(p)}.$$

If we multiply ω by a unit λ of R, then we see that $(\lambda\omega)^{(p)} = \lambda^p \omega^{(p)}$, and consequently $A(E, \lambda\omega) = \lambda^{1-p} A(E, \omega)$. Thus A is a modular form of weight $p-1$ in characteristic p.

Now suppose that R is a field of characteristic p. In this case, although the value of $A(E, \omega)$ depends on the choice of both E *and* ω, the property of $A(E, \omega)$ being zero or non-zero depends only on E, independent of the choice of ω. If it is non-zero, we say that E is *ordinary*; if it is zero, we say that E is *supersingular*. It is relatively easy to determine the supersingular j-invariants modulo any prime p. There are only finitely many isomorphisms classes of such curves; in fact their j-invariants all lie in the finite field $\mathbb{F}_{p^2}$. Furthermore, there is a precise formula for their number, as well as a precise formula for a polynomial of which their j-invariants are the solutions (see [**74**]).

The remainder of this section is devoted to calculating the q-expansion of the Hasse invariant. Recall from the preceding section that the Tate curve E_{Tate} in characteristic p is the quotient (in an appropriate sense) of the multiplicative group $\mathbb{G}_m$ over $\mathbb{F}_p((q))$ by the cyclic subgroup $q^{\mathbb{Z}}$. Since the multiplicative group is already defined over $\mathbb{F}_p$, we see that in computing $E_{\text{Tate}}^{(p)}$ we just have to raise q to the pth power, and so $E_{\text{Tate}}^{(p)}$ is isomorphic to the quotient of $\mathbb{G}_m$ by the cyclic subgroup $q^{p\mathbb{Z}}$, the Frobenius isogeny is given by $t \bmod q^{\mathbb{Z}} \mapsto t \bmod q^{p\mathbb{Z}}$, and its dual $\check{\pi}$ is just the natural map $\mathbb{G}_m/q^{p\mathbb{Z}} \longrightarrow \mathbb{G}_m/q^{\mathbb{Z}}$. Recall that ω_{Tate} equals dt/t, so that $\omega_{\text{Tate}}^{(p)}$ is again dt/t, and also $(\check{\pi})^*\omega_{\text{Tate}} = dt/t$. We conclude from this that $A(E_{\text{Tate}}, dt/t) = 1$. In other words, the q-expansion of A is just the constant 1!

Note that this would be quite impossible in characteristic zero: the only modular forms over $\mathbb{C}$ with constant q-expansions are those of weight zero. On the other hand, it is the fact that the weight $p-1$ modular form A and the weight zero modular form 1 have the same q-expansion which gives rise to the possibility of congruences of modular forms of different weights, and the beautiful theory of p-adic families of modular forms. (See the discussion and references in the guide to the literature given below.)

Return to $X_0(2)$

In this section we try to shed some theoretical light on the calculations of section 2, to pave the way for the discussion in the case of a general prime p.

Let $\mathcal{O}$ denote the ring of integers in $\mathbb{C}_2$, that is, the ring of elements r for which $|r| \leq |1|$; geometrically, $\mathcal{O}$ is the closed unit disk of $\mathbb{C}_2$. The open unit disk of elements r such that $|r| < |1|$ is the maximal ideal $\mathfrak{m}$ of the valuation ring $\mathcal{O}$, and the quotient field k of $\mathcal{O}$ by $\mathfrak{m}$ is an algebraic closure of the field $\mathbb{F}_2$, which we denote by k.

Reducing modulo $\mathfrak{m}$ extends to a map from the projective j-line over $\mathbb{C}_2$ to the projective j-line over k, which we call the *specialization map*. This map has the following modular interpretation: if E is an elliptic curve such that $j(E)$ lies in $\mathcal{O}$ then E has a model with coefficients in $\mathcal{O}$ with good reduction modulo $\mathfrak{m}$, and the reduction of $j(E)$ modulo $\mathfrak{m}$ is just the j-invariant of the reduction of this model module $\mathfrak{m}$. On the other hand, if $j(E) \in \mathbb{C}_2 \setminus \mathcal{O}$ then any model of E over $\mathcal{O}$ has singular reduction modulo $\mathfrak{m}$, and the reduction of $j(E)$ modulo $\mathfrak{m}$ is the point ∞, which as we saw in section 3 corresponds to isomorphism class of a generalized elliptic curve which is singular. Finally, the point ∞ on the j-line corresponds to the singular generalized elliptic curve over $\mathbb{C}_2$, any model over $\mathcal{O}$ of which certainly has singular reduction modulo $\mathfrak{m}$, and the point ∞ certainly reduces to the point ∞.

Let us (as is customary) extend the phrase *ordinary reduction* to include either good ordinary reduction or bad reduction. (One should really require bad *multiplicative* reduction. However, $\mathbb{C}_2$ is algebraically closed, and so any E with bad reduction has a model with multiplicative reduction.) Since $j = 0$ is the unique supersingular j-invariant modulo 2, we see that the closed disk $|j| \geq |1|$ is the set of j-invariants in $X_0(1)$ having ordinary reduction. We denote this disk by $X_0(1)^h$ (h is for Hasse). The complement of $X_0(1)^h$ in $X_0(1)$ is the open disk $|j| < |1|$, and consists of the set of j-invariants having good supersingular reduction. We call this the *supersingular disk* in $X_0(1)$.

An examination of equations (IV) and (V) shows that the preimage of $X_0(1)^h$ under both B_1 and B_2 is equal to the union of the two disks $|j_2| \geq 1$ and $|j_2| \leq |2^{12}|$. We denote the former disk by $X_0(2)^h_\infty$ (because it contains the cusp ∞) and the latter by $X_0(2)^h_0$ (because it contains the cusp 0); their union we denote simply by $X_0(2)^h$. Note that $X_0(2)^h$ is preserved by w_2, and that w_2 interchanges $X_0(2)^h_\infty$ and $X_0(2)^h_0$. The fact that $X_0(2)^h$ is preserved by w_2 (which is equivalent to the fact that $X_0(1)^h$ has the same preimage under either B_1 or B_2) reflects the fact that the property of being ordinary or supersingular is an invariant of an isogeny class of elliptic curves in characteristic two.

The complement of $X_0(2)^h$ in $X_0(2)$ is the open annulus $|1| > |j_2| > |2^{12}|$. Its points correspond to those two-isogenies whose source (or equivalently target) has good supersingular reduction. We refer to it as the *supersingular annulus* in $X_0(2)$. Here is the picture:

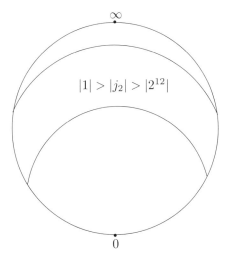

The j_2-line, drawn as a Riemann sphere, with the supersingular annulus marked

The disk $X_0(1)^h$ is contained in the disk D_1 on which B_1^{-1} is defined, and we see that B_1^{-1} restricts to yield an isomorphism $X_0(1)^h \xrightarrow{\sim} X_0(2)_\infty^h$; in particular, every elliptic curve with ordinary reduction has a canonical subgroup.

We can characterize the disk $X_0(1)^h$ as the set of j-invariants whose reduction modulo $\mathfrak{m}$ has non-zero Hasse invariant. The rest of this section is devoted to finding an analogous description of the larger disk D_1.

Let E_4 denote the weight four Eisenstein series on $\mathrm{SL}_2(\mathbb{Z})$, whose q-expansion is
$$E_4 = 1 + 240 \sum_{n=1}^\infty \sigma_3(n) q^n,$$
where $\sigma_3(n) = \sum_{d|n} d^3$. Recall that this modular form is actually well-defined in all characteristics (see [**38**], where it is denoted c_4) and that the j-invariant satisfies (indeed is defined by) the equation $j = E_4^3/\Delta$.

If E is an elliptic curve over $\mathbb{C}_2$ with good reduction, and ω is a differential on E with non-zero reduction modulo $\mathfrak{m}$, then for any modular form f over $\mathbb{C}_2$ we may compute $f(E,\omega)$, and the absolute value $|f(E,\omega)|$ is independent of the choice of ω (because any two such ω differ only by multiplication by a unit of $\mathcal{O}$). Thus we are entitled to write simply $|f(E)|$ in place of $|f(E,\omega)|$. For example, $\Delta(E,\omega)$ must be a unit in $\mathcal{O}$ (since it reduces to the discriminant of the reduction of E, which is non-zero), and so $|\Delta(E)| = |1|$, yielding the formula
$$|j(E)| = \frac{|E_4^3(E)|}{|\Delta(E)|} = |E_4^3(E)|.$$

Hence we see that $|j(E)| > |256|$ if and only if $|E_4(E)| > |2^{8/3}|$.

We now wish to relate E_4 to the Hasse invariant. The Hasse invariant A is a modular form of weight one defined modulo 2. Thus A^4 is a modular form of weight one defined modulo 8, whose q-expansion is the constant 1. (The point being that if x is a number well-defined modulo 2, than x^4 is well-defined modulo 8.) On the other hand, E_4 reduces to a modular form modulo 8 whose q-expansion is also equal to the constant 1 (because 8 divides 240). Thus by the q-expansion principle, we see that $E_4 \equiv A^4 \mod 8$. Thus $|E_4(E)| > |2^{8/3}|$ if and only if $|A(E)|^4 > |2^{8/3}|$ (because $|2^{8/3}| > |8|$, and so the second equality can be checked after reducing modulo 8, where A^4 makes sense) if and only if $|A(E)| > |2^{2/3}|$ (because $|2^{2/3}| > |2|$, and so the second inequality can be checked after reducing modulo 2, where A makes sense). Here $|A(E)|$ is denoting the valuation of the Hasse invariant of E reduced over $\mathcal{O}/2$ (which is non-Noetherian highly non-reduced local ring!), *not* E reduced over $k = \mathcal{O}/\mathfrak{m}$.

Thus we see that the annulus $|1| \geq |j| > |256|$ consists precisely of those j-invariants for which the corresponding elliptic curve E satisfies $|A(E)| > |2^{2/3}|$. We may think of these as the j-invariants of elliptic curves with "not too supersingular" reduction modulo two (in the words of [**79**]; recall that the annulus $|j| = |1|$ corresponds precisely to the elliptic curves having good ordinary reduction), and so the disk D_1 of elliptic curves consists of those elliptic curves having not too supersingular reduction (in the preceding sense) along with the elliptic curves having bad reduction.

The theory for arbitrary primes p

In this section we explain the generalization to an arbitrary prime p of the results presented above in the case $p = 2$. Let $\mathbb{C}_p$ denote the completion of the algebraic closure of $\mathbb{Q}_p$, let $\mathcal{O}$ denote its ring of integers, and let $\mathfrak{m}$ denote the maximal ideal of $\mathcal{O}$.

We let $Y_0(p)$ denote the modular curve whose points parameterize p-isogenies between elliptic curves, and $X_0(p)$ its completion. Over $\mathbb{C}$ one can construct $X_0(p)$ as the quotient $\mathcal{H}^*/\Gamma_0(p)$, and so see that it has two cusps, 0 and ∞. As p increases the genus of $X_0(p)$ also increases, and in particular $X_0(p)$ is not a rational curve unless $p = 13$ or $p < 11$. Thus one cannot in general describe $X_0(p)$ by a single parameter, as we described $X_0(2)$ by the parameter j_2.

Nevertheless, $Y_0(p)$ is an affine curve, and so it has an affine ring, which is generated by parameters which classify p-isogenies up to isomorphism (and which can be constructed explicitly via modular forms). By the Lefschetz principle, we see that we may equally regard $Y_0(p)$ as a curve over $\mathbb{C}_p$, since the moduli problem of classifying p-isogenies will depend on the same invariants whether solved over $\mathbb{C}$ or over $\mathbb{C}_p$. Then $X_0(p)$ will be the completion of $Y_0(p)$, a smooth projective curve over $\mathbb{C}_p$.

The process of passing from an isogeny to its dual yields an involution of $X_0(p)$, which we denote by w_p. Passing to the j-invariant of the source of an isogeny is a morphism $B_1 : X_0(p) \longrightarrow X_0(1)$. We let B_2 denote the composition $B_1 \circ w_p$.

We let $X_0(1)^h$ denote the set of j-invariants in $X_0(p)$ corresponding to elliptic curves with ordinary reduction (that is, either bad reduction or good ordinary reduction). The complement of $X_0(1)^h$ in $X_0(1)$ is a disjoint union of disks: each disk is a congruence class modulo $\mathfrak{m}$ of j-invariants which are congruent to a particular supersingular j-invariant in characteristic p. We refer to these disks as the supersingular disks in $X_0(1)$.

We let $X_0(p)^h$ denote the preimage of $X_0(1)^h$ under B_1^{-1}. Just as in the case of $p = 2$, and for the same reason (that being ordinary or supersingular is an isogeny class invariant in characteristic p), $X_0(p)^h$ is invariant under w_p. Furthermore, it is the union of two connected components which are interchanged by w_p, which we label $X_0(p)_\infty^h$ (the component containing ∞) and $X_0(p)_0^h$ (the component containing 0), and the restriction of B_1 to $X_0(p)_\infty^h$ is an isomorphism onto $X_0(1)^h$. (In order to interpret *connected component* in a more than intuitive sense, one must use the language of rigid analytic geometry.) The complement of $X_0(p)_\infty^h$ and $X_0(p)_0^h$ in $X_0(p)$ is a disjoint union of annuli, each the preimage under B_1 of one of the supersingular disks in $X_0(1)$. We refer to these as the *supersingular annuli* in $X_0(p)$. There is the following picture that goes with this discussion, in which one thinks of $X_0(p)$ as being obtained by gluing together two copies of the "sphere with holes" $X_0(1)^h$ (in the guise of $X_0(p)_\infty^h$ and $X_0(p)_0^h$) via the supersingular annuli:

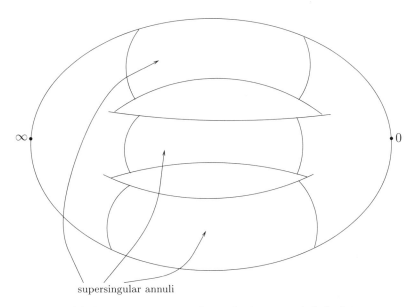

$X_0(p)$ drawn as a union of two "spheres with holes"
glued along the supersingular annuli

To prove all the facts just stated, one must study the modular curve $X_0(p)$ as a *scheme* over $\mathbb{Z}_p$, as in [**39, 85**]. The assertions then follow in a fairly standard fashion from the known structure of $X_0(p)$ over $\mathbb{Z}_p$; see the discussion and references in [**25**].

For now, note that since $B_1 : X_0(p)_\infty^h \to X_0(1)^h$ is an isomorphism, it has an inverse $B_1^{-1} : X_0(1)^h \to X_0(p)^h$, and thus any elliptic curve with ordinary reduction is equipped with a canonical subgroup of order p. Just as in the case of $p = 2$, there is a larger set D_1 containing $X_0(1)^h$ to which the map B_1^{-1} extends, and which we now describe.

Let A denote the Hasse invariant in characteristic p. If E is any elliptic curve with good reduction, we may consider its reduction modulo p (which is an elliptic curve over the ring $\mathcal{O}/p$), and the inequality $|A(E)| > |p^{p/(p+1)}|$ makes sense, since $|p^{p/(p+1)}| > |p|$. Define D_1 to the the subset of $X_0(1)$ consisting of the set of points in $X_0(1)^h$ together with those j-invariants of elliptic curves having good reduction whose Hasse invariant satisfies this inequality. Then just as in the case of $p = 2$, the map B_1^1 extends to a map $B_1^{-1}: D_1 \to X_0(p)$ which is an isomorphism between D_1 and its image D_2. If $j(E)$ lies in D_1 then the map B_1^{-1} endows E with a canonical subgroup of order p.

Let $D_1^{(n)}$ denote the disk which is the union of $X_0(1)^h$ and those j-invariants of good reduction whose corresponding elliptic curve satisfies $|A(E)| > |p^{1/p^{n-1}(p+1)}|$. Just as in the case $p = 2$, the Deligne-Tate map $\phi := B_2 \circ B_1^{-1}$ is a degree p map from $D_1^{(n)}$ onto $D_1^{(n-1)}$, and in particular if $j(E)$ lies in $D_1^{(n)}$ then E has a canonical cyclic subgroup of order p^n.

When written in terms of the uniformizing parameter q at ∞ the Deligne-Tate map has the form $q \mapsto q^p$. We will give a rather lengthy explanation of this fact, which will lead us toward an explanation of the existence of the Deligne-Tate map itself.

The Deligne-Tate map corresponds in modular terms to taking the quotient of an elliptic curve E by its canonical subgroup. Let us first try to understand why it is that the Deligne-Tate map is defined on the formal neighbourhood of the cusp ∞ on $X_0(1)$. In other words, why does E_{Tate} have a canonical subgroup? Or again, why is there a map B_1^{-1} inverse to B_1 defined on the formal neighbourhood of ∞?

This form of the question we can answer: the map B_1 is a unramified at the point ∞ on $X_0(p)$, and so induces an isomorphism of the formal neighbourhood of ∞ in $X_0(p)$ with the formal neighbourhood of ∞ in $X_0(1)$. The map B_1^{-1} is simply the inverse to this isomorphism. The existence of B_1^{-1} means that the Tate curve $E_{\text{Tate}} = \mathbb{G}_m/q^{\mathbb{Z}}$ is equipped with a canonical subgroup of order p. To compute this subgroup, we will look at the analytic picture over $\mathbb{C}$. (This is valid, since the curves $X_0(1)$ and $X_0(p)$, the projection B_1, the formal section B_1^{-1} of B_1 on the formal neighbourhood of $j = \infty$, and the Tate curve $\mathbb{G}_m/q^{\mathbb{Z}}$, are all defined over $\mathbb{Q}$, and thus so is the resulting subgroup of order p in $\mathbb{G}_m/q^{\mathbb{Z}}$. To determine this subgroup, it suffices to compute over the overfield $\mathbb{C}$ of $\mathbb{Q}$, where, as observed above, one can use analytic methods; the information we get is then perfectly applicable to the overfield $\mathbb{C}_p$ of $\mathbb{Q}$ that we are actually interested in.)

If τ is a point in the upper half-plane then the image of τ in $X_0(p)$ is the isogeny $E_\tau \longrightarrow E_{p\tau}$ given by $z \bmod \Lambda_\tau \mapsto pz \bmod \Lambda_{p\tau}$. The kernel of this isogeny is the group $\dfrac{2\pi i}{p}\mathbb{Z}/\mathbb{Z}$. Exponentiating this description of the isogeny, we see that it can also be described as the map $\mathbb{C}^\times/q^{\mathbb{Z}} \longrightarrow \mathbb{C}^\times/q^{p\mathbb{Z}}$ given by $t \mapsto t^p$, whose kernel is the image of the group μ_p of pth roots of unity in $\mathbb{G}_m/q^{\mathbb{Z}}$.

We conclude that the canonical subgroup of $E_{\text{Tate}} = \mathbb{G}_m/q^{\mathbb{Z}}$ is the image of μ_p in $\mathbb{G}_m/q^{\mathbb{Z}}$. The quotient of $\mathbb{G}_m/q^{\mathbb{Z}}$ by this group is isomorphic to $\mathbb{G}_m/q^{p\mathbb{Z}}$, and thus is a Tate elliptic curve with parameter q^p. Hence in terms of the parameter q around the cusp ∞, the Deligne-Tate map is given by the formula $q \mapsto q^p$.

The above argument actually provides the key insight as to the existence of canonical subgroups in general. Let us first explain this for elliptic curves over $\mathbb{C}_p$ having bad (multiplicative) reduction. First note that such curves are parameterized by the disk $|j| > |1|$ in $X_0(1)$. The usual power-series relating j and q (that is, $q = 1/j + 744/j^2 + \cdots$, which has integer coefficients) shows that q provides an analytic isomorphism of the disk $|j| > |1|$ in $X_0(1)$ with the open unit disk in $\mathbb{C}_p$. Since the power-series defining the Tate curve have coefficients lying in $\mathbb{Z}$, we may specialize them at any point q in this open unit disk to obtain an elliptic curve $\mathbb{C}_p^\times/q^{\mathbb{Z}}$, whose j-invariant will be that value of j corresponding to our given choice of q. Thus the elliptic curves over $\mathbb{C}_p$ having bad reduction are obtained as specializations of the Tate curve at points of the open unit q-disk (this is Tate's theory of uniformization of elliptic curves with multiplicative reduction). We may evaluate the formal section B_1^{-1} at any such point, and since this evaluation commutes with specialization, our preceding calculation with the Tate curve shows that the canonical subgroup of the curve $\mathbb{C}_p^\times/q^{\mathbb{Z}}$ will simply be the image of the subgroup μ_p of $\mathbb{C}_p^\times$.

If E is an elliptic curve over $\mathbb{C}_p$ with good ordinary reduction, then E does not admit a description in terms of the Tate curve, but the situation is almost as good. Let E denote a model for with good reduction modulo $\mathfrak{m}$. Then one finds that the formal group of E (thought of as a formal group over $\mathcal{O}$) is isomorphic to the formal multiplicative group $\hat{\mathbb{G}}_m$. (This is a consequence of the fact that the reduction modulo $\mathfrak{m}$ of each of these formal groups is of height one, so that they are isomorphic over k, together with a Newton's method/Hensel's lemma argument, which allows one to lift this isomorphism to an isomorphism over $\mathcal{O}$.) Now the p-torsion in $\hat{\mathbb{G}}_m$ is just the group μ_p, and so this isomorphism yields a copy of μ_p as a subgroup of E. This is the canonical subgroup of E.

To describe the canonical subgroup of a non-ordinary elliptic curve E whose j-invariant lies in D_1 is more difficult. One examines the formal group of E (whose reduction modulo $\mathfrak{m}$ is now of height two) and uses the bound on the Hasse invariant of E (which one knows by virtue of the assumption that $j(E)$ lies in D_1) to construct a canonically determined subgroup of this formal group of order p. The details can be found in [**79**].

p-adic modular forms

Recall again the algebraic definition of a modular form of weight k and of level one defined over the field $\mathbb{C}_p$: it is a rule that attaches to a pair (E,ω) consisting of an elliptic curve and a non-zero regular differential defined over $\mathbb{C}_p$ a number $f(E,\omega)$ which satisfies the weight k transformation rule and which behaves well in families, including those that include generalized elliptic curves among their fibres. From now on we will refer to such forms as *classical* modular forms.

A p-adic modular form of weight k is such a rule which is defined only on those pairs (E,ω) for which E has ordinary reduction (that is, as we said above, has either bad or good ordinary reduction modulo $\mathfrak{m}$). Thus any classical modular form also gives rise to a p-adic modular form, but there are also many p-adic modular forms which are not classical modular forms. For example, if $p=2$ and E has ordinary reduction, then $E_4(E,\omega) \neq 0$, and so $E_4(E,\omega)^{-1}$ is well-defined. Thus E_4^{-1} is a 2-adic modular form of weight -4. Another way to construct p-adic modular forms for any prime p is to notice that ordinary elliptic curves have canonical subgroups of order p^n for all n, and so any classical modular form on $\Gamma_0(p^n)$ gives rise to a p-adic modular form of level one.

The space of all p-adic modular forms of weight k (for some fixed p and k) is a p-adic Banach space, and just as in the classical theory this space has Hecke operators acting on it. Unfortunately, the action of these operators is rather hard to control, and one does not get, and cannot expect to get, a very good spectral theory. Thus there is not a very good theory of p-adic Hecke eigenforms.

To deal with this, one looks at a more refined type of p-adic modular form. An *overconvergent* p-adic modular form is a rule of the usual type which is defined on those pairs (E,ω) for which $j(E)$ lies in $D_1^{(n)}$ for some n. (The n is fixed for any particular overconvergent form, but may vary from form to form.) All our examples of p-adic modular forms given above are actually overconvergent modular forms: for example, we saw that E_4 is never zero on the disc D_1 in $X_0(2)$, and so E_4^{-1} is defined on that disc. Also, any classical modular form on $\Gamma_0(p^n)$ for some n yields an overconvergent modular form defined on $D_1^{(n)}$, since the elliptic curves with j-invariant in this region are precisely those which have a canonical subgroup of order p^n. Here is an example of a p-adic modular form which is *not* overconvergent: the "weight two Eisenstein series of level one" is the q-expansion $E_2 = 1 + 24\sum_{n=1}^{\infty}\sigma(n)q^n$, where $\sigma(n) = \sum_{d|n} d$. This is not a classical modular form at all (it doesn't transform correctly under the substitution $\tau \mapsto -1/\tau$), but *is* the q-expansion of a p-adic modular form for every p (see [**79**], in which it is denoted by P), which is *not* overconvergent for any prime p [**31**].

One can again define Hecke operators acting on overconvergent modular forms, and one finds that something special happens. If one works with the Hecke operator U_p (which on q-expansions is defined by $U_p(\sum_n a_n q^n) = \sum_n a_{np} q^n)$) rather than T_p, then there is a good theory of Hecke eigenforms, because the operator U_p is a *completely continuous* (or, in alternative language, *compact*) operator on the space of overconvergent modular forms, and so has a good spectral theory. More precisely, for any non-zero $\lambda \in \mathbb{C}_p$, there is a finite-dimensional space of overconvergent modular forms on which U_p has eigenvalue λ. This space will be preserved by the Hecke operators T_ℓ (as ℓ ranges over all primes different from p), and to construct Hecke eigenforms with U_p-eigenvalue λ it suffices to diagonalize these commuting operators on this finite-dimensional vector space.

The reason that U_p is completely continuous is the following: recall that the Deligne-Tate map is given in terms of the parameter q by the formula $\phi(q) = q^p$. Thus one sees that the operator U_p can be described geometrically via the *trace* of the Deligne-Tate map. More precisely, it is $1/p$ times this trace. It is simplest to explain this for U_p acting on modular forms of weight zero, which is to say modular functions. Then an overconvergent modular function is simply an analytic function f on one of the regions $D_1^{(n)}$. We claim that

$$U_p(f)(j) = \frac{1}{p}\sum_{\phi(j')=j} f(j')$$

for any point $j \in D_1^{(n)}$. To see this, it suffices to verify the formula for those j lying in the residue disc about ∞, where we can compute in terms of q; it will follow in general by analytic continuation. The claim then follows from the above formula for the action of U_p on q-expansions, together with the following simple piece of algebra:

$$\frac{1}{p}\sum_{q'^p=q}\sum_n a_n q'^n = \sum_n a_{np} q^n.$$

(See [**79**] for a precise form of this argument.) Why does this formula imply that U_p is completely continuous? Well, note that if $j \in D_1^{(n)}$ and $\phi(j') = j$, then $j' \in D_1^{(n+1)}$. Thus computing $U_p(f)$ involves restricting f from the region $D_1^{(n)}$ to the proper subregion $D_1^{(n+1)}$, and such restriction operators are always completely continuous. (This is Montel's theorem from complex analysis being applied in the (simpler) p-adic setting.)

The preceding discussion shows that the fact that the Deligne-Tate map is defined on the regions $D_1^{(n)}$ which extend some way into the supersingular annuli is fundamental to obtaining a good spectral theory for overconvergent p-adic modular forms. This explains the importance of the Deligne-Tate map in the theory of p-adic modular forms, and hence why we have devoted our efforts in this appendix to describing the ideas behind the construction of this map.

Guide to the literature

This final section contains a brief review of the literature on the p-adic theory of modular curves and modular forms, which I hope will be helpful to someone trying to enter the field. I have tried to present as accurate an account of the development and current state of the field as I can, within the limits of my own understanding of these matters. I apologize in advance for any omissions or oversights that I may inadvertently have made.

The p-adic aspects of the theory of modular curves that we have discussed in this appendix seem to make their first appearance in the literature in [**43**], in which Dwork constructs and studies "by hand" the Deligne-Tate map on the region D_1. He states that the existence of this map was conjectured by Tate (based on a calculation for $p = 2$; I don't know if it was a similar calculation to that of section 2 above) and first proved in general by Deligne. In subsequent articles [**44, 45**] Dwork studied the spectral theory of the completely continuous operator U_p on the space of overconvergent modular forms of level zero. One of the points emphasized in [**45**] in particular is the importance of restricting ones attention to overconvergent forms if one hopes to obtain a reasonable spectral theory.

In an independent line of research, Swinnerton-Dyer [**150**] began the systematization of the theory of congruences of q-expansions of modular forms (a subject which seems to have begun with the work of Ramanujan), by studying the ring of modular forms modulo p. This work is also reported on in [**131**]. In [**134**] Serre extends these results to develop a theory of congruences of modular forms modulo arbitrary powers of p, and introduces the notion of p-adic modular forms.

In [**79**] Katz gives a systematic presentation of the p-adic geometry of the modular curves, including the construction of canonical subgroups of "not too supersingular elliptic curves" (which construction he attributes to Lubin) and the consequent construction of B_1^{-1} and the Deligne-Tate map, as well as of the theory of congruences of modular forms modulo p. The results of this article, and the later article [**81**] (which extends the techniques of [**79**] to define *generalized p-adic modular functions*, a notion that includes as a special case the p-adic modular forms of [**79**] and [**134**]), subsumed most of those of Dwork, Serre and Swinnerton-Dyer mentioned above, and also generalized them to modular forms of arbitrary level.

However, although [**79**] in some sense unified the various existing p-adic theories, the applications of these theories were in two different directions. On the one hand the papers [**80, 82, 83, 134**] used the theory of congruences of modular forms to construct p-adic analytic families of modular forms (essentially families of Eisenstein series) and hence to construct p-adic L-functions (which appeared as the special values of these Eisenstein series, either at the cusp ∞ or at certain special j-values corresponding to complex-multiplication elliptic curves). On the other hand, the papers [**150, 151**] were concerned with using the same theory of congruences not to construct families of modular forms, but rather to understand the image of the Galois representations attached to Hecke eigenforms (as constructed in [**37**]). Thus their main concern was not congruences between arbitrary modular forms, but the possible congruences that could arise between Hecke eigenforms. This was something that was not dealt with in the theory of [**81**], and this may go some way to explaining the remark on this paper made in the introduction of [**151**].

Studying properties of Galois representations via congruences of modular forms turned out to be a very fruitful idea. One key direction of research was the investigation of congruences between cuspidal Hecke eigenforms and Eisenstein series. This is the main theme of the seminal paper [**94**], and formed the basis for a successful attack on the so-called *main conjecture of Iwasawa theory* over the field $\mathbb{Q}$, beginning in [**118**], continuing in [**159**], and culminating with the proof of the main conjecture in [**105**]. The investigation of possible congruences among the cuspidal eigenforms themselves proved to be an equally important topic, a discussion of which would unfortunately take us too far afield, and which is provided by Ribet's article in this volume.

In Hida's papers [**71, 70**] the properly p-adic aspect of the theory of congruences made its resurgence. In these pivotal papers Hida connected the study of congruences of eigenforms and the associated Galois representations with the study of p-adic families of modular forms by constructing *p-adic analytic families of p-adic Hecke eigenforms* and attaching to them p-adic analytic families of Galois representations. (See also [**106**] for a discussion of Hida's construction and a more refined analysis of these Galois representations.) The main technical constraint on Hida's results is that they only construct families of *ordinary* eigenforms; that is, eigenforms whose U_p eigenvalues are p-adic units.

The influence of Hida's theory was enormous. It allowed a simplification of the proof of the main conjecture of Iwasawa theory, by rephrasing it as the question of analyzing the intersection locus of the Eisenstein family with the cuspidal part of Hida's family. By extending Hida's theory to the context of Hilbert modular forms, Wiles was able to prove the main conjecture for arbitrary totally real fields [**160**]; this harks back to Serre's study of p-adic L-functions as the constant term of families of Eisenstein series. Other developments included the construction of families of p-adic L-functions attached to Hida families of cuspforms [**64, 88, 96**]. Greenberg and Stevens [**64**] used these L-functions to prove the weight two case of the conjecture of Mazur, Tate and Teitelbaum [**104**]. (See also [**65**], which presents the main ideas of their argument in a simplified setting.) of p-adic modular forms, Hida's work also motivated Mazur to develop his theory of deformations of Galois representations [**97**] (for more on which, see the main body of this article!), which proved decisive for the further development of the theory of p-adic modular forms.

In [**57**] Gouvêa used Mazur's theory in order to associate a p-adic Galois representation to each p-adic modular form, whether ordinary or not. This work marked the beginning of line of research aimed at constructing p-adic analytic families of p-adic Hecke eigenforms and Galois representations, analogous to those constructed by Hida, but in the non-ordinary situation. That such families should exist was the principal conjecture of [**61**]. This work also raised a number of questions related to overconvergent p-adic modular forms, refocusing attention on an aspect of the p-adic theory that had languished since the appearance of papers of Dwork cited above. The p-adic analytic viewpoint favoured by Dwork reemerged in a series of papers by Coleman [**25, 26, 27, 28**]. The first of these papers relates to the conjecture of [**104**], while the second and third show that an overconvergent p-adic eigenform whose U_p-eigenvalue is not too divisible by p (in a sense depending on the weight of the form in question) is necessarily classical (extending the analogous result in the ordinary case, due to Hida [**70**], and hence answering one of the questions of [**57**]).

The current course of research in the theory of p-adic modular forms has been set by the papers [**28**] and [**23**]. In the first of these papers, Coleman analyzes the variation of the spectral theory of the U_p operator on overconvergent p-adic modular forms as a function of the weight and hence constructs p-adic analytic families of Hecke eigenforms, in particular proving in a qualitative form the conjectures of [**61**]. In [**23**] this construction is globalized to construct for each prime p a p-adic rigid analytic curve called the *eigencurve*. This is, roughly speaking, the universal parameter space for p-adic analytic families of overconvergent Hecke eigenforms of finite slope (that is, whose U_p-eigenvalue is non-zero; the slope of a Hecke eigenform is the p-adic valuation of its U_p-eigenvalue). The deformation theory of Galois representations plays a key role in the construction of the eigencurve, and the universal family of finite-slope eigenforms comes equipped with a p-adic analytic family of p-adic Galois representations.

With the construction of the eigencurve, it appears that the various directions of research that instigated the development of the theory of p-adic modular forms have achieved synthesis. The overconvergent forms are understood as being the necessary "glue" that binds classical Hecke eigenforms together into analytic families. Furthermore, the existence of positive slope families will certainly have many number-theoretic applications. One such example is [**148**], in which Stevens develops a theory of p-adic L-functions in the context of Coleman's families, and uses them [**149**] to prove the higher-weight case of the conjecture of [**104**] (in the more precise form stated in [**25**]).

Yet the construction of the eigencurve raises as many questions as it answers. For example, while the structure of Hida's families of ordinary forms is very precisely understood, the structure of the higher-slope parts of the eigencurve is essentially a mystery (the references [**24, 47, 147**] provide some information on a small part of this structure for certain small values of p). Relatedly, the quantitative aspects of the conjectures of [**61**] remain unresolved; all that is known are some very special cases dealt with in [**24, 47, 147**].

Another question is that of the existence of p-adic analytic families of eigenforms of infinite slope. Such forms, whose U_p-eigenvalue vanishes, aren't accessible via the standard spectral theory techniques available for the study of the completely continuous operator U_p. One natural question that arises in this context is: are there infinite slope forms that can be written as the limit of finite slope forms? Such forms would correspond to "punctures" in the eigencurve, which would be filled in by the limiting infinite slope form. Coleman [**29**] has shown that forms obtained by twisting finite slope forms with the Teichmüller character are of this type. In the same reference, Coleman reports on a computation of Stein suggesting that some non-twist infinite slope forms might also be of this type. As to whether infinite slope forms move in p-adic families, some computations for forms on $\Gamma_0(4)$ [**46**] suggest that the answer could be yes, but essentially nothing is known in this direction.

One of the most intriguing and general questions raised by the construction of the eigencurve is that of the intrinsic Galois-theoretic interpretation of the Galois representations that it parameterizes (these are just the Galois representations attached to overconvergent p-adic Hecke eigenforms, and this problem was already posed in [**57**]). In [**23**] it is shown that the eigencurve is the rigid analytic Zariski closure (in an appropriate ambient space, which is essentially a deformation space of Galois representations) of those points corresponding to classical modular forms. The Galois representations associated to such forms are conjecturally characterized as being those which are *potentially semistable* (see [**53**] and the references contained therein for a discussion of this conjecture and the notion of potential semistability). However no analogous intrinsic characterization of the non-classical points on the eigencurve is known, even on the conjectural level (see [**86, 87**] for some investigations in this direction). Finding such a description of the eigencurve, which would not depend on the "extrinsic" construction via the theory of overconvergent p-adic modular forms, is related to the problem of generalizing the construction of the eigencurve to other settings. For a discussion of what such settings might be, and the possible relations to Iwasawa theory and a general theory of p-adic L-functions, see [**102**].

MATTHEW EMERTON
DEPARTMENT OF MATHEMATICS
UNIVERSITY OF MICHIGAN
ANN ARBOR, MI 48109
emerton@math.lsa.umich.edu

BIBLIOGRAPHY

1. V. Arnold, M. Atiyah, P. Lax, and B. Mazur (eds.), *Mathematics: Frontiers and perspectives*, American Mathematical Society, 2000.
2. W. Barth and H. Lange (eds.), *Arithmetic of complex manifolds (Erlangen, 1988)*, Lecture Notes in Mathematics, no. 1399, Springer-Verlag, 1989.
3. B. Birch and W. Kuijk (eds.), *Modular functions of one variable IV*, Lecture Notes in Mathematics, vol. 476, Springer-Verlag, Berlin, Heidelberg, New York, 1975.
4. S. Bloch and K. Kato, *L-functions and Tamagawa numbers of motives*, In Cartier et al. [**17**], pp. 333–400.
5. G. Böckle, *Demuškin groups with group actions and applications to deformations of Galois representations*, Preprint, 1998.
6. _____, *The generic fiber of the universal deformation space associated to a tame Galois representation*, Manuscripta Math. **96** (1998), 231–246.
7. G. Böckle and A. Mézard, *The prime-to-adjoint principle and unobstructed Galois deformations in the Borel case*, J. Number Theory **78** (1999), 167–203.
8. N. Boston, *Deformation theory of Galois representations*, Ph.D. thesis, Harvard University, 1987.
9. _____, *Explicit deformation of Galois representations*, Invent. Math. **103** (1991), 181–196.
10. N. Boston and B. Mazur, *Explicit universal deformations of Galois representations*, Advanced Studies in Pure Mathematics **17** (1989), 1–21.
11. N. Boston and Stephen V. Ullom, *Representations related to CM elliptic curves*, Math. Proc. Cambridge Philos. Soc. **113** (1993), 71–85.
12. N. Bourbaki, *Éléments de Mathématique: Théorie des ensembles*, Hermann, Paris, 1968.
13. J. Buhler, *Elliptic curves and modular forms*, in this volume.
14. H. Carayol, *Formes automorphes et représentations galoisiennes*, Seminar on Number Theory, 1981/1982, Univ. Bordeaux I, Talence, 1982, pp. Exp. No. 31, 20.
15. H. Carayol, *Sur les représentations ℓ-adiques associées aux formes modulaires de Hilbert*, Ann. Sci. École Norm. Sup. (4) **19** (1986), 409–468.
16. _____, *Formes modulaires et représentations Galoisiennes à valeurs dans un anneau local complet*, In Mazur and Stevens [**103**], pp. 213–237.

17. P. Cartier, L. Illusie, N. M. Katz, G. Laumon, Y. I. Manin, and K. A. Ribet (eds.), *The Grothendieck Festschrift I*, Birkhauser, 1990.
18. J. W. S. Cassels and J. Fröhlich (eds.), *Algebraic number theory (Brighton, 1965)*, Academic press, 1967.
19. J. Coates and C.-G. Schmidt, *Iwasawa theory for the symmetric square of an elliptic curve*, J. Reine Angew. Math. **375/376** (1987), 104–156.
20. J. Coates and A. Sydenham, *On the symmetric square of a modular elliptic curve*, In Coates and Yau [**22**], pp. 152–171.
21. J. Coates and M. J. Taylor (eds.), *L-functions and arithmetic*, London Mathematical Society Lecture Notes, vol. 153, Cambridge University Press, Cambridge, 1991.
22. J. Coates and S.-T. Yau (eds.), *Elliptic curves, modular forms and Fermat's last theorem (Hong Kong, 1993)*, Cambridge, MA, International Press, 1997.
23. R. Coleman and B. Mazur, *The eigencurve*, (A. J. Scholl and R. L. Taylor, eds.), London Mathematical Society Lecture Note Series, vol. 254, Cambridge University Press, 1998, pp. 1–113.
24. R. Coleman, G. Stevens, and J. Teitelbaum, *Numerical experiments on families of p-adic modular forms*, preprint.
25. R. F. Coleman, *A p-adic Shimura isomorphism and p-adic periods of modular forms*, In Mazur and Stevens [**103**], pp. 21–51.
26. _____, *Classical and overconvergent modular forms*, Invent. Math. **124** (1996), 215–241.
27. _____, *Classical and overconvergent modular forms of higher level*, J. Théor. Nombres Bordeaux **9** (1997), no. 2, 395–403.
28. _____, *p-adic Banach spaces and families of modular forms*, Invent. Math. **127** (1997), 417–479.
29. _____, *Conjectures and rumours of conjectures*, notes to a talk posted on http://www.math.berkeley.edu/ coleman/Sems/Sp00/coleman.html, 2000.
30. R. F. Coleman and B. Edixhoven, *On the semi-simplicity of the U_p-operator on modular forms*, Math. Ann. **310** (1998), 119–127.
31. R. F. Coleman, F. Q. Gouvêa, and N. Jochnowitz, *E_2, Θ, and overconvergence*, Internat. Math. Res. Notices (1995), no. 1, 23–41.
32. B. Conrad, *The flat deformation functor*, In Cornell et al. [**33**], Papers from the Instructional Conference on Number Theory and Arithmetic Geometry held at Boston University, Boston, MA, August 9–18, 1995, pp. 373–420.
33. G. Cornell, J. Silverman, and G. Stevens (eds.), *Modular forms and Fermat's last theorem*, Springer-Verlag, Berlin, Heidelberg,New York, 1997, Papers from the Instructional Conference on Number Theory and Arithmetic Geometry held at Boston University, Boston, MA, August 9–18, 1995.
34. H. Darmon, F. Diamond, and R. Taylor, *Fermat's last theorem*, In Coates and Yau [**22**].
35. A. J. de Jong, *Crystalline Dieudonné module theory via formal and rigid geometry*, Inst. Hautes Études Sci. Publ. Math. (1995), no. 82, 5–96.
36. B de Smit and H. W. Lenstra, *Explicit construction of universal deformation rings*, In Cornell et al. [**33**], Papers from the Instructional Conference on Number Theory and Arithmetic Geometry held at Boston University, Boston, MA, August 9–18, 1995, pp. 313–326.
37. P. Deligne, *Formes modulaires et représentations ℓ-adiques, Exposé 355*, Séminaire Bourbaki 1968/1969, Springer-Verlag, 1971, pp. 139–172.

38. _____, *Courbes elliptiques: formulaire d'après j. tate*, In Birch and Kuijk [**3**], pp. 53–73.
39. P. Deligne and M. Rapoport, *Les schémas de modules de courbes elliptiques*, (Berlin, Heidelberg, New York) (P. Deligne and W. Kuijk, eds.), Lecture Notes in Mathematics, vol. 349, Springer-Verlag, 1973.
40. P. Deligne and J.-P. Serre, *Formes modulaires de poids* 1, Ann. Sci. École Norm. Sup. (4) **7** (1974), 507–530, In J.-P. Serre, *Œuvres*, volume III.
41. F. Diamond and J. Im, *Modular forms and modular curves*, In Murty [**110**], Papers from the seminar held at the Fields Institute for Research in Mathematical Sciences, Toronto, Ontario, 1993–1994, pp. 39–133.
42. C. Doran and S. Wong, *Deformations of Galois representations and modular forms*, Based on lectures of Barry Mazur at Harvard University, Fall 1993.
43. B. Dwork, *p-adic cycles*, Inst. Hautes Études Sci. Publ. Math. **37** (1969), 27–115.
44. _____, *On Hecke polynomials*, Invent. Math. **12** (1971), 249–256.
45. _____, *The U_p operator of Atkin on modular functions of level 2 with growth conditions*, In Kuijk and Serre [**89**].
46. M. Emerton, Unpublished computations, 1998.
47. _____, *2-adic modular forms of minimal slope*, Ph.D. thesis, Harvard University, 1998.
48. M. Flach, *A generalisation of the Cassels-Tate pairing*, J. Reine Angew. Math. **412** (1990), 113–127.
49. _____, *Selmer groups for the symmetric square of an elliptic curve*, Ph.D. thesis, University of Cambridge, 1990.
50. _____, *A finiteness theorem for the symmetric square of an elliptic curve*, Invent. Math. **109** (1992), 307–327.
51. _____, *Annihilation of Selmer groups for the adjoint representation of a modular form*, In Murty [**110**], Papers from the seminar held at the Fields Institute for Research in Mathematical Sciences, Toronto, Ontario, 1993–1994, pp. 249–265.
52. J.-M. Fontaine and B. Mazur, *Geometric galois representations*, In Coates and Yau [**22**], pp. 41–78.
53. _____, *Geometric Galois representations*, In Coates and Yau [**22**], pp. 41–78.
54. J. Fröhlich, *Local fields*, In Cassels and Fröhlich [**18**], pp. 1–41.
55. W. Fulton and J. Harris, *Representation theory: a first course*, Springer-Verlag, 1991.
56. P. Gabriel, *Groupes formels*, Schémas en Groupes (Sém. Géometrie Algébrique, Inst. Hautes Études Sci., 1963/64), vol. Fasc. 2b, Exposé 7b, Inst. Hautes Études Sci., Paris, 1965, pp. 66–152+3.
57. F. Q. Gouvêa, *Arithmetic of p-adic modular forms*, Lecture Notes in Mathematics, vol. 1304, Springer-Verlag, Berlin, Heidelberg, New York, 1988.
58. _____, *Deforming Galois representations: controlling the conductor*, J. Number Theory **34** (1990), 95–113.
59. _____, *p-adic numbers: an introduction*, Springer-Verlag, Berlin, Heidelberg, New York, 1993.
60. _____, *Deforming galois representations: a survey*, In Murty [**110**], Papers from the seminar held at the Fields Institute for Research in Mathematical Sciences, Toronto, Ontario, 1993–1994, pp. 179–207.

61. F. Q. Gouvêa and B. Mazur, *Families of modular eigenforms*, Math. Comp. **58** (1992), 793–806.
62. _____, *On the density of modular representations*, Computational perspectives on number theory (Chicago, IL, 1995), American Mathematical Society, Providence, RI, 1998, pp. 127–142.
63. R. Greenberg, *Iwasawa theory for elliptic curves*, in this volume.
64. R. Greenberg and G. H. Stevens, *p-adic L-functions and p-adic periods of modular forms*, Invent. Math. **111** (1993), 407–447.
65. _____, *On the conjecture of Mazur, Tate and Teitelbaum*, In Mazur and Stevens [**103**], pp. 183–211.
66. B. H. Gross, *Kolyvagin's work on modular elliptic curves*, In Coates and Taylor [**21**], pp. 235–256.
67. Alexander Grothendieck, *Technique de descente et théorèmes d'existence en géométrie algébrique II. Le théorème d'existence en théorie formelle des modules*, Séminaire Bourbaki, Vol. 5 (Paris), Soc. Math. France, 1995, Exp. No. 195, pp. 369–390.
68. K. Haberland, *Galois cohomology of algebraic number fields*, VEB Deutscher Verlag, 1978.
69. Robin Hartshorne (ed.), *Algebraic geometry: Arcata 1974*, American Mathematical Society, 1975.
70. H. Hida, *Galois representations into $\mathrm{GL}_2(\mathbb{Z}_p[[X]])$ attached to ordinary cusp forms*, Invent. Math. **85** (1986), 545–613.
71. _____, *Iwasawa modules attached to congruences of cusp forms*, Ann. Sci. École Norm. Sup. (4) **19** (1986), 231–273.
72. _____, *Theory of p-adic Hecke algebras and Galois representations*, Sugaku Expositions **2** (1989), 75–102.
73. _____, *Nearly ordinary Hecke algebras and Galois representations of several variables*, JAMI Inaugural Conference Proceedings, supplement to Amer. J. Math (1990), 115–134.
74. J.-I. Igusa, *Class number of a definite quaternion with prime discriminant*, Proc. Nat. Acad. Sci. USA **44** (1958), 312–314.
75. Y. Ihara, K. A. Ribet, and J.-P. Serre (eds.), *Galois groups over $\mathbb{Q}$ (Berkeley 1990)*, Mathematical Sciences Research Institute Publications, no. 16, Springer-Verlag, 1990.
76. U. Jannsen, *Über Galoisgruppen lokaler Körper*, Invent. Math. **70** (1982/83), 53–69.
77. U. Jannsen and K. Wingberg, *Die Struktur der absoluten Galoisgruppe $\mathfrak{p}$-adischer Zahlkörper*, Invent. Math. **70** (1982/83), 71–98.
78. K. Joshi and C. Khare, *On ordinary forms and ordinary Galois representations*, Preprint, 1995.
79. N. M. Katz, *p-adic properties of modular schemes and modular forms*, In Kuijk and Serre [**89**].
80. _____, *p-adic L-functions via moduli of elliptic curves*, In Hartshorne [**69**].
81. _____, *Higher congruences between modular forms*, Ann. of Math. (2) **101** (1975), 332–367.
82. _____, *p-adic interpolation of real analytic Eisenstein series*, Ann. of Math. (2) **104** (1976), 459–571.
83. _____, *The Eisenstein measure and p-adic interpolation*, Am. Journ. Math. **99** (1977), 238–311.

84. N. M. Katz and S. Lang, *Finiteness theorems in geometric class field theory*, L'Enseignement Mathématique **XVIII** (1981), 285–319.
85. N. M. Katz and B. Mazur, *Arithmetic moduli of elliptic curves*, Annals of Mathematics Studies, vol. 108, Princeton University Press, Princeton, New Jersey, 1985.
86. M. Kisin, *Periods for p-adic modular forms*, Preprint, 1999.
87. _____, *p-adic modular forms and the Fontaine-Mazur conjecture*, Preprint, 2000.
88. K. Kitigawa, *On standard p-adic l-functions of families of elliptic cusp forms*, In Mazur and Stevens [**103**], pp. 81–110.
89. W. Kuijk and J.-P. Serre (eds.), *Modular functions of one variable III*, Lecture Notes in Mathematics, vol. 350, Springer-Verlag, Berlin, Heidelberg, New York, 1973.
90. S. Lang, *Algebra*, third ed., Addison-Wesley, 1993.
91. W. Li, *Newforms and functional equations*, Math. Ann. **212** (1975), 285–315.
92. Saunders MacLane, *Categories for the working mathematician*, Springer-Verlag, 1971.
93. P. A. Martin, *Deformações de representações Galoisianas ordinárias e de representações não ramificadas*, Ph.D. thesis, Universidade de São Paulo, 1991.
94. B. Mazur, *Modular curves and the Eisenstein ideal*, Inst. Hautes Études Sci. Publ. Math. **47** (1977), 33–186.
95. _____, *Rational isogenies of prime degree*, Invent. Math. **44** (1978), 129–162.
96. _____, *Two-variable p-adic L-functions*, unpublished manuscript, 1985.
97. _____, *Deforming Galois representations*, In Ihara et al. [**75**], pp. 385–437.
98. _____, *Two-dimensional p-adic Galois representations unramified away from p*, Compositio Math. **74** (1990), 115–133.
99. _____, *Galois deformations and Hecke curves*, Harvard University course notes, 1994.
100. _____, *An "infinite fern" in the universal deformation space of galois representations*, Collect. Math. **48** (1997), 155–193, Journées Arithmétiques (Barcelona, 1995).
101. _____, *An introduction to the deformation theory of Galois representations*, In Cornell et al. [**33**], Papers from the Instructional Conference on Number Theory and Arithmetic Geometry held at Boston University, Boston, MA, August 9–18, 1995, pp. 243–311.
102. _____, *The theme of p-adic variation*, In Arnold et al. [**1**], pp. 433–459.
103. B. Mazur and G. Stevens (eds.), *p-adic monodromy and the birch-swinnerton-dyer conjecture*, Contemporary Mathematics, vol. 165, American Mathematical Society, 1994.
104. B. Mazur, J. Tate, and J. Teitelbaum, *On p-adic analogues of the conjectures of Birch and Swinnerton-Dyer*, Invent. Math. **84** (1986), 1–48.
105. B. Mazur and A. Wiles, *Class fields of abelian extensions of* $\mathbb{Q}$, Invent. Math. **76** (1984), 179–330.
106. _____, *On p-adic analytic families of Galois representations*, Compositio Math. **59** (1986), 231–264.
107. J. S. Milne, *Arithmetic duality theorems*, Academic Press, 1986.
108. T. Miyake, *Modular forms*, Springer-Verlag, 1989.
109. P. Morandi, *Field and Galois theory*, Springer-Verlag, Berlin, Heidelberg, New York, 1996.

110. V. K. Murty (ed.), *Seminar on Fermat's Last Theorem*, Providence, RI, American Mathematical Society, 1995, Papers from the seminar held at the Fields Institute for Research in Mathematical Sciences, Toronto, Ontario, 1993–1994.
111. J. Neukirch, A. Schmidt, and K. Wingberg, *Cohomology of number fields*, Springer-Verlag, Berlin, Heidelberg, New York, 2000.
112. J. Neukrich, *Class field theory*, Springer-Verlag, Berlin, Heidelberg, New York, 1986.
113. L. Nyssen, *Pseudo-représentations*, Math. Ann. **306** (1996), 257–283.
114. Joseph Oesterlé, *Travaux de Wiles (et Taylor, ...). II*, Astérisque (1996), no. 237, Exp. No. 804, 5, 333–355, Séminaire Bourbaki, Vol. 1994/95.
115. R. Ramakrishna, *On a variation of Mazur's deformation functor*, Compositio Math. **87** (1993), 269–286.
116. M. Raynaud, *Schémas en groupes de type $(p, p, \ldots, p)$*, Bull. Soc. Math. France **102** (1974), 241–280.
117. K. A. Ribet, *Lectures on modular forms*, in this volume.
118. _____, *A modular construction of unramified p-extensions of $\mathbb{Q}(\mu_p)$*, Invent. Math. **34** (1976), 151–162.
119. _____, *Galois representations and modular forms*, Bull. Amer. Math. Soc. (N.S.) **32** (1995), 375–402.
120. D. Robinson, *A course in the theory of groups*, Springer-Verlag, 1982.
121. R. Rouquier, *Caractérisation des caractères et pseudo-caractères*, J. Algebra **180** (1996), 571–586.
122. K. Rubin, *The work of Kolyvagin on the arithmetic of elliptic curves*, In Barth and Lange [2], pp. 128–136.
123. _____, *Kolyvagin's system of Gauss sums*, In van der Geer et al. [154], pp. 309–324.
124. _____, *Euler systems*, Princeton University Press, Princeton, New Jersey, 1999.
125. M. Schlessinger, *Functors of Artin rings*, Trans. A. M. S. **130** (1968), 208–222.
126. L. Schneps (ed.), *The Grothendieck theory of dessins d'enfants*, London Mathematical Society Lecture Note Series, vol. 200, Cambridge, Cambridge University Press, 1994.
127. A. Scholl, *An introduction to Kato's Euler system*, In Scholl and Taylor [128], pp. 379–460.
128. A. Scholl and R. Taylor (eds.), *Galois representations in arithmetic algebraic geometry (Durham, 1996)*, Cambridge, Cambridge University Press, 1998.
129. S. Sen, *Continuous cohomology and p-adic Galois representations*, Invent. Math. **62** (1980/81), 89–116.
130. _____, *An infinite-dimensional Hodge-Tate theory*, Bull. Soc. Math. France **121** (1993), 13–34.
131. J.-P. Serre, *Congurences et formes modulaires (d'après H. P. F. Swinnerton-Dyer)*, Seminaire Bourbaki (1971/72), no. 416.
132. _____, *Propriétés galoisiennes des points d'ordre fini des courbes elliptiques*, Invent. Math. **15** (1972), 259–331.
133. _____, *A course in arithmetic*, Springer-Verlag, Berlin, Heidelberg, New York, 1973.
134. _____, *Formes modulaires et fonctions zêta p-adiques*, In Kuijk and Serre [89], In J.-P. Serre, *Œuvres*, vol III.

135. _____, *Local fields*, Springer-Verlag, Berlin, Heidelberg, New York, 1974.
136. _____, *Quelques applications du théorème de densité de Chebotarev*, Inst. Hautes Études Sci. Publ. Math. **54** (1981), 323–401.
137. _____, *Sur les représentations modulaires de degré* 2 *de* $\mathrm{Gal}(\overline{\mathbf{Q}}/\mathbf{Q})$, Duke Math. J. **54** (1987), 179–230.
138. _____, *Topics in Galois theory*, Jones and Bartlett Publishers, Boston, MA, 1992, Lecture notes prepared by Henri Damon [Henri Darmon], With a foreword by Darmon and the author.
139. _____, *Cohomologie Galoisienne*, fifth ed., Springer-Verlag, Berlin, 1994.
140. _____, *Galois cohomology*, Springer-Verlag, Berlin, 1997, Translated from the French by Patrick Ion and revised by the author.
141. J.-P. Serre and D. B. Zagier (eds.), *Modular functions of one variable V*, Lecture Notes in Mathematics, vol. 601, Springer-Verlag, Berlin, Heidelberg, New York, 1977.
142. I. R. Shafarevich, *Algebraic number fields*, Proceedings of the International Congress of Mathematicians, Stockholm 1962 (Djursholm), Inst. Mittag-Leffler, 1963, Translated version reprinted in I. R. Shafarevich, *Collected Mathematical Papers* (Springer-Verlag, 1989), pp. 283–294, pp. 163–176.
143. S. S. Shatz, *Profinite groups, arithmetic and geometry*, Annals of Mathematics Studies, vol. 67, Princeton University Press, Princeton, NJ, 1972.
144. G. Shimura, *Introduction to the arithmetic theory of automorphic forms*, Princeton University Press, 1971.
145. _____, *The special values of zeta functions associated with cusp forms*, Comm. Pure Appl. Math. **6** (1976), 783–804.
146. J. H. Silverman, *The arithmetic of elliptic curves*, Springer-Verlag, Berlin, Heidelberg, New York, 1986.
147. L. Smithline, *Exploring slopes of p-adic modular forms*, Ph.D. thesis, University of California at Berkeley, 2000.
148. G. Stevens, *Rigid analytic modular symbols*, unpublished manuscript, 1994.
149. _____, *Coleman's $\mathcal{L}$-invariant and families of modular forms*, unpublished manuscript, 1996.
150. H. P. F. Swinnerton-Dyer, *On ℓ-adic representaions and congruences for coefficients of modular forms*, In Kuijk and Serre [**89**].
151. _____, *On ℓ-adic representations and congruences for coefficients of modular forms*, In Serre and Zagier [**141**], pp. 63–90.
152. J. Tate, *Galois cohomology*, in this volume.
153. D. L. Ulmer, *A construction of local points on elliptic curves over modular curves*, Internat. Math. Res. Notices (1995), 349–363.
154. G. van der Geer, F. Oort, and J. Steenbrink (eds.), *Arithmetic algebraic geometry (Texel, 1989)*, Birkhauser, 1991.
155. L. C. Washington, *Introduction to cyclotomic fields*, Springer-Verlag, Berlin, Heidelberg, New York, 1982.
156. _____, *Galois cohomology*, In Cornell et al. [**33**], Papers from the Instructional Conference on Number Theory and Arithmetic Geometry held at Boston University, Boston, MA, August 9–18, 1995, pp. 101–120.
157. T. Weston, *Euler systems and arithmetic geometry*, Notes from a course given by Barry Mazur at Harvard University, available at http://www.math.harvard.edu/weston/mazur.html, 1998.

158. _____, *On Selmer groups of geometric Galois representations*, Ph.D. thesis, Harvard University, 2000.
159. A. Wiles, *Modular curves and the class group of* $\mathbb{Q}(\zeta_p)$, Invent. Math. **58** (1980), 1–35.
160. _____, *The Iwasawa conjecture for totally real fields*, Ann. of Math. **131** (1990), 493–540.
161. K. Wingberg, *Der Eindeutigkeitssatz für Demuškinformationen*, Invent. Math. **70** (1982/83), 99–113.

Introduction to Iwasawa Theory for Elliptic Curves

Ralph Greenberg

Introduction to Iwasawa Theory for Elliptic Curves

Ralph Greenberg[1]

Preface

This article is intended to be a fairly self-contained introduction to some of the ideas encompassed in the phrase "Iwasawa Theory for Elliptic Curves." Starting in the late 1950s, Iwasawa proved a number of results and formulated some important conjectures concerning the behavior of ideal class groups in a certain kind of tower of number fields - the tower of subfields of a $\mathbb{Z}_p$-extension. Inspired by Iwasawa's ideas, Mazur developed an analogous theory in the late 1960s and early 1970s. Mazur's ideas are contained in his article *Rational points of abelian varieties with values in towers of number fields* [**Ma1**]. Around the same time, Manin wrote an article entitled *Cyclotomic fields and modular curves* [**Man**] which contains a simplified account of some of Mazur's ideas, using the language and techniques of Galois cohomology. All along, one of the motivations was to provide an approach to studying the behavior of the Mordell-Weil group of elliptic curves (and, more generally, abelian varieties) and the conjecture of Birch and Swinnerton-Dyer.

The first really general result concerning the Birch and Swinnerton-Dyer conjecture was the famous Coates-Wiles theorem for elliptic curves with complex multiplication, also proved in the 1970s (in [**C-W**]). If E is such an elliptic curve defined over $\mathbb{Q}$, Coates and Wiles prove that the Mordell-Weil group $E(\mathbb{Q})$ must be finite if the Hasse-Weil L-function $L(E/\mathbb{Q}, s)$ does not vanish at s=1. Their approach was also inspired by results and conjectures of Iwasawa, but involves a rather different set of ideas than in Mazur's theory. Through the 1980s, the special case of elliptic curves with complex multiplication was pursued by a number of people culminating in theorems of Rubin, the so-called "main conjectures" of Iwasawa Theory for imaginary quadratic fields, proved in [**Ru1**]. Several of Mazur's conjectures are consequences of Rubin's theorems, including conjectures 3.2 and 3.3 (stated in Chapter 3) if $F/\mathbb{Q}$ is abelian and conjecture 4.16 stated in Chapter 4, all under the assumption that the elliptic curve E has complex multiplication, is defined over $\mathbb{Q}$, and has ordinary reduction at p. A more complete account of

---

[1]Department of Mathematics, Box 354350, University of Washington, Seattle, Washington, USA 98195–4350.
E-mail address: greenber@math.washington.edu.
The research behind this article has been supported by grants from the National Science Foundation.

the history of these ideas can be found in my article *Iwasawa Theory - Past and Present* [**Gr3**].

One of the central theorems proven in Mazur's article [**Ma1**] is his Control theorem, which asserts that the Selmer group for an abelian variety behaves well Galois-theoretically in a $\mathbb{Z}_p$-extension for any prime p where the abelian variety has good, ordinary reduction. This is also proven in Manin's article [**Man**]. The goal of the present article is to give a somewhat different, and perhaps simpler proof of this theorem and to deduce various corollaries concerning the behavior of the Mordell-Weil group and the Tate-Shafarevich group in a $\mathbb{Z}_p$-extension. We will restrict attention to the case of elliptic curves, although all of the arguments do work for abelian varieties. These results will be the subject of Chapter 4. The proof of the Control theorem (theorem 4.1) presented here is based on Galois cohomology, but differs from the proofs in the articles of Mazur and of Manin in that we avoid the use of a certain result concerning the behavior of the norm map on formal groups of height 1. Instead we use the fact that if E is an elliptic curve with good, ordinary reduction at a prime p, then the p-primary subgroup of the Selmer group for E over finite extensions of $\mathbb{Q}$ (and certain infinite extensions) has a very simple and elegant description in terms of the Galois cohomology for the group $E[p^\infty]$ of p-power torsion points on E. This description is essentially contained in theorems 2.4, 2.6, and 2.9 from Chapter 2. A proof of the Control theorem along the same lines can be found in my article *Iwasawa Theory for Elliptic Curves* [**Gr2**], which goes much further in discussing this whole topic. One also finds a variety of interesting examples in that article. In the present article, my intention is to provide a more approachable introduction than [**Gr2**].

Chapter 1 is a survey of results concerning the behavior of the Mordell-Weil group of an elliptic curve in various towers of number fields, including the tower of subfields of a $\mathbb{Z}_p$-extension, but also some interesting towers of nonabelian extensions. The theme is closely related to the topic of Mazur's ICM talk [**Ma2**]. But we also discuss one rather puzzling example concerning the Mordell-Weil groups of an elliptic curve of conductor 37 in the tower of subfields of a certain Galois extension with Galois group $PGL_2(\mathbb{Z}_p)$.

This article is an expanded form of the lectures that I gave at the Park City Mathematics Institute during the summer of 1999. I would like to thank the organizers of PCMI for the invitation to give those lectures. I am also grateful to many of the participants who provided interesting questions and comments. These definitely helped me in writing this article. I also would like to express my thanks to Adam Logan for his assistance at Park City in getting notes to the participants and handling the problem sessions and to Alexandra Nichifor, Jim Mailhot and Brian Conrad for carefully reading the manuscript and pointing out many inaccuracies.

CHAPTER 1
Mordell-Weil Groups

Let E be an elliptic curve defined over $\mathbb{Q}$. Suppose that K is a finite extension of $\mathbb{Q}$. The famous Mordell-Weil theorem states that the abelian group $E(K)$ of K-rational points on E is finitely generated. That is,

$$(1.1) \qquad E(K) \cong \mathbb{Z}^r \times T$$

where $r = \mathrm{rank}_{\mathbb{Z}}(E(K))$ is a nonnegative integer and $T = E(K)_{\mathrm{tors}}$ is finite. Of course, this may fail to be true if K is an infinite extension of $\mathbb{Q}$. For example, let $\overline{\mathbb{Q}}$ denote the algebraic closure of $\mathbb{Q}$ in the complex numbers $\mathbb{C}$. Then

$$(1.2) \qquad E(\overline{\mathbb{Q}})_{\mathrm{tors}} = E(\mathbb{C})_{\mathrm{tors}} \cong (\mathbb{Q}/\mathbb{Z})^2$$

which is certainly not finite. Also, the quotient group $E(\overline{\mathbb{Q}})/E(\overline{\mathbb{Q}})_{\mathrm{tors}}$ has infinite rank. More precisely, $E(\overline{\mathbb{Q}})$ is a divisible group and so one can consider $E(\overline{\mathbb{Q}})/E(\overline{\mathbb{Q}})_{\mathrm{tors}}$ as a vector space over $\mathbb{Q}$. It turns out that its dimension is infinite. This is equivalent to the statement that $\mathrm{rank}_{\mathbb{Z}}(E(L))$ is unbounded as L varies over all finite extensions of $\mathbb{Q}$. A proof is outlined in exercises 1.5–1.8.

But there are infinite algebraic extensions K of $\mathbb{Q}$ such that $E(K)$ is finitely generated. The following simple theorem (pointed out by B. Mazur in [**Ma1**]) gives a sufficient (and obviously necessary) condition for Galois extensions.

Theorem 1.3. *Let K be a Galois extension of $\mathbb{Q}$. Assume that $E(K)_{\mathrm{tors}}$ is finite and that $\mathrm{rank}_{\mathbb{Z}}(E(L))$ is bounded as L varies over all finite extensions of $\mathbb{Q}$ contained in K. Then $E(K)$ is finitely generated.*

Proof. Choose a finite extension L of $\mathbb{Q}$ contained in K such that $\mathrm{rank}_{\mathbb{Z}}(E(L))$ is as large as possible. Then $E(K)/E(L)$ must be a torsion group. Suppose that $P \in E(K)$. Then there exists an integer $m \geq 1$ such that $mP \in E(L)$. Let $\sigma \in \mathrm{Gal}(K/L)$. Then

$$m(\sigma(P) - P) = \sigma(mP) - mP = O_E.$$

Therefore, $\sigma(P) - P \in E(K)_{\mathrm{tors}}$. If $t = |E(K)_{\mathrm{tors}}|$, then $t(\sigma(P) - P) = O_E$. That is, $\sigma(tP) = tP$ for all $\sigma \in \mathrm{Gal}(K/L)$. This implies that $tP \in E(L)$ for all $P \in E(K)$. Hence we can define a homomorphism $\varphi : E(K) \to E(L)$ by $\varphi(P) = tP$. The image of φ is finitely generated since it is a subgroup of $E(L)$. The kernel of φ is just $E(K)_{\mathrm{tors}}$. It follows that $E(K)$ is indeed finitely generated. $\square$

Verifying the hypotheses of the above theorem is hard, especially the hypothesis concerning the behavior of $\operatorname{rank}_{\mathbb{Z}}(E(L))$. Here is one interesting and rather deep result, which is a consequence of theorems of K. Kato and D. Rohrlich.

Theorem 1.4. *Suppose that E is a modular elliptic curve. Let Σ be a finite set of primes. Then $\operatorname{rank}_{\mathbb{Z}}(E(L))$ is bounded as L varies over all finite, abelian extensions of $\mathbb{Q}$ which are unramified outside Σ.*

Suppose that $\Sigma = \{p_1, \ldots, p_n\}$. By the Kronecker-Weber theorem, the fields L in the above theorem are just the finite extensions of $\mathbb{Q}$ contained in $K = \mathbb{Q}(\mu_{p_1^\infty}, \ldots, \mu_{p_n^\infty})$, where, for any prime p, we let μ_{p^∞} denote the group of p-power roots of unity in $\overline{\mathbb{Q}}$. This field K is just $\mathbb{Q}_\Sigma^{\mathrm{ab}}$, the maximal abelian extension of $\mathbb{Q}$ unramified outside Σ. A theorem of K. Ribet [**Ri**] implies that $E(\mathbb{Q}_\Sigma^{\mathrm{ab}})_{\mathrm{tors}}$ is finite. (His theorem is much stronger than this. It states that $A(F\mathbb{Q}^{\mathrm{ab}})_{\mathrm{tors}}$ is finite, where F is any number field, A is an abelian variety defined over F, and $\mathbb{Q}^{\mathrm{ab}}$ is the maximal abelian extension of $\mathbb{Q}$.) Thus the hypotheses in theorem 1.3 are satisfied for $K = \mathbb{Q}_\Sigma^{\mathrm{ab}}$ and one obtains the following consequence:

Theorem 1.5. *Suppose that E is modular and that Σ is any finite set of primes. Then $E(\mathbb{Q}_\Sigma^{\mathrm{ab}})$ is finitely generated.*

One result which we will prove completely is the following theorem of B. Mazur (from [**Ma1**]). We will assume that K/F is a $\mathbb{Z}_p$-extension. That is, K/F is a Galois extension and $\operatorname{Gal}(K/F) \cong \mathbb{Z}_p$, the additive group of p-adic integers. Since $\mathbb{Z}_p = \varprojlim \mathbb{Z}/p^n\mathbb{Z}$, one could equivalently require that $K = \bigcup_n F_n$, where F_n is a cyclic extension of F of degree p^n and $F_n \subset F_{n+1}$ for all $n \geq 0$. If E is an elliptic curve defined over a number field F, we let $\mathrm{III}_E(F)$ denote the corresponding Tate-Shafarevich group (which will be defined in chapter 2), and $\mathrm{III}_E(F)_p$ denotes its p-primary subgroup. Here is Mazur's theorem.

Theorem 1.6. *Suppose that E is an elliptic curve defined over a number field F. Let p be a prime such that E has good, ordinary reduction at all primes of F lying over p. Assume that both $E(F)$ and $\mathrm{III}_E(F)_p$ are finite. Let $K = \bigcup_n F_n$ be a $\mathbb{Z}_p$-extension of F. Then $\operatorname{rank}_{\mathbb{Z}}(E(F_n))$ is bounded for $n \geq 0$.*

The proof of this theorem is a good illustration of the ideas of Iwasawa theory, and will be one of the main goals of this article. The theorem will be a consequence (corollary 4.9) of Mazur's "Control Theorem" (which is the main result of chapter 4) together with basic results about the structure of modules over a certain ring Λ (which will be the subject of chapter 3). But instead of studying Mordell-Weil groups directly, we study the behavior of Selmer groups which are introduced in chapter 2. The assumption that E has good, ordinary reduction at primes over p is crucial. To briefly recall the definition, suppose that v is a prime of F where E has good reduction and let $\widetilde{E}_v$ denote the reduction of E modulo v (so that $\widetilde{E}_v$ is an elliptic curve over the residue field f_v of v). Let $p = \mathrm{char}(f_v)$ and let $\overline{f}_v$ denote an algebraic closure of f_v. If $\widetilde{E}_v(\overline{f}_v)$ has points of order p, then we say that $\widetilde{E}_v$ is ordinary and that E has ordinary reduction at v. Otherwise, we say that $\widetilde{E}_v$ is supersingular and that E has supersingular reduction at v.

If F is any number field and p is any prime, then there is one $\mathbb{Z}_p$-extension of F which can be described explicitly. It is a subfield of $F(\mu_{p^\infty})$. There is a canonical homomorphism

$$\chi : \operatorname{Gal}(F(\mu_{p^\infty})/F) \to \mathbb{Z}_p^\times$$

defined in the following way: Let $\sigma \in \text{Gal}(F(\mu_{p^\infty})/F)$. Suppose that ζ_{p^n} is a primitive p^n-th root of unity. Then $\sigma(\zeta_{p^n}) = \zeta_{p^n}^{u_n}$, where $u_n \in \mathbb{Z}$ and $(u_n, p) = 1$. (Each u_n is only well-defined modulo p^n.) It is easy to see that $\{u_n\}$ is a Cauchy sequence in $\mathbb{Z}_p$ converging to a well-defined element $u \in \mathbb{Z}_p^\times$. We define $\chi(\sigma) = u$. Then $\sigma(\zeta) = \zeta^{\chi(\sigma)}$ for all $\zeta \in \mu_{p^\infty}$. (See exercises 1.1 and 1.2.) The homomorphism χ is continuous, injective, and its image has finite index in $\mathbb{Z}_p^\times$. Using the p-adic log, one sees that $\mathbb{Z}_p^\times \cong (\mathbb{Z}/p\mathbb{Z})^\times \times \mathbb{Z}_p$ if p is odd (or $\{\pm 1\} \times \mathbb{Z}_2$ if $p = 2$). One can then show that $\text{im}(\chi) \cong \Delta \times \Gamma$, where Δ is a finite group and $\Gamma \cong \mathbb{Z}_p$. Therefore, there is a unique subfield F_∞ of $F(\mu_{p^\infty})$ such that $\text{Gal}(F_\infty/F) \cong \Gamma \cong \mathbb{Z}_p$. F_∞ is called the "cyclotomic $\mathbb{Z}_p$-extension" of F.

If F is totally real, then it is conjectured that the only $\mathbb{Z}_p$-extension of F is the one constructed above. This is known as Leopoldt's conjecture. But if F is not totally real, then it turns out that there are infinitely many different $\mathbb{Z}_p$-extensions of F. For example, if F is an imaginary quadratic field, then, for any prime p, there exists a Galois extension $\widetilde{F}$ of F which contains all $\mathbb{Z}_p$-extensions of F and such that $\text{Gal}(\widetilde{F}/F) \cong \mathbb{Z}_p^2$. Thus $\text{Gal}(\widetilde{F}/F)$ has infinitely many (in fact uncountably many) quotients isomorphic to $\mathbb{Z}_p$, and each such quotient corresponds to $\text{Gal}(F_\infty/F)$ for some F_∞. All of this can be proved by class field theory. There are two such $\mathbb{Z}_p$-extensions F_∞/F which have the property that $F_\infty/\mathbb{Q}$ is Galois. One of them is the cyclotomic $\mathbb{Z}_p$-extension F_∞^c/F, and then $\text{Gal}(F_\infty^c/\mathbb{Q}) \cong (\mathbb{Z}/2\mathbb{Z}) \times \mathbb{Z}_p$. In fact, $F_\infty^c = F\mathbb{Q}_\infty$, where $\mathbb{Q}_\infty \subset \mathbb{Q}(\mu_{p^\infty})$ is the unique $\mathbb{Z}_p$-extension of $\mathbb{Q}$. The other is the so-called anticyclotomic $\mathbb{Z}_p$-extension F_∞^{ac}/F which is characterized by $F_\infty^{\text{ac}} = \bigcup_n F_n^{\text{ac}}$, $F_n^{\text{ac}} \subset F_{n+1}^{\text{ac}}$, where F_n^{ac} is Galois over $\mathbb{Q}$ and $\text{Gal}(F_n^{\text{ac}}/\mathbb{Q})$ is isomorphic to the dihedral group of order $2p^n$.

We can now state two more theorems, both of which are rather deep and will not be proven. (Their proofs are a combination of results of K. Rubin, D. Rohrlich, Gross-Zagier and myself.)

Theorem 1.7. *Suppose that F is an imaginary quadratic field and that E is an elliptic curve defined over $\mathbb{Q}$ such that $\text{End}(E) \otimes_{\mathbb{Z}} \mathbb{Q} \cong F$. Suppose that p is any prime and that $F_\infty = \bigcup_n F_n$ is a $\mathbb{Z}_p$-extension of F. Assume that $F_\infty \neq F_\infty^{\text{ac}}$. Then $\text{rank}_{\mathbb{Z}}(E(F_n))$ is bounded for $n \geq 0$.*

Theorem 1.8. *Suppose that F and E are as in theorem 1.7. Let p be a prime where E has good reduction. Let $F_\infty^{\text{ac}} = \bigcup_n F_n^{\text{ac}}$ be the anticyclotomic $\mathbb{Z}_p$-extension.*

  (i) *If E has supersingular reduction at p, then $\text{rank}_{\mathbb{Z}}(E(F_n^{\text{ac}}))$ is unbounded as n varies.*
  (ii) *Assume that E has ordinary reduction at p. If the Hasse-Weil L-series $L(E/\mathbb{Q}, s)$ has an odd order zero at $s = 1$, then $\text{rank}_{\mathbb{Z}}(E(F_n^{\text{ac}}))$ is unbounded as n varies. But if $L(E/\mathbb{Q}, s)$ has an even order zero at $s = 1$, then $\text{rank}_{\mathbb{Z}}(E(F_n^{\text{ac}}))$ is bounded for $n \geq 0$.*

What one can prove is actually quite a bit more precise. Assume first that E has ordinary reduction at p and that $L(E/\mathbb{Q}, s)$ has an odd order zero at $s = 1$. Then the rank of $E(F_n^{\text{ac}})$ grows quite regularly and is given by the formula

$$(1.9) \qquad \text{rank}_{\mathbb{Z}}(E(F_n^{\text{ac}})) = 2p^n + c.$$

for all sufficiently large n, where c is some constant. That is, the rank increases by exactly $2\varphi(p^n)$ at the n-th layer for all $n \gg 0$. (Compare this with exercise 1.9.)

On the other hand, if E has supersingular reduction, one has

$$(1.10) \qquad \mathrm{rank}_{\mathbb{Z}}(E(F_n^{\mathrm{ac}})) - \mathrm{rank}_{\mathbb{Z}}(E(F_{n-1}^{\mathrm{ac}})) = \varepsilon_n \varphi(p^n)$$

for all sufficiently large n, where ε_n is either 0 or 2, depending just on the parity of n. That is, new points of infinite order appear in every other layer of the anticyclotomic $\mathbb{Z}_p$-extension of F. More precisely, if $L(E/\mathbb{Q}, s)$ has an even order zero at $s = 1$, then $\varepsilon_n = 0$ for odd n and $\varepsilon_n = 2$ for even n (and $n \gg 0$). The reverse holds when $\mathrm{ord}_{s=1}(L(E/\mathbb{Q}, s))$ is odd. We should remark that $\mathrm{rank}_{\mathbb{Z}}(E(L))$ must be even if L is any finite extension of F. This is simply because $E(L)$ is a module over the ring $R = \mathrm{End}(E)$, this ring is an order in F and hence $\mathrm{rank}_{\mathbb{Z}}(R) = 2$, and therefore $\mathrm{rank}_{\mathbb{Z}}(E(L)) = 2 \, \mathrm{rank}_R(E(L))$.

The above theorems suggest that there are some systematic patterns influencing the behavior of $\mathrm{rank}_{\mathbb{Z}}(E(L))$ as L varies over finite extensions of $\mathbb{Q}$. The anticyclotomic $\mathbb{Z}_p$-extension of an imaginary quadratic field seems to play a special role. This turns out to be related to the following group theoretic fact. If G is a dihedral group, then g and g^{-1} are conjugate elements for all $g \in G$. The same statement is true for projective limits of dihedral groups such as $\mathrm{Gal}(F_\infty^{\mathrm{ac}}/\mathbb{Q})$, where F_∞^{ac} is the anticyclotomic $\mathbb{Z}_p$-extension of any imaginary quadratic field F (and for any prime p).

Another profinite group with the property that g and g^{-1} are always conjugate is $G = \mathrm{PGL}_2(\mathbb{Z}_p)$. (See exercise 1.11.) This group can also be realized as a Galois group. For example, let E be an elliptic curve defined over $\mathbb{Q}$. For any prime p, let $T_p(E)$ denote the Tate module for E and p. Thus, $T_p(E) \cong \mathbb{Z}_p^2$ and $\mathrm{Gal}(\overline{\mathbb{Q}}/\mathbb{Q})$ acts continuously and $\mathbb{Z}_p$-linearly on $T_p(E)$. This gives a homomorphism

$$\rho = \rho_{E,p} : \mathrm{Gal}(\overline{\mathbb{Q}}/\mathbb{Q}) \to \mathrm{Aut}_{\mathbb{Z}_p}(T_p(E)) \cong \mathrm{GL}_2(\mathbb{Z}_p).$$

There is also an induced homomorphism $\widetilde{\rho} : \mathrm{Gal}(\overline{\mathbb{Q}}/\mathbb{Q}) \to \mathrm{PGL}_2(\mathbb{Z}_p)$. If we let $M = \overline{\mathbb{Q}}^{\ker(\rho)}$, then $\mathrm{Gal}(M/\mathbb{Q}) \cong \mathrm{im}(\rho)$. If E does not have complex multiplication (i.e., if $\mathrm{End}(E) = \mathbb{Z}$), then a theorem of Serre states that $\mathrm{im}(\rho)$ is of finite index in $\mathrm{GL}_2(\mathbb{Z}_p)$ for all p, with equality for all but finitely many p. Thus, for sufficiently large p, $\mathrm{Gal}(M/\mathbb{Q}) \cong \mathrm{GL}_2(\mathbb{Z}_p)$, and M contains a subfield $K = \overline{\mathbb{Q}}^{\ker(\widetilde{\rho})}$ such that $\mathrm{Gal}(K/\mathbb{Q}) \cong \mathrm{PGL}_2(\mathbb{Z}_p)$.

The symmetric groups S_n for $n \geq 1$ also have the above conjugacy property, as well as any group of exponent 2. Suppose that G is any group with this conjugacy property. Then it is easy to see that the following property holds: *If χ is the character of any finite dimension representation $\sigma : G \to \mathrm{GL}_n(\mathbb{C})$, then χ is real-valued.* Conversely, if G is a profinite group with this property, then G also has the property that g and g^{-1} are conjugate for all $g \in G$.

If K is any finite Galois extension of $\mathbb{Q}$, then there is a well-known factorization of the Dedekind zeta function $\zeta_K(s)$:

$$\zeta_K(s) = \prod_\chi L(\chi, s)^{d_\chi}$$

where χ runs over the characters of all irreducible representations of $\mathrm{Gal}(K/\mathbb{Q})$ over $\mathbb{C}$ and d_χ denotes the dimension of the corresponding representation. The factor $L(\chi, s)$ is the Artin L-function associated to χ (or to the $\mathbb{C}$-representation it determines), which is defined by a certain Euler product. The well-known functional equation for $\zeta_K(s)$ relates the values $\zeta_K(s)$ and $\zeta_K(1-s)$. The functional equation for $L(\chi, s)$ relates the values $L(\chi, s)$ and $L(\overline{\chi}, 1-s)$. Here $\overline{\chi}$ is the character of

Gal($K/\mathbb{Q}$) obtained by complex conjugation, which is also irreducible. If E is an elliptic curve defined over $\mathbb{Q}$, then the Hasse-Weil L-function $L(E/K, s)$ is conjectured to have an analytic continuation to the entire complex plane and to satisfy a functional equation relating the values $L(E/K, s)$ to $L(E/K, 2-s)$. But one can define L-functions $L(E/\mathbb{Q}, \chi, s)$ for every irreducible character χ of Gal($K/\mathbb{Q}$), again by a certain Euler product. One then has a factorization (which is proved by a formal calculation with the Euler factors):

$$(1.11) \qquad L(E/K, s) = \prod_\chi L(E/\mathbb{Q}, \chi, s)^{d_\chi}$$

where χ runs over the characters of the irreducible $\mathbb{C}$-representations of Gal($K/\mathbb{Q}$) as before. There is a conjectural functional equation which relates $L(E/\mathbb{Q}, \chi, s)$ to $L(E/\mathbb{Q}, \overline{\chi}, 2-s)$. Let

$$\Lambda(E/\mathbb{Q}, \chi, s) = A_\chi^s \Gamma(s)^{d_\chi} L(E/\mathbb{Q}, \chi, s),$$

where A_χ is a certain positive constant. Then the conjectured functional equation is

$$(1.12) \qquad \Lambda(E/\mathbb{Q}, \chi, s) = w(E, \chi)\Lambda(E/\mathbb{Q}, \overline{\chi}, 2-s)$$

where the so-called root number $w(E, \chi)$ is a complex number of absolute value 1. If χ is real-valued, then $\overline{\chi} = \chi$, and $w(E, \chi) = \pm 1$. If one considers the power series expansion of $\Lambda(E/\mathbb{Q}, \chi, s)$ about $s=1$ (assuming the conjectured analyticity), it is easy to see that

$$w(E, \chi) = (-1)^{\mathrm{ord}_{s=1}(L(E/\mathbb{Q}, \chi, s)}.$$

In particular, if $w(E, \chi) = -1$, then $L(E/\mathbb{Q}, \chi, 1) = 0$, and this would contribute at least d_χ to the order of vanishing of $L(E/K, s)$ at $s=1$.

If σ is the irreducible representation with character χ, then we can regard σ as a continuous, finite-dimensional, complex representation of $G_\mathbb{Q}$. Every such representation σ factors through Gal($K/\mathbb{Q}$) for some finite, Galois extension $K/\mathbb{Q}$. The definition of $L(E/\mathbb{Q}, \chi, s)$ does not depend on the choice of $K/\mathbb{Q}$.

The conjectures about L-functions and their analytic continuation even provide conjectural values for the root numbers. We will mention one useful formula. Let w_E denote the root number for $L(E/\mathbb{Q}, s)$. Thus $w_E = (-1)^{\mathrm{ord}_{s=1}(L(E/\mathbb{Q}, s))}$. Let N_E denote the conductor of the elliptic curve E (which is a product of certain powers of the primes where E has bad reduction). If σ is the representation of $G_\mathbb{Q}$ with character χ, let $K_\chi = \overline{\mathbb{Q}}^{\ker(\sigma)}$. Thus σ can be regarded as a faithful representation of Gal($K_\chi/\mathbb{Q}$). We let N_χ denote the conductor of χ, which is a certain product of the primes ramified in $K_\chi/\mathbb{Q}$. Assume that $(N_E, N_\chi) = 1$. Let $\delta_\chi = \det(\sigma)$, which can be regarded as a Dirichlet character. Then it turns out that

$$(1.13) \qquad w(E, \chi) = \delta_\chi(-N_E) w_E^{d_\chi}.$$

For a proof that the conjectural root number turns out to be given by (1.13), see proposition 10 in [**Ro4**]. It follows that under the hypothesis that $(N_E, N_\chi) = 1$, the value of $w(E, \chi)$ depends only on δ_χ, N_E, w_E, and the parity of d_χ.

Suppose for example that $d_\chi = 1$ so that χ can be regarded as a Dirichlet character whose conductor is relatively prime to N_E. The L-function $L(E/\mathbb{Q}, \chi, s)$

then has a simple description in terms of $L(E/\mathbb{Q}, s)$. If we write $L(E/\mathbb{Q}, s)$ as the Dirichlet series $\sum_{n=1}^{\infty} a_n n^{-s}$, then

$$L(E/\mathbb{Q}, \chi, s) = \sum_{n=1}^{\infty} a_n \chi(n) n^{-s}.$$

It is only when χ is real-valued that the functional equation can force the function $L(E/\mathbb{Q}, \chi, s)$ to vanish at $s = 1$. According to (1.13), the sign in the functional equation is then $w(E, \chi) = \chi(-N_E) w_E$.

Suppose that K is Galois over $\mathbb{Q}$ and $G = \text{Gal}(K/\mathbb{Q})$ is dihedral of order $2n$, where n is odd. The irreducible representations of G over $\mathbb{C}$ are all 2-dimensional except for the two which factor through $G/G' \cong \mathbb{Z}/2\mathbb{Z}$. Let σ be any such 2-dimensional representation and let χ denote its character. Then χ is real-valued and δ_χ is the nontrivial character φ of $\text{Gal}(F/\mathbb{Q})$, where $F = K^{G'}$ is the unique quadratic field contained in K. Assuming that the primes dividing N_E are unramified in $K/\mathbb{Q}$, the sign in the functional equation for $L(E/\mathbb{Q}, \chi, s)$ is $\varphi(-N_E)$.

As another example, suppose that E' is an elliptic curve without complex multiplication whose conductor is relatively prime to the conductor of E. Let p be a prime. By Serre's theorem, the representation $\rho_{E'} : G_\mathbb{Q} \to \text{GL}_2(\mathbb{Z}_p)$ giving the action of $G_\mathbb{Q}$ on $E'[p^\infty]$ is surjective for all but finitely many primes p. Let p be any such prime. Thus $\text{Gal}(\mathbb{Q}(E'[p^\infty])/\mathbb{Q}) \cong \text{GL}_2(\mathbb{Z}_p)$ and $\mathbb{Q}(E'[p^\infty])$ contains a subfield K such that $G = \text{Gal}(K/\mathbb{Q}) \cong \text{PGL}_2(\mathbb{Z}_p)$. ($K$ is the field $\overline{\mathbb{Q}}^{\ker(\widetilde{\rho}_{E',p})}$.) If χ is the character of an irreducible representation of G, then χ is real-valued. This follows from exercise 1.11. Also d_χ is even for all but finitely many χ's, and δ_χ is one of the two characters of $G/G' \cong \mathbb{Z}/2\mathbb{Z}$. The field $F = K^{G'}$ is the unique quadratic subfield of $\mathbb{Q}(\mu_{p^\infty}) : F = \mathbb{Q}\left(\sqrt{(-1)^{\frac{p-1}{2}} p}\right)$. Each of these two characters occurs as δ_χ for infinitely many χ's. If d_χ is even, then $w(E, \chi) = \delta_\chi(-N_E)$. In particular, if $\left(\frac{-N_E}{p}\right) = -1$ then both signs ± 1 occur infinitely often in the functional equations for $L(E/\mathbb{Q}, \chi, s)$, but if $\left(\frac{-N_E}{p}\right) = +1$, then the signs are $+1$ (with finitely many exceptions) and so the functional equation will not force $L(E/\mathbb{Q}, \chi, 1)$ to vanish.

All of these remarks are relevant to the behavior of the ranks of Mordell-Weil groups because of the Birch and Swinnerton-Dyer conjecture which states in part that, for any finite extension K of $\mathbb{Q}$, we should have

(1.14) $$\text{rank}_\mathbb{Z}(E(K)) = \text{ord}_{s=1}(L(E/K, s)).$$

There is a refined version of this conjecture which is suggested by the factorization (1.11). Assume that $K/\mathbb{Q}$ is Galois and let $G = \text{Gal}(K/\mathbb{Q})$. Then G acts on $E(K)$ and one can consider the finite-dimensional representation space $E(K) \otimes_\mathbb{Z} \mathbb{C}$. Its dimension is $\text{rank}_\mathbb{Z}(E(K))$. We have a decomposition as a direct sum of irreducible representations

(1.15) $$E(K) \otimes_\mathbb{Z} \mathbb{C} \cong \bigoplus_\chi V_\chi^{m_\chi(E)}$$

where χ varies over the irreducible characters of G and V_χ denotes the underlying vector space for the corresponding representation of G. The multiplicities $m_\chi(E)$ are nonnegative integers which depend only on E and χ, and not on the choice of Galois extension $K/\mathbb{Q}$. (See exercise 1.3.) If χ_0 is the character of the trivial representation of G, then $m_{\chi_0}(E) = \text{rank}_{\mathbb{Z}}(E(\mathbb{Q}))$. Also, if τ is any automorphism of the field $\mathbb{Q}(\chi)$ generated by the values of χ, then the conjugate character χ^τ is the character of another irreducible representation of G and one can easily show that $m_{\chi^\tau}(E) = m_\chi(E)$ for all such τ. (This is exercise 1.4.)

If $K/\mathbb{Q}$ is a finite Galois extension, then it is clear that

(1.16) $$\text{rank}_{\mathbb{Z}}(E(K)) = \sum_\chi d_\chi m_\chi(E)$$

where χ runs over the irreducible character of $G = \text{Gal}(K/\mathbb{Q})$. Assuming analytic continuation, it is also clear from (1.11) that

(1.17) $$\text{ord}_{s=1}(L(E/K, s)) = \sum_\chi d_\chi \, \text{ord}_{s=1}(L(E/\mathbb{Q}, \chi, s)).$$

The refined version of the Birch and Swinnerton-Dyer conjecture states that

(1.18) $$m_\chi(E) = \text{ord}_{s=1}(L(E/\mathbb{Q}, \chi, s))$$

for every irreducible character χ of G.

We can now say a little more about the proofs of some of the theorems stated earlier. If we use (1.16) for $K = L$, where L is any finite, abelian extension of $\mathbb{Q}$ unramified outside Σ, then theorem 1.4 amounts to the assertion that $m_\chi(E) = 0$ for all but finitely many one-dimensional characters χ such that $K_\chi/\mathbb{Q}$ is unramified outside Σ. Such χ's can be identified with Dirichlet characters whose conductor is divisible only by primes in Σ. Therefore, theorem 1.4 is a consequence of the following two theorems.

Theorem 1.19 (Rohrlich [**Ro1, Ro3**]). *Suppose that E is a modular elliptic curve and let Σ be a finite set of primes. Then $L(E/\mathbb{Q}, \chi, 1) \neq 0$ for all but finitely many Dirichlet character χ whose conductors are divisible only by primes in Σ.*

Theorem 1.20 (Kato [**Ka**]). *Suppose that E is a modular elliptic curve and that χ is a one-dimensional character of $\text{Gal}(K/\mathbb{Q})$ for some abelian extension $K/\mathbb{Q}$. Assume that $L(E/\mathbb{Q}, \chi, 1) \neq 0$. Then $m_\chi(E) = 0$.*

As a special case, let $\mathbb{Q}_\infty = \bigcup_n \mathbb{Q}_n$ be the cyclotomic $\mathbb{Z}_p$-extension of $\mathbb{Q}$, where p is any prime. Then theorem 1.4 implies that $\text{rank}_{\mathbb{Z}}(E(\mathbb{Q}_n))$ is bounded, which means that there is an integer $n_0 = n_0(E, p)$ such that

$$\text{rank}(E(\mathbb{Q}_n)) = \text{rank}(E(\mathbb{Q}_{n_0}))$$

for all $n \geq n_0$. Can $n_0(E, p)$ be chosen independently of E and/or p? We know nothing about this question. But here are some examples.

(1) $E: y^2 + xy = x^3 - 3x + 1$ and $p = 3$. Then $N_E = 34$. It turns out that $\text{rank}_{\mathbb{Z}}(E(\mathbb{Q})) = 0$, but $\text{rank}_{\mathbb{Z}}(E(\mathbb{Q}_n)) = 2$ for all $n \geq 1$. For each of the two faithful characters χ of $\text{Gal}(\mathbb{Q}_1/\mathbb{Q})$, we have $m_\chi(E) = 1$. A point of infinite order on $E(\mathbb{Q}_1)$ is $P = (\beta, -\beta)$ where $\beta = \zeta_9 + \zeta_9^{-1}$ and ζ_9 is a primitive 9-th root of unity. (Note: $\mathbb{Q}_1$ is the maximal real subfield of $\mathbb{Q}(\zeta_9)$ and is generated over $\mathbb{Q}$ by β.) It is easy to check that $P \in E(\mathbb{Q}_1)$. To see that it has infinite order, it is enough to verify that $E(\mathbb{Q}_1)_{\text{tors}} = E(\mathbb{Q})_{\text{tors}}$. Now $E(\mathbb{Q})_{\text{tors}}$ also has order 2. (Cremona's tables [**Cr**] give all of the needed data about this and other examples.) We must show that $E(\mathbb{Q}_1)_{\text{tors}}$ also has order 2. Let η denote the unique prime of $\mathbb{Q}_1$ lying above 3. The residue field of η is $\mathbb{F}_3$ and one can verify that $\widetilde{E}(\mathbb{F}_3)$ has order 6. But the reduction map $E(\mathbb{Q}_1)_{\text{tors}} \to \widetilde{E}(\mathbb{F}_3)$ is injective on the prime-to-3 torsion. (See proposition 3.1

in [**Si1**]. This standard result will also be very helpful in several exercises.) Also by exercise 1.13, one sees that the 3-primary subgroup of $E(\mathbb{Q}_1)_{\text{tors}}$ is trivial and so $E(\mathbb{Q}_1)_{\text{tors}}$ does indeed have order 2.

(2) $E : y^2 + xy = x^3 - 115x + 392$ and $p = 2$. Then $N_E = 195$. It turns out that $E(\mathbb{Q})$ is finite (of order 8). Now $\mathbb{Q}_1 = \mathbb{Q}(\sqrt{2})$ and $P = (0, 14\sqrt{2}) \in E(\mathbb{Q}_1)$. But one can check that $E(\mathbb{Q}_1)_{\text{tors}}$ still has order 8. (Consider the reduction of E at the primes of $\mathbb{Q}_1$ lying over 2 and over 7.) Hence P must have infinite order. It turns out that $\text{rank}(E(\mathbb{Q}_1)) = 1$. Now $\mathbb{Q}_2 = \mathbb{Q}_1(\sqrt{1+\sqrt{2}})$ and, with the assistance of a computer, one can explicitly find a point $Q \in E(\mathbb{Q}_2)$ such that $Tr_{\mathbb{Q}_2/\mathbb{Q}_1}(Q) = O_E$. One can also verify that $E(\mathbb{Q}_2)_{\text{tors}} = E(\mathbb{Q})_{\text{tors}}$. (See exercise 3.11, where a proof that $E(\mathbb{Q}_\infty)_{\text{tors}} = E(\mathbb{Q})_{\text{tors}}$ is outlined.) Using these facts, it is not hard to show that Q and P are independent and that $\text{rank}(E(\mathbb{Q}_2)) \geq 3$. For a more detailed discussion of this example, see [**Gr2**], pages 137-8, where it is shown that $\text{rank}(E(\mathbb{Q}_2)) = 3$ and that there is no further increase in the rank, i.e., $\text{rank}(E(\mathbb{Q}_n)) = 3$ for all $n \geq 2$.

Remark. According to a theorem of Rohrlich [**Ro5**], one can find examples of elliptic curves E such that $\text{rank}(E(\mathbb{Q}_n)) > \text{rank}(E(\mathbb{Q}_{n-1}))$ if $[\mathbb{Q}_n : \mathbb{Q}] \leq 9$, i.e., if $p = 2$, $n = 1, 2$, or 3, $p = 3$, $n = 1$ or 2, and $p = 5$ or 7, $n = 1$.

Let Σ be a fixed finite set of primes, E a fixed elliptic curve defined over $\mathbb{Q}$, and d a fixed positive integer. Theorem 1.4 suggests the following question. Let χ vary over all irreducible characters of $\text{Gal}(\mathbb{Q}_\Sigma/\mathbb{Q})$ such that $d_\chi = d$. What can one say about the behavior of $m_\chi(E)$? (Note: $\mathbb{Q}_\Sigma$ denotes the compositum of all finite extensions of $\mathbb{Q}$ unramified outside Σ. A continuous finite-dimensional representation $\sigma : \text{Gal}(\mathbb{Q}_\Sigma/\mathbb{Q}) \to \text{GL}_d(\mathbb{C})$ must factor through a quotient group $\text{Gal}(K/\mathbb{Q})$ for some finite Galois extension K of $\mathbb{Q}$. We are letting χ vary over the characters of all such irreducible representations σ.)

Theorems 1.4, 1.19, and 1.20 concern the case where $d = 1$. Only finitely many χ's will be real-valued. Thus theorem 1.19 really concerns the values $L(E/\mathbb{Q}, \chi, 1)$ where $\chi \neq \overline{\chi}$ and these values *cannot* be forced to vanish by the functional equations. If χ is real-valued, then theorem 1.20 is a result of Kolyvagin. In the special case where E has complex multiplication, then theorem 1.20 is a consequence of a theorem of Rubin (generalizing the Coates-Wiles Theorem proved in [**C-W**].

Let $d = 2$. There may be infinitely many χ's such that $d_\chi = 2$ and $\chi = \overline{\chi}$. Suppose for example that $\mathbb{Q}_\Sigma$ contains an imaginary quadratic field F and let F_∞^{ac} be the anticyclotomic $\mathbb{Z}_p$-extension of F for some $p \in \Sigma$. Then $F_\infty^{\text{ac}} \subset \mathbb{Q}_\Sigma$. (See exercise 1.14.) All the irreducible characters of $\text{Gal}(F_\infty^{\text{ac}}/\mathbb{Q})$ will be real-valued and all but finitely many will be 2-dimensional. In light of theorems 1.19 and 1.20, it is natural to make the following *tentative* conjecture.

(1) *Let χ vary over all irreducible 2-dimensional characters of $\text{Gal}(\mathbb{Q}_\Sigma/\mathbb{Q})$ such that $\chi \neq \overline{\chi}$. Then $L(E/\mathbb{Q}, \chi, 1) \neq 0$ and $m_\chi(E) = 0$ for all but finitely many such χ's.*

As for the real-valued characters, the following conjecture seems natural (but also very tentative.)

(2) *Assume that E does not have complex multiplication. Let χ vary over all irreducible 2-dimensional characters of $\text{Gal}(\mathbb{Q}_\Sigma/\mathbb{Q})$ such that $\chi = \overline{\chi}$. Then, for all but finitely many such χ's, (a) $L(E/\mathbb{Q}, \chi, 1) \neq 0$ and $m_\chi(E) = 0$ if $w(E, \chi) = +1$, and (b) $L'(E/\mathbb{Q}, \chi, 1) \neq 0$ and $m_\chi(E) = 1$ if $w(E, \chi) = -1$.*

Virtually nothing is known to support these conjectures. It may suffice to allow χ to vary over irreducible representations of any fixed dimension d, but as a later example will show, some restriction on d_χ is necessary at least for the second conjecture.

Theorem 1.7 is a consequence of the first conjecture (and is really the only supporting evidence we have). Suppose that E satisfies the hypotheses in that theorem. If F_∞ is the cyclotomic $\mathbb{Z}_p$-extension of the imaginary quadratic field F, then the conclusion of theorem 1.7 follows from theorem 1.4 and is valid for any modular elliptic curve. If F_∞ is any other $\mathbb{Z}_p$-extension of F (except F_∞^{ac}), then F_n will not be Galois over $\mathbb{Q}$ for $n \gg 0$. Nevertheless it is possible to factor $L(E/F_n, s)$ as a certain product of L-functions of the form $L(E/\mathbb{Q}, \chi, s)$ for suitable irreducible characters χ of $\mathrm{Gal}(K/\mathbb{Q})$, where K is a Galois extension of $\mathbb{Q}$ containing F_n. Also, $\mathrm{rank}_{\mathbb{Z}}(E(F_n))$ can be expressed as the corresponding sum of the quantities $m_\chi(E)$. The χ's which occur here will mostly be the characters of the irreducible 2-dimensional representations of $G_{\mathbb{Q}}$ induced form the 1-dimensional representations ρ of G_F which factor through $\mathrm{Gal}(F_n/F)$. With finitely many exceptions, we will have $\chi \neq \overline{\chi}$ and so the first conjecture above would imply that $L(E/\mathbb{Q}, \chi, 1)$ is nonzero and $m_\chi(E) = 0$ for all but finitely many such χ's. One can actually prove these statements (assuming E has complex multiplication by F) and deduce that $\mathrm{rank}(E(F_n))$ is indeed bounded.

Theorem 1.8 is somewhat different, but is still an illustration of the same philosophy. The difference is that if χ is an irreducible character of dimension 2 for the dihedral group $\mathrm{Gal}(F_n^{\mathrm{ac}}/\mathbb{Q})$ and if E has complex multiplication by F, then $L(E/\mathbb{Q}, \chi, s)$ can itself be factored in a natural way as a product of two L-functions and it is the signs in the functional equations for these L-functions which explain the results stated in theorem 1.8. This is the so-called "exceptional case" in the terminology of Mazur's ICM talk [**Ma2**] on this topic.

The next example was a real surprise to me. The possibility of such an example was suggested by L. Howe. Let E be the elliptic curve defined by the equation

(1.21) $$E : y^2 + y = x^3 - x.$$

Then E has conductor 37. (There are four elliptic curves of conductor 37, falling into two isogeny classes. We could take E to be any of these curves for this example.) Let E' be the elliptic curve defined by the equation

(1.22) $$E' : y^2 + xy + y = x^3 + x^2 - 8x + 6.$$

The conductor of E' is 1225. The special property that this elliptic curve has is that it admits an isogeny defined over $\mathbb{Q}$ of degree 37. That is, there is another elliptic curve E'' defined over $\mathbb{Q}$ (whose equation we won't bother to write down) and a $\mathbb{Q}$-isogeny $\varphi : E' \to E''$ such that $\Phi = \ker(\varphi)$ has order 37. The elliptic curve E' doesn't have complex multiplication and so Serre's theorem implies that $\mathrm{Gal}(\mathbb{Q}(E'[p^\infty])/\mathbb{Q}) \cong \mathrm{GL}_2(\mathbb{Z}_p)$ for all but finitely many primes p. Pick one such prime (except $p = 2$ or $p = 37$). As mentioned earlier, $\mathbb{Q}(E'[p^\infty])$ then contains a subfield K such that $\mathrm{Gal}(K/\mathbb{Q}) \cong \mathrm{PGL}_2(\mathbb{Z}_p)$ and every irreducible character χ of $\mathrm{Gal}(K/\mathbb{Q})$ is real-valued. Note that $N_E = 37$ and N_χ is a product of the primes 5, 7, and p. Hence $(N_E, N_\chi) = 1$ and so (as discussed earlier), we have $w(E, \chi) = \delta_\chi(-37) w_E^{d_\chi}$. (For the above choice of E, we have $w_E = -1$.) Assume now that $(\frac{-37}{p}) = +1$. Then $w(E, \chi) = +1$ for all irreducible characters χ of $\mathrm{Gal}(K/\mathbb{Q})$ which have even dimension (which means all but four). Thus one might guess that $L(E/\mathbb{Q}, \chi, 1) \neq 0$ and $m_\chi(E) = 0$ for all but finitely many such χ's and therefore that rank($E(L)$) is bounded as L varies over the finite extensions of $\mathbb{Q}$ contained in K. Using exercise 1.12 and theorem 1.3, one would then conclude that $E(K)$ is finitely generated. But this turns out not to be so. One can in fact construct many points on $E(K)$ by using the projection map $\pi_0 : X_0(37) \to E$.

Suppose more generally that E is any elliptic curve defined over $\mathbb{Q}$ and that there is a curve X defined over $\mathbb{Q}$ of genus at least 2 together with a nonconstant rational map $\pi : X \to E$, also defined over $\mathbb{Q}$. The
following lemma is quite useful for systematically constructing points on E.

Lemma 1.23. *Let K be a Galois extension of $\mathbb{Q}$ such that only finitely many primes of $\mathbb{Q}$ are ramified in $K/\mathbb{Q}$. Assume that $E(K)_{\text{tors}}$ is finite and that $X(K)$ is infinite. Then* rank($E(L)$) *is unbounded as L varies over the finite extensions of $\mathbb{Q}$ contained in K.*

Proof. Assume to the contrary that $E(L)$ has bounded rank. Then, by theorem 1.3, there is a finite extension L of $\mathbb{Q}$ contained in K such that $E(K) = E(L)$. Now by a theorem of Faltings (formerly known as the Mordell Conjecture), $X(L)$ is finite. But $X(K)$ is infinite and so we can pick a point $x \in X(K)$ such that $x \notin X(L)$. Then $\pi(x) \in E(K) = E(L)$. Also, for any $g \in \text{Gal}(K/L)$, we have $\pi(g(x)) = g(\pi(x)) = \pi(x)$. That is, if we let $P = \pi(x)$, then
$$\{g(x)\}_{g \in \text{Gal}(K/L)} \subset \pi^{-1}(P).$$
Let $n = \deg(\pi)$. Then $\pi^{-1}(P)$ has at most n elements. (One can think of the map $\pi : X(\mathbb{C}) \to E(\mathbb{C})$ as an n-sheeted covering of Riemann surfaces.) It follows that, for every $x \in X(K)$, $[L(x) : L] \leq n$. Thus $[L(x) : \mathbb{Q}]$ is bounded and $L(x)/\mathbb{Q}$ can be ramified only at primes in the fixed finite set of primes ramified in $K/\mathbb{Q}$. By a well-known theorem of Hermite, there can only be finitely many possibilities for the field $L(x)$ as x varies over $X(K)$. Hence there must be a finite extension L' of L contained in K such that $X(K) = X(L')$. But this contradicts the theorem of Faltings. $\square$

To apply this lemma to the example, we take X to be the curve defined by the modular equation $f_{37}(x,y) = 0$. For any N, the modular polynomial $f_N(x,y)$ is an element of $\mathbb{Z}[x,y]$, irreducible over $\mathbb{C}$, such that the solutions over $\mathbb{C}$ of the equation $f_N(x,y) = 0$ are the points of the form $(j(z), j(Nz))$. Recall that $j(z)$ is defined for all z in the upper half-plane $\mathcal{H}$ and is just the j-invariant of the elliptic curve $E' = \mathbb{C}/(\mathbb{Z}1 + \mathbb{Z}z)\Omega$, where Ω is any nonzero complex number. The isomorphism class of an elliptic curve over $\mathbb{C}$ is determined by its j-invariant. Now $E'' = \mathbb{C}/(\mathbb{Z}1 + \mathbb{Z}Nz)\Omega$ has j-invariant $j(Nz)$, and the pair (E', E'') could be any pair of elliptic curves connected by a cyclic isogeny of degree N. Thus if X is defined by $f_N(x,y) = 0$, then the points on $X(K)$ correspond to such pairs where E' and E'' are both defined over K. The curve X is birationally equivalent over $\mathbb{Q}$ to the curve $X_0(N)$ and hence has the same genus.

Let $N = 37$ and let E', E'' be as specified earlier. (E' is defined by the equation (1.22) and $E'' = E'/\Phi$ where Φ has order 37.) Both E' and E'' are defined over $\mathbb{Q}$. Now let Π be any cyclic subgroup of $E'[p^\infty]$ of order p^m where $m \geq 1$. Let E'_Π denote the elliptic curve E'/Π which is defined over the field $\mathbb{Q}(E'[p^\infty])$. Let $g \in \text{Gal}(E'[p^\infty]/K)$. Then g acts on $E'[p^\infty]$ as a scalar and so $g(\Pi) = \Pi$. That means that the j-invariant of E'_Π is fixed by g and so belongs to K. Since $p \neq 37$, $\varphi(\Pi)$ is a cyclic subgroup of $E''[p^\infty]$ of order p^m. Letting $E''_\Pi = E''/\varphi(\Pi)$, we have that E''_Π is also defined over K. Obviously E'_Π and E''_Π are related by an isogeny of degree 37 and so the pair (E'_Π, E''_Π) corresponds to a point on X defined over K. These are all distinct points for different choices of Π. (Here is a sketch of the reason. Suppose that Π_1 and Π_2 are cyclic subgroups of $E'[p^\infty]$ of order p^m. If $(E'_{\Pi_1}, E''_{\Pi_1})$ and $(E'_{\Pi_2}, E''_{\Pi_2})$ correspond to the same point of X, then E'_{Π_1} and E'_{Π_2} have the same j-invariant and so there would be an isomorphism $\varepsilon : E'_{\Pi_1} \overset{\sim}{\to} E'_{\Pi_2}$ defined over $\mathbb{C}$. For $i = 1, 2$, let $\varphi_i : E' \to E'_{\Pi_i}$ be the natural isogeny with kernel Π_i and let $\widehat{\varphi}_i$ denote the dual isogeny. Then $\widehat{\varphi}_2 \circ \varphi_2$ and $\widehat{\varphi}_2 \circ \varepsilon \circ \varphi_1$ are both in $\mathrm{End}(E')$ and have the same degree (namely p^{2m}). Since $\mathrm{End}(E') = \mathbb{Z}$, we have $\widehat{\varphi}_2 \circ \varphi_2 = p^m$ and $\widehat{\varphi}_2 \circ \varepsilon \circ \varphi_1 = \pm p^m$, which implies that $\varphi_2 = \pm \varepsilon \circ \varphi_1$ and hence $\Pi_1 = \Pi_2$.) Therefore, $X(K)$ is indeed infinite.

Now only 5, 7, and p are ramified in $K/\mathbb{Q}$. By exercise 1.12, $E(K)_{\mathrm{tors}}$ is finite. Also, as we mentioned above, X is birationally equivalent to $X_0(37)$ over $\mathbb{Q}$, which is known to have genus 2, and composing a rational map from X to $X_0(37)$ with π_0 gives a nonconstant rational map $\pi : X \to E$ defined over $\mathbb{Q}$. Hence lemma 1.23 implies that $E(K)$ has infinite rank. Thus, if E, K, and p are as above (including the hypothesis that $(\frac{-37}{p}) = +1$), then as χ varies over the irreducible characters of $\mathrm{Gal}(K/\mathbb{Q})$ such that d_χ is even (all but finitely many), the root numbers $w(E, \chi)$ should all be $+1$, $\mathrm{ord}_{s=1}(L(E, \chi, s))$ should be even, $m_\chi(E)$ should be positive for infinitely many χ's, and therefore the conjecture in (1.18) asserts that infinitely many of the functions $L(E, \chi, s)$ should have at least a double zero at $s = 1$. For each such χ, $m_\chi(E)$ should be at least 2 (and even).

In the case where $(\frac{-37}{p}) = -1$, L. Howe has systematically calculated the root numbers $w(E, \chi)$ in [**Ho**]. Interestingly, it turns out that these roots numbers are still $+1$ for those characters χ where $m_\chi(E)$ is positive because of the contribution of the points constructed above. But those points only contribute 1 to $m_\chi(E)$, although this multiplicity should again be even. Furthermore, assuming just that $m_\chi(E) \geq 1$ when $w(E, \chi) = -1$, he finds the following lower bound for $\mathrm{rank}(E(L_m))$, where L_m is the subfield of K (or of $\mathbb{Q}(E'[p^m])$) such that $\mathrm{Gal}(L_m/\mathbb{Q}) \cong \mathrm{PGL}_2(\mathbb{Z}/p^m\mathbb{Z})$:

$$\mathrm{rank}_\mathbb{Z}(E(L_m)) \geq p^{2m} - p^{2m-1} - p - 1.$$

Note for comparison that $[L_m : \mathbb{Q}] = (p^2 - 1)p^{3m-2}$. The above method for constructing points on $E(K)$ was first exploited by M. Harris (in [**Ha**]) who proves that those points will make a contribution to $\mathrm{rank}_\mathbb{Z}(E(L_m))$ of $(p+1)p^{m-1} + O(1)$ as $n \to \infty$. All of these remarks and results show that there is a serious discrepancy between what the root number calculations might suggest for the ranks and what explicit, systematic constructions give.

Another interesting example to consider is the following. Let E be the same elliptic curve as before (defined in (1.21)), but now we take E' to be defined by the equation $y^2 = x^3 - x$. Then E' has conductor 32 and is an elliptic curve with complex multiplication. In fact, $\mathrm{End}(E') \cong \mathbb{Z}[i]$, where i corresponds to the automorphism $(x, y) \to (-x, iy)$ of E'. Now $37 = (6+i)(6-i)$ in $\mathbb{Z}[i]$. Let $\varphi \in \mathrm{End}(E')$ correspond to $6 + i$. Then φ is an isogeny of E' to itself defined over $F = \mathbb{Q}(i)$ and its kernel is a subgroup of $E'(\overline{\mathbb{Q}})$ of order 37. Thus $(j(E'), j(E'))$ is a point on $X(\mathbb{Q})$, where X is again the curve defined by $f_{37}(x, y) = 0$. If $p \neq 37$, then just as before, for every cyclic subgroup Π of $E'[p^\infty]$ of order p^m, we obtain a pair of elliptic curves $E'_\Pi = E'/\Pi$ and $E''_\Pi = E'/\varphi(\Pi)$. (Note that E'_Π and E''_Π are not necessarily isomorphic.) Then $(j(E'_\Pi), j(E''_\Pi))$ is a point on X

defined over $\mathbb{Q}(E'[p^\infty])$ and, for the same reason as before, is actually defined over $K = \overline{\mathbb{Q}}^{\ker(\widetilde{\rho}_{E'})}$, where $\rho_{E'} : G_\mathbb{Q} \to \mathrm{Aut}(T_p(E'))$ gives the action of $G_\mathbb{Q}$ on the Tate module $T_p(E')$ (and equivalently on $E'[p^\infty]$). As before, $\widetilde{\rho}_{E'}$ is the corresponding projective representation. We have $\mathrm{Gal}(K/\mathbb{Q}) \cong \mathrm{im}(\widetilde{\rho}_{E'})$, which is isomorphic to a subgroup of $PGL_2(\mathbb{Z}_p)$. This time $\mathrm{im}(\widetilde{\rho}_{E'})$ is much smaller and is not even of finite index in $PGL_2(\mathbb{Z}_p)$.

We can consider $T_p(E')$ as a module over the ring $R_p = \mathrm{End}(E') \otimes_\mathbb{Z} \mathbb{Z}_p$ and the action of $\rho_{E'}(g)$ on $T_p(E')$ is R_p-linear for every $g \in G_F$. Now $\mathrm{rank}_{R_p}(T_p(E')) = 1$. From these facts, one can deduce that $\rho_{E'}(G_F)$ is isomorphic to a subgroup of $R_p^\times$ and hence is abelian. That is, $\mathrm{Gal}(F(E'[p^\infty])/F)$ is abelian. According to exercise 1.16, we have $F \subset \widetilde{F} \subset \mathbb{Q}(E'[p^\infty])$ and $F \subset F_\infty^{\mathrm{ac}} \subset K$, where $\widetilde{F}$ is the compositum of all $\mathbb{Z}_p$-extensions of F (so that $\mathrm{Gal}(\widetilde{F}/F) \cong \mathbb{Z}_p^2$) and F_∞^{ac} is the anticyclotomic $\mathbb{Z}_p$-extension of F. Furthermore, $[K : F_\infty^{\mathrm{ac}}] < \infty$. K is a union of dihedral extensions of $\mathbb{Q}$ and if χ is the character of an irreducible 2-dimensional representation of $\mathrm{Gal}(K/\mathbb{Q})$, then δ_χ is the nontrivial character of $\mathrm{Gal}(F/\mathbb{Q})$ and so $w(E, \chi) = \delta_\chi(-37) = -1$. That is, $\mathrm{ord}_{s=1}(L(E, \chi, s))$ is odd and $m_\chi(E)$ should be positive for all such χ. Using lemma 1.23, we can at least say that $E(K)$ has infinite rank. (Exercise 1.10 implies that $E(K)_{\mathrm{tors}}$ is finite. It is not hard to show that there are infinitely many distinct j-invariants $j(E'_\Pi)$ as Π varies (although it is possible for E'_{Π_1} and E'_{Π_2} to be isomorphic over $\mathbb{C}$ for distinct Π_1, Π_2). It follows that $X(K)$ is infinite.) Therefore, $m_\chi(E)$ is positive for infinitely many χ's.

It is interesting to apply lemma 1.23 to $K = \mathbb{Q}_\Sigma^{\mathrm{ab}}$, where Σ is any finite set of primes. Combined with theorem 1.5, we get the following result.

Theorem 1.24. *Suppose that X is a curve defined over $\mathbb{Q}$ with genus at least 2. Suppose also that there is a nonconstant rational map $\pi : X \to E$ defined over $\mathbb{Q}$, where E is a modular elliptic curve. Then $X(\mathbb{Q}_\Sigma^{\mathrm{ab}})$ is finite.*

The hypotheses are satisfied for example when X is defined by $x^4 + y^4 = 1$, which has genus three and is a two-sheeted covering of the elliptic curve E defined by $y^2 = 1 - x^4$.

Exercises.

In the following exercises, E denotes a fixed elliptic curve defined over $\mathbb{Q}$.

1.1. Suppose that A is any p-primary, abelian group. Let $z \in \mathbb{Z}_p$. Then $z = \lim_{n \to \infty} z_n$ for some sequence $\{z_n\}$ of integers. Show that if $a \in A$, then there exists an integer n_0 such that $a^{z_n} = a^{z_{n_0}}$ for all $n \geq n_0$. Show that if we define $a^z = a^{z_{n_0}}$, then A becomes a $\mathbb{Z}_p$-module. If A is given the discrete topology, show that the map $\mathbb{Z}_p \times A \to A$ defined by $(z, a) \to a^z$ is continuous. (If we use additive notation for A, then we would of course write za.)

1.2. Suppose that $A \cong (\mathbb{Q}_p/\mathbb{Z}_p)^r$ for some $r \geq 1$. Prove that the automorphism group of A is isomorphic to $\mathrm{GL}_r(\mathbb{Z}_p)$. (Two interesting examples are $A = \mu_{p^\infty}$ and $A = E[p^\infty]$.)

1.3. Let $K/\mathbb{Q}$ be a finite Galois extension. Let H be a subgroup of $G = \mathrm{Gal}(K/\mathbb{Q})$ and let $L = K^H$. The inclusion $E(L) \subset E(K)$ induces a $\mathbb{C}$-linear map

$$E(L) \otimes_\mathbb{Z} \mathbb{C} \to (E(K) \otimes_\mathbb{Z} \mathbb{C})^H.$$

Prove that this map is an isomorphism.

Suppose now that H is a normal subgroup of G and that $\sigma : G \to \mathrm{GL}_d(\mathbb{C})$ is an irreducible representation such that $H \subset \ker(\sigma)$. Then σ can also be regarded as an irreducible representation σ' of $G/H = \mathrm{Gal}(L/\mathbb{Q})$. Show that the multiplicity of σ' in the (G/H)-representation space $E(L) \otimes_\mathbb{Z} \mathbb{C}$ is the same as the multiplicity of σ in the G-representation space $E(K) \otimes_\mathbb{Z} \mathbb{C}$.

1.4. Let $K/\mathbb{Q}$ be a finite Galois extension. For every irreducible character χ of $\mathrm{Gal}(K/\mathbb{Q})$, define $m_\chi(E)$ as in the text. (See (1.15).) Prove that if τ is any automorphism of the field $\mathbb{Q}(\chi)$ generated by the values of χ, then χ^τ is another irreducible character of $\mathrm{Gal}(K/\mathbb{Q})$ and $m_{\chi^\tau}(E) = m_\chi(E)$. (Note: It is conjectured that $\mathrm{ord}_{s=1}(L(E/\mathbb{Q}, \chi^\tau, s)) = \mathrm{ord}_{s=1}(L(E/\mathbb{Q}, \chi, s))$. Even in the case that $d_\chi = 1$, this is only known if $\mathrm{ord}_{s=1}(L(E/\mathbb{Q}, \chi, s)) \le 1$.)

1.5. Let K denote the compositum of all quadratic extensions of $\mathbb{Q}$. Prove that $E(K)_\mathrm{tors}$ is finite.

1.6. Prove that there exist infinitely many quadratic fields F with the property that $\mathrm{rank}_\mathbb{Z}(E(F)) > \mathrm{rank}_\mathbb{Z}(E(\mathbb{Q}))$. (Hint: E is defined by an equation of the form $y^2 = f(x)$, where $f(x) \in \mathbb{Q}[x]$ and has degree 3. If p is any sufficiently large prime, show that p is ramified in $F/\mathbb{Q}$, where $F = \mathbb{Q}\left(\sqrt{f(\frac{1}{p})}\right)$, that $E(F)_\mathrm{tors} = E(\mathbb{Q})_\mathrm{tors}$, and that $\mathrm{rank}_\mathbb{Z}(E(F)) > \mathrm{rank}_\mathbb{Z}(E(\mathbb{Q}))$.)

1.7. Let K be a Galois extension of $\mathbb{Q}$ such that $\mathrm{Gal}(K/\mathbb{Q}) \cong (\mathbb{Z}/2\mathbb{Z})^n$ for some $n \ge 1$. Prove that
$$\mathrm{rank}_\mathbb{Z}(E(K)) = \mathrm{rank}_\mathbb{Z}(E(\mathbb{Q})) + \sum_F \mathrm{rank}_\mathbb{Z}\left(E(F)/E(\mathbb{Q})\right)$$
where F varies over all quadratic subfields of K.

1.8. Let K denote the composition of all quadratic extensions of $\mathbb{Q}$. Show that $E(K)$ has infinite rank.

1.9. Let F_∞ be any $\mathbb{Z}_p$-extension of a number field F. Let F_n denote the n-th layer (so that $\mathrm{Gal}(F_n/F)$ is cyclic of order p^n). Prove the congruence
$$\mathrm{rank}_\mathbb{Z}(E(F_n)) \equiv \mathrm{rank}_\mathbb{Z}(E(F_{n-1})) \pmod{\varphi(p^n)}.$$

1.10. Let F be an imaginary quadratic field. Let L be any finite extension of F and let $K = F_\infty^\mathrm{ac} L$, where F_∞^ac is the anticyclotomic $\mathbb{Z}_p$-extension of F for any prime p. Prove that $E(K)_\mathrm{tors}$ is finite. (Hint: First show that any prime of F which is inert in $F/\mathbb{Q}$ must split completely in F_∞^ac/F.)

1.11. Let $G = \mathrm{PGL}_2(\mathbb{Z}_p)$ where p is any prime. Show that for all $g \in G$, g and g^{-1} are conjugate in G. This implies that the character of any finite dimensional representation of G over $\mathbb{C}$ is real-valued. (Hint: If $A \in \mathrm{GL}_2(\mathbb{Z}_p)$, show that A and $B = \det(A)A^{-1}$ are conjugate in $\mathrm{GL}_2(\mathbb{Z}_p)$ by considering the equation $XA = BX$ for $X = \begin{bmatrix} x & y \\ z & -x \end{bmatrix}$.)

1.12. Suppose that E' is an elliptic curve defined over $\mathbb{Q}$. Let p be any prime. Suppose that $\widetilde{\rho}_{E'} : G_\mathbb{Q} \to \mathrm{PGL}_2(\mathbb{Z}_p)$ is as described earlier. Let $K = \overline{\mathbb{Q}}^{\ker(\widetilde{\rho}_{E'})}$. Show that if E' has good, supersingular reduction at a prime $l \ne p$, then the residue field for any prime of K lying above l is finite. N. Elkies has proven that infinitely many such primes exist. Let E be any elliptic curve/$\mathbb{Q}$ (or even any abelian variety). Deduce that $E(K)_{\text{tors}}$ is finite.

1.13. Let F_∞/F be a $\mathbb{Z}_p$-extension. Assume that E is an elliptic curve defined over F. Prove that if $E(F)_{p\text{-prim}} = 0$, then $E(F_\infty)_{p\text{-prim}} = 0$.

1.14. Let F_∞/F be a $\mathbb{Z}_p$-extension. Suppose that λ is a prime of F and $\lambda \nmid p$. Show that λ is unramified in F_∞/F. (Hint: Let L be any finite abelian extension of F in which λ is tamely ramified. Show that the ramification index for λ in L/F divides $N(\lambda) - 1$. What can the corresponding inertia subgroup of $\text{Gal}(F_\infty/F)$ be?)

1.15. Suppose that F_∞/F is a $\mathbb{Z}_p$-extension. We can write $E(F_\infty)_{\text{tors}} = A \times B$, where A is the subgroup of elements whose order is relatively prime to p and B is $E(F_\infty)_{p\text{-prim}}$. Assume that E has good reduction at p. Prove that A is finite. Show by example that B can be infinite. (Hint: For the first assertion, show that at least one prime of F lying over p is ramified in F_∞/F. For the second assertion, try $E: y^2 = x^3 - x$, $p \equiv 1 \pmod{4}$, and $F = \mathbb{Q}(i)(E[p])$.

1.16. Let $R_p = \text{End}(E') \otimes_{\mathbb{Z}} \mathbb{Z}_p$, where E' is the elliptic curve $y^2 = x^3 - x$ considered at the end of this section and p is any prime. Show that $R_p^\times \cong \mathbb{Z}_p^2 \times \Delta$, where Δ is a finite group. Show that $M = F(E'[p^\infty])$ cannot be just a finite extension of $F(\mu_{p^\infty})$ (which is a subfield). Deduce that $\widetilde{F}$ is a subfield of M and $[M : \widetilde{F}] < \infty$. Deduce that $\mathbb{Q}(E'[p^\infty])$ contains F and that $M = \mathbb{Q}(E'[p^\infty])$. If $K = \overline{\mathbb{Q}}^{\ker(\widetilde{\rho}_{E'})}$, show that K contains F_∞^{ac} and that $[K : F_\infty^{\text{ac}}] < \infty$.

CHAPTER 2
Selmer groups.

Let K be any algebraic extension of $\mathbb{Q}$. Suppose that E is an elliptic curve defined over K. One of the principal ways in which one studies the Mordell-Weil group $E(K)$ is by using Galois cohomology. Fix $n \geq 2$. Then there is a natural injective homomorphism of $E(K)/nE(K)$ into $H^1(G_K, E[n])$ which is called the Kummer map. If K is a finite extension of $\mathbb{Q}$, then $H^1(G_K, E[n])$ turns out to be an infinite group. (See exercise 2.10.) But one can show that the image of $E(K)/nE(K)$ under the Kummer map is contained in a certain finite subgroup of $H^1(G_K, E[n])$ (called the n-Selmer group). In that way, one proves the weak Mordell-Weil theorem which asserts that $E(K)/nE(K)$ is finite when $[K:\mathbb{Q}] < \infty$. Using the theory of heights one then proves the Mordell-Weil theorem:

$$E(K) \cong \mathbb{Z}^r \times T \text{ for some } r \geq 0 \text{ and some finite group } T.$$

So then $E(K)/nE(K) \cong (\mathbb{Z}/n\mathbb{Z})^r \times T/nT$ and, if one knows the structure of the n-Selmer group, one would be able to give an upper bound on r.

In these notes we will consider all n's at once, replacing $E(K)/nE(K) = E(K) \otimes_{\mathbb{Z}} (\mathbb{Z}/n\mathbb{Z})$ by the direct limit $E(K) \otimes_{\mathbb{Z}} (\mathbb{Q}/\mathbb{Z})$ and the n-Selmer group by a certain subgroup $\mathrm{Sel}_E(K)$ of $H^1(G_K, E(\overline{\mathbb{Q}})_{\mathrm{tors}})$ which is called the Selmer group for E over K. It fits into an exact sequence

$$0 \to E(K) \otimes_{\mathbb{Z}} (\mathbb{Q}/\mathbb{Z}) \to \mathrm{Sel}_E(K) \to \text{Ш}_E(K) \to 0$$

where $\text{Ш}_E(K)$ is the so-called Tate-Shafarevich group for E over K. ($\text{Ш}_E(K)$ is a certain subgroup of $H^1(G_K, E(\overline{\mathbb{Q}}))$ which we will also soon define.) Now

$$E(K) \otimes_{\mathbb{Z}} (\mathbb{Q}/\mathbb{Z}) \cong (\mathbb{Q}/\mathbb{Z})^r$$

and therefore knowing the structure of $\mathrm{Sel}_E(K)$ would give an upper bound on r.

We will give two equivalent definitions of $\mathrm{Sel}_E(K)$. After that, the main objective of this chapter will be to describe an alternative definition for the p-primary subgroup of $\mathrm{Sel}_E(K)$ which involves only the Galois module $E[p^\infty]$. (For any prime p, $E[p^\infty] = \bigcup_m E[p^m]$ is the p-primary subgroup of $E(\overline{\mathbb{Q}})_{\mathrm{tors}}$.) This alternative definition will be valid under the hypothesis that E has good, ordinary reduction at all primes of K lying over p. We will adopt the more usual notation for Galois cohomology groups, writing $H^i(L, \cdot)$ in place of $H^i(G_L, \cdot)$ and $H^i(K/L, \cdot)$ in place of $H^i(\mathrm{Gal}(K/L), \cdot)$ if K/L is a Galois extension.

First Definition: We define the Tate-Shafarevich group $\Sha_E(K)$ by

$$\Sha_E(K) = \ker\left(H^1(K, E(\overline{K})) \longrightarrow \prod_v H^1(K_v, E(\overline{K}_v)) \right)$$

where v runs over all primes of K, archimedean and nonarchimedean. If K is a finite extension of $\mathbb{Q}$, then K_v is the completion of K at v. If K is an infinite algebraic extension of $\mathbb{Q}$, then K_v denotes the union of the completions at v of all finite extensions of $\mathbb{Q}$ contained in K. Thus K_v is always either $\mathbb{R}$ or $\mathbb{C}$ or an algebraic extension of $\mathbb{Q}_l$ (for some prime l). The map defining $\Sha_E(K)$ is defined by using the restriction maps

$$H^1(K, E(\overline{K})) \longrightarrow H^1(K_v, E(\overline{K}_v))$$

noting that $E(\overline{K}) \subset E(\overline{K}_v)$ and G_{K_v} can be identified with a subgroup of G_K. (These identifications depend on choosing an embedding $\overline{K} \hookrightarrow \overline{K}_v$ consistent with the natural map $K \hookrightarrow K_v$. But the choice doesn't matter.) To define $\mathrm{Sel}_E(K)$, consider the map

$$\lambda : H^1(K, E(\overline{K})_{\mathrm{tors}}) \longrightarrow H^1(K, E(\overline{K}))$$

arising from the inclusion $E(\overline{K})_{\mathrm{tors}} \subset E(\overline{K})$. This map is surjective. (The reason for this is that the cokernel of the map λ is isomorphic to a subgroup of $H^1(K, E(\overline{K})/E(\overline{K})_{\mathrm{tors}})$. This group is trivial because $E(\overline{K})/E(\overline{K})_{\mathrm{tors}}$ is a uniquely divisible group.) Finally, the Selmer group is defined by

$$(2.1) \qquad \mathrm{Sel}_E(K) = \lambda^{-1}(\Sha_E(K)).$$

Second Definition: This is based on Kummer theory for E. Let L be a field of characteristic zero. Classical Kummer theory gives an isomorphism

$$L^\times / (L^\times)^n \xrightarrow{\sim} H^1(L, \mu_n)$$

for any $n \geq 1$, where μ_n denotes the group of n-th roots of unity in $\overline{L}$. Taking the direct limit, one obtains an isomorphism

$$\kappa : L^\times \otimes_\mathbb{Z} (\mathbb{Q}/\mathbb{Z}) \xrightarrow{\sim} H^1(L, (\overline{L}^\times)_{\mathrm{tors}}).$$

Of course, $(\overline{L}^\times)_{\mathrm{tors}}$ is just the group of roots of unity in $\overline{L}^\times$. The map is defined as follows. Suppose that $\alpha = a \otimes r \in L^\times \otimes_\mathbb{Z} (\mathbb{Q}/\mathbb{Z})$, where $a \in L^\times$ and $r = \frac{m}{n} + \mathbb{Z} \in \mathbb{Q}/\mathbb{Z}$. Since $\overline{L}^\times$ is a divisible group, we can choose $b \in \overline{L}^\times$ such that $b^n = a^m$. Define $\varphi_\alpha : G_L \to (\overline{L}^\times)_{\mathrm{tors}}$ by $\varphi_\alpha(g) = g(b)/b$ for all $g \in G_L$. Then the cohomology class $[\varphi_\alpha]$ in $H^1(L, (\overline{L}^\times)_{\mathrm{tors}})$ is well-defined and one defines

$$\kappa(\alpha) = [\varphi_\alpha].$$

The injectivity of κ is not hard to prove. The surjectivity is a consequence of Hilbert's Theorem 90: $H^1(L, \overline{L}^\times) = 0$. If L is any finite extension of $\mathbb{Q}$, then it is easy to see $L^\times \otimes_\mathbb{Z} (\mathbb{Q}/\mathbb{Z})$ is isomorphic to a countably infinite direct sum of copies of $\mathbb{Q}/\mathbb{Z}$. Thus $H^1(L, (\overline{L}^\times)_{\mathrm{tors}})$ is quite big.

If E is an elliptic curve defined over L, then $E(\overline{L})$ is also a divisible group. Imitating Kummer theory for the multiplicative group $L^\times$, one defines the Kummer homomorphism
$$\kappa : E(L) \otimes_{\mathbb{Z}} (\mathbb{Q}/\mathbb{Z}) \longrightarrow H^1(L, E(\overline{L})_{\text{tors}})$$
in the same way: Let $\alpha = P \otimes r \in E(L) \otimes_{\mathbb{Z}} (\mathbb{Q}/\mathbb{Z})$, where $P \in E(L)$ and $r = \frac{m}{n} + \mathbb{Z}$. Choose $Q \in E(\overline{L})$ such that $nQ = mP$. Define a 1-cocycle $\varphi_\alpha : G_L \to E(\overline{L})_{\text{tors}}$ by $\varphi_\alpha(g) = g(Q) - Q$ for all $g \in G_L$. Then $[\varphi_\alpha]$ is well-defined and one defines $\kappa(\alpha) = [\varphi_\alpha]$. It is not hard to verify that the following sequence (which is called the Kummer sequence) is exact:

$$(2.2) \qquad 0 \longrightarrow E(L) \otimes_{\mathbb{Z}} (\mathbb{Q}/\mathbb{Z}) \stackrel{\kappa}{\longrightarrow} H^1(L, E(\overline{L})_{\text{tors}}) \stackrel{\lambda}{\longrightarrow} H^1(L, E(\overline{L})) \longrightarrow 0.$$

If K is an algebraic extension of $\mathbb{Q}$, we will use the Kummer sequence for $L = K$ and also for the completions $L = K_v$ where v is any prime of K. In the following commutative diagram, the vertical maps are defined in obvious ways and the rows are exact.

$$\begin{array}{ccccccccc}
0 & \longrightarrow & E(K) \otimes (\mathbb{Q}/\mathbb{Z}) & \stackrel{\kappa}{\longrightarrow} & H^1(K, E(\overline{K})_{\text{tors}}) & \stackrel{\lambda}{\longrightarrow} & H^1(K, E(\overline{K})) & \longrightarrow & 0 \\
& & \Big\downarrow a_v & & \Big\downarrow b_v & & \Big\downarrow c_v & & \\
0 & \longrightarrow & E(K_v) \otimes (\mathbb{Q}/\mathbb{Z}) & \stackrel{\kappa_v}{\longrightarrow} & H^1(K_v, E(\overline{K}_v)_{\text{tors}}) & \stackrel{\lambda_v}{\longrightarrow} & H^1(K_v, E(\overline{K}_v)) & \longrightarrow & 0
\end{array}$$

Suppose that $[\varphi] \in H^1(K, E(\overline{K})_{\text{tors}})$. According to the first definition, $[\varphi] \in \text{Sel}_E(K)$ if and only if $c_v \circ \lambda([\varphi]) = 0$ for every prime v of K. Clearly, this is equivalent to requiring that $b_v([\varphi]) \in \text{im}(\kappa_v)$ for all v. Therefore, we obtain the second definition of the Selmer group:

$$\text{Sel}_E(K) = \ker\left(H^1(K, E(\overline{K})_{\text{tors}}) \to \prod_v (H^1(K_v, E(\overline{K}_v)_{\text{tors}})/\text{im}(\kappa_v))\right).$$

That is, if $[\varphi] \in H^1(K, E(\overline{K})_{\text{tors}})$, then

$$[\varphi] \in \text{Sel}_E(K) \text{ if and only if } [\varphi|_{G_{K_v}}] \in \text{im}(\kappa_v)$$

for every prime v of K.

Let p be a prime. The p-primary subgroup of $\mathbb{Q}/\mathbb{Z}$ is isomorphic to $\mathbb{Q}_p/\mathbb{Z}_p$. Therefore, for any field L, the p-primary subgroup of $E(L) \otimes (\mathbb{Q}/\mathbb{Z})$ can be identified with $E(L) \otimes (\mathbb{Q}_p/\mathbb{Z}_p)$. We will now let κ and κ_v denote the global and local Kummer maps restricted to the p-primary subgroups: If K is any algebraic extension of $\mathbb{Q}$, we have

$$\kappa : E(K) \otimes (\mathbb{Q}_p/\mathbb{Z}_p) \to H^1(K, E[p^\infty]),$$
$$\kappa_v : E(K_v) \otimes (\mathbb{Q}_p/\mathbb{Z}_p) \to H^1(K_v, E[p^\infty]).$$

The p-primary subgroup of $\text{Sel}_E(K)$ is

$$(2.3) \qquad \text{Sel}_E(K)_p = \ker\left(H^1(K, E[p^\infty]) \to \prod_v H^1(K_v, E[p^\infty])/\text{im}(\kappa_v)\right).$$

Here v runs over all primes of K. Everything in this definition, except possibly $\text{im}(\kappa_v)$, depends just on the G_K-module $E[p^\infty]$. Most of the rest of this chapter concerns describing $\text{im}(\kappa_v)$ in a way which also involves just the Galois module $E[p^\infty]$.

The first remark to be made is that $\mathrm{im}(\kappa_v)$ is divisible. This is obvious since $E(K_v) \otimes (\mathbb{Q}_p/\mathbb{Z}_p)$ is divisible. If $v \nmid p$, we will show that $E(K_v) \otimes (\mathbb{Q}_p/\mathbb{Z}_p) = 0$ and so $\mathrm{im}(\kappa_v) = 0$. That is, if $[\varphi] \in H^1(K_v, E[p^\infty])$, then the local condition at v defining $\mathrm{Sel}_E(K)_p$ is simply that $[\varphi|_{G_{K_v}}]$ is trivial in $H^1(K_v, E[p^\infty])$.

Theorem 2.4. *(i) Suppose that E is an elliptic curve defined over an algebraic extension K_v of $\mathbb{Q}_l$ where l is a prime, $l \neq p$. Then $E(K_v) \otimes (\mathbb{Q}_p/\mathbb{Z}_p) = 0$.*
(ii) Suppose that E is an elliptic curve defined over $K_v = \mathbb{C}$ or $K_v = \mathbb{R}$. Then $E(K_v) \otimes (\mathbb{Q}_p/\mathbb{Z}_p) = 0$.
Thus, whenever $v \nmid p$, we have $\mathrm{im}(\kappa_v) = 0$.

Proof. Part (ii) is rather easy. Since $K_v = \mathbb{R}$, $E(K_v) \cong S$ or $S \times \mathbb{Z}/2\mathbb{Z}$, where $S = \mathbb{R}/\mathbb{Z}$. Now S is divisible by p and so $S \otimes (\mathbb{Q}_p/\mathbb{Z}_p) = 0$. Also, if T is any finite group, then $T \otimes \mathbb{Q}_p/\mathbb{Z}_p$ is easily seen to be trivial. Hence $E(K_v) \otimes (\mathbb{Q}_p/\mathbb{Z}_p) = 0$. If $K_v = \mathbb{C}$, then $E(K_v) \cong S \times S$ and so again we have $E(K_v) \otimes (\mathbb{Q}_p/\mathbb{Z}_p) = 0$.

Now assume that K_v is a finite extension of $\mathbb{Q}_l$, where l is a prime, $l \neq p$. There is a theorem of E. Lutz (Proposition 6.3 of [**Si1**]) which states that
$$E(K_v) \cong \mathbb{Z}_l^{[K_v:\mathbb{Q}_l]} \times T$$
where T is a finite group. Since $l \neq p$, $\mathbb{Z}_l$ is p-divisible, and therefore we have $\mathbb{Z}_l \otimes (\mathbb{Q}_p/\mathbb{Z}_p) = 0$. Again, $T \otimes (\mathbb{Q}_p/\mathbb{Z}_p) = 0$, and it follows that $E(K_v) \otimes (\mathbb{Q}_p/\mathbb{Z}_p) = 0$. If K_v is an infinite algebraic extension of $\mathbb{Q}_l$, then we have $K_v = \bigcup L_v$, where L_v runs over all finite extensions of $\mathbb{Q}_l$ contained in K_v. Since $E(L_v) \otimes (\mathbb{Q}_p/\mathbb{Z}_p) = 0$ for each such L_v, it clearly follows that $E(K_v) \otimes (\mathbb{Q}_p/\mathbb{Z}_p) = 0$ too. $\square$

If K_v is an algebraic extension of $\mathbb{Q}_p$, then the situation is considerably more subtle. Suppose first that K_v is a finite extension of $\mathbb{Q}_p$. Then the theorem of Lutz states that
$$E(K_v) \cong \mathbb{Z}_p^{[K_v:\mathbb{Q}_p]} \times T,$$
where T is a finite group. Since $\mathbb{Z}_p \otimes (\mathbb{Q}_p/\mathbb{Z}_p) \cong \mathbb{Q}_p/\mathbb{Z}_p$, we have
$$(2.5) \qquad \mathrm{im}(\kappa_v) \cong E(K_v) \otimes (\mathbb{Q}_p/\mathbb{Z}_p) \cong (\mathbb{Q}_p/\mathbb{Z}_p)^{[K_v:\mathbb{Q}_p]}.$$
In contrast, it turns out that
$$H^1(K_v, E[p^\infty]) \cong (\mathbb{Q}_p/\mathbb{Z}_p)^{2[K_v:\mathbb{Q}_p]} \times \text{ (a finite group)}.$$
Using a terminology that we define later, we would say that $\mathrm{im}(\kappa_v)$ has $\mathbb{Z}_p$-corank $[K_v : \mathbb{Q}_p]$ and that $H^1(K_v, E[p^\infty])$ has $\mathbb{Z}_p$-corank $2[K_v : \mathbb{Q}_p]$.

Assume that E has good, ordinary reduction at v. Thus, the reduction of E at v is an elliptic curve $\widetilde{E}$ defined over the residue field k_v for v and $\widetilde{E}(\overline{k}_v)$ has elements of order p. More precisely, it turns out that $\widetilde{E}[p^\infty] \cong \mathbb{Q}_p/\mathbb{Z}_p$. There is a reduction map $E(\overline{K}_v) \to \widetilde{E}(\overline{k}_v)$ which is equivariant for the action of the Galois groups: G_{K_v} acts on $E(\overline{K}_v)$ in the obvious way. There is a natural homomorphism $G_{K_v} \to G_{k_v}$ whose kernel is the inertia subgroup I_{K_v} of G_{K_v}, and G_{K_v} acts on $\widetilde{E}(\overline{k}_v)$ through the above homomorphism together with the obvious action of G_{k_v} on $\widetilde{E}(\overline{k}_v)$. In particular, there is a natural G_{K_v}-equivariant homomorphism
$$\pi : E[p^\infty] \to \widetilde{E}[p^\infty].$$
It is known that π is surjective and $\ker(\pi) \cong \mathbb{Q}_p/\mathbb{Z}_p$. The action of G_{k_v} on $\widetilde{E}[p^\infty]$ is given by a character $\psi : G_{k_v} \to \mathbb{Z}_p^\times$. We can regard ψ as a character of G_{K_v} which is trivial on I_{K_v}. The action of G_{K_v} on $\ker(\pi)$ is given by a character $\varphi : G_{K_v} \to \mathbb{Z}_p^\times$.

Thus the action ρ_E of G_{K_v} on $E[p^\infty]$ is "triangular", that is $\rho_E = \begin{bmatrix} \varphi & * \\ 0 & \psi \end{bmatrix}$. The action of G_{K_v} on μ_{p^∞} is also given by a character $\chi : G_{K_v} \to \mathbb{Z}_p^\times$ and the Weil pairing implies that $\det(\rho_E) = \chi$. Hence $\varphi\psi = \chi$.

We will denote $\ker(\pi)$ by $\mathcal{F}[p^\infty]$. The motivation for this notation is that there is a certain formal group $\mathcal{F}$ associated to E, the kernel of the reduction map $E(\overline{K}_v) \to \widetilde{E}(\overline{k}_v)$ coincides with the group $\mathcal{F}(\overline{\mathfrak{m}}_v)$ of the formal group (where $\overline{\mathfrak{m}}_v$ is the maximal ideal in the ring of integers of $\overline{K}_v$), and $\ker(\pi)$ is just the p-primary subgroup of $\mathcal{F}(\overline{\mathfrak{m}}_v)$. In fact, for the proof of Lutz' theorem, one uses the logarithm for the formal group $\mathcal{F}$ to compare $\mathcal{F}(\mathfrak{m}_v)$ with a compact subgroup of the additive group of K_v.

Consider the map $\varepsilon_v : H^1(K_v, \mathcal{F}[p^\infty]) \to H^1(K_v, E[p^\infty])$ which is induced by the inclusion $\mathcal{F}[p^\infty] \subset E[p^\infty]$. Here is the promised description of $\operatorname{im}(\kappa_v)$.

Theorem 2.6. *Assume that K_v is a finite extension of $\mathbb{Q}_p$ and that E has good, ordinary reduction at v. Then $\operatorname{im}(\kappa_v) = \operatorname{im}(\varepsilon_v)_{\operatorname{div}}$.*

Proof. Recall that $\operatorname{im}(\kappa_v)$ is isomorphic to $(\mathbb{Q}_p/\mathbb{Z}_p)^{[K_v:\mathbb{Q}_p]}$. To prove theorem 2.6, all we have to show is that $\operatorname{im}(\kappa_v) \subseteq \operatorname{im}(\varepsilon_v)$ and that $\operatorname{im}(\varepsilon_v)$ is isomorphic to $(\mathbb{Q}_p/\mathbb{Z}_p)^{[K_v:\mathbb{Q}_p]} \times$ (a finite group). The first statement is not hard to prove. The exact sequence

$$0 \to \mathcal{F}[p^\infty] \to E[p^\infty] \to \widetilde{E}[p^\infty] \to 0$$

induces an exact sequence of cohomology groups

$$H^0(K_v, E[p^\infty]) \longrightarrow H^0(K_v, \widetilde{E}[p^\infty])$$
$$\longrightarrow H^1(K_v, \mathcal{F}[p^\infty]) \xrightarrow{\varepsilon_v} H^1(K_v, E[p^\infty]) \xrightarrow{\pi_v} H^1(K_v, \widetilde{E}[p^\infty])$$
$$\longrightarrow H^2(K_v, \mathcal{F}[p^\infty]).$$

Thus $\operatorname{im}(\varepsilon_v) = \ker(\pi_v)$ and so the inclusion $\operatorname{im}(\kappa_v) \subset \operatorname{im}(\varepsilon_v)$ would follow if we show that $\pi_v \circ \kappa_v$ is the zero map. Let $[\varphi] \in \operatorname{im}(\kappa_v)$. Then there exists a $Q \in E(\overline{K}_v)$ so that $\varphi(g) = g(Q) - Q$ for all $g \in G_{K_v}$. Denote the image of Q under the reduction map $E(\overline{K}_v) \to \widetilde{E}(\overline{k}_v)$ by $\widetilde{Q}$, we have $\pi_v([\varphi]) = [\widetilde{\varphi}] \in H^1(K_v, \widetilde{E}[p^\infty])$ where $\widetilde{\varphi}(g) = g(\widetilde{Q}) - \widetilde{Q}$ for all $g \in G_{K_v}$. That is $\widetilde{\varphi}$ is a 1-coboundary for $\widetilde{E}(\overline{k}_v)$; its cohomology class is trivial in $H^1(K_v, \widetilde{E}(\overline{k}_v))$. But $\widetilde{E}(\overline{k}_v)$ is a torsion group and therefore its p-primary subgroup $\widetilde{E}[p^\infty]$ is a direct summand. Hence it is clear that $\widetilde{\varphi}$ is a 1-coboundary for $\widetilde{E}[p^\infty]$, and so $[\widetilde{\varphi}] = \pi_v([\varphi])$ is indeed trivial. This proves that $\operatorname{im}(\kappa_v) \subset \operatorname{im}(\varepsilon_v)$.

Since $\operatorname{im}(\kappa_v)$ is divisible, we must just show that $[\operatorname{im}(\varepsilon_v) : \operatorname{im}(\kappa_v)]$ is finite to prove theorem 2.6. Some terminology will be helpful. Let A be any p-primary abelian group. We can regard A as a $\mathbb{Z}_p$-module. (See exercise 1.1.) We put the discrete topology on A. Then its Pontryagin dual $\widehat{A} = \operatorname{Hom}_{\operatorname{cont}}(A, \mathbb{Q}_p/\mathbb{Z}_p)$ is a compact $\mathbb{Z}_p$-module. (See exercise 2.1.) We say that A is a cofinitely generated $\mathbb{Z}_p$-module if $\widehat{A}$ is finitely generated as a $\mathbb{Z}_p$-module. We then define $\operatorname{corank}_{\mathbb{Z}_p}(A)$ to be the rank of $\widehat{A}$ as a $\mathbb{Z}_p$-module. Thus, if A is cofinitely generated as a $\mathbb{Z}_p$-module, then $\widehat{A} \cong \mathbb{Z}_p^r \times T$ for some $r \geq 0$ and some finite group T. Therefore, $A \cong \widehat{\widehat{A}}$ must be isomorphic to $(\mathbb{Q}_p/\mathbb{Z}_p)^r \times \widehat{T}$ and has $\mathbb{Z}_p$-corank r. Then $A_{\text{div}} \cong (\mathbb{Q}_p/\mathbb{Z}_p)^r$ (which we then say is $\mathbb{Z}_p$-cofree) and $[A : A_{\text{div}}] < \infty$.

By (2.5), it follows that $\text{im}(\kappa_v)$ is $\mathbb{Z}_p$-cofree and has $\mathbb{Z}_p$-corank $[K_v : \mathbb{Q}_p]$. We must explain why $\text{im}(\varepsilon_v)$ is $\mathbb{Z}_p$-cofinitely generated and has the same $\mathbb{Z}_p$-corank. Now the order of $\ker(\varepsilon_v)$ is bounded above by the order of the finite group $H^0(K_v, \widetilde{E}[p^\infty]) = \widetilde{E}(k_v)_p$. Hence it suffices to show that $H^1(K_v, \mathcal{F}[p^\infty])$ is $\mathbb{Z}_p$-cofinitely generated and has $\mathbb{Z}_p$-corank $[K_v : \mathbb{Q}_p]$. We will base the proof of this on some theorems of Tate concerning Galois cohomology over local fields. However, in chapter 3 we will give a self-contained proof of what we need by using standard techniques of Iwasawa theory.

Suppose that A is a discrete G_{K_v}-module and that $A \cong (\mathbb{Q}_p/\mathbb{Z}_p)^r$ as a $\mathbb{Z}_p$-module. Here is a helpful but nontrivial result. We include also the case where K_v is a finite extension of $\mathbb{Q}_l$ with $l \neq p$. The notation $\widehat{A}(1)$ refers to the Tate twist:

$$\widehat{A}(1) = \text{Hom}(A, \mu_{p^\infty}).$$

Note that $\widehat{A}(1)$ is a free $\mathbb{Z}_p$-module of rank r.

Corank Lemma. *(1) Let K_v be a finite extension of $\mathbb{Q}_p$. Then $H^1(K_v, A)$ is $\mathbb{Z}_p$-cofinitely generated and its corank is equal to*

$$r[K_v : \mathbb{Q}_p] + \text{corank}_{\mathbb{Z}_p}(H^0(K_v, A)) + \text{rank}_{\mathbb{Z}_p}(H^0(K_v, \widehat{A}(1))).$$

(2) Assume that K_v is a finite extension of $\mathbb{Q}_l$ where $l \neq p$. Then $H^1(K_v, A)$ is $\mathbb{Z}_p$-cofinitely generated and its corank is equal to

$$\text{corank}_{\mathbb{Z}_p}(H^0(K_v, A)) + \text{rank}_{\mathbb{Z}_p}(H^0(K_v, \widehat{A}(1))).$$

Applying this to $A = \mathcal{F}[p^\infty] \cong \mathbb{Q}_p/\mathbb{Z}_p$ will give the desired result. We then have $r = 1$, and $H^0(K_v, \mathcal{F}[p^\infty])$ is a subgroup of $H^0(K_v, E[p^\infty]) = E(K_v)_p$, which is finite. Using the notation introduced earlier, the action of G_{K_v} on $\mathcal{F}[p^\infty]$ is by a character $\varphi : G_{K_v} \to \mathbb{Z}_p^\times$. Therefore $\widehat{A}(1)$ is isomorphic to $\mathbb{Z}_p$ and G_{K_v} acts by $\chi\varphi^{-1} = \psi : G_{K_v} \to \mathbb{Z}_p^\times$. The character ψ, which gives the action of G_{K_v} on $\widetilde{E}[p^\infty]$ is clearly nontrivial. (The inertia group I_{K_v} acts trivially, but a Frobenius automorphism acts nontrivially.) It follows that if $A = \mathcal{F}[p^\infty]$, then $H^0(K_v, \widehat{A}(1)) = 0$. The first part of the Corank Lemma implies that $H^1(K_v, \mathcal{F}[p^\infty])$, and hence $\text{im}(\varepsilon_v)$, has $\mathbb{Z}_p$-corank $[K_v : \mathbb{Q}_p]$ as needed. Theorem 2.6 follows.

Right now we will justify the Corank Lemma by using results of Tate about Galois cohomology over local fields. We let A be any G_{K_v}-module which is isomorphic to $(\mathbb{Q}_p/\mathbb{Z}_p)^r$ for some $r \geq 1$. First we show that $H^1(K_v, A)$ is $\mathbb{Z}_p$-cofinitely generated. The exact sequence

$$0 \to A[p] \to A \xrightarrow{p} A \to 0$$

induces a surjective map $H^1(K_v, A[p]) \to H^1(K_v, A)[p]$. Let L_v be a finite Galois extension of K_v such that G_{L_v} acts trivially on $A[p]$. The restriction map $H^1(K_v, A[p]) \to H^1(L_v, A[p])$ has a finite kernel. Also,

$$H^1(L_v, A[p]) = \text{Hom}(\text{Gal}(M_v/L_v), A[p])$$

where M_v is the maximal extension of L_v such that $\text{Gal}(M_v/L_v)$ is an elementary abelian p-group. Since L_v has only finitely many extensions of degree p, it is not hard to see that M_v/L_v is a finite extension, and hence $H^1(L_v, A[p])$ must be finite. So $H^1(K_v, A[p])$ and therefore $H^1(K_v, A)[p]$ are finite. This implies that $H^1(K_v, A)$ is cofinitely generated as a $\mathbb{Z}_p$-module. The same statement is obviously true for $H^0(K_v, A)$. The results stated in the following paragraph imply that it is true for $H^2(K_v, A)$.

Let M be a finite G_{K_v}-module. Assume that $|M| = p^a$. Let $\widehat{M}(1)$ denote the G_{K_v}-module $\operatorname{Hom}(M, \mu_{p^\infty})$, which also has order p^a. The results of Tate are

(i) $\prod_{j=0}^{2} |H^j(K_v, M)|^{(-1)^j} = \begin{cases} p^{-a[K_v:\mathbb{Q}_p]} & \text{if } v | p \\ 1 & \text{if } v \nmid p \end{cases}$

(ii) $H^2(K_v, M)$ is the Pontryagin dual of $H^0(K_v, \widehat{M}(1))$ and hence has the same order.

The cohomology groups occurring above are all finite. One can extend these results to infinite G_{K_v}-modules such as A by writing $A = \bigcup_n A[p^n]$ and applying the above results to $M = A[p^n]$ for all $n \geq 0$, which are of course finite. One obtains

(i) $\sum_{j=0}^{2} (-1)^j \operatorname{corank}_{\mathbb{Z}_p}(H^j(K_v, A)) = \begin{cases} -[K_v : \mathbb{Q}_p] \operatorname{corank}_{\mathbb{Z}_p}(A) & \text{if } v|p \\ 0 & \text{if } v \nmid p \end{cases}$

(ii) $H^2(K_v, A)$ is the Pontryagin dual of $H^0(K_v, \widehat{A}(1))$ and hence we have

$$\operatorname{corank}_{\mathbb{Z}_p}(H^2(K_v, A)) = \operatorname{rank}_{\mathbb{Z}_p}(H^0(K_v, \widehat{A}(1))).$$

The Corank Lemma follows immediately from these statements. Later (in theorem 3.11) we will give a self-contained proof for the case $A \cong \mathbb{Q}_p/\mathbb{Z}_p$ needed in the proof of theorem 2.6 by using an "Iwasawa-theoretic" approach. It is based on the fact that G_{K_v} acts on A by a character $\alpha : G_{K_v} \to \mathbb{Z}_p^\times$. Thus one can "trivialize" the Galois module A by taking an extension L_v/K_v such that the action of G_{K_v} on A factors through $\operatorname{Gal}(L_v/K_v)$ and such that L_v is a $\mathbb{Z}_p$-extension of some finite extension of K_v. Then we can use the restriction map and the fact that $H^1(L_v, A) = \operatorname{Hom}(\operatorname{Gal}(L_v^{\text{ab}}/L_v), A)$.

It will be important to have information about $[\operatorname{im}(\varepsilon_v) : \operatorname{im}(\kappa_v)]$. This is not hard to obtain. It is clear that an upper bound on this index is given by $|H^1(K_v, \mathcal{F}[p^\infty])/H^1(K_v, \mathcal{F}[p^\infty])_{\text{div}}|$. Now for $m \gg 0$, we have

$$H^1(K_v, \mathcal{F}[p^\infty])_{\text{div}} = p^m H^1(K_v, \mathcal{F}[p^\infty]).$$

Choose such an m and consider the exact sequence

$$0 \longrightarrow A[p^m] \longrightarrow A \xrightarrow{p^m} A \longrightarrow 0$$

where $A = \mathcal{F}[p^\infty]$. This induces an exact sequence of cohomology groups

$$H^1(K_v, A) \xrightarrow{p^m} H^1(K_v, A) \longrightarrow H^2(K_v, A[p^m])$$

and hence $[\operatorname{im}(\varepsilon_v) : \operatorname{im}(\kappa_v)]$ is bounded above by $|H^2(K_v, A[p^m])|$ for any sufficiently large m. The order of this group equals $|H^0(K_v, \widehat{A[p^m]}(1))|$. The Weil pairing $E[p^m] \times E[p^m] \to \mu_{p^m}$ induces another nondegenerate pairing

$$A[p^m] \times \widetilde{E}[p^m] \to \mu_{p^m}.$$

That is, $\operatorname{Hom}(A[p^m], \mu_{p^m}) \cong \widetilde{E}[p^m]$ as G_{K_v}-modules. It follows that

$$H^0(K_v, \widehat{A[p^m]}(1)) \cong \widetilde{E}(k_v)_p$$

for $m \gg 0$. Therefore, $\operatorname{im}(\varepsilon_v)/\operatorname{im}(\kappa_v)$ is cyclic and we have

(2.7) $$|\operatorname{im}(\varepsilon_v)/\operatorname{im}(\kappa_v)| \leq |\widetilde{E}(k_v)_p|.$$

It turns out that equality holds and the two groups are even isomorphic, but we will not need this.

We have proved the following result.

Theorem 2.8. *If K_v is a finite extension of $\mathbb{Q}_p$ and if E is an elliptic curve/K_v with good, ordinary reduction, then $\operatorname{im}(\kappa_v)$ has finite index in $\operatorname{im}(\varepsilon_v)$ and the quotient $\operatorname{im}(\varepsilon_v)/\operatorname{im}(\kappa_v)$ is a cyclic group whose order divides $|\widetilde{E}(k_v)_p|$, where k_v is the residue field of v. In particular, if $p \nmid |\widetilde{E}(k_v)|$, then $\operatorname{im}(\kappa_v) = \operatorname{im}(\varepsilon_v)$.*

If v is a prime of K where E has good reduction and if the characteristic of the residue field k_v divides $|\widetilde{E}(k_v)|$, then v is called an anomalous prime for E. For a fixed E, anomalous primes seem rather rare, although it seems likely that infinitely many such primes should exist. There is a nice discussion of this question for the curves $y^2 = x^3 + D$ in the introduction of [**Ma1**]. Obviously, E has ordinary reduction at any anomalous prime v for E. See exercise 2.12 for more about this property.

We will now compare the images of κ_v and ε_v for infinite extensions of $\mathbb{Q}_p$ under certain restrictive hypotheses. One hypothesis concerns the so-called profinite degree of an infinite extension K/F. This is defined as the least common multiple of the degrees $[L:F]$, where L varies over all finite extensions of F contained in K. This should be interpreted as a formal product Πl^{a_l} over all primes l, where $0 \leq a_l \leq \infty$. If l^∞ "divides" this product, then the power of l dividing $[L:F]$ is unbounded as L varies. The main result is the following.

Theorem 2.9. *Assume that K_v is an extension of $\mathbb{Q}_p$ with finite residue field k_v. Assume also that the profinite degree of $K_v/\mathbb{Q}_p$ is divisible by p^∞. Then $\operatorname{im}(\kappa_v) = \operatorname{im}(\varepsilon_v)$. In particular, this is true if K_v is a ramified $\mathbb{Z}_p$-extension of F_v, where F_v is a finite extension of $\mathbb{Q}_p$.*

Proof. By the same argument as in the proof of theorem 2.6, we still have the inclusion $\operatorname{im}(\kappa_v) \subset \operatorname{im}(\varepsilon_v)$. We will show that $\operatorname{im}(\varepsilon_v)$ is a divisible group and that $\operatorname{im}(\varepsilon_v)/\operatorname{im}(\kappa_v)$ is finite. This clearly implies the equality. The fact that $\operatorname{im}(\varepsilon_v)/\operatorname{im}(\kappa_v)$ is finite follows from the observations in the previous paragraph. For we can write K_v as a union of finite extensions of $\mathbb{Q}_p : K_v = \bigcup_n F_v^{(n)}$. We let $\kappa_v^{(n)}$, $\varepsilon_v^{(n)}$ denote the two maps to $H^1(F_v^{(n)}, E[p^\infty])$ that we are considering, but over the ground field $F_v^{(n)}$. We have
$$H^1(K_v, E[p^\infty]) = \varinjlim_n H^1(F_v^{(n)}, E[p^\infty])$$
where the direct limit is defined by the natural restriction maps. We then have
$$\operatorname{im}(\varepsilon_v) = \varinjlim \operatorname{im}(\varepsilon_v^{(n)}), \quad \operatorname{im}(\kappa_v) = \varinjlim \operatorname{im}(\kappa_v^{(n)}),$$
and therefore $\operatorname{im}(\varepsilon_v)/\operatorname{im}(\kappa_v)$ is finite because it is a direct limit of the finite groups $\operatorname{im}(\varepsilon_v^{(n)})/\operatorname{im}(\kappa_v^{(n)})$ whose orders are uniformly bounded by $|\widetilde{E}(k_v)_p|$.

We will show that $H^1(K_v, \mathcal{F}[p^\infty])$, is divisible, from which it follows that $\operatorname{im}(\varepsilon_v)$ is divisible too. It suffices to prove that $H^1(K_v, A)$ is divisible whenever A is a divisible p-primary G_{K_v}-module. This follows from the fact that G_{K_v} has p-cohomological dimension 1. That is, if M is any finite p-primary G_{K_v}-module,

then $H^n(K_v, M) = 0$ for all $n \geq 2$. To deduce that $H^1(K_v, A)$ is divisible, consider the exact sequence
$$0 \to A[p] \to A \to A \to 0$$
which induces an exact sequence of cohomology groups
$$H^1(K_v, A) \xrightarrow{p} H^1(K_v, A) \to H^2(K_v, A[p]).$$
Since the last group is zero, $H^1(K_v, A)$ is p-divisible. It must then be divisible since it is a p-primary group. $\square$

It is worth reviewing why G_{K_v} has p-cohomological dimension 1. The only hypothesis needed for this is that K_v is an extension of $\mathbb{Q}_l$ such that $[K_v : \mathbb{Q}_l]$ is divisible by p^∞ and l is any prime. This hypothesis implies that the p-primary subgroup of the Brauer group of K_v is zero. (The reason for this is that $\mathrm{Br}(K_v) = \varinjlim \mathrm{Br}(F_v^{(n)})$ where as before the $F_v^{(n)}$'s are finite extensions of $\mathbb{Q}_l$ such that $K_v = \bigcup_n F_v^{(n)}$. But in most treatments of local class field theory one proves that any element of $\mathrm{Br}(F_v^{(n)})$ of order p^m becomes trivial over any finite extension of degree divisible by p^m. Since the profinite degree of K_v over $F_v^{(n)}$ is divisible by p^∞, it follows that the p-primary subgroup of the direct limit is zero.) Therefore, $H^2(K_v, \overline{K}_v^\times)_p = 0$. Of course, the same argument applies to any algebraic extension of K_v and therefore $H^2(H, \overline{K}_v^\times) = 0$ for every closed subgroup H of G_{K_v}. Now consider the exact sequence of G_{K_v}-modules
$$1 \to \mu_p \to \overline{K}_v^\times \xrightarrow{p} \overline{K}_v^\times \to 1.$$
This induces an exact sequence
$$H^1(H, \overline{K}_v^\times) \to H^2(H, \mu_p) \to H^2(H, \overline{K}_v^\times)$$
for any closed subgroup H of G_{K_v}. The first term is zero by Hilbert's Theorem 90 and the third term is zero by the above remark. Hence $H^2(H, \mu_p) = 0$. Letting H be a Sylow pro-p subgroup P of G_{K_v}, it follows that $H^2(P, \mathbb{Z}/p\mathbb{Z}) = 0$. This implies that $H^2(P, M) = 0$ for any finite P-module M of p-power order (since the composition factors as a P-module must be isomorphic to $\mathbb{Z}/p\mathbb{Z}$). Since the restriction map $H^2(K_v, M) \to H^2(P, M)$ is injective, we have $H^2(K_v, M) = 0$ for any G_{K_v}-module M of p-power order, which suffices to show that G_{K_v} has p-cohomological dimension 1.

Theorems 2.6 and 2.9 can be extended to cover elliptic curves with multiplicative reduction at a prime v of K lying above p. If E is such an elliptic curve, then E/K_v is a Tate curve. Assume for simplicity that E has split, multiplicative reduction at v. Then one has Tate's parametrization
$$\overline{K}_v^\times / \langle q_E \rangle \xrightarrow{\sim} E(\overline{K}_v)$$
where $\langle q_E \rangle$ is the infinite cyclic subgroup of $\overline{K}_v^\times$ generated by the Tate period q_E for E. (For a brief introduction, see appendix C in [**Si1**]. A more detailed treatment is given in chapter V of [**Si2**].) Since q_E has positive valuation, we have $\mu_{p^\infty} \cap \langle q_E \rangle = 1$ and so the Tate parametrization (which is G_{K_v}-equivariant) induces a natural map $\mu_{p^\infty} \to E(\overline{K}_v)$. One then obtains an exact sequence
$$0 \to \mu_{p^\infty} \to E[p^\infty] \to \mathbb{Q}_p/\mathbb{Z}_p \to 0$$

of G_{K_v}-modules. The action of G_{K_v} on $\mathbb{Q}_p/\mathbb{Z}_p$ is trivial. Thus, just as in the case of good, ordinary reduction, the action ρ_E of G_{K_v} on $E[p^\infty]$ is triangular. We have $\rho_E = \begin{bmatrix} \varphi & * \\ 0 & \psi \end{bmatrix}$, where $\varphi = \chi$ (the cyclotomic character) and ψ is trivial. If E has nonsplit, multiplicative reduction at v, then the situation is similar. One just twists the above exact sequence by the unique unramified character of G_{K_v} of order 2.

In both cases, $E[p^\infty]$ contains a canonical G_{K_v}-invariant subgroup of $\mathbb{Z}_p$-corank 1, which is again $\mathcal{F}[p^\infty]$, where $\mathcal{F}$ is the formal group associated to E. The following result can then be proved either by adapting our earlier arguments to this case or by using classical Kummer theory for $K_v^\times$ together with the Tate parametrization.

Theorem 2.10. *Suppose that E has multiplicative reduction at v and that K_v is any algebraic extension of $\mathbb{Q}_p$. Then* $\mathrm{im}(\kappa_v) = \mathrm{im}(\epsilon_v)$, *where*

$$\epsilon_v : H^1(K_v, \mathcal{F}[p^\infty]) \to H^1(K_v, E[p^\infty])$$

is the map induced by the inclusion $\mathcal{F}[p^\infty]$

Thus the result is actually somewhat better than in the earlier case where E was assumed to have good, ordinary reduction: $\mathrm{im}(\kappa_v)$ and $\mathrm{im}(\epsilon_v)$ coincide for *all* algebraic extensions of $\mathbb{Q}_p$.

If E has good, supersingular reduction at v, it is still possible to describe $\mathrm{im}(\kappa_v)$ in a way which depends only on the G_{K_v}-module $E[p^\infty]$. This description is considerably more subtle and it involves the ring B_{cris} introduced by Fontaine. We assume that K_v is a finite extension of $\mathbb{Q}_p$. Let $V_p(E) = T_p(E) \otimes_{\mathbb{Z}_p} \mathbb{Q}_p$, a two-dimensional $\mathbb{Q}_p$-representation space for G_{K_v}. The ring B_{cris} is also a $\mathbb{Q}_p$-algebra (of infinite dimension) on which G_{K_v} acts. Define a subgroup

$$H^1_f(K_v, V_p(E)) = \ker\left(H^1(K_v, V_p(E)) \to H^1(K_v, V_p(E) \otimes_{\mathbb{Q}_p} B_{\mathrm{cris}})\right).$$

Here $H^1(K_v, V_p(E))$ consists of the cocycle classes of continuous cocycles, where $V_p(E)$ has the topology of a $\mathbb{Q}_p$-vector space (rather than the discrete topology). Also, $H^1(K_v, V_p(E))$ is a $\mathbb{Q}_p$-vector space of dimension $2[K_v : \mathbb{Q}_p]$. The $\mathbb{Q}_p$-subspace $H^1_f(K_v, V_p(E))$ has dimension $[K_v : \mathbb{Q}_p]$. (Any line in this subspace determines a three-dimensional representation space for G_{K_v}—an extension of $\mathbb{Q}_p$ by $V_p(E)$. This representation space turns out to be "crystalline.") Now $V_p(E)$ contains $T_p(E)$ as a G_{K_v}-invariant $\mathbb{Z}_p$-lattice and $V_p(E)/T_p(E) \cong E[p^\infty]$. The map $V_p(E) \to E[p^\infty]$ induces a homomorphism

$$H^1(K_v, V_p(E)) \to H^1(K_v, E[p^\infty]).$$

The image of $H^1_f(K_v, V_p(E))$ under this map is denoted by $H^1_f(K_v, E[p^\infty])$ and is a divisible group of $\mathbb{Z}_p$-corank $[K_v : \mathbb{Q}_p]$. According to a result of Bloch and Kato [**B-K**], one has the following description of the image of the Kummer map κ_v

$$\mathrm{im}(\kappa_v) = H^1_f(K_v, E[p^\infty]).$$

One still can consider the map $\epsilon_v : H^1(K_v, \mathcal{F}[p^\infty]) \to H^1(K_v, E[p^\infty])$ when E has good, supersingular reduction at v. Then $\widetilde{E}_v[p^\infty] = 0$, the formal group $\mathcal{F}$ has height 2 (rather than 1 as in the case of good, ordinary or multiplicative reduction), and $\mathcal{F}[p^\infty] = E[p^\infty]$. Thus the map ϵ_v is an isomorphism and the equality $\mathrm{im}(\kappa_v) = \mathrm{im}(\epsilon_v)_{\mathrm{div}}$ is certainly impossible. But for infinite extensions K_v of $\mathbb{Q}_p$ which are sufficiently ramified ("deeply ramified" in the sense of [**C-G**]), something quite interesting happens. It then turns out that $\mathrm{im}(\kappa_v) = \mathrm{im}(\epsilon_v)$ (just as in theorem 2.9 for the case of good, ordinary reduction). This is true, in particular, if K_v contains a ramified $\mathbb{Z}_p$-extension of a finite extension of $\mathbb{Q}_p$. Thus, we then have $\operatorname{im}(\kappa_v) = H^1(K_v, E[p^\infty])$.

Exercises.

2.1. Let A be a p-primary abelian group which is given the discrete topology. The Pontryagin dual of A is defined by
$$\widehat{A} = \operatorname{Hom}(A, \mathbb{Q}_p/\mathbb{Z}_p)$$
which is topologized with the "compact-open" topology (where $\mathbb{Q}_p/\mathbb{Z}_p$ is given the discrete topology). Show that $\widehat{A}$ is a pro-p group (i.e. a projective limit of finite p-groups) and hence is compact. Similarly, if X is an abelian pro-p group, show that its Pontryagin dual $\widehat{X} = \operatorname{Hom}_{\text{cont}}(X, \mathbb{Q}_p/\mathbb{Z}_p)$ is a discrete p-primary group (again with the "compact-open" topology). Show that $\widehat{\mathbb{Z}_p} \cong \mathbb{Q}_p/\mathbb{Z}_p$ and $(\widehat{\mathbb{Q}_p/\mathbb{Z}_p}) \cong \mathbb{Z}_p$ as topological groups.

2.2. Suppose that $G = \langle \sigma \rangle$ is a finite, cyclic group of order m and that A is an abelian group (written with additive notation) on which G acts. Then $H^1(G, A) = \ker(N)/\operatorname{im}(\sigma - 1)$, where $N: A \to A$ is the norm map defined by $N(a) = \sum_{i=0}^{m-1} \sigma^i(a)$ and $\sigma - 1: A \to A$ is defined by $(\sigma - 1)(a) = \sigma(a) - a$ (both for all $a \in A$).

(a) Suppose that $\Gamma \cong \mathbb{Z}_p$ and let $\gamma \in \Gamma$ be chosen so that $\langle \gamma \rangle$ is a dense subgroup of Γ. (We say that γ is a topological generator of Γ.) Suppose that A is a finite, abelian p-group on which Γ acts continuously. (We put the discrete topology on A and require that the map $\Gamma \times A \to A$ defined by $(\gamma, a) \to \gamma(a)$ be continuous.) Let $\Gamma_n = \Gamma^{p^n}$ so that Γ/Γ_n is cyclic of order p^n. Then $H^1(\Gamma, A)$ can be defined as $\varinjlim_n H^1(\Gamma/\Gamma_n, A^{\Gamma_n})$. Show that
$$H^1(\Gamma, A) = A/(\gamma - 1)A.$$

(b) Suppose that A is a discrete, p-primary abelian group on which Γ acts continuously. Prove that $A = \bigcup_n A^{\Gamma_n}$. Defining $H^1(\Gamma, A)$ as above, show that $H^1(\Gamma, A) = A/(\gamma - 1)A$.

(c) Suppose that $A \cong (\mathbb{Q}_p/\mathbb{Z}_p)^r$ as a group and that Γ acts continuously on A. Prove that $H^0(\Gamma, A)$ and $H^1(\Gamma, A)$ have the same $\mathbb{Z}_p$-corank and that if $H^0(\Gamma, A)$ is finite, then $H^1(\Gamma, A) = 0$.

2.3. Let K_v be a finite extension of $\mathbb{Q}_l$ for some prime l. Let $A = \mathbb{Q}_p/\mathbb{Z}_p$ with a trivial action of G_{K_v}. Calculate $\operatorname{corank}_{\mathbb{Z}_p}(H^1(K_v, A))$ by using local class field theory. Verify the corank lemma in this case.

More generally, verify the corank lemma if $A \cong \mathbb{Q}_p/\mathbb{Z}_p$ as a $\mathbb{Z}_p$-module and G_{K_v} acts on A by a character $\psi: G_{K_v} \to \mathbb{Z}_p^\times$ of finite order. (If ψ is nontrivial, the inflation-restriction sequence will help.)

2.4. Now take $A = \mu_{p^\infty}$. Calculate $\operatorname{corank}_{\mathbb{Z}_p}(H^1(K_v, A))$ by using Kummer theory. Verify the corank lemma in this case.

2.5. Let K_v be a finite extension of $\mathbb{Q}_l$, where l is a prime and $l \neq p$. Suppose that E is an elliptic curve defined over K_v which has good reduction. Then it is known that

$K_v(E[p^\infty])/K_v$ is an unramified extension. Prove that $H^1(K_v^{\text{unr}}/K_v, E[p^\infty]) = 0$, where K_v^{unr} denotes the maximal unramified extension of K_v. (Note that

$$\text{Gal}(K_v^{\text{unr}}/K_v) \cong \widehat{\mathbb{Z}} = \prod_q \mathbb{Z}_q$$

where q varies over all primes. Use an argument involving the inflation-restriction sequence. Exercise 2.2 will be helpful.) Using the inflation-restriction sequence, deduce that if L_v is any unramified extension of K_v, then $H^1(L_v/K_v, E(L_v)_p) = 0$. Is this necessarily true if L_v/K_v is allowed to be a ramified extension?

2.6. With the same assumptions as in exercise 2.5, show that $H^1(K_v, E[p^\infty])$ is finite, as predicted by the corank lemma. (Hint: Use the inflation-restriction sequence. It will be useful to know the structure of $\text{Gal}(K_v^{\text{tame}}/K_v)$, where K_v^{tame} is the maximal tamely ramified extension of K_v. This can be described precisely by local class field theory.)

2.7. Using the corank lemma, determine $\text{corank}_{\mathbb{Z}_p}(H^1(K_v, E[p^\infty]))$ if E is an elliptic curve defined over a finite extension K_v of $\mathbb{Q}_p$.

2.8. Suppose that E is an elliptic curve defined over $\mathbb{Q}$ and that $\text{rank}_{\mathbb{Z}}(E(\mathbb{Q})) \geq 3$. Prove that $\ker(H^1(\mathbb{Q}, E[p^\infty]) \to \prod_l H^1(\mathbb{Q}_l, E[p^\infty]))$ is infinite for every prime p. (Hint: $H^1(\mathbb{Q}, E[p^\infty])$ contains a subgroup isomorphic to $E(\mathbb{Q}) \otimes (\mathbb{Q}_p/\mathbb{Z}_p)$. Study its image under the above global-to-local map. For $l \neq p$, $H^1(\mathbb{Q}_l, E[p^\infty])$ is finite by the corank lemma or by exercise 2.6. Note: It turns out that $\text{rank}_{\mathbb{Z}}(E(\mathbb{Q})) \geq 2$ suffices to give the same conclusion.)

2.9. Let E be an elliptic curve defined over a number field F. Let Σ be a finite set of primes of F containing all the archimedean primes, the primes dividing p, and the primes where E has bad reduction. Let F_Σ denote the maximal algebraic extension of F which is unramified outside Σ. Let K be any intermediate field: $F \subset K \subset F_\Sigma$. Let Σ_K denote the primes of K lying above those in Σ. Prove that

$$\text{Sel}_E(K)_p = \ker(H^1(F_\Sigma/K, E[p^\infty]) \to \prod_{\eta \in \Sigma_K} H^1(K_\eta, E[p^\infty])/\text{im}(\kappa_\eta)).$$

The maps come from restriction maps when one chooses an embedding $F_\Sigma \hookrightarrow \overline{K}_\eta$ for each η. (Hint: Exercise 2.5 is relevant.)

Assume that K is a finite extension of F. Prove that $H^1(F_\Sigma/K, E[p^\infty])$ and hence $\text{Sel}_E(K)_p$ are cofinitely generated $\mathbb{Z}_p$-modules.

2.10. Let E be an elliptic curve defined over a finite extension K of $\mathbb{Q}$. Suppose that $n \geq 2$ and that p is any prime dividing n. Prove that $H^1(K, E[n])$ has infinitely many elements of order p. (Hint: Let $M = K(E[n])$, so that G_M acts trivially on $E[n]$. Use the inflation-restriction sequence to compare $H^1(K, E[n])$ with $H^1(M, E[n])^{\text{Gal}(M/K)}$.)

2.11. Let Σ be a finite set of primes of K containing all the primes where E has bad reduction and all the prime divisors of n, where $n \geq 2$. Let K_Σ be the maximal extension of K unramified outside the set Σ. Prove that $H^1(K_\Sigma/K, E[n])$ is finite.

2.12. Let $\widetilde{E}$ be an elliptic curve defined over a finite field k_v of characteristic p. Define $a_v = 1 + |k_v| - |\widetilde{E}(k_v)|$. A theorem of Hasse (the Riemann hypothesis for the zeta function of $\widetilde{E}$ over k_v) states that $|a_v| \leq 2\sqrt{|k_v|}$. This means that the two roots α_v and β_v of the quadratic polynomial $X^2 - a_v X + |k_v|$ have absolute value $\sqrt{|k_v|}$. It is known that if $k_v^{(n)}$ denotes the extension of k_v of degree n, then $|\widetilde{E}(k_v^{(n)})| = 1 + |k_v^{(n)}| - (\alpha_v^n + \beta_v^n)$. (If $n = 1$, this is true by definition.)

(a) Prove that $\widetilde{E}$ is ordinary if and only if $p \nmid a_v$.
(b) Prove that if $\widetilde{E}$ is supersingular, $k_v = \mathbb{Z}/p\mathbb{Z}$, and $p \geq 5$, then $a_v = 0$.
(c) Prove that $\widetilde{E}(k_v)$ has an element of order p if and only if $a_v \equiv 1 \pmod{p}$. If $k_v = \mathbb{Z}/p\mathbb{Z}$ and $p \geq 7$, show that $a_v \equiv 1 \pmod{p} \iff a_v = 1$.

2.13. Fill in the details for the proof of theorem 2.10.

2.14. Suppose that K_v is a finite extension of $\mathbb{Q}_p$ and that E/K_v has either potentially good, ordinary reduction or potentially multiplicative reduction. Show that $E[p^\infty]$ contains a unique G_{K_v}-invariant subgroup C_v of $\mathbb{Z}_p$-corank 1 which is characterized as follows: There is a subgroup of finite index in the inertia group $I_{K_v} = G_{K_v^{\mathrm{unr}}}$ whose action on C_v is given by the cyclotomic character χ.

Show that $\mathrm{im}(\kappa_v) = \mathrm{im}(\epsilon_v)_{\mathrm{div}}$, where $\epsilon_v : H^1(K_v, C_v) \to H^1(K_v, E[p^\infty])$ is the map induced by the inclusion $C_v \subset E[p^\infty]$.

CHAPTER 3
Λ-Modules.

Let $\Lambda = \mathbb{Z}_p[[T]]$, the ring of formal power series over $\mathbb{Z}_p$ in one variable. Λ is a Noetherian ring. If $g(T) = \sum_{i=0}^{\infty} b_i T^i \in \Lambda$, then $g(T)$ is invertible in Λ if and only if $b_0 \in \mathbb{Z}_p^{\times}$. This implies that Λ is a local ring with maximal ideal $\mathfrak{m} = (p, T)$. We have $\Lambda/\mathfrak{m} \cong \mathbb{Z}/p\mathbb{Z}$. In fact, $\Lambda/\mathfrak{m}^k$ is finite for every $k \geq 1$. We consider Λ as a topological ring by giving it the $\mathfrak{m}$-adic topology. (A base of neighborhoods of 0 is $\{\mathfrak{m}^k\}_{k \geq 1}$.) Then $\Lambda \cong \varprojlim \Lambda/\mathfrak{m}^k$ which makes Λ a compact topological ring. Λ is obviously not a PID, but nevertheless there is the following theorem reminiscent of the structure theory of modules over a PID.

Theorem 3.1. *Suppose that X is a finitely generated Λ-module. Then there exists a Λ-module homomorphism*

$$\varphi : X \to \Lambda^r \times \prod_{i=1}^{t} \Lambda/(f_i(T)^{e_i})$$

with a finite kernel and cokernel, where $r \geq 0$, $f_1(T), \ldots, f_t(T)$ are irreducible elements of Λ, and $e_1, \ldots, e_t$ are positive integers. The value of r, the prime ideals $(f_i(T))$, and the corresponding exponents e_i are uniquely determined by X.

Some of the Λ-modules that we will be interested in arise in the following way. Let F_∞/F be a $\mathbb{Z}_p$-extension, so that $\Gamma = \text{Gal}(F_\infty/F)$ is isomorphic to $\mathbb{Z}_p$. Choose $\gamma_0 \in \Gamma$ such that $\gamma_0|_{F_1}$ is nontrivial. Then γ_0 generates a dense subgroup of Γ. Suppose that A is a p-primary, abelian group with the discrete topology and that Γ acts continuously on A. That is, there is a homomorphism $\Gamma \to \text{Aut}(A)$ and the map $\Gamma \times A \to A$ defined by $(\gamma, a) \to \gamma(a)$ for all $\gamma \in \Gamma$, $a \in A$ is continuous. Then we can make both A and $\widehat{A} = \text{Hom}(A, \mathbb{Q}_p/\mathbb{Z}_p)$ into Λ-modules. We already know how to make A and $\widehat{A}$ into $\mathbb{Z}_p$-modules. (See exercise 1.1.) We make them into $\mathbb{Z}_p[T]$-modules by defining $Ta = \gamma_0(a) - a$ for all $a \in A$. T is an endomorphism of A. In exercise 3.3, it is shown that T is "topologically nilpotent"' and so both A and $\widehat{A}$ can be made into Λ-modules.

Let E be an elliptic curve defined over a field F. Then $A = H^1(F_\infty, E[p^\infty])$ is a p-primary abelian group and has a natural action of Γ. We can then regard A

as a Λ-module (with the discrete topology). If F is a finite extension of $\mathbb{Q}_l$ for any prime l, then $\widehat{A}$ is a finitely generated Λ-module. If F is a number field, then let Σ be a finite set of primes of F containing all archimedean primes, all primes dividing p, and all primes where E has bad reduction. Then $E[p^\infty]$ is a $\mathrm{Gal}(F_\Sigma/F)$-module, and $A = H^1(F_\Sigma/F_\infty, E[p^\infty])$ also has a natural action of Γ and can be regarded as a Λ-module. Again, $\widehat{A}$ turns out to be a finitely generated Λ-module. Also, $\mathrm{Sel}_E(F_\infty)_p$ is a Λ-submodule of $H^1(F_\Sigma/F_\infty, E[p^\infty])$ (just because it is Γ-invariant). There are many difficult, unsolved questions concerning the structure of these Λ-modules. We will state a few of them, using the following convenient terminology. If A is a discrete Λ-module as above, then we say that A is Λ-cotorsion if $\widehat{A}$ is Λ-torsion, A is Λ-cofinitely generated if $\widehat{A}$ is Λ-finitely generated, and we define $\mathrm{corank}_\Lambda(A)$ to be $\mathrm{rank}_\Lambda(\widehat{A})$.

Conjecture 3.2. *Suppose that F is a finite extension of $\mathbb{Q}$ and that E is any elliptic curve defined over F. Let Σ be chosen as above. Then*
$$\mathrm{corank}_\Lambda(H^1(F_\Sigma/F_\infty, E[p^\infty])) = [F : \mathbb{Q}].$$

In general, all that is known about this is that
$$\mathrm{corank}_\Lambda(H^1(F_\Sigma/F_\infty, E[p^\infty])) \geq [F : \mathbb{Q}]$$
and that the equality is equivalent to the assertion that $H^2(F_\Sigma/F_\infty, E[p^\infty])$, which can also be considered as a Λ-module, is Λ-cotorsion.

Conjecture 3.3. *Suppose that F is a finite extension of $\mathbb{Q}$ and that E is an elliptic curve defined over F which has good, ordinary reduction at all primes of F lying over p. Suppose that F_∞/F is the cyclotomic $\mathbb{Z}_p$-extension. Then $\mathrm{Sel}_E(F_\infty)_p$ is Λ-cotorsion.*

This was conjectured by Mazur in [**Ma1**]. One case in which it is known to be true is when F is an abelian extension of $\mathbb{Q}$ and E is a modular elliptic curve defined over $\mathbb{Q}$. In this case, both conjectures 3.2 and 3.3 have been proven by Kato (and earlier by Rubin if E has complex multiplication). Conjecture 3.3 should be true under the weaker hypothesis that E has either potentially ordinary or potentially multiplicative reduction at all primes of F (lying over p.) In general, the following conjecture (due to P. Schneider) gives the Λ-corank of $\mathrm{Sel}_E(F_\infty)_p$.

Conjecture 3.4. *Let F_∞/F be the cyclotomic $\mathbb{Z}_p$-extension of a number field F. Assume that E is an elliptic curve defined over F. Then*
$$\mathrm{corank}_\Lambda(\mathrm{Sel}_E(F_\infty)_p) = \sum_{v\,pss} [F_v : \mathbb{Q}_p]$$
where the sum is over all primes of F lying over p where E has potentially supersingular reduction.

This conjecture turns out to imply conjecture 3.2 (for the cyclotomic $\mathbb{Z}_p$-extension F_∞/F). One can prove that $\sum_{v\,pss} [F_v : \mathbb{Q}_p]$ is a lower bound on $\mathrm{corank}_\Lambda(\mathrm{Sel}_E(F_\infty)_p)$, and therefore $\mathrm{Sel}_E(F_\infty)_p$ cannot be Λ-cotorsion if there is at least one prime of F lying over p where E has potentially supersingular reduction.

We will return now to various basic theorems concerning Λ and its module theory, partially proving theorem 3.1. We start with the following result.

CHAPTER 3. Λ-MODULES.

Theorem 3.5. *Suppose that $g(T) \in \Lambda$, but $g(T) \notin p\Lambda$. If $g(T) = \sum_{i=0}^{\infty} b_i T^i$, let d be the smallest integer such that $b_d \in \mathbb{Z}_p^{\times}$. Let $f(T) \in \Lambda$. Then there exists a polynomial $r(T) \in \mathbb{Z}_p[T]$ of degree $< d$, and an element $h(T) \in \Lambda$ such that*

$$f(T) = g(T)h(T) + r(T).$$

Furthermore, $h(T)$ and $r(T)$ are uniquely determined by $f(T)$ and $g(T)$.

Proof. The uniqueness is easy. If $g(T)h(T) + r(T) = g(T)h'(T) + r'(T)$, then $r'(T) - r(T) = s(T)$ is divisible by $g(T)$ in Λ. Let $\widetilde{\Lambda} = \Lambda/p\Lambda$ which can be thought of as $\mathbb{F}_p[[T]]$, where $\mathbb{F}_p = \mathbb{Z}/p\mathbb{Z}$ is the field with p elements. We will denote the image of an element $f(T) \in \Lambda$ under the canonical homomorphism $\Lambda \to \widetilde{\Lambda}$ by $\widetilde{f}(T)$. Assume that $s(T) \neq 0$. Then we can write $s(T) = p^m s_0(T)$, where $s_0(T) \notin p\Lambda$ and $m \geq 0$. Now p is a prime element of Λ because $\widetilde{\Lambda}$ is an integral domain. Since $g(T)$ divides $s(T)$, it then follows that $g(T)$ divides $s_0(T)$. But $\widetilde{s}_0(T)$ is a nonzero polynomial in $\widetilde{\Lambda}$ of degree $< d$ which is divisible by $\widetilde{g}(T)$ and therefore by T^d in $\widetilde{\Lambda}$. This is clearly impossible. Hence $s(T) = 0$ and so $r(T) = r'(T)$. This implies that $h(T) = h'(T)$ too.

We prove the existence by using the fact that $\widetilde{\Lambda}$ is a discrete valuation ring and has a simple division algorithm. If $\widetilde{f}(T) = \sum_{i=0}^{\infty} \widetilde{a}_i T^i$ (where $\widetilde{a}_i \in \mathbb{F}_p$ for all i), then we can write $\widetilde{f}(T) = T^d(\sum_{i=d}^{\infty} \widetilde{a}_i T^{i-d}) + \sum_{i=0}^{d-1} \widetilde{a}_i T^i$. Since $\widetilde{g}(T)$ and T^d generate the same ideal in $\widetilde{\Lambda}$, we can write $\widetilde{f}(T) = \widetilde{g}(T)\widetilde{h}(T) + \widetilde{r}(T)$, where $\widetilde{h}(T) \in \widetilde{\Lambda}$ and $\widetilde{r}(T)$ is a polynomial in $\widetilde{\Lambda}$ of degree $< d$. Lifting back to Λ, we get a division algorithm modulo $p\Lambda$. That is, we can write

$$f(T) = g(T)h_1(T) + r_1(T) + pf_1(T)$$

where $h_1(T), f_1(T) \in \Lambda$ and $r_1(T)$ is a polynomial of degree $< d$. Applying this to $f_1(T)$ instead of $f(T)$, we then obtain

$$f(T) = g(T)h_2(T) + r_2(T) + p^2 f_2(T)$$

where again $h_2(T), f_2(T) \in \Lambda$, $r_2(T)$ is a polynomial of degree $< d$, and $h_2(T) \equiv h_1(T) \pmod{p\Lambda}$, $r_2(T) \equiv r_1(T) \pmod{p\mathbb{Z}_p[T]}$. Continuing, we obtain an equation for each $n \geq 1$:

$$f(T) = g(T)h_n(T) + r_n(T) + p^n f_n(T)$$

where $h_n(T) \in \Lambda$ and $h_{n+1}(T) \equiv h_n(T) \pmod{p^n \Lambda}$, $r_n(T) \in \mathbb{Z}_p[T]$ with degree $< d$ and $r_{n+1}(T) \equiv r_n(T) \pmod{p^n \mathbb{Z}_p[T]}$. Clearly, the limits exist in Λ and we can write $f(T) = g(T)h(T) + r(T)$ where $h(T) = \lim_{n \to \infty} h_n(T)$, $r(T) = \lim_{n \to \infty} r_n(T)$. Here $h(T) \in \Lambda$ and $r(T)$ is a polynomial of degree $< d$. This proves the existence. $\square$

Corollary 3.6. *Let $g(T)$ be as in theorem 3.5. Then $\Lambda/(g(T))$ is isomorphic to $\mathbb{Z}_p^d$ as a $\mathbb{Z}_p$-module.*

Proof. This is clear because the map $f(T) + (g(T)) \to r(T)$ defines a $\mathbb{Z}_p$-module isomorphism of $\Lambda/(g(T))$ to $\{r(T) | r(T) \in \mathbb{Z}_p[T], \deg(r(T)) < d\}$. $\square$

A polynomial $\sum_{i=0}^{d} c_i T^i \in \mathbb{Z}_p[T]$ is called a "distinguished polynomial" if $c_d = 1$ and $c_i \in p\mathbb{Z}_p$ for all $i < d$. (An equivalent condition is that the polynomial be monic and that all of its roots in $\overline{\mathbb{Q}}_p$ have positive valuation.)

Corollary 3.7. *Let $g(T)$ be as in theorem 3.5. Then there exists a unique distinguished polynomial $g_0(T)$ such that $g(T) = u(T)g_0(T)$, where $u(T) \in \Lambda^\times$. The degree of $g_0(T)$ is d.*

Proof. Multiplication by T defines an endomorphism of the $\mathbb{Z}_p$-module $\Lambda/(g(T))$. Since this is free with $\mathbb{Z}_p$-rank d, there is a monic polynomial $g_0(T)$ of degree d such that $g_0(T)$ annihilates $\Lambda/(g(T))$. Namely, take $g_0(T)$ to be the characteristic polynomial of T acting on this $\mathbb{Z}_p$-module. Thus, $(g_0(T)) \subset (g(T))$. Applying corollary 3.6 to $g_0(T)$, we see that $\Lambda/(g_0(T))$ is isomorphic to $\mathbb{Z}_p^{d_0}$ where $d_0 \leq d$. But there is a natural surjective homomorphism $\Lambda/(g_0(T)) \to \Lambda/(g(T))$ of $\mathbb{Z}_p$-modules. Hence $d_0 = d$ and this homomorphism must be an isomorphism. That is, $g_0(T)$ is distinguished of degree d and $(g_0(T)) = (g(T))$. Thus, $g(T) = u(T)g_0(T)$ as stated. It is clear that $\Lambda/(g(T))$ can be identified with $\mathbb{Z}_p[T]/(g_0(T))$ and that $g_0(T)$ as defined above is the minimal polynomial of the endomorphism T of $\Lambda/(g(T))$. This implies uniqueness because if $g(T) = u'(T)g_0'(T)$ where $g_0'(T)$ is a distinguished polynomial, then $g_0'(T)$ annihilates $\Lambda/(g(T))$ and hence $g_0(T)|g_0'(T)$ in $\mathbb{Z}_p[T]$. It is also clear from the above that $\deg(g_0(T)) = \deg(g_0'(T))$ and so $g_0(T) = g_0'(T)$. (Alternatively, one could see that $g_0(T)$ and $g_0'(T)$ have the same degree by reducing modulo $p\Lambda$.) □

Theorem 3.8. *Let $g(T)$ be a nonzero element of Λ. Then we can write $g(T)$ uniquely in the form*

$$g(T) = p^m u(T) g_0(T)$$

where $m \geq 0$, $u(T) \in \Lambda^\times$, and $g_0(T)$ is a distinguished polynomial. Also, $g(T)$ is an irreducible element of Λ if and only if either $m = 1$ and $g_0(T) = 1$ or $m = 0$ and $g_0(T)$ is irreducible as an element of $\mathbb{Q}_p[T]$.

Proof. We can obviously write $g(T)$ as $g(T) = p^m g_1(T)$ where $g_1(T) \in \Lambda$, but $\notin p\Lambda$. Here m is unique. We can then apply corollary 3.7 to $g_1(T)$, obtaining the first assertion. Now obviously p is irreducible in Λ. On the other hand, if $g_0(T)$ is reducible in Λ, then by applying corollary 3.7, one would find that $g_0(T)$ can be factored as a product of distinguished polynomials of degree $< \deg(g_0(T))$. Hence $g_0(T)$ would be reducible in $\mathbb{Q}_p[T]$. Conversely, if $g_0(T)$ is reducible in $\mathbb{Q}_p[T]$, then it is reducible in $\mathbb{Z}_p[T]$ and its monic, irreducible factors are also distinguished. One can see this by reducing modulo $p\Lambda$. It is then clear that $g(T)$ is reducible in Λ. □

Thus, up to multiplication by elements of $\Lambda^\times$, the irreducible elements of Λ are either p or any distinguished polynomial in $\mathbb{Z}_p[T]$ which is irreducible over $\mathbb{Z}_p$ or, equivalently, over $\mathbb{Q}_p$. Since $\mathbb{Z}_p[T]$ is a UFD, it is easy to deduce that Λ is also a UFD.

Now let X be any finitely generated, torsion Λ-module. We will give a partial proof of theorem 3.1. We let $Y = X_{\mathbb{Z}_p\text{-tors}}$. It is easy to see that Y is a Λ-submodule of X, and therefore is also finitely generated. This implies that Y has finite exponent, i.e., $p^m Y = 0$ for some $m \geq 0$. Also, X/Y is torsion-free as a $\mathbb{Z}_p$-module. Assume first that $Y = 0$. Let $g(T)$ be a nonzero annihilator of X. If we write $g(T) = p^m g'(T)$ where $m \geq 1$, then it is clear that $g'(T)$ annihilates X. Hence we can assume that $g(T) \notin p\Lambda$. Now if X has n generators, X will be a homomorphic image of $(\Lambda/(g(T)))^n$. It follows from corollary 3.6 that X is finitely generated as a $\mathbb{Z}_p$-module and, since $Y = 0$, X must be a free $\mathbb{Z}_p$-module: $X \cong \mathbb{Z}_p^\lambda$ for some $\lambda \geq 0$.

Consider the $\mathbb{Q}_p$-vector space $V = X \otimes_{\mathbb{Z}_p} \mathbb{Q}_p$ of dimension λ together with the endomorphism T. We can consider V as a module over $\mathbb{Q}_p[T]$, and since this ring is a PID, we have

$$V \cong \prod_{i=1}^{t} \mathbb{Q}_p[T]/(f_i(T)^{e_i})$$

where the $f_i(T)$'s are uniquely determined, monic irreducible polynomials over $\mathbb{Q}_p$ and the e_i's are uniquely determined nonnegative integers (everything up to order, of course). However, if we write $g(T) = u(T)g_0(T)$ where $g_0(T)$ is a distinguished polynomial, then $g_0(T)$ annihilates V and it follows that $f_i(T)|g_0(T)$ in $\mathbb{Q}_p[T]$ for each i. But this implies that each $f_i(T)$ is a distinguished polynomial in $\mathbb{Z}_p[T]$. Now X is a $\mathbb{Z}_p$-lattice in V and so is $X' = \prod_{i=1}^{t} \mathbb{Z}_p[T]/(f_i(T)^{e_i})$. It follows that for some $k \geq 0$, $p^k X \subset X'$. Clearly $p^k X$ is isomorphic to X. The index $[X' : p^k X]$ is obviously finite. Identifying $\mathbb{Z}_p[T]/(f_i(T)^{e_i})$ with $\Lambda/(f_i(T)^{e_i})$, we then obtain an injective Λ-module homomorphism

$$\varphi : X \to \prod_{i=1}^{t} \Lambda/(f_i(T)^{e_i})$$

where the $f_i(T)$'s are irreducible distinguished polynomials, and hence irreducible elements of Λ. The cokernel of φ is finite. This proves theorem 3.1 in the important special case where X is assumed to be $\mathbb{Z}_p$-torsion free. Note that

$$\text{rank}_{\mathbb{Z}_p}(X) = \sum_{i=1}^{t} e_i \deg(f_i(T))$$

since, as we have seen, $f(T) = f_i(T)^{e_i}$ is a distinguished polynomial and $\Lambda/(f(T))$ is isomorphic to $\mathbb{Z}_p^{\deg(f)}$ as a $\mathbb{Z}_p$-module.

Two finitely generated, torsion Λ-modules X_1 and X_2 are said to be pseudo-isomorphic if there is a Λ-module homomorphism $\psi : X_1 \to X_2$ such that $\ker(\psi)$ and $\text{coker}(\psi)$ are both finite. We then write $X_1 \sim X_2$. It turns out that this is an equivalence relation. Also, if X is any finitely generated, torsion Λ-module, then it is not hard to show that $X \sim Y \times X/Y$, where $Y = X_{\mathbb{Z}_p\text{-tors}}$ as before. Thus, to prove theorem 3.1 for X, it would suffice to show that Y is pseudo-isomorphic to a direct sum of finitely many Λ-modules of the form $\Lambda/(p^{\mu_j})$, where $\mu_j \geq 1$. For a complete proof of theorem 3.1, including the cases where $Y \neq 0$ or where $\text{rank}_\Lambda(X) > 0$, we refer the reader to [**Wa**] or to [**Bo**], Chapter VII, §4. The proofs found in those references are quite different—one involving a matrix reduction argument, the other a much more conceptual and general module-theoretic approach. A more succinct argument (along the lines of the more general theory given in [**Bo**]) can be found in [**Se**]. This is based on the module theory of complete local rings of Krull dimension 2. Λ is such a ring.

Suppose now that X is an arbitrary finitely generated, torsion Λ-module. In theorem 3.1, we would have $r = 0$ and we can assume that each $f_i(T)$ is taken to be either p or an irreducible, distinguished polynomial. Then the characteristic ideal of X is defined to be the ideal generated by the polynomial $f_X(T) = \prod_{i=1}^{t} f_i(T)^{e_i}$. Letting $Y = X_{\mathbb{Z}_p\text{-tors}}$ as before, we can apply our previous observations to the Λ-module X/Y. If we let $\lambda_X = \deg(f_X(T))$, then the invariant λ_X also has the following interpretations:

$$\lambda_X = \text{rank}_{\mathbb{Z}_p}(X/Y) = \dim_{\mathbb{Q}_p}(X \otimes_{\mathbb{Z}_p} \mathbb{Q}_p).$$

We also define an invariant $\mu = \mu_X$ by $f_X(T) \in p^\mu \Lambda$, but $\notin p^{\mu+1}\Lambda$. That is, μ_X is the sum of the e_i's over all i's such that $f_i(T) = p$. In fact, $\mu_X = \mu_Y$ depends only on the structure of $Y = X_{\mathbb{Z}_p\text{-tors}}$ as a Λ-module. Assume that $p^m Y = 0$. For each $j, 0 \leq j \leq m$, let $Y[p^j]$ denote the kernel of the map $Y \xrightarrow{p^j} Y$, multiplication by p^j. Then for $1 \leq j \leq m$, $Y[p^j]/Y[p^{j-1}]$ can be considered as a $\widetilde{\Lambda}$-module, where recall that $\widetilde{\Lambda} = \Lambda/p\Lambda$ is the ring $\mathbb{F}_p[[T]]$. The above $\widetilde{\Lambda}$-modules are finitely generated. Then μ_X has the following interpretation:

$$\mu_X = \sum_{j=1}^{m} \text{rank}_{\widetilde{\Lambda}}(Y[p^j]/Y[p^{j-1}]).$$

Since a finitely generated, torsion $\widetilde{\Lambda}$-module is finite, it is clear that $\mu_X > 0$ if and only if $X_{\mathbb{Z}_p\text{-tors}}$ is infinite. Other equivalent statements are that $X[p]$ is infinite or that X/pX is infinite. If $\mu_X = 0$, then X is a finitely generated $\mathbb{Z}_p$-module.

Our last general theorem concerning Λ-modules is the following very useful result which allows us to obtain information about X from certain quotients.

Theorem 3.9. *Let X be an abelian pro-p group on which Γ acts continuously. We regard X as a Λ-module. Then*
  (1) $X = 0 \iff X/TX = 0 \iff X/\mathfrak{m}X = 0$.
  (2) *X is finitely generated as a Λ-module if and only if $X/\mathfrak{m}X$ is finite dimensional as an $\mathbb{F}_p$-vector space. The minimal number of generators of X as a Λ-module is $\dim_{\mathbb{F}_p}(X/\mathfrak{m}X)$.*
  (3) *If X/TX is finite, then X is a torsion Λ-module.*

Proof. Part (1) is the subject of exercise 3.2. For part (2), it is obvious that if X is finitely generated, then $\dim_{\mathbb{F}_p}(X/\mathfrak{m}X)$ is finite. To prove the converse, let $d = \dim_{\mathbb{F}_p}(X/\mathfrak{m}X)$ and pick $x_1, \ldots, x_d \in X$ so that their images in $X/\mathfrak{m}X$ form a basis over $\mathbb{F}_p$. Let $Y = \Lambda x_1 + \cdots + \Lambda x_d$, the Λ-submodule generated by $x_1, \ldots, x_d$. Since Λ is compact, Y is a closed subgroup of X. Hence X/Y is a pro-p group. But one can easily verify that $\mathfrak{m}(X/Y) = X/Y$. By part (1), this implies that $X/Y = 0$ and so $X = Y$. That is, X is generated as a Λ-module by the d elements $x_1, \ldots, x_d$.

Now assume that X/TX is finite. Then $X/\mathfrak{m}X$ is finite and so X is finitely generated as a Λ-module. Let $x_1, \ldots, x_d$ be a set of generators. Suppose that X/TX has exponent p^k. Then $p^k x_i \in TX$ for $1 \leq i \leq d$. Therefore we can write

$$p^k x_i = \sum_{j=1}^{d} T a_{ij}(T) x_j$$

for $1 \leq i \leq d$, where the $a_{ij}(T)$'s are elements of Λ. That is,

$$\sum_{j=1}^{d} (p^k \delta_{ij} - T a_{ij}(T)) x_j = 0$$

for each i. Let A be the $d \times d$ matrix $[p^k \delta_{ij} - T a_{ij}(T)]$. Let A^* be its adjoint matrix so that $A^* A = \det(A) I_d$. Let $g(T) = \det(A)$, which must be a nonzero element of Λ because $g(0) = p^{dk}$. We have $g(T) x_i = 0$ for each i and therefore $g(T)$ annihilates X. This proves (3). $\square$

Suppose now that G is any profinite group. That is, $G = \varprojlim G_n$, where the G_n's are finite groups. For simplicity, we will take the indexing set to be the set of nonnegative integers, although this is not essential. We also may assume that the homomorphisms $\varphi_{m,n} : G_m \to G_n$ for $m \geq n \geq 0$ which define the inverse limit are surjective. Then one obtains induced $\mathbb{Z}_p$-algebra homomorphisms

$$\psi_{m,n} : \mathbb{Z}_p[G_m] \to \mathbb{Z}_p[G_n].$$

Here, if $\mathcal{G}$ is any group, then $\mathbb{Z}_p[\mathcal{G}]$ denotes the group ring for $\mathcal{G}$ over $\mathbb{Z}_p$. The maps $\psi_{m,n}$ are surjective. One then defines the completed group ring for G over $\mathbb{Z}_p$ as

$$\mathbb{Z}_p[[G]] = \varprojlim \mathbb{Z}_p[G_n],$$

where the inverse limit is defined by the maps $\psi_{m,n}$. Now $\mathbb{Z}_p[G_n]$ is a compact topological ring, a $\mathbb{Z}_p$-algebra isomorphic to $\mathbb{Z}_p^{|G_n|}$ as a $\mathbb{Z}_p$-module. It follows that $\mathbb{Z}_p[[G]]$ is also a compact topological ring. Alternatively, one can define $\mathbb{Z}_p[[G]]$ as an inverse limit of the finite rings $(\mathbb{Z}/p^k\mathbb{Z})[G_n]$, each with the discrete topology, where $k \geq 1$, $n \geq 0$. For each n, there is a continuous, surjective, $\mathbb{Z}_p$-algebra homomorphism $\psi_n : \mathbb{Z}_p[[G]] \to \mathbb{Z}_p[G_n]$. Also, the ordinary group algebra $\mathbb{Z}_p[G]$ for G over $\mathbb{Z}_p$ can be identified with a subring of $\mathbb{Z}_p[[G]]$ in a natural way and turns out to be dense. This and other elementary properties of $\mathbb{Z}_p[[G]]$ are the subject of exercise 3.12.

We are interested in the special case $G = \Gamma \cong \mathbb{Z}_p$. Then we can take $G_n = \Gamma/\Gamma^{p^n}$, which is cyclic of order p^n. If γ_0 is a topological generator of Γ, then $\gamma_0^{(n)} = \gamma_0 \Gamma^{p^n}$ generates G_n for each $n \geq 0$. The next result shows that the rings $\mathbb{Z}_p[[\Gamma]]$ and $\Lambda = \mathbb{Z}_p[[T]]$ can be identified. The identification is not canonical, but depends on the choice of γ_0.

Theorem 3.10. *There is a unique isomorphism*

$$\varepsilon : \Lambda \to \mathbb{Z}_p[[\Gamma]]$$

of topological $\mathbb{Z}_p$-algebras such that $\varepsilon(T) = \gamma_0 - 1$.

Proof. $\Lambda = \mathbb{Z}_p[[T]]$ is a compact $\mathbb{Z}_p$-algebra. For each $n \geq 0$, let $\omega_n = (1+T)^{p^n} - 1$, which is easily seen to be a distinguished polynomial of degree p^n. By corollary 3.6, $\Lambda/(\omega_n)$ is isomorphic to $\mathbb{Z}_p^{p^n}$ as a $\mathbb{Z}_p$-module. More precisely, by theorem 3.5, it is clear that $\{(1+T)^i + (\omega_n)\}_{0 \leq i < p^n}$ is a $\mathbb{Z}_p$-module basis for $\Lambda/(\omega_n)$. Also $(1+T)^{p^n} \equiv 1 \pmod{(\omega_n)}$. Hence we can define an isomorphism

$$\Lambda/(\omega_n) \to \mathbb{Z}_p[\Gamma/\Gamma^{p^n}]$$

of $\mathbb{Z}_p$-algebras by mapping $1 + T + (\omega_n)$ to $\gamma_0^{(n)}$. Thus we obtain continuous, surjective, $\mathbb{Z}_p$-algebra homomorphisms $\Lambda \to \mathbb{Z}_p[\Gamma/\Gamma^{p^n}]$ for all $n \geq 0$ which map T to $\gamma_0^{(n)} - 1$. This induces a continuous, $\mathbb{Z}_p$-algebra homomorphism

$$\varepsilon : \Lambda \to \mathbb{Z}_p[[\Gamma]].$$

The image of ε is dense, but since Λ is compact, this image is also closed. Hence ε is surjective. It is clear that $\varepsilon(T) = \gamma_0 - 1$ and that this uniquely determines the map ε

(assuming it is a continuous, $\mathbb{Z}_p$-algebra homomorphism). Also, $\ker(\varepsilon) \subset \bigcap_n(\omega_n)$, which is trivial. Hence ε is injective. The compactness of Λ implies that ε is a homeomorphism too. $\square$

The final topic of this section is to study the structure of certain specific Λ-modules in order to prove the following special case of the corank lemma of chapter 2. The argument is somewhat long, but is a good illustration of standard techniques of Iwasawa theory.

Theorem 3.11. *Let K_v be a finite extension of $\mathbb{Q}_p$. Suppose that A is a G_{K_v}-module and that $A \cong \mathbb{Q}_p/\mathbb{Z}_p$ as a group. Then $H^1(K_v, A)$ is a cofinitely generated $\mathbb{Z}_p$-module and $\mathrm{corank}_{\mathbb{Z}_p}(H^1(K_v, A)) = [K_v : \mathbb{Q}_p] + \delta_A(K_v)$, where $\delta_A(K_v) = 1$ if $A \cong \mathbb{Q}_p/\mathbb{Z}_p$ or $A \cong \mu_{p^\infty}$ as G_{K_v}-modules and $\delta_A(K_v) = 0$ otherwise.*

Note that $H^0(K_v, A)$ is either finite or all of A. The latter occurs only when $A \cong \mathbb{Q}_p/\mathbb{Z}_p$ as G_{K_v}-modules (i.e., the action of G_{K_v} on A is trivial). Also, $\widehat{A}(1) \cong \mathrm{Hom}(A, \mu_{p^\infty})$ and $H^0(K_v, \widehat{A}(1))$ is either trivial or all of $\widehat{A}(1)$, which is isomorphic to $\mathbb{Z}_p$ as a group. The latter occurs only when $A \cong \mu_{p^\infty}$ as G_{K_v}-modules. Thus

$$\delta_A(K_v) = \mathrm{corank}_{\mathbb{Z}_p}(H^0(K_v, A)) + \mathrm{rank}_{\mathbb{Z}_p}(H^0(K_v, \widehat{A}(1)))$$

and so the above theorem is just the content of the corank lemma when $r = 1$ and $v | p$. The two cases where $\delta_A(K_v) = 1$ are rather straightforward and are the subject of exercises 2.3 and 2.4. Therefore, we will assume that $\delta_A(K_v) = 0$ in the following proof.

Proof. The action of G_{K_v} on A is given by a homomorphism $\psi : G_{K_v} \to \mathbb{Z}_p^\times$. If ψ has finite order, then it factors through $\Delta = \mathrm{Gal}(F/K_v)$ where F is a finite extension of K_v. The corank lemma in this case is the subject of exercise 2.3. The argument just depends on local class field theory over F and studying the action of Δ on $F^\times$. It is straightforward and so we will now assume that $\mathrm{im}(\psi)$ is infinite. Let $F_\infty = \overline{K}_v^{\ker(\psi)}$ so that $G = \mathrm{Gal}(F_\infty/K_v)$ acts faithfully on A. Also, $G \cong \mathrm{im}(\psi)$, which is a subgroup of $\mathbb{Z}_p^\times$. Hence $G \cong \Delta \times \Gamma$, where $\Gamma \cong \mathbb{Z}_p$ and Δ is a finite group of order dividing $p - 1$ if p is odd and of order 1 or 2 if $p = 2$. Let $F_0 = F_\infty^\Gamma$ so that $\Delta = \mathrm{Gal}(F_0/K_v)$ and $\Gamma = \mathrm{Gal}(F_\infty/F_0)$. That is, F_∞/F_0 is a $\mathbb{Z}_p$-extension. For $n \geq 0$, we let $F_n = F_\infty^{\Gamma^{p^n}}$. Then $F_\infty = \bigcup_n F_n$ and F_n/F_0 is cyclic of degree p^n. The inflation-restriction sequence

$$0 \to H^1(G, A) \xrightarrow{i} H^1(K_v, A) \xrightarrow{r} H^1(F_\infty, A)^G \to H^2(G, A)$$

is exact and will allow us to study $H^1(K_v, A)$ by studying $H^1(F_\infty, A)^G$ instead.

If p is odd, then $H^2(G, A) = 0$ and so r is surjective. This follows easily from the facts that $|\Delta|$ is not divisible by p and Γ is a free pro-p group and hence has p-cohomological dimension 1. If $p = 2$ and $|\Delta| = 1$, then, for the same reason, $H^2(G, A) = 0$ and so r is surjective in this case too. But if $|\Delta| = 2$, then it is not hard to verify that $H^2(G, A) \cong \mathbb{Z}/2\mathbb{Z}$. Thus, $\mathrm{coker}(r)$ is finite of order ≤ 2 and is usually trivial. Now $\ker(r) = \mathrm{im}(i)$ also turns out to be finite of order ≤ 2 and is usually trivial. If $|\Delta| = 1$, then $G = \Gamma$ and, since we are assuming that $H^0(K_v, A)$ is finite, one can apply exercise 2.2(c) to see that $H^1(G, A) = 0$. If p is odd and $|\Delta| > 1$, then $A^\Delta = 0$. It is then even easier to see that $H^1(G, A) = 0$ (using a simple inflation-restriction argument). But if $p = 2$ and $|\Delta| = 2$, then A^Δ has order

2 and one finds that $H^1(G, A) \cong \mathbb{Z}/2\mathbb{Z}$. The above remarks show that
$$\operatorname{corank}_{\mathbb{Z}_p}(H^1(K_v, A)) = \operatorname{corank}_{\mathbb{Z}_p}(H^1(F_\infty, A)^G).$$
But $H^1(F_\infty, A) = \operatorname{Hom}_{\operatorname{cont}}(G_{F_\infty}, A)$ since G_{F_∞} acts trivially on A. Let M_∞ denote the maximal, abelian pro-p extension of F_∞. That is, M_∞ is the compositum of all finite, abelian p-extensions of F_∞. Let $X = \operatorname{Gal}(M_\infty/F_\infty)$. Then we have $H^1(F_\infty, A) = \operatorname{Hom}_{\operatorname{cont}}(X, A)$. From here on, we will omit the subscript cont, taking the requirement of continuity as understood.

Note that M_∞ is a Galois extension of K_v and that $\operatorname{Gal}(M_\infty/K_v)$ can be regarded as a group extension of the quotient group $\operatorname{Gal}(F_\infty/K_v) = G$ by the closed, normal subgroup $X = \operatorname{Gal}(M_\infty/F_\infty)$. Hence there is a natural action of G on X (by inner automorphisms). Then we have $H^1(F_\infty, A)^G = \operatorname{Hom}_G(X, A)$. Now X is a $\mathbb{Z}_p$-module on which G acts $\mathbb{Z}_p$-linearly and continuously. Let X^ψ denote the maximal quotient of X on which G acts by ψ. Then we have
$$\operatorname{corank}_{\mathbb{Z}_p}(H^1(K_v, A)) = \operatorname{corank}_{\mathbb{Z}_p}(\operatorname{Hom}(X^\psi, A)) = \operatorname{rank}_{\mathbb{Z}_p}(X^\psi)$$
since $A \cong \mathbb{Q}_p/\mathbb{Z}_p$ as a group. The rest of the argument involves studying X as a G-module.

All of the characters of Δ have values in $\mathbb{Z}_p^\times$. If p is odd, so that $p \nmid |\Delta|$, then one has a decomposition of X by the characters of Δ (often referred to as the Δ-decomposition):
$$X = \bigoplus_{\chi \in \widehat{\Delta}} X^\chi.$$
Here $X^\chi = \{x \in X | \delta(x) = \chi(\delta)x \text{ for all } \delta \in \Delta\} = e_\chi X$, where e_χ is the idempotent for χ in $\mathbb{Z}_p[\Delta]$. If $p = 2$ and $|\Delta| = 2$, which is the only case where $|\Delta|$ is divisible by p, then we define X^χ as the maximal quotient of X on which Δ acts by χ. Now Δ acts on A by the character $\chi = \psi|_\Delta$, which we denote by ψ_Δ. Then
$$\operatorname{Hom}_\Delta(X, A) = \operatorname{Hom}(X^{\psi_\Delta}, A).$$
Furthermore, Γ acts on A by the character $\psi_\Gamma : \Gamma \to 1 + p\mathbb{Z}_p$, where $\psi_\Gamma = \psi|_\Gamma$. If Y is any $\mathbb{Z}_p$-module on which Γ acts and if $\kappa : \Gamma \to 1 + p\mathbb{Z}_p$ is a continuous homomorphism, then we let Y^κ denote $Y/(\gamma_0 - \kappa(\gamma_0))Y$, which is the maximal quotient of Y on which Γ acts by the character κ. Here γ_0 is a fixed topological generator of Γ. In fact, we have $X^\psi = (X^{\psi_\Delta})^{\psi_\Gamma}$ and so it follows that
$$\operatorname{corank}_{\mathbb{Z}_p}(H^1(K_v, A)) = \operatorname{rank}_{\mathbb{Z}_p}(X^{\psi_\Delta}/(\gamma_0 - \psi_\Gamma(\gamma_0))X^{\psi_\Delta}).$$
We determine this rank by considering X^{ψ_Δ} as a Λ-module.

The Galois group X is an abelian, pro-p group. Applying exercise 3.3 to the discrete, p-primary Γ-module $\operatorname{Hom}(X, \mathbb{Q}_p/\mathbb{Z}_p)$, one can regard X as a Λ-module. We will prove the following statements about X which were proved by Iwasawa in [**Iw**].

(a) X is a finitely generated Λ-module.
(b) X has Λ-rank equal to $[K_v : \mathbb{Q}_p]|\Delta|$. More precisely, for each character χ of Δ, X^χ has Λ-rank equal to $[K_v : \mathbb{Q}_p]$.
(c) If F_∞ contains the group μ_{p^∞} of p-power roots of unity, then the Λ-torsion submodule $X_{\Lambda\text{-tors}}$ is isomorphic to $T_p(\mu_{p^\infty})$, the Tate module of μ_{p^∞}. Otherwise, $X_{\Lambda\text{-tors}} = 0$.

The key ingredient in proving these statements is local class field theory which gives the structure of $\text{Gal}(M_n/F_n)$, where M_n denotes the maximal abelian pro-p extension of F_n. We have

$$\text{Gal}(M_n/F_n) \cong \mathbb{Z}_p^{[F_n:\mathbb{Q}_p]+1} \times \mu_{F_n}$$

where μ_{F_n} denotes the group of p-power roots of unity in F_n. This comes from the reciprocity map $F_n^\times \to \text{Gal}(F_n^{\text{ab}}/F_n)$, which is injective and has dense image, together with the well-understood structure of $F_n^\times$ (which can be studied by using the p-adic log map). Note that $[F_n : \mathbb{Q}_p] = [K_v : \mathbb{Q}_p]|\Delta|p^n$ for $n \geq 0$. Furthermore, Δ can be identified with a subgroup of $\text{Gal}(F_n/K_v)$ and so acts on $\text{Gal}(M_n/F_n)$. We can consider the Δ-decomposition

$$\text{Gal}(M_n/F_n) = \bigoplus_{\chi \in \widehat{\Delta}} \text{Gal}(M_n/F_n)^\chi$$

and, since the reciprocity map is Δ-equivariant, one finds that

$$\text{rank}_{\mathbb{Z}_p}(\text{Gal}(M_n/F_n)^\chi) = \begin{cases} [K_v:\mathbb{Q}_p]p^n & \text{if } \chi \text{ is nontrivial.} \\ [K_v:\mathbb{Q}_p]p^n + 1 & \text{if } \chi \text{ is trivial.} \end{cases}$$

The extra 1 if χ is trivial could be thought of as corresponding to $\text{Gal}(F_\infty/F_n) \cong \mathbb{Z}_p$ since $F_\infty \subset M_n$ and Δ acts trivially on $\text{Gal}(F_\infty/F_n)$. Thus, we actually have that $\text{rank}_{\mathbb{Z}_p}(\text{Gal}(M_n/F_\infty)^\chi) = [K_v:\mathbb{Q}_p]p^n$ for all $n \geq 0$ and all characters χ of Δ.

The field M_n is just the maximal abelian extension of F_n contained in M_∞ and therefore $\text{Gal}(M_\infty/M_n)$ is the closure of the commutator subgroup $\text{Gal}(M_\infty/F_n)'$ of $\text{Gal}(M_\infty/F_n)$. Consider the exact sequence

$$1 \to X \to \text{Gal}(M_\infty/F_n) \to \text{Gal}(F_\infty/F_n) \to 1.$$

If $g \in \text{Gal}(F_\infty/F_n)$ and $x \in X$, recall that the action of g on x is defined by $g(x) = \widetilde{g}x\widetilde{g}^{-1}$, where $\widetilde{g} \in \text{Gal}(M_\infty/F_n)$ is chosen so that $\widetilde{g}|_{F_\infty} = g$. Thus, $g(x)x^{-1} = \widetilde{g}x\widetilde{g}^{-1}x^{-1}$ is a commutator in $\text{Gal}(M_\infty/F_n)$. We will leave it to the reader (in exercise 3.13) to prove the useful fact that

$$\text{Gal}(M_\infty/F_n)' = \{\gamma_0^{p^n}(x)x^{-1} | x \in X\}$$

and that this is a closed subgroup of $\text{Gal}(M_\infty/F_n)$. Switching to the usual additive notation for the Λ-module X, we therefore have

$$\text{Gal}(M_\infty/M_n) = (\gamma_0^{p^n} - 1)X = \omega_n X \quad \text{and} \quad \text{Gal}(M_n/F_\infty) = X/\omega_n X$$

for all $n \geq 0$, where, as previously, $\omega_n = (1+T)^{p^n} - 1$.

Statement (a) is now an immediate consequence of theorem 3.9, part 2 because $\omega_0 = T$ and X/TX is isomorphic to the finitely generated $\mathbb{Z}_p$-module $\text{Gal}(M_0/F_\infty)$. This implies that $X/\mathfrak{m}X$ is finite. Statement (b) is a consequence of exercise 3.6. For the first part, we note that the above remarks imply that $\text{rank}_{\mathbb{Z}_p}(X/\omega_n X) = [F_n : \mathbb{Q}_p] = rp^n$, where $r = [K_v : \mathbb{Q}_p]|\Delta|$. To obtain the second part, note that the isomorphism $X/\omega_n X \cong \text{Gal}(M_n/F_\infty)$ is Δ-equivariant. It follows that $(X/\omega_n X)^\chi \cong X^\chi/\omega_n X^\chi$ is isomorphic to $\text{Gal}(M_n/F_\infty)^\chi$ and so, for each $\chi \in \widehat{\Delta}$ and $n \geq 0$, we have

$$\text{rank}_{\mathbb{Z}_p}(X^\chi/\omega_n X^\chi) = [K_v : \mathbb{Q}_p]p^n.$$

Using exercise 3.6 again, we have $\text{rank}_\Lambda(X^\chi) = [K_v : \mathbb{Q}_p]$ as stated.

The Λ-torsion submodule of X is a little more subtle to study. First we recall that G_{F_∞} has p-cohomological dimension 1, as was explained at the end of chapter 2. Consequently, as also explained at the end of chapter 2, it follows that $H^1(F_\infty, \mathbb{Q}_p/\mathbb{Z}_p)$ is a divisible group. That is, $\widehat{X} = \text{Hom}(X, \mathbb{Q}_p/\mathbb{Z}_p)$ is divisible and so X is torsion-free as a $\mathbb{Z}_p$-module. In particular, it follows that X has no nonzero, finite Λ-submodules. Let $Y = X_{\Lambda\text{-tors}}$ and let $W = X/Y$. Then W is a torsion-free Λ-module. This easily implies that we have an exact sequence

$$0 \to Y/\omega_n Y \to X/\omega_n X \to W/\omega_n W \to 0$$

for all $n \geq 0$. Now $\text{rank}_\Lambda(W) = [K_v : \mathbb{Q}_p]|\Delta|$ and so it follows that $W/\omega_n W$ has $\mathbb{Z}_p$-rank $[K_v : \mathbb{Q}_p]|\Delta|p^n$, which is the same as the $\mathbb{Z}_p$-rank of $X/\omega_n X$. This implies that $Y/\omega_n Y$ must be finite and isomorphic to a subgroup of $(X/\omega_n X)_{\mathbb{Z}_p\text{-tors}} = \mu_{F_n}$. If μ_{F_n} has bounded order as $n \to \infty$ (i.e., if μ_{p^∞} is not contained in F_∞), then so does $Y/\omega_n Y$. It follows that $Y = \varprojlim Y/\omega_n Y$ is itself finite and therefore we must have $Y = 0$. This proves the second part of statement (c).

For the first part of (c), note again that $Y = \varprojlim Y/\omega_n Y$. Since $Y/\omega_n Y$ is cyclic for all n, it is clear that if Y is infinite, then $Y \cong \mathbb{Z}_p$ as a $\mathbb{Z}_p$-module. Now, by exercise 3.5, it follows that the $\mathbb{Z}_p$-torsion subgroup of $W/\omega_n W$ has bounded order as $n \to \infty$. Hence if μ_{F_n} has unbounded order (i.e., if $\mu_{p^\infty} \subset F_\infty$), then $Y/\omega_n Y$ must have unbounded order too. Since $Y/\omega_n Y$ is isomorphic to a subgroup of μ_{F_n}, it follows that $Y \cong \varprojlim \mu_{F_n} = T_p(\mu_{p^\infty})$ as stated.

It is now easy to complete the proof of theorem 3.11. We proved before that $\text{corank}_{\mathbb{Z}_p}(H^1(K_v, A)) = \text{rank}_{\mathbb{Z}_p}(X^{\psi_\Delta}/g(T)X^{\psi_\Delta})$, where $g(T) = T - b$ with $b = \psi_\Gamma(\gamma_0) - 1 \in p\mathbb{Z}_p$. Thus $g(T)$ is a distinguished polynomial of degree 1. Statements (a), (b), (c) imply that X^{ψ_Δ} is pseudo-isomorphic to $\Lambda^{[K_v:\mathbb{Q}_p]}$ or to $Y \times \Lambda^{[K_v:\mathbb{Q}_p]}$, where $Y = T_p(\mu_{p^\infty})$. The latter occurs if $\mu_{p^\infty} \subset F_\infty$ and ψ_Δ is the character giving the action of Δ on μ_{p^∞}. Now $\Lambda/g(T)\Lambda$ has $\mathbb{Z}_p$-rank 1 by corollary 3.6 (which is rather obvious in this case). If $\mu_{p^\infty} \not\subset F_\infty$, it then follows that $H^1(K_v, A)$ indeed has $\mathbb{Z}_p$-corank equal to $[K_v : \mathbb{Q}_p]$. If $\mu_{p^\infty} \subset F_\infty$, then $G = \text{Gal}(F_\infty/K_v)$ acts on $T_p(\mu_{p^\infty})$ by a character χ. We are assuming $\psi \neq \chi$. If $\psi_\Delta \neq \chi_\Delta$, then $Y = 0$. If $\psi_\Delta = \chi_\Delta$, then $Y = T_p(\mu_{p^\infty})$, but $\psi_\Gamma \neq \chi_\Gamma$ and so it follows that $Y/g(T)Y$ is finite. In either case, we again find that $X^{\psi_\Delta}/g(T)X^{\psi_\Delta}$ has $\mathbb{Z}_p$-rank $[K_v : \mathbb{Q}_p]$. $\square$

Exercises

3.1. Let A be a finite, abelian p-group on which Γ acts continuously. Prove that $T^n A = 0$ for $n \gg 0$. Prove that $\mathfrak{m}^n A = 0$ when $n \gg 0$, where $\mathfrak{m}$ denotes the maximal ideal of Λ.

3.2. Let A be a discrete p-primary, abelian group on which Γ acts continuously. Prove that $A = \bigcup_n A^{\Gamma_n}$ (where $\Gamma_n = \Gamma^{p^n}$ for all $n \geq 0$). Prove that

$$A = 0 \iff A^\Gamma = 0 \iff A[p]^\Gamma = 0.$$

Use this to prove part (1) of theorem 3.9 by taking $A = \text{Hom}_{\text{cont}}(X, \mathbb{Q}_p/\mathbb{Z}_p)$.

3.3. Let A be as in exercise 3.2. We know how to make A into a $\mathbb{Z}_p$-module. Prove that if $a \in A$, then $T^n a = 0$ for $n \gg 0$. Using this, show that A can be made into a Λ-module. Let $X = \widehat{A}$. Show that if $x \in X$, then $T^n x \to 0$ in X. Using this, make X into a Λ-module.

3.4. Let Z be a finitely generated Λ-module. Show that Z is finite if and only if $\mathfrak{m}^n Z = 0$ for some $n \geq 0$. Prove that every finitely generated Λ-module X contains a Λ-submodule Z which is finite and contains all other finite Λ-submodules of X. Prove that if $f(T) \in \Lambda$, then $\Lambda/(f(T))$ contains no nonzero, finite Λ-submodules. Deduce that the kernel of the map occurring in theorem 3.1 must be the maximal finite Λ-submodule Z of X.

3.5. Suppose that X is a Λ-submodule of Λ^r and that $H = \Lambda^r/X$ is finite. For $n \geq 1$, let $\omega_n = (1+T)^{p^n} - 1$. Prove that $H \cong (X/\omega_n X)_{\mathbb{Z}_p\text{-tors}}$ as a Λ-module if $n \gg 0$. It follows that the finite Λ-module H is uniquely determined by the isomorphism class of X.

3.6. Suppose that X is a finitely generated Λ-module and that $\text{rank}_{\mathbb{Z}_p}(X/\omega_n X) = rp^n + O(1)$ as $n \to \infty$. Prove that $\text{rank}_\Lambda(X) = r$. Now assume in addition that $X/TX \cong \mathbb{Z}_p^r$. Prove that $X \cong \Lambda^r$ as a Λ-module.

3.7. In this problem, let $X = \Lambda/(f(T))$, where $f(T)$ is a nonzero element of Λ.
  (a) Suppose that $X/\omega_n X$ is finite for all $n \geq 0$. Prove that
  $$|X/\omega_n X| = u \prod_{\zeta^{p^n}=1} f(\zeta - 1).$$
  where $u \in \mathbb{Z}_p^\times$. (Hint: Multiplication by $f(T)$ defines a $\mathbb{Z}_p$-linear map $\Lambda/(\omega_n) \to \Lambda/(\omega_n)$. The cokernel of this map is $X/\omega_n X$ and its order is the determinant, up to a p-adic unit.)
  (b) If $f(T) = p^m$, then part (a) implies that $(X/\omega_n X) = p^{mp^n}$ for all $n \geq 0$. Assume that $f(T)$ is a distinguished polynomial of degree d. Assume that $X/\omega_n X$ is finite for all $n \geq 0$. Prove that $|X/\omega_n X| = p^{dn+c}$ for all $n \gg 0$, where c is some constant.

3.8. (a) Assume that X is a finitely generated torsion Λ-module and that $X/\omega_n X$ is finite for all n. Let $\mu = \mu_X$, $\lambda = \lambda_X$ as defined in this chapter. Using theorem 3.1, prove that $|X/\omega_n X| = p^{\mu p^n + \lambda n + O(1)}$ as $n \to \infty$.
  (b) Assume that X is a finitely generated, torsion Λ-module. Let t_n denote the number of zeros of $f_X(t)$ of the form $T = \zeta - 1$, where $\zeta^{p^n} = 1$ (counting multiplicities). Prove that $X/\omega_n X$ is infinite if and only if $t_n \geq 1$. Prove that $\text{rank}_{\mathbb{Z}_p}(X/\omega_n X) \leq t_n$.

3.9. Assume that A is a discrete, p-primary abelian group on which Γ acts continuously. Assume also that A^Γ has order p. Prove that either A is finite or that $\widehat{A} \cong \Lambda/(f(T))$, where $f(T)$ is an irreducible element of Λ. Deduce that if A is infinite, then either A has exponent p or A is $\mathbb{Z}_p$-cofree of finite $\mathbb{Z}_p$-corank.

3.10. Assume that A is a discrete, p-primary abelian group on which Γ acts continuously. Assume that A^Γ is finite and that $A^{\Gamma_{n+1}} = A^{\Gamma_n}$ for some $n \geq 0$. Prove that $A = A^{\Gamma_n}$.

3.11. (a) Let F_∞/F be a $\mathbb{Z}_p$-extension. Suppose that E is an elliptic curve defined over F and that $E(F_1)_p = E(F)_p$. Prove that $E(F_\infty)_p = E(F)_p$.
  (b) Let E be an elliptic curve defined over $\mathbb{Q}$ with the following properties: E has good reduction at 2 and $\widetilde{E}(\mathbb{F}_2)$ has order 4, E has good reduction at 7, and $E(\mathbb{Q})_{\text{tors}}$ has order 8. Let $\mathbb{Q}_\infty$ be the cyclotomic $\mathbb{Z}_2$-extension of $\mathbb{Q}$. Show that $E(\mathbb{Q}_\infty)_{\text{tors}} = E(\mathbb{Q})_{\text{tors}}$. (Note: The hypotheses are satisfied by the elliptic curve $y^2 + xy = y^3 - 115x + 392$ considered in chapter 1.)

CHAPTER 3. Λ-MODULES.

3.12. (a) Let G be a profinite group. Consider the natural map
$$\mathbb{Z}_p[G] \to \mathbb{Z}_p[[G]].$$
Prove that this map is injective and that its image is dense.

(b) Suppose that G and H are profinite groups and that there is a continuous, surjective homomorphism $\varphi : G \to H$. Prove that the induced homomorphism $\mathbb{Z}_p[[G]] \to \mathbb{Z}_p[[H]]$ is also surjective.

(c) Assume that G is a pro-p group (i.e., an inverse limit of finite p-groups). Prove that $\mathbb{Z}_p[[G]]$ is a local ring.

3.13. In the notation of the proof of theorem 3.11, show that
$$\operatorname{Gal}(M_\infty/F_n)' = \{\gamma_0^{p^n}(x)x^{-1} | x \in X\}$$
and that this is a closed subgroup of $\operatorname{Gal}(M_\infty/F_n)$.

CHAPTER 4
Mazur's Control Theorem.

The main result of this section is the following theorem of Mazur concerning the Galois theoretic behavior of Selmer groups. We suppose that F is a finite extension of $\mathbb{Q}$ and that E is an elliptic curve defined over F.

Theorem 4.1. *Assume that p is a prime and that E has good, ordinary reduction at all primes of F lying over p. Assume that $F_\infty = \bigcup_n F_n$ is a $\mathbb{Z}_p$-extension of F. Then the natural maps*

$$\operatorname{Sel}_E(F_n)_p \to \operatorname{Sel}_E(F_\infty)_p^{\operatorname{Gal}(F_\infty/F_n)}$$

have finite kernels and cokernels. Their orders are bounded as $n \to \infty$.

Of course, as before, F_n denotes the unique subfield of F_∞ containing F such that $[F_n : F] = p^n$. After proving this theorem, we will derive various corollaries, including theorem 1.6.

Our proof of theorem 4.1 is based on the description of the images of the local Kummer homomorphisms presented in chapter 2, specifically theorems 2.4, 2.6, 2.8, and 2.9. We will also prove a special case of conjecture 3.3. Let E be any elliptic curve defined over F. Let K be an algebraic extension of F. For every prime η of K, we let

$$\mathcal{H}_E(K_\eta) = H^1(K_\eta, E[p^\infty])/\operatorname{im}(\kappa_\eta).$$

Let $\mathcal{P}_E(K) = \prod_\eta \mathcal{H}_E(K_\eta)$, where η runs over all primes of K. Thus,

$$\operatorname{Sel}_E(K)_p = \ker\left(H^1(K, E[p^\infty]) \to \mathcal{P}_E(K)\right),$$

where the map is induced by restricting cocycles to decomposition groups. Also, we put

$$\mathcal{G}_E(K) = \operatorname{im}\left(H^1(K, E[p^\infty]) \to \mathcal{P}_E(K)\right).$$

Let $F_\infty = \bigcup_n F_n$ be an arbitrary $\mathbb{Z}_p$-extension of F. Consider the following commutative diagram with exact rows.

$$\begin{array}{ccccccccc} 0 & \to & \mathrm{Sel}_E(F_n)_p & \to & H^1(F_n, E[p^\infty]) & \to & \mathcal{G}_E(F_n) & \to & 0 \\ & & \downarrow s_n & & \downarrow h_n & & \downarrow g_n & & \\ 0 & \to & \mathrm{Sel}_E(F_\infty)_p^{\Gamma_n} & \to & H^1(F_\infty, E[p^\infty])^{\Gamma_n} & \to & \mathcal{G}_E(F_\infty)^{\Gamma_n}. & & \end{array}$$

Here $\Gamma_n = \mathrm{Gal}(F_\infty/F_n) = \Gamma^{p^n}$. The maps s_n, h_n, and g_n are the natural restriction maps. The snake lemma then gives the exact sequence

$$0 \to \ker(s_n) \to \ker(h_n) \to \ker(g_n) \to \mathrm{coker}(s_n) \to \mathrm{coker}(h_n).$$

Therefore, we must study $\ker(h_n)$, $\mathrm{coker}(h_n)$, and $\ker(g_n)$, which we do in a sequence of lemmas.

Lemma 4.2. $\mathrm{Ker}(h_n)$ *is finite and has bounded order as n varies.*

Proof. By the inflation-restriction sequence, $\ker(h_n) \cong H^1(\Gamma_n, B)$, where B is the p-primary subgroup of $E(F_\infty)$. If B is finite, which is usually the case, then we will have $H^1(\Gamma_n, B) = \mathrm{Hom}(\Gamma_n, B)$ for $n \gg 0$. Lemma 4.2 follows immediately. But it is not necessary to assume the finiteness of B. If γ denotes a topological generator of Γ, then $H^1(\Gamma_n, B) = B/(\gamma^{p^n} - 1)B$. (See exercise 2.2.) Since $E(F_n)$ is finitely generated, the kernel of $\gamma^{p^n} - 1$ acting on B is finite. Now B_{div} has finite $\mathbb{Z}_p$-corank. Hence (by exercise 2.2(c)), it follows that $(\gamma^{p^n} - 1)B_{\mathrm{div}} = B_{\mathrm{div}}$ and so it is clear that

$$B_{\mathrm{div}} \subseteq (\gamma^{p^n} - 1)B \subseteq B.$$

Since B is a cofinitely generated $\mathbb{Z}_p$-module, B_{div} has finite index in B. Obviously, $H^1(\Gamma_n, B)$ has order bounded by $[B:B_{\mathrm{div}}]$, a bound which is independent of n. If B is finite, then $\ker(h_n)$ has the same order as $H^0(\Gamma_n, B)$, namely $|E(F_n)_p|$. □

Lemma 4.3. $\mathrm{Coker}(h_n) = 0$.

Proof. By the inflation-restriction exact sequence, the sequence

$$H^1(F_n, E[p^\infty]) \to H^1(F_\infty, E[p^\infty])^{\Gamma_n} \to H^2(\Gamma_n, B)$$

is exact, where $B = H^0(F_\infty, E[p^\infty])$ again. But $\Gamma_n \cong \mathbb{Z}_p$ is a free pro-p group. Therefore Γ_n has p-cohomological dimension 1. Hence $H^2(\Gamma_n, B) = 0$ and so h_n is surjective as claimed. □

Let v be any prime of F. We will let v_n denote any prime of F_n lying over v. To study $\ker(g_n)$, we focus on each factor in $\mathcal{P}_E(F_n)$ by considering

$$r_{v_n} : \mathcal{H}_E((F_n)_{v_n}) \to \mathcal{H}_E((F_\infty)_\eta)$$

where η is any prime of F_∞ lying above v_n. ($\mathcal{P}_E(F_\infty)$ has a factor for all such η's, but the kernels will be the same.) If v is archimedean, then v splits completely in F_∞/F, i.e., $F_v = (F_\infty)_\eta$ for all $\eta|v$. Thus, $\ker(r_{v_n}) = 0$. For nonarchimedean v, we consider separately $v \nmid p$ and $v|p$.

Lemma 4.4. *Suppose that v is a nonarchimedean prime not dividing p. Then $\ker(r_{v_n})$ is finite and has bounded order as n varies. If E has good reduction at v or if v splits completely in F_∞/F, then $\ker(r_{v_n}) = 0$ for all n.*

Proof. Let Γ_v denote the decomposition subgroup of Γ for any nonarchimedean prime v. If Γ_v is trivial, then v splits completely in F_∞/F and hence the lemma is obvious because $F_v = (F_\infty)_\eta$ and $\ker(r_{v_n}) = 0$ for all n. Otherwise, Γ_v has finite index in Γ and so v is finitely decomposed in F_∞/F. We now assume this. By theorem 2.4, $\mathcal{H}_E(M_\eta) = H^1(M_\eta, E[p^\infty])$ for every algebraic extension M_η of F_v. Let $B_v = H^0(K, E[p^\infty])$, where $K = (F_\infty)_\eta$. Since v is unramified and finitely decomposed in F_∞/F, K is the unramified $\mathbb{Z}_p$-extension of F_v (in fact, the only $\mathbb{Z}_p$-extension of F_v). The group B_v is isomorphic to $(\mathbb{Q}_p/\mathbb{Z}_p)^e \times$ (a finite group), where $0 \leq e \leq 2$. Let $\Gamma_{v_n} = \mathrm{Gal}(K/(F_n)_{v_n})$, which is isomorphic to $\mathbb{Z}_p$, topologically generated by γ_{v_n}, say. Then, using exercise 2.2 again, we have

$$\ker(r_{v_n}) \cong H^1(\Gamma_{v_n}, B_v) \cong B_v/(\gamma_{v_n} - 1)B_v.$$

Since $E((F_n)_{v_n})$ has a finite p-primary subgroup, it is clear that $(\gamma_{v_n}-1)B_v$ contains $(B_v)_{\mathrm{div}}$ (just as in the proof of lemma 4.2) and hence

(4.5) $$|\ker(r_{v_n})| \leq |B_v/(B_v)_{\mathrm{div}}|.$$

This bound is independent of n and v_n. We have equality if $n \gg 0$. Now assume that E has good reduction at v. Then, since $v \nmid p$, $F_v(E[p^\infty])/F_v$ is unramified. It is clear that $K \subseteq F_v(E[p^\infty])$ and that $\Delta = \mathrm{Gal}(F_v(E[p^\infty])/K)$ is a finite, cyclic group of order prime to p. One can then prove that $B_v = E[p^\infty]^\Delta$ is divisible. Therefore, $\ker(r_{v_n}) = 0$ as stated. $\square$

Now assume that $v|p$. For each n, we let f_{v_n} denote the residue field for $(F_n)_{v_n}$. It doesn't depend on the choice of v_n. Now either v splits completely in F_∞/F, v is ramified in F_∞/F, or v is unramified but finitely decomposed in F_∞/F. In the first two cases, the finite field f_{v_n} stabilizes and so f_η is finite, where η is any prime of F_∞ lying over v and f_η is the corresponding residue field. Let $\widetilde{E}$ denote the reduction of E at v. Then we have

Lemma 4.6. *Suppose that v is a prime dividing p. Assume that E has good, ordinary reduction at v. Then $\ker(r_{v_n})$ is finite and has bounded order as n varies.*

Proof. If v splits completely in F_∞/F, then $\ker(r_{v_n}) = 0$ for all n. Assume that v is ramified in F_∞/F. In this case, theorem 2.9 is applicable. Then r_{v_n} can be viewed as the composition of the following two maps:

$$a_{v_n} : H^1((F_n)_{v_n}, E[p^\infty])/\mathrm{im}(\kappa_{v_n}) \longrightarrow H^1((F_n)_{v_n}, E[p^\infty])/\mathrm{im}(\varepsilon_{v_n})$$
$$b_{v_n} : H^1((F_n)_{v_n}, E[p^\infty])/\mathrm{im}(\varepsilon_{v_n}) \longrightarrow H^1((F_\infty)_\eta, E[p^\infty])/\mathrm{im}(\varepsilon_\eta).$$

Here we are using the notation of chapter 2. Now a_{v_n} is clearly surjective. Hence

$$|\ker(r_{v_n})| = |\ker(a_{v_n})| \cdot |\ker(b_{v_n})|.$$

But $\ker(a_{v_n}) = \mathrm{im}(\varepsilon_{v_n})/\mathrm{im}(\kappa_{v_n})$ and, by theorem 2.8, the order of this group is bounded by $|\widetilde{E}(f_\eta)_p|$, which is finite. On the other hand, the exact sequence

$$0 \to \mathcal{F}[p^\infty] \to E[p^\infty] \to \widetilde{E}[p^\infty] \to 0$$

of Galois modules gives us the following commutative diagram:

$$\begin{array}{ccc}
0 \longrightarrow H^1((F_n)_{v_n}, E[p^\infty])/\mathrm{im}(\varepsilon_{v_n}) & \xrightarrow{\pi_{v_n}} & H^1((F_n)_{v_n}, \widetilde{E}[p^\infty]) \\
\downarrow {b_{v_n}} & & \downarrow {c_{v_n}} \\
0 \longrightarrow H^1((F_\infty)_\eta, E[p^\infty])/\mathrm{im}(\varepsilon_\eta) & \xrightarrow{\pi_\eta} & H^1((F_\infty)_\eta, \widetilde{E}[p^\infty]).
\end{array}$$

Therefore, $|\ker(b_{v_n})| \leq |\ker(c_{v_n})|$. Now
$$\ker(c_{v_n}) \cong H^1((F_\infty)_\eta/(F_n)_{v_n}, \widetilde{E}(f_\eta)_p) \cong \widetilde{E}(f_\eta)_p/(\gamma_{v_n}-1)\widetilde{E}(f_\eta)_p$$
where γ_{v_n} is a topological generator of $\mathrm{Gal}((F_\infty)_\eta/(F_n)_{v_n})$. Thus $|\ker(c_{v_n})|$ is bounded above by $|\widetilde{E}(f_\eta)_p|$ and so we obtain

(4.7) $$|\ker(r_{v_n})| \leq |\widetilde{E}(f_\eta)_p|^2$$

if v is ramified in F_∞/F.

Finally, assume that v is unramified but finitely decomposed in F_∞/F. Then $(F_\infty)_\eta$ is the unramified $\mathbb{Z}_p$-extension of F_v. In this case, $\ker(r_{v_n}) = 0$ for all n. We will sketch the argument. (Since f_η is infinite, we can't apply theorem 2.9.) The Kummer sequence (2.2) reduces this to showing that $H^1(L/M, E(L))$ is trivial whenever M is a finite extension of F_v and L/M is a finite, unramified p-extension. Let l and m denote the residue fields of L and M, respectively. Then we have an exact sequence
$$0 \to \mathcal{F}(\mathfrak{m}_L) \to E(L) \to \widetilde{E}(l) \to 0,$$
where $\mathfrak{m}_L$ denotes the maximal ideal of L and $\mathcal{F}(\mathfrak{m}_L)$ the group of points on the formal group $\mathcal{F}$. It is enough to verify that $H^1(L/M, \mathcal{F}(\mathfrak{m}_L))$ and $H^1(L/M, \widetilde{E}(l))$ are both trivial. Now $\mathrm{Gal}(L/M) = \mathrm{Gal}(l/m)$. The vanishing of $H^1(L/M, \mathcal{F}(\mathfrak{m}_L))$ can be proved by using the fact that $\mathcal{F}(\mathfrak{m}_L)$ has a filtration where the successive quotients are isomorphic as Galois modules to those for the formal additive group. There the filtration is defined by powers of the maximal ideal of L. Multiplication by a uniformizing parameter for M induces isomorphisms (as Galois modules) to the additive group l. One then uses the fact that $H^1(l/m, l) = 0$. The vanishing of $H^1(L/M, \widetilde{E}(l)) = H^1(l/m, \widetilde{E}(l))$ follows from that of $H^1(m, \widetilde{E}(\overline{m}))$, where $\overline{m}$ denotes the algebraic closure of m, and this can be shown just as in exercise 2.5. $\square$

Let v be any prime of F. If v splits completely in F_∞/F, then $\ker(r_{v_n}) = 0$ and so these primes contribute nothing to $\ker(g_n)$. For all other primes v of F, the number of primes v_n of F_n lying above v is bounded as $n \to \infty$. If E has good reduction at v and $v \nmid p$, then $\ker(r_{v_n}) = 0$, and so these primes contribute nothing to $\ker(g_n)$. Thus there are only finitely many v's that need to be considered and, for each such v, there are only a bounded number of v_n's. For those v_n's, lemmas 4.4 and 4.6 imply that $|\ker(r_{v_n})|$ is bounded. Combining these observations, we have proved the following lemma.

Lemma 4.8. *The order of* $\ker(g_n)$ *is bounded as n varies.*

Lemma 4.2 implies that $\ker(s_n)$ is finite and has bounded order no matter what type of reduction E has at $v|p$. Lemmas 4.3 and 4.8 show that $\mathrm{coker}(s_n)$ is finite and of bounded order, assuming that E has good, ordinary reduction at all $v|p$. Thus, the proof of theorem 4.1 is now complete.

Here is one corollary of theorem 4.1. It implies theorem 1.6 since the finiteness of $\mathrm{Sel}_E(F)_p$ is equivalent to that of $E(F)$ and $\Sha_E(F)_p$.

Corollary 4.9. *Suppose that E is an elliptic curve defined over F and that p is a prime such that E has good, ordinary reduction at all primes of F lying above p. Assume also that $\mathrm{Sel}_E(F)_p$ is finite. Then $\mathrm{Sel}_E(F_\infty)_p$ is Λ-cotorsion. Consequently, $\mathrm{rank}_\mathbb{Z}(E(F_n))$ is bounded as n varies.*

Proof. Under the above hypotheses, theorem 4.1 implies that $\text{Sel}_E(F_\infty)_p^\Gamma$ is finite. Let $X = \text{Hom}(\text{Sel}_E(F_\infty)_p, \mathbb{Q}_p/\mathbb{Z}_p)$. Then X is a Λ-module and X/TX is finite. (Note that X/TX is the maximal quotient of X on which Γ acts trivially. Hence X/TX is the Pontryagin dual of $\text{Sel}_E(F_\infty)_p^\Gamma$.) By theorem 3.9, it follows that X is a finitely generated, torsion Λ-module. Hence $\text{Sel}_E(F_\infty)_p$ is indeed Λ-cotorsion. As we saw in chapter 3, $X/X_{\mathbb{Z}_p\text{-tors}} \cong \mathbb{Z}_p^\lambda$ for some $\lambda \geq 0$. This implies that

$$(\text{Sel}_E(F_\infty)_p)_{\text{div}} \cong (\mathbb{Q}_p/\mathbb{Z}_p)^\lambda.$$

Using theorem 4.1 again, it follows that $(\text{Sel}_E(F_n)_p)_{\text{div}} \cong (\mathbb{Q}_p/\mathbb{Z}_p)^{t_n}$ where $t_n \leq \lambda$. (One just needs the finiteness of the kernels for this.) Now

$$E(F_n) \otimes (\mathbb{Q}_p/\mathbb{Z}_p) \cong (\mathbb{Q}_p/\mathbb{Z}_p)^{\text{rank}_\mathbb{Z}(E(F_n))}$$

is a subgroup of $(\text{Sel}_E(F_n)_p)_{\text{div}}$ and therefore we obtain the inequality $\text{rank}_\mathbb{Z}(E(F_n)) \leq \lambda$ for all $n \geq 0$. □

The next result gives sufficient conditions for $\text{Sel}_E(F_\infty)_p$ to be trivial.

Corollary 4.10. *Suppose that E is an elliptic curve defined over F and that p is a prime such that E has good, ordinary reduction at all primes of F lying over p. Assume that $\text{Sel}_E(F)_p = 0$, that none of the primes of F over p are anomalous for E, and that $E(F_v)_p = 0$ for all primes v of F where E has bad reduction. Then $\text{Sel}_E(F_\infty)_p = 0$.*

Proof. By lemma 4.3, $\text{coker}(h_0) = 0$. Assume that $v|p$. If v is unramified in F_∞/F, then $\ker(r_v) = 0$. If v is ramified in F_∞/F, then the residue field f_η for a prime η of F_∞ above v is a finite p-extension of f_v, the residue field for v. Since $\widetilde{E}(f_\eta)_p^{\text{Gal}(f_\eta/f_v)} = \widetilde{E}(f_v)_p = 0$, it follows that $\widetilde{E}(f_\eta)_p = 0$. Hence, by (4.7), $\ker(r_v) = 0$. Assume now that v is a prime of F where E has bad reduction (and so $v \nmid p$). If η is a prime of F_∞ lying above v, then $\text{Gal}((F_\infty)_\eta/F_v)$ is either trivial or isomorphic to $\mathbb{Z}_p$. Since $E((F_\infty)_\eta)_p^{\text{Gal}((F_\infty)_\eta/F_v)} = E(F_v)_p = 0$, it follows that $E((F_\infty)_\eta)_p = 0$. Thus the group B_v occurring in the proof of lemma 4.4 is trivial. Therefore, it is clear that $\ker(r_v) = 0$. Combining these remarks, we see that $\ker(g_0) = 0$. Therefore, $\text{coker}(s_0) = 0$. That is, the map $\text{Sel}_E(F)_p \to \text{Sel}_E(F_\infty)_p^\Gamma$ is surjective. The hypothesis that $\text{Sel}_E(F)_p = 0$ implies that $\text{Sel}_E(F_\infty)_p^\Gamma = 0$. Therefore $X/TX = 0$, where $X = \text{Hom}(\text{Sel}_E(F_\infty)_p, \mathbb{Q}_p/\mathbb{Z}_p)$. By theorem 3.9, $X = 0$ and so $\text{Sel}_E(F_\infty)_p = 0$ too. □

Remark. The hypothesis for the primes v where E has bad reduction can be weakened. It suffices to assume that $B_v = H^0((F_\infty)_\eta, E[p^\infty])$ is divisible. This is equivalent to assuming that the Tamagawa factor c_v for E/F_v is not divisible by p. (See exercise 4.6).

The next result concerns the growth of the order of $\text{III}_E(F_n)_p$, assuming that this group is finite for all n. The hypothesis that $E(F_n)$ is finite for all n is not necessary, but it simplifies the argument. For a more general result (not assuming the finiteness of $E(F_n)$), see theorem 1.10 in [**Gr2**], which is proved at the end of section 3 of that article.

Corollary 4.11. *Assume that E is an elliptic curve defined over F and that E has good, ordinary reduction at all primes of F lying above p. Let F_∞/F be a $\mathbb{Z}_p$-extension. Assume that both $E(F_n)$ and $\text{III}_E(F_n)_p$ are finite for all n. Then there* *exist integers $\lambda, \mu \geq 0$ depending only on E and F_∞/F such that*

$$|\text{III}_E(F_n)_p| = p^{\lambda n + \mu p^n + O(1)} \text{ as } n \to \infty.$$

Proof. Let $X = \text{Hom}(\text{Sel}_E(F_\infty)_p, \mathbb{Q}_p/\mathbb{Z}_p)$. Then $\text{Sel}_E(F_n)_p$ is finite for all $n \geq 0$ and therefore so is $\text{Sel}_E(F_\infty)_p^{\Gamma_n}$. Now $|\text{Sel}_E(F_\infty)_p^{\Gamma_n}| = |X/\omega_n X|$, where, as in chapter 3, $\omega_n = (1+T)^{p^n} - 1$. This is because $X/\omega_n X$ is the maximal quotient of X on which Γ_n acts trivially. Hence $X/\omega_n X \cong \text{Hom}(\text{Sel}_E(F_\infty)_p^{\Gamma_n}, \mathbb{Q}_p/\mathbb{Z}_p)$ is just the Pontryagin dual of $\text{Sel}_E(F_\infty)_p^{\Gamma_n}$ and so has the same order. By theorem 4.1, $|\text{Sel}_E(F_n)_p|/|X/\omega_n X| = p^{e_n}$, where $|e_n|$ is bounded as n varies. Corollary 4.9 implies that X is a finitely generated, torsion Λ-module. The stated result then follows from exercise 3.8(a). The integers λ and μ are just the Iwasawa invariants of the Λ-module X: $\lambda = \lambda_X$, $\mu = \mu_X$. □

Remark. Under the hypotheses of corollary 4.11, the integers λ and/or μ can be positive. Examples where $\mu > 0$ were found by Mazur and are discussed in his paper [**Ma1**]. One such example is outlined in exercises 4.1–4.5. Other examples and results about λ and μ can be found in section 5 of [**Gr2**], e.g. corollary 5.6, propositions 5.7, 5.10, and various examples where λ and μ can be calculated precisely.

Theorem 1.8 shows that it is possible for $\text{rank}(E(F_n))$ to be unbounded in a $\mathbb{Z}_p$-extension $F_\infty = \bigcup_n F_n$ even if E has good, ordinary reduction at the primes over p. But, as suggested by the comment after that theorem, there is some regularity to the growth of the rank. This is explained by the following result.

Corollary 4.12. *Suppose again that E is an elliptic curve defined over F which has good, ordinary reduction at all primes of F lying over p. Let F_∞/F be a $\mathbb{Z}_p$-extension. Let $r = \text{corank}_\Lambda(\text{Sel}_E(F_\infty)_p)$. Then*

$$\text{corank}_{\mathbb{Z}_p}(\text{Sel}_E(F_n)_p) = rp^n + O(1)$$

as $n \to \infty$. In particular, if $\text{III}_E(F_n)_p$ is finite for all n, then $\text{rank}(E(F_n)) = rp^n + O(1)$ as $n \to \infty$.

Proof. Let $X = \text{Hom}(\text{Sel}_E(F_\infty)_p, \mathbb{Q}_p/\mathbb{Z}_p)$. We know that X is a finitely generated Λ-module. According to theorem 3.1 and corollary 3.6, X is pseudo-isomorphic to $\Lambda^r \times Y \times Z$, where Y is a free $\mathbb{Z}_p$-module of finite rank and Z is a torsion group of bounded exponent. As in the proof of corollary 4.11, $X/\omega_n X$ is the Pontryagin dual of $\text{Sel}_E(F_\infty)_p^{\Gamma_n}$ and so

$$\text{rank}_{\mathbb{Z}_p}(X/\omega_n X) = \text{corank}_{\mathbb{Z}_p}(\text{Sel}_E(F_\infty)_p^{\Gamma_n}).$$

By theorem 4.1, $\text{corank}_{\mathbb{Z}_p}(\text{Sel}_E(F_\infty)_p^{\Gamma_n}) = \text{corank}_{\mathbb{Z}_p}(\text{Sel}_E(F_n)_p)$. Now $\Lambda/\omega_n \Lambda$ has $\mathbb{Z}_p$-rank p^n. (It can be identified with the group ring $\mathbb{Z}_p[\Gamma/\Gamma_n]$.) Also, $Y/\omega_n Y$ has bounded $\mathbb{Z}_p$-rank and $Z/\omega_n Z$ is finite. It is then clear that $\text{rank}_{\mathbb{Z}_p}(X/\omega_n X) = rp^n + O(1)$ as $n \to \infty$ and this implies the first statement. Under the assumption that $\text{III}_E(F_n)_p$ is finite, $E(F_n) \otimes (\mathbb{Q}_p/\mathbb{Z}_p)$ has finite index in $\text{Sel}_E(F_n)_p$ and so has the same $\mathbb{Z}_p$-corank. The $\mathbb{Z}_p$-corank of $E(F_n) \otimes (\mathbb{Q}_p/\mathbb{Z}_p)$ is equal to $\text{rank}(E(F_n))$, giving the second statement. □

What happens if E doesn't have good, ordinary reduction at the primes of F lying over p? If E has potentially ordinary reduction at all primes of F over p (i.e., good, ordinary reduction at those primes over some finite extension of F), then theorem 4.1 is still valid. The proof requires only some small modifications. The situation is more interesting if E is only assumed to have (potentially) ordinary or multiplicative reduction at the primes of F above p. The conclusion of theorem 4.1 should conjecturally still hold if F_∞ is the cyclotomic $\mathbb{Z}_p$-extension of F. This can be proved if $F = \mathbb{Q}$, but the proof needs the rather deep result that the Tate period q_E for E (assuming that E has multiplicative reduction at p) is not of the form $q_E = \zeta p^a$, where ζ is a root of unity. This was only established recently by Barré-Sirieix, Diaz, Gramain and Philibert who proved that q_E is actually transcendental if E has an algebraic j-invariant. (See section 3 of [**Gr2**] for a more thorough discussion.) For arbitrary $\mathbb{Z}_p$-extensions, theorem 4.1 would not in general be true under the above assumption.

Now let's assume that E has supersingular (or potentially supersingular) reduction for at least one prime of F lying over p. If F_∞/F is a $\mathbb{Z}_p$-extension which is ramified at such a prime, then one can show that $\text{Sel}_E(F_\infty)_p$ has positive Λ-corank. If $r = \text{corank}_\Lambda(\text{Sel}_E(F_\infty)_p)$, then, as shown in the proof of corollary 4.12, $\text{corank}_{\mathbb{Z}_p}(\text{Sel}_E(F_\infty)_p^{\Gamma_n}) = rp^n + O(1)$ as $n \to \infty$. Assume that $\text{rank}(E(F_n))$ is bounded and that $\text{III}_E(F_n)_p$ is finite for all n. (Theorems 1.7 and 1.20 provide many cases where this assumption is known to hold.) Then $\text{Sel}_E(F_n)_p$ would have bounded $\mathbb{Z}_p$-corank as $n \to \infty$ and so the conclusion in theorem 4.1 would fail to be true. In fact, the cokernel of the map $\text{Sel}_E(F_n)_p \to \text{Sel}_E(F_\infty)_p^{\Gamma_n}$ would have unbounded $\mathbb{Z}_p$-corank. (The kernel would still be finite and of bounded order by lemma 4.2.)

It is lemma 4.6 which is not valid if E has supersingular reduction at a prime v of F lying over p. For any $n \geq 0$, consider the map

$$r_{v_n} : \mathcal{H}_E((F_n)_{v_n}) \to \mathcal{H}_E((F_\infty)_\eta).$$

As discussed in chapter 2, if $d_n = [(F_n)_{v_n} : \mathbb{Q}_p]$, then $H^1((F_n)_{v_n}, E[p^\infty])$ has $\mathbb{Z}_p$-corank $2d_n$ and $\text{im}(\kappa_{v_n})$ has $\mathbb{Z}_p$-corank d_n. Therefore, $\mathcal{H}_E((F_n)_{v_n})$ has $\mathbb{Z}_p$-corank equal to d_n. This is true no matter what reduction type E has at the prime v. Assume that E has supersingular (or potentially supersingular) reduction at v. Assume also that $(F_\infty)_\eta/F_v$ is a ramified $\mathbb{Z}_p$-extension (i.e., the inertia subgroup of $\Gamma = \text{Gal}(F_\infty/F)$ for v is nontrivial). Under these assumptions, it turns out surprisingly that $\text{im}(\kappa_\eta) = H^1((F_\infty)_\eta, E[p^\infty])$. That is, $\mathcal{H}_E((F_\infty)_\eta) = 0$. Thus, even though the $\mathbb{Z}_p$-corank of $\mathcal{H}_E((F_n)_{v_n})$ is unbounded, the direct limit of these groups under the restriction maps is zero. Hence $\ker(r_{v_n}) = \mathcal{H}_E((F_n)_{v_n})$ for all $n \geq 0$.

The difference between the ordinary and supersingular cases shows up rather clearly in theorem 1.8, especially in the more precise statements (1.9) and (1.10). According to (1.10), under the assumption that F is an imaginary quadratic field and that $\text{End}(E) \otimes \mathbb{Q} \cong F$, we have

$$\text{rank}(E(F_n^{\text{ac}})) = 2 \sum_{j=1}^{n}{'} \varphi(p^j) + O(1).$$

if E has supersingular reduction at p. Here $\sum'$ denotes the sum over those j's with a fixed parity (depending on the parity of $\text{ord}_{s=1}(L(E/\mathbb{Q}, s))$. Compare this with corollary 4.12. Another interesting result is the following recent theorem of Kurihara which should be compared with corollary 4.11.

Theorem 4.13. *Suppose that E is an elliptic curve defined over $\mathbb{Q}$ which has good, supersingular reduction at a prime $p \geq 5$. Assume that $L(E, 1)/\Omega_E$ and the Tamagawa factors for all primes where E has bad reduction are not divisible by p. Assume also that $\operatorname{Gal}(\mathbb{Q}(E[p])/\mathbb{Q}) \cong GL_2(\mathbb{Z}/p\mathbb{Z})$. Let $\mathbb{Q}_\infty = \bigcup_n \mathbb{Q}_n$ be the cyclotomic $\mathbb{Z}_p$-extension of $\mathbb{Q}$. Then $\text{III}_E(\mathbb{Q}_n)_p$ is finite and its order is p^{t_n} where*

$$t_n = \sum_{j=1}^{n-1}{}' p^j - \left[\frac{n}{2}\right]$$

for $n > 0$. Here $\sum'$ denotes the sum over integers $j \equiv n - 1 \pmod 2$. Also, the Mordell-Weil group $E(\mathbb{Q}_n)$ is finite for all $n \geq 0$.

We will close this article with a brief discussion of the so-called "Main Conjecture." This conjecture has its roots in a conjecture of Iwasawa which gives an interpretation of the p-adic L-functions constructed by Kubota and Leopoldt in terms of the structure of certain Galois groups as Λ-modules. Iwasawa's conjecture was proved by Mazur and Wiles in 1983. But in the early 1970s, Mazur formulated a similar conjecture for a p-adic analogue of the Hasse-Weil L-function of an elliptic curve $E/\mathbb{Q}$ under the assumption that E has good, ordinary reduction at p. We refer the reader to [**Gr3**] for a more thorough account of the history of these ideas.

Let $\Gamma = \operatorname{Gal}(\mathbb{Q}_\infty/\mathbb{Q})$, where $\mathbb{Q}_\infty$ denotes the cyclotomic $\mathbb{Z}_p$-extension of $\mathbb{Q}$. Let $X = \operatorname{Hom}(\operatorname{Sel}_E(\mathbb{Q}_\infty)_p, \mathbb{Q}_p/\mathbb{Z}_p)$ which is known to be a finitely generated, torsion (due to Kato) Λ-module. This is just the Pontryagin dual of the discrete, abelian group $\operatorname{Sel}_E(\mathbb{Q}_\infty)_p$ and so we denote it by $\widehat{\operatorname{Sel}_E(\mathbb{Q}_\infty)_p}$. Let $f_E(T)$ denote a generator of the characteristic ideal of X. It should be clear that the behavior of $\operatorname{Sel}_E(\mathbb{Q}_n)_p$ as n varies is somehow closely related to the valuation of the numbers $f_E(\zeta - 1)$, where ζ varies over the p^n-th roots of unity. (See exercises 3.7, 3.8, and corollary 4.11, for example.) In turn, the Birch and Swinnerton-Dyer conjecture suggests that there should be a close link with the numbers $L(E/\mathbb{Q}, \chi, 1)$, where χ varies over the characters of $\operatorname{Gal}(\mathbb{Q}_n/\mathbb{Q})$, $n \geq 0$.

If E is an elliptic curve$/\mathbb{Q}$, then it is now known that E is modular. One consequence is that the L-function $L(E/\mathbb{Q}, \chi, s)$ can be analytically continued to the entire complex plane, where χ is any Dirichlet character. In particular, $L(E/\mathbb{Q}, \chi, 1)$ is defined. A theorem of Shimura implies that $L(E/\mathbb{Q}, \chi, 1)/\Omega_E$ is algebraic, where $\Omega_E = \int_{E(\mathbb{R})} \frac{dx}{y}$ is the so-called "real period" for E. Also, let $\alpha_p, \beta_p \in \overline{\mathbb{Q}}$ be defined by $\alpha_p + \beta_p = a_p$, $\alpha_p \beta_p = p$, where $a_p = 1 + p - \widetilde{E}(\mathbb{F}_p)$. Assume that E has good, ordinary reduction at p. Then $p \nmid a_p$ (exercise 2.12). This means that p splits in the imaginary quadratic field $\mathbb{Q}(\alpha_p, \beta_p)$. Let $\tau(\chi) \in \overline{\mathbb{Q}}$ be the Gauss sum for the Dirichlet character χ. As in chapter 1, we will regard a Dirichlet character χ as a 1-dimensional representation of a Galois group $\operatorname{Gal}(K_\chi/\mathbb{Q})$. The p-adic L-function $L_p(E/\mathbb{Q}, s)$ is a $\mathbb{Q}_p$-valued analytic function of a p-adic variable s which is completely determined by the collection of numbers $L(E/\mathbb{Q}, \chi, 1)/\Omega_E$, where χ varies over all Dirichlet characters such that $K_\chi \subset \mathbb{Q}_\infty$. That is, χ varies over $\widehat{\Gamma}$, the Pontryagin dual of Γ.

Here is the precise definition. Fix an embedding of $\overline{\mathbb{Q}}$ into $\overline{\mathbb{Q}}_p$. Then we can consider χ as having values in $\overline{\mathbb{Q}}_p$. The numbers $\alpha_p, \beta_p, \tau(\chi)$, and $L(E/\mathbb{Q}, \chi, 1)/\Omega_E$ can be regarded as elements of $\overline{\mathbb{Q}}_p$. It is clear that $\alpha_p, \beta_p \in \mathbb{Q}_p$ and that one of these numbers (say α_p) is a p-adic unit. Fix a topological generator γ_0 of Γ as in chapter 3. Then $\chi \in \widehat{\Gamma}$ is determined by $\zeta = \chi(\gamma_0)$, which is a p^n-th root of unity. If $T = \gamma_0 - 1$, then we define $\chi(T) = \chi(\gamma_0) - 1 = \zeta - 1$. Note that if $g(T) = \sum_{i=0}^{\infty} b_i T^i \in \Lambda$, then it makes sense to write $g(\chi(T))$ since the infinite series $g(\zeta - 1) = \sum b_i(\zeta - 1)^i$ converges in $\overline{\mathbb{Q}}_p$. This also makes sense if $g(T) \in \Lambda \otimes_{\mathbb{Z}_p} \mathbb{Q}_p$, the ring of power series with bounded denominators. The existence of $L_p(E/\mathbb{Q}, s)$ is a consequence of a theorem of Mazur and Swinnerton-Dyer which states that there exists a power series $g_E(T) \in \Lambda \otimes_{\mathbb{Z}_p} \mathbb{Q}_p$ satisfying the following interpolation property.

(4.14)
$$\begin{aligned}(1) \quad g_E(0) &= (1 - \beta_p p^{-1})^2 L(E/\mathbb{Q}, 1)/\Omega_E \\ (2) \quad g_E(\chi(T)) &= (\beta_p)^m L(E/\mathbb{Q}, \chi, 1)/\Omega_E \tau(\chi)\end{aligned}$$

if χ has conductor $p^m > 1$.

Here, if χ is a faithful character of $\text{Gal}(\mathbb{Q}_n/\mathbb{Q})$, where $n \geq 1$, then the conductor of χ is p^m where $m = n + 1$ if p is odd, $m = n + 2$ if $p = 2$. Since a nonzero element of $\Lambda \otimes_{\mathbb{Z}_p} \mathbb{Q}_p$ can have only finitely many zeros, $g_E(T)$ is clearly uniquely determined by the above interpolation property. It also does not depend on the choice of embedding $\overline{\mathbb{Q}} \to \overline{\mathbb{Q}}_p$. The proof of the existence of $g_E(T)$ can be found in [**M-T-T**]. There, p-adic L-functions are associated to modular forms under very general hypotheses.

Finally, here is the definition of $L_p(E/\mathbb{Q}, s)$. One can identify $\Gamma = \text{Gal}(\mathbb{Q}_\infty/\mathbb{Q})$ with $\text{Gal}(\mathbb{Q}(\mu_{p^\infty})/\mathbb{Q})$ if p is odd (or with $\text{Gal}(\mathbb{Q}(\mu_{2^\infty})/\mathbb{Q}(\mu_4))$ if $p = 2$). Thus, we can regard γ_0 as an element of $\text{Gal}(\mathbb{Q}(\mu_{p^\infty})/\mathbb{Q})$. Define κ_0 by $\gamma_0(\zeta) = \zeta^{\kappa_0}$ for all $\zeta \in \mu_{p^\infty}$. (We have $\kappa_0 \in 1 + p\mathbb{Z}_p$ if p is odd, $\kappa_0 \in 1 + 4\mathbb{Z}_2$ if $p = 2$.) Then, by definition,

(4.15) $$L_p(E/\mathbb{Q}, s) = g_E(\kappa_0^{s-1} - 1)$$

for all $s \in \mathbb{Z}_p$. This function is actually independent of the choice of γ_0. Note also that $L_p(E/\mathbb{Q}, 1) = g_E(0) = (1 - \beta_p p^{-1})^2 L(E/\mathbb{Q}, 1)/\Omega_E$ and hence we have $L_p(E/\mathbb{Q}, 1) \neq 0 \iff L(E/\mathbb{Q}, 1) \neq 0$.

Here then is the statement of the Main Conjecture which was formulated by Mazur.

Conjecture 4.16. *The characteristic ideal of the Λ-module $X = \widehat{\text{Sel}_E(\mathbb{Q}_\infty)}_p$ is generated by $g_E(T)$.*

Implicit in this statement is the assertion that $g_E(T) \subset \Lambda$. This is known to be true in most cases, e.g. if the $G_\mathbb{Q}$-module $E[p]$ is irreducible where p is an odd prime. The most general result concerning conjecture 4.16 is the theorem of Rubin [**Ru1**] establishing it when E is an elliptic curve/$\mathbb{Q}$ with complex multiplication and p is any odd prime where E has good, ordinary reduction. Some simple consequences of conjecture 4.16 are given in exercises 4.7, 4.8.

Exercises

In exercises 4.1–4.5, we consider the elliptic curve $E = X_0(11)$, which is defined by $y^2 + y = x^3 - x^2 - 10x - 20$. In Cremona's tables, one finds that E has two $\mathbb{Q}$-isogenies of degree 5 and that $E(\mathbb{Q}) \cong \mathbb{Z}/5\mathbb{Z}$. Also it can be shown that $\text{III}_E(\mathbb{Q})$ is trivial.

4.1. Deduce that $E[5] \cong \mu_5 \times \mathbb{Z}/5\mathbb{Z}$ as a $G_\mathbb{Q}$-module.

4.2. Let $\Sigma = \{\infty, 5, 11\}$. One can consider $H^1(\mathbb{Q}_\Sigma/\mathbb{Q}_\infty, \mu_5)$ as a discrete Λ-module, where $\mathbb{Q}_\infty$ denotes the cyclotomic $\mathbb{Z}_5$-extension of $\mathbb{Q}$. Show that the compact Λ-module $H^1(\mathbb{Q}_\Sigma/\mathbb{Q}_\infty, \mu_5)\widehat{}$ has μ-invariant equal to 1. (Hint: Use Kummer theory and consider the units of the fields $\mathbb{Q}_n$. The fact that the class number of $\mathbb{Q}_n$ is not divisible by 5 is helpful)

4.3. Show that E has good, ordinary reduction at $p = 5$. Show that
$$\mu_5 \subset \ker(E[5^\infty] \to \widetilde{E}[5^\infty]).$$
(Note: To define the reduction map, one must choose an embedding $\overline{\mathbb{Q}} \to \overline{\mathbb{Q}}_5$.)

4.4. There is a natural map
$$\varepsilon : H^1(\mathbb{Q}_\Sigma/\mathbb{Q}_\infty, \mu_5) \to H^1(\mathbb{Q}_\Sigma/\mathbb{Q}_\infty, E[5^\infty]).$$
Show that $\ker(\varepsilon)$ is finite and that $\text{im}(\varepsilon) \subset \text{Sel}_E(\mathbb{Q}_\infty)_5$. The Λ-module $\text{Sel}_E(\mathbb{Q}_\infty)\widehat{}_5$ is Λ-torsion by corollary 4.9. Show that its μ-invariant is positive.

4.5. Prove that $\dim_{\mathbb{Z}/p\mathbb{Z}}(\text{III}_E(\mathbb{Q}_n)[p]) \geq p^n - c$ for all n, where c is some constant. (Theorem 1.6 is helpful here.)

4.6. Let E be an elliptic curve defined over a finite extension F_v of $\mathbb{Q}_l$, where $l \neq p$. Let K_η denote the unramified $\mathbb{Z}_p$-extension of F_v (which is the only $\mathbb{Z}_p$-extension of F_v). Let $B_v = H^0(K_\eta, E[p^\infty])$. Prove that $H^1(K_\eta/F_v, B_v)$ has order $c_v^{(p)}$, where $c_v^{(p)}$ is the largest power of p dividing $c_v = [E(F_v) : E_0(F_v)]$. Deduce that B_v is divisible if and only if $p \nmid c_v$. (Here $E_0(F_v)$ denotes the subgroup of $E(F_v)$ consisting of the points with nonsingular reduction. See [**Si1**], Chapter 7, §6 for a discussion of the group $E(F_v)/E_0(F_v)$.)

4.7. Assume that E is an elliptic curve/$\mathbb{Q}$ with good, ordinary reduction at p. Kato has proved that the characteristic ideal of $\text{Sel}_E(\mathbb{Q}_\infty)\widehat{}_p$ contains $p^a g_E(T)$ for some $a \geq 0$. Using this, prove that $\text{rank}_\mathbb{Z}(E(\mathbb{Q})) \leq \text{ord}_{s=1}(L_p(E/\mathbb{Q}, s))$.

4.8. Assume that E is an elliptic curve/$\mathbb{Q}$. Kolyvagin has proved that if $L(E/\mathbb{Q}, 1) \neq 0$, then both $E(\mathbb{Q})$ and $\text{III}_E(\mathbb{Q})$ are finite. Assuming conjecture 4.16, prove the following partial converse: If $L(E/\mathbb{Q}, 1) = 0$, then either $E(\mathbb{Q})$ is infinite or $\text{III}_E(\mathbb{Q})_p$ is infinite for every prime p where E has good, ordinary reduction.

BIBLIOGRAPHY

[B-D] M. Bertolini, H. Darmon, Nontriviality of families of Heegner points and ranks of Selmer groups over anticyclotomic towers, *Jour. Ramanujan Math. Soc.* **13** (1998), 15–24.

[B-K] S. Bloch, K. Kato, L-functions and Tamagawa numbers of motives. In: *Grothendieck Festschrift 1, Progress in Math.*, Birkhäuser 86 (1990), 333–400.

[Bo] Bourbaki, *Commutative Algebra*, Addison-Wesley, 1972.

[C-G] J. Coates, R. Greenberg, Kummer theory for abelian varieties over local fields, *Invent. Math.* **124** (1996), 129–174.

[C-W] J. Coates, A. Wiles, On the conjecture of Birch and Swinnerton-Dyer, *Invent. Math.* **39** (1977), 223–251.

[Cr] J. E. Cremona, Algorithms for Modular Elliptic Curves, Cambridge University Press (1992).

[Gr1] R. Greenberg, On the Birch and Swinnerton-Dyer Conjecture, *Invent. Math.* **72** (1983), 241–265.

[Gr2] R. Greenberg, Iwasawa theory for elliptic curves, in *Lecture Notes in Math.*, **1716** (1999), 51–144.

[Gr3] R. Greenberg, Iwasawa Theory — Past and Present, to appear in *Advanced Studies in Pure Mathematics*.

[Ha] M. Harris, Systemic growth of Mordell-Weil groups of abelian varieties in towers of number fields, *Invent. Math.* **51** (1979), 123–141.

[Ho] L. Howe, Twisted Hasse-Weil L-functions and the result of Mordell-Weil groups, *Canadian J. Math.* **49** (1997), 749–771.

[Iw] K. Iwasawa, On $\mathbb{Z}_l$-extensions of algebraic number fields, *Ann. of Math.* **98** (1973), 246–326.

[Ka] K. Kato, p-adic Hodge theory and values of zeta functions of modular curves. Preprint. (See also [**Ru2**] and [**Sc**].)

[Ku] P. F. Kurcanov, Elliptic curves of infinite rank over Γ-extensions, *Mat. Sbornik* **90** (1973), 320–329.

[Man] Y. I. Manin, Cyclotomic fields and modular curves, *Russian Math. Surveys* **26**, no. 6 (1971), 7–78.

[Ma1] B. Mazur, Rational points of abelian varieties with values in towers of number fields, *Invent. Math.* **18** (1972), 183–266.

[Ma2]   B. Mazur, Modular curves and arithmetic, *Proceedings of the International Congress of Mathematicians*, Warsaw, 1983, 185–211.
[M-T-T] B. Mazur, J. Tate, J. Teitelbaum, On p-adic analogues of the conjectures of Birch and Swinnerton-Dyer, *Invent. Math.* **84** (1986), 1–48.
[Ri]   K. Ribet, Torsion points of abelian varieties in cyclotomic extensions, *Enseign. Math.* **27** (1981), 315–319.
[Ro1]   D. Rohrlich, On L-functions of elliptic curves and cyclotomic towers, *Invent. Math.* **75** (1984), 404–423.
[Ro2]   D. Rohrlich, On L-functions of elliptic curves and anticyclotomic towers, *Invent. Math.* **75** (1984), 383–408.
[Ro3]   D. Rohrlich, L-functions and division towers, *Math. Ann.* **281** (1988), 611–632.
[Ro4]   D. Rohrlich, Galois theory, elliptic curves, and root numbers, *Comp. Math.* **100** (1996), 311–349.
[Ro5]   D. Rohrlich, Realization of some Galois representations of low degree in Mordell-Weil groups, *Math. Research Letters* **4** (1997), 123–30.
[Ru1]   K. Rubin, The "main conjectures" of Iwasawa theory for imaginary quadratic fields, *Invent. Math.* **103** (1991), 25–68.
[Ru2]   K. Rubin, Euler systems and modular elliptic curves, *London Math. Soc. Lecture Note Series* **259** (1998), 351–368.
[Sc]   A. Scholl, An introduction to Kato's Euler systems, *London Math. Soc. Lecture Note Series* **259** (1998), 379–460.
[Se]   J-P. Serre, Classes des corps cyclotomiques (d'après K. Iwasawa), *Sém. Bourbaki* 1958/59, no. 174.
[Si1]   J. Silverman, The Arithmetic of Elliptic Curves, *Graduate Texts in Math.* **106**, Springer-Verlag 1986.
[Si2]   J. Silverman, Advanced Topics in the Arithmetic of Elliptic Curves, *Graduate Texts in Math.* **151**, Springer-Verlag 1994.
[Wa]   L. Washington, Introduction to Cyclotomic Fields, *Graduate Texts in Math.* **83**, Springer-Verlag 1982.

Galois Cohomology

John Tate

Galois Cohomology
John Tate[1]

Galois Cohomology

Preface

I thank Helena Verrill and William Stein for their help in getting this account of my talks at Park City into print. After Helena typed up her original notes of the talks, William was a great help with the editing, and put them in the canonical format for this volume.

The somewhat inefficient organization of this account is mainly a result of the fact that, after the first talk had been given with the idea that it was to be the only one, a second was later scheduled, and these are the notes of the material in the two talks in the order it was presented.

The bible for this subject is Serre [6], in conjunction with [5] or [1]. Haberland [2] is also an excellent reference.

Group modules

Consider a group G and an abelian group A equipped with a map
$$G \times A \to A,$$
$$(\sigma, a) \mapsto \sigma a.$$
We use notation $\sigma, \tau, \rho, \ldots$ for elements of G, and $a, b, a', b', \ldots$ for elements of A. To say that A is a G-set means that
$$\tau(\sigma a) = (\tau\sigma)a \quad \text{and} \quad 1a = a,$$
for all $\sigma, \tau \in G$ and $a \in A$, where 1 is the identity in G. To say that A is a G-module means that, in addition, we have
$$\sigma(a + b) = \sigma a + \sigma b,$$

---

[1]Department of Mathematics; Austin, TX.
E-mail address: tate@math.utexas.edu.
1991 *Mathematics Subject Classification*. 11.
Key words and phrases. Galois cohomology, Elliptic curves.

©2001 American Mathematical Society

for all $\sigma \in G$ and $a, b \in A$. This is all equivalent to giving A the structure of $\mathbf{Z}[G]$-module.

Given a G-module A as above, the subgroup of fixed elements of A is

$$A^G := \{a \in A \mid \sigma a = a \text{ for all } \sigma \in G\}.$$

We say G *acts trivially* on A if $\sigma a = a$ for all $a \in A$; thus $A^G = A$ if and only if the action is trivial. When $\mathbf{Z}$, $\mathbf{Q}$, $\mathbf{Q}/\mathbf{Z}$ are considered as G-modules, this is with the trivial action, unless stated otherwise.

If we take $G = \mathrm{Gal}(K/k)$, with K a Galois extension of k of possibly infinite degree, then we have the following examples of fixed subgroups of G-modules:

A	A^G
K^+ as an additive group	k^+
K^* as a multiplicative group	k^*
$E(K)$, where E/k is an elliptic curve	$E(k)$.

The action on $E(K)$ above is given by $\sigma(x, y) = (\sigma x, \sigma y)$ for a point $P = (x, y)$, if E is given as a plane cubic. In general, if C is a commutative algebraic group over K, we can take $A = C(K)$, and then $A^G = C(k)$.

Cohomology

We now define the cohomology groups $H^r(G, A)$, for $r \in \mathbf{Z}$. Abstractly, these are the right derived functors of the left exact functor

$$\{G\text{-modules}\} \to \{\text{abelian groups}\}$$

that sends $A \mapsto A^G$. Since $A^G = \mathrm{Hom}_{\mathbf{Z}[G]}(\mathbf{Z}, A)$, we have a canonical isomorphism

$$H^r(G, A) = \mathrm{Ext}^r_{\mathbf{Z}[G]}(\mathbf{Z}, A).$$

More concretely, the cohomology groups $H^r(G, A)$ can be computed using the "standard cochain complex" (see, e.g., [1, pg. 96]). Let

$$C^r(G, A) := \mathrm{Maps}(G^r, A);$$

an element of $C^r(G, A)$ is a function f of r variables in G,

$$f(\sigma_1, \ldots, \sigma_r) \in A,$$

and is called an *r-cochain*. (If, in addition, A and G have a topological structure, then we instead consider continuous cochains.) There is a sequence

$$\cdots \to 0 \to 0 \to C^0(G, A) \xrightarrow{\delta} C^1(G, A) \xrightarrow{\delta} C^2(G, A) \xrightarrow{\delta} \cdots$$

Here $C^0(G, A) = A$, since an element f of $C^0(G, A)$ is given by the single element $f_0(\bullet) \in A$, its value at the unique element $\bullet \in G^0$. The maps δ are defined by

$$\begin{aligned}(\delta f_0)(\sigma) &= \sigma f_0(\bullet) - f_0(\bullet), \\ (\delta f_1)(\sigma, \tau) &= \sigma f_1(\tau) - f_1(\sigma \tau) + f_1(\sigma), \\ (\delta f_2)(\sigma, \tau, \rho) &= \sigma f_2(\tau, \rho) - f_2(\sigma \tau, \rho) + f_2(\sigma, \tau \rho) - f_2(\sigma, \tau),\end{aligned}$$

and so on. Note that $\delta \circ \delta = 0$. The cohomology groups are given by

$$H^r(G, A) = \ker \delta / \mathrm{im}\, \delta \subset C^r(G, A) / \mathrm{im}\, \delta.$$

Cocycles are elements of the kernel of δ, and coboundaries are elements of the image of δ. We have

$$H^0(G, A) = A^G,$$
$$H^1(G, A) = \frac{\text{crossed-homomorphisms}}{\text{principal crossed-homomorphisms}}$$
$$= \text{Hom}(G, A), \text{ if action is trivial,}$$
$$H^2(G, A) = \text{classes of "factor sets"}.$$

The groups $H^2(G, A)$ and $H^1(G, A)$ arise in many situations. Perhaps the simplest is their connection with group extensions and their automorphisms. Given a G-module A, suppose $\mathcal{G}$ is a group extension of G by A, that is, $\mathcal{G}$ is a group which contains A as a normal subgroup such that $\mathcal{G}/A \cong G$, where the given action of G on A is the same as the conjugation action induced by this isomorphism. Construct a 2-cocycle $a_{\sigma,\tau}$ as follows. For each element $\sigma \in G$, let $u_\sigma \in \mathcal{G}$ be a coset representative corresponding to σ. Then $\mathcal{G} = \coprod_\sigma A u_\sigma$, i.e., every element of $\mathcal{G}$ is uniquely of the form au_σ. Thus

$$u_\sigma u_\tau = a_{\sigma,\tau} u_{\sigma\tau}$$

for some $a_{\sigma,\tau} \in A$. The map $(\sigma, \tau) \mapsto a_{\sigma,\tau}$ is a 2-cocycle.

Exercise 0.1. Using the associative law, check that $a_{\sigma,\tau}$ is a 2-cocycle, and if $\mathcal{G}'$ is another extension of G by A, then there is an isomorphism $\mathcal{G}' \cong \mathcal{G}$ that induces the identity on A and G if and only if the corresponding 2-cocycles differ by a coboundary.

Exercise 0.2. Conversely, show that every 2-cocycle arises in this way. For example, in the trivial case, if $a_{\sigma,\tau} = 1$ for every σ and τ, then we can take $\mathcal{G}$ to be the semidirect product $G \ltimes A$.

Therefore we may view $H^2(G, A)$ as the group of isomorphism classes of extensions of G by A with a given action of G on A.

Exercise 0.3. Show that an automorphism of $\mathcal{G}$ that induces the identity on A and on $G = \mathcal{G}/A$ is of the form $au_\sigma \mapsto ab_\sigma u_\sigma$ with $\sigma \mapsto b_\sigma$ a 1-cocycle, and it is an inner automorphism induced by an element of A if and only if $\sigma \mapsto b_\sigma$ is a coboundary.

Examples

Given a finite Galois extension K/k, and a commutative algebraic group C over k, the following notation is frequently used:

$$H^r(K/k, C) := H^r(\text{Gal}(K/k), C(K)).$$

We have $H^0(K/k, C) = C(k)$ as above, and

$$H^1(K/k, \mathbf{G}_m) = H^1(K/k, K^*) = 0,$$
$$H^2(K/k, \mathbf{G}_m) = \text{Br}(K/k) \subset \text{Br}(k)$$

The first equality is Hilbert's Theorem 90. In the second equality $\text{Br}(k)$ is the Brauer group of k; this is the group of equivalence classes of central simple algebras with center k that are finite dimensional over k; two such algebras are equivalent if they are matrix algebras over k-isomorphic division algebras.

The map from $H^2(K/k, \mathbf{G}_m)$ to $\text{Br}(K/k)$ is defined as follows. Given a 2-cocycle $a_{\sigma,\tau}$, define a central simple algebra over k by $\mathcal{A} = \oplus K u_\sigma$, which is a vector spaces over K with a basis $\{u_\sigma\}$ indexed by the elements $\sigma \in G$. Multiplication is defined by the same rules as for group extensions above (with $A = K^*$), extended linearly.

Characterization of $H^r(G, -)$

For fixed G and varying A the groups $H^r(G, A)$ have the following fundamental properties:

1. $H^0(G, A) = A^G$.
2. $H^r(G, -)$ is a functor
$$\{G\text{-modules}\} \to \{\text{abelian groups}\}.$$
3. Each short exact sequence
$$0 \to A' \to A \to A'' \to 0$$
gives rise to connecting homomorphisms (see below)
$$\delta : H^r(G, A'') \to H^{r+1}(G, A')$$
from which we get a long exact sequence of cohomology groups, functorial in short exact sequences in the natural sense.
4. If A is "induced" or "injective", then $H^r(G, A) = 0$ for all $r \neq 0$.

These properties characterize the sequence of functors H^i equipped with the δ's uniquely, up to unique isomorphism.

For $c \in H^r(G, A'')$, define $\delta(c)$ as follows. Let $c_1 : G^r \to A''$ be a cocycle representing c. Lift c_1 to any map (cochain) $c_2 : G^r \to A$. Since $\delta(c_1) = 0$, the map $\delta(c_2) : G^{r+1} \to A$ has image in A', so defines a map $\delta(c_2) : G^{r+1} \to A'$, and thus represents a class $\delta(c) \in H^{r+1}(G, A')$.

For an infinite Galois extension, one uses cocycles that come by inflation from finite Galois subextensions. This amounts to using continuous cochains, where continuous means with respect to the Krull topology on G and the discrete topology on A.

Abstracting this situation leads to the notion of the cohomology of a profinite group G (i.e., a projective limit, in the category of topological groups, of finite groups G_i) operating continuously on a discrete module A. Without loss of generality the G_i can be taken to be the quotients G/U of G by its open normal subgroups U, and then A is the union of its subgroups A^U. The cohomology groups $H^r(G, A)$ computed with continuous cochains are direct limits, relative to the inflation maps (see Section), of the cohomology groups $H^r(G/U, A^U)$ of the finite quotients, because the continuous cochain complex $C^*(G, A)$ is the direct limit of the complexes $C^*(G/U, A^U)$. Also, it is easy to see that the groups $\{H^r(G, -)\}_r$ are characterized by δ-functoriality on the category of *discrete* G-modules.

Kummer theory

Let k^{sep} be a separable closure of a field k, and put $G_k = \text{Gal}(k^{\text{sep}}/k)$. Let $m \geq 1$ be an integer, and assume that the image of m in k is nonzero. Associated to the exact sequence
$$0 \longrightarrow \mu_m \longrightarrow (k^{\text{sep}})^* \xrightarrow{m} (k^{\text{sep}})^* \longrightarrow 0,$$

we have a long exact sequence

$$0 \longrightarrow \mu_m \cap k \longrightarrow k^* \xrightarrow{m} k^*$$
$$\longrightarrow H^1(G_k, \mu_m) \longrightarrow H^1(G_k, (k^{\text{sep}})^*) = 0,$$

where the last equality is by Hilbert's Theorem 90. Thus $H^1(G_k, \mu_m) \cong k^*/(k^*)^m$.

Now assume that the group of mth roots of unity μ_m is contained in k. Then

$$H^1(G_k, \mu_m) = \text{Hom}_{\text{cont}}(G_k, \mu_m),$$

so

$$k^*/(k^*)^m \cong \text{Hom}_{\text{cont}}(G_k, \mu_m).$$

Using duality, this isomorphism describes the finite abelian extensions of k whose Galois group is killed by m. For example, consider a Galois extension K/k such that $G = \text{Gal}(K/k)$ is a finite abelian group that is killed by m. Since G is a quotient of $G_k = \text{Gal}(k^{\text{sep}}/k)$, we have a diagram

$$\begin{array}{ccc} k^*/(k^*)^m & \xrightarrow{\cong} & \text{Hom}_{\text{cont}}(G_k, \mu_m) \\ \uparrow & & \uparrow \\ B & \xrightarrow{\cong} & \widehat{G} := \text{Hom}(G, \mu_m), \end{array}$$

where B is the subgroup of $k^*/(k^*)^m$ corresponding to $\widehat{G}$ under the isomorphism.

Exercise 0.4. Show that

$$K = k(\sqrt[m]{B}) = k(\{\sqrt[m]{b} \mid b \in B\}),$$

and $[K : k] = \#B$.

The case when G cyclic is the crucial step in showing that a polynomial with solvable Galois group can be solved by radicals.

For the rest of this section, we assume that k is a number field and continue to assume that k contains μ_m. Let S be a finite set of primes of k including all divisors of m and large enough so that the ring $\mathcal{O}_S$ of S-integers of k is a principal ideal ring.

Exercise 0.5. Show that the extension $K(\sqrt[m]{B})$ above is unramified outside S if and only if $B \subset U_S k^{*m}/k^{*m} \cong U_S/U_S^m$, where $U_S = \mathcal{O}_S^*$ is the group of S-units of k.

Exercise 0.6. Let k_S be the maximal extension of k which is unramified outside S, and let $G_S = \text{Gal}(k_S/k)$. Then $\text{Hom}_{\text{cont}}(G_S, \mu_m) = U_S/U_S^m$. It follows that $\text{Hom}_{\text{cont}}(G_S, \mu_m)$ is finite, because U_S is finitely generated.

Now let E be an elliptic curve over k. The m-torsion points of E over $\overline{k}$ form a group $E_m = E_m(\overline{k}) \approx (\mathbf{Z}/m\mathbf{Z})^2$. Suppose that, in addition to the conditions above, S also contain the places at which E has bad reduction. Then it is a fact that $E(k_S)$ is divisible by m, so we have an exact sequence

$$0 \to E_m \to E(k_S) \xrightarrow{m} E(k_S) \to 0.$$

Taking cohomology we obtain an exact sequence

$$0 \to E(k)/mE(k) \to H^1(k_S/k, E_m) \to H^1(k_S/k, E)_m \to 0,$$

where the subscript m means elements killed by m. Thus, to prove that $E(k)/mE(k)$ is finite (the "weak Mordell-Weil theorem"), it suffices to show that $H^1(k_S/k, E_m)$ is finite. Let $k' = k(E_m)$ be the extension of k obtained by adjoining the coordinates of the points of order m. Then k'/k is finite and unramified outside S. Hence $H^1(k'/k, E_m)$ is finite, and the exact inflation-restriction sequence (see Section)

$$0 \to H^1(k'/k, E_m) \to H^1(k_S/k, E_m) \to H^1(k_S/k', E_m)$$

shows that it suffices to prove $H^1(k_S/k, E_m)$ is finite when $k = k'$. But then

$$H^1(k_S/k, E_m) \cong \mathrm{Hom}_{\mathrm{cont}}(G_S, E_m) \cong \mathrm{Hom}_{\mathrm{cont}}(G_S, \mu_m)^2$$

is finite by Exercise 0.6.

Exercise 0.7. Take $k = \mathbf{Q}$ and let E be the elliptic curve $y^2 = x^3 - x$. Let $m = 2$ and $S = \{2\}$, $U_S = \langle -1, 2 \rangle$, and show $(E(\mathbf{Q}) : 2E(\mathbf{Q})) \leq 16$. (In fact, $E(\mathbf{Q}) = E_2$ is of order 4, killed by 2, but to show that we need to examine what happens over $\mathbf{R}$ and over $\mathbf{Q}_2$, not just use the lack of ramification at the other places.)

Exercise 0.8. Suppose $S' = S \cup \{P_1, P_2, \ldots, P_t\}$ is obtained by adding t new primes to S. Then $U_{S'} \cong U_S \times \mathbf{Z}^t$. Hence $H^1(k_{S'}/k, E)_m \cong H^1(k_S/k, E) \times (\mathbf{Z}/m\mathbf{Z})^{2t}$. Hence $H^1(k, E)$ contains an infinite number of independent elements of order m. Hilbert Theorem 90 is far from true for E.

Functor of pairs (G, A)

A *morphism of pairs* $(G, A) \mapsto (G', A')$ is given by a pair of maps ϕ and f,

$$G \xleftarrow{\phi} G' \quad \text{and} \quad A_\phi \xrightarrow{f} A',$$

where ϕ is a group homomorphism, and f is a homomorphism of G'-modules, and A_ϕ means A with the G' action induced by ϕ. A morphism of pairs induces a map

$$H^r(G, A) \to H^r(G', A')$$

got by composing the map $H^r(G, A) \to H^r(G', A_\phi)$ induced by ϕ with the map $H^r(G', A_\phi) \to H^r(G', A')$ induced by f. We thus consider $H^r(G, A)$ as a functor of pairs (G, A).

If G' is a subgroup of G then there are maps

$$H^r(G, A) \underset{\text{corestriction}}{\overset{\text{restriction}}{\rightleftarrows}} H^r(G', A).$$

Here the corestriction map (also called the "transfer map") is defined only if the index $[G : G']$ is finite.

When $r = 0$ the corestriction map is the trace or norm:

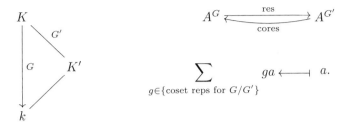

Corollary 0.9. *If G is of finite cardinality m, then*
$$mH^r(G, A) = 0 \text{ for } r \neq 0.$$

Proof. Letting $G' = \{1\}$, we have
$$(\text{corestriction}) \circ (\text{restriction}) = [G : G'] = [G : \{1\}] = m.$$
Since $H^r(\{1\}, A) = 0$ for $r \neq 0$, this composition is 0, as claimed. $\square$

Exercise 0.10. Restriction to a p-Sylow subgroup is injective on the p-primary component of $H^r(G, A)$.

The Shafarevich group

Let k be a number field, ν a place of k, and k_ν the completion of k at ν. Let $\overline{k}_\nu$ be an algebraic closure of k_ν and let $\overline{k}$ be the algebraic closure of k in $\overline{k}_\nu$. These four fields are illustrated in the following diagram.

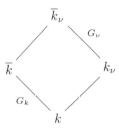

Let E be an elliptic curve over k. We have natural morphisms of pairs
$$(G_k, E(\overline{k})) \to (G_\nu, E(\overline{k}_\nu)),$$
for each place ν, hence a homomorphism
$$H^1(k, E) \to \prod_\nu H^1(k_\nu, E),$$
where the product is taken over all places of k. The kernel of this map is the Shafarevich group $\text{III}(k, E)$, which is conjectured to be finite.

If you can prove that III is finite, then you will be famous, and you will have shown that the descent algorithm to compute the Mordell-Weil group, which seems to work in practice, will always work. Until 1986, there was no single instance where it was known that III was finite! Now much is known for $k = \mathbf{Q}$ if the rank of $E(\mathbf{Q})$ is 0 or 1; see [**3**] and [**4**] for results in this direction. Almost nothing is known for higher ranks.

The inflation-restriction sequence

Recall that a morphism of pairs
$$(G, A) \to (G', A')$$
is a map $G' \to G$ and a G'-homomorphism $A \to A'$, where G' acts on A via $G' \to G$. In particular, we can take G' to be a subgroup H of G. Here are three special instances of the above map:

1) restriction $H^r(G, A) \to H^r(H, A)$
2) inflation $H^r(G/H, A^H) \to H^r(G, A)$
   (for $H \triangleleft G$, $G \to G/H$, $A^H \subset A$)
3) conjugation $H^r(H, A) \xrightarrow{\tilde{\sigma}} H^r(\sigma H \sigma^{-1}, A)$, $\sigma \in G$
   (for $\sigma h \sigma^{-1} \mapsto h$ and $a \mapsto \sigma a$)

Theorem 0.11. *If $\sigma \in H$, then the conjugation map $\tilde{\sigma}$ is the identity.*

Exercise 0.12. Given a commutative algebraic group C defined over k one sometimes uses the notation $H^r(k, C) := H^r(k^{\text{sep}}/k, C)$, where k is a separable algebraic closure of k. Show that this makes sense, in the sense that if k_1^s and k_2^s are two separable closures of k, then the isomorphism $H^r(\text{Gal}(k_1^s/k), C(k_1^s)) \cong H^r(\text{Gal}(k_2^s/k), C(k_2^s))$ induced by a k-isomorphism $\varphi : k_1^s \to k_2^s$ is independent of the choice of φ.

Theorem 0.13. *If H is a normal subgroup of G, then there is a "Hochschild-Serre" spectral sequence*
$$E_2^{rs} = H^r(G/H, H^s(H, A)) \Rightarrow H^{r+s}(G, A)$$

By Theorem 0.11, G acts on $H^r(H, A)$ and H acts trivially, so this makes sense. (The profinite case follows immediately from the finite one by direct limit; cf. the end of Section .) The low dimensional corner of the spectral sequence can be pictured as follows.

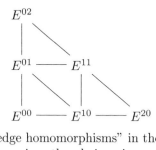

Inflation and restriction are "edge homomorphisms" in the spectral sequence. The lower left corner pictured above gives the obvious isomorphism $A^G \cong (A^H)^{G/H}$, and the exact sequence

$$0 \longrightarrow H^1(G/H, A^H) \xrightarrow{\text{inf}} H^1(G, A) \xrightarrow{\text{res}} H^1(H, A)^{G/H}$$
$$\xrightarrow{d} H^2(G/H, A^H) \xrightarrow{\text{inf}} H^2(G, A).$$

The map d is the "transgression" and is induced by $d_2 : E_2^{01} \to E_2^{20}$.

Exercise 0.14.

1. Show that this last sequence, or at least the first line, is exact by using standard 1-cocycles.
2. If $H^1(H, A) = 0$, so that $E_2^{r1} = 0$ for all r, then the sequence obtained by increasing the superscripts on the H's by 1 is exact.

Consider a subfield K of k^{sep} that is Galois over k, and let C be a commutative algebraic group over k.

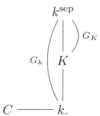

The inflation-restriction sequence is

$$0 \longrightarrow H^1(K/k, C(K)) \longrightarrow H^1(k, C(k^{\text{sep}})) \longrightarrow H^1(K, C(k^{\text{sep}}))^{\text{Gal}(K/k)}$$
$$\longrightarrow H^2(K/k, C(K)) \longrightarrow H^2(k, C(k^{\text{sep}})).$$

If $C = \mathbf{G}_m$, then $H^1(K, C(k^{\text{sep}})) = 0$, and there is an inflation-restriction sequence with $(1, 2)$ replaced by $(2, 3)$:

$$0 \longrightarrow H^2(K/k, K^*) \longrightarrow H^2(k, (k^{\text{sep}})^*) \longrightarrow H^2(K, (k^{\text{sep}})^*)^{\text{Gal}(K/k)}$$
$$\longrightarrow H^3(K/k, K^*) \longrightarrow H^3(k, (k^{\text{sep}})^*).$$

An element $\alpha \in H^2(K, (k^{\text{sep}})^*)^{\text{Gal}(K/k)}$ represents a central simple algebra A over K which is isomorphic to all of its conjugates by $\text{Gal}(K/k)$. As the diagram indicates, the image α in $H^3(K/k, K^*)$ is the "obstruction" whose vanishing is the necessary and sufficient condition for such an algebra A to come by base extension from an algebra over k.

Cup products

G-pairing

If A, A', and B are G-modules, then

$$A \times A' \xrightarrow{b} B$$

is a G-pairing if it is bi-additive, and respects the action of G:

$$b(\sigma a, \sigma a') = \sigma b(a, a').$$

Such a pairing induces a map $\tilde{b}$

$$\cup : H^r(G, A) \times H^s(G, A') \xrightarrow{\tilde{b}} H^{r+s}(G, B),$$

as follows: given cochains f and f', one defines (for a given b) a cochain $f \cup f'$ by

$$(f \cup f')(\sigma_1, \ldots, \sigma_{r+s}) = b(f(\sigma_1, \ldots, \sigma_r), \sigma_1 \ldots \sigma_r f'(\sigma_{r+1}, \ldots, \sigma_{r+s})),$$

and checks the rule
$$\delta(f \cup f') = \delta f \cup f' + (-1)^r f \cup \delta f'.$$
If $\delta f = \delta f' = 0$, then also $\delta(f \cup f') = 0$; i.e., if f and f' are cocycles, so is $f \cup f'$. Similarly one checks that the cohomology class of $f \cup f'$ depends only on the classes of f and f'. Thus we obtain the desired pairing b.

If $r = 0$ and $a \in A^G$ is fixed, then $a' \mapsto b(a, a')$ defines a G-homomorphism $\varphi_a : A' \to B$, and $\alpha' \mapsto a \cup \alpha'$ is the map $H^r(G, A') \to H^r(G, B)$ induced by φ_a.

If H is a subgroup of G, and $\alpha \in H^r(G, A)$ and $\beta \in H^s(H, A')$, then we can form
$$\mathrm{res}(\alpha) \cup \beta \in H^{r+s}(H, B).$$
Suppose that the index of H in G is finite, so that corestriction is defined; then one can show that
$$\mathrm{cores}(\mathrm{res}(\alpha) \cup \beta) = \alpha \cup \mathrm{cores}(\beta) \in H^{r+s}(G, B).$$

Duality for finite modules

If A and B are G-modules, we make the group $\mathrm{Hom}_{\mathbf{Z}}(A, B)$ into a G-module by defining $(\sigma f)(a) = \sigma(f(\sigma^{-1} a))$. Note then that $\mathrm{Hom}_G(A, B) = (\mathrm{Hom}_{\mathbf{Z}}(A, B))^G$. Also, the obvious pairing $A \times \mathrm{Hom}_{\mathbf{Z}}(A, B) \to B$ is a G-pairing. The canonical map

$$(*) \qquad A \to \mathrm{Hom}_{\mathbf{Z}}(\mathrm{Hom}_{\mathbf{Z}}(A, B))$$

is a G-homomorphism. In case A is finite, killed by m, and B has a unique cyclic subgroup of order m, the map $(*)$ is an isomorphism; one can thus recover A from its "dual" $\mathrm{Hom}_{\mathbf{Z}}(A, B)$ which has the same order as A. There are two especially important such duals for finite A.

- The *Pontrjagin Dual* of A is $\mathrm{Hom}_{\mathbf{Z}}(A, \mathbf{Q}/\mathbf{Z})$; this equals $\mathrm{Hom}_{\mathbf{Z}}(A, \mathbf{Z}/m\mathbf{Z})$ if $mA = 0$.
- The *Cartier Dual* of A is $\mathrm{Hom}_{\mathbf{Z}}(A, \mu(k^{\mathrm{sep}}))$; this equals $\mathrm{Hom}_{\mathbf{Z}}(A, \mu_m(k^{\mathrm{sep}}))$ if $mA = 0$.

In the Pontrjagin case, G is an arbitrary profinite group and acts trivially on $\mathbf{Q}/\mathbf{Z}$. Taking limits, this duality extends to a perfect duality (i.e., an antiequivalence of categories) between discrete abelian torsion groups and profinite abelian groups.

In the Cartier case, $G = \mathrm{Gal}(k^{\mathrm{sep}}/k)$ or some quotient thereof, and $m \neq 0$ in k. (The Cartier dual of a p-group in characteristic p is a *group scheme*, not just a Galois module.) If E is an elliptic curve over k and the image of m in k is nonzero, the *Weil pairing* $E_m(k^{\mathrm{sep}}) \times E_m(k^{\mathrm{sep}}) \to \mu_m$ identifies E_m with its Cartier dual.

Local fields

Let k be a local field, i.e., the field of fractions of a complete discrete valuation ring with finite residue field F. Let K be a finite extension of k.

Fundamental facts:
$$H^1(K/k, K^*) = 0 \qquad \text{(Hilbert's Theorem 90)}$$
$$H^2(K/k, K^*) = \mathbf{Z}/[K:k]\mathbf{Z}$$
$$H^2(k, \mathbf{G}_m) = \mathrm{Br}(k) = \mathbf{Q}/\mathbf{Z}$$

The equality $\mathrm{Br}(k) = \mathbf{Q}/\mathbf{Z}$ is given canonically, by the Hasse invariant, as follows: The group $\mathrm{Br}(k)$ is the Brauer group, defined in §. Consider the inflation-restriction sequence for $H^2(-, \mathbf{G}_m)$ in the tower of fields

where k^{ur} is the maximal unramified extension of k. Since every central division algebra over a local field has an unramified splitting field, we have $\mathrm{Br}(k^{\mathrm{ur}}) = 0$, and hence an isomorphism

$$\mathrm{Br}(k) \cong H^2(k^{\mathrm{ur}}/k, \mathbf{G}_m) = H^2(\mathrm{Frob}^{\widehat{\mathbf{Z}}}, (k^{\mathrm{ur}})^*).$$

Using the exact sequence

$$0 \longrightarrow U(k^{\mathrm{ur}}) \longrightarrow (k^{\mathrm{ur}})^* \xrightarrow{\text{valuation}} \mathbf{Z} \longrightarrow 0$$

and the fact that the unit group of an unramified extension has trivial cohomology in dimension $\neq 0$, we find that we can replace $(k^{\mathrm{ur}})^*$ by $\mathbf{Z}$, and hence

$$\mathrm{Br}(k) \cong H^2(\widehat{\mathbf{Z}}, \mathbf{Z}) = H^1(\widehat{\mathbf{Z}}, \mathbf{Q}/\mathbf{Z}) = \mathbf{Q}/\mathbf{Z};$$

the middle equality comes from the short exact sequence

$$0 \to \mathbf{Z} \to \mathbf{Q} \to \mathbf{Q}/\mathbf{Z} \to 0$$

and the fact that $\mathbf{Q}$, being uniquely divisible, has trivial cohomology in nonzero dimensions. The resulting map

$$\mathrm{Br}(k) \to \mathbf{Q}/\mathbf{Z}$$

is the called the Hasse invariant.

Theorem 0.15. *Let A be a finite G_k-module of order prime to the characteristic of k. Let*

$$A^* = \mathrm{Hom}(A, \mathbf{G}_m) = \mathrm{Hom}(A, \mu(\overline{k}))$$

be the Cartier dual of A. Then the G-pairing

$$A \times A^* \to \overline{k}^*$$

induces a pairing

$$H^r(k, A) \times H^{2-r}(k, A^*) \to H^2(G_k, (k^{\mathrm{sep}})^*) = \mathrm{Br}(k) = \mathbf{Q}/\mathbf{Z}.$$

This is a perfect pairing of finite groups, for all $r \in \mathbf{Z}$. It is nontrivial only if $r = 0, 1, 2$, since for $r \geq 3$,

$$H^r(k, A) = 0 \quad \text{for all } A,$$

i.e., "the cohomological dimension of a non-archimedean local field is 2."

Example. By Kummer theory, we have

$$k^*/(k^*)^m = H^1(k, \mu_m(\overline{k})).$$

Thus there is a perfect pairing

$$H^1(k, \mathbf{Z}/m\mathbf{Z}) \quad \times \quad H^1(k, \mu_m(\overline{k})) \longrightarrow \mathbf{Q}/\mathbf{Z}$$
$$\| \qquad\qquad\qquad \|$$
$$\mathrm{Hom}(G_k, \mathbf{Z}/m\mathbf{Z}) \quad \times \quad k^*/(k^*)^m$$

The left hand equality is because the action is trivial. Conclusion:

$$G_k^{\mathrm{ab}}/(G_k^{\mathrm{ab}})^m \cong k^*/(k^*)^m.$$

Taking the limit gives *Artin reciprocity*:

$$k^* \hookrightarrow G_k^{ab} ;$$

the image is dense.

Let E/k be an elliptic curve. In some sense,

$$E = \mathrm{Ext}^1(E, \mathbf{G}_m)$$

in the category of algebraic groups. There is a pairing

$$H^r(k, E) \times H^s(k, E) \to H^{r+s+1}(k, \mathbf{G}_m).$$

For example, taking $r = 0$ and $s = 1$, we have the following theorem.

Theorem 0.16. *Let E be an elliptic curve over a non-archimedean local field k, then we have the following perfect pairing between Pontrjagin duals.*

$$H^0(k, E) \quad \times \quad H^1(k, E) \longrightarrow H^2(k, \mathbf{G}_m) = \mathbf{Q}/\mathbf{Z}$$
$$\| \qquad\qquad\qquad \|$$
$$\begin{array}{c} E(k) \\ \textit{profinite} \end{array} \quad \times \quad \begin{array}{c} H^1(k, E) \\ \textit{discrete, torsion} \end{array}$$

Sketch of Proof. We use the Weil pairing. Letting D denote "Pontrjagin dual", we have a diagram

$$\begin{array}{ccccccccc}
0 & \longrightarrow & E(k)/mE(k) & \longrightarrow & H^1(k, E_m) & \longrightarrow & H^1(k, E)_m & \longrightarrow & 0 \\
& & \downarrow & & \downarrow & & \downarrow & & \\
0 & \longrightarrow & H^1(k, E)_m^D & \longrightarrow & H^1(k, E_m)^D & \longrightarrow & (E(k)/mE(k))^D & \longrightarrow & 0
\end{array}$$

The rows are exact. The top one from the Kummer sequence, and the bottom is the dual of the top one. The middle vertical arrow is an isomorphism by Theorem 0.15. The outside vertical arrows are induced by the pairing of Theorem 0.16. The diagram commutes, so they are also isomorphisms, and Theorem 0.16 follows by passage to the limit with more and more divisible m. □

It was in trying to prove Theorem 0.16 that I was led to Theorem 0.15 in the late 1950's. Of course the "fundamental facts" and the Artin isomorphism are a much older story.

BIBLIOGRAPHY

1. J. W. S. Cassels and A. Fröhlich (eds.), *Algebraic number theory*, London, Academic Press Inc. [Harcourt Brace Jovanovich Publishers], 1986, Reprint of the 1967 original.
2. K. Haberland, *Galois cohomology of algebraic number fields*, VEB Deutscher Verlag der Wissenschaften, Berlin, 1978, With two appendices by Helmut Koch and Thomas Zink.
3. V. A. Kolyvagin, *On the Mordell-Weil group and the Shafarevich-Tate group of modular elliptic curves*, Proceedings of the International Congress of Mathematicians, Vol. I, II (Kyoto, 1990) (Tokyo), Math. Soc. Japan, 1991, pp. 429–436.
4. K. Rubin, *The "main conjectures" of Iwasawa theory for imaginary quadratic fields*, Invent. Math. **103** (1991), no. 1, 25–68.
5. J-P. Serre, *Local fields*, Springer-Verlag, New York, 1979, Translated from the French by Marvin Jay Greenberg.
6. _____, *Galois cohomology*, Springer-Verlag, Berlin, 1997, Translated from the French by Patrick Ion and revised by the author.

The Arithmetic of Modular Forms

Wen-Ching Winnie Li

The Arithmetic of Modular Forms
Wen-Ching Winnie Li[1]

Introduction

The 1999 IAS/Park City Mentoring Program for women took place on May 17-27 at the Institute for Advanced Study, Princeton, NJ. I was the organizer of the graduate program. In order to be consistent with the central theme "The Arithmetic of Elliptic Curves" of the graduate and research component of PCMI this year, the title of the graduate Mentoring Program was chosen to be "The Arithmetic of Elliptic Curves, Modular Forms, and Calabi-Yau Varieties". The purpose was to give a preview of the main subjects to be discussed at Park City on the one hand, and on the other hand, we diverged from their theme by discussing the higher-dimensional analogue connected to geometry and physics. The lectures were delivered by three mathematicians on the following subjects:

1. Wen-Ching W. Li: The arithmetic of modular forms (3 lectures),
2. Noriko Yui: Arithmetic of certain Calabi-Yau varieties and mirror symmetry (3 lectures),
3. Alice Silverberg: Conjectures about elliptic curves (2 lectures).

What follows is the written account of the three lectures I delivered. As they are the beginning of the eight lectures of the program, the first lecture serves as an overall introduction of the theme of the program. The second lecture surveys the theory of classical modular forms and discusses certain proved and unproved conjectures in the area. The third lecture draws connections among modular forms, elliptic curves, and representations of Galois groups with the common thread being the associated L-functions. Comparisons with progress made for automorphic forms of GL_2 over function fields are also given there.

---

[1] Department of Mathematics, Pennsylvania State University, University Park, PA 16803.
E-mail address: wli@math.psu.edu.

These lecture notes were written while the author was visiting the National Center for Theoretical Sciences, Mathematics Division, Hsinchu, Taiwan, partially supported by the NSC in Taiwan, and the Institute for Advanced Study, Princeton, NJ, partially supported by Ellentuck Fund. She would like to thank both institutions for their hospitalities. This research was supported in part by the NSF grants DMS96-22938 and DMS99-70651.

©2001 American Mathematical Society

LECTURE 1
Introduction to Elliptic Curves, Modular Forms, and Calabi-Yau Varieties

1.1. Classical Modular Forms

The group
$$SL(2,\mathbb{Z}) = \{\begin{pmatrix} a & b \\ c & d \end{pmatrix} : a,b,c,d \in \mathbb{Z}, \quad ad - bc = 1\}$$
acts on the Poincaré upper half-plane
$$\mathcal{H} = \{z \in \mathbb{C} : \Im z > 0\}$$
by fractional linear transformations, that is, $\gamma = \begin{pmatrix} a & b \\ c & d \end{pmatrix}$ sends z to $\gamma z = \frac{az+b}{cz+d}$. A typical fundamental domain $\mathcal{D}$ of $\mathcal{H}/SL(2,\mathbb{Z})$ is the part of $\mathcal{H}$ which is inside the vertical strip from $x = -\frac{1}{2}$ to $x = \frac{1}{2}$ but outside the unit circle centered at the origin; on the vertical boundaries $-\frac{1}{2} + iy$ is identified with $\frac{1}{2} + iy$ for $y \geq \sqrt{3}/2$, and on the circular boundary $e^{i\theta}$ is identified with $e^{i(\pi-\theta)}$ for $\pi/3 \leq \theta \leq \pi/2$. Under the hyperbolic metric $\frac{dxdy}{y^2}$ on $\mathcal{H}$, the distance between $-\frac{1}{2} + iy$ and $\frac{1}{2} + iy$ approaches zero as the height y tends to infinity. Hence the fundamental domain $\mathcal{D}$ looks like a sphere with the point $i\infty$ removed. We call $i\infty$ a cusp of the group $SL(2,\mathbb{Z})$. The image of $i\infty$ under the action of $SL(2,\mathbb{Z})$ consists of $i\infty$ and all rational numbers.

Let Γ be a subgroup of $SL(2,\mathbb{Z})$ of finite index. By gluing together images of $\mathcal{D}$ under suitably chosen representatives of Γ in $SL(2,\mathbb{Z})$ we obtain a fundamental domain for $\mathcal{H}/\Gamma$. We may compactify $\mathcal{H}/\Gamma$ by adding the cusps of Γ which are the orbits of Γ in $SL(2,\mathbb{Z})(i\infty)$. The compactified Riemann surface, $\widehat{\mathcal{H}/\Gamma}$, is a curve of genus g over $\mathbb{C}$. It is called a modular curve when Γ is a congruence subgroup of $SL(2,\mathbb{Z})$.

Let $f(z)dz$ be a holomorphic differential on $\widehat{\mathcal{H}/\Gamma}$. Since $f(\gamma z)d(\gamma z) = f(z)dz$ and $d(\gamma z) = (cz+d)^{-2}dz$ for $\gamma = \begin{pmatrix} a & b \\ c & d \end{pmatrix}$, we find
$$f(\gamma z) = (cz+d)^2 f(z).$$

Further, the holomorphicity of the differential implies that $f(z)$ is holomorphic on $\mathcal{H}$ and at the cusps of Γ, and, moreover, it also vanishes at each cusp of Γ. The last assertion requires some explanation. Since Γ has finite index in $SL_2(\mathbb{Z})$, let M be the smallest positive integer such that Γ contains translations by M. We may choose $q = e^{2\pi i z/M}$ to be a uniformizing element at the cusp $i\infty$. Then $dz = \frac{M}{2\pi i}\frac{dq}{q}$ and the holomorphicity of $f(z)dz = f(q)\frac{M}{2\pi i}\frac{dq}{q}$ at $i\infty$ implies that f must vanish at $i\infty$. The same happens at other cusps. We call such f a cusp form of weight 2 for Γ. In general, given an integer k, we call a complex-valued function f on $\mathcal{H}$ a *modular form* of weight k for Γ if

(1.1) f is holomorphic on $\mathcal{H}$ and at all cusps of Γ;

(1.2) $f(\gamma z) = (cz+d)^k f(z)$ for all $\gamma = \begin{pmatrix} a & b \\ c & d \end{pmatrix}$ in Γ.

If f satisfies the additional condition

(1.3) f vanishes at all cusps of Γ,

then we call f a *cusp form* of weight k for Γ. Thus cusp forms of positive even weight $2k$ can be thought of as holomorphic k-tensors on $\widehat{\mathcal{H}/\Gamma}$.

The subgroups of particular interest to us are congruence subgroups of the following type, involving a positive integer N:

$$\Gamma_0(N) = \{\begin{pmatrix} a & b \\ c & d \end{pmatrix} \in SL_2(\mathbb{Z}) : c \equiv 0 \pmod{N}\},$$

$$\Gamma_1(N) = \{\begin{pmatrix} a & b \\ c & d \end{pmatrix} \in SL_2(\mathbb{Z}) : c \equiv 0 \pmod{N}, a \equiv d \equiv 1 \pmod{N}\},$$

$$\Gamma(N) = \{\begin{pmatrix} a & b \\ c & d \end{pmatrix} \in SL_2(\mathbb{Z}) : b \equiv c \equiv 0 \pmod{N}, a \equiv d \equiv 1 \pmod{N}\}.$$

It turns out that in order to study modular forms for the last two subgroups, it suffices to study forms of weight k level N and Dirichlet character χ (cf. [**L1**]), in other words, forms f satisfying (1.1) and

(1.2)' $f(\gamma z) = \chi(d)(cz+d)^k f(z)$ for all $\gamma = \begin{pmatrix} a & b \\ c & d \end{pmatrix}$ in $\Gamma_0(N)$.

Of course cusp forms should satisfy condition (1.3). Note that if χ is the trivial character, then these are forms of weight k for $\Gamma_0(N)$. Denote by $\mathcal{M}(k, N, \chi)$ (resp. $\mathcal{C}(k, N, \chi)$) the space of such modular (resp. cusp) forms. It is a finite-dimensional vector space over $\mathbb{C}$. In particular, a form f in $\mathcal{M}(k, N, \chi)$ is invariant under the translation $z \longmapsto z+1$, hence it has a Fourier expansion

$$f(z) = \sum_{n \geq 0} a_n e^{2\pi i n z},$$

and $a_0 = 0$ if f is a cusp form. The Dirichlet series attached to f is

$$D(f, s) = \sum_{n \geq 1} a_n n^{-s}.$$

We are interested in the arithmetic properties of a_n and the analytic properties of $D(f, s)$.

1.2. Elliptic Curves

An elliptic curve over $\mathbb{C}$ is $\mathbb{C}/L$, where L is a lattice in $\mathbb{C}$, in other words, $L = \mathbb{Z}\omega_1 + \mathbb{Z}\omega_2$ with ω_1 and ω_2 linearly independent over $\mathbb{R}$. Two elliptic curves $\mathbb{C}/L$ and $\mathbb{C}/L'$ are equivalent if there is an analytic group isomorphism g from the first to the second. From $g(z+z') = g(z) + g(z')$ for all z, z' in $\mathbb{C}$ and g being analytic and one-to-one we conclude that $g(z) = uz$ for some $u \in \mathbb{C}^\times$. Hence $\mathbb{C}/L$ and $\mathbb{C}/L'$ are equivalent if and only if $L' = uL$ for some element $u \in \mathbb{C}^\times$, which in turn means that L and L' are two equivalent lattices. In an equivalence class of a given lattice, there exists a lattice with basis $z, 1$, where $z \in \mathcal{H}$. Further, two lattices with basis $z, 1$ and $z', 1$ are equivalent if and only if $z = \gamma(z')$ for some element $\gamma \in SL_2(\mathbb{Z})$. This shows that the quotient $\mathcal{H}/SL_2(\mathbb{Z})$ parametrizes equivalence classes of elliptic curves defined over $\mathbb{C}$.

The group of points in an elliptic curve $E = \mathbb{C}/L$ with order dividing N is

$$E[N] := \frac{1}{N}L/L \cong (\mathbb{Z}/N\mathbb{Z}) \times (\mathbb{Z}/N\mathbb{Z}).$$

There is a pairing, called Weil pairing, from $E[N] \times E[N]$ to the group of N-th roots of unity. The reader is referred to Chap 3 of [**Si**] for more details. The triple $(E; P, Q)$ denotes an elliptic curve E with two chosen generators P and Q of $E[N]$. Two triples $(E; P, Q)$ and $(E'; P', Q')$ are equivalent if there is an isomorphism from E to E' mapping P to P' and Q to Q'. Similarly we define equivalence of $(E; P)$ where P is a point in E of order N, and equivalence of $(E; C)$ where C is a cyclic subgroup of E of order N. Like elliptic curves, these equivalence classes are parametrized by modular curves.

Theorem 1.1. *$\mathcal{H}/\Gamma(N)$ (resp. $\mathcal{H}/\Gamma_1(N)$, $\mathcal{H}/\Gamma_0(N)$) parametrizes the equivalence classes of $(E; P, Q)$ with a fixed value of the Weil pairing of P and Q (resp. $(E; P)$, $(E; C)$).*

Hence the modular curves $\mathcal{H}/\Gamma(N)$, $\mathcal{H}/\Gamma_1(N)$, and $\mathcal{H}/\Gamma_0(N)$ are moduli spaces of elliptic curves endowed with various level structures.

The Riemann surface $\widehat{\mathcal{H}/SL_2(\mathbb{Z})}$ has genus zero, therefore the space of meromorphic functions on this surface is a rational function field over $\mathbb{C}$. Traditionally one chooses the generator j to be the function with Fourier expansion

$$j(q) = \frac{1}{q} + 744 + 196884q + c_2 q^2 + \cdots, \qquad q = e^{2\pi i z}.$$

Thus j is a modular form for $SL_2(\mathbb{Z})$ of weight zero. More precisely, it is a holomorphic function on $\mathcal{H}$ with a simple pole at the cusp $i\infty$, and it induces a bijection from $\widehat{\mathcal{H}/SL_2(\mathbb{Z})}$ to the projective line $P^1(\mathbb{C})$. In other words, it parametrizes the equivalence classes of elliptic curves over $\mathbb{C}$; $j(z)$ is called the j-invariant of the elliptic curve arising from the lattice with basis $z, 1$.

Using the Weierstrass $\mathcal{P}$-function $\mathcal{P}_L(u)$ attached to a lattice L, one obtains an algebraic model of $E = \mathbb{C}/L$ given by

$$y^2 = 4x^3 - g_4(L)x - g_6(L)$$

with the correspondence $u \in \mathbb{C}/L$ mapping to $(x,y) = (\mathcal{P}_L(u), \mathcal{P}'_L(u))$. Here

$$g_4(L) = 60 \sum_{\alpha \in L, \alpha \neq 0} \frac{1}{\alpha^4}, \qquad g_6(L) = 140 \sum_{\alpha \in L, \alpha \neq 0} \frac{1}{\alpha^6}.$$

The j-invariant of E is equal to $1728 g_4(L)^3/\Delta(L)$, where $\Delta = g_4^3 - 27 g_6^2$ is a cusp form of weight 12 for $SL_2(\mathbb{Z})$, g_4 and g_6 are modular forms, but not cusp forms, for $SL_2(\mathbb{Z})$ of weight 4 and 6, respectively. They are called Eisenstein series. The form Δ has the following infinite product expression

$$\Delta(z) = (2\pi)^{12} q \prod_{n \geq 1} (1-q^n)^{24} = (2\pi)^{12} \eta(z)^{24},$$

with η being the Dedekind eta function.

The algebraic geometric definition of an elliptic curve defined over a field K is an irreducible nonsingular projective curve of genus 1 with a marked K-rational point. It can be described as solutions of an equation of the form

(1.2.1) $$y^2 z + a_1 xyz + a_3 yz^2 = x^3 + a_2 x^2 z + a_4 xz^2 + a_6 z^3,$$

where coefficients $a_1, ..., a_6$ are in K. In this model, $[0:1:0]$ is the marked point.

1.3. Zeta Functions

Let V be a nonsingular projective variety of dimension d defined over a finite field k of q elements. Assume further that V is geometrically irreducible, that is, viewed as a variety $\bar{V}$ defined over an algebraic closure $\bar{k}$ of k, $\bar{V}$ is irreducible. Denote by $\bar{N}_n$ the number of points of V over a degree n field extension of k. The zeta function attached to V is defined as

$$Z_V(U) = exp(\sum_{n \geq 1} \bar{N}_n \frac{U^n}{n}).$$

In 1949 Weil conjectured that $Z_V(U)$ is a quotient of two polynomials with coefficients in $\mathbb{Q}$, it satisfies the functional equation

$$Z_V(\frac{1}{q^d U}) = \pm (q^{d/2} U)^c Z_V(U),$$

for some integer c, and it satisfies the Riemann hypothesis (cf. [**W1**]). The rationality and functional equation were established in 1960 by Dwork [**DW**], while the Riemann hypothesis was not proved until 1974 by Deligne [**D1**]. Particularly, Deligne showed that, if $\Phi^{(i)}$ is the map on the ith ℓ-adic cohomology group $H^i(\bar{V}, \mathbb{Q}_\ell)$ induced from the Frobenius map on $\bar{V}$, and

$$P_i(U) = \det(1 - \Phi^{(i)} U)$$

is the characteristic polynomial of the inverse of $\Phi^{(i)}$, then

$$Z_V(U) = \frac{P_1(U) P_3(U) \cdots P_{2d-1}(U)}{P_0(U) P_2(U) \cdots P_{2d}(U)},$$

and for $0 \leq i \leq 2d$,

$$P_i(U) = \prod_{1 \leq j \leq B_i} (1 - \alpha_{ij} U)$$

with α_{ij} algebraic integers of absolute value $q^{i/2}$. For curves of genus 1, this was proved by Hasse [**Ha**], and arbitrary genus by Weil [**W2**].

Let E be an elliptic curve defined by a Weierstrass equation (1.2.1) with coefficients in $\mathbb{Z}$. Then for all except finitely many primes p, the equation (1.2.1) mod p also defines an elliptic curve E_p over the field $\mathbb{Z}/p\mathbb{Z}$. Its zeta function has the form

$$Z_{E_p}(U) = \frac{1 - a_p U + pU^2}{(1-U)(1-pU)},$$

where a_p is equal to $p+1$ minus the number of $\mathbb{Z}/p\mathbb{Z}$-rational points on E_p. The conductor N of E records the bad reductions of E so that a prime p divides N if and only if E mod p is not an elliptic curve. The Hasse-Weil zeta function of E is defined as

$$Z(E, s) = \prod_{p \nmid N} \frac{1}{1 - a_p p^{-s} + p^{1-2s}} \prod_{p | N} \frac{1}{1 - a_p p^{-s}},$$

where, at a prime p dividing N, $a_p = 0, 1$ or -1 depending on the reduction type of E at p. Express $Z(E, s)$ as a Dirichlet series

$$Z(E, s) = \sum_{n \geq 1} a_n n^{-s}.$$

The conjecture of Taniyama-Shimura-Weil asserts that for an elliptic curve E defined over $\mathbb{Q}$ with conductor N and the Hasse-Weil zeta-function $Z(E, s) = \sum_{n \geq 1} a_n n^{-s}$, the function defined by

$$f_E(z) = \sum_{n \geq 1} a_n e^{2\pi i n z}$$

is a cusp form of weight 2 for the group $\Gamma_0(N)$. Wiles [**Wi**] and Taylor-Wiles [**TW**] proved this conjecture for semi-stable elliptic curves, which, together with earlier work by Ribet [**R**] and others, enabled Wiles to prove Fermat's Last Theorem. This conjecture is now completely established via the joint efforts by Breuil, Conrad, Diamond, and Taylor [**BCDT**]. The reader can consult several articles in this volume for more details on this topic.

1.4. Calabi-Yau Varieties

As we venture geometrically into higher dimensional varieties, Calabi-Yau varieties are a natural object to study. A Calabi-Yau variety is a smooth projective variety V of dimension n with trivial canonical bundle and trivial cohomology groups $H^i(V, \mathcal{O}_V) = 0$ for $0 < i < n$. One dimensional Calabi-Yau varieties are elliptic curves; two dimensional ones are $K3$ surfaces. Calabi-Yau 3-folds have been important in physics. Noriko Yui will introduce Calabi-Yau varieties and mirror symmetry. The reader is referred to her lectures [**Yui**] for more details and references. Here we give a brief account of the connection with modular forms.

For example, equation (1.2.1) with coefficients $a_1, ..., a_6$ in $\mathbb{Z}[t]$ and with non-constant discriminant $\Delta(t)$ defines a $K3$ surface V over $\mathbb{Q}$. Clearly this $K3$ surface admits a fibration by elliptic curves. Under mild assumptions, equation (1.2.1) modulo p defines a surface over $\mathbb{Z}/p\mathbb{Z}$, which, after desingularization, is also a $K3$

surface, which we denote by V_p. The vanishing condition on cohomology implies that its zeta function has the form

$$Z_{V_p}(U) = \frac{1}{(1-U)(1-p^2U)P_2(U)},$$

where $P_2(U)$ is a polynomial of degree 22 and $P_2(0) = 1$. Using geometry, one can factor $P_2(U)$ into two pieces:

$$P_2(U) = Q_p(U)R_p(U),$$

where $R_p(U)$ arises from algebraic cycles and it satisfies $R_p(\frac{U}{p}) \in \mathbb{Z}[U]$, while $Q_p(U)$ arises from transcendental cycles. Under suitable conditions, the part $H^{1,1}$ in the Hodge decomposition of H^2 as $H^{2,0} \bigoplus H^{1,1} \bigoplus H^{0,2}$ has dimension 20 and $Q_p(U)$ has degree 2 for all primes p not dividing some integer N. The general Langlands philosophy predicts that there is a way to define the factors $Q_p(U)$ at the places p dividing N such that the zeta function attached to V

$$Z(V, s) = \prod_p \frac{1}{Q_p(p^{-s})} = \sum_{n \geq 1} a_n n^{-s}$$

is the Dirichlet series attached to a cusp form. Some cases have been worked out by various authors. Here we exhibit a few examples from the paper [**SB**] by Stienstra and Beukers, with the affine part of the $K3$ surfaces defined by

A : $\quad s^2 = xy(x-y)(y-1)(1-x),$
B : $\quad s^2 = -xy(x+y)(y+1)(1+x),$
C : $\quad s^2 = xy(x+y+1)(xy+y+x),$

respectively. In each case the weight of the cusp form f is 3, as a result of the Ramanujan-Petersson conjecture for cusp forms (proved by Deligne) and the Riemann hypothesis, and its level is determined by the lattice of the transcendental cycle. Indeed, the corresponding f for each case was computed as follows:

A : $\quad f(z) = \eta(4z)^6 \in \mathcal{C}(3, 16, \varepsilon_{16}),$
B : $\quad f(z) = \eta(z)^2\eta(2z)\eta(4z)\eta(8z)^2 \in \mathcal{C}(3, 8, \varepsilon_8),$
C : $\quad f(z) = \eta(2z)^3\eta(6z)^3 \in \mathcal{C}(3, 12, \varepsilon_{12}).$

Here ε_m is a quadratic character of $\mathbb{Z}$ mod m, and $\mathcal{C}(k, N, \varepsilon_m)$ denotes the space of cusp forms of weight k level N and character ε_m.

Similarly, for a Calabi-Yau 3-fold, the interesting part is $P_3(U)$ occurring in the zeta function of the mod p Calabi-Yau 3-fold. In special cases, one expects to find cusp forms of weight 4 related to the Hasse-Weil zeta-function of the 3-fold. These are some higher dimensional analogues of the Taniyama-Shimura-Weil conjecture.

Over $\mathbb{Q}$, elliptic curves are the best studied geometric object for the connection between global zeta functions arising from geometry and automorphic forms. More is known over a function field. This will be discussed in the third lecture.

LECTURE 2
The Arithmetic of Modular Forms

2.1. Eisenstein Series

Let Γ be $\Gamma_0(N)$, or $\Gamma_1(N)$, or $\Gamma(N)$. Hecke introduced a way to construct certain modular forms of weight $k \geq 3$ for Γ as follows. For each $\delta \in SL_2(\mathbb{Z})$, define

$$E_\delta(z) = \sum_{\begin{pmatrix} a & b \\ c & d \end{pmatrix} \in \delta(\Gamma/\Gamma_\infty)} \frac{1}{(cz+d)^k},$$

where Γ_∞ is the subgroup of translations contained in Γ. The infinite sum converges absolutely to a modular form E_δ of weight k for Γ, called an Eisenstein series. These Eisenstein series E_δ generate a subspace $\mathcal{E}(k,\Gamma)$ of the space $\mathcal{M}(k,\Gamma)$ of modular forms of weight k for Γ with dimension equal to the number of cusps of Γ. For $k=2$ or 1, some modification is needed to take care of the convergence problem, and the dimension of the space of Eisenstein series is reduced accordingly. In all cases Hecke showed that, for $k \geq 1$, the space $\mathcal{M}(k,\Gamma)$ of modular forms decomposes as a direct sum of the space of Eisenstein series and the space of cusp forms:

$$\mathcal{M}(k,\Gamma) = \mathcal{E}(k,\Gamma) \bigoplus \mathcal{C}(k,\Gamma).$$

We also have a similar decomposition of the space $\mathcal{M}(k,N,\chi)$ of modular forms of weight k level N and character χ:

$$\mathcal{M}(k,N,\chi) = \mathcal{E}(k,N,\chi) \bigoplus \mathcal{C}(k,N,\chi).$$

Here each space is finite dimensional. The space of Eisenstein series $\mathcal{E}(k,N,\chi)$ is fully understood in the following sense. Take two Dirichlet characters χ_1 and χ_2 of $\mathbb{Z}$ with conductor N_1 and N_2 respectively. Let $\chi = \chi_1\chi_2$. Using the associated Dirichlet L-functions

$$L(\chi_i, s) = \sum_{n \geq 1} \chi_i(n) n^{-s}, \quad i = 1, 2,$$

we write the product

$$L(\chi_1,s)L(\chi_2,s-k+1) = \sum_{n\geq 1} a_n n^{-s},$$

as a Dirichlet series and use the coefficients to define

$$E_{\chi_1,\chi_2}(z) = -L(\chi_1,0)L(\chi_2,-k+1) + \sum_{n\geq 1} a_n e^{2\pi i n z}.$$

It can be shown that E_{χ_1,χ_2} lies in the space of Eisenstein series of weight k level $N_1 N_2$ and character χ. Notice that the twisted Bernoulli numbers appear in the constant term of E_{χ_1,χ_2}. As (χ_1,χ_2) runs through all such pairs with $N_1 N_2$ dividing N and $\chi_1 \chi_2 = \chi$, the functions $E_{\chi_1,\chi_2}(dz)$ with d ranging through the factors of $N/N_1 N_2$ span the whole space $\mathcal{E}(k,N,\chi)$. Hence our main concern is the structure of the space of cusp forms $\mathcal{C}(k,N,\chi)$.

2.2. Hecke Operators

Given $f(z) = \sum_{n\geq 0} a_n e^{2\pi i n z} \in \mathcal{M}(k,N,\chi)$, a natural question concerning the Fourier coefficients of f is to know when the arithmetic function $n \mapsto a_n$ is multiplicative with respect to a prime p, that is, when $a_{mp^r} = a_m a_{p^r}$ for all m prime to p and r positive. To formulate this question in another way, introduce the associated Dirichlet series

$$D(f,s) = \sum_{n\geq 1} a_n n^{-s}.$$

We say that $D(f,s)$ is *Eulerian at* p if there are complex numbers $c_1, c_p, c_{p^2}, \ldots$ such that

$$D(f,s) = \left(\sum_{p\nmid n} a_n n^{-s}\right)(c_1 + c_p p^{-s} + c_{p^2} p^{-2s} + \cdots).$$

In other words, $a_{np^r} = a_n c_{p^r}$ for all n prime to p. To answer this question, Hecke introduced the operator T_p which sends $f(z) = \sum_{n\geq 0} a_n e^{2\pi i n z}$ to

$$(T_p f)(z) = \sum_{n\geq 0}(a_{np} + \chi(p)p^{k-1} a_{n/p}) e^{2\pi i n z}.$$

Here we used the convention that $a_x = 0$ if x is not an integer. It is easily shown that T_p is an operator on $\mathcal{M}(k,N,\chi)$ preserving the subspace of cusp forms.

Theorem 2.1. (Hecke) *Let f be a nonzero form in $\mathcal{M}(k,N,\chi)$. Let p be a prime not dividing N. Then $D(f,s)$ is Eulerian at p if and only if f is an eigenfunction of T_p. In this case, the Euler factor at p has the form*

$$\sum_{r\geq 0} c_{p^r} p^{-rs} = \frac{1}{1 - c_p p^{-s} + \chi(p) p^{k-1-2s}},$$

where c_p is the eigenvalue, that is, $T_p f = c_p f$.

Hecke's work on modular forms can be found in his collected papers [**H**].

Let Γ be a subgroup of $SL_2(\mathbb{Z})$ of finite index. Denote by $\overline{\Gamma}$ its image in $PSL_2(\mathbb{Z})$. Let $\mathcal{D}(\Gamma)$ be a fundamental domain for Γ. Petersson introduced a Hermitian inner product $<\ ,\ >$ on forms of weight k for Γ via

$$<f,g> = \frac{i}{2[PSL_2(\mathbb{Z}):\overline{\Gamma}]} \iint_{\mathcal{D}(\Gamma)} f(z)\overline{g(z)}(\Im z)^{k-2} dz \wedge d\bar{z},$$

which generalizes the usual integral of the exterior product of two holomorphic one forms on $\widehat{\mathcal{H}/\Gamma}$. One can show that the integral is convergent, and even absolutely convergent, if at each cusp of Γ at least one of f, g vanishes. In particular, $<\ ,\ >$ is well-defined on the subspace of cusp forms. Moreover, the value of the inner product $<f,g>$ is independent of the choice of Γ for which both f and g are modular forms.

Theorem 2.2. (Petersson) *Let f and g be two forms in $\mathcal{C}(k, N, \chi)$, and let p be a prime not dividing N. Then*

$$<T_p f, g> = \chi(p) <f, T_p g>.$$

Thus the Hecke operator T_p is normal on $\mathcal{C}(k, N, \chi)$, and hence is diagonalizable. An eigenvalue c_p of T_p satisfies $c_p = \chi(p)\overline{c_p}$. In particular, if χ is trivial, then all eigenvalues of T_p are real. Since T_p commutes with $T_{p'}$ for any two primes p, p' not dividing N, the Hecke operators T_p, $p \nmid N$, on $\mathcal{C}(k, N, \chi)$ can be simultaneously diagonalized. This proves

Corollary 2.3. *The space $\mathcal{C}(k, N, \chi)$ has an orthogonal basis consisting of common eigenfunctions of T_p for all primes $p \nmid N$.*

2.3. The Structure of $\mathcal{C}(k, N, \chi)$– the Theory of Newforms

Observe that if M is a proper divisor of N and g is a form in $\mathcal{C}(k, M, \chi)$, then g is also a form in $\mathcal{C}(k, N, \chi)$ since $\Gamma_0(N)$ is a subgroup of $\Gamma_0(M)$. Moreover, if d divides N/M, then $g(dz)$ lies in $\mathcal{C}(k, dM, \chi)$ and hence is in $\mathcal{C}(k, N, \chi)$. We call $g(dz)$ a "push-up" of g. Denote by $\mathcal{C}^-(k, N, \chi)$ the space generated by functions in $\mathcal{C}(k, M, \chi)$ as M runs through proper divisors of N, as well as their push-ups. A form in $\mathcal{C}^-(k, N, \chi)$ is called an oldform. The space of oldforms is invariant under the Hecke operators T_p with $p \nmid N$. Note that if $g \in \mathcal{C}(k, M, \chi)$ is an eigenfunction of T_p, then so is its push-up $g(dz)$, and both have the same eigenvalue with respect to T_p. Therefore when the space $\mathcal{C}^-(k, N, \chi)$ is decomposed as a direct sum of common eigenspaces of T_p, $p \nmid N$, each common eigenspace has dimension at least two. The orthogonal complement, $\mathcal{C}^+(k, N, \chi)$, of $\mathcal{C}^-(k, N, \chi)$ in $\mathcal{C}(k, N, \chi)$ is invariant under all T_p, $p \nmid N$, and each nonzero form in a common eigenspace of $\mathcal{C}^+(k, N, \chi)$ is called a *newform* of weight k level N and character χ since it genuinely has level N. Thus $\mathcal{C}(k, N, \chi)$ has a basis consisting of newforms of weight k level factors of N and character χ, as well as their push-ups (to level dividing N). The study of cusp forms is now reduced to the study of newforms.

The theory of (classical) newforms is developed by Atkin-Lehner [**AL**], Miyake [**M**], and Li [**L1**]. Here we summarize certain main results below. Refer to chap 7 of [**L2**] for more details.

Theorem 2.4. (2.4.1) Let $f(z) = \sum_{n \geq 1} a_n e^{2\pi i n z}$ be a newform of weight k level N and character χ. Then $a_1 \neq 0$. So we may assume f normalized, that is, $a_1 = 1$. As such, we have $T_p f = a_p f$ for all primes $p \nmid N$.

(2.4.2) (Strong multiplicity one) Let

$$f(z) = \sum_{n \geq 1} a_n e^{2\pi i n z} \quad \text{and} \quad g(z) = \sum_{n \geq 1} b_n e^{2\pi i n z}$$

be two normalized newforms. If $a_p = b_p$ for all except finitely many primes p, then $f = g$.

In particular, each common eigenspace in $\mathcal{C}^+(k, N, \chi)$ is 1-dimensional. This gives another description of newforms of level N.

At each prime q dividing N, there is the U_q operator which sends $f(z) = \sum_{n \geq 1} a_n e^{2\pi i n z} \in \mathcal{C}(k, N, \chi)$ to $\sum_{n \geq 1} a_{nq} e^{2\pi i n z}$ in the same space. Further, U_q's commute with themselves and with all T_p's. Therefore each normalized newform of level N is also an eigenfunction of U_q, with eigenvalue the q-th Fourier coefficient.

Theorem 2.5. Let $f(z) = \sum_{n \geq 1} a_n e^{2\pi i n z}$ be a normalized newform of weight k level N and character χ. Then

(2.5.1) $T_p f = a_p f$ for all primes $p \nmid N$, $U_q f = a_q f$ for all primes $q | N$. In other words, the associated Dirichlet series of f has the form

$$D(f, s) = \sum_{n \geq 1} a_n n^{-s}$$

$$= \prod_{p \nmid N} \frac{1}{1 - a_p p^{-s} + \chi(p) p^{k-1-2s}} \prod_{q | N} \frac{1}{1 - a_q q^{-s}};$$

(2.5.2) $|a_q| = q^{(k-1)/2}$ if χ is not a character of $\mathbb{Z}$ mod N/q;

(2.5.3) If χ is a character of $\mathbb{Z}$ mod N/q, then $a_q = 0$ if $q^2 | N$; and $a_q^2 = \chi(q) q^{k-2}$ if $q^2 \nmid N$.

It is worth remarking that in case $k = 1$, the last case in (2.5.3) never occurs.

The estimate of the size of a_p is a much deeper matter. The Ramanujan-Petersson conjecture asserts that

$$|a_p| \leq 2 p^{(k-1)/2} \quad \text{for} \quad p \nmid N.$$

This was proved by Eichler and Shimura [**S**] for weight $k = 2$, Deligne [**D2**] for $k \geq 2$, and Deligne-Serre [**DS**] for $k = 1$. We state this fact in another way as follows.

Theorem 2.6. Let $f(z) = \sum_{n \geq 1} a_n e^{2\pi i n z}$ be a normalized newform of weight k level N and character χ. Then, for $p \nmid N$, in writing

$$1 - a_p p^{-s} + \chi(p) p^{k-1-2s} = (1 - \alpha_p p^{-s})(1 - \beta_p p^{-s}),$$

we have $|\alpha_p| = |\beta_p| = p^{(k-1)/2}$.

Example. Suppose $\Gamma_0(N)$ has genus 1; for example, $N = 11$. This means that the modular curve $\widehat{\mathcal{H}/\Gamma_0(N)}$ has genus 1, in other words, it is an elliptic curve E. The space $\mathcal{C}(2, \Gamma_0(N))$ is hence 1-dimensional. By our analysis above, the nonzero functions in this space are all newforms of weight 2 and level N and the space is generalized by the normalized newform $f(z) = \sum_{n\geq 1} a_n e^{2\pi i n z}$. The associated Dirichlet series of f

$$D(f,s) = \sum_{n\geq 1} a_n n^{-s}$$

$$= \prod_{p\nmid N} \frac{1}{1 - a_p p^{-s} + p^{k-1-2s}} \prod_{q\mid N} \frac{1}{1 - a_q q^{-s}}$$

has the same Euler factors as the function $Z(E,s)$ of the elliptic curve E as proved by Eichler and Shimura, and the Ramanujan-Petersson conjecture for f in this case is nothing but the Riemann hypothesis for E modulo the primes not dividing N.

LECTURE 3
Connections among Modular Forms, Elliptic Curves, and Representations of Galois Groups

3.1. Functional Equations

The L-function associated to a modular form $f(z) = \sum_{n \geq 0} a_n e^{2\pi i n z}$ is the associated Dirichlet series multiplied by a suitable Γ-factor to account for the factor at the infinite place:

$$L(f, s) = (2\pi)^{-s} \Gamma(s) \sum_{n \geq 1} a_n n^{-s}.$$

Here $\Gamma(s)$ is the Gamma function given by the integral form for $\Re(s) > 0$ as

$$\Gamma(s) = \int_0^\infty e^{-t} t^{s-1} dt.$$

The analytic behavior of $L(f, s)$ was first considered by Hecke [**H**].

Theorem 3.1. *Suppose $f \in \mathcal{M}(k, N, \chi)$. Then $L(f, s)$ defined above converges absolutely for $\Re(s) > k$. It has an analytic continuation to the whole s-plane, holomorphic everywhere except possibly for a simple pole at $s = k$ and at $s = 0$ with the residue at $s = 0$ being $-a_0$, and it is holomorphic if f is a cusp form. It is bounded at infinity in each vertical strip of finite width and satisfies the functional equation*

$$L(f, s) = i^k N^{\frac{k}{2}-s} L(g_f, k - s),$$

where

$$g_f(z) = N^{-k/2} z^{-k} f\left(\frac{-1}{Nz}\right) \in \mathcal{M}(k, N, \bar{\chi}).$$

In particular, if f is a normalized newform of level N, then $g_f = \lambda_f \bar{f}$ for a nonzero constant λ_f and

$$\bar{f}(z) = \sum_{n \geq 0} \overline{a_n} e^{2\pi i n z}$$

so that the functional equation becomes

$$L(f, s) = \lambda_f i^k N^{\frac{k}{2}-s} L(\bar{f}, k - s) = \varepsilon(f, s) L(\bar{f}, k - s).$$

Remark. As proved in [**L1**], a normalized newform f is characterized analytically by $D(f,s)$ having Euler product at all primes and satisfying the functional equation of the form

$$L(f,s) = C_1 N^{\frac{k}{2}-s} L(\bar{f}, k-s)$$

for some nonzero constant C_1.

Let f be a normalized newform of weight k level N and character χ. Let η be a Dirichlet character with conductor m prime to N. The twist of f by η is

$$f_\eta(z) = \sum_{n \geq 1} a_n \eta(n) e^{2\pi i n z},$$

which can be shown to be a newform of weight k level Nm^2 and character $\chi\eta^2$. Its L-function satisfies the functional equation

$$L(f_\eta, s) = \varepsilon(f_\eta, s) L(\overline{f_\eta}, k-s),$$

where

(3.1.1) $$\varepsilon(f_\eta, s) = \varepsilon(f,s) \frac{g(\eta)}{g(\bar{\eta})} \eta(-N) \chi(m) m^{k-2s}$$

and $g(\eta)$ is the Gauss sum attached to the character η (cf. [**ALi**]).

Hecke was the first to consider when a holomorphic function on the Poincaré upper half-plane with prescribed Fourier coefficients is a modular form. He gave a criterion for the full modular group $SL_2(\mathbb{Z})$. Weil [**W3**] extended it to congruence subgroups.

Theorem 3.2 (Converse theorem for modular forms). *Given a sequence of complex numbers $a_0, a_1, a_2, \ldots$ satisfying the bound $|a_n| = O(n^c)$ for some positive constant c and all $n \geq 1$, form the function $f(z) = \sum_{n \geq 0} a_n e^{2\pi i n z}$, which is a holomorphic function on $\mathcal{H}$. Let χ be a character modulo N. If for all Dirichlet characters η with conductor m prime to N (including η trivial), $L(f_\eta, s)$ has analytic continuation to the whole s-plane, holomorphic everywhere except possibly with two simple poles at $s = k$ and $s = 0$, bounded at each vertical strip of finite width, and satisfies the functional equation*

$$L(f_\eta, s) = \varepsilon(f_\eta, s) L(\bar{f}_{\bar\eta}, k-s),$$

where $\varepsilon(f_\eta, s)$ is related to $\varepsilon(f,s)$ as in (1.2), then f lies in $\mathcal{M}(k, N, \chi)$. Further, if $D(f,s)$ converges absolutely at $s = k - \delta$ for some $k > \delta > 0$, then f is a cusp form.

3.2. Connections with Elliptic Curves

The converse theorem plays a very important role in drawing connections between modular forms and other objects via associated L-functions. We exhibit a few examples. More details can be found in chap 7 of [**L2**].

Let K be an imaginary quadratic extension of $\mathbb{Q}$. Let χ be a Grossencharacter (or idele class character) of K, algebraic of type k at the infinite place. Hecke showed that the L-function attached to χ and its twists by Dirichlet characters of $\mathbb{Z}$ have the analytic properties as stated in Theorem 3.2., hence there is a normalized newform f of weight k such that $L(f,s) = L(\chi, s)$. The level of f is equal to the norm from K to $\mathbb{Q}$ of the product of the different $\mathcal{D}_{K/\mathbb{Q}}$ and the conductor of χ, and the character of f is equal to the restriction of χ to ideles of $\mathbb{Q}$ times the quadratic character $\eta_{K/\mathbb{Q}}$ attached to K. Such cusp forms are called *cusp forms of CM type*.

Another tie is with elliptic curves. Let E be an elliptic curve defined over $\mathbb{Q}$ of conductor N and with Hasse-Weil zeta function

$$(3.2.1) \qquad Z(E,s) = \prod_{p \nmid N} \frac{1}{1 - a_p p^{-s} + p^{1-2s}} \prod_{p \mid N} \frac{1}{1 - a_p p^{-s}}.$$

Put

$$L(E,s) = (2\pi)^{-s} \Gamma(s) Z(E,s).$$

When E has complex multiplications, $L(E,s)$ is equal to $L(\chi, s)$ for some idele class character χ of an imaginary quadratic extension of $\mathbb{Q}$, and hence it is an L-function attached to a cusp form, as explained above.

The Taniyama-Shimura-Weil conjecture, first stated in 1950's in a more vague form, asserts that the Euler factors of the L-function attached to an elliptic curve defined over $\mathbb{Q}$ of conductor N agree with those of the L-function attached to a certain cusp form of weight 2 for the congruence subgroup $\Gamma_0(N)$. Weil's converse theorem (Theorem 3.2) provided strong theoretical support and made the conductor precise. To date, this conjecture is completely settled by the joint effort of Breuil, Conrad, Diamond, and Taylor [**BCDT**], along the line of Wiles's seminal work [**Wi**].

Now we return to the elliptic E over $\mathbb{Q}$ with Hasse-Weil function $Z(E,s)$ given by (3.2.1). Because the Riemann hypothesis holds, we may write

$$1 - a_p p^{-s} + p^{1-2s} = (1 - \alpha_p p^{-s})(1 - \beta_p p^{-s}), \quad \text{where} \quad |\alpha_p| = |\beta_p| = p^{1/2}.$$

Write $\alpha_p = e^{i\theta_p}\sqrt{p}$ so that $a_p/\sqrt{p} = e^{i\theta_p} + e^{-i\theta_p}$ is a real number between -2 and 2. We are interested in the distribution of the angles θ_p, or equivalently, $a_p/\sqrt{p} = 2\cos\theta_p$, as p varies.

The Sato-Tate conjecture for elliptic curves. Suppose E does not have CM. Then the set $\{a_p/\sqrt{p} : p \nmid \text{cond } E\}$ is uniformly distributed with respect to the Sato-Tate measure

$$\mu_{ST} = \frac{1}{\pi}\sqrt{1 - \frac{x^2}{4}}\, dx \quad \text{on} \quad [-2, 2].$$

In other words, the sum of the Dirac measure supported at $a_p/\sqrt{p}$ for primes $p \leq x$ divided by the number of such primes converges weakly to μ_{ST} as x tends to infinity.

This conjecture is extended to cusp forms. More precisely, let $f(z)$ be a normalized newform of weight k for the group $\Gamma_0(N)$ with Fourier coefficients a_n, $n \geq 1$. At a prime $p \nmid N$, the Ramanujan-Petersson conjecture holds, namely,

$$1 - a_p p^{-s} + p^{k-1-2s} = (1 - \alpha_p p^{-s})(1 - \overline{\alpha_p} p^{-s}).$$

Thus $a_p/p^{(k-1)/2}$ lies in the interval $[-2, 2]$. The Sato-Tate conjecture asserts that if f is not of CM type, then the set $\{a_p/p^{(k-1)/2} : p \nmid N\}$ is uniformly distributed with respect to the Sato-Tate measure μ_{ST}.

So far, there is good numerical evidence to support both conjectures, but the conjecture is not proved to hold for a single elliptic curve or cusp form.

3.3. Connections with Representations of the Galois Group over $\mathbb{Q}$

Let K be a finite Galois extension of $\mathbb{Q}$. The elements in K integral over $\mathbb{Z}$ form a ring $\mathcal{O}_K$, called the ring of integers of K. Given a prime p in $\mathbb{Z}$, the principal ideal $p\mathcal{O}_K$ decomposes as a product

$$p\mathcal{O}_K = (\mathcal{P}_1 \cdots \mathcal{P}_r)^e,$$

where each $\mathcal{P}_i$ is a maximal ideal of $\mathcal{O}_K$. The Galois group $Gal(K/\mathbb{Q})$ permutes the ideals $\mathcal{P}_i$, and the subgroup of elements in $Gal(K/\mathbb{Q})$ leaving $\mathcal{P}_i$ fixed is called the decomposition group of $\mathcal{P}_i$, denoted by $D_{\mathcal{P}_i}$. There is a natural surjection $\phi_{\mathcal{P}_i}$ from $D_{\mathcal{P}_i}$ to the Galois group of the residue field of $\mathcal{P}_i$, which is cyclic of order the degree of K over $\mathbb{Q}$ divided by re, generated by the Frobenius automorphism. A pull back to $D_{\mathcal{P}_i}$ of the Frobenius automorphism of the residue field is called a Frobenius element at $\mathcal{P}_i$. The kernel of $\phi_{\mathcal{P}_i}$ is called the inertia subgroup at $\mathcal{P}_i$.

Since the Galois group acts transitively on the $\mathcal{P}_i$'s, the decomposition groups $D_{\mathcal{P}_i}$ are conjugate for $1 \leq i \leq r$, and so are the inertia subgroups. If p does not divide the discriminant of K over $\mathbb{Q}$, then the ramification index $e = 1$. In this case the inertia subgroup at $\mathcal{P}_i$ is trivial so that the decomposition group $D_{\mathcal{P}_i}$ is cyclic, generated by the Frobenius element. Thus the Frobenius elements coming from the different decomposition groups $D_{\mathcal{P}_i}$, $1 \leq i \leq r$, are conjugate to each other. They determine a unique conjugacy class $Frob_p$, called the *Frobenius (conjugacy class) at p*.

We shall be concerned with continuous representations ρ of $G := Gal(\bar{\mathbb{Q}}/\mathbb{Q})$ acting on a finite-dimensional vector space V over different kinds of fields. The group G is endowed with the pro-finite topology such that for every finite Galois extension K of $\mathbb{Q}$, the subgroup $Gal(\bar{\mathbb{Q}}/K)$ is open in G.

The first kind are complex representations; that is, the underlying space V is a vector space over $\mathbb{C}$ endowed with the usual topology. Let ρ be a continuous representation acting on V. The profinite topology on G forces ρ to have an open kernel and hence there is a finite Galois extension K of $\mathbb{Q}$ so that the kernel of ρ is $Gal(\bar{\mathbb{Q}}/K)$ and ρ is a faithful representation of $Gal(K/\mathbb{Q})$. To ρ Artin introduced the L-function :

$$L(\rho, s) = \text{some } \Gamma\text{-factor} \prod_p \frac{1}{\det(1 - \rho(Frob_p)p^{-s}|V^{I_p})}.$$

Here the Γ-factor depends on the action of ρ at the local Galois group at each infinite place of K. In particular, if $K = \mathbb{Q}$, there is only one infinite place ∞, and the Galois group at ∞ is generated by the complex conjugation c. In the case that ρ is two dimensional, we call ρ an *odd* or *even* representation according to its determinant at c being -1 or 1. The Γ-factor for an odd degree two representation of G is the same as the Γ-factor attached to a holomorphic modular form, introduced at the beginning of this lecture. At each finite place p, we choose a decomposition group $\mathcal{P}_i$, and denote its inertia subgroup by I_p. The space V^{I_p} is the subspace of V fixed by I_p, so the restriction to V^{I_p} of ρ is the same for all choices of Frobenius elements in $\mathcal{P}_i$. While this restricted action depends on the choice of $\mathcal{P}_i$, its characteristic polynomial does not, hence the above L-function is well-defined and converges absolutely on the half-plane $\Re(s)$ large.

LECTURE 3. CONNECTIONS AND COMPARISONS

Artin has shown that degree 1 complex representations and idele class characters of $\mathbb{Q}$ with finite order are in one-to-one correspondence such that the L-functions attached to both representations have the same Euler factors. Hence one derives from Hecke's work that the L-function attached to nontrivial characters are holomorphic. Artin conjectured that this should be true in general:

Artin conjecture. Let ρ be an irreducible complex representation of $Gal(\bar{\mathbb{Q}}/\mathbb{Q})$ of degree greater than 1. Then $L(\rho, s)$ has holomorphic continuation to the whole s-plane, is bounded in each vertical strip of finite width, and satisfies the functional equation

$$L(\rho, s) = \varepsilon(\rho, s) L(\check{\rho}, 1-s),$$

where $\check{\rho}$ is the contragredient of ρ and $\varepsilon(\rho, s)$ is the product of a nonzero constant with an exponential function N^{-s}.

Brauer proved the conjecture with meromorphicity replacing holomorphicity. Further, if ρ is replaced by $\rho \otimes \eta$, the twist of ρ by an idele class character η of conductor relatively prime to the conductor of ρ, then $\varepsilon(\rho \otimes \eta, s)$ is related to $\varepsilon(\rho, s)$ just like (3.1.1). In view of the converse theorem for modular forms, Artin's conjecture for degree two odd irreducible complex representations ρ of G is equivalent to the existence of a normalized newform f_ρ of weight 1 such that $L(\rho, s)$ and $L(f_\rho, s)$ have the same Euler factors. More precisely, if $f_\rho(z) = \sum_{n \geq 1} a_n e^{2\pi i n z}$ has level N and character χ, then at primes $p \nmid N$ we have $Tr\rho(Frob_p) = a_p$ and $det\rho(Frob_p) = \chi(p)$. Deligne and Serre [**DS**] proved the converse, namely, to each weight 1 normalized newform f, there is an odd degree two irreducible complex representation ρ_f of G such that the Euler factors of $L(f, s)$ and $L(\rho_f, s)$ agree. It remains to show that the ρ_f's exhaust all degree two odd irreducible representations of G.

Another kind of representations of G is a compatible system of finite dimensional ℓ-adic representations which arise from geometry. The representation space V is finite-dimensional over a finite extension of $\mathbb{Q}_\ell$ endowed with the usual ℓ-adic topology. To each normalized newform f of weight $k \geq 2$ and level N, there is an ℓ-adic representation ρ_f of G so that the local factors of $L(f, s)$ and $L(\rho_f, s)$ agree at the primes p not dividing $N\ell$. This fact has numerous important consequences.

The third kind are modular representations of G, that is, the representation space V is a finite-dimensional vector space over a finite field or its algebraic closure, endowed with discrete topology. For odd irreducible degree 2 modular representations of G, Serre published a conjecture [**Se**] in 1987 predicting that they all arise from modular forms. More precisely, if ρ is such a representation acting on a vector space of characteristic p, then there is a normalized newform $f = \sum_{n \geq 1} a_n e^{2\pi i n z}$ of weight k level N and character χ whose Fourier coefficients a_n lie in the ring of integers $\mathcal{O}_K$ of a number field K and there is a prime ideal $\mathcal{P}$ of $\mathcal{O}_K$ of residual characteristic p such that $Tr\rho(Frob_p) \equiv a_p \pmod{\mathcal{P}}$ and $det\rho(Frob_p) \equiv p^{k-1}\chi(p) \pmod{\mathcal{P}}$. Such f, if exists, is nonunique. Serre also gave a recipe for the smallest possible weight and level of such a form.

3.4. Comparisons with Automorphic Forms of GL_2 over Function Fields

The Poincaré upper half-plane can be identified with the homogeneous space $GL_2(\mathbb{R})/O_2(\mathbb{R})Z(\mathbb{R})$, where $Z(\mathbb{R})$ is the center of $GL_2(\mathbb{R})$. Just like $\mathbb{R}$ is the completion of $\mathbb{Q}$ at the infinite place ∞, the group $GL_2(\mathbb{R})$ is the component at ∞ of $GL_2(A_\mathbb{Q})$ over the ring of adeles $A_\mathbb{Q}$ of $\mathbb{Q}$. By the strong approximation theorem for SL_2, we have

$$GL_2(A_\mathbb{Q}) = GL_2(\mathbb{Q}) \cdot GL_2(\mathbb{R}) \prod_p GL_2(\mathbb{Z}_p).$$

Thus a modular form can be viewed as a function on $GL_2(A_\mathbb{Q})$, left invariant by $GL_2(\mathbb{Q})$ and right invariant by a subgroup of $\prod_p GL_2(\mathbb{Z}_p)$ of finite index. Moreover, it satisfies some growth condition and behaves in a prescribed way under the action of the group $O_2(\mathbb{R})$ and the center of $GL_2(A_\mathbb{Q})$. The reader is referred to chap 8 of [**L2**] for more details. This viewpoint allows one to replace $\mathbb{Q}$ by other number fields or function fields, and GL_2 by other algebraic groups. Here we keep the group to be GL_2 and consider it over a function field K, which is a finite extension of $F(t)$ with the field of constants being the finite field F. We compare the results for this case with what we discussed above for classical modular forms, which are (part of) automorphic forms for GL_2 over $\mathbb{Q}$.

First, the Ramanujan-Petersson conjecture holds, which was proved by Drinfeld [**Dr**] using his work on Drinfeld modules to establish a correspondence from automorphic representations of GL_2 over K to ℓ-adic representations of Galois group $Gal(K^{sep}/K)$, as conjectured by Langlands.

The Sato-Tate conjecture for elliptic curves over K holds. More precisely, Yoshida [**Y**] proved that an elliptic curve defined over K with nonconstant j-invariant satisfies the Sato-Tate conjecture. By the way, the K3-surface in the first lecture can be regarded as an elliptic curve defined over $\mathbb{Q}(t)$, a characteristic zero function field.

The Taniyama-Shimura-Weil conjecture holds for elliptic curves defined over K. This follows from the work of Grothendieck, who proved the analytic continuation and functional equation for the L-function attached to an elliptic curve as well as twists by idele class characters, the work of Deligne, who showed that the global ε-factor occurring in the functional equations satisfy the relation (3.1.1) [**D3**], and the work of Jacquet and Langlands, who proved the converse theorem for GL_2 over K [**JL**].

Hence the automorphic forms for GL_2 over K arising from elliptic curves with nonconstant j-invariant satisfy the Sato-Tate conjecture. We exhibit more examples of automorphic forms satisfying the Sato-Tate conjecture below.

Let k be a finite field of characteristic p. Let ψ be the additive character $\psi(x) = e^{2\pi i x/p}$ of the prime field $F_p = \mathbb{Z}/p\mathbb{Z}$. Thus ψ composed with the trace from k to F_p yields a nontrivial additive character of k. For an nonzero element a in a finite field k, define the Kloosterman sum

$$Kl(k;a) = \sum_{x \in k^\times} \psi \circ Tr_{k/F_p}(x + \frac{a}{x}).$$

A. Weil showed that $|Kl(k;a)| \le 2\sqrt{|k|}$. Hence the normalized Kloosterman sum $Kl(k;a)/\sqrt{|k|}$ lies in the interval $[-2,2]$. The theorem below, proved by Chai and

Li [**CL**], asserts the existence of an automorphic form for GL_2 over a function field which, at all but finitely many places, is an eigenfunction of the Hecke operator with eigenvalue given by a Kloosterman sum arising from a global element b; and furthermore, the Sato-Tate conjecture holds for such a form if b is not a constant.

Theorem 3.3 (Kloosterman sum conjecture over a function field). *Let K be a function field with the field of constants F as above. Let b be a nonzero element in K. Then there exists an automorphic form f of GL_2 over K such that at every place v of K, which is neither a zero nor a pole of b, f is an eigenform of the Hecke operator at v with eigenvalue $-Kl(F_v; b)$. Here F_v denotes the residue field of K at v and b is viewed as a nonzero element in the residue field F_v. Moreover, if b is not in F, then f satisfies the Sato-Tate conjecture.*

We remark that the analogous question over $\mathbb{Q}$ is still unsloved. Refer to [**CL**] for more detailed discussion.

Another example from [**CL**] is similar to the above, with the Kloosterman sum replaced by the character sum

$$\lambda_\chi(F_v; b) = \sum_{\substack{(x,y) \in F_v \times F_v \\ y^2 = (x-1)^2 + 4bx}} \chi \circ N_{F_v/F}(x),$$

in which χ is a nontrivial character of $F^\times$.

As a consequence of [**CL**], one knows that the Hasse-Weil zeta function attached to a hyperelliptic curve of genus 2 over a function field K is an automorphic L-function of GL_4 over K, unlike the case of $\mathbb{Q}$, where elliptic curves are the only case systematically studied, as discussed in the last section of Lecture 1.

As for the Ramanujan-Petersson conjecture for automorphic L-functions of GL_n over a function field, L. Lafforgue proved in [**La**] that the conjecture always holds when n is odd, and there may be an additional possibility when n is even.

BIBLIOGRAPHY

[AL] A. O. L. Atkin and J. Lehner, *Hecke operators on $\Gamma_0(m)$)*, Math. Ann. **185** (1970), 134–160.

[ALi] A. O. L. Atkin and W.-C. W. Li, *Twists of newforms and pseudo- eigenvalues of W-operators*, Invent. Math. **48** (1978), 221–243.

[BCDT] C. Breuil, B. Conrad, F. Diamond, and R. Taylor, *On the modularity of elliptic curves over $\mathbb{Q}$*, preprint (1999).

[CL] C.-L. Chai and W.-C. W. Li, *Character sums and automorphic forms*, preprint (1999).

[D1] P. Deligne, *La conjecture de Weil I*, Inst. Hautes Etudes Sci. Publ. Math. **43** (1974), 273–307.

[D2] P. Deligne, *Formes modulaires et représentations ℓ-adiques*, Lecture Notes in Math. vol. 179, Springer-Verlag, Berlin, Heidelberg, New York, 1971.

[D3] P. Deligne, *Les constantes des équations fonctionelles des fonctions L*, Modular Functions of One Variable. II, vol. Lecture Notes in Math. **349** Springer-Verlag, Berlin-Heidelberg-New York, 1973, pp. 501–595.

[DS] P. Deligne and J.-P. Serre, *Formes modulaires de poids 1*, Ann. Sci. E. N. S. **7** (1974), 507–530.

[Dr] V. G. Drinfel'd, *The proof of Petersson's conjecture for $GL(2)$ over a global field of characteristic p*, Functional Anal. Appl. **22** (1988), 28–43.

[DW] B. Dwork, *On the rationality of the zeta function of an algebraic variety*, Amer. J. Math. **82** (1960), 631–648.

[Ha] H. Hasse, *Beweis des Analogons der Riemannschen Vermutung für die Artinschen und F. K. Schmidtschen Kongruenzzetafunktionen in gewissen elliptischen Fällen*, Ges. d. Wiss. Nachrichten, Math. Phys. Klasse **Heft 3** (1933), 253–262.

[H] E. Hecke, *Mathematische Werke*, Vandenhoeck und Ruprecht, Göttingen, 1959.

[JL] H. Jacquet and R. Langlands, *Automorphic Forms on $GL(2)$*, Lecture Notes in Math., vol 114, Springer-Verlag Berlin, Heidelberg, New York, 1970.

[La] L. Lafforgue, *Chtoucas de Drinfeld et conjecture de Ramanujan-Petersson*, Asterisque **243** (1997).

[L1] W.-C. W. Li, *Newforms and functional equations*, Math. Ann. **212** (1975), 285–315.

[L2] W.-C. W. Li, *Number Theory with Applications*, World Scientific, Singapore, New Jersey, London, Hong Kong, 1996.

[M] T. Miyake, *On automorphic forms on GL_2 and Hecke operators*, Annals of Math. **94** (1971), 174–189.

[R] K. Ribet, *On modular representations of $Gal(\bar{\mathbb{Q}}/\mathbb{Q})$ arising from modular forms*, Invent. Math. **100** (1990), 431–476.

[Se] J.-P. Serre, *Sur les représentations modulaires de degré 2 de $Gal(\bar{\mathbb{Q}}/\mathbb{Q})$*, Duke Math. J. **54** (1987), 179–230.

[S] G. Shimura, *Introduction to the Arithmetic Theory of Automorphic Forms*, Iwanami Shoten and Princeton Univ. Press, Princeton, 1971,

[Si] J. Silverman, *The Arithmetic of Elliptic Curves*, Springer, New York, 1992.

[SB] J. Stienstra and F. Beukers, *On the Picard-Fuchs equation and the formal Brauer group of certain elliptic K3-surfaces*, Math. Ann. **271** (1985), 269–304.

[W1] A. Weil, *Numbers of solutions of equations in finite fields*, Bulletin of Amer. Math. Soc. **55** (1949), 497–508.

[W2] A. Weil, *Sur les Courbes Algébriques et les Variétés Qui s'en Deduisent*, Hermann, Paris, 1948.

[W3] A. Weil, *Über die Bestimmung Dirichletsher Reihen durch Funktionalgleichungen*, Math. Ann. **168** (1967), 149–156.

[Wi] A. Wiles, *Modular elliptic curves and Fermat's Last Theorem*, Annals of Math. **141** (1995), 443–551.

[TW] R. Taylor and A. Wiles, *Ring-theoretic properties of certain Hecke algebras*, Annals of Math. **141** (1995), 553–572.

[Y] H. Yoshida, *On an analogue of the Sato-Tate conjecture*, Invent. Math. **19** (1973), 261–277.

[Yui] N. Yui, *Arithmetic of certain Calabi-Yau varieties and mirror symmetry*, this volume

Arithmetic of Certain Calabi–Yau Varieties and Mirror Symmetry

Noriko Yui

Arithmetic of Certain Calabi–Yau Varieties and Mirror Symmetry

Noriko Yui[1]

Introduction

These notes were prepared for a series of lectures which I was invited to give for the Mentoring Program for Women at the Institute for Advanced Study, Princeton in the spring of 1999. The Program was a part of the 1999 Park City Mathematics Institutes on *Arithmetic Algebraic Geometry* which took place in the summer of 1999.

In recent years, Calabi–Yau manifolds have been investigated enthusiastically in connection with superstring theory and mirror symmetry. The research activities have mostly been focused on the aspects of geometry, topology and physics around Calabi–Yau manifolds. The article *Mathematical Aspects of Mirror Symmetry* by David Morrison in the IAS/Park City Mathematics Series Volume **3** gives an excellent exposition of mirror symmetry from mathematical (i.e., geometrical and topological) points of view.

On the other hands, we may consider Calabi–Yau manifolds as algebraic objects, that is, we may consider Calabi–Yau varieties, say, defined over the field $\mathbb{Q}$ of rational numbers, or more generally, over number fields. Calabi–Yau varieties of dimension one are nothing but elliptic curves, and those of dimension two are $K3$ surfaces. Calabi–Yau varieties of dimension three are the so-called Calabi–Yau threefolds, which may be regarded as three-dimensional generalizations of elliptic curves. Being higher dimensional analogues of elliptic curves, Calabi–Yau varieties defined over number fields are therefore expected to be endowed with very rich arithmetic properties, and this is what I wish to convey to the reader of these notes.

---

[1]Department of Mathematics and Statistics, Queen's University, Kingston, Ontario CANADA K7L 3N6 .
E-mail address: yui@mast.queensu.ca.
Received by the editors June 14, 2000.
1991 *Mathematics Subject Classification*. Primary 11G40, 14G10, 14J30; Secondary 14J10, 14J20, 14J28, 14J30, 14J32.
Key words and phrases. Elliptic curves, $K3$ surfaces, Calabi–Yau threefolds, rigid Calabi–Yau threefolds, Modularity conjectures, Mirror symmetry, Mirror maps, Mirror–Moonshine.
The author was supported in part by a Research Grant from the Natural Sciences and Engineering Research Council of Canada.

©2001 American Mathematical Society

After Wiles's celebrated work which established the modularity (the conjecture of Shimura–Taniyama–Weil) for elliptic curves defined over $\mathbb{Q}$, it seems very natural to consider the modularity conjectures for higher dimensional analogue of elliptic curves, namely, for special classes of Calabi–Yau varieties of dimension 2 and 3 defined over $\mathbb{Q}$ (or number fields).

Indeed, in Lecture 1, we will discuss the modularity conjecture for certain Calabi–Yau threefolds defined over $\mathbb{Q}$. I will first formulate the modularity conjecture for *rigid* Calabi–Yau threefolds over $\mathbb{Q}$. Then I will give some evidence in support of the conjecture. More precisely, I will first define L–series of rigid Calabi–Yau threefolds defined over $\mathbb{Q}$, and then compute the order of vanishing of L–series at $s = 2$. Further we recapitulate these results in the framework of the conjecture of Beilinson and Bloch. The intermediate Jacobians of rigid Calabi–Yau varieties are elliptic curves (a priori defined over $\mathbb{C}$). We will relate the conjecture of Beilinson and Bloch for rigid Calabi–Yau threefolds to that of Birch and Swinnerton-Dyer for elliptic curves. The mirror symmetry conjecture fails to hold for the class of rigid Calabi–Yau threefolds in the sense that a mirror pair of a rigid Calabi–Yau threefold is no longer a Calabi–Yau threefold.

In Lecture 2, our discussions are concentrated on certain classes of Calabi–Yau varieties, so-called *orbifold Calabi–Yau varieties*, or in short *Calabi–Yau orbifolds*. For these Calabi–Yau orbifolds, the mirror symmetry conjecture hold true. Our goals are to describe the orbifold construction for Calabi–Yau hypersurfaces of Fermat type of dimension 2 and 3, and then deduce their arithmetic properties from the dominant Fermat hypersurfaces. We will determine L–series, and will discuss how to compute the order of vanishing of these L–series. Again we recapitulate their L–series in the context of the conjectures of Tate, and of Beilinson and Bloch.

In Lecture 3, we will look into $K3$ surfaces (Calabi–Yau varieties of dimension 2). Our aim in this lecture has somewhat different flavor from the first two lectures. We will discuss yet another modularity conjecture, namely, the modularity of the mirror maps of certain families of $K3$ surfaces. Lian and Yau [**LY2**] *Arithmetic properties of mirror maps and quantum coupling*, CMP **176** (1996) first observed that the mirror maps of certain one-parameter families of $K3$ surfaces with Picard number 19 (obtained via orbifolding construction) may be related (polynomially) to Hauptmoduln of genus zero subgroups of $PSL_2(\mathbb{R})$. This phenomenon is the so-called *mirror moonshine phenomenon*. We will discuss selected examples of one-parameter families of $K3$ surfaces. We compute mirror maps of such families. The mirror maps thus computed are in support of the mirror–moonshine phenomenon.

In these three lectures, I have tried to convey the excitement abound studying arithmetic properties of Calabi–Yau varieties and mirror symmetry. In fact, there are seemingly more questions and conjectures than theorems. It is a gold mine waiting to be explored!

Acknowledgments

I would like to thank the organizers of the 1999 Mentoring Program for Women in Mathematics for their invitation to deliver three lectures at the Institute for Advanced Study.

Thanks are also due to the several people who made suggestions to preliminary versions of this note. This includes Srinath Baba, Imin Chen, Igor Dolgachev, Charles Doran, Yasuhiro Goto, Shinobu Hosono, Bong Lian, Ron Livné, Ling Long, Masa-Hiko Saito, J.-P. Serre, Jan Stienstra, and Don Zagier.

Special thanks are due to Joe Buhler, Ling Long, Ken Ono, Chad Schoen, and William Stein for their help in calculating the q–expansions of modular forms and the order of vanishing of L–series discussed in Lecture 1.

Finally, my heartfelt thanks go to my former postdoctoral fellow, Helena Verrill for providing me with a texfile of the lecture notes that she took during my three lectures. The version presented here is a revised and expanded version of her lecture notes.

LECTURE 1
The Modularity Conjecture for Rigid Calabi–Yau Threefolds over the Field of Rational Numbers

In the first lecture, we shall formulate the modularity conjecture for rigid Calabi–Yau threefolds defined over the field $\mathbb{Q}$. We shall discuss illustrating examples of rigid Calabi–Yau threefolds over $\mathbb{Q}$ in support of the modularity conjecture.

1. Definition of Calabi–Yau varieties

Let X be a smooth projective variety over a field K, (e.g., $K = \mathbb{C}$, $\mathbb{Q}$ or a number field), of dimension d.

Definition 1. X is a *Calabi–Yau* variety if
(1) $H^i(X, \mathcal{O}_X) = 0$ for every $i, 0 < i < d$, and
(2) $K_X := \wedge^d \Omega^1_X \simeq \mathcal{O}_X$.

Remark 1.1. (a) The invariants defining a Calabi–Yau variety are geometric, and accordingly, they are defined for complex varieties. If a variety in question is defined over a field $\mathbb{Q}$ or a number field, the invariants are defined for its complexification.

(b) As topological manifolds, Calabi–Yau varieties are simply connected Kähler manifolds with trivial canonical bundle.

(c) The condition (2) implies that the geometric genus of X,

$$p_g(X) := \dim H^0(X, K_X) = \dim H^d(X, \mathcal{O}_X) = 1.$$

(The second equality holds because $H^0(X, K_X) \simeq H^d(X, \mathcal{O}_X)$ by the Serre duality.)

Remark 1.2. (a) If $d = 1$, the condition (1) is empty. Then the condition (2) implies that a Calabi–Yau variety of dimension 1 (furnished with at a rational point) is nothing but an elliptic curve.

(b) If $d = 2$, $H^1(X, \mathcal{O}_X) = 0$ and $p_g(X) = 1$, so Calabi–Yau varieties of dimension 2 are $K3$ surfaces. Kummer surfaces, quartics in $\mathbb{P}^3$, and double covers of $\mathbb{P}^2$ branched along a sextic are standard examples of $K3$ surfaces. Lecture 3 will be devoted to $K3$ surfaces.

(c) If $d = 3$, $H^1(X, \mathcal{O}_X) = H^2(X, \mathcal{O}_X) = 0$, and $p_g(X) = 1$. These are called Calabi–Yau threefolds, which will be the main objects of our discussions in Lectures 1 and 2.

We introduce the Hodge numbers $h^{p,q} := \dim H^q(X, \Omega_X^p)$, where $\Omega_X^p := \wedge^p \Omega_X^1$ and $H^q(X, \Omega_X^p)$ is the (p,q)-th Hodge cohomology group of X. We encode this information into a diamond, called the *Hodge* diamond. There are symmetries of the Hodge diamond:

$$h^{p,q} = h^{q,p} \quad ; \quad h^{p,q} = h^{d-p,d-q} = h^{d-q,d-p}.$$

The first identity follows from the operation of complex conjugation and the latter from the Serre duality on the Hodge cohomology groups.

Lines of symmetry are indicated below:

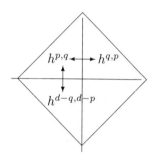

Now we shall digress.

(a) For $d = 1$ the Hodge diamond is given by:

$$\begin{array}{ccc} & h^{0,0} & \\ h^{1,0} & & h^{0,1} \\ & h^{1,1} & \end{array} \qquad \begin{array}{ccc} & 1 & \\ 1 & & 1 \\ & 1 & \end{array}$$

The Betti numbers are $B_0 = 1$, $B_1 = 2$ and $B_2 = 1$, and the Euler characteristic $E(X) := B_0 - B_1 + B_2 = 0$.

(b) For $d = 2$ the Hodge diamond is:

$$\begin{array}{ccccc} & & h^{0,0} & & \\ & h^{1,0} & & h^{0,1} & \\ h^{2,0} & & h^{1,1} & & h^{0,2} \\ & h^{2,1} & & h^{1,2} & \\ & & h^{2,2} & & \end{array}$$

The condition of being a Calabi–Yau variety in this case says that $h^{0,1} = 0 = h^{1,0}$ and $h^{0,2} = 1 = h^{2,0}$. For $K3$ surfaces, we have $h^{1,1} = 20$, and so the Hodge diamond is of the form:

$$\begin{array}{ccccc} & & 1 & & \\ & 0 & & 0 & \\ 1 & & 20 & & 1 \\ & 0 & & 0 & \\ & & 1 & & \end{array}$$

Adding the quantities horizontally we get the Betti numbers:

$$\begin{array}{rcccl}
1 & & & = B_0 = & 1 \\
0 & + & 0 & = B_1 = & 0 \\
1 + 20 & + & 1 & = B_2 = & 22 \\
0 & + & 0 & = B_3 = & 0 \\
& 1 & & = B_4 = & 1
\end{array}$$

The Euler characteristic, which is the alternating sum of Betti numbers, is computed to be:

$$E(X) = \sum_{i=0}^{4} (-1)^i B_i = 1 + 22 + 1 = 24.$$

(c) For $d = 3$ the Hodge diamond is:

$$\begin{array}{ccccccc}
& & & h^{0,0} & & & \\
& & h^{1,0} & & h^{0,1} & & \\
& h^{2,0} & & h^{1,1} & & h^{0,2} & \\
h^{3,0} & & h^{2,1} & & h^{1,2} & & h^{0,3} \\
& h^{3,1} & & h^{2,2} & & h^{1,3} & \\
& & h^{3,2} & & h^{2,3} & & \\
& & & h^{3,3} & & &
\end{array}$$

Since Calabi–Yau threefolds are Kähler manifolds by Yau's theorem, $h^{1,1} > 0$. That $h^{2,0} = 0$ follows from $h^{1,0} = 0$ by the Serre duality. Indeed, we have $H^1(X, \mathcal{O}_X) \simeq H^2(X, \Omega_X^3)^*$ by the Serre duality, and the latter further is isomorphic to $H^2(X, \mathcal{O}_X)$ by the Calabi–Yau condition (2) in Definition 1.

Using the condition (1) and (2), and symmetries, we have:

$$\begin{array}{ccccccc}
& & & 1 & & & \\
& & 0 & & 0 & & \\
& 0 & & h^{1,1} & & 0 & \\
1 & & h^{2,1} & & h^{1,2} & & 1 \\
& 0 & & h^{2,2} & & 0 & \\
& & 0 & & 0 & & \\
& & & 1 & & &
\end{array}
\qquad
\begin{array}{rcl}
B_0 &=& 1 \\
B_1 &=& 0 \\
B_2 &=& h^{1,1} \\
B_3 &=& 2(1 + h^{2,1}) \\
B_4 &=& h^{2,2} \\
B_5 &=& 0 \\
B_6 &=& 1
\end{array}$$

The Betti numbers are calculated by adding the horizontal quantities. The Euler characteristic of a Calabi–Yau threefold is computed by taking the alternating sums of the Betti numbers:

$$E(X) = \sum_{i=0}^{6} (-1)^i B_i = 2(h^{1,1} - h^{2,1}).$$

In contrast to the $d = 2$ case, the Euler characteristic of a Calabi–Yau threefold depends on $h^{1,1}$ and $h^{2,1}$. It is still an open problem if there is a constant bounding the absolute value of this number for all Calabi–Yau threefolds; in fact, there is a conjecture (cf. Werner and van Geemen [**WvG**]) that the Euler characteristic of a Calabi–Yau threefold is bounded.

Conjecture 1.3. There is a constant C such that for all Calabi–Yau threefolds X, $|E(X)| < C$.

For weighted Calabi–Yau orbifolds in weighted projective spaces $\mathbb{P}^4(Q)$ with weight Q, the absolute constant was determined to be $C = 960$ from the lists complied by Klemm and Schimmrigh [**KSc**] and Kreuzer and Skarke [**KSk**].

However, this constant is not yet known to exist in general.

2. The Mirror Symmetry Conjecture

(a) For $d = 2$, there are several notions of the mirror symmetry for $K3$ surfaces. There should be relations among these variants of mirror symmetry conjecture for $K3$ surfaces. However, we are not going into discussion on this topic here but we will defer it to Lecture 3.

(b) For $d = 3$, a very "naive" (topological) version of the Mirror Symmetry Conjecture is formulated as follows:

Given a Calabi–Yau threefold X, there exists a mirror Calabi–Yau threefold $X^\vee$ such that
$$\begin{aligned} h^{1,1}(X^\vee) &= h^{2,1}(X), \\ h^{2,1}(X^\vee) &= h^{1,1}(X), \\ E(X^\vee) &= -E(X). \end{aligned}$$

In terms of the Hodge diamond, Mirror symmetry occurs here:

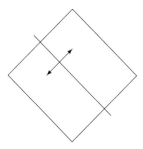

Here is a question : *Given a Calabi–Yau threefold, can one always find a naive mirror Calabi–Yau threefold?* The answer is **no**. There is a special class of Calabi–Yau threefolds for which the Mirror Symmetry Conjecture fails. On the other hand, these Calabi–Yau threefolds are indeed the objects of arithmetic interest. This is exactly what we plan to do in this lecture.

There are much more to the Mirror Symmetry Conjecture for Calabi–Yau threefolds than just exchanging the Hodge numbers of mirror pairs of Calabi–Yau threefolds. A crude form of the Mirror Symmetry Conjecture is stated as follows.

Conjecture 2.1. There exists a pair of Calabi–Yau threefolds $(X, X^\vee)$ and an isomorphism $H^{1,1}(X) \simeq H^{2,1}(X^\vee)$ such that this isomorphism should exchange the prepotential of the A–model Yukawa coupling defined on the complexified Kähler cone $\mathcal{K}_X \subset H^{1,1}(X)$ with the prepotential of the B–model Yukawa coupling defined on the complex structure moduli space whose tangent space is $H^{2,1}(X^\vee)$.

A fuller discussion on the Mirror Symmetry Conjecture will be postponed till Lecture 2. For the necessary background on physics behind the Mirror Symmetry Conjecture, the reader is referred to [**MirSym1**] and [**MirSym2**].

3. Rigid Calabi–Yau threefolds over $\mathbb{Q}$

Definition 2. A Calabi–Yau threefold X is said to be *rigid* if $h^{2,1} := h^{2,1}(X) = 0$.

The quantity $h^{2,1}$ represents the number of infinitesimal deformations of complex structures on X, so if this is zero there are no deformations. The Euler characteristic of any rigid Calabi–Yau threefold takes a positive value.

The Hodge diamond of a rigid Calabi–Yau threefold looks like:

$$\begin{array}{ccccccc} & & & 1 & & & \\ & & 0 & & 0 & & \\ & 0 & & h^{1,1} & & 0 & \\ 1 & & 0 & & 0 & & 1 \\ & 0 & & h^{2,2} & & 0 & \\ & & 0 & & 0 & & \\ & & & 1 & & & \end{array}$$

Remark 3.1. Any rigid Calabi–Yau threefold would provide a counter example to the Mirror Symmetry Conjecture. Indeed, if X is a rigid Calabi–Yau threefold, then $h^{2,1}(X) = 0$, then its naive mirror partner $X^\vee$ should have $h^{1,1}(X^\vee) = h^{2,1}(X) = 0$. This means that $X^\vee$ is not a Calabi–Yau threefold any longer. For each rigid Calabi–Yau threefold, there is a proposal what should be its mirror partner. For instance, for a rigid Calabi–Yau threefold obtained as an orbifold of a triple product of an elliptic curve with some finite group action, a mirror partner should be a Fano variety in $\mathbb{P}^8$ (see, e.g., Candelas, Derrick and Parks [**CDP**]).

Remark 3.2. Rigid Calabi–Yau varieties defined over $\mathbb{Q}$ might be considered as three–dimensional analogues of elliptic curves over $\mathbb{Q}$ (at least from cohomological points of view). In particular, rigid Calabi–Yau threefolds have $B_3 = 2$, and their third etale cohomology groups give rise to rank two motives, and they are believed to be amenable to arithmetic investigation.

More generally, we may consider any Calabi–Yau threefold X, which is not necessarily rigid, but whose third cohomology group $H^3(X, \mathbb{Q}_l)$ decomposes to motivic factors, e.g., $H^3(X, \mathbb{Q}_l) = \mathcal{I} \oplus \mathcal{M}$ where the Hodge structure of $\mathcal{M}$ is of type $(3,0) + (0,3)$ while that of $\mathcal{I}$ has $h^{3,0}(\mathcal{I}) = 0$. Then our argument can be applied to the motivic factor $\mathcal{M}$ in place of the variety X.

For instance, Bloch [**Bl1**] considered the motivic factor $\mathcal{M}$ arising from the Jacobian of the diagonal quartic curve $aT_0^4 + bT_1^4 = T_2^4$; $a, b \in \mathbb{Q}^*$, Buhler, Schoen and Top [**BST**] considered triple products of elliptic curves, and Livné [**L2**] studied motivic orthogonal two-dimensional Galois representations.

4. The modularity conjecture for rigid Calabi–Yau threefolds over $\mathbb{Q}$

The modularity conjecture for elliptic curves defined over $\mathbb{Q}$ has been established in totality by the exhaustive efforts of Wiles and his former students. For details, see Wiles [**Wi**], Taylor–Wiles [**TW**], Breuil, Conrad, Diamond and Taylor [**BCDT**].

Theorem: *Every elliptic curve E over $\mathbb{Q}$ is modular in the sense that*

$$L(E, s) = L(f, s)$$

where f is a cusp form of weight 2 for $\Gamma_0(N)$ where N is the conductor of E.

This is the theorem that we would like to generalize to rigid Calabi–Yau threefolds over $\mathbb{Q}$.

4.1. Henceforth, we assume that a rigid Calabi–Yau threefold X is defined over $\mathbb{Q}$. Let $\bar{\mathbb{Q}}$ denotes the algebraic closure of $\mathbb{Q}$. Let $\mathbb{Q}_\ell$ denote the field of ℓ-adic rationals where ℓ is a prime. We choose a suitable integral model for a rigid Calabi–Yau threefold X over $\mathbb{Q}$.

—A rational prime p is a good prime if X mod p defines a smooth variety over $\mathbb{F}_p$.
—For a good prime p, let Frob_p be the Frobenius automorphism on X mod p.

Let Frob_p act on $H^3(X_{\bar{\mathbb{Q}}}, \mathbb{Q}_\ell)$ where $p \neq \ell$. Define the polynomial $P_{p,3}(T)$ by

$$P_{p,3}(T) := \det(1 - \text{Frob}_p\, T\, |H^3_{et}(X_{\bar{\mathbb{Q}}}, \mathbb{Q}_\ell)),$$

where T is an indeterminant.

We have $P_{p,3} \in 1 + T\mathbb{Z}[T]$, and $\deg P_{p,3} = 2$. $P_{p,3}$ is given by

$$P_{p,3}(T) = 1 - t_3(p)T + p^3 T^2.$$

4.2. What is $t_3(p)$?

Recall now that for elliptic curves X, we had $t_1(p) := \text{trace}(\text{Frob}_p\, |\, H^1_{et}(X_{\bar{\mathbb{Q}}}, \mathbb{Q}_\ell)) = 1 + p - \#X(\mathbb{F}_p)$. This relation is from the Lefschetz fixed point formula. Furthermore, the Riemann Hypothesis assert that we have $|t_1(p)| \leq 2p^{1/2}$.

For a Calabi–Yau threefold X, we have:

$$\#X(\mathbb{F}_p) = \sum_{n=0}^{6} (-1)^n t_n(p)$$

where $t_n(p) := \text{trace}(\text{Frob}_p\, |H^n_{et}(X_{\bar{\mathbb{Q}}}, \mathbb{Q}_\ell))$. So

$$\#X(\mathbb{F}_p) = t_0(p) - t_1(p) + t_2(p) - t_3(p) + t_4(p) - t_5(p) + t_6(p)$$

By looking at the Hodge diamond and using various dualities and the Riemann hypothesis, we have

$$t_1(p) = t_5(p) = 0,\ t_0(p) = 1, t_6(p) = p^3,\ \text{and}\ \ t_4(p) = pt_2(p).$$

This gives rise to the formula:

$$\#X(\mathbb{F}_p) = 1 + p^3 + (1+p)t_2(p) - t_3(p).$$

Moreover, we have $|t_2(p)| \leq p h^{1,1}$, and the equality holds if and only if all $H^{1,1}$ cycles are defined over $\mathbb{Q}$. We have

$$t_3(p) = 1 + p^3 + (1+p)t_2(p) - \#X(\mathbb{F}_p).$$

By virtue of the Riemann Hypothesis, we have $|t_3(p)| \leq 2p^{3/2}$.

Definition 3. The L–series of X is

$$L(X, s) = L(H^3_{et}(X_{\bar{\mathbb{Q}}}, \mathbb{Q}_\ell), s)$$

$$= (*) \times \prod_{p\ \text{good}} \frac{1}{1 - t_3(p)p^{-s} + p^{3-2s}},$$

where $(*)$ is a factor corresponding to primes of bad reduction.

Conjecture 4.1. A rigid Calabi–Yau threefold over $\mathbb{Q}$ is **modular**, in the sense that $L(X, s) = L(f, s)$, where f is a modular form of weight 4 on $\Gamma_0(N)$ for some N divisible only by the primes of bad reduction.

Remark 4.2. This conjecture may be regarded as a concrete realization of the conjecture of Fontaine and Mazur [**FM**] that all irreducible odd 2–dimensional Galois representations "coming from geometry" should be modular, up to a Tate twist. For a recent development on the conjecture of Fontaine and Mazur, the reader is referred to the article of Richard Taylor [**Tay**]. See also Livné [**L2**] for examples of orthogonal two–dimensional Galois representations.

Remark 4.3. J.-P. Serre informed the author that the conjecture 4.1 is a consequence of Serre's conjecture on residual Galois representations of degree 2, specifically, the conjectural Theorem 6 on page 211 of his 1987 article [**Ser**] in Duke Journal of Mathematics.

5. Evidence for the modularity conjecture

We can discuss only a handful of examples here. For a more complete list of rigid Calabi–Yau threefolds, the reader is referred, for instance, to the articles of M.-H. Saito and Yui [**SY**], van Geemen and Nygaard [**vGN**], van Straten [**vS**], and Werner and van Geemen [**WvG**].

5.1. Chad Schoen's quintic

Schoen [**Sch1**] considered the Fermat hyperplane of degree 5 in $\mathbb{P}^4$ defined by the equation:
$$X_0^5 + X_1^5 + X_2^5 + X_3^5 + X_4^5 = 0 \subset \mathbb{P}^4.$$

As $h^{2,1} = 101$, this Calabi–Yau threefold is not rigid, that is, it has lots of deformations, which we have to kill to get a rigid Calabi–Yau threefold. So we add one more homogeneous term to the defining equation of the Fermat quintic hypersurface and define:
$$Y : X_0^5 + X_1^5 + X_2^5 + X_3^5 + X_4^5 - 5X_0 X_1 X_2 X_3 X_4 = 0 \subset \mathbb{P}^4.$$

The variety Y is singular having 125 nodes. We ought to resolve singularities on Y to get a smooth Calabi–Yau threefold, X. Then X is rigid, and has
$$h^{3,0} = 1, h^{2,1} = 0, h^{1,1} = 25, \quad \text{and} \quad E(X) = 50.$$

What is the L–series of X? Looking at the defining equation we would guess the bad prime is 5. In fact, 5 is the only bad prime. The modularity result is given as follows:

Theorem: *The L–series is a modular form, $L(X, s) = L(f, s)$, where f is a normalized newform of weight 4 on $\Gamma_0(25)$, the first few terms of the q–expansion are given as follows:*

$$\begin{aligned}
f(q) &= \eta(q)^4 \eta(q^5)^4 + 5\eta(q)^3 \eta(q^5)^4 \eta(q^{25}) + 20\eta(q)^2 \eta(q^5)^4 \eta(q^{25})^2 \\
&\quad + 25\eta(q)\eta(q^5)^4 \eta(q^{25})^3 + 25\eta(q^5)^4 \eta(q^{25})^4 \\
&= q + q^2 + 7q^3 - 7q^4 + 7q^6 + 6q^7 - 15q^8 + 22q^9 - 43q^{11} - 49q^{12} \\
&\quad - 28q^{13} + 6q^{14} + 41q^{16} + 91q^{17} + 22q^{18} - 35q^{19} + 42q^{21} \\
&\quad - 43q^{22} + 162q^{23} - 105q^{24} - 28q^{26} - 35q^{27} - 42q^{28} + 160q^{29} \\
&\quad + 42q^{31} + 161q^{32} - 301q^{33} + 91q^{34} - 154q^{36} - 314q^{37} + O(q^{38})
\end{aligned}$$

5.2. Hirzebruch's Quintic

This quintic threefold is constructed from a regular pentagon in $\mathbb{R}^2$ with the following vertices $(-\frac{1}{2}, \pm\frac{u\sqrt{2-u}}{2})$, $(\frac{1-u}{2}, \pm\frac{\sqrt{2-u}}{2})$, $(u, 0)$ where $u = \frac{1+\sqrt{5}}{2}$:

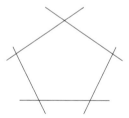

Take the product of the equations of the five lines in the pentagon to get:

$$F(x, y) = \left(x + \frac{1}{2}\right)\left(y^4 - y^2(2x^2 - 2x + 1) + \frac{1}{5}(x^2 + x - 1)^2\right).$$

Let Y be given by the following affine equation:

$$Y : F(x, y) - F(u, w) = 0.$$

Then Y has 126 nodes. Resolving singularities, we obtain a smooth Calabi–Yau threefold, X, with

$$h^{3,0} = 1, h^{2,1} = 0, h^{1,1} = 152, \quad \text{and} \quad E(X) = 304.$$

This was constructed by Hirzebruch in [**Hir**].

The bad primes are 2 and 5, and there are no other bad primes. What is the L–series of X? The result is formulated as follows:

Theorem: The L–series is $L(X, s) = L(f, s)$ where f is a modular form of weight 4 for $\Gamma_0(2 \cdot 5^2) = \Gamma_0(50)$ with the q–expansion:

$$\begin{aligned}f(q) = {}& q - 2q^2 - 2q^3 + 4q^4 + 4q^6 - 26q^7 - 8q^8 - 23q^9 - 28q^{11} - 8q^{12} - 12q^{13} \\ &+ 52q^{14} + 16q^{16} + 64q^{17} + 46q^{18} - 60q^{19} + 52q^{21} + 56q^{22} + 58q^{23} + 16q^{24} \\ &+ 24q^{26} + 100q^{27} - 104q^{28} + 90q^{29} - 128q^{31} - 32q^{32} + 56q^{33} - 128q^{34} \\ &- 92q^{36} - 236q^{37} + 120q^{38} + 24q^{39} + 242q^{41} - 104q^{42} - 362q^{43} - 112q^{44} \\ &- 116q^{46} - 226q^{47} - 32q^{48} + 333q^{49} - 128q^{51} - 48q^{52} + 108q^{53} + O(q^{51}).\end{aligned}$$

5.3. A complete intersection Calabi–Yau threefold of degree $2, 2, 2, 2$ in $\mathbb{P}^7$

van Geemen and Nyggard [**vGN**] considered a complete intersection Y of degree $2, 2, 2, 2$ in $\mathbb{P}^7$ defined by the equations:

$$\begin{aligned}Y_0^2 &= X_0^2 + X_1^2 + X_2^2 + X_3^2, \\ Y_1^2 &= X_0^2 - X_1^2 + X_2^2 - X_3^2, \\ Y_2^2 &= X_0^2 + X_1^2 - X_2^2 - X_3^2, \\ Y_3^2 &= X_0^2 - X_1^2 - X_2^2 + X_3^2.\end{aligned}$$

Let X be a smooth resolution of Y. Then

$$h^{3,0} = 1, h^{2,1} = 0, h^{1,1} = 128, \quad \text{and} \quad E(X) = 256.$$

Judging from the defining equations, we see that 2 is the only bad prime. What is the L–series of X in this case?

LECTURE 1. THE MODULARITY FOR RIGID CALABI–YAU THREEFOLDS

Theorem: The L–series is computed to be $L(X,s) = L(f,s)$ where f is a modular form of weight 4 on $\Gamma_0(8)$, which is expressed by the η–products:

$$f(q) = \eta(q^2)^4 \eta(q^4)^4$$

and its q–expansion is given by:

$$\begin{aligned} f(q) = & \; q - 4q^3 - 2q^5 + 24q^7 - 11q^9 - 44q^{11} + 22q^{13} + 22q^{13} + 8q^{15} \\ & + 50q^{17} + 44q^{19} - 96q^{21} - 56q^{23} - 121q^{25} + 152q^{27} + 198q^{29} \\ & - 160q^{31} + O(q^{33}) \end{aligned}$$

5.4. Verrill's rigid Calabi–Yau threefold

The root lattice A_3 has the Dynkin diagram:

$$\underset{v_1}{\bullet} - \underset{v_2}{\bullet} - \underset{v_3}{\bullet}$$

The roots of A_3 can be given as $\{e_i - e_j \mid 1 \leq i, j \leq 4, \, i \neq j\}$, where $\{e_i \mid i = 1, 2, 3, 4\}$ is the standard basis of $\mathbb{R}^4$. For this root lattice do a bit of toric geometry and combinatorics. To each $e_i - e_j$ associate the monomial $X_i X_j^{-1}$ to get a polynomial. The characteristic polynomial of the adjoint representation is

$$\begin{aligned} \chi_{A_3} &= \sum_{1 \leq i,j \leq 4, i \neq j} X_i X_j^{-1} \\ &= (X_1 + X_2 + X_3 + X_4)(X_1^{-1} + X_2^{-1} + X_3^{-1} + X_4^{-1}) - 4 \end{aligned}$$

and we set

$$\chi_{A_3} = \lambda \in \mathbb{P}^1.$$

This construction gives rise to a toric variety, which is rational. We take a double cover of the base, replacing λ by $(t-1)^2/t$, to get

$$Y : (X_1 + X_2 + X_3 + X_4)(X_1^{-1} + X_2^{-1} + X_3^{-1} + X_4^{-1}) - 4 = \frac{(t-1)^2}{t} \subset \mathbb{P}^3 \times \mathbb{P}^1.$$

Let X be a smooth resolution of Y. Then X is a Calabi–Yau threefold with

$$h^{3,0} = 1, h^{2,1} = 0, h^{1,1} = 50, \quad \text{and} \quad E(X) = 100.$$

This rigid Calabi–Yau threefold was constructed by Verrill [**V**].

The bad primes are 2 and 3. What is the L–series of X?

Theorem: The L–series is $L(X,s) = L(f,s)$ where f is a modular form of weight 4 on $\Gamma_0(6)$, expressed as the η-products:

$$f(q) = \eta(q)^2 \eta(q^2)^2 \eta(q^3)^2 \eta(q^6)^2$$

The q–expansion of $f(q)$ is given as follows:

$$\begin{aligned} f(q) = & \; q - 2q^2 - 3q^3 + 4q^4 + 6q^5 + 6q^6 - 16q^7 - 8q^8 + 9q^9 - 12q^{10} + 12q^{11} \\ & - 12q^{12} + 38q^{13} + 32q^{14} - 18q^{15} + 16q^{16} - 126q^{17} - 18q^{18} + 20q^{19} \\ & + 24q^{20} + 48q^{21} + O(q^{22}). \end{aligned}$$

Exercise: Let Y be a complete intersection in $\mathbb{P}^5$ defined by the equations:

$$X_1 + X_2 + X_3 + X_4 + X_5 + X_6 = 0,$$
$$\frac{1}{X_1} + \frac{1}{X_2} + \frac{1}{X_3} + \frac{1}{X_4} + \frac{1}{X_5} + \frac{1}{X_6} = 0.$$

Let X be a smooth resolution of Y. This is known as the *Barth–Nieto* quintic. Gritsenko and Hulek [**GH**] have considered modular forms attached to this quintic.

We can compute its L-series: $L(X,s) = L(f,s)$ where f is a weight 4 modular form for $\Gamma_0(6)$.

Is there any relation to Verrill's Calabi–Yau threefold? Recently, Hulek, et. al [HSvGvS] have answered this question.

6. Strategy for establishing the modularity conjecture

How can one establish this kind of modularity results? What kind of methods are available for this purpose? We shall describe below three approaches toward this end.

(A) Geometric structures: Firstly it will be helpful if we know the structure of Calabi–Yau threefolds under consideration.

There is a distinguished class of rigid Calabi–Yau threefolds which are endowed with the structure of product varieties.

Theorem 6.1. (Schoen [**Sch2**]) *Let $r : S \to \mathbb{P}^1$ and $r' : S' \to \mathbb{P}^1$ be relatively minimal rational elliptic surfaces with sections. Let $\Sigma(S)$ and $\Sigma(S')$ denote the images of the singular fibers of S and S', respectively. Let $Y := S \times_{\mathbb{P}^1} S'$ be the fiber product. Suppose that the fibers of r and r' over all points in $\Sigma(S) \cap \Sigma(S')$ are all semi-stable, i.e., of type I_b with $b > 0$. Then Y has only ordinary double points as singularities. Let X be the projective variety obtained after small resolutions of these singularities. Then X is a rigid Calabi–Yau threefold exactly when all fibers of r and r' are semi-stable, and, in addition, $Y, Y', \Sigma(S), \Sigma(S')$ and $\Sigma(S) \cap \Sigma(S')$ satisfy certain conditions.*

One special case of the above theorem is especially relevant to our discussion.

Example 6.2. Suppose that S is a rational elliptic modular surface with exactly four semi-stable singular fibers. Beauville [**Bea**] and Schoen [**Sch2**] found all stable elliptic families with four semi-stable singular fibers (in fact, six in all). And then these turn out all to be modular. (This list was also rediscovered by Sebbar [**Seb**] as the torsion-free genus 0 congruence subgroups of $SL(2,\mathbb{Z})$ of index 12 with exactly 4 cusps.) Let S' be a rational elliptic modular surface isogenous to S. Then the fiber product $S \times_{\mathbb{P}^1} S'$ has only nodal singularities. So by Schoen's theorem, we obtain six rigid Calabi–Yau threefolds. These six rigid Calabi–Yau threefolds are all modular from the construction. The L-series of these rigid Calabi–Yau threefolds are expressed in terms of η products.

Proposition 6.3. (M.-H. Saito and Yui [**SY**], and [**Y2**]) *If X is the rigid Calabi–Yau threefold constructed by Verrill, then X is birationally equivalent to a smooth resolution $\widetilde{S \times_{\mathbb{P}^1} S}$ of the fiber product $S \times_{\mathbb{P}^1} S$ where S is an elliptic modular surface of level 6 with (projective) congruence subgroup $\Gamma_1(6) \simeq \Gamma_0(6)$. Then we have the Shimura isomorphism:*

$$H^{3,0}(X) = H^{3,0}(\widetilde{S \times_{\mathbb{P}^1} S}) \simeq \mathcal{S}_4(\Gamma_1(6)) = \mathcal{S}_4(\Gamma_0(6))$$

where $\mathcal{S}_4(G)$ denotes the vector space of cusp forms of weight 4 on a group G. It then follows that L-series $L(X,s) = L(f,s)$ where f is a cusp form of weight 4 on $\Gamma_0(6)$.

Proof: In fact, the elliptic surface S in question is defined by the equation
$$F(X_1, X_2, X_3)\\ = (s+1)X_1X_2X_3 - (X_1 + X_2 + X_3)(X_2X_3 + X_3X_1 + X_1X_2) = 0 \subset \mathbb{P}^2 \times \mathbb{P}^1.$$
This hypersurface has singularities. After resolving them, we obtain a smooth surface. (In fact, this is the universal family of elliptic curves corresponding to the modular curves $X_1(6)$.) Then we see that Verrill's rigid Calabi–Yau threefold defined in 5.4 is birationally equivalent to a smooth resolution of the fiber product $S \times_{\mathbb{P}^1} S$.

(We thank A. Sebbar for pointing out that there is only one projective congruence subgroup of index 6 in $\mathrm{SL}_2(\mathbb{Z})$, so that $\Gamma_1(6) \simeq \Gamma_0(6)$) (projectively).) □

This raises a natural question:

Problems: (1) What is the structure of a rigid Calabi–Yau threefold in general?

(2) Construct rigid Calabi–Yau threefolds which are not birationally equivalent to self-fiber products of elliptic modular surfaces, or of elliptic surfaces.

(B) Serre's criterion: Secondly, we will make use of the criterion of Serre (based on "Faltings' method" [**F**], and formulated by Livné [**L1**]) to establish the equivalence of two 2–dimensional ℓ–adic Galois representations.

First we need to introduce the notion of *non-quadratic* (resp. *non-cubic*) for a set. Let T be a subset of a (finite dimensional) vector space V. Then T is said to be is *non-quadratic* (resp. *non-cubic*) if every homogeneous polynomial of degree $d = 2$ (resp. $d = 3$) on V which vanishes on T vanishes on V. Note that the image of a non-quadratic (resp. non-cubic) set $\subset V$ under a surjective linear map is again non-quadratic (resp. non-cubic).

Serre's criterion: *Let $\mathbb{Q}_2$ be the field of 2-adic rational numbers. Pick a finite set, S, of rational primes, and let $\mathbb{Q}_S$ denote the compositum of all quadratic extensions of $\mathbb{Q}$ unramified outside S. Suppose*
$$\rho_1, \rho_2 : \mathrm{Gal}(\overline{\mathbb{Q}}/\mathbb{Q}) \to GL_2(\mathbb{Q}_2)$$
are continuous Galois representations, unramified outside S.

Furthermore, suppose that ρ_1, ρ_2 satisfy the following conditions:

(1)
$$\mathrm{trace}\,\rho_1 \equiv \mathrm{trace}\,\rho_2 \equiv 0\,(\mathrm{mod}\,2),$$
$$\det \rho_1 \equiv \det \rho_2\,(\mathrm{mod}\,2).$$

(2) There exists a finite set T of rational primes, with $T \cap S = \emptyset$, such that

(i) the image of the set $\{\mathrm{Frob}_p\}_{p \in T}$ in (the $(\mathbb{Z}/2\mathbb{Z})$–vector space) $\mathrm{Gal}(\mathbb{Q}_S/\mathbb{Q})$ is non-cubic, and

(ii) for all $p \in T$,
$$\mathrm{trace}\,\rho_1(\mathrm{Frob}_p) = \mathrm{trace}\,\rho_2(\mathrm{Frob}_p),$$
$$\det \rho_1(\mathrm{Frob}_p) = \det \rho_2(\mathrm{Frob}_p).$$

Then the two Galois representations ρ_1 and ρ_2 have isomorphic semi-simplifications, so that
$$L(\rho_1, s) = L(\rho_2, s).$$

The modularity of the rigid Calabi–Yau threefolds discussed in the previous section can be established using the Serre criterion, namely, by comparing the two

2–adic Galois representations arising from a rigid Calabi–Yau threefold in question. They are:
$$\rho_1 : \mathrm{Gal}(\overline{\mathbb{Q}}/\mathbb{Q}) \to \mathrm{Aut}(H^3_{et}(X_{\overline{\mathbb{Q}}}, \mathbb{Q}_2)) \cong GL_2(\mathbb{Q}_2)$$
and
$$\rho_2 : \mathrm{Gal}(\overline{\mathbb{Q}}/\mathbb{Q}) \to GL_2(\mathbb{Q}_2)$$
attached to a modular form of weight 4 on some $\Gamma_0(N)$. For details, see Schoen[**Sch1**], van Geemen and Nyggard[**vGN**], Werner and van Geemen [**WvG**], and Verrill[**V**].

(**C**) **Wiles' method**: We may invoke Wiles' technique and its far-reaching generalizations on the modularity of 2–dimensional Galois representations. We follow the exposition of Diamond [**Di**], and Diamond's private note on the Fontaine–Mazur conjecture for continuous 2–dimensional Galois representations. (See also a recent preprint of Richard Taylor [**Tay**].) For this, we ought to study continuous 2–dimensional Galois representations arising from rigid Calabi–Yau threefolds over $\mathbb{Q}$. Suppose that
$$\rho : \mathrm{Gal}(\overline{\mathbb{Q}}/\mathbb{Q}) \to \mathrm{Aut}(H^3(X_{\overline{\mathbb{Q}}}, \mathbb{Q}_\ell)) \simeq GL_2(\mathbb{Q}_\ell)$$
is a continuous 2–dimensional Galois representation associated to a rigid Calabi–Yau threefold over $\mathbb{Q}$, where ℓ is some rational prime. Impose some hypothesis on the ρ, such as, ρ is unramified outside a finite set S of primes, and has a crystalline restriction to $\mathrm{Gal}(\overline{\mathbb{Q}}_\ell/\mathbb{Q}_\ell)$ of Hodge–Tate type $(0,3)$ with $3 < \ell - 1$. Suppose that we can find just one rational ℓ such that
$$\bar{\rho} : \mathrm{Gal}(\overline{\mathbb{Q}}/\mathbb{Q}) \to GL_2(\mathbb{F}_\ell)$$
is modular and has absolutely irreducible restriction to $\mathrm{Gal}_L = \mathrm{Gal}(\overline{\mathbb{Q}}/L)$ where $L = \mathbb{Q}(\sqrt{(-1)^{(\ell-1)/2}\ell})$. Then ρ itself is modular.

With Wiles' method, one has only to find one good prime $\ell \geq 5$ and then establish the modularity for the residual mod ℓ Galois representation $\bar{\rho}$ satisfying a certain absolute irreducibility restriction. Then the modularity property is contagious and spread to all primes establishing the modularity of ρ itself. We may test the criteria with $\ell = 5$ provided that 5 is a good prime (if not try $\ell = 7$). However, the real challenge is to prove the mod 5 representation is modular, for instance, the image may not be solvable!

7. The intermediate Jacobian of a Calabi–Yau threefold

Now we use the Hodge theory to define the intermediate Jacobian of a Calabi–Yau threefold. Let X be a rigid Calabi–Yau threefold, which we consider over $\mathbb{C}$. Then there is the Hodge decomposition:
$$H^3(X, \mathbb{C}) = H^{3,0}(X) \oplus H^{2,1}(X) \oplus H^{1,2}(X) \oplus H^{0,3}(X) = H^{3,0}(X) \oplus H^{0,3}(X).$$
The Hodge filtration is given by
$$H^3(X, \mathbb{C}) = F^0 \supset F^1 \supset F^2 \supset F^3$$
where
$$\begin{aligned} F^1 &= H^{3,0}(X) \oplus H^{2,1}(X) \oplus H^{1,2}(X) = H^{3,0}(X), \\ F^2 &= H^{3,0}(X) \oplus H^{2,1}(X) = H^{3,0}(X), \\ F^3 &= H^{3,0}(X). \end{aligned}$$

The intermediate Jacobian $J^2(X)$ of X is defined by

$$J^2(X) = \frac{H^3(X, \mathbb{C})}{F^2 H^3(X, \mathbb{C}) + H^3(X, \mathbb{Z})} \cong H^{3,0}(X)^\vee / H_3(X, \mathbb{Z})$$

where $H^{3,0}(X)^\vee$ denotes the dual of $H^{3,0}(X)$. Then $J^2(X)$ is a complex torus $E(\mathbb{C})$ of dimension 1 such that $J^2(X) \simeq E(\mathbb{C})$.

In what follows we consider rigid Calabi–Yau threefolds defined over a number field K, and assume that

The intermediate Jacobian $J^2(X)$ is equipped with a K–rational point, and that there is a K–isomorphism $J^2(X) \simeq E$ where K is a number field, and E is an elliptic curve defined over K.

For instance, this assumption may be fulfilled if $J^2(X)$ is of CM-type.

Conjecture 7.1. A rigid Calabi–Yau threefold X over $\mathbb{Q}$ is modular if and only if the intermediate Jacobian $J^2(X)$ is defined over $\mathbb{Q}$ and hence is modular.

Remark 7.2. The conjecture, however, is not claiming that $L(X_K, s) = L(E_K, s)$. Rather there should be some relations between the two L–series, e.g., if $L(E_K, s)$ has a q–expansion : $L(E_K, s) = \mathcal{G}(q)$, then there should be a polynomial $\mathcal{F} = \mathcal{F}(\mathcal{G})$ depending on $\mathcal{G}$ and a positive integer k (e.g., $k = 2$) such that $L(X_K, s) = \mathcal{F}(\mathcal{G})(q^k)$.

Remark 7.3. Now that the conjecture of Shimura–Taniyama–Weil is settled in its totality, all elliptic curves over $\mathbb{Q}$ are modular. Therefore, to prove the conjecture 7.1, it is sufficient to show that the intermediate Jacobian $J^2(X)$ is defined over $\mathbb{Q}$.

Problem: We know that an elliptic curve (over $\mathbb{C}$) can be defined by one complex number $z \in \mathfrak{H}$, which is defined by the quotient of period integrals. In the same vein, we may be able to define the intermediate Jacobian $J^2(X)$ of a rigid Calabi–Yau threefold X by a single complex number $z \in \mathfrak{H}$. We know that $H^3(X, \mathbb{C})/H^3(X, \mathbb{Z})$ is a dimension one complex torus so it should be an elliptic curve. We define the map

$$H_3(X, \mathbb{Z}) \to H^0(X, \Omega^3)^\vee, \quad \gamma \mapsto \int_\gamma \omega$$

where ω is a unique holomorphic 3–form on X and $\gamma \in H_3(X, \mathbb{Z})$ is a topological cycle. Then the image of this map is a lattice in $H^0(X, \Omega^3) \simeq \mathbb{C}$. Suppose that $H_3(X, \mathbb{Z}) \simeq \mathbb{Z}\gamma_1 + \mathbb{Z}\gamma_2$. We ought to find the images $\int_{\gamma_1} \omega$ and $\int_{\gamma_2} \omega$ and then their quotient to obtain a complex number.

Characterize complex numbers $z \in \mathfrak{H}$ obtained above so that the associated elliptic curves are defined over $\mathbb{Q}$.

Some initial attempt for calculating the intermediate Jacobians of the rigid Calabi–Yau threefolds in Example 6.2 has been made by William Stein and Helena Verrill. For instance, let $S : x^3 + y^3 + z^3 - \lambda xyz = 0 \subset \mathbb{P}^2 \times \mathbb{P}^1$ be a pencil of elliptic curves with parameter λ. Then S has exactly four semi-stable singular fibers all of type I_b where b is the number of $\mathbb{P}^1$ in the singular fiber. Take the fiber product over the base to obtain a variety in $\mathbb{P}^2 \times \mathbb{P}^2$. For instance, the fiber product $S \times_{\mathbb{P}^1} S$ is defined by the hypersurface:

$$rst(x^3 + y^3 + z^3) = (r^3 + s^3 + t^3)xyz \subset \mathbb{P}^2 \times \mathbb{P}^2$$

where $[x:y:z]$ and $[r:s:t]$ are homogeneous coordinates for $\mathbb{P}^2$. Then the intermediate Jacobian of the rigid Calabi–Yau threefold thus obtained has j–invariant 0.

8. The order of vanishing of L–series at $s=2$

In this section, we will confine ourselves to rigid Calabi–Yau threefolds and their L–series. We will postpone our discussions on non-rigid Calabi–Yau threefolds (orbifolds) defined over number fields and their L–series to Lecture 2.

Let X be a rigid Calabi–Yau threefold defined over $\mathbb{Q}$ or a number field K. Suppose that we can determine their L–series. Now we may ask

Problem: What is the order of vanishing of the L–series $L(X_\mathbb{Q}, s)$ (or $L(X_K, s)$) at $s=2$?

There is a fairly good algorithm to do this, for instance, if one has the functional equation for the L–series in question, with explicit knowledge of the conductor, the order of vanishing may be empirically determined with high degree of accuracy. See for instance, Buhler, Schoen and Top [**BST**]. For the examples discussed below (8.1–8.4), the actual calculation of the central critical values of L–series at $s=2$ have been carried out by Ling Long (a graduate student at Penn State).

8.1. Chad Schoen's quintic. We know that the L–series is the Mellin transform of a modular form of weight 4 on $\Gamma_0(25)$. In fact, the L–series is expressed in terms of Jacobi sums.

Theorem: (Bloch [**Bl2**]). $L(X_\mathbb{Q}, 2) \neq 0$. But, over the quadratic extension $\mathbb{Q}(\sqrt{5})$, we have $L(X_{\mathbb{Q}(\sqrt{5})}, s)$ vanishes to the order 1 at $s=2$.

Remark 8.1. The above theorem may be reformulated as follows: Let ψ be the quadratic character associated to the field $\mathbb{Q}(\sqrt{5})$. Then the (twisted) L–series $L(\psi \otimes H^3(\overline{X}, \mathbb{Q}_l, s))$ vanishes to order 1 at $s=2$. Numerically we compute that the central critical value of $L(X_\mathbb{Q}, 2)$ is approximately $1.565244414\cdots$, from which we conclude that $L(X_\mathbb{Q}, 2) \neq 0$. The sign of the functional equation is -1. This implies that the L–series is *odd* in the sense that if L–series vanishes at $s=2$ over some finite extension of $\mathbb{Q}$, then the order of vanishing ought to be odd. That over the field extension $\mathbb{Q}(\sqrt{5})$, the L–series vanishes to the order 1 at $s=2$ is the main result of Bloch [**Bl2**]

8.2. Hirzebruch's quintic. We know that the L–series is the Mellin transform of a modular form of weight 4 on $\Gamma_0(50)$.

Theorem: $L(X_\mathbb{Q}, 2) = 0$, but $L'(X_\mathbb{Q}, 2) \neq 0$. Furthermore, $\mathrm{ord}_{s=2} L(X_\mathbb{Q}, s) = 1$.

Proof: The central critical value is 0. The sign of the functional equation is -1, so that the order of $L(X_\mathbb{Q}, s)$ at $s=2$ is odd. Since the first derivative $L'(X_\mathbb{Q}, s)$ does not vanish at $s=2$, the order of vanishing of $L(X_\mathbb{Q}, s)$ at $s=2$ must be 1. $\square$

8.3. A complete intersection Calabi–Yau threefold of degree $(2,2,2,2)$. We know that the L–series is the Mellin transform of a modular form of weight 4 on $\Gamma_0(8)$. This modular form is expressed in terms of the η-functions: $\eta(q^2)^4 \eta(q^4)^4$.

Again computing the central critical value, and the sign of the functional equation, we reach our conclusion:

Proposition: $L(X_\mathbb{Q}, 2) \neq 0$.

8.4. Verrill's rigid Calabi–Yau threefold. We know that the L–series is the Mellin transform of a modular form $f(q) = \eta(q)^2 \eta(q^2)^2 \eta(q^3)^3 \eta(q^6)^2$ of weight 4 on $\Gamma_0(6)$. The central critical value is approximately $.5097104242 \cdots$ and the sign of the functional equation is $+1$. Therefore, we may conclude:

Proposition: $L(X_\mathbb{Q}, 2) \neq 0$.

Problem: When $L(X_\mathbb{Q}, 2) \neq 0$, for which algebraic extension of $\mathbb{Q}$, the L–series will vanish at $s = 2$?

9. The conjecture of Beilinson and Bloch

Now we interpret the order of vanishing at $s = 2$ of the L–series of rigid Calabi–Yau threefolds over $\mathbb{Q}$ or a number field K in the context of the conjecture of Beilinson and Bloch, that is, in connection with the rank of the Chow group $CH^2(X)_{hom}$ of X (see Beilinson [**Be**], and Bloch [**Bl1**], [**Bl2**]).

9.1. Dimension One Case. First we recall the structure of the Mordell–Weil group on an elliptic curve E defined over a number field K. By Mordell's theorem, the Mordell–Weil group $E(K) = \mathrm{Pic}^0(E_K)$ of K–rational points on E is a finitely generated Abelian group. The conjecture of Birch and Swinnerton-Dyer [**BS-D**] predicts a relation between the rank of $E(K)$ and the order of vanishing of the L–series at $s = 1$:

$$\mathrm{rank}_\mathbb{Z} E(K) = \mathrm{order}_{s=1} L(E_K, s).$$

There is ample evidence in support of the conjecture of Birch and Swinnerton-Dyer for elliptic curves over $\mathbb{Q}$. Moreover, now that the validity of the conjecture of Shimura–Taniyama–Weil has been established for elliptic curves over $\mathbb{Q}$ in totality, it might be said that a complete proof for the conjecture of Birch and Swinnerton-Dyer would be in sight.

9.2. Dimension Three Case. For any smooth projective threefold (e.g., Calabi–Yau threefold) X defined over a number field K, the Chow group of null homologous cycles of co-dimension 2 on X, $CH^2(X_K)_{hom}$, is the object what replaces the Mordell–Weil group $E(K)$ of an elliptic curve E over K. This group is defined to be the kernel of the cycle class map:

$$CH^2(X_K)_{hom} := \mathrm{Ker}\left[CH^2(X_K) \to H^4(X_{\bar{K}}, \mathbb{Z}_l(2))\right] \quad (l : \text{prime})$$

and it is an Abelian group. However, it is not known whether it is finitely generated or not. Beilinson [**Be**] and Bloch [**Bl1**],[**Bl2**] formulated the following conjecture which generalizes the conjecture of Birch and Swinnerton-Dyer to smooth projective threefolds defined over number fields.

Conjecture 9.1. (Beilinson–Bloch) Let X be a smooth projective threefold defined over a number field K. Then X satisfies the following properties:

(1) $CH^2(X_K)_{hom}$ is finitely generated, and

(2) the rank of $CH^2(X_K)_{hom}$ is equal to the order of the L–series of X_K at $s = 2$:
$$\text{rank}_\mathbb{Z} CH^2(X_K)_{hom} = \text{ord}_{s=2} L(X_K, s).$$

The Chow group $CH^2(X_K)$ is a notoriously difficult object to compute. Consequently, there is very little evidence to support the conjecture of Beilinson and Bloch, though there are several results along this line.

Example 9.2. As we discussed in 8.1, Bloch [**Bl2**] computed the L–series and its order at $s = 2$ for Chad Schoen's quintic. Also Bloch has conjectured about the filtration of the Chow groups to establish the validity of the above conjecture:
$$F^0 CH^2(X_\mathbb{Q})_\mathbb{Q} = 0, \quad \text{but}$$
$$F^0 CH^2(X_{\mathbb{Q}(\sqrt{5})})_\mathbb{Q} \cong gr^0 CH_2(X_{\mathbb{Q}(\sqrt{5})})_\mathbb{Q} \cong \mathbb{Q}$$
If this is true, then
$$\text{rank}_\mathbb{Z} CH^2(X_{\mathbb{Q}(\sqrt{5})})_{hom} = 1 \quad \text{and} \quad \text{ord}_{s=2} L(X_{\mathbb{Q}(\sqrt{5})}, s) = 1$$
establishing the validity of the conjecture of Beilinson and Bloch.

Theorem 9.3. *Let X be Hirzebruch's rigid Calabi–Yau threefold defined over $\mathbb{Q}$ introduced in* **5.2**. *Then we have a partial support of the conjecture of Beilinson and Bloch for $X_\mathbb{Q}$. That is,*
$$\text{ord}_{s=2} L(X_\mathbb{Q}, s) = 1 \leq \text{rank}_\mathbb{Z} CH^2(X_\mathbb{Q})_{hom}.$$

Proof: By the discussions in 8.2, we know that the L–series of Hirzebruch's rigid Calabi–Yau threefold $X_\mathbb{Q}$ vanishes to the order 1 at $s = 2$. Kimura [**Ki**] has constructed explicitly a codimension 2 algebraic cycles in $CH^2(X_\mathbb{Q})_{hom}$ with non-trivial Abel–Jacobi image. This gives rise to the inequality:
$$\text{rank}_\mathbb{Z} CH^2(X)_{hom} \geq 1 = \text{ord}_{s=2} L(X_\mathbb{Q}, s).$$
□

That there are no more algebraic cycles in $CH^2(X_\mathbb{Q})_{hom}$ is still an open problem.

As the above two examples indicate, the problem seems more tractable for rigid Calabi–Yau threefolds over $\mathbb{Q}$. We shall digress and reformulate the conjecture of Beilinson and Bloch in the case of rigid Calabi–Yau threefolds over $\mathbb{Q}$, and close the first lecture by proposing the following yet another conjecture.

Conjecture 9.4. Let X be a rigid Calabi–Yau threefold defined over $\mathbb{Q}$. Then the following properties hold:

(1) The intermediate Jacobian $J^2(X)$ is also defined over $\mathbb{Q}$, and has a $\mathbb{Q}$–rational point.

(2) The Abel-Jacobi map (modulo torsion)
$$\Phi_\mathbb{Q} : CH^2(X_\mathbb{Q})_{hom} \otimes \mathbb{Q} \to J^2(X_\mathbb{Q}) \otimes \mathbb{Q}$$
is injective.

(3) We have the inequalities:
$$\begin{array}{ccc} \text{rank}_\mathbb{Z} CH^2(X_\mathbb{Q})_{hom} & \leq & \text{rank}_\mathbb{Z} E(\mathbb{Q}) \\ \| & & \| \\ \text{ord}_{s=2} L(X_\mathbb{Q}, s) & \leq & \text{ord}_{s=1} L(E_\mathbb{Q}, s) \end{array}$$

where the equality on the left (resp. right) is the conjecture of Beilinson and Bloch (resp. the conjecture of Birch and Swinnerton-Dyer).

(4) $CH^2(X_\mathbb{Q})_{hom}$ is finitely generated.

Remark 9.5. In order to establish the injectivity of the Abel–Jacobi map $\Phi_\mathbb{Q}$ modulo torsion, it suffices to prove the second Bloch–Beilinson filtration

$$F^2CH^2(X)_\mathbb{Q} := Ker\,\Phi_\mathbb{Q} = 0.$$

LECTURE 2
Arithmetic of Orbifold Calabi–Yau varieties over Number Fields

Let us recall from the Lecture 1 that a smooth projective variety X of dimension d defined over a field K is a Calabi–Yau variety if
(1) $H^i(X, \mathcal{O}_X) = 0$ for every i, $0 < i < d$,
(2) $K_X \simeq \mathcal{O}_X$ ($\Rightarrow p_g(X) = 1$).

In this lecture we shall describe a specific construction of Calabi–Yau varieties of dimension $d \leq 3$. For the necessary background on geometry and physics of mirror symmetry, the reader is referred to the excellent survey articles by Morrison [**M2**], and Cox and Katz [**CK**]. For fuller details of arithmetic nature, the reader may consult the articles by Yui [**Y1**], [**Y2**] and [**Y3**].

1. Fermat hypersurfaces and their deformations

Let K be a field (e.g., $K = \mathbb{Q}$, or a number field). Let $\bar{K}$ denote its algebraic closure. Let $F(X_0, X_1, \ldots X_{d+1}) \in K[X_0, X_1, \ldots X_{d+1}]$ be a homogeneous polynomial of degree $m \geq d+1$. For instance, $F(X_0, X_1, \cdots, X_{d+1})$ may be a Fermat hypersurface:

$$F(X_0, X_1, \cdots, X_{d+1}) = X_0^m + X_1^m + \cdots + X_{d+1}^m = 0 \subset \mathbb{P}^{d+1}$$

or its deformations, or more generally non-Fermat hype hypersurfaces.

Question: *When is the hypersurface* $V : F(X_0, X_1, \ldots X_{d+1}) = 0 \subset \mathbb{P}^{d+1}$ *a Calabi–Yau variety?*

We have the canonical bundle $K_V \simeq \mathcal{O}_V(m - (d+2))$, so that $K_V \simeq \mathcal{O}_V$ if and only if $m = d + 2$.

For instance, this gives rise to Calabi–Yau Fermat hypersurfaces and their deformations as in the following examples:

Example 1.1. In all the examples below, λ denotes a parameter.
(a) If $d = 1$, then $m = 3$, and we get the Fermat cubic:

$$X_0^3 + X_1^3 + X_2^3 = 0 \subset \mathbb{P}^2,$$

or its one-parameter deformation:
$$X_0^3 + X_1^3 + X_2^3 + 3\lambda X_0 X_1 X_2 = 0 \subset \mathbb{P}^2 \times \mathbb{P}^1.$$

(b) If $d = 2$, then $m = 4$, and we get the Fermat quartic:
$$X_0^4 + X_1^4 + X_2^4 + X_3^4 = 0 \subset \mathbb{P}^3,$$

or its one-parameter deformation:
$$X_0^4 + X_1^4 + X_2^4 + X_3^4 + 4\lambda X_0 X_1 X_2 X_3 = 0 \subset \mathbb{P}^3 \times \mathbb{P}^1.$$

(c) If $d = 3$, then $m = 5$, and we get the Fermat quintic:
$$X_0^5 + X_1^5 + X_2^5 + X_3^5 + X_4^5 = 0 \subset \mathbb{P}^4,$$

or its one-parameter deformation:
$$X_0^5 + X_1^5 + X_2^5 + X_3^5 + X_4^5 + 5\lambda X_0 X_1 X_2 X_3 X_4 = 0 \subset \mathbb{P}^4 \times \mathbb{P}^1.$$

Remark 1.2. For proof of the fact that deformations of Calabi–Yau varieties are also Calabi–Yau can be found, for instance, in Cox and Katz [**CK**], Prop. (4.1.3). There is a subtle difference between Fermat varieties and deformations of Fermat varieties. Fermat varieties are dominated by product varieties, while their deformations are not. (See Schoen [**Sch3**].)

These examples represent too small classes of Calabi–Yau varieties. Henceforth our next endeavor is to construct more Calabi–Yau varieties, by passing to weighted projective varieties. For generalities on weighted projective varieties, see, for instance, Dolgachev [**Dol1**].

2. Orbifold Calabi–Yau varieties

The starting variety is the Fermat hypersurface of degree m and dimension d,
$$V : X_0^m + X_1^m + \cdots + X_{d+1}^m = 0 \subset \mathbb{P}^{d+1}.$$
The geometric genus of V is $p_g(V) = \binom{m-1}{d+1}$. If $m > d+2$ then $p_g > 1$ and so V is not a Calabi–Yau variety if $m > d+2$. The Hodge numbers of V are $h^{ij} = 0$ if $i + j \neq d$ and $i \neq j$, and $= 1$ if $i + j \neq d$ but $i = j$.

In order to construct more examples of Calabi–Yau varieties, we shall pick a *weight*: $Q = (q_0, q_1, \cdots, q_{d+1}) \in \mathbb{N}^{d+2}$, satisfying

(1) $\gcd(q_0, \cdots, \hat{q}_j, \cdots, q_{d+1}) = 1$ for every j, and

(2) q_i divides m for each i.

For each i define
$$\mu_{q_i} := \mathrm{Spec}(K[T]/(T^{q_i} - 1)).$$
This is a finite group scheme. In fact, passing onto the closure $\bar{\mathbb{Q}}$, we have the realization: $\mu_{q_i}(\bar{\mathbb{Q}}) \cong \mathbb{Z}/q_i\mathbb{Z}$ which is a finite group of order q_i. Take the direct product of μ_{q_i}, $i = 0, 1, \cdots, n+1$, to define
$$\mu_Q := \mu_{q_0} \times \mu_{q_1} \cdots \times \mu_{q_{d+1}}.$$
This finite group scheme acts on V. In fact, over the algebraic closure $\bar{\mathbb{Q}}$ the action is given by
$$\mu_Q : (X_0, X_1, \cdots, X_{d+1}) \mapsto (\xi_{q_0}^{e_0} X_0, \cdots, \xi_{q_{d+1}}^{e_{d+1}} X_{d+1})$$

where $e_i \in \mathbb{Z}/q_i\mathbb{Z}$ for each i. Define the quotient

$$Y := V/\mu_Q$$

Then Y is a singular variety with abelian quotient singularities.

To resolve the singularities on Y we use the following result.

Theorem 2.1. (Greene, Roan and S.-T. Yau [**GRY**]) *Let W be a singular algebraic variety of dimension $d \leq 3$ with at most abelian quotient singularities with trivial canonical sheaf ω_W. Then there exists a resolution Z of W, which is smooth, with trivial canonical bundle K_Z. (Such a resolution may be called a crepant resolution.)*

Problem (1) Determine all pairs $< m, Q = (q_0, q_1, \ldots q_{d+1}) >$ giving rise to a singular Calabi–Yau varieties Y, that is, $\omega_Y \simeq \mathcal{O}_Y$.

(2) Once we have a singular Calabi–Yau variety Y, then we wish to have a resolution, X, of Y which is a crepant resolution, i.e.,

$$K_X = \pi^* \omega_Y = \pi^* \mathcal{O}_Y = \mathcal{O}_X.$$

We call smooth Calabi–Yau varieties obtained via the above procedure *orbifold* Calabi–Yau varieties.

The solutions to the above problem are collected in the following theorem.

Theorem 2.2. (Yui [**Y1**] and [**Y2**]; cf. Goto [**G**] for $d = 2$)
(a) *If $d = 1$, there exist three pairs:*

$$< 3, (1, 1, 1) >, \; < 4, (1, 1, 2) >, \; < 6, (1, 2, 3) > .$$

(b) *If $d = 2$, there exist 14 pairs all of which are listed as below:*

$$< 4, (1, 1, 1, 1) >, \quad < 6, (1, 1, 1, 3) >,$$
$$< 6, (1, 1, 2, 2) >, \quad < 8, (1, 1, 2, 4) >,$$
$$< 10, (1, 2, 2, 5) >, \quad < 12, (2, 3, 3, 4) >,$$
$$< 12, (1, 3, 4, 4) >, \quad < 12, (1, 2, 3, 6) >,$$
$$< 12, (1, 1, 4, 6) >, \quad < 18, (1, 2, 6, 9) >,$$
$$< 20, (1, 4, 5, 10) >, \quad < 24, (1, 3, 8, 12) >,$$
$$< 30, (2, 3, 10, 15) >, \quad < 42, (1, 6, 14, 21) > .$$

(c) If $d = 3$, there exist 147 pairs, some of which are listed below:

$$< 5, (1,1,1,1,1) >, \qquad < 6, (1,1,1,1,2) >,$$
$$< 8, (1,1,1,1,4) >, \qquad < 8, (1,1,2,2,2) >,$$
$$< 9, (1,1,1,3,3) >, \qquad < 10, (1,1,1,2,5) >,$$
$$\ldots \qquad \ldots$$
$$< 12, (2,2,2,3,3) >, \qquad < 15, (1,3,3,3,5) >,$$
$$\ldots \qquad \ldots$$
$$< 60, (1,12,12,1,5,20) >, \qquad < 60, (3,5,12,20,20) >,$$
$$\ldots \qquad \ldots$$
$$< 714, (3,14,102,238,357) >, \quad < 924(1,21,102,238,357) >,$$
$$< 966, (2,21,138,322,483) >, \quad < 1806, (1,42,258,602,903) > .$$

In particular, none of these 147 orbifold Calabi–Yau threefolds are rigid.

Remark 2.3. In Theorem 2.2, one notes that

If dim=1, then the largest degree = 6.
If dim=2, then the largest degree = 42.
If dim=3, then the largest degree = 1806.

Consider the sequence

$$1, 2, 6, 42, 1806$$

Note that $1, 2 = 1 \cdot 2, 6 = 2 \cdot 3, 42 = 6 \cdot 7, 1806 = 42 \cdot 43$, so the next term in the sequence should be $1806 \cdot 1807$. But whereas $2, 3, 7, 43$ are prime, 1807 is not, in fact, $1807 = 13 \cdot 139$. This is related to "fake Fermat's Little theorem": The congruence

$$a^{\phi(m)} \equiv 1 \pmod{m} \; \forall m$$

is Fermat's Little Theorem. Replace this by the "false" statement:

$$a^{m+1} \equiv a \pmod{m} \; \forall a$$

The entire list of solutions m to this congruence is given by the above sequence. In fact, a solution m must have the following properties: (1) m is square-free, and (2) if $m = \prod_{i=1}^{r} p_i$ is a product of distinct primes p_i, then $p_i - 1 | m$ for every $i < r$ and $m = p_r(p_r - 1)$ with $p_r = p_1 \cdots p_{r-1} + 1$.

Perhaps this provides a number theoretic reason for why we have to take dimension ≤ 3 in Theorem 2.1.

Remark 2.4. For $d = 2$, a list of complete intersection Calabi–Yau orbifolds in weighted projective 3–spaces was obtained by Reid [**Re**] and later rediscovered by Yonemura [**Yon**]. There are in total 95 admissible weights giving rise to $K3$ surfaces. Among those 95 of Reid–Yonemura, the above 14 pairs are realized as orbifolds of Fermat hypersurfaces.

Remark 2.5. For $d = 3$, Klemm and Schimmrigk [**KSc**] and also Kreuzer and Skarke [**KSk**] have compiled complete list of transversal Calabi–Yau hypersurfaces in weighted projective 4–spaces. There are in total 7,555 weights which admit Calabi–Yau hypersurfaces. Our 147 pairs of Fermat type are among them. There is also a sublist of Calabi–Yau hypersurfaces in weighted projective 4–spaces which carry $K3$ fibrations by Hosono, Lian and S.-T. Yau [**HLY1**], and Hunt and Schimmrigk [**HS**] give thorough discussion about $K3$ fiber structures. The above 147

Calabi–Yau of Fermat type do not necessarily carry $K3$ fibrations. For further discussions on $K3$ fibered Calabi–Yau threefolds, the reader should consult the articles [**HLY1**] and [**HS**].

Remark 2.6. For $d \geq 2$, the Calabi–Yau orbifolds in Theorem 2.2 are again covered by product varieties in weighted projective spaces. For fuller discussion, the reader is recommended to the articles by Shioda and Katsura [**SK**], and by Hunt and Schimmrigk [**HS**].

Further we consider deformations of the Calabi–Yau orbifolds obtained in Theorem 2.2. Since deformation process will not change the triviality of the canonical bundle, the deformed varieties will be again Calabi–Yau varieties. However, deformations are no longer realized as quotients of product varieties, nor dominated by product varieties (see Schoen [**Sch3**]). Details will be discussed in §8 below where we will construct mirror Calabi–Yau varieties.

Example 2.7. One– or multi–parameter deformations of the Calabi–Yau orbifolds in Theorem 2.2 are again Calabi–Yau varieties.

(a) If $d = 1$, we obtain three one–parameter deformations

$$\tilde{E}_6 \quad Y_0^3 + Y_1^3 + Y_2^3 + \lambda Y_0 Y_1 Y_2 = 0 \quad \subset \mathbb{P}^2((1,1,1)),$$
$$\tilde{E}_7 \quad Y_0^4 + Y_1^4 + Y_2^2 + \lambda Y_0 Y_1 Y_2 = 0 \quad \subset \mathbb{P}^2((1,1,2)),$$
$$\tilde{E}_8 \quad Y_0^6 + Y_1^3 + y_2^2 + \lambda Y_0 Y_1 Y_2 = 0 \quad \subset \mathbb{P}^2((1,2,3)).$$

(b) If $d = 2$, the 14 pairs give rise to multi–parameter deformations. For instance, for $<m, Q> = <10, (1, 2, 2, 5)>$, we get

$$Y_0^{10} + Y_1^5 + Y_2^5 + Y_3^2 + \lambda Y_0 Y_1 Y_2 Y_3 + \phi Y_0^5 Y_1^5 = 0 \subset \mathbb{P}^3((1,1,2,5)).$$

For $<m, Q> = <42, (1, 6, 14, 21)>$, we get

$$Y_0^{42} + Y_1^7 + Y_2^3 + Y_3^2 + \lambda Y_0 Y_1 Y_2 Y_3 = 0 \subset \mathbb{P}^3((1,6,14,21)).$$

(c) Let $d = 3, m = 12$ and $Q = (1, 1, 2, 2, 6)$. We get a Calabi–Yau threefold defined by the equation:

$$Y_0^{12} + Y_1^{12} + Y_2^6 + Y_3^6 + Y_4^2 + \lambda Y_0 Y_1 Y_2 Y_3 Y_4 + \phi Y_2^3 Y_3^3 = 0 \subset \mathbb{P}^4((1,1,2,2,6)).$$

Let $d = 3, m = 24$ and $Q = (1, 1, 2, 8, 12)$. We get a Calabi–Yau threefold defined by the equation:

$$Y_0^{24} + Y_1^{24} + Y_2^{12} + Y_3^3 + Y_4^2 + \lambda Y_0 Y_1 Y_2 Y_3 Y_4 + \phi Y_0^{12} Y_1^{12} = 0 \subset \mathbb{P}^4((1,1,2,8,12)).$$

3. Sketch of proof of Theorem 2.2

We know by a general theorem in weighted projective geometry that

$$\omega_Y \simeq \mathcal{O}_Y(m - q_0 - q_1 \cdots - q_{d-1})$$

so it follows that $\omega_Y \simeq \mathcal{O}_Y$ when

(3) $m = q_0 + \cdots + q_{d+1}$.

Therefore to answer the above **Problem** (1), we look for pairs $<m, Q>$ satisfying the conditions (1), (2) in §2, and the condition (3) above.

(If dimension $d > 3$, the condition (3) is *sufficient*, but not *necessary* as was pointed out by Greene, Roan and S.-T. Yau [**GRY**].)

To answer the above **Problem** (2), we need to have explicit knowledge about singularities on the quotient Y. Then we look for a crepant resolution so that the desingularization process will not change the triviality of the canonical bundle. The Hodge number $h^{1,0}(X) = 0$ for both dimensions $d = 2$ and $d = 3$. By the Lefschetz hyperplane theorem, we know that our Fermat hypersurfaces V have $h^{1,0}(V) = 0$ for both dimensions $d = 2$ and $d = 3$. By taking orbifolds of V by finite automorphism groups of V, the Hodge numbers $h^{1,0}(X) = 0$ for both dimensions $d = 2$ and $d = 3$ as the exceptional sets of singularities arise all from the singularities of weighted projective spaces. For $d = 3$, $h^{2,0}(X) = h^{1,0}(X)$ by Serre's duality. The Hodge number $h^{1,1}$ is in general non-zero, and may be calculated using Roan's formula in [**Ro3**].

Let $\Sigma(Y)$ denote the singular locus of Y.

Theorem 3.1. (Goto [**G**]) *Let $d = 2$. If $P \in \Sigma(Y)$ then P is a cyclic quotient singularity of type $A_{n,n-1}$ for $n|m$. This means that*

$$\pi^{-1}(P) \sim E_1 \cup E_2 \cup \cdots \cup E_r$$
where E_i are exceptional curves $E_i \cong \mathbb{P}^1$

The intersection numbers of the exceptional curves are given by

$$E_i \cdot E_j = \begin{cases} 1 & |i-j| = 1 \\ 0 & |i-j| \geq 2 \\ -2 & |i-j| = 0 \end{cases}$$

The number r is given by the number of terms in the "minus continued fraction":

$$\frac{n}{n-1} = 2 - \cfrac{1}{2 - \cfrac{1}{2 - \cfrac{1}{2-\cdots}}}$$

Note that we use the notation of Hirzebruch to describe type A singularities. There is another notation; for instance, $A_{n,n-1}$ is the same as A_{n-1} in other notations.

Example 3.2. Let $d = 2$, $m = 30$, and $Q = (2, 3, 10, 15)$. There are singularities of type $A_{5,4}$, $2\, A_{3,2}$, and of type $3\, A_{2,1}$. Then

$$\frac{5}{4} = 2 - \cfrac{1}{2 - \cfrac{1}{2-\frac{1}{2}}}$$

There are 4 copies of 2 in the minus continued fraction arising from $A_{5,4}$, so we get 4 exceptional curves. Similarly, we get $2 \cdot 2$ (resp. $3 \cdot 1$) exceptional curves from $2\, A_{3,2}$ (resp. $3\, A_{2,1}$). Therefore, we have in total 11 exceptional curves obtained from blowing up singularities.

For $d = 3$, there is an algorithm due to Roan ([**Ro1**],[**Ro2**]) which describes and then resolve singularities on orbifold Calabi–Yau threefolds.

Theorem 3.3. Let $d = 3$. Let $< m, Q = (q_0, q_1, q_2, q_3, q_4) >$ be a pair in Theorem 2.2. Let Y be the quotient variety. Then in the weighted projective space $\mathbb{P}^4(Q)$, Y is defined by the following equation:
$$Y : Y_0^{m/q_0} + Y_1^{m/q_1} + Y_2^{m/q_2} + Y_3^{m/q_3} + Y_4^{m/q_4} = 0 \subset \mathbb{P}^4(\mathbb{Q})$$
where $\deg(Y_i) = q_i$ for each i, $0 \leq i \leq 4$. For each $l \in \{0, 1, \ldots, m-1\}$, let
$$I_l := \left\{ i, 0 \leq i \leq 4 \,\middle|\, l\frac{q_i}{m} \notin \mathbb{Z} \right\}.$$
Define
$$Y_{I_l} := Y \cap \{Y_j = 0 | j \in I_l\}.$$
Then $Y_{I_l} \in \Sigma(Y)$.

Digression:

(i) First note that if $\#I_l \neq 2, 3$, there are no singularities.

(ii) If $\#I_l = 2$, then Y_{I_l} defines a curve. If C is an irreducible component of I_l, then C is smooth and its blowing-up, $\pi^{-1}(C) \sim C \times \mathbb{P}^1$, is a ruled surface. The multiplicity of C, denoted $m_l(C)$, is given by $m_l(C) = c_{I_l} - 1$ where $c_{I_l} =: \gcd(q_i \,|\, i \notin I_l)$, which is ≥ 2.

Moreover, an irreducible component C is again a smooth weighted curve in a weighted projective space $\mathbb{P}^2(Q')$ of degree $m' := m/c_{I_l}$ with a reduced weight $Q' = (q'_i, q'_j, q'_k)$ where $i, j, k \notin I_l$, and $q'_\bullet := q_\bullet/c_{I_l}$. The geometric genus of C is given by the coefficient of the term $t^{m'-(q'_i+q'_j+q'_k)}$ in the formal power series expansion
$$\sum_{i=0}^{\infty} \beta_i t^i = \frac{1 - t^{m'}}{(1 - t^{q'_i})(1 - t^{q'_j})(1 - t^{q'_k})}.$$

(iii) If $\#I_l = 3$, Y_{I_l} defines a point, and it is blown up to $\mathbb{P}^2$.

Example 3.4. Let $m = 8$, $Q = (1, 1, 2, 2, 2)$. There is only one integer of interest, namely, $l = 4$. Since $4\frac{2}{8} \in \mathbb{Z}$, $I_4 = \{0, 1\}$. Then
$$Y_{I_4} = Y \cap \{Y_0 = Y_1 = 0\}, \quad Y_2^4 + Y_3^4 + Y_4^4 = 0$$
is an irreducible curve of genus 3. The multiplicity is $\gcd(2, 2, 2) - 1 = 2 - 1 = 1$.

Example 3.5. Let $m = 24$, $Q = (1, 2, 3, 6, 12)$. For $l \in \{8, 16\}$, $I_l = \{0, 1\}$, $c_{I_l} = \gcd(3, 6, 12) = 3$ so that $m_{I_l} = 2$. A smooth irreducible curve is a weighted projective curve of degree 8 with weight $Q' = (3/3, 6/3, 12/3) = (1, 2, 4)$:
$$Y_2^8 + Y_3^4 + Y_4^2 = 0 \subset \mathbb{P}^2((1, 2, 4))$$
and its genus is computed to be 1.

For $l = 12$, we have $I_{12} = \{0, 1\}$, and $c_{I_{12}} = \gcd(2, 6, 12) = 2$ so that $m_{I_{12}} = 1$. A smooth irreducible curve is a weighted projective curve with weight $Q' = (2/2, 6/2, 12/2) = (1, 3, 6)$:
$$Y_1^{12} + Y_3^4 + Y_4^2 = 0 \subset \mathbb{P}^2((1, 3, 6))$$
which is of genus 1.

For $l = 4, 20$, $I_l = \{0, 1, 2\}$, $c_{I_l} = \gcd(6, 12) = 6$ and a singular point is defined by
$$Y_3^4 + Y_4^2 = 0 \subset \mathbb{P}^1((1, 2)).$$

4. The L–series of orbifold Calabi–Yau varieties

Let X be one of the Calabi–Yau orbifold obtained above. When the field of definition for X, e.g., a number field K, or the algebraic closure $\bar{\mathbb{Q}}$, needs to be specified, we will put a subindex to X, e.g., X_K, $X_{\bar{\mathbb{Q}}}$.

Now we recall the definition of the L–series of Lecture 1, taking the field of definition for X to be $\mathbb{Q}$. Let $\bar{\mathbb{Q}}$ be its algebraic closure. For a good finite prime p of $\mathbb{Q}$, let Frob_p be the arithmetic Frobenius morphism on X mod p.

Definition 4. The L–series of X is defined as the following Euler product:

$$L(X,s) = L(H^d_{et}(X_{\bar{\mathbb{Q}}}, \mathbb{Q}_l), s) = (*) \prod_{p:good} P_{p,d}(p^{-s})^{-1},$$

where for good prime p,

$$P_{p,d}(T) = \det(1 - \mathrm{Frob}_p\, T | H^d_{et}(X_{\bar{\mathbb{Q}}}, \mathbb{Q}_l)) \in 1 + T\,\mathbb{Z}[T]$$

is the characteristic polynomial of Frob_p on the middle l-adic étale cohomology group $H^d(X_{\bar{\mathbb{Q}}}, \mathbb{Q}_l)$. Here again $(*)$ denotes the factors corresponding to bad primes.

Conjecturally, the product converges for $Re(s) > d + \frac{1}{2}$ and should be continued analytically to the whole complex plane.

Remark 4.1. The L–series for a Calabi–Yau variety defined over a number field K can be defined similarly: Replace rational primes p by finite prime ideals $\mathfrak{p}$ of K, and let

$$L(X_K, s) := (*) \prod_{\mathfrak{p}:good} P_{\mathfrak{p},d}(N\mathfrak{p}^{-s})^{-1}$$

where $\mathfrak{p}$ runs over all finite primes of K, and $N\mathfrak{p}$ denotes the norm of $\mathfrak{p}$.

The degree of the polynomial $P_{p,d}$ coincides with the d–th Betti number, B_d. For example, for $d = 2$, $\deg P_{p,2} = 22$, and for $d = 3$, $\deg P_{p,3} = B_3$. Note that for rigid Calabi–Yau threefolds given in Lecture 1, $B_3 = 2$, but for any orbifold Calabi–Yau threefold constructed above, it never has $B_3 = 2$.

Passing to the closure $\bar{\mathbb{Q}}$, we have

$$P_{p,d}(T) = \prod_{j=1}^{B_d}(1 - \alpha_{d,j}T) \in \bar{\mathbb{Q}}[T].$$

The reciprocal roots $\alpha_{d,j}$ are algebraic integers with complex absolute value

$$|\alpha_{d,j}| = p^{d/2} \qquad \text{(the Riemann Hypothesis)}.$$

In order to describe the reciprocal roots $\alpha_{d,j}$ we now invoke a classical result of Weil [**We**].

Proposition 4.2. *Let V be the Fermat hypersurface of degree m and dimension d defined over $\mathbb{Q}$. Then the eigenvalues of the Frobenius endomorphism are expressed in terms of Jacobi sums, which are algebraic integers in the m–th cyclotomic field $\mathbb{Q}(\zeta_m)$ over $\mathbb{Q}$ with complex absolute value $p^{d/2}$. The L–series of V, $L(V_{\mathbb{Q}}, s)$, is a Hecke L–series with Jacobi sum Grossencharacters of $\mathbb{Q}(\zeta_m)$. Here ζ_m denotes a primitive m–th root of unity.*

Confer Gouvêa and Yui [**GY**] for algorithmic approach to the computation of the local L–series of Fermat hypersurfaces via Fermat motives. In fact, Jacobi sums are in one-to-one correspondence with the set of vectors

$$\mathfrak{B}_d := \left\{ \mathbf{a} = (a_0, a_1, \cdots, a_{d+1}) \mid a_i \in (\mathbb{Z}/m\mathbb{Z}), a_i \neq 0, \sum_{i=0}^{d+1} a_i = 0 \right\}.$$

Now we can determine the L–series of our orbifold Calabi–Yau varieties over $\mathbb{Q}$.

Theorem 4.3. (Goto [**G**] for $d = 2$; Yui [**Y1**], [**Y2**] and [**Y3**].) *Let X be an orbifold Calabi–Yau variety of dimension d defined over $\mathbb{Q}$ constructed in Theorem 2.2.*

(a) *Let $d = 2$. Then*

$$L(X_\mathbb{Q}, s) \doteq L^Q_{Hecke,2}(s) \times L_\Sigma(s-1).$$

Here $L^Q_{Hecke,2}(s)$ is the Hecke L–series with Jacobi sum Grossencharacter of $\mathbb{Q}(\zeta_m)$ of dimension 2 compatible with the weight Q. $L_\Sigma(s)$ is the L–series associated to the singular locus $\Sigma(Y)$. Over the subfield $\mathbb{Q}(\zeta_{2m}) \subset \mathbb{Q}(\zeta_m)$, $L_\Sigma(s)$ is expressed as $L_\Sigma(s) = \zeta_K(s)^{1+e}$ where $\zeta_K(s)$ is the Dedekind zeta function of K, and e is the number of exceptional curves obtained after blowing up singularities. If all the exceptional curves are defined over $\mathbb{Q}$, $\zeta_K(s)$ coincides with the Riemann zeta-function.

(b) *Let $d = 3$. Then*

$$L(X_\mathbb{Q}, s) \doteq L^Q_{Hecke,3}(s) \times \prod_{\substack{C \in \Sigma(Y) \text{ smooth irred.}}} \zeta(C, s-1)^{m_l(C)}$$

where $L^Q_{Hecke,3}(s)$ is the Hecke L–series with Jacobi sum Grossencharacter of $\mathbb{Q}(\zeta_m)$ of dimension 3, compatible with the weight Q, and $\zeta(C, s)$ is the Hasse–Weil zeta function of a smooth irreducible curve $C \in \Sigma(Y)$ defined over $\mathbb{Q}$ with multiplicity $m_l(C)$. Moreover, $\zeta(C, s)$ is the Hecke L–series with Jacobi sum Grossencharacter of $\mathbb{Q}(\zeta_{m'})$ of degree m' compatible with the reduced weight Q'. Here m' and Q' are defined as in Theorem 3.3.

5. Sketch of proof of Theorem 4.2

We shall derive a formula for the Betti number B_d of the orbifold Calabi–Yau variety of dimension d. This boils down to the computation of the d-th cohomology group of the variety in question. For $d = 2$ (resp. $d = 3$), we follow the argument of Goto [**G**] (resp. Roan [**Ro1**],[**Ro2**]).

Proposition 5.1. *Let X be an orbifold Calabi–Yau variety of dimension d, degree m and weight $Q = (q_0, q_1, \cdots, q_{d+1})$ where $< m, Q >$ is one the pairs in Theorem 2.2.*

(a) *For $d = 2$, we have*

$$B_2(X) = \#\mathfrak{B}_2(Q) + e$$

where

$$\mathfrak{B}_2(Q) := \{\mathbf{a} = (a_0, a_1, a_2, a_3) \in \mathfrak{B}_2 \mid a_i \in (q_i\mathbb{Z}/m\mathbb{Z})\}.$$

(b) For $d = 3$, the notations of Theorem 3.3 remaining in force, we have

$$B_3(X) = \#\mathfrak{B}_3(Q) + \sum_C m_l(C) B_1(C)$$

where

$$\mathfrak{B}_3(Q) := \{\mathbf{a} = (a_0, a_1, a_2, a_3, a_4) \in \mathfrak{B}_3 \,|\, a_i \in (q_i\mathbb{Z}/m\mathbb{Z})\}.$$

Here the sum runs over all irreducible curves in the singular locus $\Sigma(Y)$ with multiplicity $m_l(C)$ and $B_1(C)$ is equal to twice the genus of C.

This follows from the construction of our orbifold Calabi–Yau varieties, in particular, from the structure of the d-th cohomology group with a coefficient $*$:

$$H^d(X, *) \otimes \mathbb{Q} \simeq (H^d(V, *)^{\mu_Q} \otimes \mathbb{Q}) \oplus (H^d(\pi^{-1}(\Sigma(Y)), *) \otimes \mathbb{Q}).$$

Then the L–series is obtained by determining the characteristic polynomial of the Frobenius endomorphism Frob_p for each good prime p. We consider the l–adic etale cohomoloogy groups. The first factor $H^d(V, \mathbb{Q}_l)^{\mu_Q}$ gives rise to the Hecke L–series with Jacobi sum Grossencharacters compatible with weight Q, and the factor $H^d(\pi^{-1}(\Sigma(Y)), \mathbb{Q}_l)$ contribute to the Dedekind (or Riemann) zeta-function or the Hasse–Weil zeta-functions.

6. The Tate conjecture

Here we consider orbifold Calabi–Yau varieties of dimension 2, namely orbifold $K3$ surfaces. We wish to compute the order of vanishing of the L–series of our orbifold $K3$ surfaces over number fields at $s = 1$, and then relate it to a certain group of algebraic cycles on X.

Conjecture 6.1. $(d = 2)$ Let K be a number field. Let X be a $K3$ surface defined over K. Then the conjecture of Tate asserts that:

$$\text{ord}_{s=1} L(X_K, s) = \text{rank}_{\mathbb{Z}} \text{NS}(X_K)$$

where $\text{NS}(X_K)$ denotes the Néron–Severi group of X over K. (Note that for $K3$ surfaces, $\text{NS}(X_K) = \text{Pic}(X_K)$, the Picard group of X over K.)

Results on the Tate conjecture for our orbifold $K3$ surfaces are collected in the following theorem.

Theorem 6.2. The Tate conjecture holds true for any $K3$ surface in characteristic zero (Tate [**T**]). In particular, for the $K3$ surfaces corresponding to the pairs $<m, Q>$ in Theorem 2.2, we have

$$\text{ord}_{s=1} L(X_K, s) = \text{rank}_{\mathbb{Z}} \text{NS}(X_K) \leq 22 - \varphi(m)$$

where φ is the Euler function. Furthermore, over the closure $\bar{\mathbb{Q}}$, all these quantities coincide.

Proposition 6.3. Let X be the weighted projective $K3$ surfaces in Theorem 2.2. Then over $\mathbb{Q}$, the rank of $NS(X_{\mathbb{Q}})$ is computed as follows:

$<m,Q>$	$e+1$	$\rho(X_{\mathbb{Q}})$	$\bar{\rho}(X)$	$\varphi(m)$
$<4,(1,1,1,1)>$	1	8	20	2
$<6,(1,1,1,3)>$	1	8	20	2
$<6,(1,1,2,2)>$	4	7	20	2
$<8,(1,1,2,4)>$	3	12	18	4
$<10,(1,2,2,5)>$	6	1	18	4
$<12,(2,3,3,4)>$	12	3	18	4
$<12,(1,3,4,4)>$	10	3	18	4
$<12,(1,2,3,6)>$	7	6	18	4
$<12,(1,1,4,6)>$	2	6	18	4
$<18,(1,2,6,9)>$	6	3	16	6
$<20,(1,4,5,10)>$	10	4	14	8
$<24,(1,3,8,12)>$	8	6	14	8
$<30,(2,3,10,15)>$	12	5	14	8
$<42,(1,6,14,21)>$	10	10	10	12

The maximum rank of $NS(X)$, denoted $\bar{\rho}(X)$ was computed by Goto [**G**]. To determine the actual rank of $NS(X_{\mathbb{Q}})$, one needs to know the exact field of definition of each exceptional curve (see Goto [**G**]), as well as the number of Jacobi sums which are equal to p at every reduction modulo p.

7. The conjecture of Beilinson and Bloch

Now we shall consider orbifold Calabi–Yau threefolds. In this case there is the conjecture of Beilinson and Bloch, which we now formulate.

Conjecture 7.1. Let $d = 3$. Let X be a Calabi–Yau threefold defined over a number field K. Let $CH^2(X_K)$ denote the Chow group of classes of algebraic cycles of codimension 2 on X defined over K modulo rational equivalence, and let $CH^2(X_K)_{hom}$ be the subgroup of null-homologous cycles in $CH^2(X_K)$, that is,

$$CH^2(X_K)_{hom} := Ker\,[CH^2(X_K) \to H^4_{et}(\overline{X}, \mathbb{Q}_l)].$$

Then the conjecture of Beilinson and Bloch [**Be**], [**Bl1**], [**Bl2**] claims that the group $CH^2(X_K)_{hom}$ is finitely generated, and

$$ord_{s=2}L(X_K, s) = rank_{\mathbb{Z}}CH^2(X_K)_{hom}.$$

Here is our strategy for tackling the conjecture for our orbifold Calabi–Yau threefolds.

First we recall a classical result that the Hecke L–series with Jacobi sum Grossencharacter satisfies the standard conjecture for L–series, in particular, it can be meromorphically continued to the whole complex plane $\mathbb{C}$ with at most one simple pole (and a point where it has a simple pole can be explicitly determined). Also it satisfies a functional equation, and it gives a relation between the two L–series under the change of variables s to $4 - s$, that is, if $\Lambda(X_{\mathbb{Q}}, s)$ denotes the extended L–series, then $\Lambda(X_{\mathbb{Q}}, s) = w\Lambda(X_{\mathbb{Q}}, 4 - s)$ with a root number $w \in \{\pm 1\}$. At $s = 2$, if $w = -1$, the L–series vanishes. Its order is 1 if $L'(X_{\mathbb{Q}}, 2) \neq 0$. However, the functional equation does not give any information about the vanishing of the L–series when $w = 1$.

Step 1: Since the third Betti number $B_3(X)$ is rather large, the actual calculations are done passing onto Fermat motives, as in Gouvêa and Yui [**GY**]. We first obtain the motivic decomposition of the quotient variety V/μ_Q under the action of $Gal(\mathbb{Q}(\zeta_m)/\mathbb{Q}) \simeq (\mathbb{Z}/m\mathbb{Z})^\times$. Then we realize the motivic decomposition on the cohomology group $H^3(V, \mathbb{Q}_l)^{\mu_Q}$. For each Fermat motive $\mathcal{M}$, we can define the L–series $L(\mathcal{M}, s)$, and compute its L–series. (Confer Yui [**Y2**] for some of these calculations.) Then putting together all motivic data, we get

$$L^Q_{Hecke,3}(s) = \prod L(\mathcal{M}, s)$$

where the product runs over the set of $(\mathbb{Z}/m\mathbb{Z})^\times$-orbits in $\mathfrak{B}_3(Q)$. Since each Fermat motive $\mathcal{M}$ consists of conjugate Jacobi sums under the action of $(\mathbb{Z}/m\mathbb{Z})^\times$, $\mathcal{M}$ is defined over $\mathbb{Q}$. Then we determine the functional equation for each $L(\mathcal{M}, s)$ and its root number. Then $L^Q_{Hecke,3}(2) \neq 0$ if and only if $L(\mathcal{M}, 2) \neq 0$ for every Fermat motive $\mathcal{M}$, and $L^Q_{Hecke,3}(2) = 0$ if there is one Fermat motive $\mathcal{M}$ for which $L(\mathcal{M}, 2) = 0$.

Among these Fermat motives, there is a unique motive associated to the weight Q, and we know that this motive gives rise to the transcendental cycles, so that its L–series does not vanish at $s = 2$.

Step 2: Now we compute the L–series associated to $H^3(\pi^{-1}(\Sigma(Y)), \mathbb{Q}_l)$. By Theorem 4.2, each factor is the Hasse–Weil zeta-function $\zeta(C, s)$ of a smooth irreducible curve in $\Sigma(Y)$. It is again a Hecke L–series with Jacobi sum Grossencharacter of dimension 1 compatible with a reduced weight Q'. If it does not vanish at $s = 1$, then the shifted Hasse–Weil zeta-function $\zeta(C, s - 1)$ does not vanish at $s = 2$.

Step3: How do we show that L–series of a Fermat motive $\mathcal{M}$ vanishes or does not vanish at $s = 2$? For this we ought to compute the leading term of the Taylor expansion of the L–series at $s = 2$. Buhler has explained to me how to do this, but I have not yet mastered the method.

Now we collect our results in support of the conjecture.

Proposition 7.2. *There are examples of orbifold Calabi–Yau threefolds X in Theorem 2.2 for which $L(X_\mathbb{Q}, 2) \neq 0$. In this case, the conjecture of Beilinson and Bloch is valid if and only if $CH^2(X_\mathbb{Q})_{hom}$ is torsion.*

However, to establish the fact that this group is actually torsion remains a challenging problem.

Example 7.3. Let $m = 8, Q = (1, 1, 2, 2, 2)$. Then $\mathfrak{B}_3(Q)$ consists of the following vectors with multiplicities given in the parenthesis.

a	*multiplicity*	$h^{3,0}$	$h^{2,1}$
(1,1,2,2,2)	4	1	1
(5,1,2,2,2)	12	0	2
(3,7,2,2,2)	4	0	2
(5,1,4,2,4)	12	0	2
(7,1,2,2,4)	12	0	2
(2,4,2,2,6)	12	0	2
(4,4,2,2,4)	6	0	2
(3,3,2,2,6)	6	0	2
(3,3,4,2,4)	6	0	2
(3,1,4,2,6)	24	0	2
(3,1,4,4,4)	4	0	2

There is only one smooth irreducible curve in the singular locus. That is, $C : Y_2^4 + Y_3^4 + Y_4^4 = 0$ which is of genus 3, and the Hasse–Weil zeta-function of C is given by the Hecke L–series with Jacobi sum Grossencharacter of $\mathbb{Q}(e^{2\pi/8})$.

Conjecture 7.4. There are examples of orbifold Calabi–Yau threefold over $\mathbb{Q}$ such that $L(X_{\mathbb{Q}}, 2) = 0$. There is a smooth irreducible curve C defined over $\mathbb{Q}$ in the singular locus of X which gives rise to an algebraic cycle in $CH^2(X)_{hom}$ of infinite order.

Example 7.5. The curve C should give rise to algebraic cycles of codimension 2 on the Calabi–Yau threefold X. This is exactly what Bloch discussed in [**Bl1**] among other things. Let $J(C)$ denote the Jacobian of C. Bloch [**Bl1**] showed that there is a submotive $\mathcal{M}$ of $H^3(J(C))$ of Hodge structure $(3,0) + (0,3)$ such that $L(\mathcal{M}, s)$ vanishes to order 1 at $s = 2$.

Conjecture 7.6. There exists a character ρ associated to an abelian number field K for which the L–series $L(X_K, s) := L(\rho \otimes H^3(X_{\mathbb{Q}}, s))$ vanishes at $s = 2$. In this case, there is a smooth irreducible curve C which gives rise to an algebraic cycle of codimension 2 of infinite order.

8. Construction of Mirror Calabi–Yau varieties

In this section, we shall describe how to construct mirror Calabi–Yau varieties of our orbifold Calabi–Yau varieties. This construction is due to Greene and Plesser [**GP**], and is known as *Greene–Plesser orbifolding construction*. (See also an article by Roan [**Ro2**] for the mirrors of Calabi–Yau orbifolds.) The Greene–Plesser construction are done again in weighted projective spaces. This construction is based on the following fact, which can be found in Cox and Katz [**CK**], Prop. (4.2.3).

Proposition 8.1. *Let X be a Calabi–Yau variety. Let G be a discrete group of symmetries on X. Then the smooth resolution of the orbifold X/G as well as its deformations are again Calabi–Yau varieties.*

Remark 8.2. There is a generalization of this construction by Batyrev [**Ba**] which works for any family of Calabi–Yau hypersurfaces in toric varieties. Batyrev describes mirror symmetry in terms of reflexive polytopes. Batyrev has constructed mirror pairs of Calabi–Yau varieties starting with reflexive polytopes. In essence, a dual pair of reflexive polytopes gives rise to a mirror pair of Calabi–Yau hypersurfaces. To describe Batyrev's construction, we need at least two more lectures. Accordingly, Batyrev's theory cannot not be covered in these lectures.

For $K3$ surfaces X, Mirror Symmetry takes place in the Hodge cohomology group $H^{1,1}(X)$. There are several variants of Mirror Symmetry for $K3$ surfaces. Here we employ the notion of Mirror Symmetry for $K3$ surfaces due to Dolgachev [**Dol2**]. According to Dolgachev, this mirror symmetry has its origin in Arnold's strange duality. The mirror symmetry for families of lattice polarized $K3$ surfaces will be discussed in Lecture 3, §8.

Conjecture 8.3. ($d = 2$) Let X be a $K3$ surface. Then there is a $K3$ surface $X^\vee$ which is a mirror partner of X if there is an orthogonal splitting:

$$\text{Pic}(X)^\perp = U_2 \perp \text{Pic}(X^\vee)$$

where U_2 is the hyperbolic lattice of rank 2.

Arithmetic digression: Let $(X, X^\vee)$ be a mirror pair of $K3$ surfaces defined over a number field K. Let $\bar{K}$ be the algebraic closure of K. Then a mirror pair satisfies

$$T(X_{\bar{\mathbb{Q}}}) = \text{Pic}(X)^\perp_{H^2(X_{\bar{\mathbb{Q}}}, \mathbb{Z})} = U_2 \oplus \text{Pic}(X^\vee_{\bar{\mathbb{Q}}})$$

where $T(X)$ is the lattice of transcendental cycles on X. This implies that

$$\text{Pic}(X_{\bar{\mathbb{Q}}}) \oplus U_2 \oplus \text{Pic}(X^\vee_{\bar{\mathbb{Q}}}) = H^2(X_{\bar{\mathbb{Q}}}, \mathbb{Z})$$

so that a mirror pair satisfies the identity:

$$\rho(X_{\bar{\mathbb{Q}}}) + \rho(X^\vee_{\bar{\mathbb{Q}}}) = 20.$$

Example 8.4. Let X be the $K3$ surface with $m = 4$ and weight $Q = (1, 1, 1, 1)$ defined by the equation

$$p := X_0^4 + X_1^4 + X_2^4 + X_3^4 = 0 \subset \mathbb{P}^3.$$

X has discrete symmetry $G = (\mathbb{Z}/4\mathbb{Z})^3 = <g_1> \times <g_2> \times <g_3>$ with

$$g_1 = (1, 0, 0, 3), \, g_2 = (0, 1, 0, 3), \, g_3 = (0, 0, 1, 3)$$

and the relation $g_1 g_2 g_3 = (1, 1, 1, 1) = (0, 0, 0, 0)$. Here (d_0, d_1, d_2, d_3) stands for the action

$$(X_0, X_1, X_2, X_3) \mapsto (\zeta^{d_0} X_0, \zeta^{d_1} X_1, \zeta^{d_2} X_2, \zeta^{d_3} X_3)$$

where $\zeta = e^{2\pi i}/4$. We consider a smooth 1–parameter deformation of the G–invariant variety defined by the equation with a parameter $\phi \in \mathbb{C}$:

$$p_\phi := X_0^4 + X_1^4 + X_2^4 + X_3^4 - 4\phi X_0 X_1 X_2 X_3.$$

Then a smooth resolution of $\{p_\phi = 0\}/G$, denoted by X^*, has the Picard group:

$$\text{Pic}(X^*) = U_2 \perp (-E_8)^2 \perp <-4>.$$

Then the mirror partner $X^{*\vee}$ of X^* is a generic quartic surface $X^{*\vee} \subset \mathbb{P}^4$ in the sense of Dolgachev, that is, $\text{Pic}(X^{*\vee}) = <4>$, and $T(X^*) = U_2 \perp <4>$.

Now we recall a "naive" Mirror Symmetry Conjecture for Calabi–Yau threefolds.

LECTURE 2. ORBIFOLD CALABI–YAU VARIETIES

Conjecture 8.5. ($d = 3$) For a given non-rigid Calabi–Yau threefold X, there is a mirror Calabi–Yau threefold $X^\vee$ such that

$$h^{1,1}(X^\vee) = h^{2,1}(X),\ h^{2,1}(X^\vee) = h^{1,1}(X)$$

so that

$$E(X^\vee) = -E(X).$$

Example 8.6. We consider Calabi–Yau threefolds X of Fermat type in weighted projective 4–space such that $h^{1,1}(X) = 1$. There are only four such Calabi–Yau threefolds given by a pair $<m, Q>$ where m is the degree and $Q = (q_0, q_1, q_2, q_3, q_4)$ denotes a weight.

		$h^{1,1}$	$h^{2,1}$	E
X_I	$<5, (1,1,1,1,1)>$	1	101	-200
X_{II}	$<6, (1,1,1,1,2)>$	1	103	-204
X_{III}	$<8, (1,1,1,1,4)>$	1	149	-296
X_{IV}	$<10, (1,1,1,2,5)>$	1	145	-288

Now we shall construct mirror Calabi–Yau threefolds by Greene–Plesser orbifolding construction. Namely, we consider orbifolds of these Calabi–Yau threefolds by taking the quotients by discrete groups of isometries. We ought to choose groups G of isometries so that

(a) G leaves the unique (up to a scalar multiplication) holomorphic 3-form invarinat,

(b) singularities can be resolved to give rise to smooth resolutions which is again Calabi–Yau threefold, and

(c) the Hodge numbers $h^{1,1}$ and $h^{2,1}$ are interchanged.

Let X be a Calabi–Yau threefold and G a group of discrete symmetries. Consider the quotient X/G, which has singularities since G does not in general act freely. Resolving singularities, let $X^\vee = \widehat{X/G}$ be a smooth resolution, which is again a Calabi–Yau variety. We impose the condition on G so that we have $h^{1,1}(X^\vee) = h^{2,1}(X)$ and $h^{2,1}(X^\vee) = h^{1,1}(X)$.

Theorem 8.7. *Let X be one of the 147 Calabi–Yau threefolds corresponding to a pair $<m, Q>$ in Theorem 2,2. Then there exists a mirror Calabi–Yau threefold $X^\vee$ of X, by the orbifolding construction.*

Example 8.8. For X_I in Example 8.6, take

$$G = (\mathbb{Z}/5\mathbb{Z})^3 = <g_1> \times <g_2> \times <g_3>$$

where the action of g_i on X_I is defined by

$$g_1 = (1, 0, 0, 4, 0): (X_0, X_1, X_2, X_3, X_4) \mapsto (\xi_5 X_0, X_1, X_2, \xi_5^4 X_3, X_4),$$
$$g_2 = (1, 0, 4, 0, 0): (X_0, X_1, X_2, X_3, X_4) \mapsto (\xi_5 X_0, X_1, \xi_5^4 X_2, X_3, X_4),$$
$$g_3 = (1, 4, 0, 0, 0): (X_0, X_1, X_2, X_3, X_4) \mapsto (\xi_5 X_0, \xi_5^4 X_1, X_2, X_3, X_4).$$

The orbifold X_I/G has the crepant resolution $X_I^\vee = \widehat{X_I/G}$, whose Hodge numbers are given by

$$h^{1,1}(X_I^\vee) = 101, h^{2,1}(X_I^\vee) = 1, E(X_I^\vee) = 200 = -E(X_I).$$

The mirror partner of X_I may be defined by a 1–parameter deformation
$$p_\phi : X_0^5 + X_1^5 + X_2^5 + X_3^5 + X_4^5 - 5\phi X_0 X_1 X_2 X_3 X_4 = 0,$$
that is $X_I^\vee$ is a smooth resolution of $\{p_\phi = 0\}/G$.

Example 8.9. The following table lists mirror partners of the Calabi–Yau threefolds in Example 8.5 in terms of the groups of isometry together with generators, and Hodge numbers and Euler characteristics.

	Group G	Generators	$h^{2,1}$	$h^{1,1}$	E
$X_I^\vee$	$(\mathbb{Z}/5\mathbb{Z})^3$	$\{(1,0,0,4,0),(1,0,4,0,0),(1,4,0,0,0)\}$	1	101	200
$X_{II}^\vee$	$(\mathbb{Z}/3\mathbb{Z}) \times (\mathbb{Z}/6\mathbb{Z})^2$	$\{(0,2,2,2,0),(0,1,0,0,5),(0,1,0,5,0)\}$	1	103	204
$X_{III}^\vee$	$(\mathbb{Z}/8\mathbb{Z})^3$	$\{(0,1,0,0,7),(0,1,0,7,0),(0,1,7,0,0)\}$	1	149	296
$X_{IV}^\vee$	$(\mathbb{Z}/10\mathbb{Z})^2$	$\{(0,0,1,0,9),(0,0,1,9,0)\}$	1	145	288

The mirror partner may be defined by 1-parameter deformations of the orbifolds $\{p_\phi = 0\}/G$ where p_ϕ are given as follows.

$$\begin{array}{ll}
X_{II} & X_0^6 + X_1^6 + X_2^6 + X_3^6 + X_4^3 - 6\phi X_0 X_1 X_2 X_3 X_4 = 0, \\
X_{III} & X_0^8 + X_1^8 + X_2^8 + X_3^8 + X_4^2 - 4\phi X_0 X_1 X_2 X_3 X_4 = 0, \\
X_{IV} & X_0^{10} + X_1^{10} + X_2^{10} + X_3^5 + X_4^2 - 5\phi X_0 X_1 X_2 X_3 X_4 = 0.
\end{array}$$

The above construction can be generalized, and indeed there is a list of Calabi–Yau orbifolds of Fermat hypersurface Calabi–Yau threefolds and their mirror partners in Klemm and Theisen [**KT**].

Example 8.10. These examples were first obtained by Greene and Plesser [**GP**]. Take a Calabi–Yau threefold X on which a group G acts freely. Then the Euler characteristic of the quotient is given by
$$E(X/G) = E(X)/\#G.$$

For instance, let X_I be the Fermat quintic, and let $G = (\mathbb{Z}/5\mathbb{Z})^2$. Then G acts freely on X_I so that its quotient X_I/G is a smooth variety with Euler characteristic $E(X_I/G) = -200/25 = -8$. We can take different groups acting on X_I as well. In the following table, we list in the first column the group G of discrete symmetries acting on X_I, the second column a generator of the group, the third and the fourth columns list the Hodge numbers $h^{2,1}$ and $h^{1,1}$ of the quotient X_I/G and the last the Euler characteristic.

Group G	Generators	$h^{2,1}$	$h^{1,1}$	E
$(\mathbb{Z}/5\mathbb{Z})$	$(1,2,3,4,0)$	21	1	-40
$(\mathbb{Z}/5\mathbb{Z})^2$	$\{(1,2,3,4,0),(1,0,2,2,0)\}$	1	21	40
$(\mathbb{Z}/5\mathbb{Z})^2$	$\{(1,0,0,4,0),(1,2,3,4,0)\}$	21	17	-8
$(\mathbb{Z}/5\mathbb{Z})$	$(1,2,2,0,0)$	17	21	8

9. Problems

In this section, we will collect open problems.

Problem (A): The Mirror Symmetry Conjecture

There are much more to the Mirror Symmetry Conjecture for Calabi–Yau threefolds than just exchanging the Hodge numbers of mirror pairs of Calabi–Yau threefolds. Now we will illustrate the Mirror Symmetry Conjecture (**Conjecture 2.1**) in Lecture 1 by an example.

Example 9.1. Consider the Calabi–Yau threefold X_I and its mirror partner in Examples 8.6 and 8.8. We write simply X for X_I and $X^\vee$ for $X_I^\vee$. For this mirror pair, the prepotentials of the A-model Yukawa coupling $\Psi_{X,A}(q)$ and of the B-model Yukawa coupling $\Psi_{B,X^\vee}(q)$ are computed (see for instance, Candelas, de la Ossa, Greene and Parkes [**COGP**], or Morrison [**M2**]):

$$\Psi_{A,X}(q) = 5 + \sum_{d=1}^{\infty} n_d d^3 \frac{q^d}{1-q^d}$$

where n_d denotes the number of rational curves of degree d contained in X, and

$$\Psi_{B,X^\vee}(q) = 5 + 2,875q + 609,250q^2 + 317,206,375q^3 + O(q^4).$$

The linear isomorphism $\phi : H^{1,1}(X) \simeq H^{2,1}(X^\vee)$ gives rise to the identity

$$\phi^*(\Psi_{B,X^\vee}(q)) = \Psi_{A,X}(q)$$

up to $d = 4$. The mirror symmetry conjecture claims the equality in all degree.

The ultimate problem is to establish the Mirror Symmetry Conjecture for our mirror pairs of orbifold Calabi–Yau threefolds.

Problem (B): The L–series of mirror Calabi–Yau threefolds

This problem is divided into several subproblems.

Subproblem B.1: The L–series of the orbifold Calabi–Yau threefolds

The determination of the L–series of the orbifold Calabi–Yau threefolds constructed in Theorem 2.2 remains a challenging problem.

Example 9.2. We first consider the Fermat quintic Calabi–Yau threefold. Let

$$X_I : X_0^5 + X_1^5 + X_2^5 + X_3^5 + X_4^5 = 0 \subset \mathbb{P}^4.$$

We look at Fermat motives. The following table lists all Fermat motives for X_I.

a	*multiplicities*	*dimension*	$h^{3,0}$	$h^{2,1}$
$(1,1,1,1,1)$	1	4	1	1
$(1,1,1,3,4)$	20	4	0	40
$(1,1,2,2,4)$	30	4	0	60

We see that $h^{3,0} = 1$, $h^{2,1} = 101$ so that $B_3 = 204$. Further $h^{1,1} = 1$ and hence $\chi(X_I) = -200$. There is a unique motive corresponding to the weight $Q = (1,1,1,1,1)$. Now we look at the action of the group $G = (\mathbb{Z}/5\mathbb{Z})^3 = <g_1> \times <g_2> \times <g_3>$ in Example 8.7 on these Fermat motives. The action is given by $\mathbf{d}(g_1) = \zeta_5^{d_0} 1^{d_1} 1^{d_2} \zeta_5^{4d_3} 1^{d_4}$, and similarly for g_2 and g_3. Then only the motive corresponding to the weight $Q = (1,1,1,1,1)$ is left invariant under the G–action. This motive gives rise to the invariant monomial $X_0 X_1 X_2 X_3 X_4$, which is on the mirror $X_I^\vee$ and in fact, in $H^{3,0}(X_I^\vee)$. Further, $(3,3,3,3,3) \in H^{2,1}(X_I^\vee)$ and hence $B_3(X_I^\vee) = 4$. The L-series of $X_I^\vee$ at $\phi = 1$ coincides with the L-series of the motive corresponding to the weight $Q = (1,1,1,1,1)$, which is a Hecke L-series with Jacobi sum Grossencharacter of $\mathbb{Q}(\zeta_5)$.

Subproblem B.2: The L–series of one–parameter deformations of the orbifold Calabi–Yau threefolds

For instance, determine the L–series of a 1–parameter deformation of a Calabi–Yau threefold X_λ of degree m and weight $Q = (q_0, q_1, \cdots, q_4)$:

$$X_0^{m/q_0} + X_1^{m/q_1} + \cdots + X_4^{m/q_4} + \lambda X_0 X_1 \cdots X_4 = 0$$

in $\mathbb{P}^4(Q)$. How can one determine the L–series of X_λ?

The same question can be addressed to multi–parameter deformations of Calabi–Yau threefolds in Theorem 2.2.

Subproblem B.3: The L–series of Calabi–Yau threefolds which are mirrors of Calabi–Yau threefolds considered in Subproblem B.2

For instance, let G be the discrete group of symmetries on X_λ and let X_λ^* be a smooth resolution of the quotient X_λ/G. What is the L–series of X_λ^*? Are there any relation between the L–series of X_λ and the L–series of X_λ^*?

Subproblem B.4: Relations between the L–series of a mirror pair of Calabi–Yau threefolds

Can one detect the mirror symmetry phenomenon for a mirror pair of Calabi–Yau threefolds in their L–series?

LECTURE 3
$K3$ Surfaces, Mirror Moonshine Phenomenon

In the third lecture, we shall give a detailed discussion on dimension two Calabi–Yau varieties, namely, $K3$ surfaces. In particular, we shall study the so-called *mirror maps* of certain families of lattice polarized $K3$ surfaces.

1. Mirror Moonshine phenomena

The starting point of our discussion is the *Moonshine Conjecture for the Monster* of Conway and Norton [**CN**]. The conjecture was proved in the affirmative by Borcherds [**Bo**] in 1992.

Let $\mathbb{M}$ denote the monster, that is, the largest sporadic finite simple group. The order of $\mathbb{M}$ is about 8×10^{58}, and the the number of conjugacy classes in $\mathbb{M}$ is 194.

Conjecture 1.1. (The Moonshine Conjecture=Theorem of Borcherds) Let $g \in \mathbb{M}$ be any element. Then associated to a rational conjugacy class of g, there is a McKay–Thompson series $T_g(q)$ of the form:

$$T_g(q) = \frac{1}{q} + \sum_{n \geq 1} c_g(n) q^n, \quad c_g(n) \in \mathbb{Z} \quad \forall n$$

where $q = e^{2\pi i z}$, $z \in \mathcal{H} = \{z = x + iy | y > 0\}$: the upper-half complex plane, satisfying the following properties:

(1) $g \mapsto c_g(n)$ is a (reducible) character of $\mathbb{M}$.

(2) There exists an integer h and $h | gcd(24, o(g))$, where $o(g)$ is the order of g, and a congruence subgroup $\Delta(g) \subset \mathrm{PSL}_2(\mathbb{R})$ such that $\Gamma_0(ho(g)) \subseteq \Delta(g) \subseteq$ the normalizer of $\Gamma_0(ho(g))$ in $\mathrm{PSL}_2(\mathbb{R})$.

(3) The group $\Delta(g)$ acts on $\mathfrak{H}$ by the linear fractional transformation, and the compactification of the quotient space, $(\mathfrak{H}/\Delta(g))^*$, is a genus zero compact Riemann surface, and $T_g(q)$ is a Hauptmodul for $\Delta(g)$.

Remark 1.2. (a) McKay–Thompson series depend on the rational conjugacy class of an element g, so there are at most 194 McKay–Thompson series associated to the monster $\mathbb{M}$. Indeed, the actual number of McKay–Thompson series arising from $\mathbb{M}$ is 174.

(b) If g is the identity element **1** of $\mathbb{M}$, then $T_\mathbf{1}(q) = j(q) - 744$ where $j(q)$ is the elliptic modular j–function for the group $\Delta(\mathbf{1}) = \mathrm{PSL}_2(\mathbb{Z})$.

In 1994, B. Lian and S.-T. Yau [**LY1**] (see also the subsequent paper [**LY2**]) observed that McKay–Thompson series are appearing as mirror maps of certain pencils of $K3$ surfaces.

Conjecture 1.3. (Mirror Moonshine conjecture) The mirror map $z(q)$ of a certain one–parameter family of $K3$ surfaces is related to McKay–Thompson series $T_g(q)$ for some $g \in \mathbb{M}$ in the following sense: there exist $a, b \in \mathbb{Q}$, $a \neq 0$ such that

$$\frac{a}{z(aq)} + b = T_g(q) \text{ for some } g \in \mathbb{M}.$$

In what follows, I will explain the mirror moonshine conjecture, and give some examples in support of the conjecture. I will also discuss some generalizations of the conjecture, for instance, McKay–Thompson series may be replaced by Hauptmoduln, and one–parameter families of $K3$ surfaces by multi–parameter families. For fuller discussions, the reader is referred to the article by Verrill and Yui [**VY1**] and a monograph [**VY2**]. Also confer the article of Doran [**Dor**] where he addresses the commensurability and the modularity questions of the mirror maps in connection with the mirror moonshine conjecture.

2. $K3$ surfaces

All surfaces are assumed to be defined over $\mathbb{C}$, and all surfaces are assumed to be algebraic.

Definition 5. A surface X is a called a $K3$ *surface* if
(1) $H^1(X, \mathcal{O}_X) = 0$, and
(2) $K_X \equiv 0$ (so that $p_g(X) = 1$).

The Hodge diamond and Betti numbers of a $K3$ surface are given by

$$\begin{array}{ccccc} & & 1 & & \\ & 0 & & 0 & \\ 1 & & 20 & & 1 \\ & 0 & & 0 & \\ & & 1 & & \end{array}$$

and

$$B_0 = B_4 = 1,\ B_1 = B_3 = 0,\ B_2 = 22.$$

There is the Hodge decomposition:

$$H^2(X, \mathbb{C}) = H^{2,0}(X) \oplus H^{1,1}(X) \oplus H^{0,2}(X)$$

where $H^{i,j}(X) := H^j(X, \Omega_X^i)$. There is a nowhere vanishing holomorphic 2–form, ω_X, on X which is a basis for the one–dimensional subspace space $H^{2,0}(X)$ of $H^2(X, \mathbb{C})$.

We shall list several examples of $K3$ surfaces.

Example 2.1. (a) In Lecture 2, we listed 14 examples of $K3$ surfaces obtained from Fermat surfaces by orbifolding construction. More examples of $K3$ surfaces can be found in Yonemura [**Yon**] where he listed 95 hypersurface simple $K3$ singularities. The construction is done in weighted projective spaces.

(b) Let $F(X_0, X_1, X_2, X_3)$ be a homogeneous polynomial of degree 4 in $\mathbb{P}^3$ with no multiple roots, and let X denote the zero locus of this polynomial. Then X is a $K3$ surface with a holomorphic 2–form is given by the residue of $dx_1 \wedge dx_2 \wedge dx_3/F(x_1, x_2, x_3)$ where $x_i := X_i/X_0$.

(c) Let $F(X_0, X_1, X_2)$ be a homogeneous polynomial of degree 6 in $\mathbb{P}^2$ with no multiple roots, and let $Z^2 = F(X_0, X_1, X_2)$ be a double cover of $\mathbb{P}^2$ branched along a sextic curve $F(X_0, X_1, X_2) = 0$. When the sextic curve has singularities, they are of type A, D or E. Then the minimal resolution is a $K3$ surface. A holomorphic 2–form is of the form $dx_1 \wedge dx_2/Z$ where $x_i = X_i/X_0$.

(d) Let A be an abelian surface. Let $\iota : A \to A$ be the involution $\iota(a) = -a$. Then the quotient $A/<\iota>$ surface is a singular surface with 16 singular points which are all rational double points. Resolving them, we obtain a smooth $K3$ surface, $X = Kum(A/<\iota>)$, called the Kummer surface.

We shall introduce a special class of $K3$ surfaces called elliptic modular $K3$ surfaces. We follow the exposition of Shioda [**Sh**].

Example 2.2. Let G be a subgroup of finite index of $\mathrm{PSL}_2(\mathbb{Z})$. The group G acts on the upper-half complex plane $\mathfrak{H}$ by the linear fractional transformation. The quotient $\mathfrak{H}/G$ together with a finite number of cusps form a compact Riemann surface, $(\mathfrak{H}/G)^* =: \Delta_G$, of genus g. Then there is a holomorphic map

$$J_G : \Delta_G \to \mathbb{P}^1$$

after identifying $\Delta_{\mathrm{PSL}_2(\mathbb{Z})}$ with $\mathbb{P}^1$ by the elliptic modular j–function. Let Σ be the set of cusps and elliptic points of G, and put $\Delta' = \Delta_G \setminus \Sigma$. Let U' be the universal covering of Δ' with the projection $\pi : U' \to \Delta'$. Then there is a holomorphic map $\omega : U' \to \mathfrak{H}$ such that

$$J_G(\pi(\tilde{u})) = j(\omega(\tilde{u})) \quad (\tilde{u} \in U').$$

Moreover, there is a representation φ of the fundamental group $\pi_1(\Delta')$ into $\mathrm{PSL}_2(\mathbb{Z})$:

$$\varphi : \pi_1(\Delta') \to G$$

such that

$$\omega(\gamma \cdot \tilde{u}) = \varphi(\gamma) \cdot \omega(\tilde{u}) \quad \tilde{u} \in U', \gamma \in \pi_1(\Delta')$$

where γ acts on $\mathfrak{H}$ by the linear fractional transformation. The representation φ determines a sheaf $\mathfrak{G}_\Delta$ over Δ_G, locally constant over Δ' with the general stalk $\mathbb{Z} \oplus \mathbb{Z}$. Under this situation, attached to the group G, there exists an elliptic surface B_G over Δ_G with a global section, called the elliptic modular surface.

Now we choose G to be torsion-free genus zero congruence subgroups of of level N of $\mathrm{PSL}_2(\mathbb{Z})$. The numerical characters of B_G are computed as follows. The irregularity index of B_G is 0 as it is equal to the genus of Δ_G, the geometric genus is $p_g(B_G) = -2 + t/2 = \mu/12 - 1$ where t is the number of cusps of G, and μ denotes the index of G in $\mathrm{PSL}_2(\mathbb{Z})$.

Sebbar[**Seb**] has classified all such subgroups $G \subset \mathrm{PSL}_2(\mathbb{Z})$ which give rise to elliptic modular $K3$ surfaces. There are 9 elliptic modular $K3$ surfaces. In order to get $K3$ surfaces, we see immediately that the index must be $\mu = 24$ and the number of cusps ought to be $t = 6$. All possible groups which give rise to $K3$ surfaces are $\Gamma(4)$ (level 4), $\Gamma_0(3) \cap \Gamma(2)$ (level 6), $\Gamma_1(7)$ (level 7), $\Gamma_1(8)$, $\Gamma_0(8) \cap \Gamma(2)$ (level 8), $\left\{ \pm \begin{pmatrix} 1+4a & 2b \\ 4c & 1+4d \end{pmatrix} a \equiv c \pmod{2} \right\}$ (level 8), $\Gamma_0(12)$ (level 12), $\Gamma_0(16)$ (level 16), and $\left\{ \pm \begin{pmatrix} 1+4a & b \\ 8c & 1+4d \end{pmatrix} a \equiv c \pmod{2} \right\}$ (level 16).

We may also consider pencils of $K3$ surfaces. This will be discussed below in §6.

3. The $K3$ lattice

In this section we shall discuss the lattice structures endowed on $K3$ surfaces.

Definition 6. A free $\mathbb{Z}$-module L of finite rank n with an integral symmetric bilinear form

$$< \ , \ > : L \times L \to \mathbb{Z}$$

is called a *lattice* of rank n. There is a matrix representation of L with respect to a $\mathbb{Z}$-basis for L, which is denoted again by L. A lattice L is *unimodular* if $det(L) = \pm 1$, and *indefinite* if the matrix L is indefinite. The signature of L is that of the quadratic form $L \otimes \mathbb{R}$ over $\mathbb{R}$.

A lattice L is *even* if $< x, x > \equiv 0 \pmod{2}$ for every $x \in L$.

An *isometry* of lattices is a $\mathbb{Z}$-module isomorphism preserving the symmetric bilinear forms. The group of isometries of a lattice L is denoted by $O(L)$.

Example 3.1. (a) A hyperbolic lattice U_2 is a lattice of rank 2 with the matrix $\begin{pmatrix} 0 & 1 \\ 1 & 0 \end{pmatrix}$, which is even, unimodular of signature $(1,1)$.

(b) The lattice E_8 is the unique positive definite even unimodular lattice of rank 8.

Indefinite unimodular even lattices are completely classified up to isometry by the following theorem.

Theorem 3.2. (Milnor [Mil]). *Let L be an indefinite even unimodular lattice of signature (r_+, r_-). Then*

(a) $r_+ - r_- \equiv 0 \pmod{8}$, *and*

(b) $L \simeq U_2^s \oplus (-E_8)^t$ *where $s = min\{r_+, r_-\}$ and $t = |r_+ - r_-|/8$.*

Now let X be a $K3$ surface. It is known that $H^2(X, \mathbb{Z})$ (resp. $H_2(X, \mathbb{Z})$) is a free $\mathbb{Z}$-module of rank 22. There is the intersection paring (cup product) on X:

$$< \ , \ > : H^2(X, \mathbb{Z}) \times H^2(X, \mathbb{Z}) \to \mathbb{Z}$$

which gives rise to a quadratic form on $H^2(X, \mathbb{Z})$, so that $H^2(X, \mathbb{Z})$ captures the structure of a lattice of rank 22. This lattice has the following properties:

— unimodular (by Poincáre duality),
— even (by Wu's formula),
— indefinite, and
— of signature $(3, 19)$ (by Hirzebruch's signature theorem).

Applying Milnor's classification theorem, we see that there is a unique lattice up to isometry satisfying the above properties: namely,

$$\mathcal{L} := U_2^3 \perp (-E_8)^2.$$

We call $\mathcal{L}$ the $K3$ *lattice*.

Theorem 3.3. *Let X be an algebraic $K3$ surface. Then there is an isometry of lattices:*

$$\phi : H^2(X, \mathbb{Z}) \simeq \mathcal{L}.$$

An isometry ϕ is called a *marking* and a pair (X, ϕ) a *marked $K3$* surface.

Now we shall consider sublattices of $H^2(X, \mathbb{Z})$ and $\mathcal{L}$.

Definition 7. Let M and L be lattices. An embedding of lattice $M \hookrightarrow L$ is said to be *primitive* if L/M is torsion-free.

Proposition 3.4. *Let X be a $K3$ surface, and let ω_X be a nowhere vanishing holomorphic 2–form on X. Let*

$$\mathrm{Pic}(X) := \{\, x \in H^2(X, \mathbb{Z}) \,|\, <x, \omega_X> = 0 \,\},$$

and

$$T(X) := \{\, x \in H^2(X, \mathbb{Z}) \,|\, <x, y> = 0 \text{ for any } y \in \mathrm{Pic}(X) \,\}.$$

Then the following assertions hold:

(a) *Both $\mathrm{Pic}(X)$ and $T(X)$ are independent of the choice of ω_X, and are primitive sublattices of $H^2(X, \mathbb{Z})$.*

(b) *$\mathrm{Pic}(X)$ has rank $\rho(X)$ and signature $(1, \rho(X) - 1)$, where $\rho(X)$ is called the Picard number of X and may take any integer value between 1 and 20.*
$T(X)$ has rank $22 - \rho(X)$ and signature $(2, 19 - (\rho(X) - 1))$.

(c) *$H^{2,0}(X) \subset T(X) \otimes \mathbb{C}$ and $\mathrm{Pic}(X)$ is parameterized by algebraic cycles on X, and that $T(X) = \mathrm{Pic}(X)^\perp$ in $H^2(X, \mathbb{Z})$ with respect to $<\ ,\ >$.*

We call $\mathrm{Pic}(X)$ the Picard lattice and $T(X)$ the transcendental lattice of X.

Now we start out with primitive sublattices of $\mathcal{L}$. Then we can realize them as geometric objects as follows.

Proposition 3.5. (a) *If $\mathcal{M}$ is a primitive sublattice of $\mathcal{L}$ of signature $(1, \rho - 1)$, then there exists an algebraic $K3$ surface X and an isometry of lattices $\mathrm{Pic}(X) \simeq \mathcal{M}$.*

(b) *If $\mathcal{T}$ is a primitive sublattice of $\mathcal{L}$ of signature $(2, 20 - \rho)$, then there exists an algebraic $K3$ surface X and an isometry $T(X) \simeq \mathcal{T}$.*

4. Lattice polarized $K3$ surfaces

Let X be a $K3$ surface. Let

$$H^2(X, \mathbb{C}) = H^{2,0}(X) \oplus H^{1,1}(X) \oplus H^{0,2}(X)$$

be the Hodge decomposition, and put $H^{1,1}(X)_\mathbb{R} = H^{1,1}(X) \cap H^2(X, \mathbb{R})$. Then the subspace $H^{1,1}(X)_\mathbb{R}$ has signature $(1, 19)$, and the cone $\{x \in H^{1,1}(X)_\mathbb{R} \,|\, <x, x> > 0\}$ consists of two connected components. We denote by $V(X)^+$ the component which contains the class of $(1,1)$-forms associated to Kähler metrics. Let $\Delta(X) :=$

$\{\delta \in \text{Pic}(X) \mid <\delta, \delta> = -2\}$ be the roots of $\text{Pic} Pic(X)$. Then by the Riemann-Roch theorem, $\Delta(X) = \Delta^+(X) \coprod -\Delta^+(X)$, where $\Delta^+(X)$ consists of effective divisor classes. Let $C(X)^+ := \{x \in V^+(X) \mid <x, \delta> \gg 0 \; for \; any \; \delta \in \Delta^+(X)\}$ be the *Kähler cone* of X and $C(X)$ its closure in $H^{1,1}(X)_{\mathbb{R}}$. We set

$$\text{Pic}(X)^+ := C(X) \cap H^2(X, \mathbb{Z}), \quad \text{Pic}(X)^{++} := C(X)^+ \cap H^2(X, \mathbb{Z}).$$

The elements of $\text{Pic}(X)^+$ (resp. $\text{Pic}(X)^{++}$) are pseudo-ample divisor classes, i.e., numerically effective divisor classes with positive self-intersection (resp. ample divisor classes).

The following theorem is the most fundamental in the theory of $K3$ surfaces.

Theorem 4.1. *Let X, and X' be two $K3$ surfaces. Then the following statements are equivalent:*

(i) *There is an isometry $g : H^2(X', \mathbb{Z}) \to H^2(X, \mathbb{Z})$, satisfying the following properties:*

(a) $g(\otimes \mathbb{C})(H^0(X, \Omega_X^2)) = H^0(X', \Omega_{X'}^2)$, *and*
(b) $(g \otimes \mathbb{R})$ *preserves the Kähler cone.*
(Such an isometry is called an effective Hodge isometry.)

(ii) *there is an isomorphism of K3 surfaces $f : X \to X'$ such that $f^* = g$.*

In other words, an effective Hodge isometry g arises as $f^* : H^2(X', \mathbb{Z}) \to h^2(X, \mathbb{Z})$ of some isomorphism $f : X \to X'$ of K3 surfaces X, X'.

Now let $\mathcal{M}$ be an even non-degenerate lattice of signature $(1, t)$ with integral symmetric bilinear form $<\;,\;>$. We can define the cone:

$$V(\mathcal{M}) = \{x \in \mathcal{M}_{\mathbb{R}} \mid <x, x> \gg 0\},$$

which has two components. Fix one of the components and call it $V(\mathcal{M})^+$. Let

$$\Delta(\mathcal{M}) = \{\delta \in \mathcal{M} \mid <\delta, \delta> = -2\}$$

and fix a subset $\Delta(\mathcal{M})^+$ such that $\Delta(\mathcal{M}) = \Delta(\mathcal{M})^+ \coprod (-\Delta(\mathcal{M})^+)$, and if $\delta_i \in \Delta(\mathcal{M})^+$ and $\delta = \sum n_i \delta_i$ with $n_i \geq 0$, then $\delta \in \Delta(\mathcal{M})^+$. Define

$$C(\mathcal{M})^+ = \{h \in V(\mathcal{M})^+ \mid <h, \delta> \gg 0 \; \text{for all} \; \delta \in \Delta(\mathcal{M})^+\}.$$

Definition 8. Let $\mathcal{M}$ be an even non-degenerate lattice of signature $(1, t)$.

(a) An $\mathcal{M}$–polarized $K3$ surface is a pair (X, j) where X is a $K3$ surface and $j : \mathcal{M} \hookrightarrow \text{Pic}(X)$ is a primitive lattice embedding.

(b) An $\mathcal{M}$–polarized $K3$ surface (X, j) is called *pseudo-ample* (resp. *ample*) if $j(\mathcal{M})$ contains a pseudo-ample (resp. ample) divisor class.

(c) Let $\mathcal{M}$ be a sublattice of $\mathcal{L}$ of signature $(1, t)$. A *marked $\mathcal{M}$–polarized $K3$ surface* is a triplet (X, ϕ, j_ϕ) where X is a $K3$ surface, $\phi : H^2(X, \mathbb{Z}) \to \mathcal{L}$ is the isometry of lattices and $j_\phi := \phi^{-1} \mid \mathcal{M} : \mathcal{M} \hookrightarrow \text{Pic}(X)$ is a primitive embedding.

5. Moduli space of lattice polarized $K3$ surfaces

There exists the fine moduli space for marked lattice polarized $K3$ surfaces. We shall recall this fact from Dolgachev [**Dol2**].

Let X be a $K3$ surface, and let ω_X be a holomorphic 2-form on X which spans $H^{2,0}(X)$. Let $\{\gamma_1, \gamma_2, \cdots, \gamma_{22}\}$ be a $\mathbb{Z}$–basis for $H_2(X, \mathbb{Z})$.

LECTURE 3. K3 SURFACES, AND MIRROR MOONSHINE

Definition 9. The *period map* of X is
$$p(X) := [\int_{\gamma_1} \omega_X : \int_{\gamma_2} \omega_X : \cdots : \int_{\gamma_{22}} \omega_X] \in \mathbb{P}^{21}$$
and the *period domain* is the set of images of the period map, and denoted by Ω.

If $\gamma_i \in \text{Pic}(X)$, then $\int_{\gamma_i} \omega_X = 0$, therefore, the period domain Ω is a subset of $\mathbb{P}(T(X) \otimes \mathbb{C})$, and satisfies $\Omega \wedge \Omega = 0$ and $\Omega \wedge \overline{\Omega} > 0$.

Now we shall describe the period domains of lattice polarized $K3$ surfaces. Let $\mathcal{M} \subset \mathcal{L}$ be a primitive sublattice of signature $(1,t)$. Let $\mathcal{N} = \mathcal{M}_\mathcal{L}^\perp$ be the orthogonal complement of $\mathcal{M}$ in $\mathcal{L}$. It is a lattice of signature $(2, 19-t)$. Let (X, ϕ, j_ϕ) be a marked $\mathcal{M}$–polarized $K3$ surface, where $\phi : H^2(X, \mathbb{Z}) \to \mathcal{L}$ is a marking, and $j_\phi := \phi^{-1}|_\mathcal{M} : \mathcal{M} \hookrightarrow \text{Pic}(X)$ be a primitive lattice embedding.

The marking ϕ defines a point $\phi(\omega_X)$ in $\mathbb{P}(\mathcal{L} \otimes \mathbb{C})$, which lives in $\mathbb{P}(\mathcal{N} \otimes \mathbb{C})$. Let $\mathcal{Q}$ denote the quadric in $\mathbb{P}(\mathcal{N} \otimes \mathbb{C})$ corresponding to the quadratic form $b(x, y)$ on $\mathcal{N} \otimes \mathbb{C}$ defined by $\mathcal{N}$. Let
$$\mathfrak{D}_\mathcal{M} := \{[\eta] \in \mathbb{P}(\mathcal{N} \otimes \mathbb{C}) \mid b([\eta], [\eta]) = 0,\ b([\eta], [\bar{\eta}]) > 0\}$$
be the subset of the quadric $\mathcal{Q}$ defined by the inequality $\Omega \wedge \overline{\Omega} > 0$.

Theorem 5.1. *Let $\mathcal{M} \subset \mathcal{L}$ be a primitive sublattice of signature $(1,t)$. Then the period map:*
$$p : \{(X, \phi, j_\phi)\}/\sim\ \longrightarrow \mathfrak{D}_\mathcal{M}$$
gives rise to a bijection between the isomorphism classes of marked pseudo-ample $\mathcal{M}$–polarized $K3$ surfaces and the domain $\mathfrak{D}_\mathcal{M}$.

Furthermore, the domain $\mathfrak{D}_\mathcal{M}$ is isomorphic to two copies of the symmetric homogeneous domain of type IV of the form:
$$O(2, 19-t)/SO(2) \times O(19-t).$$

Next we wish to get rid of markings of $\mathcal{M}$–polarized $K3$ surfaces. For this we introduce the group
$$\Gamma(\mathcal{M}) := \{\sigma \in O(\mathcal{L}) \mid o(m) - m \quad \text{for any} \quad m \in \mathcal{M}\}$$
and let $\Gamma_\mathcal{M}$ be the image of $\Gamma(\mathcal{M})$ under the injective map $\Gamma(\mathcal{M}) \to O(\mathcal{N})$.

Theorem 5.2. (Dolgachev [**Dol2**].) *Let $\mathcal{M} \subset \mathcal{L}$ be a primitive sublattice of signature $(1,t)$. Then the moduli space, $\mathfrak{M}_\mathcal{M}$, of pseudo-ample (or ample) $\mathcal{M}$–polarized $K3$ surfaces exists, and in fact, there is a bijection:*
$$\mathfrak{M}_\mathcal{M} \to \mathfrak{D}_\mathcal{M}/\Gamma_\mathcal{M}.$$

Then the coarse moduli space $\mathfrak{M}_\mathcal{M}$ defines a quasi-projective algebraic variety of dimension $20 - (t+1) = 19 - t$.

Example 5.3. Let $\mathcal{M} \subset \mathcal{L}$ be of signature $(1, 19)$, so $\text{rank}\,\mathcal{M} = 20$. Then $\mathcal{M}$-polarized $K3$ surfaces have the maximum Picard number, .i.e., $\rho(X) = 20$. These $K3$ surfaces are called *singular $K3$ surfaces*, and by Shioda and Inose [**SI**], we have

$$\{\text{ singular K3 surfaces }\}/\text{isom} \overset{1-1}{\longleftrightarrow} \left\{\begin{array}{c}\text{positive deftinite even}\\\text{bilinear quadratic forms}\end{array}\right\}/SL_2(\mathbb{Z})$$

We will digress.

Fix a discriminant d of a quadratic form. Then the class number of the quadratic form $h(d) < \infty$. So there are only finitely many isomorphism classes of singular $K3$ surfaces with a fixed discriminant d (which is equivalent to fixing the Picard lattice $\mathcal{M}$).

On the other hand, for a fixed Picard lattice $\mathcal{M}$, the coarse moduli space $\mathfrak{M}_\mathcal{M}$ is a dimension 0 quasi-projective algebraic variety. So $\mathfrak{M}_\mathcal{M}$ is Noetherian, and hence it consists of a finite number of points. Therefore, the isomorphism classes of singular $K3$ surfaces with a fixed Picard lattice $\mathcal{M}$ is finite.

Now consider sublattices $\mathcal{M}$ of signature $(1,18)$, so rank $\mathcal{M} = 19$. E.g.,
$$\mathcal{M} = U_2 \perp (-E_8)^2 \perp <-2N>, \quad N \in \mathbb{Z}, N \geq 1.$$

Theorem 5.4. (Dolgachev [**Dol2**].) *Suppose that $\mathcal{M} = U_2 \perp (-E_8)^2 \perp <-2N>$, $N \in \mathbb{Z}, N \geq 1$, then the coarse moduli space $\mathfrak{M}_{K3,\mathcal{M}}$ has dimension 1. In fact*
$$\mathfrak{M}_\mathcal{M} \cong \mathfrak{H}/\Gamma_0(N)^*.$$

Here
$$\Gamma_0(N)^* = <\Gamma_0(N), w_N>$$
where w_N is the Fricke involution
$$w_N = \begin{pmatrix} 0 & \frac{-1}{\sqrt{N}} \\ \sqrt{N} & 0 \end{pmatrix}.$$

Remark 5.5. Recall that $X_0(N)$ is the compactification of the open Riemann surface $\mathfrak{H}/\Gamma_0(N)$. We know that $X_0(N)$ is the moduli space of elliptic curves E together with cyclic N-isogeny
$$E[N] \cong (\mathbb{Z}/N\mathbb{Z})^2, \quad C[N] \cong \mathbb{Z}/N\mathbb{Z}.$$
Here E and E' are elliptic curves, and $E' = E/C(N)$. The Fricke involution is the map which exchanges E and E'. Therefore the product $E \times E'$ is fixed under the Fricke involution, and
$$\mathfrak{H}/\Gamma_0(N)^* = \text{ the moduli space of abelian surfaces } E \times E'$$

Definition 10. A $K3$ surface X *admits a Shioda–Inose structure* if there is an abelian surface A and rational maps of degree 2 and a diagram

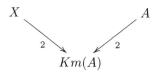

such that $T(X) = T(A)$. Here $Km(A)$ is the Kummer surface defined in Example 2.1. (d), i.e., $Km(A) = \widetilde{A/<\iota>}$ is the resolution of the quotient by the involution $\iota : A \to A$, $a \mapsto -a$.

Theorem 5.6. (Morrison) [**M1**] *If X is a $K3$ surface with Picard rank $19, 18$, then X always has a Shioda Inose structure. If the Picard rank is 17, then with some additional condition on the transcendental lattice, X admits a Shioda–Inose structure.*

6. Picard–Fuchs differential equations for one–parameter families of $K3$ surfaces

In this section we shall consider pencils of $K3$ surfaces, varying continuously with parameters. We shall discuss one–parameter families of $K3$ surfaces and determine the Picard–Fuchs differential equations for the families. For the basic theory of Picard–Fuchs differential equation, the reader is referred to Yoshida's monograph [**Yos**].

Let $B = \mathbb{P}^1 \setminus \{\text{finite number of points}\}$, and let X be a certain variety with a map $\pi : X \to B$. We assume that for all $z \in B$ the fiber X_z, given by $X_z := \pi^{-1}(z)$ is a $K3$ surface, so that $\{X_z\}_{z \in B}$ is a one–parameter family of $K3$ surfaces with a parameter z.

Here are some examples of one–parameter families of $K3$ surfaces.

Example 6.1. (a) Let
$$1 - (1 - XY)Z - tXY(1 - X)(1 - Y)(1 - Z) = 0$$
be the affine equation of the family of algebraic $K3$ surface with parameter t. This family of $K3$ surfaces arises naturally in connection with the irrationality of $\zeta(3)$, and has been first considered by Beukers and Peters [**BP**]. In this case, $B = \mathbb{P}^1 \setminus \{0, \infty, (\sqrt{2} \pm 1)^4\}$, and $\{X_z\}_{z \in B}$ is a one–parameter family of $K3$ surfaces.

(b) Let X be one of the 14 $K3$ surfaces listed in Lecture II obtained from Fermat surfaces by orbifolding construction. We can get one–parameter families of $K3$ surfaces by taking their deformations. The typical examples are:
$$X_0^4 + X_1^4 + X_2^4 + X_3^4 + 4\lambda X_0 X_1 X_2 X_3 = 0 \subset \mathbb{P}^3 \times \mathbb{P}^1$$
and
$$X_0^{42} + X_1^7 + X_2^3 + X_3^2 + \lambda X_0 X_1 X_2 X_3 = 0 \subset \mathbb{P}^3 \times \mathbb{P}^1$$
where $\lambda \in \mathbb{P}^1$ is a parameter.

Definition 11. Let $\{X_z\}_{z \in B}$ be a one–parameter family of $K3$ surfaces. For each fiber X_z, $z \in B$, there is a unique holomorphic 2–form ω_z which spans $H^2(X_z, \mathbb{C})$.

(a) For a fixed value z_0 of z, we have the monodromy representation
$$\pi_1(B, z_0) \to Aut(\mathbb{P} H_2(X_z, \mathbb{Z}))$$
and the image of this map, denoted $\mathcal{G}$, is called the *monodromy group* of the family $\{X_z\}_{z \in B}$.

(b) Let $\{\gamma_1, \gamma_2, \cdots, \gamma_{22}\}$ be a basis for $H_2(X_z, \mathbb{Z})$, which is flat with respect to parameter z. The *period map* is defined by
$$B \to \mathbb{P}^{21}/\mathcal{G} \quad ; \quad z \mapsto \left[\int_{\gamma_1} \omega_z : \cdots : \int_{\gamma_{22}} \omega_z\right]$$
where each function $\int_{\gamma_i} \omega_z$ is a *period*.

(c) The *Picard–Fuchs* differential equation is a differential equation satisfied by the periods, with the same monodromy group.

Now the existence of a Picard–Fuchs differential equation for one–parameter family of $K3$ surfaces follows from the proposition below.

Proposition 6.2. Let $\{X_z\}_{z\in B}$ be a one–parameter family of $K3$ surfaces with generic Picard number $\rho(X_z)$. Let $r = 22 - \rho(X_z)$. Then the periods $\int_{\gamma_i} \omega_z$ satisfy the Picard-Fuchs differential equation, of order r:

$$\left(\frac{d^r}{dz^r} + P_{r-1}(z)\frac{d^{r-1}}{dz^{r-1}} + \cdots + P_o(z)\right)f = 0$$

The idea of the proof is to look at the $r+1$ classes $[\omega_z], [\frac{\partial \omega_z}{\partial z}], \ldots, [\frac{\partial^r \omega_z}{\partial z^r}]$ in the r–dimensional vector space $H_{DR}(X_z, \mathbb{C})/\text{Pic}(X_z) \otimes \mathbb{C}$, which must be linearly dependent, and the linear relation gives rise to the differential equation.

Now we shall consider a special case.

Theorem 6.3. Suppose $\mathcal{M} = U_2 \perp (-E_8)^2 \perp <-2N>$, $N \geq 1$ be a primitive sublattice of $\mathcal{L}$, so $\text{rk}\,\mathcal{M} = 19$. Let $\{X_z\}_{z\in B}$ be a one–parameter family of $\mathcal{M}$-polarized $K3$ surfaces with generic Picard number $\rho(X_z) = 19$. Then $r = 22 - \rho(X_z) = 3$, and the family $\{X_z\}_{z\in B}$ satisfies an order 3 Picard-Fuchs differential equation.

Furthermore this order 3 differential equation is the symmetric square of an order two differential equation for a family of elliptic curves.

This follows from the fact that X_z admits a Shioda–Inose structure with $E_z \times E'_z$ as a underlying abelian surface. In fact, $E_z = \mathbb{C}/\mathbb{Z} + z\mathbb{Z}$ and $E'_z = \mathbb{Z}/\mathbb{Z} + (-1/Nz)\mathbb{Z}$. Therefore, by Theorem 5.6, $T(X_z) = T(E_z \times E'_z)$ and the Picard–Fuchs differential equation for X_z is a symmetric square of the Picard–Fuchs differential equation for E_z.

7. Mirror maps

We consider one–parameter families of $K3$ surfaces. The notations and hypothesis of §6 remain in force.

Definition 12. A regular singular point of a Picard–Fuchs differential equation is called a *point of maximal unipotent monodromy* if the local monodromy group matrix $\mathcal{G}$ is such that $\mathcal{G} - Id_k$ is nilpotent to the maximum order k.

Remark 7.1. It is not always true that every Picard–Fuchs differential equation is equipped with a point of maximal unipotent monodromy.

Now we shall discuss solutions of the Picard–Fuchs differential equations. Let $\{X_z\}_{z\in B}$ be a one–parameter family of $K3$ surfaces with generic Picard number $\rho(X_z) = 19$. Then the Picard–Fuchs differential equation of the family has order 3.

Definition 13. Suppose that $z = 0$ is a point of maximal unipotent monodromy for the Picard-Fuchs differential equation of the family. At $z = 0$, the Frobenius method guarantees that it has a holomorphic solution :

$$\omega_0(z) = 1 + \sum_{n\geq 1} a_n z^n$$

and a logarithmic solution:

$$\omega_1(z) = log(z)\omega_0(z) + \sum_{n\geq 1} b_n z^n.$$

Put
$$t = \frac{\omega_1(z)}{2\pi i \omega_0(z)} \quad \text{and} \quad q = e^{2\pi it} = ze^{\sum_{n\geq 1} b_n z^n/\omega_0(z)}.$$
Invert z to express it as a q–series:
$$z = z(q) = q + O(q^2).$$
The function $z(q)$ is called the mirror map of the family $\{X_z\}_{z \in B}$.

Lian and Yau ([**LY1**] and [**LY2**]) have formulated the following conjecture, so-called the *Mirror Moonshine Conjecture* for $K3$ surfaces constructed via orbifolding construction.

Conjecture 7.2. Let $\{X_z\}_{z \in B}$ be a one–parameter family of $K3$ surfaces (obtained from orbifolding construction) such that
 (a) its Picard–Fuchs differential equation is of Fuchsian type of order 3,
 (b) $z = 0$ is a point of maximal unipotent monodromy, and
 (c) there is a unique holomorphic solution $\omega_0(z)$ such that $\omega_0(0) = 1$ and $\omega_0'(0) \neq 0$.
Let $z(q)$ be the mirror map of the family $\{X_z\}_{z \in B}$. Then
$$\frac{\pm a}{z(aq)} + b = T_g(q) \quad \text{for some} \quad a, b \in \mathbb{Q}, \, a \neq 0$$
where $T_g(q)$ is the McKay–Thompson series corresponding to an element g of the Monster $\mathbb{M}$.

Now we give some examples in support of the Mirror Moonshine Conjecture. We first consider the one–parameter family of $K3$ surfaces in Example 6.1. This is the family of $K3$ surfaces considered by Beukers and Peters [**BP**] and Peters [**P**].

Theorem 7.3. Let $\{X_t\}$ be a one–parameter family of $K3$ surfaces with a parameter t in Example 6.1. For $t \neq 0, 1, (\sqrt{2} \pm 1)^4$, the surface X_t defines a smooth $K3$ surface, having the following properties:

(a) (Peters [**P**]). Let $\mathcal{M} = U_2 \perp (-E_8)^2 \perp <-12>$. Then $\{X_t\}$ is an $\mathcal{M}$–polarized family of $K3$ surface. In fact,
$$\text{Pic}(X_t) \simeq \mathcal{M} \quad \text{and} \quad T(X_t) \simeq U_2 \perp <12>.$$

(b) (Beukers and Peters [**BP**]). The Picard–Fuchs differential equation is given by
$$(t^4 - 34t^2 + t^2)(\frac{d}{dt})^3 + (6t^3 - 153t^2 + 3t)(\frac{d}{dt})^2 + (7t^2 - 112t + 1)(\frac{d}{dt}) + (t - 5).$$
This is a symmetric square of the order 2 differential equation:
$$(t^3 - 34t^2 + t)(\frac{d}{dt})^2 + (2t^2 - 51t + 1)(\frac{d}{dt}) + \frac{1}{4}(t - 10).$$

(c) (Beukers and Peters [**BP**]). The solutions of the Picard–Fuchs differential equation at $t = 0$ are given by
$$\omega_0(t) = \sum_{n=0}^{\infty} a_n t^n, \quad a_n = \sum_{k=0}^{n} \binom{n}{k}^2 \binom{n+k}{k}^2,$$
$$\omega_1(t) = \omega_0(t) \log t - 3 - 3t - 9t^2 + 103t^3 + \cdots$$

and
$$\omega_2(t) = -\frac{1}{2}\omega_0(t)\log^2 t + \omega_1(t)\log t - 36t - 558t^2 + \cdots.$$

(c) (Verrill and Yui [**VY1**]). We have
$$t(q) = \left(\frac{\eta(q)\eta(q^6)}{\eta(q^2)\eta(q^3)}\right)^{12} = q - 12q^2 + 66q^3 - 220q^4 + \cdots$$
and the mirror map is given by
$$\frac{1}{t(q)} + 12 = T_{6B}(q) \quad \text{with symbol} \quad 6+6.$$

We consider a family of quartic hypersurfaces in $\mathbb{P}^3$. Let $[X_0 : X_1 : X_2 : X_3]$ denote the projective coordinate of $\mathbb{P}^3$. Let $II = (\mathbb{Z}/4\mathbb{Z})^2 = <g_1> \times <g_2>$, where g_1 and g_2 act as $(1,0,3,0)$ and $(0,1,0,3)$ respectively:
$$g_1 : (X_0, X_1, X_2, X_3) \mapsto (\zeta_4 X_0, X_1, \zeta_4^3 X_3),$$
$$g_2 : (X_0, X_1, X_2, X_3) \mapsto (X_0, \zeta_4 X_1, X_2, \zeta_4^3 X_3)$$
where $\zeta_4 = e^{2\pi i/4}$. Then the II–invariant monomials are
$$X_0^4, X_1^4, X_2^4, X_3^4, X_0 X_1 X_2 X_3.$$
This gives rise to the II–invariant one–parameter family of quartics
$$X(z) : X_0^4 + X_1^4 + X_2^4 + X_3^4 + 4zX_0 X_1 X_2 X_3 = 0.$$
If $z^4 \neq 1$, the quotient X_z/II has six rational double points of type A_3. Resolving singularities, we obtain a smooth family of $K3$ surfaces, denoted by $\{X_z\}_{z\in B}$ where $B = \mathbb{P}^1 \setminus \{z \mid z^4 = 1\}$. This is an example of orbifolding construction discussed in Lecture 2.

Theorem 7.4. *The family $\{X_z\}_{z\in B}$ of quartic $K3$ surfaces constructed above has the following properties:*

(a) Let $\mathcal{M} = U_2 \perp (-E_8)^2 \perp <-4>$. Then $\{X_z\}$ is an $\mathcal{M}$-polarized family of $K3$ surfaces. In fact,
$$\text{Pic}(X_z) \simeq \mathcal{M} \quad \text{and} \quad T(X_z) \simeq U_2 \perp <4>.$$

(b) (Lian and Yau [**LY1**]). The Picard–Fuchs differential equation is given by
$$(\Theta^3 - 4z(4\Theta + 1)(4\Theta + 2)(4\Theta + 3))\omega(z) = 0$$
where $\Theta = z\frac{d}{dz}$.

(c) (Lian and Yau [**LY1**]). The solutions to the Picard–Fuchs differential equation are given by
$$\omega_0(z) = \sum_{n=0}^{\infty} \frac{(4n)!}{(n!)^4} z^n$$
and
$$\omega_1(z) = \omega_0(z)\log z + O(z).$$

(d) We have
$$z(q) = q - 104q^2 + 6444q^3 - 311744q^4 + \cdots$$

and the mirror map is given by
$$\frac{1}{z(q)} - 96 = T_{2A}(q)$$
is the McKay–Thompson series corresponding to $\Gamma_0(2)+ = \Gamma_0(2)^+$.

In the same vein as Theorem 7.2, we obtain a family of one–parameter $\mathcal{M}$–polarized $K3$ surfaces for $\mathcal{M} = U_2 \perp (-E_8)^2 \perp <-2>$.

Theorem 7.5. Let $II = (\mathbb{Z}/6\mathbb{Z}) \times (\mathbb{Z}/2\mathbb{Z}) = <g_1> \times <g_2>$ where
$$g_1 : (X), X_1, X_2, X_3 \mapsto (\zeta_6 X_0, X_1, \zeta_6^5 X_2, X_3),$$
$$g_2 : (X_0, X_1, X_2, X_3) \mapsto (X_0, X_1, X_2, -X_3)$$
where $\zeta_6 = e^{2\pi i/6}$. Let
$$X(z) : X_0^6 + X_1^6 + X_2^6 + X_3^2 + 6zX_0X_1X_2X_3 = 0$$
be the II–invariant one–parameter family of sextic surfaces in weighted projective space $\mathbb{P}^3((1,1,1,3))$. If $z^6 \neq 1$, the quotient $X(z)/II$ has a smooth $K3$ resolution, denoted by X_z. Let $B = \mathbb{P}^1 \setminus \{z \mid z^6 = 1\}$.

(a) $\{X_z\}_{z \in B}$ is a family of $\mathcal{M}$-polarized $K3$ surfaces and in fact,
$$\text{Pic}(X_z) \simeq \mathcal{M} \quad \text{and} \quad T(X_z) \simeq U_2 \perp <2>.$$

(b) The Picard–Fuchs differential equation is given by
$$(\Theta^3 - 8z(6\Theta + 1)(6\Theta + 2)(6\Theta + 3))\,\omega(z) = 0$$
where $\Theta = z \frac{d}{dz}$.

(c) The solutions to the Picard–Fuchs differential equation are given by
$$\omega_0(z) = \sum_{n=0}^{\infty} \frac{(6n)!}{(n!)^3(3n)!} z^n$$
and
$$\omega_1(z) = \omega_0(z) \log z + O(q).$$

(d) We have
$$z(q) = q + 744q^2 + 356652q^3 - 140361152q^4 + \cdots$$
and he mirror map is given by $\frac{1}{z(q)} = j(q)$, which is the McKay–Thompson series T_{1A} corresponding to the full modular group $\Gamma = \text{PSL}_2(\mathbb{Z})$.

Remark 7.6. Dolgachev [Dol2] gives another one–parameter family of $K3$ surfaces lattice polarized by the lattice $\mathcal{M} = U_2 \perp (-E_8)^2 \perp <-2N>$ for $N \in \{1, 2, 3\}$.

Dolgachev's examples are all elliptic $K3$ surfaces with sections. For instance, for $N = 1$, Dolgachev looks for a $K3$ surface with an elliptic fibration $f : X \to \mathbb{P}^1$ with a section and two reducible fibers of type $\tilde{E}_8$ and one reducible fiber of type $\tilde{A}_1$. Similarly, for $N = 2, 3$, Dolgachev specifies elliptic fibrations, i.e., types of reducible fibers, to obtain $K3$ surfaces, which are elliptic surfaces with sections.

For Dolgachev's families, the mirror maps are again expressed in terms of McKay–Thompson series associated to genus zero congruence subgroups of level N.

Remark 7.7. In Examples 7.4 (resp. Example 7.5), it should be pointed out that the transcendental lattice of $X(z)$ remains unchanged under orbifolds construction. Therefore, the orbifold X_z has the same transcendental lattice as that of $X(z)$. This implies that the one–parameter families $X(z)$ and X_z satisfy the same Picard–Fuchs differential equation.

Remark 7.8. In a later article, Lian and Yau [**LY3**] have computed more examples of mirror maps for $K3$ surfaces which are constructed torically, and have conjectured the monstrous moonshine phenomenon there as well. Verrill and Yui [**VY1**] have provided a counter-example to the conjecture for torically constructed one–parameter families. To remedy the situation, a revised monstrous moonshine phenomenon involving the notion of commensurability had to be brought into the picture. Subsequently, Doran [**Dor**] has answered the question of commensurablity of mirror maps to McKay–Thompson series, as well as the modularity of mirror maps of $K3$ surfaces lattice polarized by lattices of the form $\mathcal{M} = U_2 \perp (-E_8)^2 \perp <-2N>$ where $N \in \mathbb{Z}$ with $N \geq 1$.

8. Generalizations and open problems

There are several natural questions to be investigated.

Problem 8.1. Given a primitive sublattice $\mathcal{M} \subset \mathcal{L}$, construct $K3$ surfaces polarized by $\mathcal{M}$. Is there any algorithmic procedure for this construction? This problem should be considered in conjunction with the construction of mirror $K3$ surfaces.

For instance, let $\mathcal{M} = U_2 \perp (-E_8)$. A one–parameter family of the weighted Fermat surface of degree 42 and weight $Q = (1,6,14,21)$ defined by

$$X_0^{42} + X_1^7 + X_2^3 + X_3^2 + \lambda X_0 X_1 X_2 X_3 = 0$$

is an $\mathcal{M}$-polarized $K3$ surfaces for generic λ. (Incidentally, this family of $K3$ surfaces is a self-mirror family.)

Problem 8.2. Compute and analyze mirror maps of one–or multi–parameter families of $K3$ surfaces with generic Picard number ≤ 18.

For instance, we consider orbifolds of the deformations of weighted Fermat surfaces of degree m and weight $Q = (q_0, q_1, q_2, q_3)$ in $\mathbb{P}^3(Q)$.

m	Q	$\mathcal{M}^*$	$Picard\,rank$
6	$(1,1,2,2)$	$U_2 \perp (-E_8) \perp D_4 \perp A_2$	16
8	$(1,1,2,4)$	$U_2 \perp (-E_8) \perp D_7$	17
12	$(1,1,4,6)$	$U_2 \perp (-E_8)^2$	18

These $K3$ surfaces may be obtained by orbifolding construction. For instance, start with a $K3$ surface with a defining equation of the form:

$$X_{\lambda,\phi} : X_0^{m/q_0} + X_1^{m/q_1} + X_2^{m/q_2} + X_3^{m/q_3} + \lambda X_0 X_1 X_2 X_3 + \phi X_0^{m/2} X_1^{m/2} = 0$$

with parameters λ, ϕ. Let $\mathcal{G}$ be a discrete group of symmetries leaving the holomorphic 2–form invariant. Take the orbifold $X_{\lambda,\phi}/\mathcal{G}$ and then its smooth resolution, which we denote by X^*. Then X^* is a $K3$ surface, lattice polarized by $\mathcal{M}^*$.

The Picard–Fuchs differential equations and their solutions may be determined using the GKZ hypergeometric system as in Hosono, Lian and S.-T. Yau [**HLY2**]. According to Hosono (private communication), for each family, the period domain is a subset of $\mathfrak{H} \times \mathfrak{H}$, and the mirror map is expressed in terms of two McKay–Thompson series.

To provide rigorous proofs is one of our problems.

Problem 8.3. There are several variants of mirror symmetry for $K3$ surfaces. Here we will describe the versions of mirror symmetry due to Dolgachev via Arnold's strange duality [**Dol2**] for lattice polarized $K3$ surfaces.

Definition 14. Let $\mathcal{M} \subset \mathcal{L}$ be a primitive sublattice of signature $(1, t)$. Assume that the orthogonal complement $\mathcal{M}^\perp$ admits an orthogonal splitting $\mathcal{M}^\perp = U_2 \perp \mathcal{M}^*$. Let $\mathcal{F}(\mathcal{M})$ be a complete family of pseudo-ample $\mathcal{M}$-polarized $K3$ surfaces. The *mirror family* $\mathcal{F}^*(\mathcal{M}^*)$ is defined to be any family of pseudo-ample $\mathcal{M}^*$-polarized $K3$ surfaces.

Belcastro [**Bel**] has computed the Picard lattice for the 95 families of $K3$ surfaces in weighted projective 3-spaces determined by M. Reid and rediscovered by Yonemura [**Yon**]. As an application, she has determined whether the mirror family for each one is in the list of Reid–Yonemura. The conclusion is that the 95 families of $K3$ surfaces are not closed under mirror symmetry.

Among 95 families, there are 11 weights that give rise to $K3$ surfaces with $\mathcal{M} = -E_8 \perp U_2$ so that $\mathcal{M}^* = \mathcal{M}$. These $K3$ surfaces are self-mirrors.

The construction of mirror families of $K3$ surfaces in general is still an open question.

Problem 8.4. There is the notion of mirror symmetry due to Batyrev[**Ba**] based on reflexive polytopes and toric geometry. However, we have no time nor space to go into Batyrev's mirror symmetry here. We close this lecture by giving one example due to Narumiya and Shiga [**NS**].

Recently, Narumiya and Shiga [**NS**] a family of $K3$ surfaces associated to the simplest reflexive polytope. Let

$$P = \begin{pmatrix} 1 & 0 & 0 & -1 \\ 0 & 1 & 0 & -1 \\ 0 & 0 & 1 & -1 \end{pmatrix}$$

and

$$P^* = \begin{pmatrix} 3 & -1 & -1 & -1 \\ -1 & 3 & -1 & -1 \\ -1 & -1 & 3 & -1 \end{pmatrix}$$

be the dual polytope of P. Then both P and P^* are reflexive polytopes.

The toric variety associated to P^* gives rise to a generic family

$$c_1 xy^2 z + c_2 xy^2 z + c_3 xyz^2 + c_4 xyz + c_5 = 0 \quad c_i \in \mathbb{C}.$$

Specializing this family, we obtain the following one-parameter family of $K3$ surfaces:

$$\mathcal{F}(P^*) = \{ S(\lambda) : xyz(x + y + z + \lambda) + 1 = 0 \}$$

by setting $c_1 = c_2 = c_3 = c_5 = 1$, $c_4 = \lambda$, or another family:

$$\mathcal{F}(P^*) = \{ S(\mu) : xyz(x + y + z + 1) + \mu^4 = 0 \}$$

by setting $c_1 = c_2 = c_3 = c_4 = 1, c_5 = \mu$. Then both $S(\lambda)$ and $S(\mu)$ are families of $K3$ surfaces. One can see that $S(\lambda)$ has the transcendental lattice

$$T_{S(\lambda)} = U_2 \oplus <4> \quad \text{and} \quad \text{Pic}(S(\lambda)) = U_2 \oplus (-E_8)^2 \oplus <-4>.$$

In fact, the surface $S(\lambda)$ admits a Shioda–Inose structure and hence it is birationally equivalent a Kummer surface of the product of two isogenous rational elliptic surfaces.

The mirror partner of $\mathcal{F}(P^*)$ is $\mathcal{F}(P)$ which contains all quartic hypersurfaces in $\mathbb{P}^3$.

To compute the mirror map (or rather the reciprocal of the mirror map) of the family $\mathcal{F}(P^*)$, one uses the family with parameter μ and compute the period integral $I(S(\mu))$. The result of their calculation is:

$$I(S(\mu)) = (2\pi i)^2 F_{3,2}(\frac{1}{4}, \frac{2}{4}, \frac{3}{4}; 1, 1, ; (4\mu)^4) = \{2\pi i F_{2,1}(\frac{1}{8}, \frac{3}{8}; 1; (4\mu)^4)\}^2.$$

Here $F_{3,2}$ is the generalized hypergeometric function

$$F_{3,2}(a, b, b'; c, c', x) = \sum_n \frac{(a)_n (b)_n (b')_n}{(c)_n (c')_n (1)_n} x^n$$

where $(a)_n = a(a+1)\cdots(a+n-1)$ and $F_{2,1}$ is the Gauss hypergeometric function.

Then one has:

$$\mu(q) = \sqrt{-1} q^{-1/4}(1 + 26q + 79q^2 + 326q^3 + 755q^4 + 2106q^5 + \cdots)$$

So we get

$$\mu^4(q) - 96 = T_{2A}(q)$$

which is the McKay–Thompson series associated to $\Gamma_0(2)^+$.

The function μ of the Narumiya–Shiga family is related to the mirror map of the Lian–Yau family in Example 7.4 by the equation: $\mu^4 = \frac{1}{z}$.

(Narumiya–Shiga called $\mu(q)$ as the mirror map. However, according to our definition, a mirror map ought to be holomorphic. Therefore, the inverse of $\mu(q)$ should be the mirror map of the family.)

Problem 8.5. There are dualities due to Kyoji Saito [S] on characteristic polynomials of weight systems, and Kobayashi [Kob] on weight systems. Ebeling[E] has generalized Kobayahsi's duality and discussed its connection to mirror moonshine. A natural question is: Are they any possible relations among all these dualities and mirror symmetry?

BIBLIOGRAPHY

[Ba] Batyrev, V., *Dual polyhedra and mirror symmetry for Calabi–Yau hypersurfaces in toric varieties*, J. Alg. Geometry **3** (1994), 493–535.

[Bea] Beauville, A., *Les familles stables de courbes elliptiques sur $\mathbb{P}^1$ admettant quatre fibres singulariéres*, C.R. Acad. Sci. Paris Série I **294** (1982), 657–660.

[Be] Beilinson, A., *Higher regulators and values of L–functions*, J. Soviet Math. **30** (1985), 2036–2070.

[Bel] Belcastro, S.-M., *Picard lattices of families of K3 surfaces*, math.AG/9809008, 1998.

[BP] Beukers, F., and Peters, C., *A family if K3 surfaces and $\zeta(3)$*, J. Reine Angew. Math. **351** (1984), 42–54.

[BS-D] Birch, B., and Swinnerton-Dyer, H.P.F., *Notes on elliptic curves I*, J. Reine Angew. Math. **212** (1963), 7–25.

[Bl1] Bloch, S., *Algebraic cycles and values of L-functions*, J. Reine Angew. Math. **350** (1984), 94–108.

[Bl2] Bloch, S., *Algebraic cycles and values of L-functions II*, Duke J. Math. **52** (1985), 379–397.

[Bo] Borcherdes, R., *Monstrous moonshine and monstrous Lie superalgebras* Invent. Math. **109** (1992), 405–444.

[BCDT] Breuil, C., Conrad, B., Diamond, F., and Taylor, R., *On the modularity of elliptic curves over $\mathbb{Q}$*, preprint 1999.

[BST] Buhler, J., Schoen, C., and Top, J., *Cycles, L-functions and triple products of elliptic curves*, J. Reine Angew. Math. **492** (1997), 93–133.

[CDP] Candelas, P., Derrick, E., and Parkes, L., *Generalized Calabi–Yau manifolds and the mirror of a rigid manifold*, Nucl. Phys. B**407** (1993), 115–154.

[CK] Cox, D., and Katz, S., *Mirror Symmetry and Algebraic Geometry*, Math. Surveys and Monographs **68**, Amer. Math. Soc., 1999.

[CN] Conway, J., and Norton, S., *Monstrous moonshine*, London Math. Soc. **11** (1979), 308–339,

[COGP] Candelas, P., de la Ossa, C., Green, P.S., and Parkes, L., *A pair of Calabi–Yau manifolds as an exactly solvable superconformal theory*, Nuclear Phys. B **359** (1991), 21–74.

[D] Deligne, P., *Formes modulaires et représentations ℓ-adiques*, Sem. Bourbaki **355** 1968/69, Lecture Notes Math. **179**, Springer–Verlag 1979.

[Di] Diamond, F., *An extension of Wiles's results*, in *Modular Forms and Fermat's*

Last Theorem, edited by G. Cornell, J.H. Silverman and G. Stevens (1997), Springer–Verlag, 475–489.

[Dol1] Dolgachev, I., *Weighted projective varieties*, in Lecture Notes in Math. **956** (1982), Springer–Verlag, 24–71.

[Dol2] Dolgachev, I., *Mirror symmetry for lattice polarized K3 surfaces*, J. Math. Sci. **81** (1996), 2599–2630.

[Dor] Doran, C., *Picard–Fuchs uniformization: Modularity of the mirror map and mirror-moonshine*, Proc. NATO ASI/CRM Summer School Banff 1998, "The Arithmetic and Geometry of Algebraic Cycles", CRM Proc. & Lecture Notes Series **24** (2000), Amer. Math. Soc., 257–281.

[E] Ebeling, W., *Strange duality, mirror symmetry, and the Leech lattice*, preprint alg-geom/9612010, 1996.

[F] Faltings, G., *Endlichkeitssätze für abelsche Varietäten über Zahlkörpern*, Invent. Math. **73** (1983), 349–366.

[FM] Fontaine, M., and Mazur, B., *Geometric Galois representations*, in *Elliptic Curves, Modular Forms and Fermat's Last Theorem*, Hong Kong 1993, World Scientific, 41–78.

[G] Goto, Y., *Arithmetic of certain weighted diagonal surfaces over finite fields*, J. Number Theory **59** (1996), 37–81.

[GP] Greene, B., and Plesser, R., *Duality in Calabi–Yau moduli spaces*, Nuclear Phys. B **338** (1990), 15–37.

[GRY] Greene, B., Roan, S.-S., and Yau, S.-T., *Geometric singularities and spectra of Landau–Ginsburg models*, Comm. Math. Phys. **142** (1991), 245–259.

[GY] Gouvêa, F. Q., and Yui, N., *Arithmetic of Diagonal Hypersurfaces over Finite Fields*, London Math. Soc. Lecture Notes Series **209**, Cambridge University Press 1995.

[GH] Gritsenko, V., and Hulek, K., *The modular form of the Barth–Nieto quintic*, Internl. Math. Res. Notices (1999), 915–937.

[Hir] Hirzebruch, F., *Some examples of threefolds with trivial canonical bundle*, Collected Works **II** (1995), Springer-Verlag, 757–770.

[HLY1] Hosono, S., Lian, B., and Yau, S.-T., *Galabi–Yau varieties and pencils of K3 surfaces*, alg-geom/9603020.

[HLY2] Hosono, S., Lian, B. and Yau, S.-T.,*GKZ hypergeometric systems in mirror symmetry of Calabi–Yau hypersurfaces*, Commun. Math. Phys. **182** (1996), 535–577.

[HSvGvS] Hulek, K., Spandaw, J., van Geemen, B., and van Straten, D., *The modularity of the Barth–Nieto quintic and its relatives*, preprint 2000.

[HS] Hunt, B., and Schimmrigk, R., *K3 fibered Calabi–Yau threefolds. I. The twist map*, Intrnat. J. Math. **10** (1999), No. 7, 833–869.

[Ki] Kimura, K., *An example of algebraic cycles with non-trivial Abel–Jacobi image*, J. Algebra **222** (1999), 129–145.

[KSc] Klemm, A., and Schimmrigk, R., *Landau–Ginsburg string vacua*, Nuclear Phys. B **411** (1994), 559–583.

[KT] Klemm, A., and Theisen, S., *Considerations of one-modulus Galabi–Yau compactifications: Picard–Fuchs equations, Kähler potentials and mirror maps*, Nuclear Phys. B. **389** (1993), 153–180.

[Kob] Kobayashi, M., *Duality of weights, mirror symmetry and Arnold's strange duality*, preprint alg-geom/9502004. 1995.

[Kon] Kondo, S., *Quadratic forms and K3, Enriques surfaces*, Sugaku Expositions **6** (1993), No.1, 53–72.

[KSk] Kreuzer, M., and Skarke, H., *No mirror symmetry in Landau–Ginsburg spectra!*, Nuclear Phys. B. **388** (1993), 113–130.

[LY1] Lian, B., and Yau, S.-T., *Mirror symmetry, rational curves on algebraic manifolds and hypergeometric series*, Proc. International Congress of Mathematical Physics 1994 .

[LY2] Lian, B., and Yau, S.-T., *Arithmetic properties of mirror maps and quantum coupling*, Commu. Math. Phy. **176** (1996), No.1., 163–191.

[LY3] Lian, B., and Yau, S.-T., *Mirror maps, modular relations and hypergeometric series II*, S-Dulaity and Mirror Symmetry (Trieste, 1995), Nuclear Phys. B. Proc. Suppl. **46** (1996), 248–262.

[L1] Livné, R., *Cubic exponential sums and Galois representation*, in *Current Trends in Arithmetic Algebraic Geometry*, Contemporary Math. **67** (1987), Amer. Math. Soc., 187–201.

[L2] Livné, R., *Motivic orthogonal two-dimensional representations of $Gal(\bar{\mathbb{Q}})/\mathbb{Q}$)*, Israel J. Math. **92** (1995), 149–156.

[Mil] Milnor, J., and Husemoller, D., *symmetric Bilinear Forms*, Springer–Verlag 1971.

[M1] Morrison, D., *On K3 surfaces with large Picard number*, Invent. Math. **75** (1984), 105–121.

[M2] Morrison, D., *Mathematical aspects of mirror symmetry*, in *Complex Algebraic Geometry*, J. Kollár ed., IAS/Park City Math. Series **3** (1997), Amer. Math. Soc., 265–340.

[NS] Narumiya, N. and Shiga, H., *The mirror map for a family of K3 surfaces induced from the simplest 3 dimensional reflexive polytope*, Proc. the Montreal Moonshine Conference, CRM Proc. & Lecture Notes, Amer. Math. Soc. (to appear).

[P] Peters, C., *Monodromy and Picard–Fuchs equations for families of K3 surfaces and elliptic curves*, Ann. scient. Ec. Norm. Sup. 4e série **19** (1986), 583–607.

[Re] Reid, M., *Canonical 3-folds*, in *Algebraic Geometry Angers 1979* (1980), Sijthoff and Noordhoff, 273–310.

[Ro1] Roan, S.-S., *On Calabi–Yau orbifolds in weighted projective spaces*, Intern. J. Math. **1** (1990), 273–310.

[Ro2] Roan, S.-S., *The mirror of Calabi–Yau orbifolds*, Intern. J. Math. **2** (1991), 439–455.

[Ro3] Roan, S.-S., *On $c_1 = 0$ resolution of quotient singularity*, Intern. J. Math. **5** (1994), 523–536.

[S] Saito, Kyoji, *Duality of regular systems of weights*, Asian J. Math. **2** (1998), 983–1047.

[SY] Saito, Masa-Hiko and Yui, N., *The modularity conjecture for rigid Calabi–Yau threefolds over $\mathbb{Q}$*, preprint 2000.

[Sch1] Schoen, C., *On the geometry of a special determinantal hypersurface associated to the Mumford–Horrocks vector bundle*, J. Reine Angew. Math. **364** (1986), 85–111.

[Sch2] Schoen, C., *On fiber products of rational elliptic surfaces with section*, Math. Z. **197** (1988), 177–199.

[Sch3] Schoen, C., *Varieties dominated by product varieties*, Intern. J. Math. **7** (1996), 541–571.

[Seb] Sebbar, A., *Classification of torsion-free genus zero congruence subgroups of $PSL(2,\mathbb{Z})$*, preprint 1999.

[Ser] Serre, J.-P., *Sur les représentations modulaires de degré 2 de Gal($\overline{\mathbb{Q}}/\mathbb{Q}$)*, Duke J. Math. **54** (1987), 179–230.

[Sh] Shioda, T., *On elliptic modular surfaces*, J. Math. Soc. Japan **24** (1972), No. 1, 20–59.

[SI] Shioda, T., and Inose, K., *On singular K3 surfaces*, in "Complex Analysis and Algebraic Geometry", eds. Baily, W., and Shioda, T., (1997), Kinokuniya Shoten and Cambridge University Press, 119–136.

[SK] Shioda, T., and Katusra, T., *On Fermat varieties*, Tôhoku J. Math. **31** (1979), 97–115.

[T] Tate, J., *Conjectures on algebraic cycles in ℓ-adic cohomology*, in Motives, Part I, Proc. Sympo. Pure Math. **55** (1994), Amer. Math. Soc., 71–83.

[Tay] Taylor, R., *Remarks on a conjecture of Fontaine and Mazur*, preprint 2000.

[TW] Taylor, R., and Wiles, A., *Ring-theoretic properties of certain Hecke algebras*, Ann. Math. **141** (1995), 553–572.

[vGN] van Geemen, B., and Nygaard, N., *On the geometry and arithmetic of some Siegel modular threefolds*, J. Number Theory **53** (1995), 45–87.

[vS] van Straten, D., *A quintic hypersurface in $\mathbb{P}^4$ with 130 nodes*, Topology **32** (1993), 857–864.

[V] Verrill, H., *The L-series of certain Calabi-Yau threefolds*, J. Number Theory **81** (2000), 310–334.

[VY1] Verrill, H., and Yui, N., *Thompson series, and the mirror maps of pencils of K3 surfaces*, Proc. 1998 CRM Summer School Banff "Arithmetic and Geometry of Algebraic Cycles", CRM Proc. & Lecture Notes Series **24** (2000), Amer. Math. Soc., 399–432.

[VY2] Verrill, H., and Yui, N., *The mirror maps of certain pencils of K3 surfaces, and Hauptmoduln*, research monograph, in preparation.

[We] Weil, A., *Jacobi sums as "Grossenkaraktere"*, Trans. Amer. Math. Soc. **73** (1952), 487–495.

[WvG] Werner, J., and van Geemen, B., *New examples of threefolds with $c_1 = 0$*, Math. Z. **203** (1990), 211–225.

[Wi] Wiles, A., *Modular elliptic curves and Fermat's Last Theorem*, Ann. Math. **141** (1995), 443–551.

[Yon] Yonemura, T., *Hypersurface simple K3 singularities*, Tôhoku J. Math. **42** (1990), 351–380.

[Yos] Yoshida, M., *Fuchsian Differential Equations. With special emphasis on the Gauss-Schwarz theory*, Aspects of Mathematics, **E11**, Vieweg & Sohn, Braunschweig 1987.

[Y1] Yui, N., *The L-series of certain Calabi-Yau orbifolds over number fields*, preprint 1997.

[Y2] Yui, N., *Arithmetic of certain Calabi-Yau varieties over number fields*, in the Proc. NATO ASI Banff 1998 "Arithmetic and Geometry of Algebraic Cycles", NATO Science Series **548** (2000), Kluwer Academic Publisher, 519–565.

[Y3] Yui, N., *Arithmetic of Orbifold Calabi-Yau Varieties over Number Fields*, Research Monograph, in preparation.

[MirSym1] Yau, S.-T. (editor), *Mirror Symmetry I*, AMS/IP Studies in Advanced

Math. **9**, Amer. Math. Soc. and International Press 1998. (Previously, *Essays on Mirror Manifolds*, International Press 1992).

[MirSym2] Yau, S.-T.(editor), *Mirror Symmetry Ii*, AMS/IP Studies in Advanced Mathematics **1**, Amer. Math. Soc. and International Press 1997.